Achieving sustainable turfgrass management

It is widely recognised that agriculture is a significant contributor to global warming and climate change. Agriculture needs to reduce its environmental impact and adapt to current climate change whilst still feeding a growing population, i.e. become more 'climate-smart'. Burleigh Dodds Science Publishing is playing its part in achieving this by bringing together key research on making the production of the world's most important crops and livestock products more sustainable.

Based on extensive research, our publications specifically target the challenge of climate-smart agriculture. In this way we are using 'smart publishing' to help achieve climate-smart agriculture.

Burleigh Dodds Science Publishing is an independent and innovative publisher delivering high quality customer-focused agricultural science content in both print and online formats for the academic and research communities. Our aim is to build a foundation of knowledge on which researchers can build to meet the challenge of climate-smart agriculture.

For more information about Burleigh Dodds Science Publishing simply call us on +44 (0) 1223 839365, email info@bdspublishing.com or alternatively please visit our website at www.bdspublishing.com.

Related titles:

Managing soil health for sustainable agriculture Volume 1: Fundamentals
Print (ISBN 978-1-78676-188-0); Online (ISBN 978-1-78676-191-0, 978-1-78676-190-3)

Managing soil health for sustainable agriculture Volume 2: Monitoring and management
Print (ISBN 978-1-78676-192-7); Online (ISBN 978-1-78676-195-8, 978-1-78676-196-1)

Achieving sustainable crop nutrition
Print (ISBN 978-1-78676-312-9); Online (ISBN 978-1-78676-314-3, 978-1-78676-315-0)

Advances in measuring soil health
Print (ISBN 978-1-78676-426-3); Online (ISBN 978-1-78676-428-7, 978-1-78676-429-4)

Climate change and agriculture
Print (ISBN 978-1-78676-320-4); Online (ISBN 978-1-78676-322-8, 978-1-78676-323-5)

Water management for sustainable agriculture
Print (ISBN 978-1-78676-176-7); Online (ISBN 978-1-78676-178-1, 978-1-78676-179-8)

Understanding and minimising fungicide resistance
Print (ISBN 978-1-80146-198-6); Online (ISBN 978-1-80146-200-6, 978-1-80146-201-3)

Chapters are available individually from our online bookshop:
https://shop.bdspublishing.com

climate•SMART•publishing
IN AGRICULTURAL SCIENCE

BURLEIGH DODDS SERIES IN AGRICULTURAL SCIENCE
NUMBER 125

Achieving sustainable turfgrass management

Edited by Professor Michael Fidanza, Pennsylvania State University, USA

burleigh dodds
SCIENCE PUBLISHING

Published by Burleigh Dodds Science Publishing Limited
82 High Street, Sawston, Cambridge CB22 3HJ, UK
www.bdspublishing.com

Burleigh Dodds Science Publishing, 1518 Walnut Street, Suite 900, Philadelphia, PA 19102-3406, USA

First published 2023 by Burleigh Dodds Science Publishing Limited

Library of Congress Control Number: 2022949196

British Library Cataloguing in Publication Data
A catalogue record for this book is available from the British Library

ISBN 978-1-80146-019-4 (print)
ISBN 978-1-80146-022-4 (PDF)
ISBN 978-1-80146-021-7 (ePub)
ISSN 2059-6936 (print)
ISSN 2059-6944 (online)

DOI 10.19103/AS.2022.0110

Typeset by Deanta Global Publishing Services, Dublin, Ireland

Contents

Part 3 Case studies

Series list

Title	Series number
Achieving sustainable cultivation of maize - Vol 1 From improved varieties to local applications Edited by: Dr Dave Watson, CGIAR Maize Research Program Manager, CIMMYT, Mexico	001
Achieving sustainable cultivation of maize - Vol 2 Cultivation techniques, pest and disease control Edited by: Dr Dave Watson, CGIAR Maize Research Program Manager, CIMMYT, Mexico	002
Achieving sustainable cultivation of rice - Vol 1 Breeding for higher yield and quality Edited by: Prof. Takuji Sasaki, Tokyo University of Agriculture, Japan	003
Achieving sustainable cultivation of rice - Vol 2 Cultivation, pest and disease management Edited by: Prof. Takuji Sasaki, Tokyo University of Agriculture, Japan	004
Achieving sustainable cultivation of wheat - Vol 1 Breeding, quality traits, pests and diseases Edited by: Prof. Peter Langridge, The University of Adelaide, Australia	005
Achieving sustainable cultivation of wheat - Vol 2 Cultivation techniques Edited by: Prof. Peter Langridge, The University of Adelaide, Australia	006
Achieving sustainable cultivation of tomatoes Edited by: Dr Autar Mattoo, USDA-ARS, USA and Prof. Avtar Handa, Purdue University, USA	007
Achieving sustainable production of milk - Vol 1 Milk composition, genetics and breeding Edited by: Dr Nico van Belzen, International Dairy Federation (IDF), Belgium	008
Achieving sustainable production of milk - Vol 2 Safety, quality and sustainability Edited by: Dr Nico van Belzen, International Dairy Federation (IDF), Belgium	009
Achieving sustainable production of milk - Vol 3 Dairy herd management and welfare Edited by: Prof. John Webster, University of Bristol, UK	010
Ensuring safety and quality in the production of beef - Vol 1 Safety Edited by: Prof. Gary Acuff, Texas A&M University, USA and Prof. James Dickson, Iowa State University, USA	011
Ensuring safety and quality in the production of beef - Vol 2 Quality Edited by: Prof. Michael Dikeman, Kansas State University, USA	012
Achieving sustainable production of poultry meat - Vol 1 Safety, quality and sustainability Edited by: Prof. Steven C. Ricke, University of Arkansas, USA	013
Achieving sustainable production of poultry meat - Vol 2 Breeding and nutrition Edited by: Prof. Todd Applegate, University of Georgia, USA	014

Introduction

Turfgrass is required to meet a challenging range of aesthetic, functional and environmental requirements, whilst also mitigating the threat of abiotic and biotic stresses which are being accentuated by climate change. The turfgrass industry is also facing increasing pressure to reduce its environmental impact and advance more sustainable maintenance practices that utilise and/or optimise fewer agronomic-related resources. This volume summarises the wealth of recent research that addresses these challenges, whilst also identifying potential mitigation strategies to reduce the sector's contribution to climate change, such as reduced fertiliser use and water conservation.

The chapters are split into three parts: Part 1 chapters focus on the physiology, breeding and cultivation of turfgrass species, specifically focusing on breeding techniques, soil management, phosphite utilisation and irrigation and water management. Chapters in this part also draw attention to maintenance practices of turfgrass as well as advances in using turfgrass for athletics fields, sports pitches and ornamental lawns. Chapters in Part 2 review various biotic and abiotic stresses that affect turfgrass management, such as diseases, insect pests and weeds. Chapters also look at the importance of plant growth regulation, abiotic stress management, organic matter management and the use of biostimulants in turfgrass. Part 3 chapters provide case studies on various management practices for turfgrass, such as the use of unmanned aircraft systems, methods for selecting turfgrass varieties and cultivars as well as water management and soil testing.

Part 1 Physiology, breeding and cultivation

The volume opens with a chapter that focuses on understanding turfgrass physiology. Chapter 1 first begins by discussing physiological responses to four of the most common abiotic stress conditions: drought, heat, freezing and salinity stresses. The chapter reviews the importance of understanding how turfgrasses respond to drought, heat, freezing and salt stress, highlighting the historical perspectives and recent advances in each form of stress. The chapter also emphasises how these stresses damage plants and how in turn plants respond to tolerate these conditions. Sections on emerging issues and future directions for research for each form of stress are also provided.

The next chapter focuses on advances in breeding for improved cultivars of turfgrass. Chapter 2 draws attention to major cool-season species such as bentgrass, fine and tall fescue, Kentucky bluegrass and perennial ryegrass. It focuses on major warm-season species such as bermudagrass, buffalograss, seashore paspalum, St. Augustinegrass and Zoysiagrass. For each species, the chapter discusses biological characteristics, traditional breeding strategies, current breeding objectives and advanced breeding methods for the major cool- and warm-season turfgrass species. Emerging challenges and future potential areas to develop research in are also highlighted.

The subject of Chapter 3 is advances in soil management for successful establishment and maintenance of turfgrass. It first explores soil physical properties relevant to turfgrass management, focusing specifically on soil texture, compaction, water-holding capacity and structure. The chapter then moves on to examine soil chemical properties relevant to turfgrass management, such as sodicity and plant-available nutrient content. This is then followed by an analysis of some of the most well-known soil biological properties crucial for turfgrass management, drawing specific attention to nitrogen availability and the importance of the soil microbiome. The chapter concludes by identifying areas for future study in soil management for turfgrass species.

Chapter 4 focuses on advances in phosphite utilisation for turfgrass. The chapter begins by first describing the chemical structure of both phosphate and phosphite and how the differences between them can affect their use as a nutrient source in plants. It moves on to review the use of phosphite in turfgrass and how it is an emerging nutrient source and is considered in some cases, controversial. The chapter goes on to examine how phosphite can be used as a source of phosphorus nutrition and highlights how it can be incorporated into turfgrass nutrition strategies. A section on disease control in turfgrass via phosphite is also included, before concluding with an overview of the importance of expanding currently available research in phosphite use in turfgrass management.

The next chapter looks at advances in irrigation and water management of turfgrass. Chapter 5 reviews the importance of estimating turfgrass water needs. With this in mind, the chapter moves on to examine advances with sprinkler devices and technologies as well as the importance of using subsurface drip irrigation as an alternative way to provide turfgrass with water. The chapter also provides alternatives to potable water and highlights several turfgrass management practices that can also reduce water needs, such as mowing, fertilisation, soil surfactants and plant growth regulators. This then leads into the next section, which focuses on the use of plant growth regulators in turfgrass management. The chapter also highlights the use of soil surfactants in turfgrass to manage soil hydrophobicity.

The subject of Chapter 6 is advances in maintenance practices of turfgrass. It begins by highlighting the importance of maintaining turfgrass quality through the use of mowing, drawing specific attention to various technologies and maintenance strategies such as autonomous mowers, vertical mowing and grooming. The chapter moves on to examine putting green management, looking at aspects such as cultivation practices, rolling and firmness measurement. A section on turfgrass renovation is also provided, focusing on elements such as weed control, seed bed preparation, non-chemical control and other plant and soil issues that can occur during turfgrass renovation. The chapter also reviews the use of turf colourants and highlights how their application can be a useful tool for maintaining turfgrass.

Chapter 7 examines advances in turfgrass breeding and genetics for the purpose of athletic fields and sports pitches. The chapter first discusses the most appropriate types of species and cultivar for this, focusing on how improvements can be made to both warm-season and cool-season turfgrass. The chapter then discusses the different methods that can be used in order to improve the sports field's quality, such as alternative root zone amendments, seed coating and seed germination covers, hybrid turfgrass systems and thick cut sod. A section on sports field maintenance is also provided, drawing attention to aspects such as cultivation, topdressing, irrigation, grow lights and cooling fans as well as painting techniques. The chapter also reviews performance testing and site specific management.

The final chapter of Part 1 addresses the advancements in turfgrass for ornamental lawns. It first focuses on breeding and selection advances, drawing attention to consumer lawn preferences and the implications this can lead to in terms of contemporary breeding and production, as well as discussing the current breeding and selection trends for ornamental turfgrass lawns. The chapter then moves on to review various management practices, such as soil preparation, planting and establishment, fertilisation and nutrient management, mowing and cultivation and irrigation and water management. A section on non-traditional or alternative lawns is also provided, which is then followed by a review of socioenvironmental research practices to advance sustainable lawn management.

Part 2 Biotic and abiotic stresses

Part 2 opens with a chapter that covers advances in turfgrass disease management. Chapter 9 begins by first introducing one of the most destructive diseases of cool-season turfgrasses, anthracnose. The chapter first highlights how this disease's severity is influenced by plant nutrition, mowing, irrigation and sand topdressing and how it is crucial to develop these management practices to decrease the effect and occurrence of anthracnose in cool-season turfgrass. The chapter moves on to address dollar spot, a disease that affects all turfgrass species and how cultural management practices can provide more effect management of dollar spot. The presence of Microdochium patch and snow moulds are also touched upon, followed by brown patch, large patch and root diseases of turfgrasses. A section on host resistance is also included, a recently developed disease management strategy for turfgrass ecosystems. This is then followed by an overview of fungicide management, in particular, the previously addressed dollar spot. The chapter concludes by highlighting where future research developments should turn their attention to.

Moving on from Chapter 9, Chapter 10 addresses the advances in turfgrass insect pest management. It starts by first reviewing the recent studies of major turfgrass insect pests, such as annual bluegrass weevil, bill bugs, white grubs and fall armyworm. The chapter then moves on to discuss recent studies on emerging turfgrass pests, focusing specifically on European crane flies and *Chaetocnema minuta*. A section on integrated pest management practices is also provided, drawing attention to methods such as pest detection and monitoring, chemical control, insecticide resistance and non-target issues with chemical insecticides. The chapter also discusses biological control, cultural control, host plant tolerance and resistance to pests and the use of plant growth promoting rhizobacteria.

The next chapter focuses on advances in turfgrass weed management. Chapter 11 discusses relevant and recent developments in synthetic herbicides, alternatives to synthetic herbicides and non-chemical strategies for weed management including cultural, physical, and genetic approaches. Advances in integrated weed management strategies and barriers preventing adoption of integrated weed management strategies are also discussed. Opportunities for research and development to improve integrated weed management are presented throughout.

Chapter 12 examines advances in plant growth regulation in turfgrass. It begins by providing an overview of the current use of growth regulators in both cool-season and warm-season turfgrass species. It then goes on to discuss enhancing rooting and establishment as well as turfgrass wear tolerance. A section on seed head suppression and production is also included, which is then followed by an overview of the suppression and control of annual bluegrass. Enhancing disease tolerance and tolerance to environmental extremes such as cold and drought are also discussed. Shade tolerance enhancement is also reviewed and a case study on how turfgrass species and light quality can determine the effectiveness of using plant growth regulators for shaded turf is also included.

The subject of Chapter 13 is advances in abiotic stress management in turfgrass. Commonly observed abiotic stresses in turfgrass systems include drought, high temperature, low temperature, low irradiance, traffic, salinity, and chemical phytotoxicity. Managing abiotic stresses with reduced irrigation, fertilizer, energy, and pesticides has become critical to the sustainability of the turfgrass industry. The chapter first describes how sustainable strategies and practices can be implemented to manage abiotic stress, focusing specifically on managing turfgrass growth, moderating the environmental conditions causing the stress, improving abiotic stress tolerance and masking abiotic stress through methods such as colourants. It moves on to discuss ways of detecting abiotic stress and how data-driven decision-making can be used to manage it.

Chapter 14 focuses on advances in managing organic matter in turfgrass ecosystems. The chapter begins by first discussing the effects of organic matter on turfgrass, then moves on to review how organic matter can be monitored and measured. It then discusses the relationship between organic matter and turfgrass diseases, focusing on how different levels of organic matter can increase or decrease the development of major diseases such as dollar spot. The chapter also draws attention to soil microbial populations can affect organic matter content.

The final chapter of Part 2 addresses the advances in biostimulants in turfgrass. Chapter 15 reviews the use of biostimulants in turfgrass management. It first identifies the different types of biostimulants that can be used for turfgrass, focusing on phytohormones, biopolymers, protein hydrolysates and other nitrogen-containing compounds, botanical and synthetic bioactive compounds. The chapter also highlights the possibility of using humic substances, organics, inorganics/minerals, biologicals/microbials, soil surfactants and other naturally derived or synthetic bioactive compounds.

Part 3 Case studies

The first chapter of the four case study chapters in this section focuses on the use of unmanned aircraft systems in turfgrass monitoring. Chapter 16 begins by providing a brief overview of light reflectance by plants and how this can be used by sensors to assess how well a plant can utilise this for photosynthesis. The chapter moves on to emphasise the importance of mission planning in terms of using these systems, specifically indicating that defining the target of interest, how much resolution is required to identify patterns via an image, selecting the appropriate sensor and aircraft are all crucial elements to this process. Flight planning and data collection frequency are also deemed important as well. It also reviews data processing considerations and provides examples of small unmanned aircraft system applications in turfgrass.

Chapter 17 discusses selecting turfgrass varieties and cultivars, using the National Turfgrass Evaluation Program (NTEP) as a basis for the chapter's discussion. The chapter first describes the factors that impact turfgrass sustainability, then moves on to highlight how the data collected from using the NTEP can be used to select appropriate cultivars. An overview of the updated NTEP database is also provided, which is then followed by six case studies demonstrating ways of selecting sustainable cultivars via the use of the NTEP. These case studies include: identifying a geographical region, the use of seasonal analysis, choosing appropriate site characteristics, identifying key sustainability factors, locating and comparing cultivars and finally, the advanced selection of cultivars with improved drought resistance.

Chapter 18 examines the potential relationship between turfgrasses and pollinators. The chapter begins by first considering how land use change and pollinator decline correlate with one another, specifically drawing attention to the drastic change in human population which in turn, has resulted in the need to significantly change the landscape. The chapter then considers how the presence of pollinators can be influenced by this changed landscape, followed by a section that considers using urban grasslands as sources of pollinator habitat and forage. This is then followed by a review of the various plants that can be used as part of turfgrasses for pollinator habitats, again, a case study is provided to support the discussion. The chapter also reviews the management of pollinator friendly mixtures, focusing specifically on how turfgrass management practices such as mowing and the use of pesticides need to adapt in order to support pollinators. A section on the additional considerations that also need to be taken into account when preparing turfgrass for pollinator species is also included.

The next chapter considers the management of water for turfgrass in arid environments. Chapter 19 first reviews the importance of soil moisture for turfgrass, focusing on methods to improve moisture uniformity such as deep watering, soil moisture monitoring, supplemental irrigation and the use of wetting agents. Several case studies are provided to emphasise the importance of water and irrigation management, the first three focus on how turfgrass conversion can result in significant water savings. The next two draw attention to the use of native grasses and alternative plant materials following turfgrass removal. The use of subsurface drip irrigation on golf courses is also discussed.

The final chapter of the book looks at using soil testing in turfgrass management. Chapter 20 addresses how soil testing can be used to analyse soil types, it first highlights the common soil test extractants used, such as Bray P1 and P2, Mehlich 1 and 3, Olsen (also known as sodium bicarbonate), Morgan, Lancaster, ammonium acetate and diethylenetriaminepentaacetic acid. The chapter also includes discussion on the use of monocalcium phosphate and saturated paste extract, before moving on to highlight how soil test calibration can be used as a basis for fertiliser recommendations. A case study in sustainable soil test interpretation is also included.

Acknowledgements

It is my pleasure to express my sincere gratitude and appreciation to all the authors and co-authors of each chapter in *Achieving sustainable turfgrass management*. These contributors are the leading experts in their discipline, and this textbook represents the first concentrated effort and focus on sustainability in turfgrass science applicable for academia, government, industry, and the practitioner. Also, I am grateful to Burleigh Dodds Science Publishing (Cambridge, United Kingdom) for their diligence and professionalism throughout the entire process, and for their willingness to include turfgrass science for their sustainable series in agricultural science. I am also grateful for the unconditional encouragement and support of this project from my wife, Anjanette, and my children, Victoria, Julia, Jack, Paul and Quinn.

Michael Fidanza, Ph.D.
Professor of Plant and Soil Science
Pennsylvania State University
Reading, Pennsylvania
USA

Part 1

Physiology, breeding and cultivation

Chapter 1

Advances in understanding turfgrass physiology

David Jespersen, University of Georgia, USA; Benjamin Wherley, Texas A&M University, USA; and Michelle DaCosta, University of Massachusetts Amherst, USA

1 Introduction

From a highly managed and well-manicured golf course putting green to a low-input right-of-way, turfgrasses must cope with a wide range of environmental extremes, which frequently fall outside of optimal growing conditions. These challenging environments can alter a plant's basic physiology, negatively impacting growth and development, and eventually result in the death of turf areas. Unfortunately, there is reason to believe that in the coming future environmental conditions in which turfgrasses are grown will only become more challenging. Global climate change is estimated to raise air temperatures by more than 2°C, and beyond elevated temperatures, and there is expected to be more extreme weather patterns of every sort, from floods to droughts (IPCC, 2021). Further compounding these issues is an increasing need to reduce inputs applied to turf areas for both environmental and economic reasons. This is particularly true of water resources, which are increasingly limited and in competition for demand from other users, but is also true of fertilizers, pesticides, and even the labor needed to implement cultural management practices.

http://dx.doi.org/10.19103/AS.2022.0110.01

The development of more sustainable turfgrasses relies on a better understanding of plant physiological response to stressful conditions, which includes determining how plants are damaged from specific stresses and the mechanisms employed to protect and prolong survival under stress. In turn, this information will help inform best turfgrass management practices, identify traits to integrate into advanced cultivars, and ultimately leverage new technologies that allow for the continuing use of functional green spaces despite these environmental challenges. This chapter will examine four of the most detrimental abiotic stress that turfgrasses commonly encounter (i.e. drought, heat, freezing, and salinity stresses) and which are expected to increase in duration or severity in the future.

2 Drought stress

2.1 Overview and significance of drought stress in turf

Turfgrasses are a major component of urban green spaces throughout the world. In addition to esthetic, recreational, and functional benefits, turfgrasses offer many important ecosystem services which benefit humans and the environment (Beard and Green, 1994; Chang et al., 2021). However, in regions where rainfall is not adequate to meet plant demand, turfgrasses may require supplemental irrigation to maintain acceptable quality and function. Although the exact acreage of irrigated turfgrass in the USA is not known, total turfgrass acreage has been estimated to be 30 million acres, making it three times larger than any other single irrigated crop (Milesi et al., 2005). In more arid parts of the USA, more than 50% of residential water usage is sometimes devoted to landscape irrigation (Degen, 2007; Haley et al., 2007; Mayer et al., 1999). Based on conservative estimates by Cabrera et al. (2013), total water use by golf courses combined with landscapes in Texas represented 47% of the total use within the urban/municipal water sector and 13% of the total annual demand by all activities, positioning urban irrigation as the state's third largest water user behind agriculture and other urban uses.

In recent years, water conservation has become a major priority for many municipalities, driven largely by urban population growth combined with periodic droughts (WHO, 2020). While urban landscapes have traditionally been comprised predominantly of turfgrass, municipalities and water purveyors are increasingly providing homeowners monetary incentives to remove turfgrass and replace it with alternative 'water-efficient' landscapes utilizing native or locally adapted alternative plant species (Addink, 2005; Pincetl et al., 2019). Another water conservation strategy employed by water purveyors is mandatory irrigation restrictions, which may limit residential landscape irrigation to a given day of the week or, under more extreme measures, restrict irrigation once every 2 weeks or less (Hejl et al., 2021). In addition to residential irrigation restrictions, larger turf sites such as golf courses or athletic fields are also commonly forced to deal with reduced water allocations imposed by water purveyors as a means of meeting water conservation goals (GCSAA, 2015). When this occurs, turf managers must prioritize water to critical areas such as greens and tees or impose deficit irrigation on larger acreage areas such as fairways in order to meet or remain in compliance with water allocation goals. For turfgrasses to remain a sustainable component of future urban green spaces, they must be capable of providing functional surfaces with significantly less water in the form of rain or irrigation than in the past and, at times, surviving extended periods of prolonged drought.

Considering the importance and widespread use of turfgrasses in the modern landscape and growing need for water conservation, turfgrass drought resistance has become a trait of major importance within the industry. As a result, a substantial amount of plant physiological research and breeding efforts are currently being devoted to this issue. The following is intended to provide a review of the current understanding of turfgrass drought resistance while highlighting recent advances which have improved our knowledge in this area.

2.2 Historical perspectives, recent advances and future direction with drought stress research in turf

Drought develops when root zone soil moisture becomes insufficient to support turfgrass water demand. Therefore, drought is most common during periods of high evapotranspiration and low rainfall or inadequate irrigation. The ability of turfgrass plants to resist drought, termed drought resistance, is a broad classification involving one or multiple drought-resistance mechanisms, including avoidance, tolerance, or escape.

2.3 Drought avoidance in turfgrasses

Drought avoidance is considered to be the most widely employed drought-resistance mechanism among turfgrasses (Levitt, 1980). Avoidance is characterized by the ability of a turfgrass plant to maintain favorable internal water potential and normal physiological function during drought, primarily by minimizing transpiration and/or maximizing water uptake. Turfgrass species noted for possessing superior drought-avoidance characteristics include tall fescue [*Lolium arundinaceum* (Schreb.) SJ Darbysh.], bermudagrass [*Cynodon dactylon* (L.) Pers.], buffalograss [*Bouteloua dactyloides* (Nutt.) J.T. Columbus], and bahiagrass (*Paspalum notatum* Flueggé) (Fry and Huang, 2004).

Considering nearly 98% of water uptake by turfgrass roots is lost to transpiration through leaves, species and cultivars capable of minimizing transpiration are often those best adapted to avoid drought. Although some discrepancy exists, the preponderance of scientific literature indicates that drought-avoidant turfgrasses minimize water losses primarily through shoot morphological adaptations, including compact growth habits (Kim, 1983; Kneebone and Pepper, 1982; Shearman, 1986), reduced vertical extension rates and/or horizontal orientation of leaves (Bowman and Macaulay, 1991; Kim, 1983; Kim and Beard, 1988; Shearman, 1986), and increased leaf epicuticular wax contents (Beard and Sifers, 1997; Fu and Huang, 2004). In addition to shoot attributes that lead to reduced evapotranspiration (i.e. soil evaporation plus transpiration) losses, drought-avoidant turfgrasses may also increase water uptake during drought through deep rooting (Beard, 1973, 1985; Huang et al., 1997; Huang and Fu, 2001; Stienke et al., 2011; Wherley et al., 2014, 2015), increased root biomass/density (Carrow, 1996; Huang, 1999; Huang et al., 1997; Jespersen and Schwartz, 2018), and root hydraulic conductance/ hydraulic lift (Blizzard and Boyer, 1980; Passioura, 1988; Richards and Passiouia, 1981). Enhanced water uptake through increased rooting depth/density is one of the most desirable drought-avoidance mechanisms because it enables the turf to draw upon a greater volume of soil water, delaying the need for irrigation. It is also particularly useful where sporadic rainfall is received during the summer months, aiding the plant in maintaining favorable water relations between rainfall events.

Stomatal regulation is one of the first and most sensitive physiological responses to occur during the onset of soil drying in turfgrasses. The ability of a turfgrass to regulate stomata in response to soil water stress is a primary mechanism by which it is able to maintain favorable tissue water relations during drought (Kramer and Boyer, 1995). Decreased stomatal conductance, while beneficial in terms of minimizing evapotranspiration losses, also inhibits CO_2 influx, thereby depressing photosynthesis. Although it can be influenced by many factors, stomatal closure in response to drought is thought to be conferred primarily through accumulation of the plant hormone abscisic acid (ABA) in leaves (Davies and Zhang, 1991; Zeevaart and Creelman, 1988). As soil dries, ABA production increases in roots and is transported to shoots, where it rapidly accumulates and results in guard cell turgor loss and stomatal closure (Bradford and Hsiao, 1982). In addition to reducing transpirational water loss through stomata, ABA has also been thought to increase hydraulic conductivity of roots in some species, allowing water uptake to temporarily meet evaporative demand, and thereby sustaining stomatal conductance during the onset of drought (Steudle and Peterson, 1998; Weatherley, 1982). Although the exact mechanism of this response has not been well understood, Fiscus (1981) showed that ABA application to roots caused a transient release of solutes to the xylem, which was thought to be responsible for the initial increase in hydraulic conductivity. More recently, Carminati and Javaux (2020) used a soil-plant hydraulic model to demonstrate that soil hydraulic conductivity loss, rather than xylem vulnerability, may be the main driver of stomatal closure during drought.

Although somewhat limited in turfgrass, studies examining the relationship between ABA signaling and drought resistance of turfgrasses show this relationship to be somewhat complex. During drought, lower amounts of ABA are found to accumulate in drought-resistant compared to drought-sensitive turfgrass cultivars. This was the case for Kentucky bluegrass (*Poa pratensis* L.) cultivars (Wang et al., 2003a) and was also observed when comparing ABA levels between drought-resistant velvet bentgrass (*Agrostis canina* L.) and drought-sensitive creeping (*Agrostis stolonifera* L.) and colonial bentgrass species (*Agrostis capillaris* L.) (DaCosta and Huang, 2007). In general, decreased ABA accumulation in leaves during drought has been considered advantageous during short-term drought, as it could allow for continued gas exchange, photosynthesis, and energy production. Additionally, stomata of drought-resistant Kentucky bluegrass cultivars have been shown to be more sensitive to increasing ABA levels than those of drought-sensitive cultivars (Wang et al., 2003a). Unfortunately, despite advances within other crop systems, fewer published data are available regarding endogenous ABA responses on stomatal conductivity and root hydraulic conductance in warm-season turfgrasses during drought, making this an area of needed research moving forward.

Exogenous ABA applications have also proven favorable for improving drought responses in both warm- and cool-season turf species. Application of ABA to leaves of warm-season turfgrasses carpetgrass [*Axonopus compressus* (Sw.) Beauv.] and 'TifEagle' hybrid bermudagrass [*C. dactylon* (L.) Pers. × *C. transvaalensis* Burtt-Davy] prior to drought increased the relative water content during drought stress. Furthermore, exogenous ABA promoted increased antioxidant enzyme activities of superoxide dismutase (SOD), catalase (CAT), and peroxidase (POD) during drought stress, reducing the accumulation of H_2O_2 and associated cell membrane damage (Lu et al, 2009; Mohammadi et al., 2017). Exogenous application of ABA also enhanced physiological activities and improved drought tolerance in cool-season turfgrasses tall fescue (Jiang and Huang, 2002), Kentucky bluegrass (Wang et al., 2003a), and creeping bentgrass (McCann and Huang, 2008). Despite the observed benefits of exogenous ABA application on turf health and survival under drought stress in these small pot studies, its widespread application in protecting turf from drought stress injury in field situations remains limited, mostly due to its high cost of production (Lu et al., 2003). There have been trace amounts of ABA detected within some commercial seaweed extracts (Stirk et al., 2014); however, elucidation of the direct benefits of ABA from such applications remains difficult due to the complex nature of these extracts.

While traditionally the majority of turfgrass drought studies have examined stomatal response to soil drying, a more recently observed physiological strategy for avoiding drought stress is through limitation of transpiration at high atmospheric vapor pressure deficits (VPDs). In plants demonstrating this response, transpiration rates increase linearly with increasing VPD until a break point VPD is reached, above which there is little or no further increase in transpiration rate. While the limited transpiration trait appears to be rare in commercial crop varieties, it is thought to improve water-use efficiency by limiting gas exchange during high VPD conditions and conserving soil water during early phases of drought in order to sustain plant activity later into drought (Sinclair et al., 2017). This drought-avoidance mechanism has been associated with several crop species (Bunce, 1981; Gholipoor et al., 2013; Sadok and Sinclair, 2009) and more recently in several turfgrasses (Sermons et al., 2012; Sinclair et al., 2007; Wherley and Sinclair, 2009). Sermons et al. (2012) found that the transpiration of tall fescue increased with increasing VPD up to 1.45 kPa but plateaued thereafter, suggesting enhanced stomatal control of gas exchange at higher VPD. Their work helped to explain the earlier observation of Sinclair et al. (2007), who found that tall fescue shoot growth declined with increasing VPD beyond 1.2 kPa at a constant temperature of 21°C. More recently, Wherley and Sinclair (2009) examined transpiration responses of a collection of C_3 and C_4 turfgrasses to increasing VPD. The authors showed that transpiration rates of C_4 grasses increased linearly in response to increasing VPD across the range of 0.8–3.0 kPa. In contrast, transpiration rates of C3 increased sharply with increasing VPD across the range of low VPDs but became constrained at higher VPDs (>1.35 kPa), with limited transpiration rates at higher VPD most evident in creeping bentgrass and perennial ryegrass (*L. perenne* L.). The authors concluded that assuming restricted transpiration rates reflected limitations on leaf CO_2 uptake, and the commonly observed decline in the growth of C_3 (and success of C_4) grasses at elevated temperature may partially result from VPD sensitivity. Although none of the authors of the aforementioned studies conclusively determined the physiological basis of the limited transpiration response, Sinclair et al. (2017) has suggested that it appears to be a consequence of limited hydraulic conductivity in the plant, possibly related to aquaporin expression and abundance. Given that turfgrasses are grown for esthetics and function rather than yield, the limited transpiration trait may be important and thus warrant closer examination by turfgrass breeders and physiologists in the future.

2.4 Drought tolerance in turfgrasses

Drought tolerance is defined as the ability of a plant to maintain physiological function when little to no water is available to the plant. As such, drought-tolerant plants experience but also survive drought. Turfgrass species noted for possessing superior drought-tolerance characteristics include fine fescue, zoysiagrass (*Zoysia* spp.), and buffalograss (Fry and Huang, 2004). Turfgrass drought tolerance is generally understood to be conferred through one or more mechanisms, including osmotic adjustment (OA) (DaCosta and Huang, 2006; Jiang and Huang, 2001c; Katuwal et al., 2021; Qian and Fry, 1997), cell wall elasticity (White et al., 2001), antioxidant defense systems (DaCosta and Huang, 2007; Katuwal et al., 2020b; Lu et al., 2009; Zhang and Schmidt, 2000), accumulation of stress-induced proteins (Jiang and Huang, 2002), and/or dormancy (Hejl et al., 2016; Hong and Bremer 2021; Steinke et al., 2010, 2011). Steinke et al. (2011) accurately point out that unfortunately, there is decreasing perceived importance of dormancy, as modern-day homeowners and homeowners associations visually demand fully green turfgrass stands (Steinke et al., 2011).

One of the more well-studied drought-tolerance mechanisms in the recent years has been that of OA (sometimes referred to as osmoregulation), defined as the active accumulation of solutes in plant tissues in response to dehydration (Jones and Turner, 1980). Osmoregulation lowers the osmotic potential of plant cells, thereby allowing them to maintain turgor pressure under drought stress. The role of OA in the maintenance of cell turgor and enhanced soil water uptake during drought was first demonstrated in pea (*Pisum sativum* L.) and maize (*Zea mays*) root tips (Greacen and Oh, 1972; Sharp and Davies, 1979). Despite the growing body of research surrounding the potential importance of OA as a defense mechanism, it has been suggested that the accumulation of solutes in drought-tolerant plants during soil dry down could simply be due to growth slowing faster than assimilation, resulting in solute accumulation. Leaf expansion is often more sensitive than photosynthesis to decreased leaf water potential, and this could explain the accumulation of photosynthetic products in leaves, resulting in solute accumulation and associated OA (Munns, 1988). A range of compatible solutes have been shown to accumulate during OA in plants, including amino acids (proline), ammonium compounds (glycine betaine), sugars (sucrose, glucose, and fructose), polyols (mannitol), inorganic ions (potassium, magnesium, and chloride), and organic acids (malate) (Zhang et al., 1999).

Liu et al. (2015) examined genetic variations in OA among a collection of bentgrass (*Agrostis* spp.) genotypes with differential drought tolerance and determined the major organic osmoregulants contributing to OA and their relation to drought tolerance. The authors found soluble sugars to be the predominant contributors to OA, followed by glycine betaine and proline, with polyamines contributing the least to bentgrass drought tolerance.

A considerable amount of research has examined the role of OA in a number of crops as well as warm- and cool-season turfgrasses. Studies generally show that drought-avoidant plants maintain higher leaf water potential, have lower dehydration tolerance, and exhibit less OA than their drought-tolerant counterparts in response to drought (Ludlow, 1987; Qian and Fry, 1997). For example, Qian and Fry (1997) evaluated OA in a collection of turfgrass species in the greenhouse and found OA was greatest in buffalograss and zoysiagrass followed by bermudagrass, with the lowest OA shown for tall fescue. The authors also noted a positive relationship between the extent of OA and recuperative capacity following severe drought. Within cool-season species, velvet bentgrass showed 50–60% greater OA than creeping bentgrass in response to soil dry down, contributing to its prolonged higher turf quality and leaf relative water content as drought stress ensued (DaCosta and Huang, 2006).

More recently, Katuwal et al. (2020b) evaluated drought-tolerance and drought-avoidance mechanisms of six cultivars and experimental lines of bermudagrass and found that 'TifTuf' showed up to 200% greater OA compared to poorer performing cultivars. In addition to OA, antioxidant enzyme activity, including SOD, ascorbate peroxidase, POD, CAT, and glutathione reductase, appeared to be positively correlated with drought-tolerance differences among the bermudagrasses tested. The authors emphasized the importance of cultivars possessing combined drought-avoidance and drought-tolerance mechanisms in future development of elite germplasm for drought. This appears to be particularly important in light of the variety of constructed urban soils in which turfgrasses are increasingly expected to grow. Many residential landscape soils are highly compacted during construction and/or contain very shallow topsoil, which has been shown to dramatically affect the ability of even the most drought-resistant turfgrasses to survive and recover from prolonged drought (Steinke et al., 2010, 2011). For this reason, many have suggested that topsoil quality may be of greater importance than cultivar selection when it comes to the development of water-efficient urban landscapes.

3 Heat stress

3.1 Overview and significance of heat stress in turf

Heat stress is caused by plants being exposed to temperatures above their optimal range, with bouts of heat stress ranging from short but intense periods of high heat (as might be seen during a heat wave and sometimes known as 'heat shock') to prolonged conditions of elevated temperatures which are frequently described as chronic heat stress (as occurs over a summers' growing season). Plants exposed to these conditions experience reduced growth, premature senescence and leaf yellowing, and eventual loss of turf areas due to plant death. Outside of average temperatures, other factors can ultimately influence the extent of damage at these supra-optimal temperatures. Elevated root zone temperatures are more damaging to plants than elevated air temperatures due to the sensitivity of root tissues, resulting in reduced root growth and eventual root death before aboveground tissues (Beard, 1973; Xu and Huang, 2000, 2001). Other environmental conditions such as high humidity will affect a plant's ability to undergo transpirational cooling, ultimately leading to more severe levels of heat stress (Bonos and Murphy, 1999). In other grass species, high nighttime temperatures, which are expected to increase disproportionately during climate change, have been found to be more detrimental than high daytime temperatures (Li et al., 2011; Peng et al., 2004).

Much of the research on heat stress in turfgrasses has focused on cool-season species utilizing C_3 photosynthetic pathways, and the content of the chapter will reflect this. There have been few studies that have looked at heat stress in warm-season grasses utilizing the C_4 photosynthetic pathway, with C_4 plants being documented to withstand temperatures greater than 50°C (Wallner et al., 1982). Whereas the optimal temperatures for the growth of cool-season grasses has been documented to be between 15.5°C and 25°C, an optimal temperature range of 27–35°C has been reported for warm-season grasses (DiPaola and Beard, 1992; Miller, 1960; Youngner, 1961). For both cool-season and warm-season grasses, optimal root zone temperatures are significantly lower than shoot temperatures, by at least 5°C. Even among turfgrass species within a given photosynthetic pathway (i.e. C_3 vs C_4), there are differences in heat tolerance and associated temperatures, resulting in tissue damage. For example, among C_3 grasses, tall fescue is known to be relatively heat tolerant, whereas species like perennial ryegrass and annual bluegrass (*Poa annua* L.) are considerably less heat tolerant (Fry and Huang, 2004). While heat stress comes in several different forms and severities, there are several physiological effects that are commonly seen in turfgrasses that ultimately lead to reduced performance, growth, and plant death.

3.2 Historical perspectives and recent advances with heat stress in turf

3.2.1 Changes in carbon metabolism

For a turfgrass plant to grow and maintain its cellular integrity, the plant needs to maintain a proper balance in its carbon metabolism. In general, this means the plant should be fixing more carbon dioxide into sugars via photosynthesis than the loss of sugars it is consuming during respiration. Typically, as temperatures increase, there is a reduction in photosynthesis and an increase in respiration, disrupting plant growth and metabolism (Hull, 1992; Liu and Huang, 2000a, 2001). These patterns of decreased photosynthesis and increased respiration, leading to reduced carbohydrates, have been found in a number of turfgrass species, including creeping bentgrass, Kentucky bluegrass, and perennial ryegrass (Duff and Beard, 1974; Huang and Gao, 2000; Watschke et al., 1973; Xu and Huang, 2003). Declines in photosynthesis have been associated with stomatal closure, decreased chlorophyll pigment contents, disruptions in electron transport for light harvesting (as measured by changes in chlorophyll fluorescence), as well as disruption to the carbon fixation process of the dark reactions of photosynthesis (Berry and Bjorkman, 1980; Huang et al., 2014; Zhao et al., 2014). Liu and Huang (2008) found that the activation of Rubisco (ribulose-1,5-bisphosphate carboxylase oxygenase), the enzyme responsible for the initial fixation of carbon, was a major factor responsible for photosynthetic decline (which corresponds to results found in other species; Salvucci and Crafts-Brandner, 2004).

In contrast to decreasing photosynthesis, there are increasing rates of respiration associated with more heat-sensitive plants during heat stress (DiPaola and Beard, 1992; Huang et al., 2012; Liu and Huang, 2000a). While respiration is an essential metabolic process that is needed for the growth and repair of plant cells, excessive respiration can lead to a depletion of carbohydrate stores and subsequent carbon starvation by the plant. These declines in carbohydrate reserves have been implicated as a factor in root dieback during heat stress (Sweeney et al., 2001; Xu and Huang, 2001). Further shifting the balance from carbohydrate accumulation to carbohydrate consumption is the increase in photorespiration occurring in C_3 grasses under high temperatures and high sunlight conditions (Rachmilevitch et al., 2008; Watschke et al., 1973).

One of the major visual symptoms of heat stress related to the impairment of photosynthesis is a yellowing of leaf tissues due to declines in chlorophyll content. Recent studies have started to elucidate the underling regulation of chlorophyll metabolism, which may ultimately help in the development of cultivars with 'stay-green' traits that maintain their green color better during heat stress. Zhang et al. (2016) identified a number of genes associated with chlorophyll metabolism and the degradation of pigments. These genes included *LpNYC*, *NOL*, *SGR*, and *PPH*. Additional work has identified *PPH* as a critical gene responsible for the degradation of chlorophyll under heat stress-induced chlorosis in perennial ryegrass and creeping bentgrass (Jespersen et al., 2016; Zhang et al., 2016). The activity of enzymes associated with other degradation pathways such as chlorophyll PODs and chlorophyllase have also been associated with the decline in chlorophyll under heat stress in bentgrasses (Rossi et al., 2017). Ongoing work is furthering the understanding between these genes and regulatory signals such as reactive oxygen species and hormones that play roles in signaling during stress-induced senescence (Xu et al., 2018; Zhang et al., 2019).

With the advent of rapid pulse amplitude modulation fluorometers, additional insights into changes in photochemistry, in particular during light harvesting, have been explored in response to heat stress. The use of chlorophyll fluorescence has been widely used to estimate the maximum potential photochemical efficiency of Photosystem II (Fv/Fm), which is a sensitive measure of heat stress-induced damage (Huang et al., 2001; Jespersen et al., 2013; Jiang and Huang, 2001b; Watkins et al., 2007; Xu and Huang, 2007). These Fv/Fm measurements are based on a normalized ratio of minimum and maximum chlorophyll fluorescence values to estimate the quantum efficiency of photosynthetic reaction centers when in their open state. More recently, fluorescence kinetics have been used to estimate the efficiency of electron transport during light harvesting. Cui et al. (2006) found that tall fescue lines with greater heat tolerance were better able to maintain systems involved in light harvesting and electron transport of photosynthesis. Similar results have been found in other studies, further implicating heat stress in damaging photosynthetic machinery of the thylakoids, resulting in photoinhibition and production of reactive oxygen species (Bi et al., 2016; Hu et al., 2018 Katuwal et al., 2020a).

3.2.2 Protein metabolism

Beyond disruptions to photosynthesis, turfgrass plants experience widespread changes in protein metabolism, including decreased protein synthesis and increased protein degradation during heat stress (He et al., 2005; Wehner and Watschke, 1984). As proteins become damaged and denatured during heat stress, they experience a loss of function and form toxic aggregates, ultimately resulting in damage to metabolic processes. Defensive proteins induced by heat stress include heat shock proteins, which act as chaperone proteins that help refold or recycle damaged proteins (Vierling, 1991). A number of studies have found that plants that induce heat shock proteins to a greater degree are generally more heat tolerant (DiMascio et al., 1994; He and Huang, 2007; Jespersen and Huang, 2015; Xu et al., 2011; Xu and Huang, 2012). Heat shock proteins may play important roles in not only the development of stress tolerance but also how plants activate defense in the future (Hu et al., 2015). Recently, a tall fescue heat shock protein gene FaHSP17.8-CII was demonstrated to play important roles in regulating responses to heat stress and stress memory signaling (Bi et al., 2021). This potential 'heat stress memory gene' had greater expression during subsequent exposure to repeated heat stress, as well as the ability to maintain higher expression levels for nearly 4 days, and was associated with reduced oxidative damage and maintenance of photosynthetic activity. Small heat shock proteins have shown promise as potential molecular markers, but constitutive expression in transgenic plants has led to issues in normal growth and stress signaling (Jespersen et al., 2018; Sun et al., 2016). Another family of proteins responsible for regulating heat stress responses are heat shock factors (HSFs). HSFs are transcription factors that are induced by heat stress to regulate responses to elevated temperatures, as well as other stresses. As with heat shock proteins, the induction of HSFs has been strongly associated with improved thermotolerance in turfgrasses (Liu et al., 2019; Wang et al., 2016). Perennial ryegrass HSF (LpHSFC1b) was significantly upregulated in response to heat stress and is believed to be one of the core regulators of abiotic stress responses, as it was also seen to be upregulated in response to drought and cold stress (Sun et al., 2020). The expression of LpHSFC1b in *Arabidopsis thaliana* (L.) results in greater heat stress tolerance but no greater tolerance to drought or cold stress. The accumulation of heat-induced defense proteins is an important aspect of acquired heat tolerance, but more work is needed to understand this large family of proteins, their production, and how they influence signaling and stress memory.

Metabolic dysfunction, including damage to proteins and membranes, that occurs during heat stress can ultimately have the effect of producing reactive oxygen species (Liu and Huang, 2000b; Jiang and Huang, 2001a). These damaging molecules are often produced when the reactions of photosynthesis or respiration are disturbed, resulting in a slowdown in electron transport, and reactive oxygen compounds proceed to cause oxidative damage to other cellular components, such as proteins, membranes, and nucleic acids. Under heat stress, the production of reactive oxygen species increases as does the associated oxidative damage as estimated by lipid peroxidation (Xu et al., 2006). It has been widely demonstrated in a number of turfgrass species that greater levels of antioxidant metabolism are associated with reduced heat-stress-induced oxidative damage. Tolerant cultivars in creeping bentgrass, Kentucky bluegrass, and perennial ryegrass have demonstrated improved expression or activity of a number of antioxidant enzymes, including super oxide dismutase, CAT, and POD (Du et al., 2009; Jiang and Huang, 2001a; Li et al., 2014; Wang et al., 2003b; Xu et al., 2006).

3.2.3 Fatty acid metabolism

Membranes are key cellular structures that are essential for regulating the movement of molecules across organelles and between cells. Membranes are essential for electron transport chains and the development of potential gradients needed for energy reactions in the cell. During heat stress, damage to membranes occurs either through oxidative damage from reactive oxygen species or through changes in physical properties, as elevated temperatures cause increased fluidity of membranes (by weakening the interactions between adjacent fatty acid chains). These disturbances can have wide-ranging effects on cells, from reduced photosynthetic efficiency to complete cell death. Effects of increased membrane damage are characterized by decreased membrane stability, as measured through increased electrolyte leakage and lipid peroxidation (Liu and Huang, 2000b). Understanding the composition of fatty acids comprising membranes, as well as the changes in fatty acid metabolism in response to heat stress, is an important step for improving heat tolerance and preventing the underlying collapse of cellular physiology. Decline in fatty acid biosynthesis occurs during heat stress and has been associated with declines in key fatty acids as well as expression of genes associated with fatty acid metabolism (Hu et al., 2020). When exposed to heat stress, cool-season turfgrasses such as tall fescue, Kentucky bluegrass, and hybrid bluegrass (*Poa pratensis* L. × *Poa arachnifera* Torr.) decrease their double bond index (DBI, an estimation of fatty acid saturation), resulting in greater levels of saturation of fatty acid chains (Su et al., 2009). Additionally, heat-tolerant hybrid bluegrass had a greater proportion of digalactosyldiacylglycerol (DGDG, a highly abundant glycolipid), higher ratio of DGDG:monogalactosyldiacylglycerol, and greater saturation of fatty acids than heat-sensitive tall fescue when exposed to supra-optimal growth temperatures. Changes in membrane composition in response to heat stress have also been found in tall fescue (Hu et al., 2020). Hu et al. (2018) found that the DBI and the level of unsaturation of fatty acids increased in a heat-tolerant tall fescue in comparison to a heat-sensitive line, which contrasts many other studies. In hard fescue (*Festuca trachyphylla*), improved heat tolerance was associated with a greater content of ethyl sterols, as well as a shift to more saturated fatty acids (and a reduction in unsaturation levels) (Wang et al., 2017). The increased saturation levels increased the interactions among adjacent fatty acid molecules, giving stability to membranes and counteracting the increased fluidity caused by elevated temperatures.

3.3 Emerging issues and future direction with heat stress research in turf

The ultimate goal of heat stress research in turfgrasses is to better understand the physiological responses of plants to supra-optimal temperatures and to develop cultural management practices to improve plant resilience against heat stress or breed new cultivars better able to genetically withstand heat stress. Many different approaches are being used to accomplish these goals, from biostimulants to genetic mapping. Understanding the regulation and induction of heat tolerance is a critical area of research. While not covered in this chapter, several other excellent reviews have discussed the role of hormone regulation (DaCosta and Huang, 2013; Huang et al., 2014), and there is ongoing research evaluating the role of exogenous compounds (i.e. biostimulant products) on heat tolerance. Researchers have also looked at alternative species, not widely used for turf, that may be better able to withstand elevated temperatures (Breuillin-Sessoms and Watkins, 2020). Another aspect that has received some attention but likely needs further research is the combination of heat stress with other stresses. In many environments, heat stress will co-occur with abiotic stresses, including drought or salinity, or biotic stress such as disease which takes advantage of a weakened plant (Cross et al., 2013; Jiang and Huang, 2000; Thompson et al., 2017). Much of the research has shown that while heat is a major factor, these combined factors compound damage to plants. Developing a deeper understanding of how heat stress damages plants, the mechanisms responsible for improved tolerance and how these responses are regulated is essential to advancement of turfgrasses capable of withstanding the increasing temperatures of a changing climate.

4 Freezing stress

4.1 Overview and significance of freezing stress in turf

Turfgrasses growing in northern and transition zone climates are often exposed to sub-freezing temperatures during the winter months. Most cool-season and some warm-season grasses exhibit mechanisms of cold temperature acclimation that promote freezing tolerance, although successful overwintering of these species can be impacted by many factors, including plant genetics (inter- and intraspecific differences in cold acclimation capacity and freezing tolerance), site-specific conditions (e.g. light availability, soil drainage, and traffic), and weather conditions in autumn and winter months. More generally, winterkill and winter injury are non-specific terms used to describe the damage based on the failure of turfgrasses to recover in spring months. However, the specific causes of winterkill can be attributed to different factors, including direct low-temperature kill, crown hydration, desiccation, anoxia, and low-temperature-active fungi. In many cases, it is often difficult to ascribe winterkill to a single factor, as several of these winter-related stressors may occur simultaneously over the same location in winter months. Moreover, due to variable winter weather conditions, there may be significant year-to-year variation determining the potential for mild-to-severe winterkill damage. These uncertainties in predicting and diagnosing causal factors are some of the primary constraints in developing turfgrass management recommendations to minimize winterkill.

Among the various winter-related stressors, damage from direct low-temperature kill due to freezing temperatures is considered a common component contributing to winterkill across many turfgrass species. Therefore, the focus of this review will emphasize the advances related to our current understanding of the development and maintenance of freezing tolerance of turfgrasses. Although there is an overlap in physiological mechanisms and impact of other secondary stresses on freezing tolerance, such as anoxia and infection by snow mold fungi, we refer to additional resources on ice encasement and winter diseases in other recently published reviews (Matsumoto and Hsiang, 2016; Merewitz, 2021).

4.2 Freezing stress

Freezing injury is the result of stress associated with the formation of ice crystals in plant tissues. Depending on the specific localization, the presence of ice can cause mechanical stress (e.g. rupturing of cell membranes and organelles due to intracellular ice) and osmotic stress (e.g. dehydration of cells due to low water potential gradients caused by extracellular ice). Freezing injury of crowns and other meristematic regions within rhizomes and stolons prevent turfgrass recovery in spring, thus making these tissues of prime importance in the study of turfgrass freezing tolerance. Beard and Olien (1963) were among the first to report the differential sensitivity of freezing damage for the upper vs lower meristematic regions of the crown, demonstrating greater sensitivity of the lower crown region (responsible for the initiation of roots) compared to the upper region (responsible for leaf initiation) of annual bluegrass crowns. Much of the early understanding on the mechanisms of crown injury due to freezing was adapted from research on forage grasses and winter cereals. For example, the formation of ice was shown to first occur in the vascular transition zone and below the shoot apical meristem of crowns, which then inhibited the regeneration of new roots (Olien and Marchetti, 1976). These studies helped to explain why regrowth of leaves may often be observed following freezing events, followed by eventual death of the plants due to failure to regenerate roots. Given the importance of ice formation and localization as related to freezing injury, there is actually very limited information available in the turfgrass literature. In one of the only published studies in turfgrasses, Stier et al. (2003) utilized infrared thermal imaging to monitor ice initiation and propagation in whole plants of perennial ryegrass and supina bluegrass (*Poa supina* Schrad.). The authors found that freezing initiated in roots, which spread to crown and finally leaf tissues. There were not large differences in freezing times among species or between non-acclimated vs acclimated plants; however, some limitations in the instrumentation may have contributed to the lack of separation among treatments. Given the general absence of research in this area, there is still a lack of understanding of how turfgrasses actually freeze in the natural environment, as well as how temperature fluctuations or changes in soil moisture around the crown may impact the potential for freezing injury. More detailed information on potential differences in meristem freezing behavior among species and the interaction with external temperature and moisture conditions will be important to help inform management practices to minimize freezing injury. Thus, with the availability of advanced and novel imaging methods to quantify changes in the presence of water and ice in plant tissues, particularly non-destructive methods (Kovaleski et al., 2019; Livingston et al., 2018; Villouta et al., 2021), the study of ice formation is an area of turfgrass low-temperature physiology research deserving future attention.

4.3 Historical perspectives and recent advances of freezing stress research in turf

4.3.1 Turfgrass freezing tolerance rankings

The relative rankings of freezing tolerance among turfgrass species were previously reported by Beard (1973) and DiPaola and Beard (1992), based on the survival of grasses in field trials and artificial freeze tests (Beard, 1966; Gusta et al., 1980; Hodgson, 1964; Rogler, 1943). A ranking of freezing tolerance inclusive of updated research on warm-season grasses was also presented by Bertrand et al. (2013). Among warm-season turfgrasses, zoysiagrass and buffalograss generally rank the highest in freezing tolerance, common and hybrid bermudagrass rank intermediate, and St. Augustine grass [*Stenotaphrum secundatum* (Walter) Kuntze] and centipedegrass [*Erimochloa ophiuroides* (Munro) Hack.] rank the lowest in freezing tolerance. While warm-season turfgrasses are most susceptible to freezing stress, some cool-season turfgrass species including annual bluegrass and perennial ryegrass can also experience significant freezing injury in northern climates. Gusta et al. (1980) evaluated cool-season grasses under controlled freezing conditions and determined that perennial ryegrass had the highest killing temperature (–5°C to –15°C), Kentucky bluegrass had intermediate killing temperatures (–21°C to –30°C), and creeping bentgrass was the most freezing tolerant, surviving temperatures below –30°C.

One of the primary challenges in identifying intraspecific differences in freezing tolerance in field environments is the lack of consistent winter environmental conditions, which vary from year to year. This results in changes in the prevalence of specific winter stresses and therefore impacts the ranking of individual traits related to winter survival. Consequently, many studies reporting genetic variability in freezing tolerance have used controlled freeze tests to establish differences in freezing tolerance among cultivars and ecotypes of cool-season (Dionne et al., 2010; Ebdon et al., 2002) and warm-season grasses (Anderson et al., 2002; Milla-Lewis et al., 2013; Patton and Reicher, 2007; Qian et al., 2001; Rukavina et al., 2007; Wu et al., 2017). These experiments involve allowing plants to undergo natural cold acclimation in field environments or simulated cold acclimation in controlled environment growth chambers followed by exposure to a range of freezing temperatures in a programmable chamber, followed by whole-plant survival assessments to determine the lethal temperature resulting in 50% kill. There are certain limitations to controlled freeze tests, which may not necessarily predict the same level of freezing tolerance achieved in a natural environment due to differences in cold acclimation conditions in controlled vs field conditions. Moreover, variability in freeze-test protocols (e.g. rate of temperature change and length of time at different freeze-test temperatures) can also impact the extent of survival or freezing tolerance separation among genotypes or cultivars, particularly when differences in freezing tolerance are small (Anderson et al., 2003). However, there seems to be general agreement in freezing tolerance rankings between controlled environment and field evaluations, as previously demonstrated in studies using zoysiagrass (Hinton et al., 2012; Patton and Reicher, 2007), bermudagrass (Dunne et al., 2019), buffalograss (Qian et al., 2001), and perennial ryegrass (Hulke et al., 2008). Therefore, laboratory-based freezing tests have been a valuable tool in identifying intraspecific differences in freezing tolerance and provide an opportunity to screen a larger pool of germplasm in advance of multi-year field trials.

4.3.2 Mechanisms of freezing tolerance

Extensive reviews of turfgrass responses to freezing stress were previously published by Bertrand et al. (2013) and Huang et al. (2014), and therefore the current review will serve to highlight early physiology research as well as research advances since those publications. A vast majority of investigations on the freezing tolerance of turfgrasses have focused on the physiological and biochemical changes associated with cold acclimation capacity. During cold acclimation, plants undergo significant reorganization from the molecular to whole-plant levels that aid in the protection of cells under extended freezing conditions, which include a suite of modifications, including stabilization of membranes, production of protective metabolites that minimize oxidative stress and promote stability under osmotic stress, as well as production of metabolites that directly mitigate ice crystallization. Some of the earliest low-temperature physiology research in turfgrasses reported changes in soluble protein concentration, non-structural carbohydrates, fatty acid content, and crown moisture content in response to decreasing temperatures (Beard, 1973; Davis and Gilbert, 1970; DiPaola, 1979; Rogers et al., 1975). Since then, intraspecific differences in turfgrass freezing tolerance during cold acclimation have been attributed to various categories of metabolites, including higher content of unsaturated fatty acids (Cyril et al., 2002; Hoffman et al., 2010; Samala et al., 1998; Zhang et al., 2009), higher accumulation of antifreeze proteins (Zhang et al., 2010), higher accumulation of dehydrin proteins (Patton et al., 2007a; Zhang et al., 2011, 2017), higher activity of antioxidant enzymes or reactive oxygen species scavenging capacity (Xu et al., 2015; Zhang et al., 2006), modified hormone profiles such as induction of ABA (Li et al., 2018; Xu et al., 2015; Zhang et al., 2017a), and accumulation of specific carbohydrate fractions such as sucrose and fructans (Ball et al., 2002; Fry et al., 1993; Dionne et al., 2010; Giolo et al., 2013; Hoffman et al., 2010; Patton et al., 2007b; Zhang et al., 2006).

More recently, the use of different 'omics technologies' has also contributed to a more comprehensive understanding of the integrated plant physiological changes associated with the regulation of freezing tolerance. For example, transcriptomic and proteomic approaches have helped to identify genes and diverse metabolic pathways involved in cold acclimation of species such as zoysiagrass (Brown et al., 2020; Long et al., 2020; Wei et al., 2015; Xuan et al., 2013), bermudagrass (Melmaiee et al., 2015; Zhu et al., 2015), perennial ryegrass (Bernstein et al., 2019; Zhang et al., 2017b), and velvet bentgrass (Espevig et al., 2012). Much of the results from these studies support earlier research on the importance of common processes such as signal transduction and stress defense, energy and carbohydrate metabolism, and protein and amino acid metabolism in preparing plants for freezing temperatures. These studies also highlight certain dynamics of cold acclimation that may often be overlooked, namely that cold acclimation involves differential regulation of hundreds to thousands of genes and metabolites (both induction and suppression) rather than simply the accumulation of a few protective compounds in response to cold acclimation. For example, Zhang et al. (2017b) reported that 1132 genes were upregulated in perennial ryegrass in response to 7 days of cold acclimation at 4°C, whereas 754 genes were downregulated during this time period. Similarly, Long et al. (2020) found that 2944 genes were upregulated in zoysiagrass following 72 h of 4°C exposure, while 2907 were downregulated. These studies have primarily documented general responses of plants to low temperatures using a single cultivar or genotype, inferring the association of these changes between non-acclimated and cold acclimated plants with the development of freezing tolerance.

Additional information can also be gained by comparing expression profiles between different cultivars or genotypes with distinct levels of freezing tolerance. For instance, using two perennial ryegrass genotypes with varying freezing tolerance, Bernstein (2021) found that 897 genes were differentially expressed between the two genotypes in response to prolonged cold acclimation at −2°C. These data provide a framework for additional studies to identify specific genes most highly correlated with freezing tolerance, which can then be used as markers to screen larger turfgrass populations. As such, there are some important factors to consider when comparing or interpreting results across different studies, even within the same species. These include considerations of genotype or cultivar selection, differences in cold acclimation methodologies (e.g. acclimation temperature and duration of exposure, light intensity, and photoperiod), transcript or protein quantification methods (e.g. RNAseq vs microarray analysis and two-dimensional gel electrophoresis vs shotgun proteomics approaches using isobaric labeling), and testing of different plant tissues (i.e. leaves, crowns, or roots). To this last point, the physiological changes occurring in leaves and crowns or other meristematic nodes during cold acclimation may be different, but most studies focus only on one of these tissue types. The leaves have important functions during cold acclimation, such as shifting carbohydrate and nitrogen metabolism to help support the energy requirements required for preparation for freezing. On the other hand, additional mechanisms might be needed for overwintering crowns to survive extended freezing conditions as well as additional stresses such as anoxia winter diseases. To date, published studies on zoysiagrass include analyses of leaves (Long et al., 2020; Wei et al., 2015), stolons (Xuan et al., 2013), and crown nodes (Brown et al., 2020). In turf-type perennial ryegrass, transcriptome studies have focused on leaves (Zhang et al., 2017b) and crowns (Bernstein et al., 2019). Taken together, it is evident that there are still many unknowns associated with the molecular mechanisms of freezing tolerance, therefore providing future opportunities to address challenges such as how to integrate whole-plant responses with freezing tolerance mechanisms, identifying the critical components for plants to acquire and maintain freezing tolerance, and, more broadly, how freezing tolerance mechanisms are impacted by genotype by environment interactions.

4.4 Emerging issues and future direction with freezing stress research in turf

While past research has provided us with a good understanding of the common genes, regulatory pathways, and metabolites associated with cold acclimation and freezing tolerance in grasses, the translation of information from fundamental physiology research into breeding freezing-tolerant turfgrasses has thus far been limited. One contributing factor may be the amount of research in the area of freezing stress physiology, which has generally lagged behind the efforts focused on drought and heat stress resistance of turfgrasses. Nonetheless, increased research efforts in warm-season turfgrasses have recently focused on screening of commercially available cultivars and novel germplasm for enhanced winter survival, including freezing tolerance (Dunne et al., 2019; Gopinath et al., 2021; Moseley et al., 2021; Kimball et al., 2016; Wu et al., 2017; Xiang et al., 2020). The identification of quantitative trait loci (QTL) for freezing tolerance and related traits is currently an active area of research, which could enable the use of marker-assisted selection to increase the efficiency of incorporating freezing tolerance traits into new cultivars

(Brown et al., 2021; Holloway et al., 2018; Kimball et al., 2018). These efforts are particularly important for warm-season grasses such as zoysiagrass and bermudagrass that are widely used in the transition zone and are often exposed to freezing temperatures. Increased climate warming is already extending the use of warm-season grasses on golf courses and sports fields into more northern climates, providing new opportunities for selection of grasses with traits such as later fall dormancy and improved overwintering capacity (Roberts, 2015, 2019).

Freezing tolerance has not been considered a trait of high priority in cool-season turfgrass breeding programs, since most cool-season grasses generally exhibit high freezing tolerance. For a species such as annual bluegrass, for which winterkill is considered a significant stress, breeding efforts have been unsuccessful due to the genetic instability of the more desirable greens-type phenotype (Benson et al., 2021; La Mantia and Huff, 2011). While there is significantly more information relative to breeding and genetics in perennial ryegrass, these studies have largely been focused on perennial ryegrasses used as a forage crop (Goslee et al., 2017; Humphreys and Eagles, 1988; Humphreys, 1989; Wilkins, 1991). Using a global collection of 192 perennial ryegrass accessions, Yu et al. (2015) found that winter survival was related to the geographical origin of the accessions, with accessions from Northern Africa and Southern Europe exhibiting reduced winter survival compared to other accessions. In addition, they found a consistent association between winter survival and a C-repeat-binding factor (CBF) gene (*LpCBF1b*), along with other genes correlated with regrowth in spring months (e.g. late embryogenesis-abundant and antioxidant genes). Hulke et al. (2012) also reported one major QTL in a CBF gene (*LpCBFIIIc*) that was associated with freezing tolerance in turf-type perennial ryegrasses. There is evidence that recurrent selection cycles under controlled environment conditions can result in higher freezing tolerance and regrowth capacity of perennial ryegrass (Iraba et al., 2013), and additional research is needed to enhance the efficiency of breeding for freezing tolerance in this species.

One of the most pressing issues will be understanding how climatic changes in northern regions may alter the extent of winterkill in turfgrasses, due to factors such as direct low-temperature injury as well as changes in the prevalence of certain stresses such as ice encasement, crown hydration, and snow molds. Current trends for northern USA include increasing occurrence in temperature and rainfall extremes and 20–30% heavier precipitation, particularly in winter and spring months (Easterling et al., 2017). As a consequence, plants will be exposed to increases in freeze-thaw cycles and changes in snow and ice cover, potentially leading to premature deacclimation concurrent with losses in freezing tolerance. Differences in deacclimation sensitivity among turfgrasses have recently been reported, demonstrating that even short-term exposure to warming temperatures trigger molecular and biochemical changes that result in loss of freeze tolerance (Espevig et al., 2014; Hoffman, et al., 2014a,b). Deacclimation sensitivity has more recently been reported in perennial ryegrass, with simulated winter warming events triggering increased expression of genes associated with water transport and fructan metabolism and lower expression of genes associated with ice recrystallization inhibition (Bernstein et al., 2019). This suggests that increased winter temperature and moisture fluctuations may further predispose perennial ryegrass to winterkill. Compared to cool-season grasses, there is even less research on deacclimation sensitivity of warm-season turfgrasses, with investigations only reported in bermudagrass (Zhang et al., 2011) and St. Augustine grass (Kimball et al., 2017). Therefore, future turfgrass physiology research should focus additional efforts on understanding how changes in temperature and precipitation will impact the survivability of turfgrasses, which can then be used to better guide breeding efforts as well as winter injury mitigation practices.

5 Salt stress

5.1 Overview and significance of salt stress in turf

Maintaining a balanced concentration of ions in the cell is essential for a plant to properly function. In the presence of excessive salts, this balance is disturbed, damaging the plant. Salts in the form of soluble ions most commonly include Ca^{+2}, Mg^{+2}, Na^{+}, K^{+}, Cl^{-}, SO_4^{-2}, HCO_3^{-}, NO_3^{-}, and CO_3^{-2} and can result in salt-affected soils, with saline soils with high levels of sodium being the greatest concern (Carrow and Duncan, 1998). Elevated levels of salts can occur for several reasons. Some areas have naturally saline soils. Other areas by oceans may suffer from saltwater intrusion, resulting in brackish water. Areas with extremely high evaporation rates are also prone to the buildup of salts left behind from irrigation water. Additionally, recycled or reclaimed or poor-quality water for irrigation may contain high concentrations of salts and dissolved solids. Last, excessive salt may come from typical management practices such as the application of fertilizers or road salts used to deice roads in winter. While many different salts in high concentrations can ultimately damage a plant, sodium is the most widely researched in terms of salt stress. When plants are exposed to high levels of salt, there is a typical decrease in growth and leaf elongation, followed by leaf firing and eventual plant death (Guo et al., 2016; Xu et al., 2016).

It has been proposed that plants experience salt stress in two major ways (Munns and Tester, 2008). One is the osmotic stress where the accumulation of salts in the soils lowers the osmotic potential and therefore water potential of the soil, making it more difficult for plants to uptake water. Ultimately this osmotic stress will cause similar effects as drought, and a decrease in water status has been observed in turfgrasses (Hameed and Ashraf, 2008; Liu et al., 2012). The other form of salt stress plants experience is ionic stress. This is the toxic buildup of salts within a plant's cells. The accumulation of salts has many serious implications for a plant cell's physiological functioning. Salt can lead to the misfunction of proteins, particularly those that have ion cofactors and chiefly those requiring potassium for proper functioning. Additionally high salt levels may affect the transport of other solutes, such as essential mineral nutrients. One of the most common salts that accumulates during salt stress is sodium. As a positively charged cation, sodium can move though some of the same channels as other ions important for nutrition, such as potassium, resulting in cation competition that reduces the levels of essential elements. Interfering with potassium in the cell also affects photosynthetic pathways. As with many other abiotic stresses, as the plant's normal metabolic processes become disrupted there is an increased production of reactive oxygen species, leading to further damage in the cell (Hu et al., 2011; Puyang et al., 2015).

5.2 Historical perspectives and recent advances with salt stress research in turf

Among turfgrass species, there is a range of documented salt tolerance, as well as proposed mechanisms for salt tolerance. Many of these mechanisms involve regulating the balance of ions in the plant, by reducing salt uptake, sequestering ions, or extruding those salts. Additionally, maintaining potassium uptake to prevent a catastrophic shift in Na^+/K^+ ratio may be an important indicator of salt tolerance. This relationship of higher Na^+ accumulation and reduced K^+ content being associated with greater damage and sensitivity to saline conditions among genotypes has been documented in a number of species, including zoysiagrass, seashore paspalum (*Paspalum vaginatum* Sw.), and bermudagrass (Chen et al., 2009; Hu et al., 2012; Lee et al., 2004). Differences across species have also been observed in terms of how quickly and the degree to which these ions accumulate (Marcum and Murdoch, 1990).

5.2.1 Morphological adaptations

The morphology and development of specific structures in plants can play essential roles in salt tolerance, usually by assisting in the removal or sequestration of salt ions. One of the most widely studied morphological structures associated with salt tolerance are the salt glands of zoyisagrasses that extrude sodium ions from the leave (Maeda, 2019; Qian et al., 2000). Previous research has documented differences in salt gland density and ion excretion among zoysiagrass, with generally finer-textured zoysiagrasses having greater salt tolerance compared to coarse-textured species (Marcum et al., 1998). Yamamoto et al. (2016) found that *Zoysia pacifica* and *Z. matrella* had relatively greater secretion of ions compared to *Z. japonica*. Interestingly, they observed that salt gland density and total ions secreted varied among species, with salt gland density and secretion rates not being directly correlated (in contrast to earlier findings).

The use of energy-dispersive x-ray spectroscopy combined with scanning electron microscopy has allowed researchers to quantify ion contents within specific leaf structures. Unlike the ever-present salt glands in zoysiagrass, bermudagrass only develops salt glands when exposed to higher levels of salinity (Chavarria et al., 2020). Additional structures which have been identified via high-resolution microscopy are enlarged papillae on the surface of the seashore paspalum. These papillae appear to contain a greater concentration of sodium and may act as a kind of 'salt bladder' to sequester sodium ions, ultimately reducing the toxic effects of excessive sodium in the cytoplasm of leaf cells (Chavarria et al., 2020; Spiekerman and Devos, 2020).

Among a group of cool-season turfgrass species, including perennial ryegrass, Kentucky bluegrass, two fescues (*F. ovina* and *F. rubra*), and two bentgrass species (*Agrostis alba* and *Agrostis tenuis*), a range of avoidance (reducing sodium uptake) and tolerance (being able to withstand higher concentrations of accumulated sodium in tissues) was found among the tested species (Soliman et al., 2018). Generally, the ability to withstand ionic stress was the most important for overall performance, and there was a strong relationship between accumulation of proline and overall performance, with fescues and Kentucky bluegrass having both the best overall salt resistance and greatest proline levels. Proline can act as an osmolyte to help maintain plant water status, but many solutes that accumulate during stress (including proline) have also been implicated in reducing oxidative damage and stabilizing cellular constituents.

5.2.2 Sodium transporters

Controlling the movement of ions in the plant is a critical feature for salt tolerance to avoid the accumulation of salts in toxic concentrations. Ion transporters are integral membrane proteins that regulate the movement of ions throughout the plant. These transporters are a key component of salt tolerance as they are involved in the uptake of salts as well as their transport into (or out of) various tissues. The HTK family (high-affinity potassium transporter) of proteins is involved in the movement of sodium and potassium ions and has been implicated in salinity tolerance in a number of species. An HTK1 gene (SvHTK1) was upregulated in the halophytic turfgrass saltwater couch grass [*Sporobolus virginicus* (L.) Kunth] under highly saline conditions and was demonstrated to have important roles in regulating the accumulation of sodium in shoot tissues (Kawakami et al., 2020). Other genes from halophytic turfgrasses have also been shown to help regulate the movement and accumulation of ions. These include ZmVP1, a vacuolar H+ pyrophophatase from *Z. matrella* that acts as a proton pump assisting in the transport of ions and which may play an important role in the movement of ions to survive saline conditions (Chen et al., 2015). Transgenic plants overexpressing the gene ZmVP1 have a greater accumulation of potassium as well as greater compartmentalization of Na in vacuoles. Bushman et al. (2020) identified that a number of genes associated with transmembrane transport of ions and other osmoprotectants were associated with improved salt tolerance in cool-season grasses Kentucky bluegrass and perennial ryegrass. Similar transcriptome results in *Z. macrostachya* found that ion homeostasis and osmoprotectants were important for salt tolerance (Wang et al., 2020). The accumulation of organic compounds such as osmolytes has been associated with improved salt tolerance in seashore paspalum (Lee et al., 2008). Although zoysiagrasses are best known for their salt glands which extrude salt from the leaf surface, a study by Sugiura and Takahashi (2021) proposed that *Z. matrella* plants may also accumulate sodium in specific leaves, which then senesce and die, acting to compartmentalize the ions as a secondary mechanism. The same study also demonstrated that *Z. matrella* will transport water across adjacent ramets, potentially helping to dilute ions, but will not transport sodium to adjacent ramets, potentially indicating the ability of the plant to compartmentalize ions, but further study is needed. This finding agrees with other studies in other species that indicate that once sodium leaves root tissues, it tends to move in one direction.

5.2.3 Hormone responses

Signal transduction is an essential aspect of plant responses to abiotic stresses. Phytohormones are important signaling molecules that can regulate defense responses, promote or reduce the growth of plant tissues, or affect stress-induced senescence. In a study on Kentucky bluegrass exposed to salt stress, Zhang et al. (2018) found that there was a significant increase in ABA, which is typically associated with increased stress levels, activation of defense pathways, and reductions in stomatal conductance and plant growth. Additionally, there were significant declines in the auxin indole-3-acetic acid (IAA) and cytokinins zeatin riboside (ZR) and isopentenyladenosine in response to salt stress, and while there were no significant differences in GA4, there was a trend for greater gibberellic acid levels to be associated with greater turf quality. These changes in hormone contents align with the pattern of significantly reduced growth in response to salt stress seen in other studies. In two creeping bentgrasses of contrasting salt tolerance, a significant increase in the content of many hormones [ABA, ZR, IAA, jasmonic acid (JA), salicylic acid (SA), and ethylene] was seen in response to salt stress (Krishnan and Merewitz, 2015). Compared to the sensitive line 'Penncross', the tolerant line 'Mariner' had significantly greater leaf tissue concentrations of ABA, ZR, and JA and SA, and of note, both JA and SA are associated with stress defense signaling. In perennial ryegrass, external application of the cytokinin 6-benzylamino purine was able to reduce salt-related leaf senescence via the activation of antioxidant metabolism and improved K^+/Na^+ ratios (Ma et al., 2016). Improved understanding of phytohormone responses, which are essential for the regulation of growth and activation of defense mechanisms, may help improve turfgrass performance under saline conditions either by selecting cultivars with improved physiology and signaling responses or through the use of exogenous treatments (i.e. biostimulant product applications by the turfgrass practitioner).

5.3 Emerging issues and future direction with salt stress research in turf

Fresh water has limited availability and is becoming costly for use in irrigating turf areas (Rodriguez Diaz et al., 2007; Stier et al., 2013; Strandberg et al., 2012). Therefore, future restrictions on landscape irrigation are likely to increase, leading to increasing the use of reclaimed or non-potable water sources for irrigation. As a result, turfgrass salinity tolerance will likely become even more important in the future. Furthermore, in this scenario of restricted irrigation, end users may be forced to use water with high salinity and, even then, in limited quantities. The combination of drought and salinity to turfgrasses may be particularly damaging, as the loss of soil moisture will result in increased salt concentrations and further exacerbate osmotic stresses. How to best utilize reclaimed water for turf irrigation and manage the buildup of salinity in the soil will be important for turfgrass managers in the future, particularly if utilizing irrigation for leaching salts is not practical. Another direction researchers have taken is to look at other halophytic grass species that may be alternative options for turfgrass areas. Several potential grasses that are less widely used such as mangrove grass [*Aeluropus lagopoides* (L.) Thwaites], saltwater couch grass, and fine fescues have shown promise for high levels of salt tolerance; however, additional research is needed to understand whether these species have the necessary traits to produce high-quality turfgrass (Friell and Watkins, 2021; Yamamoto et al., 2015; Zamin et al., 2019). The identification and development of improved grasses is likely an essential strategy for improving the stress tolerance of turf areas. As with many abiotic stresses, tolerance is likely to be a complex multi-genic trait. Understanding the underlying genetics of salinity tolerance will aid in the development of improved cultivars in the future.

6 Conclusion and future trends

Turfgrasses must perform under many challenging environmental conditions, of which many were not discussed in this chapter, and these challenges are likely to increase as a result of climate change, demand for higher quality turf esthetics and playing surfaces, and/or limitations imposed on resources dedicated to landscape areas. Insight into a turf plant's underlying physiology is essential to maintain these functional ground covers moving forward. This knowledge can inform the selection of improved cultivars with greater stress tolerance traits and the development of enhanced cultural management techniques such as precision irrigation or the application of exogenous compounds (i.e. phytohormones, plant growth regulators, or other biostimulant products). New technologies such as genetic engineering, remote sensing, and advanced computing to guide decision-making offer opportunities to meet expected challenges by creating more resilient turf areas that are better able to efficiently utilize resources. However, to reduce the inputs required for turfgrasses and meet sustainability goals in the future, it is essential to continue to develop our understanding of the physiology of how turfgrasses interact and respond to the environment and to cultural management practices.

7 Where to look for further information

The following resources provide a good overview of the subject:

- Fry, J. and Huang, B. (2004). *Applied Turfgrass Science and Physiology*. John Wiley and Sons, Hoboken, NJ.
- Huang, B., DaCosta, M. and Jiang, Y. (2014). Research advances in mechanisms of turfgrass tolerance to abiotic stresses: from physiology to molecular biology. *Critical Reviews in Plant Sciences* 33(2-3), 141-189. DOI: 10.1080/07352689.2014.870411.

Key research in this area is associated with the following organizations:

- Crop Science Society of America C5 Division (Crops.org).
- International Turfgrass Society (its.turfsociety.com).
- U.S. Golf Association (USGA.org).
- USDA-Specialty Crops Research Initiative (https://www.nifa.usda.gov/grants/programs/specialty-crop-research-initiative-scri).

8 References

Addink, S. (2005). Cash for grass-A cost effective method to conserve landscape water? Turfgrass Research Facility. University of California, Riverside, CA. Available at: https://turfgrass.ucr.edu/reports/topics/Cash-for-Grass.pdf.

Anderson, J., Taliaferro, C. and Martin, D. (2002). Freeze tolerance of bermudagrasses: vegetatively propagated cultivars intended for fairway and putting green use, and seed-propagated cultivars. *Crop Science* 42(3), 975–977.

Anderson, J. A., Taliaferro, C. M. and Martin, D. L. (2003). Longer exposure durations increase freeze damage to turf bermudagrasses. *Crop Science* 43(3), 973–977.

Ball, S., Qian, Y. L. and Stushnoff, C. (2002). Soluble carbohydrates in two buffalograss cultivars with contrasting freezing tolerance. *Journal of the American Society for Horticultural Science* 127(1), 45–49.

Beard, J. B. (1966). Winter injury. *The Golf Course Superintendent* 34, 24–33.

Beard, J. B. (1973). *Turfgrass: Science and Culture*. Prentice Hall, Englewood Cliffs, NJ.

Beard, J. B. (1985). An assessment of water use by turfgrasses. In: Gibeault, V. A. and Cockerham, S. T. (Eds.) *Turfgrass Water Conservation*. University of California Division of Agriculture and Natural Resources, Oakland, California, 45–60.

Beard, J. B. and Green, R. L. (1994). The role of turfgrasses in environmental protection and their benefits to humans. *Journal of Environmental Quality* 23(3), 452–460. DOI: 10.2134/jeq1994.00472425002300030007x.

Beard, J. B. and Olien, C. R. (1963). Low temperature injury in the lower portion of *Poa annua* L. Crowns 1. *Crop Science* 3(4), 362–363.

Beard, J. B. and Sifers, S. I. (1997). Genetic diversity in dehydration avoidance and drought resistance within the Cynodon and Zoysia species. *International Turfgrass Society Research Journal* 8, 603–610.

Benson, C. W., Mao, Q. and Huff, D. R. (2021). Global DNA methylation predicts epigenetic reprogramming and transgenerational plasticity in *Poa annua* L. *Crop Science* 61(5), 3011–3022.

Bernstein, R., DaCosta, M., Ebdon, J. S., Jung, G., Watkins, E. and Ma, L. J. (2019). Transcriptome changes associated with differential responses to mid-winter warming events in Lolium perenne. In ASA, CSSA and SSSA International Annual Meetings (2019). ASA-CSSA-SSSA.

Bernstein, R. P. (2021). The characterization of traits associated with freezing tolerance in perennial ryegrass. Doctoral Dissertations, 2273. https://doi.org/10.7275/24586932 Available at: https://scholarworks.umass.edu/dissertations_2/2273.

Berry, J. and Bjorkman, O. (1980). Photosynthetic response and adaptation to temperature in higher plants. *Annual Review of Plant Physiology* 31(1), 491–543.

Bertrand, A., Castonguay, Y., Azaiez, A. and Dionne, J. (2013). Low-temperature stress. In: Stier, J., Horgan, B. P. and Bonos, A. (Eds.) *Turfgrass: Biology, Use, and Management*. ASA, CSSA, and SSSA, Madison, WI, 279–318.

Bi, A., Fan, J., Hu, Z., Wang, G., Amombo, E., Fu, J. and Hu, T. (2016). Differential acclimation of enzymatic antioxidant metabolism and photosystem II photochemistry in tall fescue under drought and heat and the combined stresses. *Frontiers in Plant Science* 7, 453. DOI: 10.3389/fpls.2016.00453.

Bi, A., Wang, T., Wang, G., Zhang, L., Wassie, M., Amee, M., Xu, H., Hu, Z., Liu, A., Fu, J., Chen, L. and Hu, T. (2021). Stress memory gene FaHSP17.8-CII controls thermotolerance via remodeling PSII and ROS signaling in tall fescue. *Plant Physiology* 187(3), 1163–1176 DOI: 10.1093/plphys/kiab205.

Blizzard, W. E. and Boyer, J. S. (1980). Comparative resistance of the soil and the plant to water transport. *Plant Physiology* 66(5), 809–814. DOI: 10.1104/pp.66.5.809.

Bonos, S. A. and Murphy, J. A. (1999). Growth responses and performance of Kentucky bluegrass under summer stress. *Crop Science* 39(3), 770–774. https://doi.org/10.2135/cropsci1999.0011183X003900030026x.

Bowman, D. C. and Macaulay, L. (1991). Comparative evapotranspiration rates of tall fescue cultivars. *Hortscience* 26(2), 122–123. Available at: https://journals.ashs.org/hortsci/view/journals/hortsci/26/2/article-p122.xml.

Bradford, K. J. and Hsiao, T. C. (1982). Physiological responses to moderate water stress. In: Lange, O. L., Nobel, P. S., Osmond, C. B. and Ziegler, H. (Eds) *Physiological Plant Ecology II. Encyclopedia of Plant Physiology (New Series) (vol. 12) / B.* Springer, Berlin, Heidelberg. DOI: 10.1007/978-3-642-68150-9_10.

Breuillin-Sessoms, F. and Watkins, E. (2020). Performance of multiple turfgrass species during prolonged heat stress and recovery in a controlled environment. *Crop Science* 60(6), 3344–3361. DOI: 10.1002/csc2.20262.

Brown, J. M., Yu, X., Holloway, H. M. P., DaCosta, M., Bernstein, R. P., Lu, J., Tuong, T. D., Patton, A. J., Dunne, J. C., Arellano, C., Livingston, D. P. and Milla-Lewis, S. R. (2020). Differences in proteome response to cold acclimation in Zoysia japonica cultivars with different levels of freeze tolerance. *Crop Science* 60(5), 2744–2756.

Brown, J. M., Yu, X., Holloway, H. M. P., Tuong, T. D., Schwartz, B. M., Patton, A. J., Arellano, C., Livingston, D. P. and Milla-Lewis, S. R. (2021). Identification of QTL associated with cold acclimation and freezing tolerance in *Zoysia japonica*. *Crop Science* 61(5), 3044–3055.

Bunce, J. A. (1981). Comparative responses of leaf conductance to humidity in single attached leaves. *Journal of Experimental Botany* 32(3), 629–634.

Bushman, B. S., Robbins, M. D., Robins, J. G., Thorsted, K., Harris, P. and Johnson, P. G. (2020). Response to salt stress imposed on cultivars of three turfgrass species: Poa pratensis, Lolium perenne, and Puccinellia distans. *Crop Science* 60(3), 1648–1659. DOI: 10.1002/csc2.20014.

Cabrera, R. I., Wagner, K. L. and Wherley, B. (2013). An evaluation of urban landscape water use in Texas. *Texas Water Journal* 4(2), 14–27. DOI: 10.21423/twj.v4i2.6992.

Carminati, A. and Javaux, M. (2020). Soil rather than xylem vulnerability controls stomatal response to drought. *Trends in Plant Science* 25(9), 868–880. Available at: https://doi.org/10.1016/j.tplants.2020.04.003.

Carrow, R. N. (1996). Drought avoidance characteristics of diverse tall fescue cultivars. *Crop Science* 36(2), 371–377. DOI: 10.2135/cropsci1996.0011183X003600020026x.

Carrow, R. N. and Duncan, R. R. (1998). *Salt-Affected Turfgrass Sites: Assessment and Management*. John Wiley & Sons. Hoboken, NJ.

Chang, B., Wherley, B., Aitkenhead-Peterson, J. A. and McInnes, K. J. (2021). Effects of urban residential landscape composition on surface runoff generation. *Science of the Total Environment* 783, 146977. DOI: 10.1016/j.scitotenv.2021.146977.

Chavarria, M. R., Wherley, B., Jessup, R. and Chandra, A. (2020). Leaf anatomical responses and chemical composition of warm-season turfgrasses to increasing salinity. *Current Plant Biology* 22, 100147. DOI: 10.1016/j.cpb.2020.100147.

Chen, J., Yan, J., Qian, Y., Jiang, Y., Zhang, T., Guo, H., Guo, A. and Liu, J. (2009). Growth responses and ion regulation of four warm season turfgrasses to long-term salinity stress. *Scientia Horticulturae* 122(4), 620–625. DOI: 10.1016/j.scienta.2009.07.004.

Chen, Y., Li, L., Zong, J., Chen, J., Guo, H., Guo, A. and Liu, J. (2015). Heterologous expression of the halophyte Zoysia Matrella H+-pyrophosphatase gene improved salt tolerance in Arabidopsis thaliana. *Plant Physiology and Biochemistry* 91, 49–55. DOI: 10.1016/j.plaphy.2015.04.004.

Cross, J. W., Bonos, S. A., Huang, B. and Meyer, W. A. (2013). Evaluation of heat and drought as components of summer stress on tall fescue genotypes. *Hortscience* 48(12), 1562–1567. DOI: 10.21273/HORTSCI.48.12.1562.

Cui, L., Li, J., Fan, Y., Xu, S. and Zhang, Z. (2006). High temperature effects on photosynthesis, PSII functionality and antioxidant activity of two Festuca arundinacea cultivars with different heat susceptibility. *Botanical Studies* 47, 9.

Cyril, J., Powell, G. L., Duncan, R. R. and Baird, W. V. (2002). Changes in membrane polar lipid fatty acids of seashore paspalum in response to low temperature exposure. *Crop Science* 42(6), 2031-2037.

DaCosta, M. and Huang, B. (2013). Heat-stress physiology and management. In: *Turfgrass: Biology, Use, and Management*. John Wiley & Sons, Ltd, 249-278. DOI: 10.2134/agronmonogr56.c7.

DaCosta, M. and Huang, B. (2006). Osmotic adjustment associated with variation in bentgrass tolerance to drought stress. *Journal of the American Society for Horticultural Science* 131(3), 338-344. DOI: 10.21273/JASHS.131.3.338.

DaCosta, M. and Huang, B. (2007). Changes in antioxidant enzyme activities and lipid peroxidation for bentgrass species in response to drought stress. *Journal of the American Society for Horticultural Science* 132(3), 319-326. DOI: 10.21273/JASHS.132.3.319.

Davies, W. J. and Zhang, J. (1991). Root signals and the regulation of growth and development of plants in drying soil. *Annual Review of Plant Physiology and Plant Molecular Biology* 42(1), 55-76.

Davis, D. L. and Gilbert, W. B. (1970). Winter hardiness and changes in soluble protein fractions of Bermudagrass 1. *Crop Science* 10(1), 7-9.

Degen, D. (2007). *Water Sustainability*: City of Kelowna Water Utility. Presentation for Okanagan Water Stewardship Council.

DiMascio, J. A., Sweeney, P. M., Danneberger, T. K. and Kamalay, J. C. (1994). Analysis of heat shock response in perennial ryegrass using maize heat shock protein clones. *Crop Science* 34(3), 798-804. DOI: 10.2135/cropsci1994.0011183X003400030035x.

Dionne, J., Rochefort, S., Huff, D. R., Desjardins, Y., Bertrand, A. and Castonguay, Y. (2010). Variability for freezing tolerance among 42 ecotypes of green-type annual bluegrass. *Crop Science* 50(1), 321-336.

DiPaola, J. M. (1979). *Fatty Acid Composition of Turfgrass Crowns and Its Relationship with Direct Low Temperature Hardiness*. Texas A&M University.

DiPaola, J. M. and Beard, J. B. (1992). Physiological effects of temperature stress. In: *Agronomy Monographs*. John Wiley & Sons, Ltd, 231-267. DOI: 10.2134/agronmonogr32.c7.

Du, H., Wang, Z. and Huang, B. (2009). Differential responses of warm-season and cool-season turfgrass species to heat stress associated with antioxidant enzyme activity. *Journal of the American Society for Horticultural Science* 134(4), 417-422. DOI: 10.21273/JASHS.134.4.417.

Duff, D. T. and Beard, J. B. (1974). Supraoptimal temperature effects upon *Agrostis palustris*. Part II. Influence on carbohydrate levels, photosynthetic rate, and respiration rate. *Physiologia Plantarum* 32(1), 18-22. DOI: 10.1111/j.1399-3054.1974.tb03720.x.

Dunne, J. C., Tuong, T. D., Livingston, D. P., Reynolds, W. C. and Milla-Lewis, S. R. (2019). Field and laboratory evaluation of Bermudagrass germplasm for cold hardiness and freezing tolerance. *Crop Science* 59(1), 392-399.

Easterling, D. R., Arnold, J., Knutson, T., Kunkel, K., LeGrande, A., Leung, L. R., Vose, R., Waliser, D. and Wehner, M. (2017). Precipitation change in the United States. In: Climate Science Special Report: A Sustained Assessment Activity of the U.S. Global

Change Research Program [Wuebbles, D. J., Fahey, D. W., Hibbard, K. A., Dokken, D. J., Stewart, B. C. and Maycock, T. K. (eds.)]. *A Sustained Assessment Activity of the U.S.* Global Change Research Program. U.S. Global Change Research Program, Washington, DC, 301–335.

Ebdon, J. S., Gagne, R. A. and Manley, R. C. (2002). Comparative cold tolerance in diverse turf quality genotypes of perennial ryegrass. *Hortscience* 37(5), 826–830.

Espevig, T., Höglind, M. and Aamlid, T. S. (2014). Dehardening resistance of six turfgrasses used on golf greens. *Environmental and Experimental Botany* 106, 182–188. https://doi.org/10.1016/j.envexpbot.2014.02.006.

Espevig, T., Xu, C., Aamlid, T. S., DaCosta, M. and Huang, B. (2012). Proteomic responses during cold acclimation in association with freezing tolerance of velvet bentgrass. *Journal of the American Society for Horticultural Science* 137(6), 391–399.

Fiscus, E. L. (1981). Effects of abscisic acid on the hydraulic conductance of and the total ion transport through phaseolus root systems. *Plant Physiology* 68(1), 169–174. DOI: 10.1104/pp.68.1.169.

Friell, J. and Watkins, E. (2021). Review of cool-season turfgrasses for salt-affected roadsides in cold climates. *Crop Science* 61(5), 2893–2915. DOI: 10.1002/csc2.20326.

Fry, J. and Huang, B. (2004). *Applied Turfgrass Science and Physiology*. John Wiley and Sons, Hoboken, NJ.

Fry, J. D., Lang, N. S., Clifton, R. G. P. and Maier, F. P. (1993). Freezing tolerance and carbohydrate content of low-temperature-acclimated and nonacclimated centipedegrass. *Crop Science* 33(5), 1051–1055.

Fu, J. and Huang, B. (2004). Leaf characteristics associated with drought resistance in tall fescue cultivars. *Acta Horticulturae* 661(661), 233–239. DOI: 10.17660/ActaHortic.2004.661.29.

Gholipoor, M., Choudhary, S., Sinclair, T. R., Messina, C. D. and Cooper, M. (2013). Transpiration response of maize hybrids to atmospheric vapour pressure deficit. *Journal of Agronomy and Crop Science* 199(3), 155–160. DOI: 10.1111/jac.12010.

Giolo, M., Macolino, S., Barolo, E. and Rimi, F. (2013). Stolons reserves and spring green-up of seeded Bermudagrass cultivars in a transition zone environment. *Hortscience* 48(6), 780–784.

Golf Course Superintendents Association of America (2015). Water use and conservation practices on U.S. golf courses. phase II, volume I. Available at: https://www.gcsaa.org/docs/default-source/Environment/phase-2-water-use-survey-full-report.pdf?sfvrsn=2b39123e_4.

Gopinath, L., Moss, J. Q. and Wu, Y. (2021). Quantifying freeze tolerance of hybrid bermudagrasses adapted for golf course putting greens. *Hortscience* 56(4), 478–480.

Goslee, S. C., Gonet, J. M. and Skinner, R. H. (2017). Freeze tolerance of perennial ryegrass and implications for future species distribution. *Crop Science* 57(5), 2875–2880.

Greacen, E. L. and Oh, J. S. (1972). Physics of growth. *Nature: New Biology* 235(53), 24–25.

Guo, H., Wang, Y., Li, D., Chen, J., Zong, J., Wang, Z., Chen, X. and Liu, J. (2016). Growth response and ion regulation of seashore paspalum accessions to increasing salinity. *Environmental and Experimental Botany* 131, 137–145. DOI: 10.1016/j.envexpbot.2016.07.003.

Gusta, L. V., Butler, J. D., Rajashekar, C. and Burke, M. J. (1980). Freezing resistance of perennial turfgrasses. *HortScience* 15(4), 494–496.

Haley,M. B., Dukes,M. D. and Miller, G. L. (2007). Residential irrigation water use in Central Florida. *Journal of Irrigation and Drainage Engineering* 133(5), 427-434.

Hameed, M. and Ashraf, M. (2008). Physiological and biochemical adaptations of *Cynodon dactylon* (L.) Pers. from the Salt Range (Pakistan) to salinity stress. *Flora - Morphology, Distribution, Functional Ecology of Plants* 203(8), 683-694. DOI: 10.1016/j.flora.2007.11.005.

He, Y. and Huang, B. (2007). Protein changes during heat stress in three Kentucky bluegrass cultivars differing in heat tolerance. *Crop Science* 47(6), 2513-2520. DOI: 10.2135/cropsci2006.12.0821.

He, Y., Liu, X. and Huang, B. (2005). Changes in protein content, protease activity, and amino acid content associated with heat injury in creeping bentgrass. *Journal of the American Society for Horticultural Science* 130(6), 842-847. DOI: 10.21273/JASHS.130.6.842.

Hejl, R. W., Wherley, B. G. and Fontanier, C. H. (2021). Long-term performance of warm-season turfgrass species under municipal irrigation frequency restrictions. *Hortscience* 56(10), 1221-1225. DOI: 10.21273/HORTSCI15978-21.

Hejl, R. W., Wherley, B. G., White, R. H., Thomas, J. C. and Fontanier, C. H. (2016). Deficit irrigation and simulated traffic on 'Tifway' Bermudagrass summer performance and autumn recovery. *Crop Science* 56(2), 809-817. DOI: 10.2135/cropsci2015.03.0197.

Hinton, J. D., Livingston, D. P., Miller, G. L., Peacock, C. H. and Tuong, T. (2012). Freeze tolerance of nine zoysiagrass cultivars using natural cold acclimation and freeze chambers. *HortScience* 47(1), 112-115.

Hodgson, H. J. (1964). Performance of turfgrass varieties in the subarctic. *Agronomy Abstracts*, 56, 101.

Hoffman, L., DaCosta, M. and Ebdon, J. S. (2014a). Examination of cold deacclimation sensitivity of annual bluegrass and creeping bentgrass. *Crop Science* 54(1), 413-420.

Hoffman, L., DaCosta, M., Bertrand, A., Castonguay, Y. and Ebdon, J. S. (2014b). Comparative assessment of metabolic responses to cold acclimation and deacclimation in annual bluegrass and creeping bentgrass. *Environmental and Experimental Botany* 106, 197-206.

Hoffman, L., DaCosta, M., Ebdon, J. S. and Watkins, E. (2010). Physiological changes during cold acclimation of perennial ryegrass accessions differing in freeze tolerance. *Crop Science* 50(3), 1037-1047.

Holloway, H. M. P., Yu, X., Dunne, J. C., Schwartz, B. M., Patton, A. J., Arellano, C. and Milla-Lewis, S. R. (2018). A SNP-based high-density linkage map of zoysiagrass (Zoysia japonica Steud.) and its use for the identification of QTL associated with winter hardiness. *Molecular Breeding* 38(1), 10.

Hong, M. and Bremer, D. J. (2021). Minimum water requirements of Japanese lawngrass for survival during prolonged drought. *Crop Science* 61(5), 2978-2988 DOI: 10.1002/csc2.20404.

Hu, T., Li, H. Y., Zhang, X. Z., Luo, H. J. and Fu, J. M. (2011). Toxic effect of NaCl on ion metabolism, antioxidative enzymes and gene expression of perennial ryegrass. *Ecotoxicology and Environmental Safety* 74(7), 2050-2056. DOI: 10.1016/j.ecoenv.2011.07.013.

Hu, L., Bi, A., Hu, Z., Amombo, E., Li, H. and Fu, J. (2018). Antioxidant metabolism, photosystem II, and fatty acid composition of two tall fescue genotypes With different heat tolerance Under high temperature stress. *Frontiers in Plant Science* 9, 1242. DOI: 10.3389/fpls.2018.01242.

Hu, L., Huang, Z., Liu, S. and Fu, J. (2012). Growth response and gene expression in antioxidant-related enzymes in two Bermudagrass genotypes differing in salt tolerance. *Journal of the American Society for Horticultural Science* 137(3), 134–143. DOI: 10.21273/JASHS.137.3.134.

Hu, T., Liu, S. Q., Amombo, E. and Fu, J. M. (2015). Stress memory induced rearrangements of HSP transcription, photosystem II photochemistry and metabolism of tall fescue (Festuca arundinacea Schreb.) in response to high-temperature stress. *Frontiers in Plant Science* 6, 403. DOI: 10.3389/fpls.2015.00403.

Hu, T., Sun, X. Y., Zhao, Z. J., Amombo, E. and Fu, J. M. (2020). High temperature damage to fatty acids and carbohydrate metabolism in tall fescue by coupling deep transcriptome and metabolome analysis. *Ecotoxicology and Environmental Safety* 203, 110943. DOI: 10.1016/j.ecoenv.2020.110943.

Huang, B. (1999). Water relations and root activities of Buchloe dactyloides and Zoysia japonica in response to localized soil drying. *Plant and Soil* 208(2), 179–186. DOI: 10.1023/A:1004472314631.

Huang, B. and Gao, H. (2000). Growth and carbohydrate metabolism of creeping bentgrass cultivars in response to increasing temperatures. *Crop Science* 40(4), 1115–1120. DOI: 10.2135/cropsci2000.4041115x.

Huang, B., DaCosta, M. and Jiang, Y. (2014). Research advances in mechanisms of turfgrass tolerance to abiotic stresses: from physiology to molecular biology. *Critical Reviews in Plant Sciences* 33(2–3), 141–189. DOI: 10.1080/07352689.2014.870411.

Huang, B., Duncan, R. R. and Carrow, R. N. (1997). Drought-resistance mechanisms of seven warm-season turfgrasses under surface soil drying: II. Root aspects. *Crop Science* 37(6), 1863–1869. DOI: 10.2135/cropsci1997.0011183X003700060033x.

Huang, B. and Fu, J. (2001). Growth and physiological responses of tall fescue to surface soil drying. *International Turfgrass Society Research Journal* 9, 291–296.

Huang, B., Liu, X. and Xu, Q. (2001). Supraoptimal soil temperatures induced oxidative stress in leaves of creeping bentgrass cultivars differing in heat tolerance. *Crop Science* 41(2), 430–435. DOI: 10.2135/cropsci2001.412430x.

Huang, B., Rachmilevitch, S. and Xu, J. (2012). Root carbon and protein metabolism associated with heat tolerance. *Journal of Experimental Botany* 63(9), 3455–3465. DOI: 10.1093/jxb/ers003.

Hulke, B. S., Bushman, B. S., Watkins, E. and Ehlke, N. J. (2012). Association of freezing tolerance to LpCBFIIIb and LpCBFIIIc gene polymorphism in perennial ryegrass accessions. *Crop Science* 52(5), 2023–2029.

Hulke, B. S., Watkins, E., Wyse, D. L. and Ehlke, N. J. (2008). Freezing tolerance of selected perennial ryegrass (Lolium perenne L.) accessions and its association with field winterhardiness and turf traits. *Euphytica* 163(1), 131–141.

Hull, R. J. (1992). Energy relations and carbohydrate partitioning in turfgrasses. In: *Agronomy Monographs*. John Wiley & Sons, Ltd, 175–205. DOI: 10.2134/agronmonogr32.c5.

Humphreys, M. O. (1989). Assessment of perennial ryegrass (*Lolium perenne* L.) for breeding. II. Components of winter hardiness. *Euphytica* 41(1–2), 99–106.

Humphreys, M. O. and Eagles, C. F. (1988). Assessment of perennial ryegrass (*Lolium perenne* L.) for breeding. I Freezing tolerance. *Euphytica* 38(1), 75–84.

IPCC (2021). 2021: the physical science basis.contribution of working group I to the sixth assessment report of the intergovernmental panel on climate change. *Climate Change*. Cambridge University Press, Cambridge, UK.

Iraba, A., Castonguay, Y., Bertrand, A., Floyd, D. J., Cloutier, J. and Belzile, F. (2013). Characterization of populations of turf-type perennial ryegrass recurrently selected for superior freezing tolerance. *Crop Science* 53(5), 2225–2238.

Jespersen, D. and Huang, B. (2015). Proteins associated with heat-induced leaf senescence in creeping bentgrass as affected by foliar application of nitrogen, cytokinins, and an ethylene inhibitor. *Proteomics* 15(4), 798-812. DOI: 10.1002/pmic.201400393.

Jespersen, D. and Schwartz, B. (2018). Drought avoidance traits in a collection of zoysiagrasses. *HortScience* 53(11), 1579–1585. DOI: 10.21273/HORTSCI13427-18.

Jespersen, D., Ma, X., Bonos, S. A., Belanger, F. C., Raymer, P. and Huang, B. (2018). Association of SSR and candidate gene markers with genetic variations in summer heat and drought performance for creeping bentgrass. *Crop Science* 58(6), 2644-2656. DOI: 10.2135/cropsci2018.05.0299.

Jespersen, D., Meyer, W. and Huang, B. (2013). Physiological traits and genetic variations associated with drought and heat tolerance in creeping bentgrass. *ITSRJ* 12, 459–464.

Jespersen, D., Zhang, J. and Huang, B. (2016). Chlorophyll loss associated with heat-induced senescence in bentgrass. *Plant Science* 249, 1–12. DOI: 10.1016/j.plantsci.2016.04.016.

Jiang, Y. and Huang, B. (2000). Effects of drought or heat stress alone and in combination on Kentucky bluegrass. *Crop Science* 40(5), 1358–1362. DOI: 10.2135/cropsci2000.4051358x.

Jiang, Y. and Huang, B. (2001a). Drought and heat stress injury to two cool-season turfgrasses in relation to antioxidant metabolism and lipid peroxidation. *Crop Science* 41(2), 436–442. DOI: 10.2135/cropsci2001.412436x.

Jiang, Y. and Huang, B. (2001b). Physiological responses to heat stress alone or in combination with drought: a comparison between tall fescue and perennial ryegrass. *Hortscience* 36(4), 682–686. DOI: 10.21273/HORTSCI.36.4.682.

Jiang, Y. and Huang, B. (2001c). Osmotic adjustment and root growth associated with drought preconditioning-enhanced heat tolerance in Kentucky bluegrass. *Crop Science* 41(4), 1168–1173.

Jiang, Y. and Huang, B. (2002). Protein alterations in tall fescue in response to drought stress and abscisic acid. *Crop Science* 42(1), 202–207. DOI: 10.2135/cropsci2002.2020.

Jones, M. M. and Turner, N. C. (1980). Osmotic adjustment in expanding and fully expanded leaves of sunflower in response to water deficits. *Functional Plant Biology* 7(2), 181–192.

Katuwal, K. B., Rowe, S. and Jespersen, D. (2020a). The use of 5-aminolevulinic acid to reduce heat stress related damages in tall fescue. *Crop Science* 61(5), 3206–3218. DOI: 10.1002/csc2.20294.

Katuwal, K. B., Schwartz, B. and Jespersen, D. (2020b). Desiccation avoidance and drought tolerance strategies in bermudagrasses. *Environmental and Experimental Botany* 171. DOI: 10.1016/j.envexpbot.2019.103947.

Katuwal, K. B., Tishchenko, V. and Jespersen, D. (2021). Assessing drought resistance in seashore paspalum genotypes using leaf gas exchange, osmotic adjustment, and rooting characteristics. *Crop Science* 61(3), 2121–2134. DOI: 10.1002/csc2.20420.

Kawakami, Y., Imran, S., Katsuhara, M. and Tada, Y. (2020). Na+ transporter SvHKT1;1 from a halophytic turf grass is specifically upregulated by high Na+ concentration and regulates shoot Na+ concentration. *International Journal of Molecular Sciences* 21(17), 6100. DOI: 10.3390/ijms21176100.

Kim, K. S. (1983). Comparative evapotranspiration rates of thirteen turfgrasses grown under both non-limiting soil moisture and progressive water stress conditions. MS thesis, Texas A&M University, College Station, TX.

Kim, K. S. and Beard, J. B. (1988). Comparative turfgrass evapotranspiration rates and associated plant morphological characteristics. *Crop Science* 28(2), 328–331.

Kimball, J. A., Isleib, T. G., Reynolds, W. C., Zuleta, M. C. and Milla-Lewis, S. R. (2016). Combining ability for winter survival and turf quality traits in St. Augustinegrass. *Hortscience* 51(7), 810–815.

Kimball, J. A., Tuong, T. D., Arellano, C., Livingston, D. P. and Milla-Lewis, S. R. (2017). Assessing freeze-tolerance in St. Augustinegrass: temperature response and evaluation methods. *Euphytica* 213(5), 110. https://doi.org/10.1007/s10681-017-1899-z.

Kimball, J. A., Tuong, T. D., Carbajal, E. and Milla-Lewis, S. R. (2018). Identification of QTL for cold tolerance-related traits in St. Augustinegrass. *Molecular Breeding* 38, 67.

Kneebone, W. R. and Pepper, I. L. (1982). Consumptive water use by sub-irrigated turfgrasses under desert conditions 1. *Agronomy Journal* 74(3), 419–423. DOI: 10.2134/agronj1982.00021962007400030005x.

Kovaleski, A. P., Londo, J. P. and Finkelstein, K. D. (2019). X-ray phase contrast imaging of Vitis spp. buds shows freezing pattern and correlation between volume and cold hardiness. *Scientific Reports* 9(1), 14949.

Kramer, P. J. and Boyer, J. S. (1995). *Water Relations of Plants and Soils.*. Academic Press, San Diego, CA.

Krishnan, S. and Merewitz, E. B. (2015). Phytohormone responses and cell viability during salinity stress in two creeping bentgrass cultivars differing in salt tolerance. *Journal of the American Society for Horticultural Science* 140(4), 346–355. DOI: 10.21273/JASHS.140.4.346.

La Mantia, J. M. and Huff, D. R. (2011). Instability of the greens-type phenotype in Poa annua L. *Crop Science* 51(4), 1784–1792.

Lee, G., Carrow, R. N., Duncan, R. R., Eiteman, M. A. and Rieger, M. W. (2008). Synthesis of organic osmolytes and salt tolerance mechanisms in *Paspalum vaginatum*. *Environmental and Experimental Botany* 63(1–3), 19–27. DOI: 10.1016/j.envexpbot.2007.10.009.

Lee, G., Duncan, R. R. and Carrow, R. N. (2004). Salinity tolerance of seashore paspalum ecotypes: shoot growth responses and criteria. *HortScience* 39(5), 1138–1142. DOI: 10.21273/HORTSCI.39.5.1138.

Levitt, J. (1980). *Responses of Plants to Environmental Stresses. II. Water, Radiation, Salt, and Other Stresses*. Academic Press, New York.

Li, F., Zhan, D., Xu, L., Han, L. and Zhang, X. (2014). Antioxidant and hormone responses to heat stress in two Kentucky bluegrass cultivars contrasting in heat tolerance. *Journal of the American Society for Horticultural Science* 139(5), 587–596. DOI: 10.21273/JASHS.139.5.587.

Li, H., Chen, Z., Hu, M., Wang, Z., Hua, H., Yin, C. and Zeng, H. (2011). Different effects of night versus day high temperature on rice quality and accumulation profiling of rice grain proteins during grain filling. *Plant Cell Reports* 30(9), 1641–1659. DOI: 10.1007/s00299-011-1074-2.

Li, S., Yang, Y., Zhang, Q., Liu, N., Xu, Q. and Hu, L. (2018). Differential physiological and metabolic response to low temperature in two zoysiagrass genotypes native to high and low latitude. *PLoS ONE* 13(6), e0198885.

Liu, N., Shen, Y. and Huang, B. (2015). Osmoregulants involved in osmotic adjustment for differential drought tolerance in different bentgrass genotypes. *Journal of the American Society for Horticultural Science* 140(6), 605-613. DOI: 10.21273/JASHS.140.6.605.

Liu, T., Liu, Z., Li, Z., Peng, Y., Zhang, X., Ma, X., Huang, L., Liu, W., Nie, G. and He, L. (2019). Regulation of heat shock factor pathways by γ-aminobutyric acid (GABA) Associated with thermotolerance of creeping bentgrass. *International Journal of Molecular Sciences* 20(19), 4713. DOI: 10.3390/ijms20194713.

Liu, X. and Huang, B. (2000a). Carbohydrate accumulation in relation to heat stress tolerance in two creeping bentgrass cultivars. *Journal of the American Society for Horticultural Science* 125(4), 442-447. DOI: 10.21273/JASHS.125.4.442.

Liu, X. and Huang, B. (2000b). Heat stress injury in relation to membrane lipid peroxidation in creeping bentgrass. *Crop Science* 40(2), 503-510. DOI: 10.2135/cropsci2000.402503x.

Liu, X. and Huang, B. (2001). Seasonal changes and cultivar difference in turf quality, photosynthesis, and respiration of creeping bentgrass. *Hortscience* 36(6), 1131-1135. DOI: 10.21273/HORTSCI.36.6.1131.

Liu, X. and Huang, B. (2008). Photosynthetic acclimation to high temperatures associated with heat tolerance in creeping bentgrass. *Journal of Plant Physiology* 165(18), 1947-1953. DOI: 10.1016/j.jplph.2008.05.001.

Liu, Y., Du, H., He, X., Huang, B. and Wang, Z. (2012). Identification of differentially expressed salt-responsive proteins in roots of two perennial grass species contrasting in salinity tolerance. *Journal of Plant Physiology* 169(2), 117-126. DOI: 10.1016/j.jplph.2011.08.019.

Livingston, D. P., Tuong, T. D., Murphy, J. P., Gusta, L. V., Willick, I. and Wisniewski, M. E. (2018). High-definition infrared thermography of ice nucleation and propagation in wheat under natural frost conditions and controlled freezing. *Planta* 247(4), 791-806.

Long, S., Yan, F., Yang, L., Sun, Z. and Wei, S. (2020). Responses of Manila Grass (*Zoysia matrella*) to chilling stress: from transcriptomics to physiology. *PLoS ONE* 15(7), e0235972.

Lu, S., Guo, Z. and Peng, X. (2003). Effects of ABA and S-3307 on drought resistance and antioxidative enzyme activity of turfgrass. *Journal of Horticultural Science and Biotechnology* 78(5), 663-666.

Lu, S., Su, W., Li, H. and Guo, Z. (2009). Abscisic acid improves drought tolerance of triploid Bermudagrass and involves H2O2- and NO-induced antioxidant enzyme activities. *Plant Physiology and Biochemistry* 47(2), 132-138. DOI: 10.1016/j.plaphy.2008.10.006.

Ludlow, M. M. (1987). Strategies of response to water stress. Kreeb, K. H., Richter, H. and Hinckley, T. M. (Eds), Structural and Functional Responses to Environmental Stresses: Water Shortage. *14th International Botanical Congress*, Berlin, July 24-August 1. SPB Academic Publishers, The Hague, 269-81. Available at: http://hdl.handle.net/102.100.100/260323?index=1

Ma, X., Zhang, J. and Huang, B. (2016). Cytokinin-mitigation of salt-induced leaf senescence in perennial ryegrass involving the activation of antioxidant systems and ionic balance. *Environmental and Experimental Botany* 125, 1-11. DOI: 10.1016/j.envexpbot.2016.01.002.

Maeda, Y. (2019). Effects of calcium application on the salt tolerance and sodium excretion from salt glands in zoysiagrass (Zoysia japonica). *Grassland Science* 65(3), 189-196. DOI: 10.1111/grs.12234.

Marcum, K. B. and Murdoch, C. L. (1990). Growth responses, ion relations, and osmotic adaptations of eleven C4 turfgrasses to salinity. *Agronomy Journal* 82(5), 892–896. DOI: 10.2134/agronj1990.00021962008200050009x.

Marcum, K. B., Anderson, S. J. and Engelke, M. C. (1998). Salt gland ion secretion: a salinity tolerance mechanism among five zoysiagrass species. *Crop Science* 38(3), 806–810 DOI: 10.2135/cropsci1998.0011183X003800030031x.

Matsumoto, N. and Hsiang, T. (2016). *Snow Mold: The Battle under Snow between Fungal Pathogens and Their Plant Hosts*. Springer, Verlag.

Mayer, P. W., DeOreo, W. B., Opitz, E. M., Kiefer, J. C., Davis, W. Y., Dziegielewski, B. and Nelson, J. O. (1999). *Residential End Uses of Water*. AWWA Research Foundation, Denver, CO.

McCann, S. E. and Huang, B. (2008). Drought responses of Kentucky bluegrass and creeping bentgrass as affected by abscisic acid and trinexapac-ethyl. *Journal of the American Society for Horticultural Science* 133(1), 20–26. DOI: 10.21273/JASHS.133.1.20.

Melmaiee, K., Anderson, M., Elavarthi, S., Guenzi, A. and Canaan, P. (2015). Transcriptional analysis of resistance to low temperatures in Bermudagrass crown tissues. *PLoS ONE* 10(9), e0136433.

Merewitz, E. (2021). Current outlook on ice encasement stress and management strategies in turfgrasses. *hortTechnology* 31(5), 561–565.

Milesi, C., Running, S. W., Elvidge, C. D., Dietz, J. B., Tuttle, B. T. and Nemani, R. R. (2005). Mapping and modeling the biogeochemical cycling of turf grasses in the United States. *Environmental Management* 36(3), 426–438.

Milla-Lewis, S. R., Kimball, J. A., Claure, T. E., Tuong, T. D., Arellano, C. and Livingston III, D. P. (2013). Freezing tolerance and the histology of recovering nodes in St. Augustinegrass. *Int. Turfgrass Soc. J* 12, 523–530.

Miller, V. J. (1960). Temperature effect on the rate of apparent photosynthesis of seaside bent and Bermuda-grass. *Proceedings. American Society for Horticultural Science* 75, 700–703.

Mohammadi, M. H. S., Etemadi, N., Arab, M. M., Aalifar, M., Arab, M. and Pessarakli, M. (2017). Molecular and physiological responses of Iranian perennial ryegrass as affected by trinexapac ethyl, paclobutrazol and abscisic acid under drought stress. *Plant Physiology and Biochemistry* 111, 129–143. DOI: 10.1016/j.plaphy.2016.11.014.

Moseley, D. O., Trappe, J. M., Milla-Lewis, S. R., Chandra, A., Kenworthy, K. E., Liu, W. and Patton, A. J. (2021). Characterizing the growth and winter survival of commercially available and experimental genotypes of St. Augustinegrass. *Crop Science* 61(5), 3097–3109.

Munns, R. (1988). Why measure osmotic adjustment? *Functional Plant Biology* 15(6), 717–726. DOI: 10.1071/PP9880717.

Munns, R. and Tester, M. (2008). Mechanisms of salinity tolerance. *Annual Review of Plant Biology* 59, 651–681.

Olien, C. R. and Marchetti, B. L. (1976). Recovery of hardened barley from winter injuries 1. *Crop Science* 16(2), 201–204.

Passioura, J. B. (1988). Root signals control leaf expansion in wheat seedlings growing in drying soil. *Functional Plant Biology* 15(5), 687–693. DOI: 10.1071/PP9880687.

Patton, A. J. and Reicher, Z. J. (2007). Zoysiagrass species and genotypes differ in their winter injury and freeze tolerance. *Crop Science* 47(4), 1619–1627.

Patton, A. J., Cunningham, S. M., Volenec, J. J. and Reicher, Z. J. (2007a). Differences in freeze tolerance of zoysiagrasses: I. Role of proteins. *Crop Science* 47(5), 2162–2169.

Patton, A. J., Cunningham, S. M., Volenec, J. J. and Reicher, Z. J. (2007b). Differences in freeze tolerance of zoysiagrasses: II. Carbohydrate and proline accumulation. *Crop Science* 47(5), 2170–2181.

Peng, S., Huang, J., Sheehy, J. E., Laza, R. C., Visperas, R. M., Zhong, X., Centeno, G. S., Khush, G. S. and Cassman, K. G. (2004). Rice yields decline with higher night temperature from global warming. *Proceedings of the National Academy of Sciences of the United States of America* 101(27), 9971–9975. DOI: 10.1073/pnas.0403720101.

Pincetl, S., Gillespie, T. W., Pataki, D. E., Porse, E., Jia, S., Kidera, E., Nobles, N., Rodriguez, J. and Choi, D. A. (2019). Evaluating the effects of turf-replacement programs in los Angeles. *Landscape and Urban Planning* 185, 210–221.

Puyang, X., An, M., Han, L. and Zhang, X. (2015). Protective effect of spermidine on salt stress induced oxidative damage in two Kentucky bluegrass (Poa pratensis L.) cultivars. *Ecotoxicology and Environmental Safety* 117, 96–106. DOI: 10.1016/j.ecoenv.2015.03.023.

Qian, Y. and Fry, J. D. (1997). Water relations and drought tolerance of four turfgrasses. *Journal of the American Society for Horticultural Science* 122(1), 129–133. DOI: 10.21273/JASHS.122.1.129.

Qian, Y. L., Ball, S., Tan, Z., Koski, A. J. and Wilhelm, S. J. (2001). Freezing tolerance of six cultivars of buffalograss. *Crop Science* 41(4), 1174–1178.

Qian, Y. L., Engelke, M. C. and Foster, M. J. V. (2000). Salinity effects on zoysiagrass cultivars and experimental lines. *Crop Science* 40(2), 488–492. DOI: 10.2135/cropsci2000.402488x.

Rachmilevitch, S., Lambers, H. and Huang, B. (2008). Short-term and long-term root respiratory acclimation to elevated temperatures associated with root thermotolerance for two *Agrostis* grass species. *Journal of Experimental Botany* 59(14), 3803–3809. DOI: 10.1093/jxb/ern233.

Richards, R. A. and Passioura, J. B. (1981). Seminal root morphology and water use of wheat I. Environmental effects 1. *Crop Science* 21(2), 249–252. DOI: 10.2135/cropsci1981.0011183X002100020011x.

Roberts, S. Z. (2015). Managing sports turf in the transition zone. *Sports Field Management*, (September).

Roberts, S. Z. (2019). A new crop of warm-season turfgrass varieties is making the move north, increasing options for golf course superintendents in the traditionally tricky transition zone. *Golf Course Management* (December).

Rodriguez Diaz, J. A. R., Know, J. W. and Weatherhead, E. K. (2007). Competing demands for irrigation water: golf and agriculture in Spain. *Irrigation and Drainage* 56(5), 541–549.

Rogers, R. A., Dunn, J. H. and Nelson, C. J. (1975). Cold hardening and carbohydrate composition of Meyer zoysia 1. *Agronomy Journal* 67(6), 836–838.

Rogler, G. A. (1943). Response of geographical strains of grasses to low temperatures 1. *Agronomy Journal* 35(7), 547–559.

Rossi, S., Burgess, P., Jespersen, D. and Huang, B. (2017). Heat-induced leaf senescence associated with chlorophyll metabolism in bentgrass lines differing in heat tolerance. *Crop Science* 57(S1), S-169-S-178. DOI: 10.2135/cropsci2016.06.0542.

Rukavina, H., Hughes, H. G. and Qian, Y. (2007). Freezing tolerance of 27 saltgrass ecotypes from three cold hardiness zones. *Hortscience* 42(1), 157–160.

Sadok, W. and Sinclair, T. R. (2009). Genetic variability of transpiration response to vapor pressure deficit among soybean cultivars. *Crop Science* 49(3), 955-960. DOI: 10.2135/cropsci2008.09.0560.

Salvucci, M. E. and Crafts-Brandner, S. J. (2004). Mechanism for deactivation of RuBisCO under moderate heat stress. *Physiologia Plantarum* 122(4), 513-519. DOI: 10.1111/j.1399-3054.2004.00419.x.

Samala, S., Yan, J. and Baird, W. V. (1998). Changes in polar lipid fatty acid composition during cold acclimation in 'Midiron' and 'U3'Bermudagrass. *Crop Science* 38(1), 188-195.

Sermons, S. M., Seversike, T. M., Sinclair, T. R., Fiscus, E. L. and Rufty, T. W. (2012). Temperature influences the ability of tall fescue to control transpiration in response to atmospheric vapour pressure deficit. *Functional Plant Biology* 39(12), 979-986. DOI: 10.1071/FP12172.

Sharp, R. E. and Davies, W. J. (1979). Solute regulation and growth by roots and shoots of water-stressed maize plants. *Planta* 147(1), 43-49. DOI: 10.1007/BF00384589.

Shearman, R. C. (1986). Kentucky bluegrass cultivar evapotranspiration rates. *HortScience* 21(3), 455-457.

Sinclair, T., Fiscus, E., Wherley, B., Durham, M. and Rufty, T. (2007). Atmospheric vapor pressure deficit is critical in predicting growth response of "cool-season" grass Festuca arundinacea to temperature change. *Planta* 227(1), 273-276. DOI: 10.1007/s00425-007-0645-5.

Sinclair, T. R., Devi, J., Shekoofa, A., Choudhary, S., Sadok, W., Vadez, V., Riar, M. and Rufty, T. (2017). Limited-transpiration response to high vapor pressure deficit in crop species. *Plant Science* 260, 109-118. DOI: 10.1016/j.plantsci.2017.04.007.

Soliman, W. S., Sugiyama, S. and Abbas, A. M. (2018). Contribution of avoidance and tolerance strategies towards salinity stress resistance in eight C3 turfgrass species. *Horticulture, Environment, and Biotechnology* 59(1), 29-36. DOI: 10.1007/s13580-018-0004-4.

Spiekerman, J. J. and Devos, K. M. (2020). The halophyte seashore paspalum uses adaxial leaf papillae for sodium sequestration. *Plant Physiology* 184(4), 2107-2119. DOI: 10.1104/pp.20.00796.

Steinke, K., Chalmers, D., Thomas, J. and White, R. (2011). Bermudagrass and buffalograss drought response and recovery at two soil depths. *Crop Science* 51(3), 1215-1223. DOI: 10.2135/cropsci2010.08.0469.

Steinke, K., Chalmers, D., Thomas, J., White, R. and Fipps, G. (2010). Drought response and recovery characteristics of St. Augustinegrass cultivars. *Crop Science* 50(5), 2076-2083. DOI: 10.2135/cropsci2009.10.0635.

Steudle, E. and Peterson, C. A. (1998). How does water get through roots? *Journal of Experimental Botany* 49(322), 775-788. DOI: 10.1093/jxb/49.322.775.

Stier, J. C., Filiault, D. L., Wisniewski, M. and Palta, J. P. (2003). Visualization of freezing progression in turfgrasses using infrared video thermography. *Crop Science* 43(1), 415-420.

Stier, J. C., Steinke, K., Ervin, E. H., Higginson, F. R. and McMaugh, P. E. (2013). Turfgrass benefits and issues. In: Stier, J. C., Horgan, B. P. and Bonos, S. A. (Eds) *Turfgrass: Biology Use and Management*. John Wiley & Sons, Hoboken, NJ, 105-145.

Stirk, W. A., Tarkowská, D., Turečová, V., Strnad, M. and van Staden, J. (2014). Abscisic acid, gibberellins and brassinosteroids in Kelpak®, a commercial seaweed extract made from Ecklonia maxima. *Journal of Applied Phycology* 26(1), 561–567. DOI: 10.1007/s10811-013-0062-z.

Strandberg, M., BlomBack., K., Dahl Jensen, A. M. D. and Knox, J. W. (2012). Priorities for sustainable turfgrass management: a research and industry perspective. *Acta Agriculturae Scandinavica, Section B – Soil and Plant Science* 62(sup1), 3-9. DOI: 10.1080/09064710.2012.682163.

Su, K., Bremer, D. J., Jeannotte, R., Welti, R. and Yang, C. (2009). Membrane lipid composition and heat tolerance in cool-season turfgrasses, including a hybrid bluegrass. *Journal of the American Society for Horticultural Science* 134(5), 511–520. DOI: 10.21273/JASHS.134.5.511.

Sugiura, S. and Takahashi, S. (2021). Physiological integration for salinity stress alleviation in stoloniferous turfgrass, Zoysia Matrella in heterogeneous saline environments. *Landscape and Ecological Engineering* 17(1), 21–28. DOI: 10.1007/s11355-020-00432-9.

Sun, T., Shao, K., Huang, Y., Lei, Y., Tan, L. and Chan, Z. (2020). Natural variation analysis of perennial ryegrass in response to abiotic stress highlights LpHSFC1b as a positive regulator of heat stress. *Environmental and Experimental Botany* 179, 104192. DOI: 10.1016/j.envexpbot.2020.104192.

Sun, X., Sun, C., Li, Z., Hu, Q., Han, L. and Luo, H. (2016). AsHSP17, a creeping bentgrass small heat shock protein modulates plant photosynthesis and ABA-dependent and independent signalling to attenuate plant response to abiotic stress. *Plant, Cell and Environment* 39(6), 1320–1337. DOI: 10.1111/pce.12683.

Sweeney, P., Danneberger, K., Wang, D. and McBride, M. (2001). Root weight, nonstructural carbohydrate content, and shoot density of high-density creeping bentgrass cultivars. *Hortscience* 36(2), 368–370. DOI: 10.21273/HORTSCI.36.2.368.

Thompson, C., Kennelly, M., Fry, J., Sousek, M. and Reicher, Z. (2017). Physiological and pathogenic contributors to the summer decline of Roughstalk bluegrass. *Itsrj* 13(1), 566–574. DOI: 10.2134/itsrj2016.05.0304.

Vierling, E. (1991). The roles of heat shock proteins in plants. *Annual Review of Plant Physiology and Plant Molecular Biology* 42(1), 579–620.

Villouta, C., Cox, B. L., Rauch, B., Workmaster, B. A. A., Eliceiri, K. W. and Atucha, A. (2021). A device for the controlled cooling and freezing of excised plant specimens during magnetic resonance imaging. *Plant Methods* 17(1), 41.

Wallner, S. J., Becwar, M. R. and Butler, J. D. (1982). Measurement of turfgrass heat tolerance *in vitro*. *Journal of the American Society for Horticultural Science* 107(4), 608–613.

Wang, J., Juliani, H. R., Jespersen, D. and Huang, B. (2017). Differential profiles of membrane proteins, fatty acids and sterols associated with genetic variations in heat tolerance for a perennial grass species, hard fescue. *Environmental and Experimental Botany* 140, 65–75.

Wang, K., Zhang, X. and Ervin, E. H. (2016). Small heat shock proteins, a key player in grass plant thermotolerance. In: Asea, A. A. A., Kaur, P. and Calderwood, S. K. (Eds) *Heat Shock Proteins and Plants*. Heat Shock Proteins. Springer International Publishing, Cham, 41–64. DOI: 10.1007/978-3-319-46340-7_3.

Wang, R., Wang, X., Liu, K., Zhang, X. J., Zhang, L. Y. and Fan, S. J. (2020). Comparative transcriptome analysis of halophyte Zoysia macrostachya in response to salinity stress. *Plants* 9(4), 458. DOI: 10.3390/plants9040458.

Wang, Z., Huang, B. and Xu, Q. (2003a). Effects of abscisic acid on drought responses of Kentucky bluegrass. *Journal of the American Society for Horticultural Science* 128(1), 36–41. DOI: 10.21273/JASHS.128.1.0036.

Wang, Z., Pote, J. and Huang, B. (2003b). Responses of cytokinins, antioxidant enzymes, and lipid peroxidation in shoots of creeping bentgrass to high root-zone temperatures. *Journal of the American Society for Horticultural Science* 128(5), 648–655. DOI: 10.21273/JASHS.128.5.0648.

Watkins, E., Huang, B. and Meyer, W. A. (2007). Tufted hairgrass responses to heat and drought stress. *Journal of the American Society for Horticultural Science* 132(3), 289–293. DOI: 10.21273/JASHS.132.3.289.

Watschke, T. L., Schmidt, R. E., Carson, E. W. and Blaser, R. E. (1973). Temperature influence on the physiology of selected cool season turfgrasses and Bermudagrass1. *Agronomy Journal* 65(4), 591–594. DOI: 10.2134/agronj1973.00021962006500040019x.

Weatherley, P. E. (1982). Water uptake and flow in roots. In: Lange, O. L., Nobel, P. S., Osmond, C. B. and Ziegler, H. (Eds). *Physiological Plant Ecology II. Encyclopedia of Plant Physiology* (New Series) (vol. 12). B. Springer, Berlin, Heidelberg. DOI: 10.1007/978-3-642-68150-9_4.

Wehner, D. J. and Watschke, T. L. (1984). Heat stress effects on protein synthesis and exosmosis of cell solutes in three turfgrass species 1. *Agronomy Journal* 76(1), 16–19. DOI: 10.2134/agronj1984.00021962007600010005x.

Wei, S., Du, Z., Gao, F., Ke, X., Li, J., Liu, J. and Zhou, Y. (2015). Global transcriptome profiles of 'Meyer' Zoysiagrass in response to cold stress. *PLoS ONE* 10(6), e0131153.

Wherley, B., Dukes, M. D., Cathey, S., Miller, G. and Sinclair, T. (2015). Consumptive water use and crop coefficients for warm-season turfgrass species in the Southeastern United States. *Agricultural Water Management* 156, 10–18. DOI: 10.1016/j.agwat.2015.03.020.

Wherley, B., Heitholt, J., Chandra, A. and Skulkaew, P. (2014). Supplemental irrigation requirements of zoysiagrass and Bermudagrass cultivars. *Crop Science* 54(4), 1823–1831. DOI: 10.2135/cropsci2013.11.0753.

Wherley, B. G. and Sinclair, T. R. (2009). Differential sensitivity of C3 and C4 turfgrass species to increasing atmospheric vapor pressure deficit. *Environmental and Experimental Botany* 67(2), 372–376. DOI: 10.1016/j.envexpbot.2009.07.003.

White, R. H., Engelke, M. C., Anderson, S. J., Ruemmele, B. A., Marcum, K. B. and Taylor, G. R. (2001). Zoysiagrass water relations. *Crop Science* 41(1), 133–138. DOI: 10.2135/cropsci2001.411133x.

Wilkins, P. W. (1991). Breeding perennial ryegrass for agriculture. *Euphytica* 52(3), 201–214.

World Health Organization (2020). Global health observatory data. Available at: https://www.who.int/publications/i/item/9789241565271.

Wu, Y., Chandra, A., Genovesi, A. D. and Meeks, M. (2017). Freezing tolerance of field- and growth chamber-acclimated zoysiagrass hybrids. *International Turfgrass Society Research Journal* 13(1), 421–426.

Xiang, M., Fry, J. and Wu, Y. (2020). Winter survival of experimental bermudagrasses in the upper transition zone of the US. *International Turfgrass Society Research Journal*, 1–5.

Xu, C. and Huang, B. (2012). Comparative analysis of proteomic responses to single and simultaneous drought and heat stress for two Kentucky bluegrass cultivars. *Crop Science* 52(3), 1246–1260. DOI: 10.2135/cropsci2011.10.0551.

Xu, L., Zhang, M., Zhang, X. and Han, L. B. (2015). Cold acclimation treatment-induced changes in abscisic acid, cytokinin, and antioxidant metabolism in Zoysiagrass (*Zoysia japonica*). *HortScience* 50(7), 1075–1080.

Xu, Q. and Huang, B. (2000). Growth and physiological responses of creeping bentgrass to changes in air and soil temperatures. *Crop Science* 40(5), 1363–1368. DOI: 10.2135/cropsci2000.4051363x.

Xu, Q. and Huang, B. (2001). Lowering soil temperatures improves creeping bentgrass growth under heat stress. *Crop Science* 41(6), 1878–1883. DOI: 10.2135/cropsci2001.1878.

Xu, Q. and Huang, B. (2003). Seasonal changes in carbohydrate accumulation for two creeping bentgrass cultivars. *Crop Science* 43(1), 266–271. DOI: 10.2135/cropsci2003.2660.

Xu, Q., Burgess, P., Xu, J., Meyer, W. and Huang, B. (2016). Osmotic stress- and salt stress-inhibition and gibberellin-mitigation of leaf elongation associated with up-regulation of genes controlling cell expansion. *Environmental and Experimental Botany* 131, 101–109. DOI: 10.1016/j.envexpbot.2016.06.001.

Xu, S., Li, J., Zhang, X., Wei, H. and Cui, L. (2006). Effects of heat acclimation pretreatment on changes of membrane lipid peroxidation, antioxidant metabolites, and ultrastructure of chloroplasts in two cool-season turfgrass species under heat stress. *Environmental and Experimental Botany* 56(3), 274–285. DOI: 10.1016/j.envexpbot.2005.03.002.

Xu, Y. and Huang, B. (2007). Heat-induced leaf senescence and hormonal changes for thermal bentgrass and turf-type bentgrass species differing in heat tolerance. *Journal of the American Society for Horticultural Science* 132(2), 185–192. DOI: 10.21273/JASHS.132.2.185.

Xu, Y., Wang, J., Bonos, S. A., Meyer, W. A. and Huang, B. (2018). Candidate genes and molecular markers correlated to physiological traits for heat tolerance in fine fescue cultivars. *International Journal of Molecular Sciences* 19(1), 116. DOI: 10.3390/ijms19010116.

Xu, Y., Zhan, C. and Huang, B. (2011). Heat shock proteins in association with heat tolerance in grasses. *International Journal of Proteomics* 2011, 529648. DOI: 10.1155/2011/529648.

Xuan, J., Song, Y., Zhang, H., Liu, J., Guo, Z. and Hua, Y. (2013). Comparative proteomic analysis of the stolon cold stress response between the C4 perennial grass species Zoysia japonica and Zoysia metrella. *PLoS ONE* 8(9), e75705.

Yamamoto, A., Hashiguchi, M., Akune, R., Masumoto, T., Muguerza, M., Saeki, Y. and Akashi, R. (2016). The relationship between salt gland density and sodium accumulation/secretion in a wide selection from three Zoysia species. *Australian Journal of Botany* 64(4), 277–284. DOI: 10.1071/BT15261.

Yamamoto, N., Takano, T., Tanaka, K., Ishige, T., Terashima, S., Endo, C., Kurusu, T., Yajima, S., Yano, K. and Tada, Y. (2015). Comprehensive analysis of transcriptome response to salinity stress in the halophytic turf grass Sporobolus virginicus. *Frontiers in Plant Science* 6, 241 DOI: 10.3389/fpls.2015.00241.

Youngner, V. B. (1961). Growth and flowering of Zoysia species in response to temperatures, photoperiods, and light intensities 1. *Crop Science* 1(2), 91–93.

Yu, X., Pijut, P. M., Byrne, S., Asp, T., Bai, G. and Jiang, Y. (2015). Candidate gene association mapping for winter survival and spring regrowth in perennial ryegrass. *Plant Science* 235, 37–45.

Zamin, M., Khattak, A. M., Salim, A. M., Marcum, K. B., Shakur, M., Shah, S., Jan, I. and Fahad, S. (2019). Performance of Aeluropus lagopoides (mangrove grass) ecotypes, a potential turfgrass, under high saline conditions. *Environmental Science and Pollution Research International* 26(13), 13410–13421. DOI: 10.1007/s11356-019-04838-3.

Zeevaart, J. A. D. and Creelman, R. A. (1988). Metabolism and physiology of abscisic acid. *Annual Review of Plant Physiology and Plant Molecular Biology* 39(1), 439–473.

Zhang, C., Fei, S. Z., Arora, R. and Hannapel, D. J. (2010). Ice recrystallization inhibition proteins of perennial ryegrass enhance freezing tolerance. *Planta* 232(1), 155–164.

Zhang, J., Nguyen, H. T. and Blum, A. (1999). Genetic analysis of osmotic adjustment in crop plants. *Journal of Experimental Botany* 50(332), 291–302. DOI: 10.1093/jxb/50.332.291.

Zhang, J., Xing, J., Lu, Q., Yu, G., Xu, B. and Huang, B. (2019). Transcriptional regulation of chlorophyll-catabolic genes associated with exogenous chemical effects and genotypic variations in heat-induced leaf senescence for perennial ryegrass. *Environmental and Experimental Botany* 167, 103858. DOI: 10.1016/j.envexpbot.2019.103858.

Zhang, J., Yu, G., Wen, W., Ma, X., Xu, B. and Huang, B. (2016). Functional characterization and hormonal regulation of the PHEOPHYTINASE gene LpPPH controlling leaf senescence in perennial ryegrass. *Journal of Experimental Botany* 67(3), 935–945. DOI: 10.1093/jxb/erv509.

Zhang, Q., Fry, J., Rajashekar, C., Bremer, D. and Engelke, M. (2009). Membrane polar lipid changes in zoysiagrass rhizomes and their potential role in freezing tolerance. *Journal of the American Society for Horticultural Science* 134(3), 322–328.

Zhang, X. and Schmidt, R. E. (2000). Hormone-containing products' impact on antioxidant status of tall fescue and creeping bentgrass subjected to drought. *Crop Science* 40(5), 1344–1349. DOI: 10.2135/cropsci2000.4051344x.

Zhang, X., Ervin, E. H. and LaBranche, A. J. (2006). Metabolic defense responses of seeded Bermudagrass during acclimation to freezing stress. *Crop Science* 46(6), 2598–2605.

Zhang, X., Shang, C., Liu, Y., Hu, G., Harich, K. and Ervin, E. H. (2017a). Hormone and dehydrin expression responses to cold acclimation in two zoysiagrass cultivars with contrasting freezing tolerance. *Itsrj* 13(1), 547–555.

Zhang, C., Fei, S. Z., Liu, P., Ji, T., Peng, J., Frei, U. and Hannapel, D. J. (2017b). Transcriptome changes in response to cold acclimation in perennial ryegrass as revealed by a cross-species microarray analysis. *Crop Science* 57(S1), S-179.

Zhang, X., Wang, K., Ervin, E. H., Waltz, C. and Murphy, T. (2011). Metabolic changes during cold acclimation and deacclimation in five Bermudagrass varieties. I. Proline, total amino acid, protein, and dehydrin expression. *Crop Science* 51(2), 838–846.

Zhang, X., Wu, W., Ervin, E. H., Shang, C. and Harich, K. (2018). Salt stress-induced injury is associated with hormonal alteration in Kentucky bluegrass. *HortScience* 53(1), 97–101. DOI: 10.21273/HORTSCI12413-17.

Zhao, X. X., Huang, L. K., Zhang, X. Q., Li, Z. and Peng, Y. (2014). Effects of heat acclimation on photosynthesis, antioxidant enzyme activities, and gene expression in orchardgrass under heat stress. *Molecules* 19(9), 13564–13576. DOI: 10.3390/molecules190913564.

Zhu, H., Yu, X., Xu, T., Wang, T., Du, L., Ren, G. and Dong, K. (2015). Transcriptome profiling of cold acclimation in Bermudagrass (*Cynodon dactylon*). *Scientia Horticulturae* 194, 230–236.

Chapter 2

Advances in breeding for improved cultivars of turfgrass

Phillip L. Vines, University of Georgia, USA; Ambika Chandra, Texas A&M AgriLife Research, USA; and Trent M. Tate, GO Seed, Inc., USA

1 Introduction

The primary objectives of turfgrass breeding efforts are to develop turfgrass cultivars with excellent seed or sod production, improved stress and pest tolerance, and elite turf quality under high, medium, and low maintenance practices (Meyer and Funk, 1989; Bonos et al., 2006; Bonos and Huff, 2013; Hanna et al., 2013; Meyer et al., 2017; Baxter and Schwartz, 2018). Newly developed cultivars must be phenotypically distinct, be genetically stable, adapt to a broad range of climates, and be able to withstand rapid and extreme changes in weather (Table 1). Breeding projects typically involve multi-year efforts that require interdisciplinary collaborations with researchers from institutions of both academic and industry backgrounds.

Seed production potential greatly influences the immediate and long-term success of seed-propagated turfgrass cultivars. Simply put, if a cultivar does not have adequate seed yield, it will not be produced, regardless of its performance as a mowed turfgrass (Wilkins, 1991). Therefore, seed production is a primary focus for many breeding efforts of both cool- and warm-season species outlined in this chapter. A majority of cool-season turfgrass seed production is conducted in the Pacific Northwest region of the United States. For warm-season turfgrass commercial seed production, bermudagrass

http://dx.doi.org/10.19103/AS.2022.0110.02

Table 1 Descriptions of major cool- and warm-season turfgrass species

Common name	Scientific name	Maintenance levels	Breeding system	Mode of propagation	Ploidy (2n)	Base chromosome number
Cool season						
Colonial bentgrass	*Agrostis capillaris* L.; syn. = *A. tenuis* Sibth.	High	OP	Seed	4x	7
Creeping bentgrass	*Agrostis stolonifera* L.; syn. = *A. palustris* Huds.	High	OP	Seed and vegetative	4x	7
Velvet bentgrass	*Agrostis canina* L.	High	OP	Seed	2x	7
Hard fescue	*Festuca brevipila* Tracey	Low	OP	Seed	6x	7
Sheep fescue	*Festuca ovina* L.	Low	OP	Seed	4x	7
Chewings fescue	*Festuca rubra* ssp. *commutata* Gaudin; syn. = *F. rubra* ssp. *fallax* (Thuill.) Nyman	Low	OP	Seed	6x	7
Slender creeping red fescue	*Festuca rubra* L. subsp. *littoralis* (G. Mey.) Auquier	Low	OP	Seed	6x	7
Strong creeping red fescue	*Festuca rubra* L. ssp. *rubra* Gaudin	Low	OP	Seed	8x	7
Kentucky bluegrass	*Poa pratensis* L.	Low to high	Apomictic	Seed and vegetative	8x-14x	7
Perennial ryegrass	*Lolium perenne* L.	Moderate to high	OP	Seed	2x	7
Tall fescue	*Festuca arundinacea Schreb.; syn = Lolium arundinaceum (Schreb.) Darbysh*	Low to high	OP	Seed and vegetative	6x	7
Warm season						
Bermudagrass	Cynodon spp.	Low to high	OP	Seed and vegetative	2x, 3x, 4x	9
Buffalograss	*Bouteloua dactyloides* Columbus; syn. = *Buchloë dactyloides* (Nutt.) Engelm.	Low	OP (dioecious)	Seed and vegetative	2x, 4x, 6x	10
Seashore paspalum	*Paspalum vaginatum* Swartz	Medium to high	OP	Seed and vegetative	2x, 4x, 6x	10
St. Augustinegrass	*Stenotaphrum secundatum* (Walt.) Kuntze	Low to moderate	OP	Seed and vegetative	2x, 3x, 32x	9, 10
Zoysiagrass	*Zoysia* spp. Willd.	Moderate	OP	Seed and vegetative	4x	10

(*Cynodon* spp.) is primarily produced in Arizona and California, with some production in Oklahoma and Oregon; seashore paspalum (*Paspalum vaginatum* Swartz) is produced primarily in Oregon; and zoysiagrass (*Zoysia* spp.) is currently produced in Georgia. Thus, collaborations among university breeding programs and seed production companies are important to ensure new cultivars perform well for both seed production and turf performance characteristics across various geographic regions and climates.

Vegetatively propagated turfgrass cultivars must have appropriate traits related to sod-forming ability (Segars et al., 2022). The sod production industry is a major contributor to the national economy; there were nearly 1500 farms producing 137 411 ha of sod in the United States as of 2017 (USDA, 2017). Alabama, California, Florida, Georgia, Oklahoma, South Carolina, and Texas are the major sod-producing states, with hybrid bermudagrass (*Cynodon dactylon* [L.] Pers. × *C. transvaalensis* Burtt Davy) being the most commonly produced species (Segars et al., 2022). Other notable vegetatively produced species include buffalograss (*Bouteloua dactyloides* Columbus; synonym = *Buchloë dactyloides* [Nutt.] Engelm.), centipedegrass (*Eremochloa ophiuroides* [Munro] Hack.), Kentucky bluegrass (*Poa pratensis* L.), St. Augustinegrass (*Stenotaphrum secundatum* [Walt.] Kuntze), tall fescue (*Festuca arundinacea* Schreb.; synonym = *Lolium arundinaceum* (Schreb.) Darbysh.), and zoysiagrass (*Zoysia* spp. Willd.). For breeding purposes, cultivars that are intended for sod production should have improved sod tensile strength and sod handling quality characteristics (Krans et al., 1992; Heckman et al., 2001; Han, 2009; McCalla et al., 2009; Segars et al., 2022).

Developing sustainable turfgrass cultivars that are widely adapted to a range of geographic regions is another major objective of turfgrass breeding programs. To address this initiative, germplasm must be evaluated in different geographic regions. Cooperative and collaborative testing programs including the national turfgrass evaluation program (NTEP; Beltsville, MD, USA), the cooperative turfgrass breeders test (CTBT; Albany, OR, USA), the alliance for low input sustainable turf (A-LIST; Corvallis, OR, USA), and the turfgrass water conservation alliance (TWCA; Lebanon, OR, USA) have been developed. Additionally, breeders have been successful in acquiring funds from the United States Department of Agriculture's Specialty Crop Research Initiative (Washington, DC, USA) program to collaboratively evaluate turfgrasses for adaptation in various geographic locations. Each of these programs encourages participation among private and public turfgrass breeders and is designed to evaluate and identify turfgrass cultivars that perform well for specific applications.

Cool-season turfgrass breeders utilize the CTBT to evaluate a relatively large number of cultivars in as many as ten states across the United States. This program is appealing because of the low evaluation fees compared to other testing programs. The NTEP conducts trials across as many as 40 U.S. states and six Canadian provinces. The evaluation fees for this program are higher than those of the CTBT; however, breeders are able to evaluate their cultivars across a more diverse range of climate conditions with NTEP. Both A-LIST and TWCA are utilized to identify turfgrass cultivars that meet the standards for approval by each of these programs. As of 2022, there were 17 tall fescues, 9 Kentucky bluegrasses, 5 Chewings fescues (*Festuca rubra* ssp. *commutata* Gaudin; syn. = *F. rubra* ssp. *fallax* [Thuill.] Nyman), 3 hard fescues (*Festuca brevipila* Tracey), 3 strong creeping red fescues (*Festuca rubra* L. ssp. *rubra* Gaudin), and 1 sheep fescue (*Festuca ovina* L.) included in the A-LIST approved cultivars (www.a-listturf.org). These cultivars are recognized as having excellent turf quality characteristics and excellent performance when managed with limited inputs such as water, fertilizer, and pesticides. Also as of 2022, 20 tall fescues, 15 perennial ryegrasses (*Lolium perenne* L.), 14 Kentucky bluegrasses, 5 Chewings fescues, 5 hard fescues, 5 strong creeping red fescues, and 1 hybrid bermudagrass have been included in the TWCA-approved cultivars. These cultivars are recognized as having excellent drought tolerance.

In addition to increasing breeding efficiency and improving cultivar adaptation through the cooperative efforts outlined previously, turfgrass breeders are utilizing advanced breeding techniques to select elite germplasm and improve breeding efficiency for various traits of interest. Two general categories of advanced breeding techniques include (i) DNA-based genetics and genomics methods and (ii) high-throughput plant phenotyping methods. These advanced breeding techniques are focused on rapidly and accurately associating genotypes with phenotypes, or specific traits of interest, to identify elite genotypes for specific traits. Various molecular technologies have been studied in turfgrass breeding research since the 1990s. In recent years, however, DNA sequencing technologies have become routinely available and turfgrass breeders can generate DNA sequence data rapidly and affordably (Delseny et al., 2010; Duran et al., 2010; Edwards and Batley, 2010). In turn, the process of phenotyping turfgrass germplasm for various traits of interest has become a limiting factor in advancing the breeding process. Since the mid-2010s, researchers have been exploring methods for rapid phenotyping, a practice commonly called high-throughput plant phenotyping (Araus and Cairns, 2014; Shakoor et al., 2017).

DNA-based genetics and genomics methods have been utilized in turfgrass breeding research to assess population structures, inter- and intraspecific molecular relationships, genetic diversity within germplasm pools, and molecular phylogenies among turfgrasses. These approaches are currently based on DNA markers. Additionally, researchers have investigated relationships between traits of interest and genotyping data; depending on the approach taken, these are referred to as quantitative trait locus (QTL) mapping studies, linkage disequilibrium mapping studies, genome scans, genome-wide association studies, or genomic selection (Moose and Mumm, 2008).

These efforts have been combined with comparative genomics approaches, which capitalize on syntenic relationships with model and major agronomic species such as rice (*Oryza sativa* L.), sorghum (*Sorghum bicolor* [L.] Moench.), maize (*Zea mays* L.), and foxtail millet (*Setaria italica* L.). The ultimate goal of these efforts is to identify markers or genes of interest that can be used for marker-assisted selection and transgenic approaches to improve germplasm development.

As mentioned in the previous paragraph, advancements in next-generation DNA sequencing technologies have reduced the cost and time associated with generating DNA sequence data. Thus, the phenotyping of germplasm has become a limiting factor in advancing modern breeding approaches. High-throughput plant phenotyping is a relatively new discipline that has been explored in major crop species to improve the efficiency of generating data for various plant traits of interest such as yield (Tanger et al., 2017), stress tolerance (Bhat et al., 2020), and disease resistance (Shakoor et al., 2017). Recent advances in remote sensing have offered improved, nondestructive plant phenotyping approaches. These advancements have been coupled with improvements in robotics and unmanned aerial systems technology to provide mobile, time-efficient platforms for remote sensors that have contributed to high-throughput plant phenotyping applications across large fields. Researchers are currently investigating remote sensing tools to identify specific parameters that can be used to assess specific traits of interest in plant breeding programs. Applications of high-throughput plant phenotyping are limited in turfgrass breeding programs at present and, therefore, will not be heavily discussed in the following sections.

2 Breeding of cool-season turfgrasses

Formalized breeding of cool-season turfgrasses began in the 1940s with Professor H. B. Musser, who was a forage grass breeder at Pennsylvania State University (University Park, PA, USA). Prof. Musser's efforts led to the development of an advanced creeping bentgrass (*Agrostis stolonifera* L.; synonym = *A. palustris* Huds.) cultivar, 'Penncross', which was released in 1954 (Hein, 1958). Shortly thereafter, Dr. Joseph Duich and Dr. C. Reed Funk were hired at Pennsylvania State University and Rutgers University (New Brunswick, NJ, USA), respectively, as the first full-time turfgrass breeders. In 1970 in the United States, the Plant Variety Protection Act was passed to provide intellectual property rights to breeders for seeded cultivars of asexual and sexual plants (Plant Variety Protection Act, 1970). After this law was in place, there was a noticeable increase in turfgrass breeding efforts (Meyer, 1982). Currently, there are more than 50 turfgrass breeders employed at both public and private institutions throughout the United States.

The past 70 years of cool-season turfgrass breeding efforts have resulted in the development of several useful species for various turf applications including home lawns, golf and athletic fields, parks and other recreational areas, and roadsides (Busey, 1989; Meyer and Funk, 1989; Meyer et al., 2017). The major cool-season turfgrass species include colonial bentgrass (*Agrostis capillaris* L.; synonym = *A. tenuis* Sibth.), creeping bentgrass, velvet bentgrass (*Agrostis canina* L.), hard fescue, sheep fescue, Chewings fescue, slender creeping red fescue (*Festuca rubra* L. subsp. *Littoralis* [G. Mey.] Auquier), strong creeping red fescue, Kentucky bluegrass, perennial ryegrass, and tall fescue. There are other minor cool-season species including dryland bentgrass (*Agrostis castellana* Boiss. And Reut.), redtop bentgrass (*Agrostis gigantea* Roth), blue fescue (*Festuca glauca* Vill.), annual bluegrass (*Poa annua* L.), Texas bluegrass (*Poa arachnifera* Torr.), and roughstalk bluegrass (*Poa trivialis* L.) that are also used for certain turfgrass applications. The following sections will focus on the major cool-season species.

Aside from Kentucky bluegrass, which is an apomictic species, all major cool-season turfgrass species exhibit highly outcrossing, self-incompatible reproductive characteristics (Bonos and Huff, 2013). The most common breeding scheme for these species includes a genotypic and phenotypic recurrent selection approach. Specific workflows and breeding objectives will vary among breeding programs and between turfgrass species, but the general breeding approach for outcrossing species can be summarized as outlined in Fig. 1. For this approach, breeding programs generate a pool of germplasm, over time, as breeders strive to identify and maintain elite germplasm for various traits of interest for a given turfgrass species. This set of germplasm is composed of material that has been advanced in a breeding program as well as new germplasm sources, such as collections, that are being evaluated for incorporation into a breeding program. All sources of germplasm in a breeding program are added to this germplasm pool, which is often called a stock nursery, breeder nursery, or source nursery.

Breeders continually evaluate stock nurseries to identify clones, or genotypes, that share similar traits such as color, maturity, seed yield, growth habit, and pest and stress tolerance. These clones are transplanted into isolated polycrosses and allowed to cross pollinate. Seed is harvested from each parent clone and sown into mowed progeny plot trials. Alternatively, breeders may elect to harvest seed directly from mother plants in the source nursery that were allowed to open pollinate, and this seed also may be sown into mowed progeny plot trials. Breeders evaluate germplasm in mowed trials for various turf quality traits such as color, uniformity, texture, density, and tolerance to various abiotic and biotic stresses. Elite progeny lines are identified, and individual plants are either transplanted into mowed or non-mowed spaced-plant nurseries or added to the stock nursery.

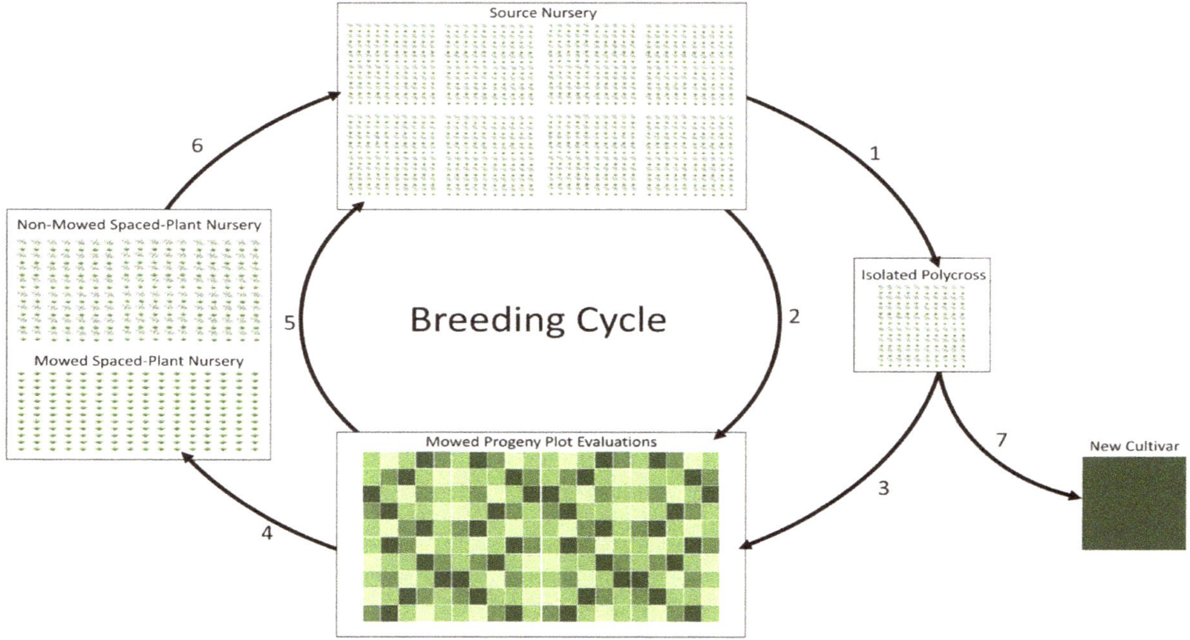

Figure 1 Phenotypic and genotypic recurrent selection breeding approach for open-pollinated turfgrass species. A source nursery is established from new germplasm collections and cycles of selection in a breeding program. Elite clones are selected from source nurseries and transplanted into isolated polycrosses (1). Seed is harvested from open-pollinated clones in a source nursery (2) or from clones in a polycross (3) and planted in mowed progeny plot evaluation trials. Individuals are selected from superior lines and further evaluated in mowed or non-mowed spaced-plant nurseries (4) or transplanted into a stock nursery (5) for subsequent breeding projects. Elite individuals are selected from spaced-plant nurseries and also transplanted into a stock nursery (6). After adequate cycles of selection and breeding, as well as consumer product testing, a new cultivar is released (7).

Mowed spaced-plant nurseries are maintained under routine mowing practices and utilized to identify individual genotypes that are tolerant to stresses such as mowing, drought, heat, cold, salinity, diseases, or insects. Spaced-plant nurseries are often inoculated with various fungi that incite important diseases on each grass species to identify specific genotypes with disease tolerance or resistance. Non-mowed spaced-plant nurseries, unlike the mowed spaced-plant nurseries, are allowed to produce seed, and these nurseries are utilized to identify specific genotypes with superior seed yield potential and tolerance to various seed production diseases. Selections from spaced-plant nurseries are clonally propagated and transplanted into stock nurseries. After adequate evaluation and selection cycles, breeders will elect to move forward with a population for cultivar development. Seed from polycrosses or vegetative clones of maternal lines from polycrosses can serve as foundation material to develop a new cultivar.

2.1 Bentgrass

2.1.1 Biological characteristics of bentgrass

Bentgrasses (*Agrostis* spp.) are a group of more than 200 species that primarily originated in western Europe (Harlan, 1992). Of the numerous species in *Agrostis*, only colonial, creeping, dryland, redtop, and velvet bentgrasses are used for turf applications, and of those, colonial, creeping, and velvet bentgrasses are most predominantly used. These three species are recognized as having fine leaf textures, high uniformities, and dense canopies while tolerating extremely low mowing heights. Accordingly, these three grasses, along with redtop bentgrass, were the first bentgrasses to be used as turf in the United States, having been introduced from Austria and Hungary as a mixture called 'South German Mixed Bentgrass' (Duich, 1985). Currently, colonial, creeping, and velvet bentgrass remains the most widely used bentgrasses for golf course tees, fairways, and putting greens.

Colonial bentgrass, also known as browntop bentgrass, is a bunch type, segmental allotetraploid ($2n = 4x = 28$), perennial grass with non-aggressive, short stolons (Jones, 1953, 1956; Chase, 1971; Hubbard, 1984; Clark et al., 1989; Brilman, 2001; Christians, 1998; Elmore et al., 1997). This species has traditionally been used as lawn and golf course grass in areas of Northern Europe and New Zealand that have mild summers (Beard, 1973; Hubbard, 1984). In the United States, colonial bentgrass is best adapted for golf course fairway or tee use (Beard, 1973; Hubbard, 1984; Elmore et al., 1997; Shildrick, 1977; Ruemmele, 2000). Compared to creeping bentgrass, colonial bentgrass exhibits less aggressive lateral growth, finer leaf texture, slightly lower canopy density, and a more vertical growth habit (Elmore et al., 1997). The genetic color of colonial bentgrass is generally brighter green than creeping bentgrass, and it also has better color retention during cool weather. Colonial bentgrass has better dollar spot (*Clarireedia jacksonii* Salgado, Beirn, Clarke, and Crouch) resistance than creeping bentgrass; however, brown patch (*Rhizoctonia solani* Kühn) is a disease of major concern on colonial bentgrass.

Creeping bentgrass is an allotetraploid (2n = 4x = 28), perennial grass that spreads by extensive stoloniferous growth (Jones, 1956). This species has a prostrate growth habit, which enables it to accommodate extremely low mowing heights; mowing heights as low as 2.5 mm are common for this species. Creeping bentgrass is by far the most widely used species for golf course putting greens in temperate regions, as it is well adapted to both warm, humid regions and cool, temperate regions of the United States. Another appealing characteristic of creeping bentgrass is that the aggressive stoloniferous growth gives it a superior ability to recover from divot damage that often occurs on golf course tees and fairways.

As mentioned previously, a landmark cultivar, Penncross, was released in 1954 as the first seeded synthetic creeping bentgrass cultivar (Musser, 1959). Modern creeping bentgrass cultivars have been bred to have superior turf quality characteristics, improved traffic tolerance, high shoot densities, and improved disease and stress tolerance, compared with older cultivars. Creeping bentgrasses are susceptible to several pathogens and pests. The most problematic and widely observed disease of creeping bentgrass is the dollar spot, but other diseases including anthracnose (*Colletotrichum cereale* Mans sensu lato Crouch, Clarke, and Hillman), brown patch, copper spot (*Gloeocercospora sorghi* Bain and Edgerton ex Deighton), and Pythium blight (*Pythium* spp.) also affect the performance of creeping bentgrass.

Velvet bentgrass is a perennial species that forms the finest-textured and densest turf of the bentgrasses (Sprague and Evaul, 1930; DeFrance et al., 1952; Bjorkman, 1954). This diploid (2n = 2x =1 4) species can tolerate very close mowing, as well as heat, cold, and shade stress and is one of the most drought tolerant of the bentgrasses used for turf (Beard, 1973; Skogley, 1973). Due to the density and vigor of this turf, even under very low mowing conditions, it is extremely effective at preventing the encroachment of annual bluegrass, which is the most prolific weed on golf courses. Velvet bentgrass is not typically considered a superior turfgrass species, but this is largely a result of mismanagement; with proper inputs it can provide an elite playing surface (Brilman and Meyer, 2000). Velvet bentgrass is susceptible to diseases such as copper spot, Pythium blight, and red thread (*Laetisaria fuciformis* [McAlpine] Burdsall). This species is typically slower to begin growing in the spring compared to other bentgrass species and tends to turn a dark purple color during cold weather in the fall. Velvet bentgrass has been used on golf courses throughout the Northeast and New England regions of the United States

(Brilman and Meyer, 2000). However, the adaptation range of velvet bentgrass has not been well documented; this is a major reason why this species has not been used as much as colonial and creeping bentgrass for high-maintenance turf.

2.1.2 Traditional breeding strategies and current breeding objectives of bentgrass

Bentgrasses are bred using a phenotypic and genotypic recurrent selection breeding scheme (Fig. 1). Although there are different traits of interest among the breeding objectives for the three major bentgrass species, traits such as excellent turf quality, improved seed yield, resistance to problematic diseases, and tolerance to stresses such as traffic, drought, and heat are shared. Many of these traits are evaluated in mowed progeny plot trials; however, noticeable improvements have been made utilizing spaced-plant nursery evaluations. With respect to disease resistance breeding, the major objectives are to continually improve dollar spot resistance in creeping bentgrass, brown patch resistance in colonial bentgrass, and copper spot resistance in velvet bentgrass.

Interspecific hybridization has also been explored in bentgrasses. For example, there have been efforts to hybridize colonial bentgrass with creeping bentgrass and creeping bentgrass with dryland bentgrass (Jones, 1956; Warnke, 2020). This is an interesting approach, as it may allow for new sources of disease resistance to be introduced within each species. For example, dollar spot disease resistance could be transferred to creeping bentgrass, and brown patch resistance could be transferred to colonial bentgrass using this approach. Moreover, an interspecific hybridization strategy could also be used to target the transfer other traits such as stolon or rhizome production among the bentgrasses.

2.1.3 Advanced breeding methods of bentgrass

Researchers have been investigating genetics and genomics approaches for bentgrasses since the late 1990s, and, as a result, DNA marker technology has been widely implemented for various breeding and research objectives in *Agrostis*. Genetic diversity has been characterized in creeping bentgrass (Caceres et al., 2000; Casler et al., 2003; Kubik et al., 2011) and colonial bentgrass (Zhao et al., 2006) using amplified fragment length polymorphism (AFLP), random amplified polymorphic DNA (RAPD), restriction fragment length polymorphism (RFLP), and simple sequence repeat (SSR) markers. Golembiewski et al. (1997) demonstrated that RAPD markers could be used to distinguish creeping bentgrass cultivars. AFLP markers have also been used to clarify taxonomic relationships in *Agrostis* (Vergara and Bughrara, 2003). Most recently, Warnke (2020) developed sequence characterized amplified region (SCAR) marker-based polymerase chain reaction (PCR) primers to identify interspecific hybrids between colonial bentgrass and creeping bentgrass.

Molecular genetics and genomics approaches have also been implemented to improve dollar spot resistance in creeping bentgrass. Multiple genetic linkage maps have been developed for creeping bentgrass and colonial bentgrass. Chakraborty et al. (2005) used RAPD, AFLP, and RFLP markers to construct a genetic linkage map of creeping bentgrass. This linkage map was subsequently utilized to identify a single major dollar spot resistance QTL on linkage group 7 (Chakraborty et al., 2006). In a separate study, dollar spot resistance QTLs were identified on creeping bentgrass linkage groups 1, 4, and 5 of a linkage map constructed based on SSR, conserved intron scanning primer, intron length polymorphism, and AFLP markers (Kubik et al., 2011; Honig et al., 2014). Lastly, a creeping × colonial bentgrass mapping population was used to derive a framework genetic linkage map based on AFLP markers (Rotter et al., 2009), which was subsequently used to identify putative dollar spot resistance loci within the colonial bentgrass genome (Rotter et al., 2009).

Although a considerable amount of genetics and genomics work has been conducted in *Agrostis*, there has been no realized implementation of these techniques in current breeding programs. Researchers are working to validate the findings of previous studies and are investigating approaches to implement marker-assisted selection for important traits such as dollar spot resistance in creeping bentgrass. There have been a few studies to investigate the usability of remote sensing-based phenotyping tools in *Agrostis* (McCall et al., 2017; Hong et al., 2019a,b), but these have not been implemented in breeding programs to date. Nonetheless, this area of research is being actively explored and is a promising means for improving breeding efficiency for bentgrass cultivar development.

2.2 Fine fescue

2.2.1 Biological characteristics of fine fescue

The fine fescues (*Festuca* spp.) are a group of cool-season grasses that have distinct, fine-textured leaves. There are many species and subspecies of fine fescue, but only five are generally considered for use as turfgrasses; these include hard fescue, sheep fescue, Chewings fescue, slender creeping red fescue, and strong creeping red fescue. Hard fescue and sheep fescue belong to the *F. ovina* complex, and Chewings fescue, slender creeping red fescue, and strong creeping red fescue belong to the *F. rubra* complex. Compared to other cool-season grasses, fine fescues are better adapted to cool, dry, and shaded environments (Carroll, 1943; Hanson et al., 1969). This group is tolerant of acidic soils and drought conditions and exhibits the best performance under lower fertility levels. These qualities have given fine fescues a reputation as low-maintenance grasses (Watkins et al., 2010; Braun et al., 2020).

Fine fescues perform best in well-drained soils and are not suited for saturated soil conditions (Murphy, 1996). In general, these grasses have poor heat and wear tolerance and lack tolerance to excessive nitrogen fertilization during periods of high temperatures (Meyer and Funk, 1989). When heavily fertilized, fine fescues can become soft, succulent, and excessive thatch producers, which make them more susceptible to diseases and summer stresses. The increasing demand for lower fertilizer and water usage makes fine fescues an option for use in certain situations to address some of these issues. Many of the newer fine fescue cultivars contain an *Epichloë festucae* Leuchtm. endophyte that improves drought tolerance, resistance to above-ground feeding insects, and, in some cases, diseases, which can reduce the need for chemical inputs normally used to treat for insects and diseases (Saha et al., 1987; Breen, 1993; Funk et al., 1993; Funk and White, 1997).

Hard fescue is a hexaploid ($2n = 6x = 42$), bunch-type grass that spreads by tillering. It has a dark green color, forms a dense cover, and grows relatively slowly. Compared to Chewings fescue, hard fescue is more tolerant of heat, drought, and low nutrient fertility. The species is widely used in many low-maintenance situations due to increased disease resistance, even under low-maintenance conditions. Hard fescue can be among the highest quality of the fine fescues on sandy, well-drained soils, and it has better tolerance to Drechslera leaf spot (*Drechslera* spp.), dollar spot, and red thread diseases than Chewings fescue and strong creeping red fescue. However, one of the major problems that severely limits hard fescue use, especially in heavier, less drained soils, is susceptibility to summer patch (*Magnaporthiopsis poae* [Landschoot and Jackson] Luo and Zhang) disease (Braun et al., 2020).

Sheep fescue is a bunch-type, tetraploid ($2n = 4x = 28$) species that has a wide color variation, ranging from blue or green to a silvery blue or silvery green. Of the fine fescue species covered here, sheep fescue is the least widely used for turfgrass applications. This species is rarely used in seed mixtures because of its color. Sheep fescue has a non-aggressive growth habit, which makes it a good addition to wildflower mixes to aid in the prevention of erosion and to add an interesting color to the mix. This species is also becoming more popular in ornamental landscapes due to its color.

Chewings fescue is a dense, bunch-type, hexaploid ($2n = 6x = 42$) perennial grass that is native to Europe (Ruemmele et al., 2003). In the 1800s, the species was introduced to New Zealand where it was first cultivated by Mr. George Chewings (Ruemmele et al., 2003). The grass was imported to North America and Europe beginning in the late 1920s and is currently used for home lawn, golf course, athletic field, and recreational turf (Braun et al., 2020). Chewings fescue belongs to the red fescue (*Festuca rubra*) complex but is different from other members in that it lacks rhizomes that the slender creeping red fescues and strong creeping red fescues have (Ruemmele et al., 2003). Even so, Chewings fescue can maintain density due to its extensive tillering capability (Beard, 1973). Chewings fescue has many desirable characteristics such as good wear tolerance, drought resistance, and shade tolerance, has low fertility and irrigation requirements, tolerates a wide range of mowing heights, and has better disease resistance than many other fine fescues (Braun et al., 2020). However, Chewings fescue is prone to excessive thatch accumulation and is susceptible to some diseases such as snow mold (*Microdochium nivale* [Fr.] Samuels and I.C. Hallett).

Both the slender creeping red fescue and strong creeping red fescue are referred to as creeping red fescues because they spread by rhizomes. As the names suggest, strong creeping red fescue has a more aggressive, spreading habit than slender creeping red fescue. Slender creeping red fescue is a hexaploid ($2n = 6x = 42$) species with weak rhizomes that extend underground. The specific epithet, 'littoralis', means 'pertaining to the seashore', which describes species associated with sand dunes, saltmarshes, or riverbanks. This accurately describes and represents the well-known salt tolerance of slender creeping red fescue; in fact, it has the best salinity tolerance of the fine fescues (Braun et al., 2020).

Strong creeping red fescue is an octoploid ($2n = 8x = 56$), rhizomatous, cool-season perennial grass that is used for recreational, athletic field, golf course, and home lawn turf (Meyer and Funk, 1989; Braun et al., 2020). The aggressive rhizomes provide a means for lateral spread and recuperative ability from diseases and summer stress. Strong creeping red fescue is desired as a turf due to its medium to dark green genetic color, wear tolerance, drought resistance, shade tolerance, and establishment rate (Meyer and Funk, 1989; Ruemmele et al., 1995; Braun et al., 2020). The relatively coarse texture and density of this species, compared with other fine fescue species, permit its use in seed mixtures with perennial ryegrass and Kentucky bluegrass (Meyer and Funk, 1989). In addition, the rhizomatous characteristic of strong creeping red fescue makes it a good option for sod mixtures, as it enhances sod strength. Like Chewings fescue, strong creeping red fescue also has low irrigation and fertility requirements; however, it prefers a taller mowing height and does not tolerate as wide of a range of mowing heights as Chewings fescue (Braun et al., 2020).

2.2.2 Traditional breeding strategies and current breeding objectives of fine fescue

Each of the fine fescue species has a unique set of breeding objectives and traits of interest; however, a phenotypic and genotypic recurrent selection breeding scheme (Fig. 1) is predominately used for all the fine fescue species discussed herein. Mowed progeny plot trials are utilized to evaluate the fine fescues for turf quality characteristics as well as performance under various abiotic and biotic stresses. Plants are identified within breeding lines that exhibit good turf quality and disease resistance and spaced-plant nurseries are established and evaluated for stem rust (*Puccinia graminis* Pers.) and crown rust (*Puccinia coronata* Corda) disease resistance, summer heat and drought stress performance, and seed yield potential to identify individual genotypes with superior performance characteristics. Elite genotypes are clonally propagated, established in a stock nursery, and utilized to develop advanced populations.

Breeding objectives for all fine fescues discussed here include improving seed yield potential, floret fertility, germination and establishment rates, traffic tolerance, shade tolerance, and performance under low-maintenance conditions (Ruemmele et al., 2003). Hard fescue breeding objectives specifically focus on improving resistance to summer patch and dollar spot diseases. Breeding objectives for sheep fescue include increased seed yield, turfgrass quality, and disease resistance. The focus of Chewings fescue breeding objectives includes improvements in resistance to leaf spot, dollar spot, and red thread diseases. Efforts are being made to improve salinity tolerance in slender creeping red fescues, as this species is the best suited for this application. In addition, slender creeping red fescue breeding objectives include improving leaf spot and summer patch disease resistance. Breeding efforts in strong creeping red fescue include improving heat and drought tolerance and resistance to stem rust, crown rust, leaf spot, dollar spot, and red thread diseases.

2.2.3 Advanced breeding methods of fine fescue

Until recently, little work had been done in developing molecular tools for fine fescue breeding applications. Inter simple sequence repeat (ISSR) markers have been used to assess genetic diversity in sheep fescue (Armoniene et al., 2010; Mohammadi et al., 2020) and hard fescue (Armoniene et al., 2010) populations. The work by Armoniene et al. (2010) also highlighted the usability of ISSR markers to aid in fine fescue species identification. Xu et al. (2018) studied hard, sheep, Chewings, slender creeping red, and strong creeping red fescues and identified single-nucleotide polymorphism (SNP) markers associated with candidate genes related to heat tolerance. Chloroplast genomes of hard, sheep, Chewings, slender creeping red, and strong creeping red fescues have been sequenced and PCR primer sets have been developed that enable distinction between these species (Qiu et al., 2019). Phylogenetic relationships among the five major fine fescue species have also been investigated in the study by Qiu et al. (2019), which corroborated prior classifications based on genome size and plant morphologies (Huff and Palazzo, 1998; Cheng et al., 2016). A hard fescue transcriptome was reported by Qiu et al. (2020); this will be useful for future molecular studies and provide a foundation for gene discovery for important traits such as summer patch resistance, for example.

2.3 Kentucky bluegrass

2.3.1 Biological characteristics of Kentucky bluegrass

Kentucky bluegrass, also known as smooth-stalked meadow grass, is a cool-season, rhizomatous grass that is widely used for golf course, athletic field, and home lawn turf throughout the northern United States and Canada (Funk, 2000; Huff, 2003). There is debate as to whether Kentucky bluegrass is indigenous to North America (Gray, 1908; Keck, 1965; Duell and Peacock, 1985; Soreng, 1985; Beard, 2012), but most agree that the species was naturalized in the United States via introductions from Europe (Bashaw and Funk, 1987). Attractive traits such as strong sod-forming ability, medium- to fine leaf texture, dark green leaf color, high shoot density, persistence, and uniformity have led to the widespread popularity and use of this species (Huff, 2003).

Kentucky bluegrass is a genetically diverse species, and it has been suggested that all traits necessary for an ideal turfgrass exist within the species; the challenge, however, is combining all the desirable traits into a single cultivar (Funk, 2000). The release of 'Merion' in the 1940s was a major advancement for Kentucky bluegrass. This cultivar was discovered by Joseph Valentine at Merion Golf Club (Ardmore, PA) and was the first Kentucky bluegrass with turf-type characteristics and tolerance to low mowing heights. Another cultivar, 'Kenblue', was released in the 1950s as a seed blend of several Kentucky bluegrass selections from old pastures in Kentucky. In 1984, a landmark cultivar, 'Midnight', was released and has remained a widely used Kentucky bluegrass cultivar nearly 40 years later (Meyer et al., 1984). Midnight proved to be a major advancement in Kentucky bluegrass because of its dark green leaf color, superior turf quality characteristics, low growing habit, and tolerance to traffic, heat, and drought stress.

2.3.2 Traditional breeding strategies and current breeding objectives of Kentucky bluegrass

Early breeding efforts in Kentucky bluegrass were based on evaluations of ecotype selections. This breeding approach was critical in the initial improvement of this species and led to landmark cultivars including Merion and Kenblue. The techniques used in Kentucky bluegrass breeding are much different from those of other cool-season turfgrass species. This is because Kentucky bluegrass primarily reproduces via an asexual reproductive mechanism called apomixis, which is a process whereby seeds are produced without the sexual union of the sperm and egg cell, resulting in an identical copy of the mother plant. This characteristic is ideal for uniformity in seed production and turf performance but limits the breeder's ability to introduce genetic variation within the species.

Researchers investigated Kentucky bluegrass for many years to identify appropriate breeding strategies to account for this facultative apomictic characteristic. In the late 1960s and early 1970s, scientists at Rutgers University found that Kentucky bluegrass hybridization events occurred at higher frequencies under greenhouse conditions during the hours from 0100 to 0400, compared to field pollinations (Funk and Han, 1967; Pepin and Funk, 1971). This method was further developed to include additional lighting for extended day lengths (Hintzen and van Wijk, 1985) and machine-assisted automated pollination (Hintzen and van Wijk, 1985; Bashaw and Funk, 1987; Riordan et al., 1988). These efforts proved to be important advancements in breeding Kentucky bluegrass cultivars, as all modern Kentucky bluegrass cultivars have been developed based on the hybridization approach.

A common breeding approach for Kentucky bluegrass hybrid development is shown in Fig. 2. The general goal for hybrid-based Kentucky bluegrass breeding is to generate as many new hybrids as possible and identify hybrids that retain high levels of apomixis, have excellent seed yield potential, and exhibit excellent performance in mowed turf plot trials. It is important to note that Kentucky bluegrass genotypes vary with respect to apomixis rates. Some genotypes are highly apomictic and exhibit very low rates of sexual fertilization, while other genotypes are less apomictic and exhibit higher rates of sexual fertilization. Research has shown that the best approach for assembling hybrid crosses is to utilize females with high rates of sexual fertilization and males with low rates of sexual fertilization (Grazi et al., 1961; Hintzen, 1979; Gates, 1997). These parents are isolated together and encouraged to cross pollinate by tapping pollen from the paternal plant onto the maternal plant between the hours of 0100 and 0400 when the mother plant is most receptive to foreign pollen.

Once the fertilization period has concluded, the resulting progeny are evaluated to identify hybrids, also known as 'off-types', that do not resemble

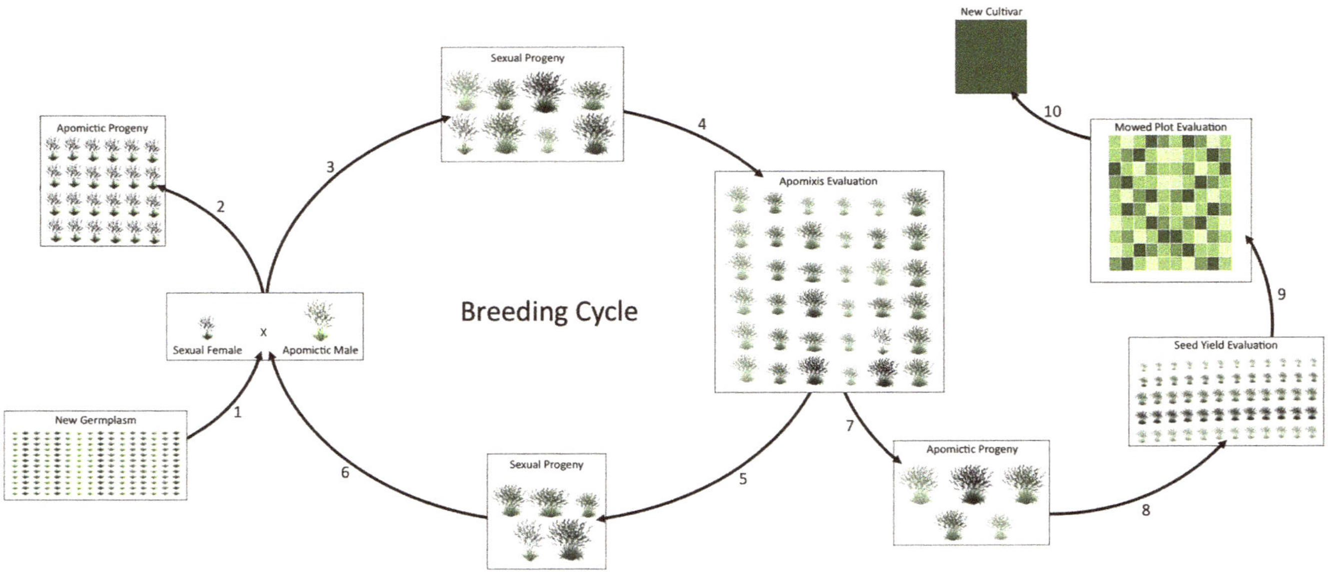

Figure 2 Breeding approach for apomictic turfgrass species. New germplasm sources are assessed and utilized for hybridization (1) using a sexual female parent and an apomictic male parent. Seed is harvested from the mother plant and progeny phenotypes are assessed; apomictic progeny are discarded (2), while sexual progeny are planted in a spaced-plant nursery (3). Seed is harvested from attractive sexual (hybrid) progeny and established in a subsequent spaced-plant nursery for assessment of apomixis levels (4). Lines with high levels of sexual progeny (5) are cycled back into the breeding program for subsequent hybridization (6), and lines with high levels of apomixis (7) are further evaluated for seed yield characteristics (8). Seed is harvested from elite lines and established in replicated mowed plot evaluation trials (9). After adequate evaluation and consumer product testing, a new cultivar is released (10).

the maternal phenotype. At this stage, progeny that resemble the maternal phenotype are presumed to have been derived from apomixis and are discarded. The hybrid, or sexual, progenies are established in spaced-plant nurseries to confirm that their phenotypes are distinct from the maternal phenotype. At this time, the suspected hybrids are also evaluated for seed yield potential. Seed is harvested from elite hybrids and established in subsequent spaced-plant nurseries to assess apomixis levels. Hybrids with apomixis levels of 85%, or higher, are further evaluated for seed yield potential and seed production traits. Seed is harvested from the lines that exhibit elite seed yield traits and established in replicated turf plot trials for turf quality evaluations. After adequate progeny testing has been completed, hybrid lines with elite turf quality and other important characteristics such as disease resistance, for example, are advanced for the development of a new cultivar.

Seed production potential is currently the major limiting factor for the development of a new Kentucky bluegrass cultivar. Therefore, as mentioned above, breeding efforts must incorporate seed yield evaluation into the breeding process. Other important foci for Kentucky bluegrass breeding programs are excellent turf quality characteristics, disease and insect resistance, and the ability to perform well under both high and low management inputs. The major diseases affecting Kentucky bluegrass include crown rust, stem rust, stripe rust (*Puccinia striiformis* Westend.), dollar spot, Bipolaris leaf spot (*Bipolaris sorokiniana* [Sacc.] Shoem.), powdery mildew (*Erysiphe graminis* D.C.), and summer patch.

2.3.3 Advanced breeding methods of Kentucky bluegrass

Several genetics and genomics studies have been conducted in Kentucky bluegrass over the last 20 years. These efforts have led to the development of markers such as AFLP (Porceddu et al., 2002; Bertoli et al., 2003; Van Treuren, 2008), selective amplification of microsatellite polymorphic loci (Porceddu et al., 2002), RAPD (Curley and Jung, 2004; Tian et al., 2006; Fard et al., 2012), SCAR (Xiaojun et al., 2015; Mingyue et al., 2020), sequence-related amplified polymorphism (SRAP) (Xiaojun et al., 2018), and SSR (Honig et al., 2010, 2012, 2018; Bushman et al., 2013; Xiaojun et al., 2018) markers. Various combinations of these marker sets have been used to assess genetic diversity within Kentucky bluegrass accessions (Bertoli et al., 2003; Tian et al., 2006; Fard et al., 2012; Bushman et al., 2013; Xiaojun et al., 2018), investigate genetic relationships among Kentucky bluegrass cultivars (Curley and Jung, 2004; van Treuren, 2008; Honig et al., 2012, 2018), construct genetic linkage maps (Porceddu et al., 2002), and identify new breeding lines (Xiaojun et al., 2015; Mingyue et al., 2020).

More recently, researchers have utilized transcriptome analyses to better understand genetic components of salt (Bushman et al., 2016) and drought (Bushman et al., 2021) stress response in Kentucky bluegrass. These types of studies enable scientists to identify candidate genes in Kentucky bluegrass, which may be further developed into breeding and selection tools or transgenic approaches for developing stress and pest tolerant Kentucky bluegrasses. Similarly, researchers are also investigating genetic components controlling apomictic processes in Kentucky bluegrass (Huff and Bara, 1993; Barcaccia et al., 1997; Porceddu et al., 2002; Patterson et al., 2005). Better understanding the genetics of apomixis and being able to manipulate this reproductive trait will be beneficial for future efforts in Kentucky bluegrass breeding.

2.4 Perennial ryegrass

2.4.1 Biological characteristics of perennial ryegrass

Perennial ryegrass is a bunch-type, obligate allogamous ($2n = 2x = 14$) perennial grass native to temperate regions of Asia, North Africa, and Europe (Terrell, 1968; Beard, 1973; Cornish, 1979). The species has been introduced to all parts of the world and is grown on every continent except Antarctica (Thorogood, 2003). Perennial ryegrass is a popular option for a broad range of turf applications because of its rapid germination and establishment, attractive turf quality characteristics, adaptability to both lower and higher mowing heights, and excellent wear tolerance (Beard, 1973; Funk and Clarke, 1989; Thorogood, 2003). In mild climates, perennial ryegrass is managed as a permanent turfgrass in parks, golf courses, athletic fields, recreational landscapes, and home lawns (Beard, 1973; Thorogood, 2003; Beard and Beard, 2005). In warmer climates, the species is primarily used to overseed warm-season turf during periods of winter dormancy (Beard and Beard, 2005; Thorogood, 2003; Turgeon, 2008).

There are several important fungal and Oomycete diseases that affect perennial ryegrass turf (Thorogood, 2003). In mowed turf situations, diseases such as Bipolaris leaf spot, brown patch, crown rust, dollar spot, gray leaf spot (*Pyricularia oryzae* Cavara), pink patch (*Limonomyces roseipellis* Stalpers and Loerakker), and red thread are primary concerns; stem rust is a devastating disease that occurs during seed production. Each of these diseases may occur at various times throughout the year and reduce the health, functionality, and aesthetic qualities of perennial ryegrass. Breeding efforts to date have improved resistance to many of these diseases in perennial ryegrass, but continued efforts are needed to ensure disease resistance is maintained in future cultivars.

2.4.2 Traditional breeding strategies and current breeding objectives of perennial ryegrass

Phenotypic and genotypic recurrent selection breeding strategies (Fig. 1) have been widely utilized in perennial ryegrass breeding programs to date. Given the widespread gray leaf spot disease issues on perennial ryegrass, most U.S. breeding programs evaluate seeded progeny plot trials for gray leaf spot disease resistance. Plants from breeding lines with gray leaf spot disease resistance and excellent turf quality are propagated, established in spaced-plant nurseries, and evaluated for additional traits of interest such as summer heat and drought stress performance, stem and crown rust resistance, and seed yield potential. Elite plants are clonally propagated and used to develop new perennial ryegrass populations. This cycle is repeated to continually improve perennial ryegrass populations for these and other traits of interest. This breeding strategy has been highly effective and continues to be the basis of many perennial ryegrass breeding programs.

Current breeding objectives for perennial ryegrass include improving seed yield potential and tolerance to abiotic and biotic stresses such as diseases, drought, heat, cold, salinity, and traffic. Concerning diseases, most breeding programs are actively focusing on the most common perennial ryegrass diseases such as Bipolaris leaf spot, crown rust, dollar spot, gray leaf spot, pink patch, red thread, and stem rust diseases. Protocols have been established to efficiently and effectively screen for resistance to many of these diseases in perennial ryegrass mowed progeny plot trials and spaced-plant nurseries. Perennial ryegrass germplasm should be continually monitored for the need to develop and implement evaluation protocols for other diseases including anthracnose, Microdochium patch (*Microdochium nivale* [Fries] Samuels and Hallett), Drechslera leaf spot, and Pythium blight.

2.4.3 Advanced breeding methods of perennial ryegrass

An immense collection of molecular genetics and genomics studies have been conducted in perennial ryegrass, and this species, within the major turfgrass species, has had by far the most attention for these types of studies. Perennial ryegrass has been the focus of this disproportionate number of studies because it is a diploid species, which makes genetic analyses more straightforward than other complex polyploids, and, in addition to being a major turfgrass species, it is also a major forage grass species (Fei, 2008). Various marker technologies including arbitrarily primed polymerase chain reaction (AP-PCR) (Huff, 1997), RAPD (Charmet et al., 1997; Kölliker et al., 1999), SSR (Kubik et al., 1999; Jones et al., 2001; Muylle et al., 2005b; Jensen et al., 2005), AFLP (Roldan-Ruiz et al., 2000; Cresswell et al., 2001; Skøt et al., 2002; Ghesquiere et al., 2003; Muylle et al., 2005a; Kalb et al., 2003), sequence-tagged site (Taylor et al., 2001), and RFLP (Yamada and Kishida, 2003) markers have been developed for perennial ryegrass. These marker sets have been implemented to differentiate ryegrass species (Charmet et al., 1997; Kölliker et al., 1999), assess genetic diversity (Cresswell et al., 2001; Skøt et al., 2002; Yamada and Kishida, 2003; Bolaric et al., 2005), identify cultivars (Huff, 1997; Kölliker et al., 1999; Roldan-Ruiz et al., 2000), and assess genetic relationships (Kubik et al., 1999; Ghesquiere et al., 2003). In addition, genetic markers have been utilized to develop genetic linkage maps (Hayward et al., 1994, 1998; Bert et al., 1999; Jones et al., 2002a,b).

Hayward et al. (1994) assembled the first genetic linkage map for ryegrass using isozyme, RAPD, and RFLP markers. This map was later refined using additional markers of the same type (Hayward et al., 1998). A second map was constructed using isozyme, AFLP, and expressed sequence tag (EST) markers (Bert et al., 1999) and improved with RFLPs (Jones et al., 2002b). This map was used to identify the locations of SSR markers from a subsequent study by Jones et al. (2002a). These and other genetic linkage maps have been used to map the perennial ryegrass self-incompatibility locus (Thorogood et al., 2002) and to map QTL for various traits including maturity-related characteristics (Hayward et al., 1994; Armstead et al., 2004; Byrne et al., 2009; King et al., 2008), vernalization response (Jensen et al., 2005), seed yield (Hayward et al., 1994; Armstead et al., 2008; Byrne et al., 2009; Sartie et al., 2018), morphological and developmental traits (Yamada et al., 2004), herbage quality traits (Cogan et al., 2005), winter hardiness (Yamada et al., 2004; Xiong et al., 2007; Paina et al., 2016), and resistance to crown rust (Thorogood et al., 2001; Dumsday et al., 2003; Muylle et al., 2005b; Sim et al., 2007; Schejbel et al., 2007), gray leaf spot (Curley et al., 2004, 2005, 2008), and powdery mildew (Schejbel et al., 2008) diseases.

Most recently, perennial ryegrass molecular research has shifted toward advanced genomics methods, which has led to the development of a transcriptome map (Studer et al., 2012), a synteny-based draft genome sequence (Byrne et al., 2015), and a reference-grade genome assembly (Frei et al., 2021). Such resources are useful tools for other approaches such as genome-wide association studies (Jaškūnė et al., 2020; Fois et al., 2022) and genomic selection (Lin et al., 2016; Faville et al., 2018, 2020). These studies will continue to be of primary interest to perennial ryegrass breeders for developing enhanced molecular breeding strategies.

2.5 Tall fescue

2.5.1 Biological characteristics of tall fescue

Tall fescue is a cool-season, bunch-type perennial grass that originated in Africa and Europe (Buckner et al., 1979). This allohexaploid ($2n = 6x = 42$)

species was introduced to the United States in the late 1800s and gained popularity in the 1940s following the release of the first commercial cultivars, 'Alta' and 'Kentucky-31', which were major forage grasses in the United States (Fergus and Buckner, 1972; Meyer and Watkins, 2003; Beard, 2012). The first turf-type tall fescue cultivar, 'Rebel', was released in 1981, and tall fescue has since become one of the major species used for home lawn, recreational, golf course, and sports field turf throughout the northern and transitional climates of the United States (Funk et al., 1981). Heat and drought tolerance, persistence under low-maintenance conditions, winter hardiness, and shade performance are examples of attributes that have resulted in the popularity of tall fescue use (Carrow, 1996; Tegg and Lane, 2004; Miller et al., 2013).

2.5.2 Traditional breeding strategies and current breeding objectives of tall fescue

Tall fescue is primarily bred using a phenotypic and genotypic recurrent selection breeding scheme as shown in Fig. 1. Typically, the breeding cycle includes evaluations of mowed turf plot trials for resistance to major diseases such as brown patch and gray leaf spot. Plants are identified within breeding lines that exhibit good disease resistance and excellent turf quality characteristics. These plants are propagated and further evaluated for resistance to seed production diseases such as stem and crown rust, summer heat and drought stress performance, and seed yield potential. Elite plants are clonally propagated and assembled into isolated crossing blocks. These individuals are allowed to cross pollinate, seed is harvested from each mother plant, and the progeny are tested for various traits of interest in turf plot evaluation trials.

Tall fescue breeding objectives have centered around developing cultivars with high seed yield potential, superior turf quality characteristics, disease resistance, traffic tolerance, improved winter performance, shade tolerance, and performance under heat and drought stress. Stem rust is a major disease that limits seed yield potential; therefore, breeding for resistance to this disease is also a primary objective of tall fescue breeding programs. Brown patch has been a problematic disease of tall fescue turf for many years, and tall fescue breeding efforts have focused heavily on improving this trait. More recently, widespread gray leaf spot epiphytotics have plagued tall fescue turf throughout the United States, and breeding for resistance to this disease has become an objective in tall fescue breeding programs.

As described earlier, selection for stem rust resistance is a major focus of tall fescue breeding programs, as this disease greatly affects seed production. In addition, breeding efforts are also focused on breeding for resistance to other important diseases that occur in mowed turf settings. Screening methods have been developed to actively evaluate for disease resistance to brown patch, crown rust, gray leaf spot, Drechslera leaf spot, pink patch, and red thread diseases. In addition to active selection for resistance to these diseases, breeders continue to monitor the need for developing and implementing evaluation protocols for other diseases of tall fescue such as dollar spot and Pythium blight.

2.5.3 Advanced breeding methods of tall fescue

Although genetics and genomics tools for tall fescue are not as far advanced as those in perennial ryegrass, there has still been good progress in developing molecular approaches for germplasm improvement in tall fescue. Some of the earliest efforts assess genetic diversity in tall fescue accessions using RFLP (Xu et al., 1994; Busti et al., 2004), AFLP (Mian et al., 2002), and expressed sequence-tagged simple sequence repeat (EST-SSR) (Mian et al., 2005) markers. Other researchers have used chloroplast, ribosomal, and nuclear DNA (Charmet et al., 1997) and RFLP markers (Xu et al., 1994) to investigate phylogenetic relationships between tall fescue and other closely related species. These studies concluded that tall fescue was more closely related to perennial ryegrass than fine fescue species, and these findings were corroborated by the work of Catalán et al. (2004).

In addition to the aforementioned markers, a public database of EST sequences for tall fescue was developed from various treatments and plant tissues in tall fescue (Saha et al., 2004; Mian et al., 2008). These resources are valuable for the development of markers for *Festuca* and other closely related species (Saha et al., 2006). Xu et al. (1995) constructed the first genetic linkage map of tall fescue using RFLP markers. A PCR-based genetic linkage map was developed shortly thereafter using AFLP, EST-SSR, and genomic SSR markers (Saha et al., 2005), and QTL for forage quality traits was mapped in a subsequent study (Saha et al., 2009). More recently, a linkage map was constructed using SSR and diversity arrays technology markers and used to map QTL for summer dormancy traits in forage tall fescue (Talukder et al., 2020).

Much of the work outlined previously has been contributions from researchers studying forage-type tall fescue. Molecular tools for turf-type tall fescue have not advanced as far as those for forage types. Nonetheless, many of the resources developed in forage-type tall fescue will apply to turf-type tall fescue breeding efforts. Another fascinating area of development in tall fescue research has been microprojectile bombardment and *Agrobacterium*-mediated transformations for resistance to gray leaf spot and brown patch diseases (Dong et al., 2008) and improved salt and drought tolerance (Tian et al., 2006; Zhao et al., 2007; Cao et al., 2009). These and additional molecular genetics and genomics tools will be valuable in improving breeding and selection efficiency for important turfgrass-related traits in modern turfgrass breeding programs.

3 Breeding of warm-season turfgrasses

Organized breeding of warm-season turfgrasses began in the 1930s with Dr. L. E. Wenger's efforts in breeding buffalograss at Kansas State University (Manhattan, KS, USA) (Wenger, 1943). Bermudagrass breeding was initiated just a few years later, in the 1940s, by Dr. G. Burton at the Coastal Plains Experiment Station (Tifton, GA, USA) (Burton, 1991). Since these early beginnings, warm-season turfgrass breeding efforts have greatly expanded and public breeding programs now exist at Mississippi State University (Starkville, MS, USA), North Carolina State University (Raleigh, NC, USA), Oklahoma State University (Stillwater, OK, USA), Texas A&M University (Dallas, TX, USA), the University of California (Riverside, CA, USA), the University of Florida (Gainesville, FL, USA), the University of Georgia (Tifton and Griffin, GA, USA), and the University of Nebraska (Lincoln, NE, USA).

The past 90 years of warm-season turfgrass breeding have led to the development of many useful species for turfgrass applications. The major warm-season turfgrass species include bermudagrass, buffalograss, seashore paspalum, St. Augustinegrass, and zoysiagrass. Other species such as bahiagrass (*Paspalum notatum* Flügge), carpetgrass (*Axonopus fissifolius* [Raddi] Kuhlm.; synonym = *Axonopus affinis* Chase), centipedegrass, and kikuyugrass (*Cenchrus clandestinus* [Hochst. ex Chiov.] Morrone; synonym = *Pennisetum clandestinum* Hochst. ex Chiov.) are minor turfgrass species and are utilized for niche and low-maintenance applications. The major warm-season species are described in the following sections.

In recent years, turfgrass breeders, physiologists, and extension specialists from North Carolina State University, Oklahoma State University, Texas A&M University, the University of Florida, and the University of Georgia have been working collaboratively to develop drought and salinity tolerance in major warm-season species including bermudagrass, seashore paspalum, St. Augustinegrass, and zoysiagrass. This work, funded by the USDA-NIFA Specialty Crops Research Initiative (Washington, DC, USA), has facilitated the exchange of germplasm between public breeding programs and the evaluation of genetic material under diverse environments. This coordinated breeding effort has supported the development of improved germplasm and release of commercially available cultivars such as 'TifTuf®' and 'Tahoma 31®' bermudagrass, 'Citrablue®' and 'Cobalt™' St. Augustinegrass, and 'CitraZoy™' and 'Lobo™' zoysiagrass.

3.1 Bermudagrass

3.1.1 Biological characteristics of bermudagrass

Bermudagrasses (*Cynodon* spp.) are a group of eight species that are native to Africa and Eurasia (de Wet and Harlan, 1970; Harlan, 1970; Harlan et al., 1970).

Of the eight species in *Cynodon*, African bermudagrass (*C. transvaalensis* Burtt-Davy) and common bermudagrass (*C. dactylon* Pers. var. *dactylon*) have had the greatest impact on the turf industry (Taliaferro, 2003; Hanna et al., 2013). More specifically, sterile, triploid interspecific hybrids produced from crosses between *C. dactylon* and *C. transvaalensis* have resulted in the most widely utilized turf-type bermudagrasses over the past 60 years (Taliaferro, 2003; Hanna et al., 2013). Bermudagrasses are warm-season, perennial grasses used for golf courses, athletic fields, parks, recreational areas, home lawns, and roadsides (Beard, 1973, 2012). Bermudagrasses are the predominant turfgrass in temperate climates with hot summers and mild winters; they are also becoming more widely used in areas of the U.S. transition zone (Beard, 1973).

All species in *Cynodon* share basic morphological characteristics; however, turf-type hybrid bermudagrasses have finer leaf and stem textures, higher shoot densities, and lower growth habits than other types (Beard, 1973; Taliaferro, 2003). Bermudagrass spreads by stolons and rhizomes, which provide good wear tolerance and recuperative capabilities and enable propagation via sprigs or sod (Beard, 1973). This grass has very good drought and heat tolerance but does not have good tolerance to shade or cold stress (McBee and Holt, 1966; Harlan and de Wet, 1969; Beard, 1973). Differences have been documented among bermudagrass cultivars with respect to cold tolerance (Harlan and de Wet, 1969). However, bermudagrass growth generally slows with a decrease in temperature, and as temperatures consistently fall below 0°C, foliage dies, and underground organs enter a state of dormancy (Taliaferro, 2003). In the spring, as temperatures begin to increase, bermudagrass growth resumes from crowns, stolons, or rhizomes that survived the winter (Taliaferro, 2003).

3.1.2 Traditional breeding strategies and current breeding objectives of bermudagrass

Bermudagrasses are highly outcrossing species because of various self-incompatibility mechanisms (Burton, 1947; Burton and Hart, 1967; Taliaferro, 2003; Hanna et al., 2013, 2016). This characteristic has been the foundation of breeding strategies for advancing *Cynodon* germplasm over the past 60 years. Bermudagrass may be propagated vegetatively or by seed, and breeding strategies have been designed for the development of each. Vegetatively propagated cultivars are produced using intra- or interspecific hybridization breeding strategies; however, because the resulting cultivars will be clonally propagated via sod or sprigs, they will be genetically and phenotypically uniform throughout a landscape. Breeding strategies for vegetative cultivars are outlined in Fig. 3. Seeded cultivars, on the other hand, are bred as populations of individuals that will share similar phenotypic traits but will be genetically diverse; this breeding strategy is based on the approach outlined

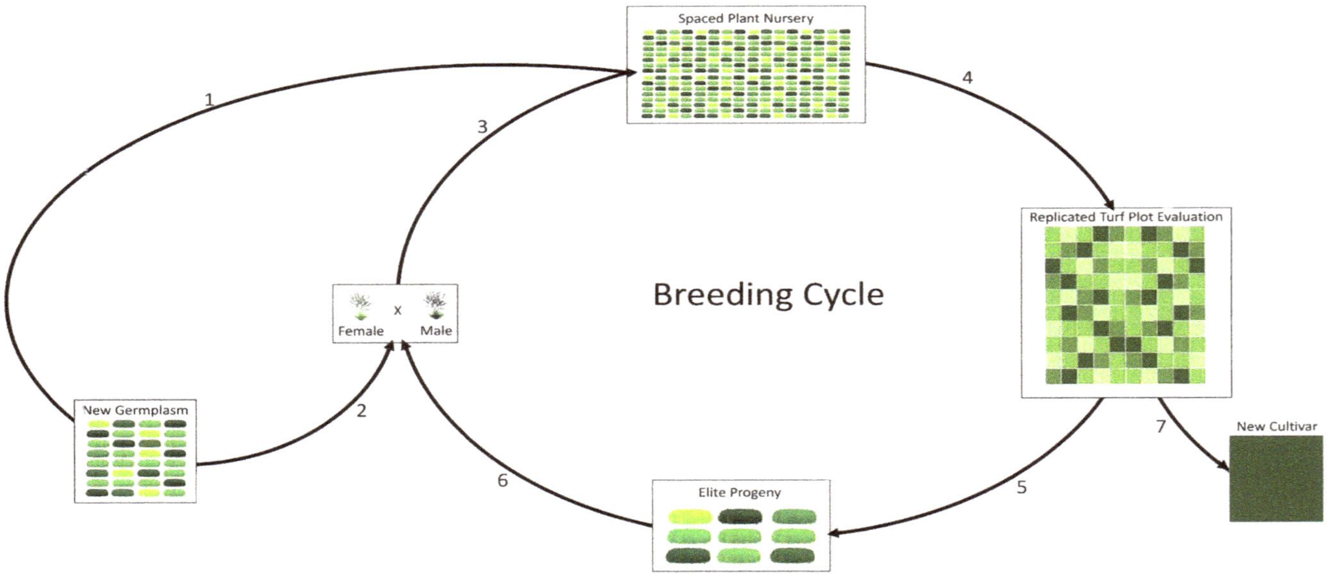

Figure 3 Breeding approach for warm-season, vegetatively propagated turfgrass species. New germplasm collections are established in spaced-plant nurseries (1) or used for crossing (2). Hybrid progeny are established in spaced-plant nurseries (3). Elite progeny are planted into replicated mowed plot trials for evaluation across different locations and various management and environmental conditions (4). Elite progeny are identified (5) and used for subsequent breeding cycles (6). After adequate evaluation and consumer product testing, a new cultivar is released (7).

in Fig. 1. There are benefits and limitations to both breeding approaches, as outlined next.

Since vegetatively propagated cultivars are established via sprigs or sod, there is no need to evaluate floret fertility, seed production potential, or progeny stability as is necessary with seed-propagated cultivars; thus, intra- and interspecific hybrids are immediately ready for turf performance evaluations (Burton, 1991). The tetraploid ($2n = 4x = 36$) *C. dactylon* var. *dactylon* has been improved for various turf characteristics using an intraspecific hybridization approach. In this method, parent clones of *C. dactylon* var. *dactylon* are paired together and allowed to interpollinate. Seed is harvested from each of the parent clones, and each seed is considered a unique progeny, which will be evaluated for turf performance characteristics. After adequate testing cycles, genotypes that have attractive characteristics may be advanced in a breeding program for subsequent breeding projects or released as an improved vegetatively propagated *C. dactylon* var. *dactylon* cultivar. 'Tiflawn' is an example of a tetraploid intraspecific hybrid bermudagrass, which was released in 1952 from the University of Georgia (Hein, 1953; Burton, 1991).

Interspecific hybridization has been the most effective means of cultivar development in bermudagrass breeding programs to date. This procedure involves pairing a parental clone of the diploid ($2n = 2x = 18$) *C. transvaalensis* with a parental clone of the tetraploid ($2n = 4x = 36$) *C. dactylon* to generate triploid ($2n = 3x = 27$) hybrids. Each resulting seed from an interspecific cross represents a potential new interspecific hybrid. When *C. transvaalensis* is used as the maternal parent, hybrids are readily identifiable, as any selfed individuals will have fine stem and leaf texture characteristics of the mother plant; these hybrids are often confirmed via chromosome counts or flow cytometry. When *C. dactylon* is used as the maternal parent, hybrids are confirmed using flow cytometry and chromosome counts. The resulting interspecific hybrid progeny are largely both seed and pollen sterile, which are desirable characteristics for the preservation of uniformity of these vegetatively propagated grasses (Hanna et al., 2016). Examples of triploid interspecific hybrid bermudagrasses include the releases of 'Tiffine' in 1953 (Hein, 1953; Robinson and Burton, 1953; Burton, 1955, 1991), 'Tifgreen' in 1956 (Hein, 1961; Burton, 1964, 1991), 'Tifway' in 1960 (Burton, 1966a, 1991), 'Tifdwarf' in 1965 (Burton, 1966b, 1991), 'TifSport' in 1995 (Hanna et al., 1997; Hanna and Elsner, 1999), 'TifEagle' in 1997 (Hanna et al., 1997; Hanna and Elsner, 1999), 'TifGrand®' in 2009 (Hanna et al., 2010), and TifTuf in 2014 (Schwartz et al., 2018) from the University of Georgia and the releases of 'Patriot' in 2004 (Taliaferro et al., 2006), 'Northbridge®' in 2012 (Wu et al., 2013), 'Latitude 36®' in 2012 (Wu et al., 2014), and Tahoma 31 in 2018 (Wu et al., 2019b) from Oklahoma State University.

Breeding efforts for seed-propagated bermudagrass cultivars do not date back as far as those for vegetatively propagated cultivars. However, notable progress has been made over the past 40 years and seed-propagated bermudagrass cultivars are becoming increasingly utilized at present time. Researchers at New Mexico State University (Las Cruces, NM, USA), Oklahoma State University, and the University of Kentucky (Lexington, KY, USA) studied various traits in bermudagrass and found that many important turf characteristics are heritable (Wofford and Baltensperger, 1985; Cluff and Baltensperger, 1991; Baltensperger et al., 1993; Kenworthy et al., 2006). Accordingly, seeded bermudagrass cultivars have been bred using phenotypic and genotypic recurrent selection techniques (Fig. 1) to improve traits such as color, density, texture, leaf length and width, internode length, spring green-up, and seed yield (Baltensperger et al., 1993). For this breeding approach, parent lines of *C. dactylon* are selected from a stock nursery and placed in isolation for cross pollination. The number of parent lines can range widely, from as few as two parents to greater than 50 parents (Taliaferro, 2003). After cross pollination, seed is harvested from each parent plant and established in mowed progeny plot trials. Germplasm is evaluated for turf quality traits and tolerance to various abiotic and biotic stresses, and superior progeny lines are identified. After adequate cycles of evaluation, breeders will elect to move forward with elite populations for cultivar development.

3.1.3 Advanced breeding methods of bermudagrass

Genetics and genomics approaches have been studied and utilized for various research and breeding objectives in *Cynodon* since the mid-1990s. DNA marker technologies including DNA amplification fingerprinting (Caetano-Anollés et al., 1995, 1997; Ho et al., 1997; Assefa et al., 1999), AFLP (Zhang et al., 1999; Wu et al., 2004, 2006; Kang et al., 2008), ISSR (Liu et al., 2007; Huang et al., 2010a; Li et al., 2011; Farsani et al., 2012; Wang et al., 2013), RAPD (Roodt et al., 2002; Zheng et al., 2005), SRAP (Yi et al., 2008; Huang et al., 2010b, 2014; Wang et al., 2011; Zheng et al., 2017), SSR (Xie et al., 2015; Yang et al., 2018), and EST-SSR (Tan et al., 2012; Khanal et al., 2017) have been developed for bermudagrass. These marker sets have been implemented to assess the origins of off-types in *Cynodon* (Caetano-Anollés et al., 1997), conduct association mapping studies for cadmium tolerance in *Cynodon dactylon* (Xie et al., 2015), characterize and identify seeded bermudagrass cultivars (Yang et al., 2018), and investigate genetic diversity (Wu et al., 2004; Liu et al., 2007; Kang et al., 2008; Yi et al., 2008; Huang et al., 2010a,b; Li et al., 2011; Farsani et al., 2012; Wang et al., 2013; Huang et al., 2014; Xie et al., 2015; Zheng et al., 2017), genetic relationships (Caetano-Anollés et al., 1995; Ho et al., 1997; Assefa et al., 1999; Zhang et al., 1999; Wu et al., 2006; Wang et al., 2011), and population structure (Zheng et al., 2017) within various *Cynodon* accessions.

In addition to the DNA marker technology described earlier, RNA-seq methods have also been utilized in *Cynodon dactylon* to conduct transcriptome analyses for low temperature tolerance (Chen et al., 2015) and cold acclimation (Zhu et al., 2015). Researchers are continuing to improve and develop these types of molecular genetics and genomics tools, and they will be valuable for future breeding and selection practices in bermudagrass breeding programs.

3.2 Buffalograss

3.2.1 Biological characteristics of buffalograss

Buffalograss is a dioecious grass with environmentally stable 1:1 male-to-female sex ratios, although monoecious forms have been reported in natural populations (Huff and Wu, 1992). Buffalograss is native to semi-arid regions of the great plains of North America and exhibits a wide range of adaptations. It is an autopolyploid ($x = 10$) where hexaploids and tetraploids are most widely distributed across the great plains and diploid and pentaploid accession have only been rarely reported (Huff et al., 1993; Johnson et al., 2001). Freezing tolerance and turfgrass performance characteristics appear to be partitioned between ploidy levels where diploids generally have lower freezing tolerance and more southern adaptation (Huff et al., 1993) but exhibit longer growing season, faster-spreading growth habit, taller seed stalks (an important trait for seed harvesting), and increased disease resistance. Hexaploids and tetraploids, on the other hand, exhibit higher freezing tolerance and greater adaptation to northern and western regions of the great plains (Huff et al., 1993; Johnson et al., 1998) and exhibit the earlier onset of fall dormancy, slower spreading growth habit, dense canopy, and shorter seed stalks. Plant breeders have been working on combining desirable characteristics into a single genotype for clonally propagated cultivar development or into populations for seeded cultivar development.

3.2.2 Traditional breeding strategies and current breeding objectives of buffalograss

Buffalograss is a stoloniferous, sod-forming perennial warm-season grass with extreme temperature tolerance, drought resistance, and low mowing and nitrogen requirements, making it a species of choice especially in low-maintenance environments. Breeding efforts in buffalograss began as early as 1936 when L. E. Wenger (1941, 1943) utilized the hybridization and selection method of plant breeding to make genetic improvements and identify transgressive segregants. Some of the earliest cultivars of buffalograss, such as 'Texoka' and 'Sharps Improved', were released in 1965 as seeded types for forage use (Ahring, 1964). With renewed interest from the turfgrass industry in cultivars requiring low management and inputs and funding support from the United States Golf Association in the 1980s, several new cultivars of buffalograss were developed and released. 'Prairie' buffalograss was the first buffalograss developed and released specifically for turfgrass use by Texas Agricultural Experiment Station in 1989, and it was also the first sodded cultivar of buffalograss. Soon thereafter, in 1991, the University of Nebraska Agricultural Experiment Station released '609', which was another sodded cultivar of buffalograss for turfgrass use, and both Prairie and 609 paved the way for the buffalograss sod industry.

Buffalograss lacks underground rhizomes and is vegetatively propagated through stolons only and therefore, important breeding objectives while developing vegetatively propagated cultivars (Fig. 3) of buffalograss is faster lateral growth for a faster grow-in time along with high sod tensile strength and handling quality. For seeded cultivar development (Fig. 1), major challenges are low bur yield, short female inflorescence height (hidden burs in the canopy) making mechanical harvest difficult, poor bur dormancy and germination, and limited herbicide options that do not suppress seed germination. Several chemical and mechanical methods, temperature treatments, and tissue culture techniques have been tested for breaking bur dormancy and improving seed germination, but all these steps add to the cost of seed production. Regardless of seeded or vegetatively propagated cultivar development, other important breeding objectives are improving shade tolerance, traffic tolerance, pest resistance, lateral spread, shoot density, and genetic color. Continued breeding effort has resulted in new and improved cultivars; for example, the seeded cultivars 'Cody' in 1997, 'Bowie' in 2001, and 'Sundancer' in 2014 and the vegetative cultivars 'Legacy' and 'Prestige' in 1997. Some of the newer cultivars address several of these challenges as continued breeding efforts have produced genetic variability for these desirable traits and characteristics.

3.3.3 Advanced breeding methods of buffalograss

Molecular characterizations of intra- and interpopulations of buffalograss have been performed using genetic markers to study genetic diversity in the germplasm (Huff et al., 1993; Budak et al., 2004; Gulsen et al., 2005; Wu et al., 2019a). Budak et al. (2004) demonstrated the utility of different genetic markers such as ISSRs, RAPDs, SRAPs, and SSRs to study the genetic relatedness among and within seeded and vegetative biotypes of buffalograss and to demonstrate the lack of genetic relationship among populations based solely on their geographic origin (Budak et al., 2005). More recently, research efforts in the past decade have been focused on developing genomic resources in buffalograss to assist in breeding and cultivar development; for example, the transcriptome analysis between '378' and 'Prestige' cultivars of buffalograss (Wachholtz et al., 2013), identification of differentially expressed genes in response to western chinch bugs (*Blissus occiduus* Barber) (Ramm et al., 2015) and the leaf spot pathogen [*Curvularia inaequalis* (Shear) Boedijn] (Amaradasa and Amundsen, 2016). Currently, a collaboration between Pennsylvania State University and the University of Nebraska is working toward developing a fully annotated and high-quality genomic sequence of a diploid buffalograss with chromosome-level resolution (Huff et al., 2021).

3.3 Seashore paspalum

3.3.1 Biological characteristics of seashore paspalum

Seashore paspalum is a warm-season perennial grass that is used for erosion control, sand dune stabilization, parks, athletic fields, and golf courses (Duncan, 2003). It is predominantly a self-incompatible, obligately outcrossing, diploid ($2n = 2x = 20$) species (Carpenter, 1958; Bashaw et al., 1970; Burson et al., 1973; Okoli, 1982; Quarin and Burson, 1983). It is heterozygous with gametophytic self-incompatibility (Espinoza and Quarín, 1997), spreads by stolons and rhizomes, and has an intermediate to fine leaf texture with a vibrant blue-green color. Seashore paspalum is considered the most salt-tolerant warm-season turfgrass, with some ecotypes tolerating up to ocean water salt levels ($ECe = 54\ dS\bullet m^{1}$) (Carrow and Duncan, 1998). It is adapted to a soil pH range of 3.6–10.2 and is tolerant of low light, waterlogged soils, and low cutting heights. The natural habitat for seashore paspalum is sandy beaches and coastal area banks. It also is found in saline environments in more inland areas (Duncan and Carrow, 2000). Seashore paspalum is native to coastal areas of tropical and subtropical regions of North and South America (Duncan, 2003). Under proper maintenance, seashore paspalum has few diseases and insect problems but can be susceptible to damage from nematodes, insects, and diseases when grown further inland, with dollar spot disease being the most problematic (Duncan and Carrow, 2000). These characteristics make seashore paspalum an important turf species due to increasing restrictions on freshwater use for turfgrass irrigation and the need to utilize secondary irrigation sources. Unpredictable rainfall patterns and drought periods which are driven by climate change make seahorse paspalum an important species to help in the reduction of potable freshwater irrigation by using gray or effluent water for irrigation (Taylor et al., 2013).

3.3.2 Traditional breeding strategies and current breeding objectives of seashore paspalum

Traditional breeding of seashore paspalum is achieved by conducting pairwise crosses of accessions and breeding lines. Each seed is then germinated, and the plants are screened for traits of interest as a selection technique. Cross pollination is conducted during the early morning hours from 0500 to 0730 at a temperature of around 18–21°C (Burson, 1985). This genetic recombination has been utilized by breeders to generate new genetic diversity to evaluate for many different traits that benefit end users such as density, color, texture, lateral spread, salt tolerance, and resistance to diseases and insects. Cultivars of seashore paspalum are predominately from a single genotype and vegetatively propagated and sold as sprigs or sod; this approach is similar to that outlined in Fig. 3. Seeded cultivars of seashore paspalum have been developed, with the first release being 'Sea Spray' (Fricker et al., 2007). One method for generating new experimental lines is to cross salt-tolerant parental lines and then subject the seedlings to saltwater flood irrigation to eliminate individuals or groups that did not inherit a high level of salt tolerance. From those highly salt-tolerant lines, plants are then selected visually for quality traits such as color, texture, and growth habit. The lines are planted in the field to be further evaluated for additional traits listed above.

3.3.3 Advanced breeding methods of seashore paspalum

Genetic and genomic research in seashore paspalum has been somewhat limited compared to other turfgrass species. Genetic diversity was evaluated for 46 seashore paspalum accessions using RAPD and SSR markers in two studies where the accessions clustered by leaf texture (Liu et al., 1994, 1995). The same SSR markers were then used to genotype ten accessions of *Paspalum* (Brown et al., 1998). In another study by Wang et al. (2006), 40 transferred polymorphic SSR markers from wheat maize and sorghum were used for characterization and evaluation of 73 *Paspalum* accessions. Their study identified 18 markers to sufficiently distinguish those accessions. Harris-Shultz et al. (2013) used a total of 80 SSR markers to assess genetic relationships among 18 *P. vaginatum* accessions. A total of 43 SSR markers were used in a set of 97 *Paspalum* accessions along with ploidy analysis to evaluate ploidy level, genetic diversity, population structure, and distribution history (Eudy et al., 2017). High-density genetic maps were constructed using genotyping by sequencing of progeny from two genetically diverse and heterozygous accessions, 509022 and HI33 (Qi et al., 2019). The work above has increased the understanding of the genetic relationships of *P. vaginatum* accessions and cultivars and has provided a basis for understanding which regions of the genome are involved with various important traits like salt tolerance.

3.4 St. Augustinegrass

3.4.1 Biological characteristics of St. Augustinegrass

St. Augustinegrass is one of the most shade-tolerant warm-season grasses widely used on lawns and landscapes in warm-humid, tropical, and subtropical regions. It is a stoloniferous turfgrass produced as sod or plugs. There are no known seeded cultivars of St. Augustinegrass that are currently commercially available. The genus *Stenotaphrum* has been reported to consist of 7 species, of which 17 St. Augustinegrass and 2 accessions of pembagrass (*S. dimidiatum* [L.] Brongn.) are available through USDA National Plant Germplasm System. St. Augustinegrass genotypes can be categorized into four groups based on their morphological characteristics, namely, the breviflorus race, the longicaudatus race, the Bitterblue race, and the Floratam group (Busey, 1986).

Overall, the genetic diversity available in St. Augustinegrass germplasm base is limited and partitioned between ploidy levels where most of the variation exists between ploidy levels. Diploid ($2n = 2x = 18$) genotypes exhibit finer leaf texture, higher shoot density, more dwarfed stature, brighter green color, and enhanced shade and cold tolerance as compared to polyploids ($2n = 27, 30, 32$), while polyploids are known to exhibit tolerance to biotic stresses like southern chinch bugs (*Blissus insularis* Barber) and sting nematodes (*Belonolaimus longicaudatus* Rau) and abiotic stress such as drought tolerance. Likewise, tetraploid or hexaploid pembagrass accessions are known to have tolerance to gray leaf spot disease [*Pyricularia grisea* (Cke.) Sacc.], southern chinch bugs, and sting nematodes. Several commercially available cultivars are either diploids or polyploids (including aneuploids) and not interploid (polyploid x diploid) hybrids because of sterility barriers between the ploidy levels.

3.4.2 Traditional breeding strategies and current breeding objectives of St. Augustinegrass

St. Augustinegrass breeding efforts have been underway since the 1940s. Breeding programs at the University of Florida, Texas A&M University, North Carolina State University, and Mississippi State University have been working on making genetic improvements in St. Augustinegrass by utilizing traditional (Fig. 3) and modern breeding methods. 'Floratam', an aneuploid cultivar of St. Augustinegrass with 32 chromosomes (Busey, 1979), was jointly released by the University of Florida and Texas A&M University in 1973 (Horn et al., 1973) for its tolerance to southern chinch bugs and the viral disease St. Augustine Decline. Floratam continues to be the predominant cultivar accounting for 81% of sod production in Florida (Satterthwaite et al., 2009) and serves as an industry standard for drought resistance in St. Augustinegrass (Steinke et al., 2010). 'Raleigh', a diploid cultivar, was released by North Carolina State University in 1980 (Bateman, 1980) and continues to serve as an industry standard for cold tolerance. Approximately 25 cultivars of St. Augustinegrass including 'Palmetto®', 'Seville', 'Bitterblue', 'Sapphire', 'Amerishade', 'DelMar', 'Mercedes', 'Classic', 'FJ Select', and Texas common have been commercialized in the past 80 years. More recently, the University of Florida released 'Captiva®' (NUF-76), a diploid cultivar (Lu et al., 2015) resistant to the polyploid-damaging populations of southern chinch bugs that overcame the resistance in Floratam.

Sterility barriers and improper pairing of chromosomes between ploidy levels have impeded the use of genetic variation present between the ploidy levels until Genovesi et al. (2009, 2017) at Texas A&M University demonstrated the use of embryo-rescue technology to break the sterility barrier and to facilitate gene exchange between ploidy levels. The first embryo-rescue-derived interploid hybrid to be commercialized was 'TamStar™' (DALSA 0605) by Texas A&M University (Chandra et al., 2015) for its superior drought resistance. Interploid hybrids have been shown to exhibit improved drought resistance (Meeks and Chandra, 2020), shade tolerance (Meeks and Chandra, 2021), and southern chinch bug resistance (van der Laat et al., 2022) as well as overall good turfgrass quality characteristics. Another embryo-rescue-derived interploid hybrid released in 2021 by Texas A&M University was Cobalt (DALSA 1618) for its superior drought resistance, shade tolerance, and dark green genetic color. In 2020, the University of Florida released Citrablue (FSA 1602), a polyploid cultivar (Kenworthy and Quesenberry, 2020) originating from a natural cross between a polyploid and a diploid St. Augustinegrass. Citrablue exhibits excellent turfgrass color and quality across multiple environments (Pereira et al., 2022).

3.4.3 Advanced breeding methods of St. Augustinegrass

Application of molecular markers in St. Augustinegrass has mostly been in the past decade for genetic diversity analysis and use in genetic linkage map construction. AFLP was used by Kimball et al. (2012) and demonstrated varying levels of genetic similarity between the original stock of vegetatively propagated Raleigh St. Augustinegrass and 46 samples of Raleigh obtained from sod farms of six different states. Milla-Lewis et al. (2013) used cytological tools, flow cytometry, and AFLP markers to characterize commercial cultivars and plant introductions belonging to different ploidy levels. Likewise, Mulkey et al. (2014) used Illumina sequencing and Wang et al. (2017) used the

454 sequencing technology to isolate microsatellite markers from the St. Augustinegrass germplasm, and Luo et al. (2020) used SNPs, ISSR, and SRAP to assess genetic diversity in the species. More recently, genomic studies have been underway to identify QTL associated with turfgrass quality (Yu et al., 2018) and agronomically important traits like freezing tolerance (Kimball et al., 2018), drought tolerance (Yu et al., 2019), and gray leaf spot resistance (Yu et al., 2020), an important step forward toward the use of these genetic markers for marker-assisted breeding.

Another recent advancement is the development of the first genetically modified St. Augustinegrass called ProVista® St. Augustinegrass by Scotts Miracle Gro. Co. (Marysville, OH, USA). Derivatives of Floratam and Raleigh were genetically modified using particle bombardment with *Arabidopsis thaliana*-derived enolpyruvylshikimate-3-phosphate synthase gene (Harriman et al., 2019) conferring resistance to glyphosate, a nonselective herbicide providing consumers with weed control option for their lawns previously not available. These derivatives were also genetically modified with spinach-derived gibberellic acid 2-oxidase gene resulting in dwarfing through reduced expression of gibberellic acid (Harriman et al., 2021) and consequently requiring reduced frequency of mowing by the consumers (Buhlman et al., 2019; Cherry et al., 2021). In collaboration with Texas A&M University's turfgrass breeding program, these genetically modified Floratam and Raleigh derivatives were hybridized with elite Texas A&M germplasm using embryo-rescue technology to develop ProVista St. Augustinegrass with overall improved stress tolerance and performance. Compared to some of the other warm-season turfgrasses, the germplasm base and available genetic variability in St. Augustinegrass is limited and therefore, use of technological advancements such as embryo-rescue, mutagenesis, transgenic approaches will continue to prove useful. Efforts to acquire new germplasm and subsequently characterizing and incorporating new genetic variation in the breeding process will enable breeders to make significant future advancements.

3.5 Zoysiagrass

3.5.1 Biological characteristics of zoysiagrass

Zoysiagrass is a perennial turfgrass that reproduces clonally through stolons and rhizomes and sexually through seed. It was introduced to the United States in the 1890s from the countries in East Asia and the Pacific Islands and is adapted to the warm, humid, and arid southern climatic region of the United States. It is a highly heterozygous and heterogeneous turfgrass with ten known species (Loch, 2015), all of which are allotetraploids ($2n = 4x = 40$) and cross-compatible promoting recombination among species (interspecific hybridization) and producing a range of phenotypic variation. Of these known species, *Zoysia japonica* Steud. (leaf width > 2.5 mm), *Z. matrella* (L.) Mar. (leaf width 1.5 mm to 2.5 mm), and *Z. pacifica* (Goudswaard) M. Hotta and S. Kuroki (leaf width < 1.5 mm) are the most commonly used species in the United States by turfgrass breeders and practitioners (Patton et al., 2017). Forbes (1955) theorized that *Zoysia* spp. are subpopulations or ecotypes within one species; this theory was later supported by Kimball et al. (2013) and Tanaka et al. (2016).

In addition to morphological differences among species, a wide range of genetic variation exists in zoysiagrass where *Z. japonica* generally exhibits good cold hardiness (e.g. clonally propagated 'Meyer' and 'Chisholm' and seed-derived 'Zenith' and 'Compadre') and therefore, adapted to the northern transitional zone, whereas *Z. matrella* and *Z. pacifica*, although lacking good cold hardiness, exhibit good shade tolerance and turfgrass quality characteristics. Zoysiagrass is a versatile grass offering a wide range of applications from roadsides, lawns, sports fields, golf course roughs, fairways, tees, and greens. Its production and use are increasing as more people are realizing its ability to produce high-quality turfgrass with fewer inputs. Its low nutritional requirements, slow growth rate, tolerance to shade, salinity, extreme temperatures, and wear make it a viable option as an economical and environmentally sustainable warm-season turfgrass species (Fry et al., 2008). Although the slow growth rate of zoysiagrass means reduced mowing requirements as compared to other warm-season turfgrasses, it also results in a slower rate of establishment and spread from sprigs or plugs or recovery from wear damage. Susceptibility to fall armyworm (*Spodoptera frugiperda* J.E. Smith), tropical sod webworm (*Herpetogramma phaeopteralis* Guenée), zoysiagrass mites (*Eriophyes zoysiae* Baker, Kono, and O'Neill), large patch disease (*Rhizoctonia solani* Kühn), early onset of wilting and leaf firing during drought stress conditions, winterkill, poor winter color retention (especially the cold-hardy germplasm), and tendency to produce thatch (especially the germplasm with finer leaf width and high shoot density) are some of the shortcomings that turfgrass breeders are working toward improving.

3.5.2 Traditional breeding strategies and current breeding objectives of zoysiagrass

Ecotype selection for cultivar development and intra- and interspecific hybridization (Fig. 3) have resulted in the release of over 50 cultivars of zoysiagrass in the past 85 years, with the majority released since the 1990s (Patton et al., 2017). Active breeding effort and interspecific hybridization between *Z. japonica* and *Z. matrella* have resulted in the development of zoysiagrasses with intermediate leaf texture (finer than *Z. japonica* and coarser than *Z. matrella*). Texas A&M University and Kansas State University have jointly released an interspecific hybrid named Innovation® (KSUZ 0802) in 2015 with cold hardiness comparable to Meyer zoysiagrass but with a faster rate of establishment, finer leaf texture, higher shoot density, and overall turfgrass quality compared to Meyer (Chandra et al., 2017a). Another cultivar released in 2019 by Texas A&M University is Lazer® (DALZ 1308) which is an interspecific hybrid between *Z. minima* (Colenso) Zotov and *Z. matrella*. Lazer is a genetically dwarf, dense, and dark green hybrid that is shade tolerant (Walton et al., 2021) and developed specifically for golf course putting surfaces offering sustainable options without sacrificing the putting green speed (Chandra et al., 2020b). Bladerunner Farms (Poteet, TX, USA) released 'Primo' (M85) and 'Prizm®' (M60) cultivars of zoysiagrass for putting green use (Doguet et al., 2016, 2017), which exhibit fine leaf texture and superior quality under putting greens management. The University of Florida released CitraZoy™ (FAES 1307) and FAES 1319 in 2020; both are interspecific hybrids between *Z. japonica* and *Z. matrella*. CitraZoy and FAES 1319 exhibit improved winter color, disease resistance and delayed wilting under drought stress. In 2021, North Carolina State University released an interspecific hybrid between *Z. japonica* and *Z. matrella* named 'Lobo' (XZ14069), which is fast to establish, drought tolerant, and maintains good color and quality under very low inputs (Braun et al., 2021).

Unlike significant improvements made in the development of clonally propagated cultivars in the past decade, the progress made in the development of seeded zoysiagrass cultivars has been slow. Economically viable seed production in zoysiagrass has been a challenge because of low and inconsistent seed yields across years, hard seed coat, poor seed germination requiring chemical or mechanical scarification, poor seedling vigor affecting initial establishment, and limited herbicide options for weed control during the establishment period. Currently, there are only two seeded cultivars (Zenith and Compadre) of zoysiagrass that are commercially available, both of which were released in 1993; however, efforts are ongoing at Texas A&M University (Chandra et al., 2020a) to identify promising germplasm and breeding methods to overcome these challenges and develop new and improved seeded zoysiagrasses.

3.5.3 Advanced breeding methods of zoysiagrass

Chandra et al. (2017b) provided a detailed overview of the molecular advances in zoysiagrass. Since this review, Holloway et al. (2018) reported the construction of SNP-based genetic linkage map and identification of QTL associated with winter hardiness for use in marker-assisted selection, and Hong et al. (2017) showed the utility of internal transcribed spacer (ITS) sequence analyses for the identification of *Z. japonica* and *Z. sinica*. Guo et al.

(2019) reported the identification of candidate genes associated with drought tolerance in zoysiagrass where nucleotide changes in these genes caused amino acid substitution suggesting such allelic diversity in these genes could contribute toward physiological variation in zoysiagrass under drought stress. Other recent advancements are the transient gene expression in zoysiagrass leaf mesophyll protoplasts as a technique to assess gene function (Kim et al., 2022) and complete chloroplast genome for *Z. japonica* and *Z. matrella* (Lee and Park, 2021a,b). Efforts are underway to breed cold tolerance in zoysiagrass as to where some of these new breeding lines will survive the cold temperature beyond the transition zone; however, because of the long winter dormancy period, these grasses are only actively growing for a much shorter duration as compared to the cool-season grasses. Efforts to select for cold-tolerant zoysiagrass with superior fall color retention and rapid spring green-up would be highly desirable by the consumers of the transition and north of the transition zone.

4 Conclusion and future trends

The many elite turfgrass cultivars available today are the result of hard work and dedication by numerous turfgrass scientists over the past 90 years. Through these efforts, there have been remarkable improvements in important traits such as turfgrass quality, disease and insect resistance, and drought, shade, and traffic tolerance. However, there remains much work to be done to continue to deliver turfgrass cultivars that are able to withstand the stresses, demands, and expectations of the future.

Global climate change will result in irregular weather patterns such as intense rainfall and drought events, extreme cold and heat conditions, and new pest issues (Luber and McGeehin, 2008; Trenberth, 2011; Mukherjee et al., 2018). These challenges will be coupled with increased restrictions on the use of inputs such as water, pesticides, and fertilizers. Therefore, turfgrass breeders will need to be forward thinking and welcoming of new and advanced breeding strategies to effectively evaluate germplasm for these and other stresses. Moreover, the continued development of genetics and genomics tools and high-throughput plant phenotyping approaches will be essential in the efficient and timely development of new turfgrass cultivars. This will require an interdisciplinary approach with researchers from backgrounds including plant breeding, plant genetics, plant physiology, plant pathology, agronomy, soil science, computer science, data science, and engineering. Collaborations among individuals with these types of expertise will allow for the development of new turfgrasses that will meet the needs of the future.

5 Where to look for further information

5.1 Textbooks

- A comprehensive introduction to plant breeding practices is found in *Breeding Field Crops*, 5th Edition, by D. A. Sleper and J. M. Poehlman.
- *Turfgrass Biology, Genetics, and Breeding*, 1st Edition, by M. D. Casler and R. R. Duncan, provides an extensive overview of the major cool- and warm-season turfgrass species, including an emphasis on breeding strategies.
- The biology and breeding of cool- and warm-season turfgrass species is summarized in *Turfgrass: Biology, Use, and Management (Agronomy Monograph No. 56)*, J. C. Stier, B. P. Horgan and S. A. Bonos, Editors.

5.2 Major scientific journals

- Plant breeding research such as trait heritability studies, plant evaluation trials, and new breeding approaches are often published in the *Journal of Agronomy and Crop Science* and *Crop Science*. Other studies of turfgrass genetics and genomics or high-throughput plant phenotyping may also be published in these journals.
- Registration of new cultivars, germplasm, and mapping populations are generally published in the *Journal of Plant Registrations* (from the Crop Science Society of America; Madison, WI, USA) and *HortScience* (from the American Society of Horticulture Science; Alexandria, VA, USA) that publishes scientific merits of new cultivar releases.
- *Theoretical and Applied Genetics* publishes many genetics and genomics studies of turfgrass species.
- High-throughput and advanced plant phenotyping studies are published in *The Plant Phenome Journal*.

5.3 Key associations and conferences

- The National Association of Plant Breeders (www.plantbreeding.org) and its associated annual meeting is a great opportunity for turfgrass breeders to interact with plant breeders of other crops.
- The Turfgrass Breeders Association (www.turfgrassbreedersassociation.com) is an organization that promotes interaction among public and private turfgrass breeders.
- Many members of the turfgrass scientific community are members of the *Crop Science Society of America* (www.crops.org) and the American Society of Agronomy (www.agronomy.org). The annual agronomy meetings (www.acsmeetings.org) are also very well attended by the turfgrass scientific community in Division C-5 (Turfgrass Science).

- The Plant and Animal Genome (www.intlpag.org) annual conference brings together researchers of both plant and animal backgrounds and fosters communication among those scientists.
- The International Turfgrass Society (www.turfsociety.com) is an organization that encourages collaborations among international turfgrass scientists. The associated International Turfgrass Research Conference is held in 4-year intervals and is well attended.
- The European Turfgrass Society (www.turfgrasssociety.eu) promotes international efforts for improved urban and environmental qualities of turfgrass.
- The American Phytopathological Society (www.apsnet.org) and its associated annual meeting is a great opportunity for turfgrass breeders to interact with turfgrass pathologists to better understand current plant pathology-related issues that may be of importance to turfgrass breeding efforts.

6 Appendix

Glossary of terms

Term	Definition
Apomixis	An asexual reproductive process that occurs in certain flowering plants and results in the production of seed without sexual fertilization.
Asexual reproduction	A type of reproduction that does not involve the union of gametes.
Breeder seed	Seed or vegetative material that is increased by the breeder to be used as the source material for the increase of foundation seed.
Certified seed	Seed that has been produced from breeder, foundation, or registered seed fields and closely maintains genetic purity or characteristics of a cultivar. An official seed certifying agency approves and certifies this type of seed.
Chromosome	A threadlike structure located in the nucleus that contains the highest level of organization of DNA and proteins. The primary function of chromosomes is to transfer genetic information to offspring from parents.
Clones	A group of plants derived by vegetative propagation from a single plant.
Cross pollination	The union of an egg from one plant with a sperm from another plant.
Cultivar	A cultivated variety that has been developed by selective breeding efforts.
Dioecious	Containing female and male reproductive organs in separate individuals.
Diploid	An organism or cell that contains two complete sets of chromosomes, one coming from each parent.
Ecotype	A group of plants with distinct characteristics that have been selected through natural environmental selection and isolation processes.

Term	Definition
Fertilization	The union of gametes (egg and sperm) to form a zygote.
Foundation seed	Seed that has been produced from breeder seed in such a way that it maintains genetic purity or characteristics of the cultivar. Foundation seed is the source for registered and certified seed.
Gene	The basic unit of inheritance that is transferred to offspring from parent and controls some characteristic of the offspring.
Genome	The complete set of genes or genetic material of an organism.
Genomics	The study of the complete set of DNA of an organism to investigate the evolution, function, structure, and mapping o genomes.
Genotype	The genetic makeup of an individual organism.
Germplasm	Living genetic resources such as seeds or plant parts that are maintained for the purpose of breeding, preservation, or research efforts.
Haploid	An organism or cell that contains a single complete set of chromosomes; the reduced number (*n*) as in egg and sperm cells.
Heritability	The quality or state of being heritable. Heritability is the measure of the proportion of observed variance in a progeny that is inherited from parents.
Hexaploid	An organism or cell that contains six complete sets of chromosomes.
Hybrid	The offspring resulting from a cross between two individuals that are different in one or more genes.
Inflorescence	The flower head of a plant including stems, stalks, bracts, and flowers.
Inherit	To receive from predecessors.
Interspecific hybridizatrion	Crossing of individuals from different species.
Intraspecific hybridization	Crossing of individuals within the same species.
Linkage	The tendency for two or more genes or segments of DNA that are closely positioned on a chromosome to be inherited together.
Linkage map	A map, or table, based on genetic recombination frequency for a population or species that illustrates the position of known genes or genetic markers relative to each other.
Monoecious	Containing both female and male reproductive organs in the same individual.
Open pollination	Also known as outcrossing, it is a form of naturally occurring cross fertilization, whereby an egg of one plant is fertilized by the sperm from another plant with the assistance of wind, insects, birds, or other natural means.
Phenotype	The external or physical appearance of an organism.
Phylogeny	Study of the evolutionary composition of an organism or group of organisms.
Pollination	Transfer of pollen from anther to stigma.
Polycross	A group of plants or clones isolated together to encourage random interpollination.

Term	Definition
Polyploid	An organism or cell with three or more sets of chromosomes.
Progeny	Descendant or offspring.
Registered seed	Seed that has been produced from breeder or foundation seed in such a way that it maintains genetic purity or characteristics of the cultivar. Registered seed is the source for certified seed.
Rogue	To remove off-type plants from a population of plants or a cultivar.
Seed	A mature ovule consisting of a seed coat, an embryo, and an endosperm.
Self pollination	The union of an egg and sperm from the same flower on the same plant or within a clone.
Synthetic cultivar	A cultivar developed through several generations of cross pollination of different strains or clones.
Tetraploid	An organism or cell that contains four complete sets of chromosomes.
Triploid	An organism or cell that contains three complete sets of chromosomes.
Variety	A group of similar plants that can be distinguished from other varieties within the same species by certain naturally occurring structural features or performance characteristics.

7 References

Ahring, R. M. (1964). The management of buffalograss for seed production in Oklahoma. *Agric. Exp. Stn. Tech. Bull. T-109*. Oklahoma State University, Stillwater.

Amaradasa, B. S. and Amundsen, K. (2016). Transcriptome profiling of buffalograss challenged with the leaf spot pathogen *Curvularia inaequalis*. *Frontiers in Plant Science* 7: 715.

Araus, J. L. and Cairns, J. E. (2014). Field high-throughput phenotyping: the new crop breeding frontier. *Trends in Plant Science* 19(1): 52–61.

Armonienė, R., Stukonis, V., Paplauskienė, V. and Brazauskas, G. (2010). The genetic diversity of fine-leaved fescue (*Festuca* L.) species in Lithuania. p. 41–45. In: Huyghe, C. (Ed.) *Sustainable Use of Genetic Diversity in Forage and Turf Breeding*. Springer, Dordrecht.

Armstead, I. P., Turner, L. B., Farrell, M., Skøt, L., Gomez, P., Montoya, T., Donnison, I. S., King, I. P. and Humphreys, M. O. (2004). Synteny between a major heading-date QTL in perennial ryegrass (*Lolium perenne* L.) and the Hd3 heading-date locus in rice. *TAG. Theoretical and Applied Genetics. Theoretische und Angewandte Genetik* 108(5): 822–828.

Armstead, I. P., Turner, L. B., Marshall, A. H., Humphreys, M. O., King, I. P. and Thorogood, D. (2008). Identifying genetic components controlling fertility in the outcrossing grass species perennial ryegrass (*Lolium perenne*) by quantitative trait loci analysis and comparative genetics. *New Phytologist* 178(3): 559–571.

Assefa, S., Taliaferro, C. M., Anderson, M. P., de Los Reyes, B. G. and Edwards, R. M. (1999). Diversity among *Cynodon* accessions and taxa based on DNA amplification fingerprinting. *Genome* 42(3): 465–474.

Baltensperger, A., Dossey, B., Taylor, L. and Klingenberg, J. (1993). Bermudagrass, *Cynodon dactylon* (L.) Pers., seed production and variety development. *International Turfgrass Society Research Journal* 7: 829–838.

Barcaccia, G., Mazzucato, A., Belardinelli, A., Pezzotti, M., Lucretti, S. and Falcinelli, M. (1997). Inheritance of parental genomes in progenies of *Poa pratensis* L. from sexual and apomictic genotypes as assessed by RAPD markers and flow cytometry. *Theoretical and Applied Genetics* 95(4): 516-524.

Bashaw, E., Hovin, A. and Holt, E. (1970). Apomixis, its evolutionary significance and utilization in plant breeding. p. 245-248. In: *Proceedings of the XI International Grassland Congress Held at Surfers Paradise*, Queensland, Australia.

Bashaw, E. G. and Funk, C. R. (1987). Apomictic grasses. p. 40-82. In: Fehr, W. E. (Ed.) *Principles of Cultivar Development (Vol. 2). Crop Species*. Macmillan Publishing Co., New York.

Bateman, D. R. (1980). Notice to sod producers and growers relative to the naming and release of the new St. Augustine cultivar 'Raleigh'. N.C. Department of Agriculture and Consumer Services, Raleigh, NC.

Baxter, L. L. and Schwartz, B. M. (2018). History of Bermudagrass turfgrass breeding research in Tifton, GA. *HortScience* 53(11): 1560-1561.

Beard, J. B. (1973). *Turfgrass: Science and Culture*. Prentice-Hall, Inc., Englewood Cliffs, NJ.

Beard, J. B. (2012). *Origin, Biogeographical Migrations and Diversifications of Turfgrasses*. Michigan State University Press, East Lansing, MI.

Beard, J. B. and Beard, J. H. (2005). *Beard's Turfgrass Encyclopedia for Golf Courses, Grounds, Lawns, Sports Fields*. Michigan State University Press, East Lansing, MI.

Bert, P. F., Charmet, G., Sourdille, P., Hayward, M. D. and Balfourier, F. (1999). A high-density molecular map for ryegrass (*Lolium perenne*) using AFLP markers. *TAG. Theoretical and Applied Genetics. Theoretische und Angewandte Genetik* 99(3-4): 445-452.

Bertoli, F. B., Johnson, R. C., Lucaroni, B., Romani, M., Russi, L., Piano, E. and Falcinelli, M. (2003). Agronomic and molecular evaluation of Kentucky bluegrass (*Poa pratensis* L.) germplasm collected in Italy. *Czech Journal of Genetics and Plant Breeding* 39: 330-332.

Bhat, J. A., Deshmukh, R., Zhao, T., Patil, G., Deokar, A., Shinde, S. and Chaudhary, J. (2020). Harnessing high-throughput phenotyping and genotyping for enhanced drought tolerance in crop plants. *Journal of Biotechnology* 324: 248-260.

Bjorkman, S. O. (1954). Chromosome studies in Agrostis. II. *Hereditas, Lund* 40: 254-258.

Bolaric, S., Barth, S., Melchinger, A. E. and Posselt, U. K. (2005). Genetic diversity in European perennial ryegrass cultivars investigated with RAPD markers. *Plant Breeding* 124(2): 161-166.

Bonos, S. A., Clarke, B. B. and Meyer, W. A. (2006). Breeding for disease resistance in the major cool-season turfgrasses. *Annual Review of Phytopathology* 44: 213-234.

Bonos, S. A. and Huff, D. R. (2013). Cool-season grasses: biology and breeding. p. 591-660. In: Stier, J. C., Horgan, B. P. and Bonos, S. A. (Eds.) *Turfgrass: Biology, Use, and Management. Agronomy Monographs* (vol. 56). CSSA, ASA, and SSSA, Madison, WI.

Braun, R. C., Milla-Lewis, S. R., Carbajal, E. M., Schwartz, B. M. and Patton, A. J. (2021). Performance and playability of experimental low-input coarse-textured zoysiagrass in multiple climates. *Grass Research* 1(1): 1-12.

Braun, R. C., Patton, A. J., Watkins, E., Koch, P. L., Anderson, N. P., Bonos, S. A. and Brilman, L. A. (2020). Fine fescues: a review of the species, their improvement, production, establishment, and management. *Crop Science* 60(3): 1142-1187.

Breen, J. P. (1993). Enhanced resistance to fall armyworm (*Lepidoptera*: Noctudiae) in *Acremonium* endophyte-infected turfgrasses. *Journal of Economic Entomology* 86(2): 621-629.

Brilman, L. A. (2001). Colonial bentgrass: an option for fairways. *Golf Course Manage* 69: 55-60.

Brilman, L. A. and Meyer, W. A. (2000). Velvet bentgrass: rediscovering a misunderstood turfgrass. *Golf Course Manage* 68: 70-75.

Brown, S., Mitchell, S., Jester, C., Liu, Z., Kresovich, S., Duncan, R., Sticklen, M. and Kenna, M. (1998). DNA typing (profiling) of seashore paspalum (*Paspalum vaginatum* Swartz) ecotypes and cultivars. P. 39-51. In: Sticklen, M. B. and Kenna, M. P. (Eds) *Turfgrass Biotechnology: Cell Molecular Genetic Approaches to Turfgrass Improvement*. Ann Arbor Press, Michigan.

Buckner, R. C., Powell, J. B. and Frakes, R. V. (1979). Historical development. *Agronomy Monographs* 20: 1-8.

Budak, H., Shearman, R. C., Gulsen, O. and Dweikat, I. (2005). Understanding ploidy complex and geographic origin of the *Buchloe dactyloides* genome using cytoplasmic and nuclear marker systems. *TAG. Theoretical and Applied Genetics. Theoretische und Angewandte Genetik* 111(8): 1545-1552.

Budak, H., Shearman, R. C., Parmaksiz, I., Gaussoin, R. E., Riordan, T. P. and Dweikat, I. (2004). Molecular characterization of buffalograss germplasm using sequence-related amplified polymorphism markers. *TAG. Theoretical and Applied Genetics. Theoretische und Angewandte Genetik* 108(2): 328-334.

Buhlman, J. L., Freshour, S. L., Harriman, R. W., Kenworthy, K. E., Unruh, J. B. and Yurisic, C. A. (2019). Measuring mowing frequency in genetically modified St. Augustinegrass. In: *International Forage and Turf Breeding Conference*. Lake Buena Vista, FL.

Burson, B. L. (1985). Cytology of *Paspalum chacoence* and *P. durifolium* and their relationship to *P. dilatatum*. *Botanical Gazette* 146(1): 124-129.

Burson, B. L., Lee, H. S. and Bennett, H. W. (1973). Genome relations Between tetraploid *Paspalum dilatatum* and four diploid *Paspalum* species 1. *Crop Science* 13(6): 739-743.

Burton, G. W. (1947). Breeding Bermudagrass for the southeastern United States. *Agronomy Journal* 39(7): 551-569.

Burton, G. W. (1955). Better turfgrasses can be produced by breeding. *USGA Journal and Turf Management* 8: 28-29.

Burton, G. W. (1964). Tifgreen (Tifton 328) Bermudagrass for golf greens. *USGA Green Section Record* 2: 11-12.

Burton, G. W. (1966a). Tifway (Tifton 419) Bermudagrass 1 (Reg. No. 7). *Crop Science* 6(1): 93-94.

Burton, G. W. (1966b). Tifdwarf Bermudagrass 1 (Reg. No. 8). *Crop Science* 6(1): 94.

Burton, G. W. (1991). *A History of Turf Research at Tifton. Green Section Record 29* (vols. 12-14). United States Golf Association, Far Hills, NJ.

Burton, G. W. and Hart, R. H. (1967). Use of self-incompatibility to produce commercial seed-propagated F_1 Bermudagrass hybrids. *Crop Science* 7(5): 524-527.

Busey, P. (1979). What is Floratam? *Florida State Horticultural Society Proceeding* 92: 228-232.

Busey, P. (1986). Morphological identification of St. Augustinegrass cultivars 1. *Crop Science* 26(1): 28-32.

Busey, P. (1989). Progress and benefits to humanity from breeding warm-season grasses for turf. *CSSA Special Publications* 15: 49-70.

Bushman, B. S., Amundsen, K. L., Warnke, S. E., Robins, J. G. and Johnson, P. G. (2016). Transcriptome profiling of Kentucky bluegrass (*Poa pratensis* L.) accessions in response to salt stress. *BMC Genomics* 17: 48.

Bushman, B. S., Robbins, M. D., Thorsted, K., Robins, J. G., Warnke, S. E., Martin, R. and Harris-Shultz, K. (2021). Transcript responses to drought in Kentucky bluegrass (*Poa pratensis* L.) germplasm varying in their tolerance to drought stress. *Environmental and Experimental Botany* 190: 104571.

Bushman, B. S., Warnke, S. E., Amundsen, K. L., Combs, K. M. and Johnson, P. G. (2013). Molecular markers highlight variation within and among Kentucky bluegrass varieties and accessions. *Crop Science* 53(5): 2245-2254.

Busti, A., Caceres, M. E., Calderini, O., Arcioni, S. and Pupilli, F. (2004). RFLP markers for cultivar identification in tall fescue (*Festuca arundinacea* Schreb.). *Genetic Resources and Crop Evolution* 51(4): 443-448.

Byrne, S., Guiney, E., Barth, S., Donnison, I., Mur, L. A. J. and Milbourne, D. (2009). Identification of coincident QTL for days to heading, spike length and spikelets per spike in *Lolium perenne* L. *Euphytica* 166(1): 61-70.

Byrne, S. L., Nagy, I., Pfeifer, M., Armstead, I., Swain, S., Studer, B., Mayer, K., Campbell, J. D., Czaban, A., Hentrup, S., Panitz, F., Bendixen, C., Hedegaard, J., Caccamo, M. and Asp, T. (2015). A synteny-based draft genome sequence of the forage grass *Lolium perenne*. *The Plant Journal: For Cell and Molecular Biology* 84(4): 816-826.

Caceres, M. E., Pupilli, F., Piano, E. and Arcioni, S. (2000). RFLP markers are an effective tool for the identification of creeping bentgrass (*Agrostis stolonifera* L.) cultivars. *Genetic Resources and Crop Evolution* 47(4): 455-459.

Caetano-Anollés, G., Callahan, L. M., Williams, P. E., Weaver, K. R. and Gresshoff, P. M. (1995). DNA amplification fingerprinting analysis of Bermudagrass (*Cynodon*): genetic relationships between species and interspecific crosses. *TAG. Theoretical and Applied Genetics. Theoretische und Angewandte Genetik* 91(2): 228-235.

Caetano-Anollés, G., Callahan, L. M. and Gresshoff, P. M. (1997). The origin of Bermudagrass (*Cynodon*) off-types inferred by DNA amplification fingerprinting. *Crop Science* 37(1): 81-87.

Cao, Y. J., Wei, Q., Liao, Y., Song, H. L., Li, X., Xiang, C. B. and Kuai, B. K. (2009). Ectopic overexpression of AtHDG11 in tall fescue resulted in enhanced tolerance to drought and salt stress. *Plant Cell Reports* 28(4): 579-588.

Carpenter, J. (1958). Production and use of seed in seashore paspalum. *Journal of the Australian Institute of Agricultural Science* 24: 252-256.

Carroll, J. (1943). Effects of drought, temperature, and nitrogen on turfgrasses. *Frontiers in Plant Science* 8: 670.

Carrow, R. N. (1996). Drought avoidance characteristics of diverse tall fescue cultivars. *Crop Science* 36(2): 371-377.

Carrow, R. N. and Duncan, R. R. (1998). *Salt-Affected Turfgrass Sites: Assessment and Management*. John Wiley and Sons, New York.

Casler, M. D., Rangel, Y., Stier, J. C. and Jung, G. (2003). RAPD marker diversity among creeping bentgrass clones. *Crop Science* 43(2): 688-693.

Catalán, P., Torrecilla, P., Rodrıguez, J. A. L. and Olmstead, R. G. (2004). Phylogeny of the festucoid grasses of subtribe Loliinae and allies (Poeae, Pooideae) inferred from ITS and trnL-F sequences. *Molecular Phylogenetics and Evolution* 31(2): 517-541.

Chakraborty, N., Bae, J., Warnke, S., Chang, T. and Jung, G. (2005). Linkage map construction in allotetraploid creeping bentgrass (*Agrostis stolonifera* L.). *TAG. Theoretical and Applied Genetics. Theoretische und Angewandte Genetik* 111(4): 795-803.

Chakraborty, N., Curley, J., Warnke, S., Casler, M. D. and Jung, G. (2006). Mapping QTL for dollar spot resistance in creeping bentgrass (*Agrostis stolonifera* L.). *TAG.*

Theoretical and Applied Genetics. Theoretische und Angewandte Genetik 113(8): 1421–1435.

Chandra, A., Genovesi, A. D. and Meeks, M. (2020a). Development of seeded zoysiagrass cultivars with improved turf quality and high seed yields. *USGA Turfgrass and Environmental Research Summary* : 54–58.

Chandra, A., Genovesi, A. D., Meeks, M., Wu, Y., Engelke, M. C., Kenworthy, K. and Schwartz, B. (2020b). Registration of 'DALZ 1308' zoysiagrass. *Journal of Plant Registrations* 14(1): 19–34.

Chandra, A., Genovesi, A. D., Wherley, B. G., Metz, S. P., Reinert, J. A., Wu, Y. Z., Skulkaew, P., Engelke, M. C., Hargey, D., Nelson, L. R., Schwartz, B. M., Raymer, P. L., Wu, Y. Q., Martin, D. L., Milla-Lewis, S. R., Miller, G., Kenworthy, K. E. and Munoz, P. (2015). Registration of 'DALSA 0605' St. Augustinegrass. *Journal of Plant Registrations* 9(1): 27–34.

Chandra, A., Fry, J. D., Genovesi, A. D., Meeks, M., Engelke, M. C., Zhang, Q., Okeyo, D., Moss, J. Q., Ervin, E., Xiong, X., Milla-Lewis, S., Brosnan, J. T., Griffin, J. and Parsons, L. (2017a). Registration of 'KSUZ 0802' zoysiagrass. *Journal of Plant Registrations* 11(2): 100–106.

Chandra, A., Milla-Lewis, S. and Yu, Q. (2017b). An overview of molecular advances in zoysiagrass. *Crop Science* 57(S1): S-73.

Charmet, G., Ravel, C. and Balfourier, F. (1997). Phylogenetic analysis in the Festuca-Lolium complex using molecular markers and ITS Rdna. *Theoretical and Applied Genetics* 94(8): 1038–1046.

Chase, A. (1971). Manual of the grasses of the United States. P. 341, 809. In: *USDA Miscellaneous Publications No. 200* (vol. 2, 2nd edn.). General publishing Co., Ltd, Toronto, ON, Canada.

Chen, L., Fan, J. B., Hu, L. X., Hu, Z., Xie, Y., Zhang, Y., Luo, Y., Nevo, E. and Fu, J. (2015). A transcriptomic analysis of Bermudagrass (*Cynodon dactylon*) provides novel insights into the basis of low temperature tolerance. *BMC Plant Biology* 15: 216, doi: 10.1186/s12870-015-0598-y.

Cheng, Y., Ma, X., Zhou, K., Humphreys, M. W. and Zhang, X. Q. (2016). Phylogenetic analysis of Festuca–Lolium complex using SRAP markers. *Genetic Resources and Crop Evolution* 63(1): 7–18.

Cherry, G., Wherley, B., Straw, C. M. and Chandra, A. (2021). St. Augustine grass cultivar performance under low input maintenance regimes. *International ASA-CSSA-SSSA Conference*. Salt Lake City, UT. Poster presentation 1174.

Christians, N. (1998). Fundamentals of turfgrass managementp. p. 46–47. Ann Arbor Press. Chelsea, MI.

Clark, K. W., Hussain, A., Bamford, K. and Bushuk, W. (1989). Identification of cultivars of *Agrostis* species by polyacrylamide gel electrophoresis of seed proteins. P. 121–125. In: *Proceedings of the 6th Int. Turfgrass Research Conference*, Tokyo, Japan.

Cluff, G. J. and Baltensperger, A. A. (1991). Heritability estimates for seed yield and seed yield components in Bermudagrass. *Agricultural Experiment Station, College of Agriculture and Home Economics*. New Mexico State University.

Cogan, N. O. I., Smith, K. F., Yamada, T., Francki, M. G., Vecchies, A. C., Jones, E. S., Spangenberg, G. C. and Forster, J. W. (2005). QTL analysis and comparative genomics of herbage quality traits in perennial ryegrass (*Lolium perenne* L.). *TAG. Theoretical and Applied Genetics. Theoretische und Angewandte Genetik* 110(2): 364–380.

Cornish, M. A. (1979). The genetics of self-incompatibility in *Lolium perenne*. Ph.D. thesis. University of Birmingham, UK.

Cresswell, A., Sackville Hamilton, N. R., Roy, A. K. and Viegas, B. M. F. (2001). Use of amplified fragment length polymorphism markers to assess genetic diversity of *Lolium* species from Portugal. *Molecular Ecology* 10(1): 229-241.

Curley, J. and Jung, G. (2004). RAPD-based genetic relationships in Kentucky bluegrass: comparison of cultivars, interspecific hybrids, and plant introductions. *Crop Science* 44(4): 1299-1306.

Curley, J., Chakraborty, N., Chang, S. and Jung, G. (2008). QTL mapping of resistance to gray leaf spot in ryegrass: consistency of QTL between two mapping populations. *Asian Journal of Turfgrass Science* 22: 85-100.

Curley, J., Sim, S. C., Jung, G., Leong, S., Warnke, S. and Barker, R. E. (2004). QTL mapping of gray leaf spot resistance in ryegrass, and synteny-based comparison with rice blast resistance genes in rice. P. 37-46. In: Hopkins, A., Wang, Z. Y., Mian, R., Sledge, M. and Barker, R. E. (Eds) *Developments in Plant Breeding*. Springer, Dordrecht.

Curley, J., Sim, S. C., Warnke, S., Leong, S., Barker, R. and Jung, G. (2005). QTL mapping of resistance to gray leaf spot in ryegrass. Theoretical and Applied Genetics 111: 1107-1117.

De Wet, J. M. J. and Harlan, J. R. (1970). Biosystematics of *Cynodon* LC rich. (Gramineae). *Taxon* 19: 565-569.

DeFrance, J. A., Odland, T. E. and Bell, R. S. (1952). Improvement of velvet bentgrass by selection 1. *Agronomy Journal* 44(7): 376-378.

Delseny, M., Han, B. and Hsing, Y. Ie (2010). High throughput DNA sequencing: the new sequencing revolution. *Plant Science* 179(5): 407-422.

Doguet, D., Doguet, D. A. and Lehman, V. G. (2016). Zoysiagrass plant named 'M85'. U.S. Plant Patent 27289. 18 October.

Doguet, D., Doguet, D. A. and Lehman, V. G. (2017). Zoysiagrass plant named 'M60'. U.S. Plant Patent 29143. 23 November.

Dong, S., Shew, H. D., Tredway, L. P., Lu, J., Sivamani, E., Miller, E. S. and Qu, R. (2008). Expression of the bacteriophage T4 lysozyme gene in tall fescue confers resistance to gray leaf spot and brown patch diseases. *Transgenic Research* 17(1): 47-57.

Duell, R. W. and Peacock, G. R. (1985). Rhizosheaths on mesophytic grasses 1. *Crop Science* 25(5): 880-883.

Duich, J. M. (1985). The bent grasses. *Weeds, Trees, and Turf* 24: 72-78.

Dumsday, J. L., Smith, K. F., Forster, J. W. and Jones, E. S. (2003). SSR-based genetic linkage analysis of resistance to crown rust (*Puccinia coronata* f. sp. *Lolii*) in perennial ryegrass (*Lolium perenne*). *Plant Pathology* 52(5): 628-637.

Duncan, R. R. (2003). Seashore paspalum. Turfgrass biology. P. 295-307. In: *Genetics, and Breeding*. John Wiley and Sons, Inc., Hoboken, NJ.

Duncan, R. R. and Carrow, R. N. (2000). *Seashore Paspalum: The Environmental Turfgrass*. John Wiley and Sons, Chelsea, MI.

Duran, C., Eales, D., Marshall, D., Imelfort, M., Stiller, J., Berkman, P. J., Clark, T., McKenzie, M., Appleby, N., Batley, J., Basford, K. and Edwards, D. (2010). Future tools for association mapping in crop plants. *Genome* 53(11): 1017-1023.

Edwards, D. and Batley, J. (2010). Plant genome sequencing: applications for crop improvement. *Plant Biotechnology Journal* 8(1): 2-9.

Elmore, C. L., Wilen, C., Cudney, D. W. and Gibeault, V. (1997). *UC IPM Pest Management Guidelines P: Turfgrass*. UC DANR Publication (vol. 3365-T).

Espinoza, F. and Quarín, C. L. (1997). Cytoembryology of *Paspalum chaseanum* and sexual diploid biotypes of two apomictic *Paspalum* species. *Australian Journal of Botany* 45(5): 871–877.

Eudy, D., Bahri, B. A., Harrison, M. L., Raymer, P. and Devos, K. M. (2017). Ploidy level and genetic diversity in the genus *Paspalum*, group *Disticha*. *Crop Science* 57(6): 3319–3332.

Fard, J. R., Zamani, Z., Moghaddam, M. R. F. and Kafi, M. (2012). Evaluation of genetic diversity among some genotypes of Kentucky bluegrass by RAPD molecular markers. *Horticulture, Environment, and Biotechnology* 53(4): 298–303.

Farsani, T. M., Etemadi, N., Sayed-Tabatabaei, B. E. and Talebi, M. (2012). Assessment of genetic diversity of Bermudagrass (*Cynodon dactylon*) using ISSR markers. *International Journal of Molecular Sciences* 13(1): 383–392.

Faville, M. J., Cao, M., Schmidt, J., Ryan, D. L., Ganesh, S., Jahufer, M. Z., Hong, S. W., George, R. and Barrett, B. A. (2020). Divergent genomic selection for herbage accumulation and days-to-heading in perennial ryegrass. *Agronomy* 10(3): 340.

Faville, M. J., Ganesh, S., Cao, M., Jahufer, M. Z. Z., Bilton, T. P., Easton, H. S., Ryan, D. L., Trethewey, J. A. K., Rolston, M. P., Griffiths, A. G., Moraga, R., Flay, C., Schmidt, J., Tan, R. and Barrett, B. A. (2018). Predictive ability of genomic selection models in a multi-population perennial ryegrass training set using genotyping-by-sequencing. *TAG. Theoretical and Applied Genetics. Theoretische und Angewandte Genetik* 131(3): 703–720.

Fei, S. Z. (2008). Recent progresses on turfgrass molecular genetics and biotechnology. *Acta Horticulturae* 783: 247.

Fergus, E. N. and Buckner, R. C. (1972). Registration of Kentucky 31 tall fescue (Reg No. 7). *Crop Science* 12: 714.

Fois, M., Bellucci, A., Malinowska, M., Greve, M., Ruud, A. K. and Asp, T. (2022). Genome-wide association mapping of crown and brown rust resistance in perennial ryegrass. *Genes* 13(1): 20.

Forbes, I. (1955). Emerald zoysia: an improved hybrid lawn grass for the south. *USGA Journal and Turf Management* 8: 23–26.

Frei, D., Veekman, E., Grogg, D., Stoffel-Studer, I., Morishima, A., Shimizu-Inatsugi, R., Yates, S., Shimizu, K. K., Frey, J. E., Studer, B. and Copetti, D. (2021). Ultralong Oxford nanopore reads enable the development of a reference-grade perennial ryegrass genome assembly. *Genome Biology and Evolution* 13(8): evab159.

Fricker, C., Wipff, J. K. and Duncan, R. R. (2007). *Hybrid Seashore Paspalum Available from Seed Called 'Sea Spray'*. U.S. Patent 7262,341 B1. Date issued: 28 August.

Fry, J., Kennelly, M. and St. John, R. (2008). Zoysiagrass: economic and environmental sense in the transition zone. *Golf Course Manage* 76: 127–132.

Funk, C. R. (2000). Long live Kentucky bluegrass, the king of grasses! Breeders strive to consolidate the desirable traits in its germplasm. *Diversity* 16: 26–28.

Funk, C. R. and Clarke, B. B. (1989). Turfgrass breeding – with special reference to turf-type perennial ryegrass, tall fescue, and endophytes. P. 3–10. In: Takotoh, H. (Ed.). *Proceedings of the 6th International Turfgrass Research Conference*, Tokyo, Japan.

Funk, C. R., Engel, R. E., Dickson, W. K. and Hurley, R. H. (1981). Registration of rebel tall fescue 1 (Reg No. 18). *Crop Science* 21: 632–632.

Funk, C. R. and Han, S. J. (1967). Recurrent intraspecific hybridization: a proposed method of breeding Kentucky bluegrass, *Poa pratensis*. *N.J. Bulletin of the Agricultural Experiment Station* 818: 3–14.

Funk, C. R. and White, J. F. (1997). *Use of Natural and Transformed Endophytes for Turf Improvement. In Neotyphodium/Grass Interactions*. Springer, Boston, MA. p. 229-239.

Funk, C. R., White, R. H. and Breen, J. P. (1993). Importance of *Acremonium* endophytes in turfgrass breeding and management. *Agriculture, Ecosystems and Environment* 44(1-4): 215-232.

Gates, M. J. (1997). Seed set variation in the presence or absence of foreign pollen in apomictic Kentucky bluegrass, Poa pratensis L. Master of sci. Doctoral dissertation, State University, University Park, PA.

Genovesi, A. D., Chandra, A. and Meeks, M. (2017). Efficiency of embryo rescue technology in St. Augustinegrass breeding. *Itsrj* 13(1): 435-442.

Genovesi, A. D., Jessup, R. W., Engelke, M. C. and Burson, B. L. (2009). Interploid St. Augustinegrass [*Stenotaphrum secundatum* (Walt.) Kuntze] hybrids recovered by embryo rescue. *In Vitro Cellular and Developmental Biology - Plant* 45(6): 659-666.

Ghesquiere, A., Calsyn, E., Baert, J. and Riek, J. D. (2003). Genetic diversity between and within ryegrass populations of the ECP/GR collection by means of AFLP markers. *Czech Journal of Genetics and Plant Breeding* 39 (Special Issue).

Golembiewski, R. C., Danneberger, T. K. and Sweeney, P. M. (1997). Potential of RAPD markers for use in the identification of creeping bentgrass cultivars. *Crop Science* 37(1): 212-214.

Gray, A. (1908). *Gray's Manual of Botany; a Handbook of the Flowering Plants and Ferns of the Central and Northeastern United States and Adjacent Canada* (7th edn.). American Book Publishing Company, New York.

Grazi, F., Umaerus, M. and Åkerberg, E. (1961). Observations on the mode of reproduction and the embryology of *Poa pratensis*. *Hereditas* 47(3-4): 489-541.

Gulsen, O., Shearman, R. C., Vogel, K. P., Lee, D. J., Baenziger, P. S., Heng-Moss, T. M. and Budak, H. (2005). Nuclear genome diversity and relationships among naturally occurring buffalograss genotypes determined by sequence-related amplified polymorphism markers. *HortScience* 40(3): 537-541.

Guo, H., Wang, Y., Zhang, B., Li, D., Chen, J., Zong, J., Li, J., Liu, J. and Jiang, Y. (2019). Association of candidate genes with drought tolerance traits in zoysiagrass germplasm. *Journal of Plant Physiology* 237: 61-71.

Han, H. R. (2009). Development of improved turf-type bermudagrasses. Thesis. Oklahoma State Univ, Stillwater, OK.

Hanna, W. W., Braman, S. K. and Schwartz, B. M. (2010). 'ST-5', a shade-tolerant turf Bermudagrass. *HortScience* 45(1): 132-134.

Hanna, W. W., Burson, B. L. and Schwartz, B. M. (2016). Polyploidy and interspecific hybridization in *Cynodon*, *Paspalum*, *Pennisetum*, and *Zoysia*. P. 318-338. In: Mason, A. S. (Ed.). *Polyploidy and Interspecific Hybridization for Crop Improvement*. CRC Press, Boca Raton, FL.

Hanna, W. W., Carrow, R. N. and Powell, A. J. (1997). Registration of 'Tift 94' Bermudagrass. *Crop Science* 37(3): 1012.

Hanna, W. W. and Elsner, J. E. (1999). Registration of 'TifEagle' Bermudagrass. *Crop Science* 39(4): 1258.

Hanna, W. W., Raymer, P. and Schwartz, B. M. (2013). Warm-season grasses: biology and breeding. P.543-590. In: Stier, J. C., Horgan, B. P. and Bonos, S. A. (Eds.) *Turfgrass: Biology, Use, and Management. Agron. Monogr.* (vol. 56). CSSA, ASA, and SSSA, Madison, WI.

Hanson, A. A., Juska, F. V. and Burton, G. W. (1969). Turfgrass science. Species and varieties. p. 370-409. In: Hanson, A. A. and Juska, F. V. (Eds.) *Turfgrass Science*. American Society of Agronomy, Madison, WI.

Harlan, J. R. (1970). *Cynodon* species and their value for grazing and hay. *Herbage Abs* 40: 233-238.

Harlan, J. R. (1992). *Crops and Man* (2nd edn.). American Society of Agronomy, Madison, WI.

Harlan, J. R. and de Wet, J. M. J. (1969). Sources of variation in *Cynodon dactylon* (L.) Pers. *Crop Science* 9(6): 774-778.

Harlan, J. R., de Wet, J. M. J., Huffine, W. W. and Deakin, J. R. (1970). A guide to the species of *Cynodon* (Gramineae). Oklahoma Agricultural Experiment Station Bulletin B 673. p. 37.

Harriman, R., Lisa, L., David, S. and Rebecca, T. (2019). Plants comprising events PP009-401, PP009-415, and PP009-469, compositions, sequences, and methods for detection thereof. U.S. Patent 10,501,753.

Harriman, R., Lisa, L. and Rebecca, T. (2021). Plants comprising a gibberellic acid 2-oxidase gene expression cassette. U.S. Patent 11 162: 113.

Harris-Shultz, K., Raymer, P., Scheffler, B. E. and Arias, R. S. (2013). Development and characterization of seashore paspalum SSR markers. *Crop Science* 53(6): 2679-2685.

Hayward, M. D., Forster, J. W., Jones, J. G., Dolstra, O., Evans, C., McAdam, N. J., Hossain, K. G., Stammers, M., Will, J., Humphreys, M. O. and Evans, G. M. (1998). Genetic analysis of *Lolium*. I. Identification of linkage groups and the establishment of a genetic map. *Plant Breeding* 117(5): 451-455.

Hayward, M. D., McAdam, N. J., Jones, J. G., Evans, C., Evans, G. M., Forster, J. W., Ustin, A., Hossain, K. G., Quader, B., Stammers, M. and Will, J. K. (1994). Genetic markers and the selection of quantitative traits in forage grasses. *Euphytica* 77(3): 269-275.

Heckman, N. L., Horst, G. L., Gaussoin, R. E. and Frank, K. W. (2001). Storage and handling characteristics of trinexapac-ethyl treated Kentucky bluegrass sod. *HortScience* 36(6): 1127-1130.

Hein, M. A. (1953). Registration of varieties and strains of Bermudagrass, II [*Cynodon dactylon* (L.) Pers.]. *Agronomy Journal* 45(11): 572-573.

Hein, M. A. (1958). Registration of varieties of strains of grasses. Bentgrasses: penncross creeping (Reg No. 1). *Agronomy Journal* 50: 399.

Hein, M. A. (1961). Registration of varieties and strains of Bermudagrass III. *Agronomy Journal* 53: 276.

Hintzen, J. J. (1979). Methods in apomictic species. p. 186-189. In: Sneep, J. and Henderiksan, A. J. T. (Eds). *Plant Breeding Perspectives*. Pudoc, Wageningen.

Hintzen, J. J. and van Wijk, A. J. P. (1985). Ecotype breeding and hybridization in Kentucky bluegrass (*Poa pratensis* L.) [turf quality, seed production, apomixy, heterozygosity]. In: *5. International Turfgrass Research Conference*. INRA, Avignon (France).

Ho, C. Y., McMaugh, S. J., Wilton, A. N., McFarlane, I. J. and Mackinlay, A. G. (1997). DNA amplification variation within cultivars of turf-type couch grasses (*Cynodon* spp.). *Plant Cell Reports* 16(11): 797-801.

Holloway, H. M. P., Yu, X., Dunne, J. C., Schwartz, B. M., Patton, A. J., Arellano, C. and Milla-Lewis, S. R. (2018). A SNP-based high-density linkage map of zoysiagrass (*Zoysia japonica* Steud.) and its use for the identification of QTL associated with winter hardiness. *Molecular Breeding* 38: 1-14.

Hong, M., Bremer, D. J. and van der Merwe, D. (2019a). Thermal imaging detects early drought stress in turfgrass utilizing small unmanned aircraft systems. *Agrosystems, Geosciences and Environment* 2(1): 1-9.

Hong, M., Bremer, D. J. and van der Merwe, D. (2019b). Using small unmanned aircraft systems for early detection of drought stress in turfgrass. *Crop Science* 59(6): 2829-2844.

Hong, M. J., Yang, D. H., Jeong, O. C., Kim, Y. J., Park, M. Y., Kang, H. G., Sun, H. J., Kwon, Y. I., Park, S. Y., Yang, P. and Song, P. S. (2017). Molecular identification of *Zoysia japonica* and *Zoysia sinica* (*Zoysia* species) based on ITS sequence analyses and CAPS. *Horticultural Science and Technology* 35: 344-360.

Honig, J. A., Averello, V., Bonos, S. A. and Meyer, W. A. (2012). Classification of Kentucky bluegrass (*Poa pratensis* L.) cultivars and accessions based on microsatellite (simple sequence repeat) markers. *HortScience* 47(9): 1356-1366.

Honig, J. A., Averello, V., Kubik, C., Vaiciunas, J., Bushman, B. S., Bonos, S. A. and Meyer, W. A. (2018). An update on the classification of Kentucky bluegrass cultivars and accessions based on microsatellite (SSR) markers. Crop *Science* 58: 1776-1787.

Honig, J. A., Bonos, S. A. and Meyer, W. A. (2010). Isolation and characterization of 88 polymorphic microsatellite markers in Kentucky bluegrass (*Poa pratensis* L.). *HortScience* 45(11): 1759-1763.

Honig, J. A., Kubik, C., Majewski, M., Poulsen, C., Weibel, E., Amundsen, K., Warnke, S. E., Meyer, W. A. and Bonos, S. A. (2014). A PCR-based linkage map of *Agrostis stolonifera* and identification of QTL markers for dollar spot resistance. *Molecular Breeding* 34(1): 185-203.

Horn, G. C., Dudeck, A. E. and Toler, R. W. (1973). *'Floratam' St. Augustinegrass: A Fast Growing New Variety for Ornamental Turf, Resistant to St. Augustine Decline and Chinch Bugs*. University of Florida, Agricultural Experiment Station.

Huang, C., Liu, G., Bai, C. and Wang, W. (2014). Genetic analysis of 430 Chinese *Cynodon dactylon* accessions using sequence-related amplified polymorphism markers. *International Journal of Molecular Sciences* 15(10): 19134-19146.

Huang, C. Q., Liu, G. D., Bai, C. J., Wang, W. Q., Zhou, S. Y. and Yu, D. Q. (2010). Estimation of genetic variation in *Cynodon dactylon* accessions using the ISSR technique. *Biochemical Systematics and Ecology* 38(5): 993-999.

Huang, C. Q., Huang, D. Y. and Zhang, Y. F. (2010). Genetic analysis for 57 accessions of *Cynodon dactylon* from 17 countries in 5 continents by SRAP markers. *Tropical Grasslands* 44: 274-281.

Hubbard, J. C. E. (1984). *Grasses: A Guide to Their Structure, Identification, Uses, and Distribution in the British Isles* (3rd edn.). Viking Penguin, Inc, New York. p. 298-299, 372.

Huff, D. R. (1997). RAPD characterization of heterogenous perennial ryegrass cultivars. *Crop Science* 37(2): 557-564.

Huff, D. R. (2003). Kentucky bluegrass. p. 27038. In: Casler, M. D. and Duncan, R. R. (Eds) *Turfgrass Biology, Genetics, and Breeding*. John Wiley and Sons, New York.

Huff, D. R. and Bara, J. M. (1993). Determining genetic origins of aberrant progeny from facultative apomictic Kentucky bluegrass using a combination of flow cytometry and silver-stained RAPD markers. *TAG. Theoretical and Applied Genetics. Theoretische und Angewandte Genetik* 87(1-2): 201-208.

Huff, D. R., Benson, C. W., Amundsen, K. and Morikone, M. (2021). A fully annotated, high quality genomic sequence of buffalograss [Abstract]. *ASA, CSSA, SSSA International Annual Meeting*. Salt Lake City, UT. Paper 135041.

Huff, D. R. and Palazzo, A. J. (1998). Fine fescue species determination by laser flow cytometry. *Crop Science* 38(2): 445–450.

Huff, D. R., Peakall, R. and Smouse, P. E. (1993). RAPD variation within and among natural populations of outcrossing buffalograss [*Buchloe dactyloides* (Nutt.) Engelm.]. *TAG. Theoretical and Applied Genetics. Theoretische und Angewandte Genetik* 86(8): 927–934.

Huff, D. R. and Wu, L. (1992). Distribution and inheritance of inconstant sex forms in natural populations of dioecious buffalograss (*Buchloe dactyloides*). *American Journal of Botany* 79(2): 207–215.

Jaškūnė, K., Aleliūnas, A., Statkevičiūtė, G., Kemešytė, V., Studer, B. and Yates, S. (2020). Genome-wide association study to identify candidate loci for biomass formation under water deficit in perennial ryegrass. *Frontiers in Plant Science* 11: 1923: 570204.

Jensen, L. B., Andersen, J. R., Frei, U., Xing, Y., Taylor, C., Holm, P. B. and Lübberstedt, T. (2005). QTL mapping of vernalization response in perennial ryegrass (*Lolium perenne* L.) reveals co-location with an orthologue of wheat VRN1. *TAG. Theoretical and Applied Genetics. Theoretische und Angewandte Genetik* 110(3): 527–536.

Johnson, P. G., Kenworthy, K. E., Auld, D. L. and Riordan, T. P. (2001). Distribution of buffalograss polyploid variation in the southern Great Plains. *Crop Science* 41(3): 909–913.

Johnson, P. G., Riordan, T. P. and Arumuganathan, K. (1998). Ploidy level determinations in buffalograss clones and populations. *Crop Science* 38(2): 478–482.

Jones, K. (1953). The cytology of some British species of agrostis and their hybrids. *British Agricultural Bulletin* 5: 316.

Jones, K. (1956). Species differentiation in *Agrostis* II. The significance of chromosome pairing in the tetraploid hybrids of *Agrostis canina* subsp. *Montana* Hartm., *A. tenuis* Sibth. *Journal of Genetics* 54(3): 377–393.

Jones, E., Dupal, M., Dumsday, J., Hughes, L. and Forster, J. (2002a). An SSR-based genetic linkage map for perennial ryegrass (*Lolium perenne* L.). *Theoretical and Applied Genetics* 105: 577–584.

Jones, E. S., Mahoney, N. L., Hayward, M. D., Armstead, I. P., Jones, J. G., Humphreys, M. O., King, I. P., Kishida, T., Yamada, T., Balfourier, F., Charmet, G. and Forster, J. W. (2002b). An enhanced molecular marker based genetic map of perennial ryegrass (*Lolium perenne*) reveals comparative relationships with other Poaceae genomes. *Genome* 45(2): 282–295.

Jones, E. S., Dupal, M. P., Kölliker, R., Drayton, M. C. and Forster, J. W. (2001). Development and characterization of simple sequence repeat (SSR) markers for perennial ryegrass (*Lolium perenne* L.). *Theoretical and Applied Genetics* 102(2–3): 405–415.

Kalb, O., Dehmer, K. J. and Weber, W. E. (2003). Classification of *Lolium multiflorum* accessions by AFLP-markers. *Czech Journal of Genetics and Plant Breeding* 39: 340–341.

Kang, S. Y., Lee, G. J., Lim, K. B., Lee, H. J., Park, I. S., Chung, S. J., Kim, J. B., Kim, D. S. and Rhee, H. K. (2008). Genetic diversity among Korean Bermudagrass (*Cynodon* spp.) ecotypes characterized by morphological, cytological and molecular approaches. *Molecules and Cells* 25(2): 163–171.

Keck, D. D. (1965). Poa. p. 1482–1490. In: Munz, P. A. (Ed.) *California Flora* (3rd edn.). University of California Press, Berkeley, CA.

Kenworthy, K. E. and Quesenberry, K. H. (2020). St. Augustinegrass plant named 'FSA1602'. U.S. Plant Patent No. 32,645. Trademark Office.

Kenworthy, K. E., Taliaferro, C. M., Carver, B. F., Martin, D. L., Anderson, J. A. and Bell, G. E. (2006). Genetic variation in *Cynodon transvaalensis* Burtt-Davy. *Crop Science* 46(6): 2376–2381.

Khanal, S., Schwartz, B. M., Kim, C., Adhikari, J., Rainville, L. K., Auckland, S. A. and Paterson, A. H. (2017). Cross-taxon application of sugarcane EST-SSR to genetic diversity analysis of Bermudagrass (*Cynodon* spp.). *Genetic Resources and Crop Evolution* 64(8): 2059-2070.

Kim, J. H., Doan, P. P. T., Lee, H. and Kim, J. (2022). Transient gene expression system in zoysiagrass leaf mesophyll protoplasts. *Plant Biotechnology Reports* 16(1): 113-121.

Kimball, J. A., Zuleta, M. C., Kenworthy, K. E., Lehman, V. G., Harris-Shultz, K. R. and Milla-Lewis, S. (2013). Genetic relationships in *Zoysia* species and the identification of putative interspecific hybrids using simple sequence repeat markers and inflorescence traits. *Crop Science* 53(1): 285–295.

Kimball, J. A., Zuleta, M. C., Martin, M. C., Kenworthy, K. E., Chandra, A. and Milla-Lewis, S. R. (2012). Assessment of molecular variation within 'Raleigh' St. Augustinegrass using amplified fragment length polymorphism markers. *Hort Science* 47(7): 839-844.

Kimball, J. A., Tuong, T. D., Arellano, C. A., Livingston, D. P., III and Milla-Lewis, S. R. (2018). Linkage analysis and the identification of quantitative trait loci associated with freeze tolerance and turf quality traits in St. Augustinegrass. *Molecular Breeding* 38(5): 67.

King, J., Thorogood, D., Edwards, K. J., Armstead, I. P., Roberts, L., Skøt, K., Hanley, Z. and King, I. P. (2008). Development of a genomic microsatellite library in perennial ryegrass (*Lolium perenne*) and its use in trait mapping. *Annals of Botany* 101(6): 845–853.

Kölliker, R., Stadelmann, F. J., Reidy, B. and Nösberger, J. (1999). Genetic variability of forage grass cultivars: A comparison of *Festuca pratensis* Huds., *Lolium perenne* L., and *Dactylis glomerata* L. *Euphytica* 106(3): 261-270.

Krans, J. V., Philley, H. W., Goatley Jr., J. M. and Tomaso-Peterson, M. (1992). The relationships of rhizome density, sod strength, and spring green-up in Bermudagrass. *Agronomy Abstracts* 84: 172.

Kubik, C., Honig, J. and Bonos, S. A. (2011). Characterization of 215 simple sequence repeat markers in creeping bentgrass (*Agrostis stolonifera* L.). *Molecular Ecology Resources* 11(5): 872–876.

Kubik, C., Meyer, W. A. and Gaut, B. S. (1999). Assessing the abundance and polymorphism of simple sequence repeats in perennial ryegrass. Crop *Science* 39: 1136–1141.

Lee, B. and Park, J. (2021). The complete chloroplast genome of Zoysia matrella (L.) Merr. Isolated in Korea (Poaceae): investigation of intraspecific variations on chloroplast genomes. Mitochondrial DNA Part B 6: 572-574.

Lee, B. and Park, J. (2021). The complete chloroplast genome of *Zoysia japonica* Steud. isolated in Korea (Poaceae): investigation of potential molecular markers on Z. japonica chloroplast genomes. *Plant Biotechnology Reports* 15(5): 707-715.

Li, H., Liu, L., Lou, Y., Hu, T. and Fu, J. (2011). Genetic diversity of Chinese natural Bermudagrass (*Cynodon dactylon*) germplasm using ISSR markers. *Scientia Horticulturae* 127(4): 555–561.

Lin, Z., Cogan, N. O., Pembleton, L. W., Spangenberg, G. C., Forster, J. W., Hayes, B. J. and Daetwyler, H. D. (2016). Genetic gain and inbreeding from genomic selection in a simulated commercial breeding program for perennial ryegrass. *The Plant Genome* 9(1). doi: 10.3835/plantgenome2015.06.0046.

Liu, W., Li, X. Q., Zhang, F., Ma, X. and Fan, Y. (2007). Genetic diversity of Bermudagrass accessions in south-west China by ISSRs molecular markers and geographic provenance. *Acta Pratacult. Sin* 16: 55-61.

Liu, Z. W., Jarret, R. L., Duncan, R. R. and Kresovich, S. (1994). Genetic relationships and variation among ecotypes of seashore paspalum (*Paspalum vaginatum*) determined by random amplified polymorphic DNA markers. *Genome* 37(6): 1011-1017.

Liu, Z. W., Jarret, R. L., Kresovich, S. and Duncan, R. R. (1995). Characterization and analysis of simple sequence repeat (SSR) loci in seashore paspalum (*Paspalum vaginatum* Swartz). *TAG. Theoretical and Applied Genetics. Theoretische und Angewandte Genetik* 91(1): 47-52.

Loch, D. S. (2015). Zoysiagrass: an overview of taxonomy, morphology, genetic variation, and environmental adaptation in relation to turf management and use. p. 30-47. In: *Proceedings of the 2015 International Zoysiagrass Symposium*, Okinawa, Japan.

Lu, H., Nagata, R., Kenworthy, K., Cherry, R., Quesenberry, K. and Busey, P. (2015). Registration of 'NUF-76'St. Augustinegrass. *Journal of Plant Registrations* 9(3): 299-303.

Luber, G. and McGeehin, M. (2008). Climate change and extreme heat events. *American Journal of Preventive Medicine* 35(5): 429-435.

Luo, Y., Zhang, X., Xu, J., Zheng, Y., Pu, S., Duan, Z., Li, Z., Liu, G., Chen, J. and Wang, Z. (2020). Phenotypic and molecular marker analysis uncovers the genetic diversity of the grass *Stenotaphrum secundatum*. *BMC Genetics* 21(1): 86.

McBee, G. G. and Holt, E. C. (1966). Shade tolerance studies on Bermudagrass and other turfgrasses 1. *Agronomy Journal* 58(5): 523-525.

McCall, D. S., Zhang, X., Sullivan, D. G., Askew, S. D. and Ervin, E. H. (2017). Enhanced soil moisture assessment using narrowband reflectance vegetation indices in creeping bentgrass. *Crop Science* 57(S1): S-161.

McCalla Jr., J., Richardson, M., Karcher, D., Landreth, J. and Patton, A. (2009). Sod production utilizing an improved seeded Bermudagrass cultivar. *Journal of Applied Turfgrass Science* 18: 1-9.

Meeks, M. and Chandra, A. (2020). Drought response and minimal water requirements of diploid and interploid St. Augustinegrass under progressive drought stress. *Crop Science* 60(2): 1048-1063.

Meeks, M. and Chandra, A. (2021). Performance of diploid and interploid hybrids of St. Augustinegrass under moderate shade. *Crop Science* 61(5): 3719-3733.

Meyer, W. A. (1982). Breeding disease-resistant cool-season turfgrass cultivars for the United States. *Plant Disease* 66(1): 341-344.

Meyer, W. A. and Funk, C. R. (1989). Progress and benefits to humanity from breeding cool-season grasses for turf. p. 31-48. In: Sleper et al. (Ed.) Contributions from *Breeding Forage* and *Turf Grasses*. CSSA Spec. Publ. 15. CSSA, Madison, WI.

Meyer, W. A., Hoffman, L. and Bonos, S. A. (2017). Breeding cool-season turfgrass cultivars for stress tolerance and sustainability in a changing environment. *Itsrj* 13(1): 3-10.

Meyer, W. A., Rose, B. L., Johnson-Cicalese, J. M. and Funk, C. R. (1984). Registration of Midnight Kentucky bluegrass. *Crop Science* 24: 822-823.

Meyer, W. A. and Watkins, E. (2003). Tall fescue (Festuca arundinacea). p. 107-127. In: M. D. Casler and R. R. Duncan (Eds.) Turfgrass Biology Genetics and Breeding. John Wiley & Sons, Hoboken, NJ.

Mian, M. A. R., Hopkins, A. A. and Zwonitzer, J. C. (2002). Determination of genetic diversity in tall fescue with AFLP markers. *Crop Science* 42(3): 944-950.

Mian, M. A. R., Saha, M. C., Hopkins, A. A. and Wang, Z. Y. (2005). Use of tall fescue EST-SSR markers in phylogenetic analysis of cool-season forage grasses. *Genome* 48(4): 637-647.

Mian, M. R., Zhang, Y., Wang, Z. Y., Zhang, J. Y., Cheng, X., Chen, L., Chekhovskiy, K., Dai, X., Mao, C., Cheung, F. and Zhao, X. (2008). Analysis of tall fescue ESTs representing different abiotic stresses, tissue types and developmental stages. *BMC Plant Biology* 8: 1-13.

Milla-Lewis, S. R., Zuleta, M. C., Van Esbroeck, G. A., Quesenberry, K. H. and Kenworthy, K. E. (2013). Cytological and molecular characterization of genetic diversity in Stenotaphrum. *Crop Science* 53(1): 296-308.

Miller, D. R., Mugaas, R. J., Meyer, M. H. and Watkins, E. (2013). Performance of low-maintenance turfgrass mixtures and blends. *hortTechnology* 23(5): 610-612.

Mingyue, T. U., Yali, H. E., Xiaoli, L. I., Ying, Z. O. U. and Xiaojun, Y. U. A. N. (2020). Development of SCAR markers related to heat tolerance in Kentucky bluegrass. *Notulae Botanicae Horti Agrobotanici Cluj-Napoca* 48(2), 509-522.

Mohammadi, R., Panahi, B. and Amiri, S. (2020). ISSR based study of fine fescue (*Festuca ovina* L.) highlighted the genetic diversity of Iranian accessions. *Cytology and Genetics* 54(3): 257-263.

Moose, S. P. and Mumm, R. H. (2008). Molecular plant breeding as the foundation for 21st century crop improvement. *Plant Physiology* 147(3): 969-977.

Mukherjee, S., Mishra, A. and Trenberth, K. E. (2018). Climate change and drought: a perspective on drought indices. *Current Climate Change Reports* 4(2): 145-163.

Mulkey, S. E., Zuleta, M. C., Keebler, J. E., Schaff, J. E. and Milla-Lewis, S. R. (2014). Development and characterization of simple sequence repeat markers for St. Augustinegrass. *Crop Science* 54(1): 401-412.

Murphy, J. A. (1996). Fine fescues: low-maintenance species for turf. *Rutgers Cooperative Research and Extension FS688*.

Musser, H. B. (1959). Turf management: grasses. *USGA Journal and Turf Management* 12: 31-32.

Muylle, H., Baert, J., Van Bockstaele, E., Moerkerke, B., Goetghebeur, E. and Roldan-Ruiz, I. (2005a). Identification of molecular markers linked with crown rust (*Puccinia coronata* f. sp. *lolii*) resistance in perennial ryegrass (*Lolium perenne*) using AFLP markers and a bulked segregant approach. *Euphytica* 143(1-2): 135-144.

Muylle, H., Baert, J., Van Bockstaele, E., Pertijs, J. and Roldan-Ruiz, I. (2005b). Four QTLs determine crown rust (*Puccinia coronata* f. sp. *lolii*) resistance in a perennial ryegrass (*Lolium perenne*) population. *Heredity* 95(5): 348-357.

Okoli, B. E. (1982). IOPB chromosome number reports LXXIV. *Taxon* 31: 127.

Paina, C., Byrne, S. L., Studer, B., Rognli, O. A. and Asp, T. (2016). Using a candidate gene-based genetic linkage map to identify QTL for winter survival in perennial ryegrass. *PLoS ONE* 11(3): e0152004.

Patterson, J. T., Larson, S. R. and Johnson, P. G. (2005). Genome relationships in polyploid *Poa pratensis* and other *Poa* species inferred from phylogenetic analysis of nuclear and chloroplast DNA sequences. *Genome* 48(1): 76-87.

Patton, A. J., Schwartz, B. M. and Kenworthy, K. E. (2017). Zoysiagrass (*Zoysia* spp.) history, utilization, and improvement in the United States: a review. *Crop Science* 57(S1): S-37.

Pepin, G. W. and Funk, C. R. (1971). Intraspecific hybridization as a method of breeding Kentucky bluegrass (*Poa pratensis* L.) for turf. *Crop Science* 11(3): 445-448.

Pereira, C. H., Paudel, D., Nunes, J. A. R., Unruh, J. B., Quesenberry, K., Kenworthy, K. and Rios, E. (2022). Multienvironment evaluation of St. Augustinegrass breeding lines for turf performance via a mixed-models approach. *International Turfgrass Society Research Journal* 14(1): 673-682.

Plant Variety Protection Act. U. S. Code, vol. 7 (1970). *Statutes at Large*, vol. 84.

Porceddu, A., Albertini, E., Barcaccia, G., Falistocco, E. and Falcinelli, M. (2002). Linkage mapping in apomictic and sexual Kentucky bluegrass (*Poa pratensis* L.) genotypes using a two way pseudo-testcross strategy based on AFLP and SAMPL markers. *TAG. Theoretical and Applied Genetics. Theoretische und Angewandte Genetik* 104(2-3): 273-280.

Qi, P., Eudy, D., Schnable, J. C., Schmutz, J., Raymer, P. L. and Devos, K. M. (2019). High density genetic maps of Seashore paspalum using genotyping-by-sequencing and their relationship to the *Sorghum bicolor* genome. *Scientific Reports* 9(1): 12183.

Qiu, Y., Hamernick, S., Ortiz, J. B. and Watkins, E. (2020). DNA content and ploidy estimation of Festuca ovina accessions by flow cytometry. *Crop Science* 60(5): 2757-2767.

Qiu, Y., Hirsch, C. D., Yang, Y. and Watkins, E. (2019). Towards improved molecular identification tools in fine fescue (*Festuca* L., Poaceae) turfgrasses: nuclear genome size, ploidy, and chloroplast genome sequencing. *Frontiers in Genetics* 10: 1223.

Quarin, C. L. and Burson, B. L. (1983). Cytogenetic relations among *Paspalum notatum* var. *saurae*, *P. pumilum*, *P. indecorum*, and *P. vaginatum*. *Botanical Gazette. Botanical Gazette* 144(3): 433-438.

Ramm, C., Wachholtz, M., Amundsen, K., Donze, T., Heng-Moss, T., Twigg, P., Palmer, N. A., Sarath, G. and Baxendale, F. (2015). Transcriptional profiling of resistant and susceptible buffalograsses in response to *Blissus occiduus* (Hemiptera: Blissidae) feeding. *Journal of Economic Entomology* 108(3): 1354-1362.

Riordan, T. P., Shearman, R. C., Watkins, J. E. and Behling, J. P. (1988). Kentucky bluegrass automatic hybridization apparatus. *Crop Science* 28(1): 183-185.

Robinson, B. P. and Burton, G. W. (1953). Tiffine (Tifton 127) turf Bermudagrass. *USGA Journal and Turfgrass Management* 6: 30.

Roldan-Ruiz, I., Calsyn, E., Gilliland, T. J., Coll, R., Van Eijk, M. J. T. and De Loose, M. (2000). Estimating genetic conformity between related ryegrass (*Lolium*) varieties. 2. AFLP characterization. *Molecular Breeding* 6(6): 593-602.

Roodt, R., Spies, J. J. and Burger, T. H. (2002). Preliminary DNA fingerprinting of the turf grass *Cynodon dactylon* (Poaceae: Chloridoideae). *Bothalia* 32(1): 117-122.

Rotter, D., Amundsen, K., Bonos, S. A., Meyer, W. A., Warnke, S. E. and Belanger, F. C. (2009). Molecular genetic linkage map for allotetraploid colonial bentgrass. *Crop Science* 49(5): 1609-1619.

Ruemmele, B. A. (2000). Breeding colonial bentgrass for drought, heat, and wear. *Diversity* 16: 34-35.

Ruemmele, B. A., Brilman, L. A. and Huff, D. R. (1995) Fine fescue germplasm diversity and vulnerability. *Crop Science* 35: 313-316.

Ruemmele, B. A., Wipff, J. K., Brilman, L. and Hignight, K. W. (2003). Fine-leaved *Festuca* species. p. 129-174. In: *Turfgrass Biology. Genetics, and Breeding*. John Wiley and Sons, Inc., Hoboken, NJ.

Saha, D. C., Johnson-Cicalese, J. M., Halisky, P. M., Van Heemstra, M. I. and Funk, C. R. (1987). Occurrence and significance of endophytic fungi in the fine fescues. *Plant Disease* 71(11): 1021-1024.

Saha, M., Kirigwi, F., Chekhovskiy, K., Black, J. and Hopkins, A. (2009). Molecular mapping of QTLs associated with important forage traits in tall fescue. p. 251-258. In: *Molecular Breeding of Forage and Turf*. Springer, New York, NY.

Saha, M. C., Cooper, J. D., Mian, M. R., Chekhovskiy, K. and May, G. D. (2006). Tall fescue genomic SSR markers: development and transferability across multiple grass species. *Theoretical and Applied Genetics* 113: 1449-1458.

Saha, M. C., Mian, M. R., Eujayl, I., Zwonitzer, J. C., Wang, L. and May, G. D. (2004). Tall fescue EST-SSR markers with transferability across several grass species. *Theoretical and Applied Genetics* 109: 783-791.

Saha, M. C., Mian, R., Zwonitzer, J. C., Chekhovskiy, K. and Hopkins, A. A. (2005). An SSR- and AFLP-based genetic linkage map of tall fescue (*Festuca arundinacea* Schreb.). *Theoretical and Applied Genetics* 110: 323-336.

Sartie, A. M., Easton, H. S., Matthew, C., Rolston, M. P. and Faville, M. J. (2018). Seed yield in perennial ryegrass (*Lolium perenne* L.): comparative importance of component traits and detection of seed-yield-related QTL. *Euphytica* 214: 1-19.

Satterthwaite, L. N., Hodges, A. W., Haydu, J. J. and Cisar, J. L. (2009). *An Agronomic and Economic Profile of Florida's Sod Industry in 2007*. Florida Cooperative Extension Service, Florida Agricultural Experiment Station.

Schejbel, B., Jensen, L. B., Asp, T., Xing, Y. and Lübberstedt, T. (2008). Mapping of QTL for resistance to powdery mildew and resistance gene analogues in perennial ryegrass. *Plant Breeding* 127(4): 368-375.

Schejbel, B., Jensen, L. B., Xing, Y. and Lübberstedt, T. (2007). QTL analysis of crown rust resistance in perennial ryegrass under conditions of natural and artificial infection. *Plant Breeding* 126(4): 347-352.

Schwartz, B. M., Hanna, W. W., Baxter, L. L., Raymer, P. L., Waltz, F. C., Kowalewski, A. R., Chandra, A., Genovesi, A. D., Wherley, B. G., Miller, G. L., Milla-Lewis, S. R., Reynolds, W. C., Wu, Y., Martin, D. L., Moss, J. Q., Kenna, M. P., Unruh, J. B., Kenworthy, K. E., Zhang, J. and Munoz, P. R. (2018). 'DT-1', a drought-tolerant triploid turf Bermudagrass. *HortScience* 53(11): 1711-1714.

Segars, C. A., Moss, J. Q., Martin, D. L. and Wu, Y. (2022). Sod production characteristics: how strong is your Bermudagrass? *International Turfgrass Society Research Journal* 14(1): 256-261, doi: 10.1002/its2.4.

Shakoor, N., Lee, S. and Mockler, T. C. (2017). High throughput phenotyping to accelerate crop breeding and monitoring of diseases in the field. *Current Opinion in Plant Biology* 38: 184-192.

Shildrick, J. P. (1977). Turfgrass seed mixtures in the United Kingdom. p. 57-64. In: Beard, J. B. (Ed.) *Proceedings of the Third International Turfgrass Conference. International Turfgrass Society and ASA and CSSA*. Madison Book Company, Madison, WI.

Sim, S., Diesburg, K., Casler, M. and Jung, G. (2007). Mapping and comparative analysis of QTL for crown rust resistance in an Italian× perennial ryegrass population. *Phytopathology* 97(6): 767-776.

Skogley, R. (1973). *Velvet bentgrass. Bull. 199*. Univ. Coop. Ext., Univ. of Rhode Island, Kingston.

Skøt, L., Sackville Hamilton, N. R., Mizen, S., Chorlton, K. H. and Thomas, I. D. (2002). Molecular genecology of temperature response in *Lolium perenne*: 2. Association of AFLP markers with ecogeography. *Molecular Ecology* 11(9): 1865-1876.

Soreng, R. J. (1985). *Poa* L. in New Mexico, with a key to middle and Southern Rocky Mountain species (Poaceae). *The Great Basin Naturalist*: 395-422.

Sprague, H. B. and Evaul, E. E. (1930). Experiments with turf grasses in New Jersey. *NJ Bulletin of the Agricultural Experiment Station* 497: 1-55.

Steinke, K., Chalmers, D., Thomas, J., White, R. and Fipps, G. (2010). Drought response and recovery characteristics of St. Augustinegrass cultivars. *Crop Science* 50: 2076-2083.

Studer, B., Byrne, S., Nielsen, R. O., Panitz, F., Bendixen, C., Islam, M. S., Pfeifer, M., Lübberstedt, T. and Asp, T. (2012). A transcriptome map of perennial ryegrass (*Lolium perenne* L.). *BMC Genomics* 13: 140.

Taliaferro, C. M. (2003). Bermudagrass. p. 235-256. In: Casler, M. D. and Duncan, R. R. (Eds) *Turfgrass Biology, Genetics, and Breeding*. Wiley, Hoboken, NJ.

Taliaferro, C. M., Martin, D. L., Anderson, J. A. and Anderson, M. P. (2006). Patriot turf bermudagrass. USPP 16,801. U.S. Patent and Trademark Office, Washington, DC.

Talukder, S. K., Bhamidimarri, S., Chekhovskiy, K. and Saha, M. C. (2020). Mapping QTL for summer dormancy related traits in tall fescue (*Festuca arundinacea* Schreb.). *Scientific Reports* 10(1): 14539.

Tan, C., Wu, Y., Taliaferro, C. M., Anderson, M. P., Tauer, C. and Samuels, T. (2012). Development of simple sequence repeat markers for Bermudagrass from its expressed sequence tag sequences and preexisting sorghum SSR markers. *Molecular Breeding* 29(1): 23-30.

Tanaka, H., Tokunaga, R., Muguerza, M., Kitazaki, Y., Hashiguchi, M., Sato, S., Tabata, S. and Akashi, R. (2016). Genetic structure and speciation of zoysiagrass ecotypes collected in Japan. *Crop Science* 56(2): 818-826.

Tanger, P., Klassen, S., Mojica, J. P., Lovell, J. T., Moyers, B. T., Baraoidan, M., Naredo, M. E. B., McNally, K. L., Poland, J., Bush, D. R., Leung, H., Leach, J. E. and McKay, J. K. (2017). Field-based high throughput phenotyping rapidly identifies genomic regions controlling yield components in rice. *Scientific Reports* 7: 42839.

Taylor, C., Madsen, K., Borg, S., Møller, M. C., Boelt, B. and Holm, P. H. (2001). The development of sequence-tagged sites (STSs) in *Lolium perenne* L.: the application of primer sets derived from other genera. *Theoretical and Applied Genetics* 103(4): 648-658.

Taylor, R. G., Scanlon, B., Döll, P., Rodell, M., Van Beek, R., Wada, Y., Longuevergne, L., Leblanc, M., Famiglietti, J. S., Edmunds, M., Konikow, L., Green, T. R., Chen, J., Taniguchi, M., Bierkens, M. F. P., MacDonald, A., Fan, Y., Maxwell, R. M., Yechieli, Y., Gurdak, J. J., Allen, D. M., Shamsudduha, M., Hiscock, K., Yeh, P. J.-F., Holman, I. and Treidel, H. (2013). Ground water and climate change. *Nature Climate Change* 3(4): 322-329.

Tegg, R. S. and Lane, P. A. (2004). A comparison of the performance and growth of a range of turfgrass species under shade. *Australian Journal of Experimental Agriculture* 44(3): 353-358.

Terrell, E. E. (1968). *A Taxonomic Revision of the Genus Lolium. Technical Bulletin No. 1392*. United States Department of Agriculture, Washington, DC.

Thorogood, D. (2003). Perennial ryegrass (*Lolium perenne* L.). p. 75-106. In: *Genetics, and Breeding* Casler, M. D. and Duncan, R. R. (Eds) *Turfgrass Biology*. John Wiley and Sons, Hoboken, NJ.

Thorogood, D., Kaiser, W. J., Jones, J. G. and Armstead, I. (2002). Self-incompatibility in ryegrass 12. Genotyping and mapping the S and Z loci of *Lolium perenne* L. *Heredity* 88(5): 385–390.

Thorogood, D., Paget, M. F., Humphreys, M. O., Turner, L. B., Armstead, I. and Roderick, H. (2001). QTL analysis of crown rust resistance in perennial ryegrass-implications for breeding. *International Turfgrass Society Research Journal* 9: 218–223.

Tian, L., Huang, C., Yu, R., Liang, R., Li, Z., Zhang, L., Wang, Y., Zhang, X. and Wu, Z. (2006). Overexpression AtNHX1 confers salt-tolerance of transgenic tall fescue. *African Journal of Biotechnology* 5.

Trenberth, K. E. (2011). Changes in precipitation with climate change. *Climate Research* 47(1): 123–138.

Turgeon, A. J. (2008). *Turfgrass Management* (8th edn.). Pearson Prentice Hall, Upper Saddle River, NJ.

USDA (2017). *Census of Agriculture. United States Summary and State Data Geographic Areas Series (vol. 1).* United States Department of Agriculture, Washington, DC.

Van der Laat, R., Dale, A. G., Arellano, C. and Milla-Lewis, S. R. (2022). Variation in southern chinch bug (*Blissus insularis*) survival and damage on St. Augustinegrass germplasm. *International Turfgrass Society Research Journal* 14(1): 642–651. doi: 10.1002/its2.67.

Van Treuren, R. (2008). AFLP fingerprinting of Kentucky bluegrass (*Poa pratensis* L.) from undisturbed Dutch grasslands: implications for conservation. *Plant Genetic Resources Newsletter* 153: 1–8.

Vergara, G. V. and Bughrara, S. S. (2003). AFLP analyses of genetic diversity in bentgrass. *Crop Science* 43(6): 2162–2171.

Wachholtz, M., Heng-Moss, T., Twigg, P., Baird, L., Lu, G. and Amundsen, K. (2013). Transcriptome analysis of two buffalograss cultivars. *BMC Genomics* 14: 613.

Walton, T., Richardson, M. D. and Karcher, D. E. (2021). Daily light integral requirements of Tifeagle bermudagrass and Lazer zoysiagrass putting greens [Abstract]. In: ASA, CSSA, SSSA International Annual Meeting. Salt Lake City, UT. Paper 134194.

Wang, M. L., Chen, Z. B., Barkley, N. A., Newman, M. L., Kim, W., Raymer, P. and Pederson, G. A. (2006). Characterization of seashore paspalum (*Paspalum vaginatum* Swartz) germplasm by transferred SSRs from wheat, maize and sorghum. *Genetic Resources and Crop Evolution* 53(4): 779–791.

Wang, Z., Liao, L., Yuan, X., Guo, A. and Liu, J. (2011). Genetic relationships of Bermudagrass (*Cynodon dactylon* var. *dactylon*) from different countries revealed by sequence-related amplified polymorphism (SRAP) analysis. *African Journal of Biotechnology* 10: 17106–17115.

Wang, Z., Liao, L., Yuan, X., Guo, H., Guo, A. and Liu, J. (2013). Genetic diversity analysis of *Cynodon dactylon* (Bermudagrass) accessions and cultivars from different countries based on ISSR and SSR markers. *Biochemical Systematics and Ecology* 46: 108–115.

Wang, Z., Raymer, P. and Chen, Z. (2017). Isolation and characterization of microsatellite markers for *Stenotaphrum* Trin. using 454 Sequencing Technology. *Hortscience* 52(1): 16–19.

Warnke, S. E. (2020). Detection of interspecific hybrids between creeping bentgrass and colonial bentgrass using DNA marker analysis. *Agrosystems, Geosciences and Environment* 3(1): e20048.

Watkins, E., Hollman, A. B. and Horgan, B. P. (2010). Evaluation of alternative turfgrass species for low-input golf course fairways. *HortScience* 45(1): 113–118.

Wenger, L. E. (1941). Soaking buffalograss (*Buchloe dactyloides*) seed to improve its germination. *Agronomy Journal* 33(2): 135–141.

Wenger, L. E. (1943). Buffalo grass. *Kansas Agricultural Experiment Station Bulletin* 321.

Wilkins, P. W. (1991). Breeding perennial ryegrass for agriculture. *Euphytica* 52(3): 201–214.

Wofford, D. S. and Baltensperger, A. A. (1985). Heritability estimates for turfgrass characteristics in bermudagrass 1. Crop *Science* 25: 133–136.

Wu, F., Chen, J., Wang, J., Wang, X., Lu, Y., Ning, Y. and Li, Y. (2019a). Intra-population genetic diversity of *Buchloe dactyloides* (Nutt.) Engelm (buffalograss) determined using morphological traits and sequence-related amplified polymorphism markers. *3 Biotech* 9(3): 97.

Wu, Y. Q., Martin, D. L., Moss, J. Q., Walker, N. R. and Fontanier, C. H. (2019b). Bermudagrass plant named 'OKC 1131'. USPP/0364715 P1. U.S. Patent and Trademark Office, Washington, DC.

Wu, Y. Q., Martin, D. L., Taliaferro, C. M., Anderson, J. A. and Moss, J. Q. (2013). Northbridge turf Bermudagrass. USPP 24,116. U.S. Patent and Trademark Office, Washington, DC.

Wu, Y. Q., Martin, D. L., Taliaferro, C. M., Anderson, J. A. and Moss, J. Q. (2014). Latitude 36 turf Bermudagrass. USPP 24,271. U.S. Patent and Trademark Office, Washington, DC.

Wu, Y. Q., Taliaferro, C. M., Bai, G. H. and Anderson, M. P. (2004). AFLP analysis of *Cynodon dactylon* (L.) Pers. var. *dactylon* genetic variation. *Genome* 47(4): 689–696.

Wu, Y. Q., Taliaferro, C. M., Bai, G. H., Martin, D. L., Anderson, J. A., Anderson, M. P. and Edwards, R. M. (2006). Genetic analyses of Chinese *Cynodon* accessions by flow cytometry and AFLP markers. *Crop Science* 46(2): 917–926.

Xiaojun, Y. U. A. N., Mingyue, T. U., Yali, H. E., Weiqian, W. A. N. G., Jun, L. I. and Shumin, Z. H. O. U. (2018). Analysis of genetic diversity in 73 Kentucky bluegrass materials by SSR and SRAP Markers. *Notulae Botanicae Horti Agrobotanici Cluj-Napoca* 46(2): 327–335.

Xiaojun, Y. U. A. N., Yali, H. E., Huang, J., Wen, H. U., Huanhuan, Z. H. O. U., Qiongyu, G. A. O. and Shumin, Z. H. O. U. (2015). Development of a SCAR marker for rapid identification of new Kentucky bluegrass breeding lines. *Notulae Botanicae Horti Agrobotanici Cluj-Napoca* 43(1), 79–85.

Xie, Y., Sun, X., Ren, J., Fan, J., Lou, Y., Fu, J. and Chen, L. (2015). Genetic diversity and association mapping of cadmium tolerance in Bermudagrass [*Cynodon dactylon* (L.) Pers.]. *Plant and Soil* 390(1–2): 307–321.

Xiong, Y., Fei, Sz, Arora, R., Brummer, E. C., Barker, R. E., Jung, G. and Warnke, S. E. (2007). Identification of quantitative trait loci controlling winter hardiness in an annual × perennial ryegrass interspecific hybrid population. *Molecular Breeding* 19(2): 125–136.

Xu, W. W., Sleper, D. A. and Chao, S. (1995). Genome mapping of polyploid tall fescue (*Festuca arundinacea* Schreb.) with RFLP markers. *TAG. Theoretical and Applied Genetics. Theoretische und Angewandte Genetik* 91(6–7): 947–955.

Xu, W. W., Sleper, D. A. and Krause, G. F. (1994). Genetic diversity of tall fescue germplasm based on RFLPs. *Crop Science* 34(1): 246–252.

Xu, Y., Wang, J., Bonos, S. A., Meyer, W. A. and Huang, B. (2018). Candidate genes and molecular markers correlated to physiological traits for heat tolerance in fine fescue cultivars. *International Journal of Molecular Sciences* 19(1): 116.

Yamada, T., Jones, E. S., Cogan, N. O. I., Vecchies, A. C., Nomura, T., Hisano, H., Shimamoto, Y., Smith, K. F., Hayward, M. D. and Forster, J. W. (2004). QTL analysis of morphological, developmental, and winter hardiness-associated traits in perennial ryegrass. *Crop Science* 44(3): 925-935.

Yamada, T. and Kishida, T. (2003). Genetic analysis of forage grasses based on heterologous RFLP markers detected by rice cDNAs. *Plant Breeding* 122(1): 57-60.

Yang, L., Wu, Y., Moss, J. Q., Zhong, S. and Yang, B. (2018). Molecular identification and characterization of seeded turf Bermudagrass cultivars using simple sequence repeat markers. *Agronomy Journal* 110(6): 2142-2150.

Yi, Y. J., Zhang, X. Q., Huang, L. K., Ling, Y., Ma, X. and Liu, W. (2008). Genetic diversity of wild *Cynodon dactylon* germplasm detected by SRAP markers. *Yichuan* 30(1): 94-100.

Yu, X., Brown, J. M., Graham, S. E., Carbajal, E. M., Zuleta, M. C. and Milla-Lewis, S. R. (2019). Detection of quantitative trait loci associated with drought tolerance in St. Augustinegrass. *PLoS ONE* 14(10): e0224620. doi: 10.1371/journal.pone.0224620.

Yu, X., Kimball, J. A. and Milla-Lewis, S. R. (2018). High density genetic maps of St. Augustinegrass and applications to comparative genomic analysis and QTL mapping for turf quality traits. *BMC Plant Biology* 18: 346.

Yu, X., Mulkey, S. E., Zuleta, M. C., Arellano, C., Ma, B. and Milla-Lewis, S. R. (2020). Quantitative trait loci associated with gray leaf spot resistance in St. Augustinegrass. *Plant Disease* 104(11): 2799-2806.

Zhang, L.-H., Ozias-Akins, P., Kochert, G., Kresovich, S., Dean, R. and Hanna, W. (1999). Differentiation of Bermudagrass (*Cynodon* spp.) genotypes by AFLP analyses. *Theoretical and Applied Genetics* 98(6-7): 895-902.

Zhao, H., Bughrara, S. S. and Oliveira, J. A. (2006). Genetic diversity in colonial bentgrass (*Agrostis capillaris* L.) revealed by *Eco*RI-*Mse*I and *Pst*I-*Mse*I AFLP markers. *Genome* 49(4): 328-335.

Zhao, J., Zhi, D., Xue, Z., Liu, H. and Xia, G. (2007). Enhanced salt tolerance of transgenic progeny of tall fescue (*Festuca arundinacea*) expressing a vacuolar Na+/H+ antiporter gene from Arabidopsis. *Journal of Plant Physiology* 164(10): 1377-1383.

Zheng, Y., Liu, J. and Chen, S. (2005). RAPD analysis of merit selections of *Cynodon dactylon* in China. *Journal of Plant Resources and Environment* 14: 6-9.

Zheng, Y., Xu, S., Liu, J., Zhao, Y. and Liu, J. (2017). Genetic diversity and population structure of Chinese natural Bermudagrass [*Cynodon dactylon* (L.) Pers.] germplasm based on SRAP markers. *PLoS ONE* 12(5): e0177508.

Zhu, H., Yu, X., Xu, T., Wang, T., Du, L., Ren, G. and Dong, K. (2015). Transcriptome profiling of cold acclimation in Bermudagrass (*Cynodon dactylon*). *Scientia Horticulturae* 194: 230-236.

Chapter 3

Advances in soil management for successful establishment and maintenance of turfgrass

Douglas J. Soldat and Paul L. Koch, University of Wisconsin, USA

1 Introduction

Soil is the foundation for turfgrass growth. Soil properties influence turfgrass appearance, function, and performance (Fig. 1). Humans have been modifying soil properties to influence turfgrass response for as long as they have been utilizing turf surfaces; and much has been written about how to assess, create, modify, or manage turfgrass root zones to achieve the desired outcome (Weil and Brady, 2017). Researchers are also interested in understanding how turfgrass management affects soil physical, chemical, and biological properties over long periods of time. Soil physical properties include density and structure, infiltration capacity, and other hydrologic properties, including water-holding characteristics. These physical properties are easy to conceptualize but often expensive or difficult to quantify. Soil chemical properties include salinity, pH, plant-available nutrient content, heavy metal content, and more. These chemical properties are easy to interpret and relatively easy to measure, which explains why chemical properties of soils are frequently or routinely managed. Finally, soil biological properties include organic matter content, microbial biomass, soil enzyme activity, various microbial community analyses (i.e. microbiome analysis), and a suite of biological indices that attempt to quantify various pools of carbon and nitrogen. These biological properties are difficult to understand and are often difficult or expensive to measure.

http://dx.doi.org/10.19103/AS.2022.0110.03

Figure 1 Example of a professional baseball game in the USA; the soil properties influence the appearance, function, and playability of this natural turfgrass sports field. Source: M. Fidanza.

In agriculture, soil health has emerged as a useful framework for helping land managers make decisions that reduce soil degradation and sustain the soil's capacity to produce (Weil and Brady, 2017). The United States Department of Agriculture's Natural Resource Conservation Service (USDA-NRCS, Beltsville, MD, USA) has defined soil health as 'the continued capacity of the soil to function as a vital living ecosystem that sustains plants, animals, and humans' (https://www.nrcs.usda.gov/wps/portal/nrcs/main/soils/health/). A healthy soil regulates the water cycle and filters pollutants; it cycles, stores, and transforms nutrients important for life and also physically supports and sustains plant and animal life. Healthy soils can produce healthy plants with fewer inputs, while unhealthy soils require more management and intervention to achieve a similar result. While the framework of soil health has not yet been applied widely to turfgrass systems, only two peer-reviewed studies were found that contained 'soil health' and 'turfgrass' in the title (Moore et al., 2019a; Slavens et al., 2009), but the concept of soil health has the potential to help turfgrass managers, policymakers, and others better understand how turfgrass management affects soil ecosystems and vice versa.

A soil health assessment includes measuring soil properties, processes, or characteristics of soils. Laboratories offering soil health assessments include university laboratories (i.e. Cornell University, Ithaca, NY, USA; Oregon State University, Corvallis, OR, USA; University of Missouri, Columbia, MO, USA) and commercial laboratories (i.e. American Agricultural Laboratory, Inc., McCook, NE, USA; Brookside Laboratories, Inc., New Bremin, OH, USA). Just as laboratories differ in interpreting soil nutrient analyses, soil health indicator selection and interpretation are not standardized. As with soil nutrient analyses, calibration and interpretation of soil health indicators will continue to change as new research adds to our understanding. Therefore, this chapter will further explore our current understanding of these three important soil properties relevant to turfgrass management and identify areas for future study.

2 Soil physical properties

Notable physical indicators of soil health include soil texture, bulk density and penetration resistance, available water capacity, and aggregate stability (Lehmann et al., 2020). These indicators attempt to illustrate the way water and nutrients are retained and transported and how suitable the root zone environment is for life (i.e. roots, microbes, and soil animals) and hence turfgrass productivity, surface characteristics, and performance.

2.1 Soil texture

Soil texture refers to the proportion of sand, silt, and clay-sized particles in a soil. Soil texture can be classified; the USDA recognizes 12 classifications that are grouped based on ranges of sand, silt, and clay (Weil and Brady, 2017) (Fig. 2). These classifications are of practical use because soil texture can be approximated by feel in field situations and a large number of properties can be inferred from textural class. For example, sand-textured soils have relatively low water-holding capacity, often good aeration, and low susceptibility to compaction. Conversely, clay-textured soils have a high water-holding capacity, poor aeration, and high susceptibility to compaction. Loam-textured soils have moderate water-holding capacity, moderate aeration, and moderate susceptibility to compaction. Soil texture is also useful because it can

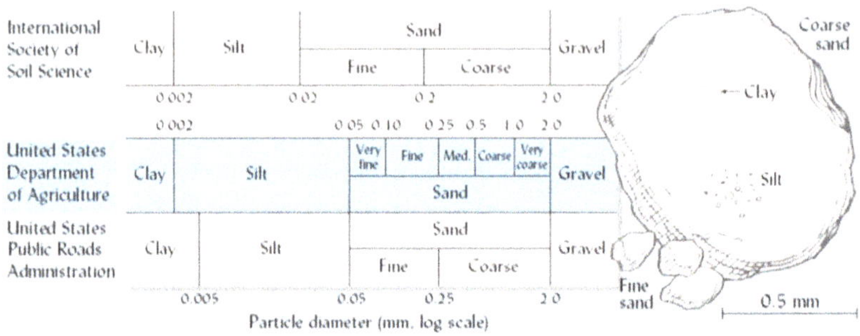

Figure 2 Soil particle size classification systems. Source: Brady and Weil (2010).

contextualize other soil properties. For example, a bulk density of 1.4 g•cm^{-3} would be considered high for a clay-textured soil but is normal for a sand-textured soil.

Soil texture is difficult to change. However, high-value turf systems often utilize root zones specially constructed with sand as the dominant component. Other areas built on non-sand soils are periodically topdressed with sand with the goal of shifting the surface soil texture to sand. Sand root zones have many properties that are desirable for high-value turfgrass systems, including rapid drainage and resistance to compaction (Turgeon and Kaminski, 2019). The primary challenges of sand root zones include poor plant-available nutrient and water-holding capacities. However, these challenges are often overcome with irrigation and fertilization inputs. Research into the optimal soil texture and soil amendments for turfgrass root zones is an ongoing endeavor. Bigelow and Soldat (2013) comprehensively reviewed the various root zone construction and modification methods of turfgrass systems.

2.2 Soil compaction

Soil compaction is commonly measured by bulk density and penetration resistance (Weil and Brady, 2017). An ideal healthy soil is comprised of approximately 50% solids (~45% mineral matter and 5% organic matter) and 50% voids or pore space. Ideally, the voids will be approximately half-filled with air and half-filled with water. When soils become compacted, the percentage of solids increases and the void space decreases. Smaller voids hold water more tightly than larger voids; therefore, compacted soils are often poorly aerated. Bulk density is a laboratory method for determining the oven-dry weight of soil per volume, while penetration resistance provides field measurements related to the compaction of the soil. Bigelow and Soldat (2013) reviewed the literature of compaction in turfgrass and Turgeon and Fidanza (2017) reviewed this history of cultivation of turfgrass soils. Soil compaction in turf systems is often confounded with traffic and wear. Dest and Ebdon (2017) recently evaluated the performance of newly sodded Kentucky bluegrass (*Poa pratensis* L.) under wear (i.e. traffic) and non-wear of compacted and non-compacted soil. Dest and Ebdon (2017) concluded that wear was the most limiting factor for Kentucky bluegrass performance, and while rooting below 7.5 cm depth was restricted in the compacted soil, the rooting in the top 7.5 cm was unaffected in the compacted soil.

2.3 Water-holding capacity

Available water-holding capacity is an especially important soil physical property for turfgrass health. Available water-holding capacity represents an amount - often reported as a root zone depth - of water that is held by the soil and available for plant uptake. It is strongly affected by soil texture, organic matter content, and condition (i.e. compaction) (Weil and Brady, 2017). For example, if a lawn uses ~20 mm water•day^{-1}, a soil with an available water-holding capacity of ~200 mm potentially represents a 10-day supply. However, a lawn growing on a soil with an available water-holding capacity of only 150 mm will 'run out' of water in less than 8 days. Managing soils to retain more plant-available water potentially would reduce irrigation water inputs for turfgrasses in areas that periodically receive rain. Soils with better available water-holding capacity will reduce the chances of excess irrigation and extend the time intervals between irrigation events, resulting in reduction in irrigation water use.

Soil amendments are known to affect available water-holding capacity. Bero and Soldat (2021) reported that topdressing a lawn with compost for three consecutive years could increase plant-available water-holding capacity by as much as 50%. However, soil texture, condition, organic matter, and amendments are not the only factors that can influence plant-available water. Researchers have demonstrated that tall fescue (*Schendonorus arundinaceus* [Schreb.] Dumort.) and perennial ryegrass (*Lolium perenne* L.) increase the available water-holding capacity of the soil compared to creeping red fescue (*Festuca rubra* L.) and Kentucky bluegrass because of the differences in root mass and soil pores created by those different grasses (Głab and Szewczyk, 2014).

2.4 Soil structure

Aggregation and aggregate stability are measures of soil structure (Weil and Brady, 2017). Soil structure represents the arrangement of soil particles and voids. Along with texture, soil structure influences air and water movements, which in turn affect biological activity in the soil. Soil structure is dynamic and affected by animals burrowing and digging, compaction, cultivation, soil moisture, shrinking and swelling of the clay components, and wetting and drying cycles. Soil structure can be described by categorizing the sizes and shapes of the structural units, commonly referred to as aggregates or peds. Another method is to test the stability of the aggregates by subjecting them to simulated rainfall. Soils that resist breaking apart under simulated rainfall are more likely to resist erosion and surface crusting in field settings. Hosseini et al. (2016) evaluated the role of endophytes in tall fescue on soil water availability and found that endophytes improved water availability to tall fescue by several mechanisms, one being improved soil aggregate stability compared to non-endophytic grass. It is well known that perennial grasses can improve or maintain soil structure by increasing soil organic carbon and producing a

Figure 3 Example of a turf profiler removing a section of root zone from a golf course putting green. Source: M. Fidanza.

network of fine roots associated with fungal hyphae (Carter et al., 1994), but few studies have evaluated the effects of turfgrass species or management practices on soil structure (Fig. 3).

3 Soil chemical properties

Chemical indicators of soil health include available plant nutrients, pH, salinity, sodicity, heavy metal content, and presence of other chemical contaminants, and more (Weil and Brady, 2017). These indicators are familiar to those who utilize traditional soil testing. Chemical indicators are useful for understanding the chemical fertility of the soil and the chemical suitability for certain microorganisms. Chemical indicators also are useful for tracking the salinity and sodium hazard (i.e. swelling potential) of soils.

Many soil chemical properties (i.e. salinity, pH, and heavy metals) influence the growth and health of turfgrass (Turgeon and Kaminski, 2019). However, the thresholds or acceptable ranges of these properties are highly species dependent. For example, annual bluegrass (*P. annua* L.) is known to be intolerant of high soil salinity, but seashore paspalum (*Paspalum vaginatum* Sw.) can be successfully irrigated with seawater in some cases (Turgeon and Kaminski, 2019). Turfgrass scientists continue to evaluate the tolerances of grasses (i.e. cultivars and species) to various soil chemical properties, and plant breeders continue to develop grasses that can potentially survive in more extreme soil chemical conditions. Successful turfgrass practitioners should understand the ideal or acceptable ranges of soil chemical properties and either choose the best adapted species or manipulate the soil profile to achieve conditions that the turfgrass stand can tolerate and thrive.

3.1 Sodicity

Sodicity is an important soil chemical property and is defined as the presence of a high proportion of sodium ions relative to other cations (Weil and Brady, 2017). While the presence of high sodium alone does not necessarily negatively impact turfgrass health, high sodium concentrations in soils can cause soil swelling which affects the balance of air and water in the soil and have a large impact on turfgrass health. Assessment of sodicity in soils involves knowing the percentage of total cation exchange sites occupied by sodium, referred to as the exchangeable sodium percentage or ESP. Sodium-induced swelling of soils can occur for sensitive soils (i.e. those containing 2:1 clays) when ESP exceeds 15%. Obear and Soldat (2014) found that a root zone containing 80% sand and 20% loam-textured soil, with a total soil clay content of 5%, showed no negative responses until ESP increased to 25%. Other sandier root zones evaluated did not show any negative responses to increasing ESP, presumably because the soil's swelling clay content was too low to change the soil's pore size distribution (Obear and Soldat, 2014).

3.1.1 Sodium adsorption ratio of irrigation water

The excess sodium that accumulates in turfgrass soils often comes from irrigation water (Turgeon and Kaminski, 2019). For this reason, chemical characterization of irrigation water is used to make predictions about the soil's potential to become sodic and to make management decisions. An early method of predicting soil sodicity was the sodium adsorption ratio (SAR), which represents the ratio of sodium to magnesium and calcium in the water. This method was found to underpredict the sodicity hazard because it does not account for calcite precipitation in the soil. When calcite precipitation occurs, the calcium of the soil solution decreases and results in a greater SAR. The adjusted SAR was then introduced to account for calcite precipitation, but field assessments found that method to overpredict the potential for sodium hazard. Suarez (1981) introduced an alternative method for estimating the SAR of water and, although the derivation contains an algebraic flaw (as described by Bleam, 2017), it has proved to be more reliable than earlier methods. The Suarez method is complex to calculate and as a result is not widely utilized by water testing laboratories. Recently, Bleam (2017) introduced a simple method which does not require water chemistry computer simulation and still accurately estimates sodium hazard. This method is based on simple best-case and worst-case assumptions regarding calcite precipitation in waters or soil solution as water evapoconcentrates. Tendency toward best-case and worst-case scenarios can be predicted by the alkalinity of the water. This method has the potential to improve management decisions regarding saline and sodic soils. However, testing, validation, and more research would be beneficial.

3.2 Plant-available nutrients

Plant-available nutrient content continues to be an area of high priority for turfgrass research. Frank and Guertal (2013) thoroughly reviewed the literature on potassium and phosphorus research in the field of turfgrass management. Since that review, studies have investigated the influence of soil potassium levels on turfgrass disease pressure and found that Microdochium patch (*Microdochium nivale* [Fr.] Samuels and Hallett) was more severe in high potassium soils (Bier et al., 2018), while anthracnose (*Colletotrichum cereale* sensu lato Crouch, Clarke, and Hillman) was reported to be more severe in soils with low soil potassium (Schmid et al., 2018). The pursuit of accurate interpretations of soil nutrient tests is complicated by these and other observations of interactions between turfgrass nutritional status and disease (Watschke et al., 2013).

4 Soil biological properties

Soil is an ecosystem that provides habitat for an enormous number of biological species that span all domains of life (Weil and Brady, 2017). Soil biological properties are soil properties that are influenced (directly or indirectly) by soil organisms. The most well-known biological property of soils is organic matter content. Soil organic matter consists of organic material in various stages of decomposition. While considered a biological property by the classification system used in this chapter, soil organic matter strongly influences soil chemical and physical properties. For this reason, soil organic matter has been widely studied. A comprehensive review of soil organic matter is beyond the scope of this chapter. However, Gaussoin et al. (2013) serves as an excellent resource regarding the characterization, development, and management of organic matter in turfgrass systems.

4.1 Nitrogen cycle

The nitrogen cycle is complex and is controlled and affected by many soil organisms (Weil and Brady, 2017). For this reason, the availability of nitrogen to turf is difficult to predict. In most other agricultural systems, this problem has been avoided by essentially flooding the system with plant-available nitrogen to avoid nutrient deficiencies. This practice has negative consequences, and now agricultural scientists are striving to better understand nitrogen cycling in soils that can potentially translate into farmers reducing the amount of fertilizer nitrogen applied without dramatically reducing yields. However, turfgrass systems are unusual because they are often maintained in a state of nitrogen deficiency (Turgeon and Kaminski, 2019). Maximizing yield is rarely a goal of turfgrass managers. Yet, nitrogen is important for helping the turfgrass recuperate from damage due to abiotic and biotic stresses. Too little nitrogen can promote turf diseases such as dollar spot (*Clarireedia* spp.) and rust (*Puccinia* spp.), among others, while excessive nitrogen is known to increase the severity of diseases such as brown patch (*Rhizoctonia solani* Kühn) and Pythium blight (*Pythium* spp.), among others (Watschke et al., 2013). Clearly, understanding the status of nitrogen availability to the turf is important (Turgeon and Kaminski, 2019), but turfgrass managers have very few tools available to them to do so.

4.2 Nitrogen availability

In pursuit of a better understanding of nitrogen availability, turfgrass scientists are investigating soil organic matter, which contains the largest pool of soil nitrogen (N) in turfgrass ecosystems (Weil and Brady, 2017). Soil organic matter is a broad category that encompasses all organic material in the soil in any state of decomposition. The carbon-to-nitrogen (C:N) ratio of bulk organic matter is often useful to predict how N may become plant available (mineralized) or made unavailable (immobilized) during the decomposition process. For example, the lower the C:N ratio, the more N will be mineralized (Janssen, 1996). Schmid et al. (2017) evaluated compost incorporation to improve a compacted, sandy soil. While the compost improved several important soil properties, Schmid et al. (2017) found that the relatively high C:N ratio of 41:1 of the compost likely hindered turfgrass performance for a short period following incorporation. Indeed, Yao and Shi (2010) reported C:N ratio of turfgrass soils to range from 10:1 to 17:1. However, the C:N ratio itself has not proven to be a good predictor of N release to turfgrass because of the complexity of the organic matter (Turgeon and Kaminski, 2019).

Methods for predicting N availability to plants from soil has proven difficult. The Illinois Soil N Test (Kahn et al., 2001) was developed to predict whether an agricultural crop would show a positive response to N fertilization. This method extracts and measures a fraction of the soil organic N pool that may be more representative of N mineralization than the bulk soil organic N pool. Gardner et al. (2013) reported that N extracted by the Illinois Soil N Test and soil organic matter content were not well correlated on the two golf courses in that study, and spatial patterns of extractable N in golf course fairways were suitable for traditional soil sampling techniques used to adequately assess the N extracted by the Illinois Soil N Test for turf soils. However, the Illinois Soil N Test has not proven to be highly predictive of N availability (Laboski et al., 2008), and given the test's complexity, it has not been widely used by agriculturalist (or turf managers) to make N fertilizer decisions. Moore et al. (2019a) was the first to evaluate usefulness of the Illinois Soil N Test in lawns. The author found that the test produced results that were linearly correlated with the N fertilizer rate (10–400 $kg \bullet ha^{-1}$ from an organic N source) on cool-season turfgrass lawns on a fine sandy-loam soil (soil organic matter 4.6–4.8%) in Connecticut, USA (Moore et al., 2019a). They also reported that another test, the Solvita Labile Available Nitrogen (SLAN) test was also linearly correlated with organic fertilization rate; however, the SLAN test was less sensitive but also more rapid than the Illinois Soil Nitrogen Test (Moore et al., 2019a). In related research, the same research group (Moore et al., 2019b) investigated the ability of the SLAN test to predict N responsiveness in cool-season lawns. They found that turfgrass response was weakly to moderately correlated with SLAN concentrations from soil samples taken in spring (Moore et al., 2019b). They also reported turfgrass response could be categorized by their probability of equaling or exceeding the response of turfgrass fertilized with urea at a particular rate (Moore et al., 2019b). Soils with high SLAN values in spring could be fertilized at lower N rates to achieve goals set by a turf manager; however, Moore et al. (2019b) cautioned that these conclusions were only valid for one soil at one location, and, as is the case with all soil testing, additional work would be necessary at other locations to increase confidence in the SLAN test. In a companion publication, Moore et al. (2019c) reported results of a similar study that investigated the Solvita CO_2-Burst test to predict turfgrass response to N fertilization. Moore et al. (2019c) reported that the test also had promise for guiding N fertilization decisions, but additional work would be required, especially in light of the large variability documented. Clearly, many questions remain to be answered before these N availability tests (i.e. https://www.paceturf.org/index.php/public/quick_test_for_soil_nitrate from PACE Turf, San Diego, CA, USA) could be confidently used to assist turfgrass managers with fertilization decisions. However, they have shown promise and deserve more attention given the importance of N management in turfgrass systems.

4.3 Soil microbiome

Scientists are also working to understand and characterize the organisms that live in the soil environment in order to improve the sustainable management of turfgrass systems (Crouch et al., 2017). The complete collection of microbial organisms (i.e. bacteria, fungi, viruses, and archaea) associated with a certain environment is called the microbiome. The turfgrass microbiome, then, is defined as the collection of all microbes associated with the leaves, stems, roots, and rhizosphere soil of a turfgrass plant and turfgrass community (York et al., 2016). Recent advances in the ability to measure the structure and function of these vast and complex microbiomes have demonstrated the enormous impact microbes can have on human health (Buffie and Pamer, 2013), plant health (Ellis, 2017), and almost every other aspect of life on Earth.

The study of turfgrass microbiomes is in relative infancy compared to many other managed crops (York et al., 2016). However, microbiome research in turfgrass is steadily increasing, especially in the soil and rhizosphere, and has consistently demonstrated that turfgrass supports a large and incredibly diverse array of microbes. For example, the soil biodiversity in New York City's Central Park (New York City, NY, USA) is immense, totaling 122 081 bacterial and 43 429 eukaryotic (mostly fungal) phylotypes, similar in biodiversity to soil collected from tropical forests, grasslands, and other natural systems (Ramirez et al., 2014). The soil microbiome on the US National Mall (Washington, DC, USA) had 751 identifiable bacterial taxa and an additional 859 that were not identifiable, and this soil biodiversity was largely unimpacted by a large renovation and installation of new sod in 2015 (Crouch et al., 2017). A global survey of 112 urban green spaces in 17 countries found that the soil microbiomes were more diverse than nearby natural ecosystems and that these urban microbiomes across the globe were more homogenous than adjacent natural systems (Delgado-Baquerizo et al., 2021).

4.3.1 Soil microbiome diversity

Soil microbial diversity also is high in intensively managed golf course turf. Annual bluegrass maintained as a putting green in New Jersey (USA) had similar soil bacterial diversity as compared to soils from tropical forests, temperate forests, and temperate grasslands despite the regular applications of synthetic fertilizers and pesticides required to maintain acceptable turfgrass quality (Beirn et al., 2017). Pesticide use was found to have limited impact on soil bacterial diversity on golf courses in Massachusetts (USA), though impacts on fungal diversity were more apparent when synthetic fungicides were used (Allan-Perkins et al., 2019). The research conducted to date on turfgrass microbiomes has clearly and consistently shown that managed turfgrass ecosystems support a large and diverse microbial community, refuting many earlier and unsupported claims that turfgrass is incompatible with a biologically healthy soil.

Turfgrass supports a diverse soil microbiome, but the importance of that diversity for sustainable turfgrass management is unclear. In grassland systems, decreasing soil biodiversity has been correlated with a decrease in numerous ecosystem functions important for the sustainable management of plants. This includes decreased soil carbon sequestration, decreased rates of decomposition, decreased nutrient retention, and reduced nutrient cycling (Wagg et al., 2014). Though turfgrass communities already have a large and diverse soil microbiome, management practices that further increase soil microbial diversity in turfgrass may increase the sustainability of turfgrass management through improved nutrient cycling and retention and greater carbon sequestration. However, research-based strategies that effectively and consistently increase soil microbial diversity in turfgrass have not yet been developed.

4.3.2 Soil microbiome and plant disease

The impact of the microbiome on plant disease management and with developing disease-suppressive soils is more nuanced. The microbes present in the soil typically dictate the ability of the soil to suppress disease, and in general there are two types of disease-suppressive soil. The first type is 'general suppression' and is defined by the inhibition of a broad number of plant pathogens through competition and antagonism from a large and diverse microbiome (Cook, 2014; Weller et al., 2002). The second type is 'specific suppression' and occurs when a specific segment of the microbial population governs suppression against a particular pathogen (Schlatter et al., 2017). Specific suppression can provide effective and long-lasting disease control but also often results in decreased microbial diversity as the specific antagonistic portion of the microbial community dominates. In turfgrass, the best example of a specific suppressive soil is for the disease take-all patch (*Gaeumannomyces graminis* [Sacc.] Arx. and D.L. Olivier) (Watschke et al., 2013). New creeping bentgrass (*Agrostis stolonifera* L.) plants are often highly susceptible to take-all patch for the first 2–4 years after seeding; however, the disease becomes less severe over time as specific antagonistic microbes like Pseudomonad bacteria grow in number (Sarniguet and Lucas, 1992). Even foliar diseases such as dollar spot can be impacted by the soil microbiome, as one study found that variations in the soil microbiome over small areas resulted in varied production of key antibiotic-producing taxa that may contribute to the highly variable nature of dollar spot symptom development (Chou et al., 2021).

4.3.3 Soil microbiome and sustainable turfgrass management

At this point, the broader ability to manipulate the turfgrass microbiome to more sustainably manage turfgrass remains more promise than reality, but the promise is immense. To date, most turfgrass microbiome research has focused on what microbes are present in various turfgrass environments. Additional research is needed to determine how these microbial communities interact with each other and how their functions related to carbon sequestration, nutrient cycling, and disease suppression can be altered to maintain healthy turfgrass while simultaneously reducing inputs of water, fertilizer, and pesticides. Additional research is also needed to determine what impacts pesticides and fertilizers have on microbiome diversity and function and what practices will increase the positive impacts of the turfgrass microbiome. Fortunately, many of these studies are already in progress and will greatly improve our knowledge of these complex and interconnected microbial communities in the coming years.

5 Conclusion and future trends

Recent advances in our understanding of physical, chemical, and biological properties of soils will help turfgrass managers work toward sustainable management of resources. However, the most important advances may come from new understandings of individual properties or the integration of sensing technology and data processing which will allow for real-time mapping of physical, chemical, and biological soil properties to assist turf managers in making informed decisions regarding the use of resources like fertilizer, water, pesticides, energy, and labor. In addition, the field of soil ecology and microbiology continues to reveal exciting new discoveries (Kostka et al., 2022). Utilizing and managing the soil microbiome to augment turfgrass growth and suppress disease may be within reach in coming decades with continued investment in research on turfgrass science, the rhizosphere of turfgrass ecosystems, and soil science.

6 Where to look for further information

An comprehensive introduction to turfgrass soil chemical properties can be found in Turfgrass Fertility and Chemical Problems by Carrow et al. (2001). Research project summaries are publicly available through the United States Golf Association Turfgrass Environmental and Research Online: https://usgatero.msu.edu/tero_browse.html. Finally, literature about how soil properties influence turfgrass establishment and management, and how establishment and management can affect soil properties can be found through the Turfgrass Information File, a computer database that provides access to all published materials about turfgrass and its maintenance. The database can be accessed here: https://tic.msu.edu/tgif.

7 References

Allan-Perkins, E., Manter, D. K. and Jung, G. (2019). Soil microbial communities on roughs, fairways, and putting greens of cool-season golf courses, *Crop Sci.* 59(4), 1753–1767.

Beirn, L. A., Hempfling, J. W., Schmid, C. J., Murphy, J. A., Clarke, B. B. and Crouch, J. A. (2017). Differences among soil-inhabiting microbial communities in *Poa annua* turf throughout the growing season, *Crop Sci.* 57(S1), S262–S273.

Bero, N. J. and Soldat, D. J. (2021). Effect of topdressing of compost, compost tea, and various fertilizers on soil and lawn characteristics, *Intl. Turfgrass Soc. Res. J.* 14(1), 110–120.

Bigelow, C. A. and Soldat, D. J. (2013). Turfgrass root zones: management, construction methods, amendment characterization, and use, In: Stier, J. C., Horgan, B. P. and Bonos, S. A. (Eds), *Turfgrass: Biology, Use, and Management*, American Society of Agronomy, Madison, WI, pp. 383–424.

Bier, P. V., Persche, M., Koch, P. and Soldat, D. J. (2018). A long term evaluation of differential potassium fertilization of a creeping bentgrass putting green, *Plant Soil* 431(1–2), 303–316.

Bleam, W. F. (2017). *Soil and Environmental Chemistry*, Academic Press, London, UK.

Brady, N. C. and Weil, R. (2010), *Elements of the Nature and Properties of Soils*, 3rd ed., Pearson Higher Education, Upper Saddle River, NJ.

Buffie, C. G. and Pamer, E. G. (2013). Microbiota-mediated colonization resistance against intestinal pathogens, *Nat. Rev. Immunol.* 13(11), 790–801.

Carrow, R. N., Waddington, D. V. and Rieke, P. E. (2001). Turfgrass Soil Fertility and Chemical Problems, John Wiley & Sons, Hoboken, NJ.

Carter, M. R., Kunelius, H. T. and Angers, D. A. (1994). Soil structural form and stability, and organic matter under cool-season perennial grasses, *Soil Sci. Soc. Am. J.* 58(4), 1194–1199.

Chou, M. Y., Shrestha, S., Rioux, R. and Koch, P. (2021). Hyperlocal variation in soil iron and the rhizosphere bacterial community determines dollar spot development in amenity turfgrass, *Appl. Environ. Microbiol.* 87(10), e00149-21.

Cook, R. J. (2014). Plant health management: pathogen suppressive soils, In: Van Alfen, N. K. (Ed.), *Encyclopedia of Agriculture and Food Systems*, Elsevier, Amsterdam, Netherlands, pp. 441–455.

Crouch, J. A., Carter, Z., Ismaiel, A. and Roberts, J. A. (2017). The US National Mall microbiome: a census of rhizosphere bacteria inhabiting landscape turf, *Crop Sci.* 57(S1), S-341-S-S-348.

Delgado-Baquerizo, M., Eldridge, D. J., Liu, Y. R., Sokoya, B., Wang, J. T., Hu, H. W., He, J. Z., Bastida, F., Moreno, J. L., Bamigboye, A. R., Blanco-Pastor, J. L., Cano-Díaz, C., Illán, J. G., Makhalanyane, T. P., Siebe, C., Trivedi, P., Zaady, E., Verma, J. P., Wang, L., Wang, J., Grebenc, T., Peñaloza-Bojacá, G. F., Nahberger, T. U., Teixido, A. L., Zhou, X. Q., Berdugo, M., Duran, J., Rodríguez, A., Zhou, X., Alfaro, F., Abades, S., Plaza, C., Rey, A., Singh, B. K., Tedersoo, L. and Fierer, N. (2021). Global homogenization of the structure and function in the soil microbiome of urban greenspaces, *Sci. Adv.* 7(28), eabg5809.

Dest, W. M. and Ebdon, J. S. (2017). The effect of wear and soil compaction on Kentucky bluegrass sod rooting and plant recovery, *Int. Turfgrass Soc. Res. J.* 13, 1–8.

Ellis, J. G. (2017). Can plant microbiome studies lead to effective biocontrol of plant diseases?, *Mol. Plant. Microbe Interact.* 30(3), 190–193.

Frank, K. W. and Guertal, E. A. (2013). Potassium and phosphorus research in turfgrass, In: Stier, J. C., Horgan, B. P. and Bonos, S. A. (Eds) *Turfgrass: Biology, Use, and Management*, American Society of Agronomy, Madison, WI.

Gardner, D. S., Horgan, B. P. and Horvath, B. J. (2013). Spatial variability of soil amino sugar nitrogen on golf course fairways, *Int. Turfgrass Soc. Res. J.* 12, 545–549.

Gaussoin, R. E., Berndt, W. L., Dockrell, C. A. and Drijber, R. A. (2013). Characterization, development, and management of organic matter in turfgrass systems, In: Stier, J. C., Horgan, B. P. and Bonos, S. A. (Eds) *Turfgrass: Biology, Use, and Management*, American Society of Agronomy, Madison, WI.

Głab, T. and Szewczyk, W. (2014). Influence of simulated traffic and roots of turfgrass species on soil pore characteristics, *Geoderma* 230–231, 221–228.

Hosseini, F., Mosaddeghi, M. R., Hajabbasi, M. A. and Sabzalian, M. R. (2016). Role of fungal endophyte of tall fescue (*Epichloë coenophiala*) on water availability, wilting point and integral energy in texturally-different soils, *Agric. Water Manag.* 163, 197–211.

Janssen, B. H. (1996). Nitrogen mineralization in relation to C:N ratio and decomposability of organic materials, *Plant Soil* 181(1), 39–45.

Kahn, S. A., Mulvaney, R. L. and Hoeft, R. G. (2001). A simple soil test for detecting sites that are nonresponsive to nitrogen fertilization, *Soil Sci. Soc. Am. J.* 65(6), 1751–1760.

Kostka, S., Fidanza, M. and Brown, W. (2022). Hey! what's going on down there? Available at: https://blog.aquaaidsolutions.com/webinar-april-14th-hey-whats-going-on-down-there?utm_content=204881949&utm_medium=social&utm_source=twitter&hss_channel=tw-209115029.

Laboski, C. A. M., Sawyer, J. E., Walters, D. T., Bundy, L. G., Hoeft, R. G., Randall, G. W. and Andraski, T. W. (2008). Evaluation of the Illinois soil nitrogen test in the North Central region of the United States, *Agron. J.* 100(4), 1070–1076.

Lehmann, J., Bossio, D. A., Kögel-Knabner, I. and Rillig, M. C. (2020). The concept and future prospects of soil health, *Nat. Rev. Earth Environ.* 1(10), 544–553. doi:10.1038/s43017-020-0080-8.

Moore, D. B., Guillard, K., Morris, T. F. and Brinton, W. F. (2019a). Correlation between Solvita labile amino-nitrogen and CO_2-burst soil health tests and response to organic fertilizer in a turfgrass soil, *Commun. Soil Sci. Plant Anal.* 50(22), 2948–2959.

Moore, D. B., Guillard, K., Geng, X., Morris, T. F. and Brinton, W. F. (2019b). Predicting cool-season turfgrass response with Solvita soil tests, Part 1: labile amino-nitrogen concentrations, *Crop Sci.* 59(4), 1779–1788.

Moore, D. B., Guillard, K., Morris, T. F. and Brinton, W. F. (2019c). Predicting cool-season turfgrass response with Solvita soil tests, Part 2: CO_2-Burst carbon concentrations, *Crop Sci.* 59(5), 2237–2248.

Obear, G. R. and Soldat, D. J. (2014). Saturated hydraulic conductivity of sand-based golf putting green root zones affected by sodium, *Soil Sci.* 179(8), 376–382.

Ramirez, K. S., Leff, J. W., Barberan, A., Bates, S. T., Betley, J., Crowther, T. W., Kelly, E. F., Oldfield, E. E., Shaw, E. A., Steenbock, C., Bradford, M. A., Wall, D. H. and Fierer, N. (2014). Biogeographic patterns in below-ground diversity in New York City's Central Park are similar to those observed globally, *Proc. R. Soc. B.* 281(1795), 20141988, doi:10.1098/rspb.2014.1988.

Sarniguet, A. and Lucas, P. (1992). Evaluation of populations of fluorescent pseudomonads related to decline of take-all patch on turfgrass, *Plant Soil* 145(1), 11–15.

Schlatter, D., Kinkel, L., Thomashow, L., Weller, D. and Paulitz, T. (2017). Disease suppressive soils: new insights from the soil microbiome, *Phytopathology* 107(11), 1284–1297.

Schmid, C. J., Murphy, J. A. and Murphy, S. (2017). Effect of tillage and compost amendment on turfgrass establishment on a compacted sandy loam, *J. Soil Water Conserv.* 72(1), 55–64.

Schmid, C. J., Clarke, B. B. and Murphy, J. A. (2018). Potassium nutrition affects anthracnose on annual bluegrass, *Agron. J.* 110(6), 2171–2179.

Slavens, M. R., Petrovic, A. M. and Barlow, J. J. (2009). The impact of recycled water irrigation on soil health and quality of newly established cool season turfgrasses during drought conditions, *Int. Turfgrass Soc. Res. J* 11(1), 471–480.

Suarez, D. L. (1981). Relation between pHc and sodium adsorption ratio (SAR) and an alternative method of estimating SAR of soil or drainage waters, *Soil Sci. Soc. Am. J.* 45(3), 469–475.

Turgeon, A. J. and Fidanza, M. A. (2017). Perspective on the history of turf cultivation, *Int. Turfgrass Soc. Res. J.* 13(1), 1–7.

Turgeon, A. J. and Kaminski, J. E. (2019). Turfgrass management, *Turfpath LLC*, State College, PA.

Wagg, C., Bender, S. F., Widmer, F. and van der Heijden, M. G. A. (2014). Soil biodiversity and soil community composition determine ecosystem multifunctionality, *Proc. Natl. Acad. Sci. U. S. A.* 111(14), 5266–5270.

Watschke, T. L., Dernoeden, P. H. and Shetlar, D. J. (2013). *Managing Turfgrass Pests*, CRC Press, Boca Raton, FL.

Weil, R. R. and Brady, N. C. (2017). *The Nature and Property of Soils* (15th edn.), Pearson, New York.

Weller, D. M., Raaijmakers, J. M., McSpadden Gardener, B. B. and Thomashow, L. S. (2002). Microbial populations responsible for specific suppressiveness to plant pathogens, *Annu. Rev. Phytopathol.* 40, 309–348.

Yao, H. and Shi, W. (2010). Soil organic matter stabilization in turfgrass ecosystems: importance of microbial processing, *Soil Biol. Biochem.* 42(4), 642–648.

York, L. M., Carminati, A., Mooney, S. J., Ritz, K. and Bennett, M. J. (2016). The holistic rhizosphere: integrating zones, processes, and semantics in the soil influenced by roots, *J. Exp. Bot.* 67(12), 3629–3643.

Chapter 4

Advances in phosphite utilization for turfgrass

John Dempsey, Independent Turfgrass Research, Ireland

1 Introduction

Phosphorus (P) is a major plant nutrient used in many metabolic processes: It is vital for cell division, early root formation, energy transfer, and it is a component of adenosine triphosphate (ATP), nucleic acids, nucleotides, coenzymes, phospholipids, and phosphoproteins (Campbell and Reece, 2002). In nature, however, because of its reactivity, P does not occur as a free element and is only found in combinations with other elements, such as oxygen (O) or hydrogen (H). The P cycle occurs by the oxidation and reduction of P compounds by electron transfer reactions and when fully oxidized the product is orthophosphate (PO_4^{3-}), also referred to as 'phosphate'. In soils at neutral pH, the phosphate ion is present as a mixture of HPO^{2-} and H_2PO_4 and because it cannot be oxidized further, it is incorporated by means of phosphorylation directly into the plant cells (McDonald et al., 2001). For many years, phosphate compounds (i.e. HPO^{2-} and H_2PO_4) were considered the only form that could supply P nutrition to plants. In fertilizer, P is normally found in the form of phosphoric acid (H_3PO_4) and its salts, such as triple super phosphate, ammonium phosphate, and potassium phosphate. All these forms readily disassociate to release hydrogen phosphate (HPO_4^{2-}) and dihydrogen phosphate (H_2PO^{4-}), which are used

http://dx.doi.org/10.19103/AS.2022.0110.04

by plants. Phosphate is the sole P-containing nutrient important for optimal plant growth and development and is required in amounts second only to nitrogen (N) (Campbell and Reece, 2002). But because of its insoluble mineral form, it can be largely unavailable to plants, leading to the widespread use of phosphate-containing fertilizers (Raghothama and Karthikeyan, 2005). Over the past number of years, however, an alternative form of P, phosphite (PO_3^{3-}) has increasingly been used, not only as a nutrient source and biostimulant but also as a plant protectant in many crop systems, including amenity turfgrass (Gómez-Merino and Trejo-Télle, 2015).

Though there is no consensus on its physiological function as a P-source for plant nutrition, experimental evidence has shown that phosphite can act as a biocide and affect plant production and productivity (Lobato et al., 2010). Positive effects of phosphate on plant metabolism are more evident when applied to the roots in hydroponic systems or to the leaves in the form of foliar sprays in the presence of sufficient phosphate. Published research conclusively indicates that phosphite functions as an effective pesticide against various species of pathogenic bacteria and oomycetes (Gómez-Merino and Trejo-Télle, 2015). Nonetheless, the use of phosphite as a sole P-source for plant nutrition is still at issue. When phosphite is applied to the soil, it comes into contact with microorganisms, which mediate the oxidation of phosphite to phosphate. Thus, by this indirect method, phosphite can become available to the plant as a P nutrient after microbial oxidative reactions (Lovatt and Mikkelsen, 2006).

This chapter will assess the sustainable use of phosphite in amenity turfgrass management, by exploring the often-contentious use of phosphite as a nutrient source, its use as a biostimulant, and its properties and efficacy as a plant protectant within many plants including turfgrasses.

2 Phosphite vs. phosphate

Phosphite for plant use is derived from the alkali metal salts of phosphorous acid, (Guest and Grant, 1991). Phosphate and phosphite are chemically very similar, as both ions are formed with a central P atom. However, with phosphate the P atom sits at the center of a tetrahedron, with oxygen atoms distributed at each point, forming a symmetrical structure, with the charge distributed evenly among these four oxygen atoms. With phosphite, one of the oxygen atoms is replaced by a hydrogen atom, and although the P atom is still at the center of the tetrahedron, the symmetry of the phosphate ion is lost. This substitution results in significant differences affecting the behavior of both molecules in plants. Both the shape of the molecule and the charge distribution influence the binding of phosphate to its interacting enzymes. For phosphate to be metabolized in any living organism, enzymatic catalyzation is required. Once phosphate has bound to an enzyme, the remaining oxygen atom becomes available to react with molecules in the reaction catalyzed by the enzyme. Phosphite however, only has one face of the tetrahedron relatively similar to all the faces of the phosphate three-dimensional structure, so if it is to bind to the surface of an enzyme that normally binds phosphate, it must bind at this face. When phosphite binds to the enzyme surface in this orientation, it is the hydrogen atom bonded to the P atom that emerges from the enzyme surface, not an oxygen atom as in phosphate. Because of this, phosphite cannot participate in the same biochemical reactions as phosphate. Due to these unique structures and considering the difference in charge distribution of the two anions, most enzymes involved with phosphoryl transfer reactions readily discriminate between phosphite and phosphate, ensuring that phosphite cannot be metabolized in plants as with phosphate. Despite this, phosphite is often promoted as a form of P nutrition, but, as discussed in the next section, the efficacy of phosphite in this regard is inconclusive and often controversial.

3 Phosphite use in turfgrass

Prior to use on any plant system the pH of phosphorous acid (H_3PO_3) needs to be modified to prevent phytotoxicity (Ouimette and Coffey, 1988). In the 1980s, the first marketable phosphite product was produced by reacting phosphonic acid with ethanol, forming ethyl-phosphonate, and then combining with aluminum ions. The resulting product is referred to as Fosetyl-Al or aluminum tris (O-ethyl phosphonate). Following uptake by plants, ethyl phosphonate is hydrolyzed in the plant to phosphorous acid, and then to phosphite. In turfgrass, this was successful as a preventative treatment in controlling oomycete pathogens such as *Pythium* spp. (Cook et al., 2009; Guest and Grant, 1991). In the past number of years, phosphorous acid is mostly neutralized by combining with an alkali salt, typically potassium hydroxide (KOH), although other alkali salts can be used (Ouimette and Coffey, 1990). The resulting solution contains mono- and di-potassium salts of phosphorous acid, forming potassium dihydrogen phosphite (KH_2PO_3) or dipotassium hydrogen phosphite (K_2HPO_3), both commonly referred to as 'potassium phosphite'. In turfgrass management, phosphite was first marketed as a plant protectant, but since the early 2000s, most of the phosphite-containing products are marketed as fertilizers or as biostimulants (Gómez-Merino and Trejo-Télle, 2015). Phosphite-containing products can be promoted legally as fertilizers, because after pH neutralization they contain cations usable as plant nutrients, such as K^+, NH^{4+}, Ca^{2+}, Mg^{2+}, Cu^{2+}, or Zn^{2+}, thus resulting in numerous phosphite-based P fertilizers currently being marketed worldwide. The use of phosphite as a plant nutrient, however, has been the subject of much discussion and even controversy (Thao and Yamakawa, 2009).

4 Phosphite as a source of phosphorus nutrition

While phosphite and phosphate are chemically very similar, their properties have very dissimilar effects on plant growth. Results of studies researching the value of phosphite as a supplier of P nutrition to many plant species range between negative to positive, with many inconclusive results in between. The inability of phosphite to provide P in a metabolically usable form was first reported in a 1950s study, which concluded that phosphite was a very poor source of P for crops although a favorable growth response was observed the year following the application (MacIntire et al., 1950). This late response was explained by Adams and Conrad (1953), who determined that oxidation of phosphite to phosphate had occurred. Phosphite undergoes a gradual transformation after addition to soil, either by biological or non-biological oxidation. Soil microorganisms (i.e. bacteria and cyanobacteria) can assimilate phosphite and release phosphate, thus gaining energy and nutrients during this biological conversion (Varadarajan et al., 2002). Therefore, since soil microorganisms can assimilate and metabolize phosphite, the impact on plant physiology of phosphite applied to the soil solution in field crops is affected by such biological processes and phosphite may not directly impact plant metabolism as a proper P-source.

Interest in phosphite as a nutrient source was revisited in the 1990s when Lovatt (1990) reported that application of phosphite, in the form of K_3PO_3, to P-deficient citrus plants restored normal plant growth, and that soil or foliar applications of phosphite could replace phosphate as a source of P in avocado. Since then, several studies have reported positive physiological responses from a wide range of phosphite-treated plants, including turfgrass (Albrigo, 1999; Cook et al., 2006; Dempsey and Owen, 2010; Lovatt, 1999; Rickard, 2000; Vincelli and Dixon, 2005). However, despite these positive correlations of phosphite application and growth, other studies as detailed in the review by Thao and Yamakawa (2009) have determined phosphite is not metabolized by plants despite its rapid uptake and translocation. While the evidence for phosphite as a source of P is inconclusive, there are positive data that indicate phosphite may have positive biostimulant-related properties when applied to turfgrass (Bigelow et al., 2021; Fidanza et al., 2015).

5 Phosphite and turfgrass nutrition

Research into phosphite specifically as a turfgrass fertilizer/biostimulant and its effects on turfgrass quality were published during the mid-2000s, Horvath et al. (2007) conducted field trials at several locations in the USA, assessing the impact of a range of phosphite products on creeping bentgrass (*Agrostis stolonifera* L.), and results showed that while there were some improvements, no phosphite product *consistently* provided a significant increase in turf quality or color. Butler et al. (2009) examined the effects of phosphite and phosphate treatments on newly seeded creeping bentgrass in a greenhouse study, and concluded that phosphite applications had limited influence on turfgrass growth and development (i.e. leaf tissue dry weight, leaf tissue P content, and root dry weight). Despite these results, there have been several published works showing positive results from phosphite treatments. Improved turfgrass quality following sequential treatment with phosphite was reported by Cook et al. (2006) on a mixed sward of creeping bentgrass and annual bluegrass (*Poa annua* L.) and by Dempsey and Owen (2010) on a creeping bentgrass sward. Potassium phosphite treatments, applied sequentially over 2 years to creeping bentgrass, velvet bentgrass (*A. canina* L.), and annual bluegrass swards, produced significantly better turfgrass quality when compared to untreated controls and plots treated with identical nutritional inputs minus the phosphite component (Dempsey et al, 2012).

Most of the above results were secondary to the focus of the studies from which they were cited, which were assessment of the disease suppressive properties of phosphite. So, while improved turfgrass quality was observed, the question arises, was this because of phosphite acting as a P nutrient source or as a biostimulant? To attempt to answer that question, research was conducted by Dempsey et al. (2022a) to investigate the rate of uptake, translocation, and fate of phosphite when applied as a foliar treatment to amenity turfgrasses, to assess phosphite's role as a source of P nutrition, and to determine the effect phosphite treatments may have on turfgrass growth, P-deficiency responses, and plant tissue and soil P accumulations.

5.1 Uptake, translocation, and accumulation of phosphite in turfgrass

In turfgrass management, phosphite is applied as a foliar treatment. Therefore, to assess the role of phosphite as a nutrient source, it was important to assess the rate of uptake, the translocation, accumulation, and fate of these applications (Dempsey et al., 2022a). To do this, phosphite was applied as a foliar treatment to creeping bentgrass and annual bluegrass in February 2011 and July 2012, at a rate of 0.35 g $PO_3^{3-} \bullet m^{-2}$. Harvesting of leaf and root tissues was carried out at regular intervals over a 6-week period. Tissue content of phosphite and phosphate was measured by high performance ion chromatography, as described in Dempsey et al. (2022a).

The results with both creeping bentgrass and annual bluegrass revealed that phosphite is rapidly taken up into leaf tissues and within hours following application to the leaf it is detectable in roots (Dempsey et al., 2022a), thus confirming phloem mobility and suggests full symplastic ambimobility. The first set of treatments and analyses were carried out in February 2011, during a period of low turfgrass growth and metabolism. Uptake into the leaf tissues during this period was rapid, with 65% in creeping bentgrass and 61% in annual bluegrass of the maximum accumulation achieved within 6 h of application. Phosphite concentration within leaf tissues attained a maximum level at 48 h post-application, and by 96 h post-application had begun to decline in both turfgrass species. Over the full study period, it was shown that phosphite levels progressively declined and by 6 weeks post-application had reduced to 10% of the maximum accumulation in creeping bentgrass and 17% of the maximum in annual bluegrass leaf tissues (Fig. 1).

Following the first series of studies, it was thought that phosphite uptake could be significantly affected by growth conditions; therefore, the second series of experiments was conducted during a period of increased metabolic activity during July 2012. The results of this second study were similar to the first with regard to rapid phosphite uptake and translocation rates, but confirmed

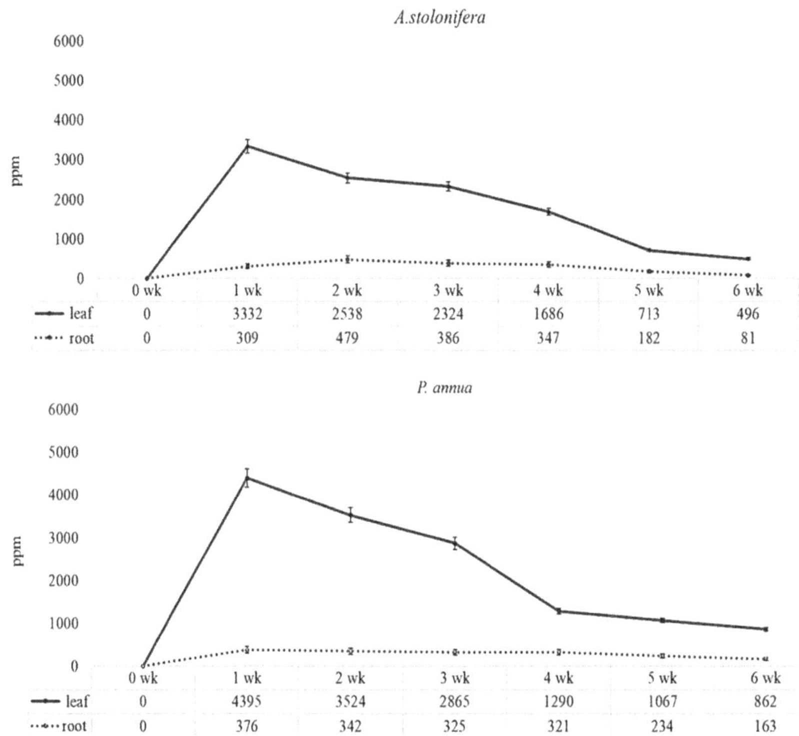

A.stolonifera

	0 wk	1 wk	2 wk	3 wk	4 wk	5 wk	6 wk
leaf	0	3332	2538	2324	1686	713	496
root	0	309	479	386	347	182	81

P. annua

	0 wk	1 wk	2 wk	3 wk	4 wk	5 wk	6 wk
leaf	0	4395	3524	2865	1290	1067	862
root	0	376	342	325	321	234	163

Figure 1 Uptake and accumulation of phosphite in creeping bentgrass (*Agrostis stolonifera*) and annual bluegrass (*Poa annua*) leaf and root tissues, from 0 weeks to 6 weeks post-application of phosphite at a rate of 0.35 g PO_3^{3-}•m^{-2}, in February 2011. Bars indicate 95% confidence intervals, $n = 6$; ppm = mg•kg^{-1}. Source: Dempsey et al. (2022a).

that phosphite uptake was increased during periods of greater growth potential (Dempsey et al., 2022a). In creeping bentgrass, for example, leaf accumulations 48 h post-application were 5520 mg•kg^{-1} in July (Fig. 2) compared to 4886 mg•kg^{-1} in the February assessments (Fig. 1). Similarly, in annual bluegrass, the phosphite leaf accumulations at 48 h post-application were 5418 mg•kg^{-1} in July (Fig. 2) compared to 5071 mg•kg^{-1} in February (Fig. 1). In creeping bentgrass, phosphite amounts in leaf tissues 6 weeks post-application had decreased to 5% of the maximum accumulation in July, compared to 10% of the maximum accumulation in February. Similarly, in annual bluegrass, amounts in leaf tissues 6 weeks post-application had dropped to 5% of the maximum accumulation in July, compared to 17% in February.

Demonstration of symplastic mobility, in that the foliar-applied phosphite translocated and was detected at 6 h post-application in the roots of treated turfgrasses, was an important outcome of this study (Dempsey et al., 2022a). Although the maximum root accumulations, 479 mg•kg^{-1} in creeping bentgrass

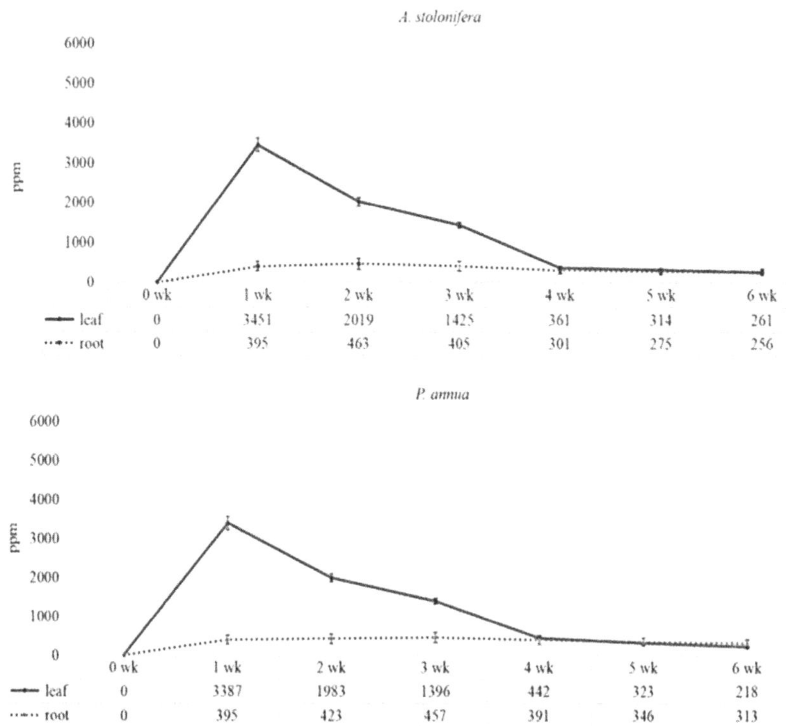

Figure 2 Uptake and accumulation of phosphite in creeping bentgrass (*Agrostis stolonifera*) and annual bluegrass (*Poa annua*) leaf and root tissues, from 0 to 6 weeks post-application of phosphite at a rate of 0.35 g PO_3^{3-}•m^{-2}, in July 2012. Bars indicate 95% confidence intervals, *n* = 6; ppm = mg•kg^{-1}. Source: Dempsey et al. (2022a).

in February and 457 mg•kg^{-1} in annual bluegrass in July (Figs 1 and 2), were much less than measured in leaf tissues, it remains a significant result, as no other compound used for pathogen suppression in turfgrasses demonstrates symplastic ambimobility, and this is the first time that this mobility of phosphite has been reported in these turfgrass species (Dempsey et al., 2022a).

5.2 Phosphite to phosphate conversion *in planta*

Determination of phosphate concentration following phosphite treatment was an interesting component of the study (Dempsey et al., 2022a), as the question of *in planta* conversion of phosphite (PO_3^{3-}) to phosphate (PO_4^{3-}) is often raised with numerous commercial suppliers claiming phosphite as a source of P nutrition following *in planta* conversion of phosphite to plant-usable phosphate. In both creeping bentgrass and annual bluegrass leaf and root tissues, during both the 2011 and 2012 assessment periods, there was not an *in planta* conversion of phosphite to phosphate, as there were no significant increases in phosphate concentration during the 6 weeks following phosphite treatment (Fig. 3). Therefore, foliar application of phosphite did not lead to *in planta* conversion to phosphate in creeping bentgrass or annual bluegrass.

5.2.1 Accumulation of phosphite in turfgrass tissues following sequential treatments over 2 years

Following a phosphite application, there is rapid uptake and accumulation of phosphite in creeping bentgrass and annual bluegrass leaf and root tissues, however, the persistence and fate of foliar-applied phosphite following long-term applications was unknown (Dempsey et al., 2022a). Therefore, evaluation of phosphite and phosphate treatments applied monthly over a two-year period for turfgrass tissue and root zone P concentrations were determined by analyzing tissue and root zone samples prior to, and at regular intervals, over the entire study period (Dempsey et al., 2022a). Over 2 years of repeated and sequential applications, phosphite in leaf tissues remained at a constant concentration, varying only with time post-application and the plant's metabolic rate as governed by seasonal growth potential, thus showing no evidence of a systemic accumulation of phosphite in leaf tissues (Fig. 4). This does not infer that phosphite is metabolized, degraded biochemically, or converted to phosphate, but rather is physically removed (i.e. removing leaf tissue by the on-going mowing regime typical for amenity turf maintenance). There was, however, a cumulative increase of phosphite concentration in roots of creeping bentgrass, and to a lesser extent, in the roots of annual bluegrass (Fig. 4). This confirms that long-term sequential foliar-applied phosphite treatments can accumulate phosphite in leaf tissues, but it can also accumulate

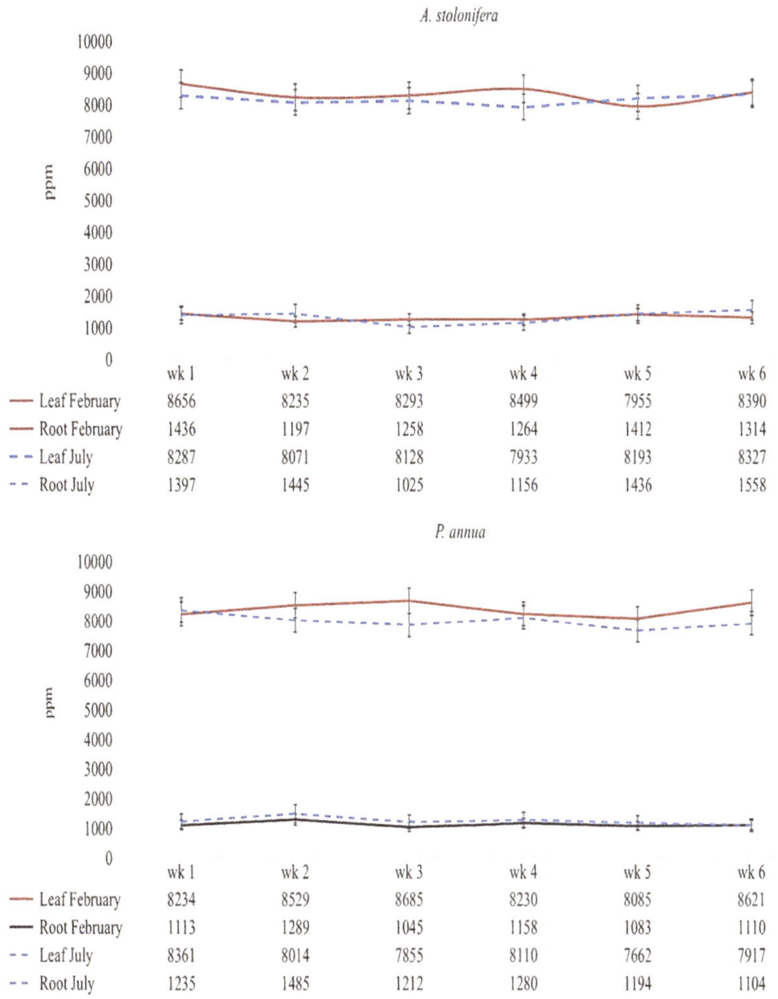

A. stolonifera

	wk 1	wk 2	wk 3	wk 4	wk 5	wk 6
Leaf February	8656	8235	8293	8499	7955	8390
Root February	1436	1197	1258	1264	1412	1314
Leaf July	8287	8071	8128	7933	8193	8327
Root July	1397	1445	1025	1156	1436	1558

P. annua

	wk 1	wk 2	wk 3	wk 4	wk 5	wk 6
Leaf February	8234	8529	8685	8230	8085	8621
Root February	1113	1289	1045	1158	1083	1110
Leaf July	8361	8014	7855	8110	7662	7917
Root July	1235	1485	1212	1280	1194	1104

Figure 3 Phosphate concentration in leaf and root tissues creeping bentgrass (*Agrostis stolonifera*) and annual bluegrass (*Poa annua*) 6 weeks post-application of phosphite at rate 0.35 g PO_3^{3-}•m^{-2}, in February 2011 and July 2012. Bars indicate 95% confidence intervals, n = 6; ppm = mg•kg^{-1}. Source: Dempsey et al. (2022a).

phosphite in roots. This lower accumulation of phosphite in annual bluegrass following long-term applications could be due to the shorter lifespan of annual bluegrass, compared to the perennial creeping bentgrass, with roots senescing more rapidly in annual bluegrass. This is evidenced further by the increased P measured in the root zone annual bluegrass compared to creeping bentgrass, as discussed below. The senescence of any turfgrass tissues that contained phosphite would most likely contribute, over time, to soil P.

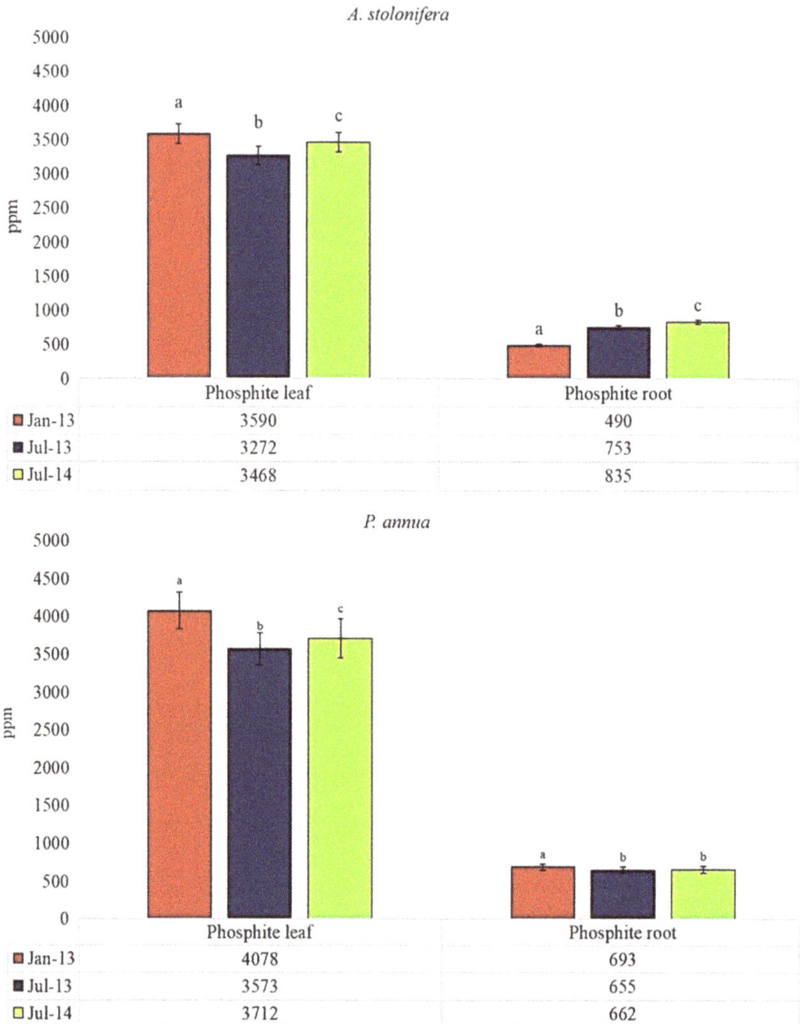

Figure 4 Phosphite accumulations in leaf and root tissues of creeping bentgrass (*Agrostis stolonifera*) and annual bluegrass (*Poa annua*) following sequential monthly applications of phosphite, at a rate of 0.35 g $PO_3^{3-}\bullet m^{-2}$, between July 2012 and July 2014. Data were recorded 1 week post-application at 6 months, 12 months, and 24 months from commencement of the applications in July 2012. Bars indicate 95% confidence limits; data means (n = 6) with different letters indicate significant difference as determined by Tukey's Honest Significance Test at $P \leq 0.05$; ppm = $mg\bullet kg^{-1}$ (Dempsey et al., 2022a).

5.2.2 The effect of sequential phosphite applications on root zone phosphorus accumulation

The use of P-containing fertilizers is a contentious issue worldwide, with some regions allowing P applications only after confirmation of P deficiency from soil test analyses (Bigelow et al., 2012). Therefore, the effect of long-term sequential applications of phosphite on P status in the root zone was also investigated (Dempsey et al., 2022a). Over 2 years of sequential treatments, phosphite and phosphate applications supplied equivalent amounts of P in leaf tissues of creeping bentgrass and annual bluegrass; however, P concentration in root zones of both turfgrass species receiving phosphite increased significantly compared to the phosphate applications. Phosphite applied to creeping bentgrass resulted in a 38% increase in root zone P concentration from 37 $mg•kg^{-1}$ at the start to 51 $mg•kg^{-1}$ at the end of the study. The P concentration increase in the annual bluegrass root zone was even greater at 54%, from 37 $mg•kg^{-1}$ at the start to 57 $mg•kg^{-1}$ at the end of the study. Over the same period, root zone P concentration following sequential phosphate applications also increased, but by a lesser extent. The root zone of creeping bentgrass receiving phosphate had an 8% P increase, from 37 $mg•kg^{-1}$ at the start to 40 $mg•kg^{-1}$ at the end, while in the annual bluegrass root zone the P increase was 19%, from 37 $mg•kg^{-1}$ at the start to 44 $mg•kg^{-1}$ at the end of the study.

This higher level of root zone P accumulation in creeping bentgrass and annual bluegrass following phosphite treatments is of interest. It could be due to phosphite being 'locked' into the rhizosphere by soil microorganisms (Dempsey et al., 2022a). Oxidation of phosphite to phosphate in soil relies on microbial activity, requiring the absorption and uptake of phosphite by soil bacteria and subsequent oxidation to phosphate. This is a slow process, however, with a half-life of several months (McDonald et al., 2001). Root zone P following phosphate treatment would be less persistent and more easily leached, considering the cation exchange capacity of those root zones were extremely low with mean values of 8.0 $meq•100\ g^{-1}$ (Dempsey et al., 2022a). Although not fully understood or explained, the steady increase of soil/ root zone P concentration following sequential phosphite treatments may become a problem for turfgrass management. Higher concentration of soil P is often attributed to increased proclivity of annual bluegrass, a species, which although dominant in many golf course putting greens, is widely regarded as an undesirable weed species (Dest and Guillard, 1987), and therefore is an area that requires more extensive study.

5.2.3 Phosphite as a source of phosphorus nutrition and effects on phosphorus-deficiency responses

An additional component of Dempsey et al. (2022a) research was to assess the properties of phosphite as a source of P nutrition for turfgrass, and its possible effects on P-deficiency response for turfgrass. Foliar treatments of phosphite and phosphate were applied at 0.35 $g•m^{-2}$ of PO_3^{3-} or PO_4^{3-}, and KCl (as the control) were applied to perennial ryegrass (*Lolium perenne* L.) and annual bluegrass bi-weekly, over a 6-month period. Two soil P concentrations were > 38 mg P•kg^{-1} (P sufficient) and < 6 mg P•kg^{-1} (P deficient). Treatment effect on turfgrass shoot growth was determined by cumulative dry leaf weights. Crowns and roots were collected at the end of the study and weighed for dry mass determination and calculation of root-to-shoot ratios. The results from Dempsey et al. (2022a) revealed in both perennial ryegrass and annual bluegrass grown in P-sufficient root zones, that phosphite-treated plants had a significant increase in oven-dried weight of leaf, crown, and root tissues compared with phosphate- or KCl-treated plants (Fig. 5). Conversely, phosphite-treated perennial ryegrass and annual bluegrass grown in P-deficient root zones had a significant reduction in oven-dried weight of leaf, crown, and root tissues compared with phosphate- and KCl-treated plants (Fig. 6).

As indicated in Figs 5 and 6, phosphite is not converted into phosphate in plants, and therefore it cannot enter the biochemical pathways in which phosphate is assimilated and metabolized (Dempsey et al., 2022a). Also, phosphite competes with phosphate for uptake via the same plant transport system, which would lead to a reduction of usable P and further phosphate depletion (Carswell et al., 1996; Danova-Alt et al., 2008; Jost et al., 2015; Varadarajan et al., 2002). Therefore, in P-limited soil, a phosphite application will not supply P in a form that can be metabolized by plants and would inhibit plant growth as P-deficiency responses would not be initiated as confirmed when phosphite applied to P-limited soil inhibited turfgrass growth (Dempsey et al., 2022a). Furthermore, since the KCl treatment resulted in increased turfgrass growth compared to the phosphite treatment, it is evidence that phosphite also suppressed the P-deficiency responses in both species (Dempsey et al., 2022a). These results agree with the findings of Ticconi et al. (2001) who concluded that phosphite inhibited P-deficiency compensatory responses in *Arabidopsis thaliana* (L.) Heynh., and Fabricio et al. (2012) who determined foliar-applied phosphite caused harmful effects to plants growing in P-limited soils. The conclusion that phosphite suppressed deficiency responses is further supported by the root oven dry weight (Fig. 6), and the root-to-shoot ratio (Fig. 7), which showed phosphite treatments produced the lowest mean ratio of roots-to-shoots in both species (Dempsey et al., 2022a).

Varadarajan et al. (2002) determined that phosphite suppressed many of the definitive responses to P limitation, such as enhanced root growth and increased root-to-shoot ratios. The results from Dempsey et al. (2022a) showed significant differences in root mass and root-to-shoot ratios between the KCl and phosphate treatments, however, significantly less root growth and reduced root-to-shoot ratios were observed in phosphite-treated plants. These results provide supportive evidence that in amenity turfgrass management, phosphite should only be applied under conditions of sufficient P soil or root zone levels for the target species involved.

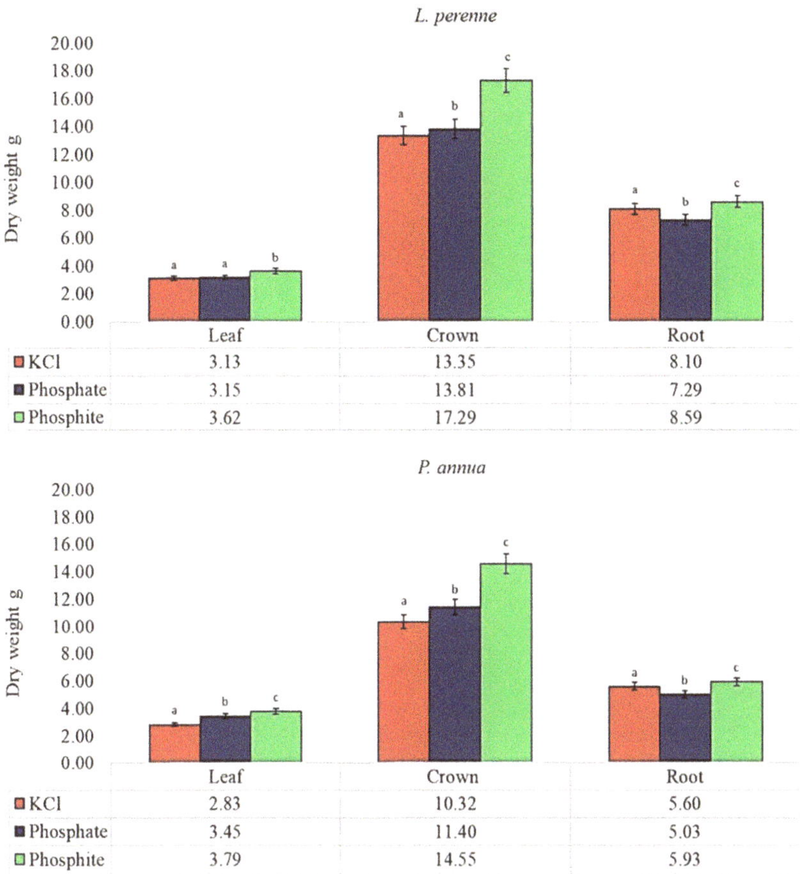

Figure 5 Effect on the growth of leaf, crown, and root tissues of perennial ryegrass (*Lolium perenne*) and annual bluegrass (*P. annua*), growing in a P-sufficient rootzone (P > 38 ppm), following sequential treatments over a 6-month period, of phosphate, phosphite, and potassium chloride (KCl; control). Bars are 95% confidence intervals, n = 6. Letters indicate significant differences within tissue type as determined by Tukey Honest Significant Different Test post-hoc analyses at $P = 0.05$; ppm = mg•kg^{-1}. Source: Dempsey et al. (2022a).

There is no evidence in the literature to support the metabolism of phosphite as a P-source or its' *in planta* conversion to a plant-usable form of P, and this is strongly supported by Dempsey et al. (2022a) following phosphite treatment in the P-deficient root zones. In Dempsey et al. (2022a), the phosphite treatment significantly increased leaf, crown, and root biomass, compared with phosphate and KCl-treated plants. There are reports of improved turfgrass quality following sequential applications of phosphite (Dempsey and Owen, 2010; Dempsey et al., 2012). Research with plant systems other than turfgrass have also reported enhanced growth responses following phosphite treatment (Albrigo, 1999;

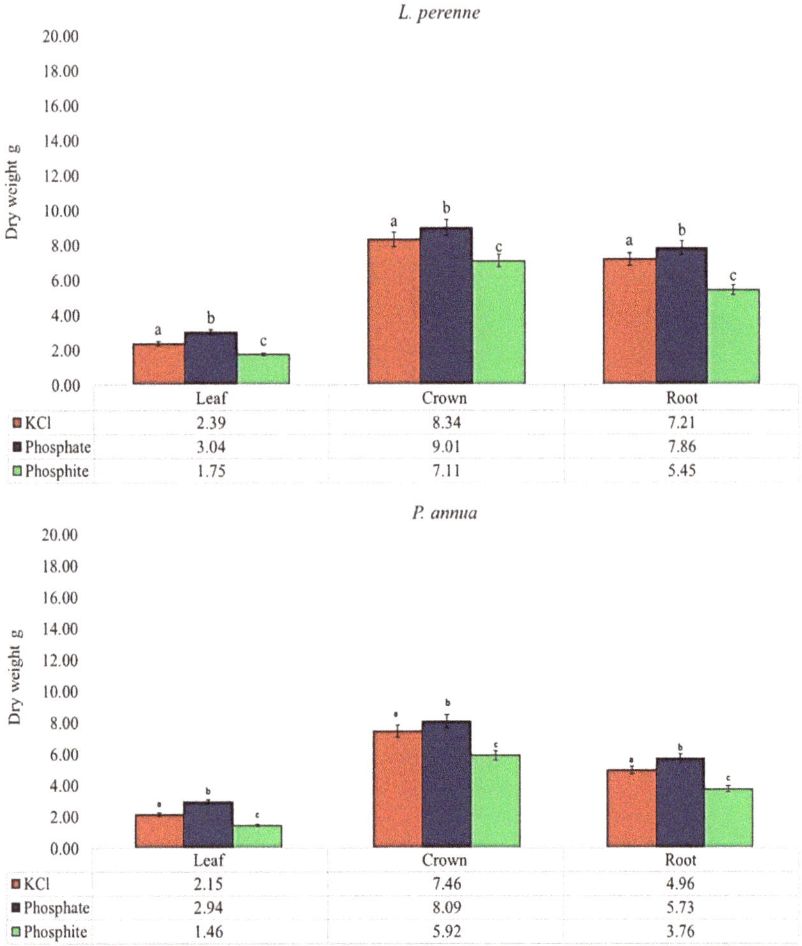

Figure 6 Effect on the growth of leaf, crown, and root tissues of perennial ryegrass (*Lolium perenne*) and annual bluegrass (*P. annua*), growing in a P-deficient root zone (P < 6 ppm), following sequential treatments over a 6-month period, of phosphate, phosphite, and potassium chloride (KCl; control). Bars are 95% confidence intervals, n = 6. Letters indicate significant differences within tissue type as determined by Tukey Honest Significant Different Test post-hoc analyses at *P* = 0.05; ppm = mg•kg^{-1}. Source: Dempsey et al. (2022a).

Lovatt, 1990b; Rickard, 2000), but the reasons for enhanced growth responses are not explained. Lovatt and Mikkelsen (2006) suggest phosphite enhanced growth may be a growth-regulatory or phytohormonal factor, affecting sugar metabolism, stimulation of the shikimic acid pathway, or internal hormonal and chemical changes. Zhang et al. (2011) concluded that while *Microcystis aeruginosa* (Kützing) Kützing could not utilize phosphite as a sole P nutrient

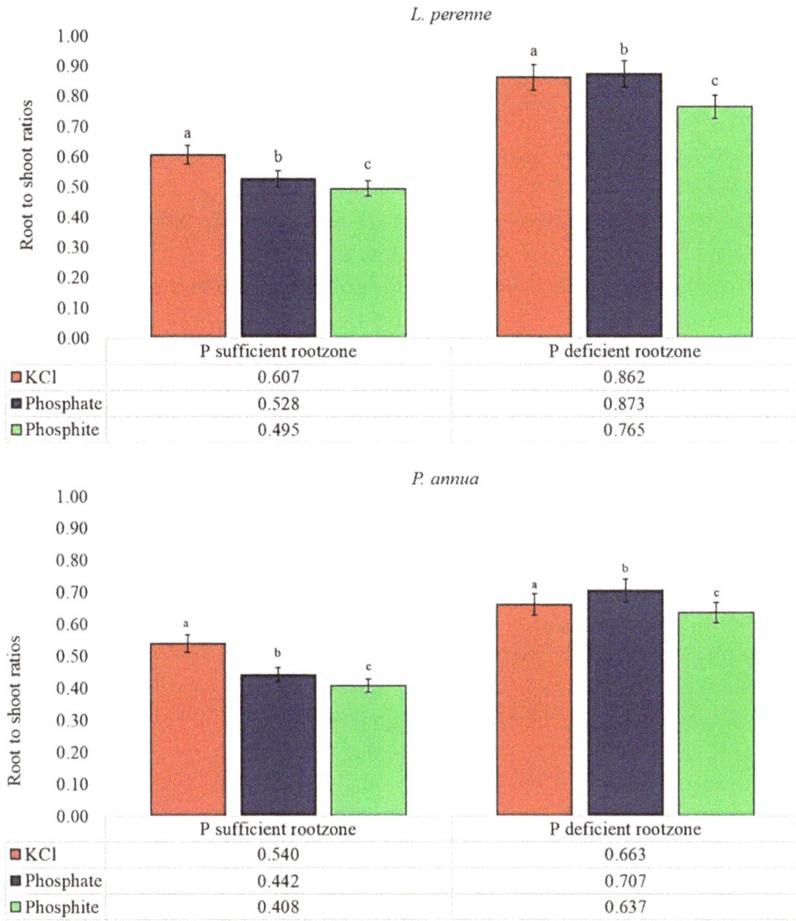

Figure 7 Effect on root-to-shoot ratios of perennial ryegrass (*Lolium perenne*) and annual bluegrass (*P. annua*), growing in P-sufficient (P > 38 ppm) and P-deficient (P < 6 ppm) root zones, following sequential treatments over a 6-month period, of phosphate, phosphite and potassium chloride (KCl; control). Bars are 95% confidence intervals, n = 6. Letters indicate significant differences within tissue type as determined by Tukey Honest Significant Different Test post-hoc analyses at $P = 0.05$; ppm = mg•kg^{-1} (Dempsey et al., 2022a).

at any concentration, when phosphite was supplied simultaneously with phosphate it increased cell numbers and chlorophyll content. In their review of the biostimulant activities of phosphite. Gómez-Merino and Trejo-Téllez (2015) concluded that phosphite can be used as a biostimulant that will enhance plant growth via activation of molecular, biochemical, and physiological responses, but these positive responses require and are attenuated in the presence of phosphate.

Root growth and development is a crucial component of all plants, but can be especially so for turfgrass, which in golf course putting greens is maintained under highly stressed situations (Smiley et al., 1992). Root development can determine how the turfgrass plant reacts in situations that can seriously impact on the viability and even survival of the sward. Abiotic and biotic challenges, such as drought, traffic related wear, and disease pressure, are constantly endangering the turfgrasses and a well-developed root system can often be the major influencing factor in the turfgrass plants success (Dernoeden, 2002). In Dempsey et al. (2022a), phosphite treatments produced the lowest mean ratio of roots-to-shoots in both creeping bentgrass and annual bluegrass in either the P-sufficient or P-deficient root zones (Fig. 7).

These root-to-shoot ratios are a direct indication of the number of roots per shoot, with the higher ratios showing the greater volume of root growth per plant. What this indicates is that in a P-sufficient situation, while phosphite treatment gave rise to increased growth of shoots, crowns, and roots, compared with phosphate and KCL treatments, the enhanced above ground growth was at the expense of the development of the root systems (Dempsey et al., 2022a). The reason for this is not clear and would require further study. In P-deficient root zones, root-to-shoot ratios also were significantly reduced by phosphite treatments. This was to be expected and is consistent with the research by Carswell et al. (1996), which concluded that phosphite treatments to P-limited plants decreased root-to-shoot ratios significantly.

Dempsey et al. (2022a) documented that phosphite is rapidly taken up and translocated by turfgrass, and that sequential applications would maintain leaf tissue accumulations of phosphite at approximately 3000 $mg\bullet kg^{-1}$. Long-term sequential phosphite applications should maintain leaf tissue accumulations at this approximate concentration, but can also lead to cumulative phosphite increases in meristematic tissues such as roots and subsequently contribute to increases in soil P. In P-sufficient root zones, foliar-applied phosphite increased biomass in turfgrass, but also led to a reduction in root-to-shoot ratios. In P-deficient root zones, foliar-applied phosphite does not supply a usable form of P in turfgrasses, and furthermore P-deficiency responses were repressed in turfgrasses.

The findings of Dempsey et al. (2022a) are of particular significance to turfgrass managers, particularly those that apply phosphite on 2–3-week intervals. This method of sequential phosphite applications would maintain phosphite levels in turfgrass leaf tissues within the range of 3000 to 3500 $mg\bullet kg^{-1}$, and should allow for improved turfgrass quality and some growth enhancement. While phosphite is included in many turfgrass nutritional programs as a biostimulant, the majority of turfgrass professionals, however, incorporate phosphite into turfgrass management programs as a means to prevent and/or suppress disease (Dempsey et al., 2012).

6 Disease control in turfgrass with phosphite

The fungicidal properties of phosphite were discovered at Rhone-Poulenc Agrochemical Laboratories in France during the 1970s (Guest and Grant, 1991), when they observed that phosphite salts were effective in controlling diseases caused by a group of oomycete fungi in the *Peronosporales* order (i.e. *Phytophthora*, *Plasmopara*, *Pythium*, and others). Fosetyl-Al, which was formulated under the trade name Aliette®, was the first phosphite-based product marketed specifically for turfgrass use. Initially, Aliette® was introduced to control *Pythium* spp., but was later combined with the mancozeb, a dithiocarbamate fungicide, for treating summer decline (i.e. abiotic and biotic stress) of creeping bentgrass (Cook et al., 2006). As a result, phosphite was used in combination with other fungicides to control yellow tuft (*Sclerophthora macrospora* [Sacc.] Thirum.) and summer stress complex of turfgrass (Cook et al., 2006; Schroetter et al., 2006). Numerous *Pythium* spp. are the causal agents of Pythium diseases of turfgrass (Dernoeden, 2002). Phosphite-based fungicides proved effective at suppressing Pythium when applied preventatively; however, the efficacy of these products varied with rates and timing of applications, and host species (Dernoeden, 2002).

The efficacy of potassium phosphite products and fosetyl-Al in controlling Pythium blight in turfgrass was studied by Cook et al. (2009), where they assessed inhibitory effects of phosphorous acid on *Pythium* spp. in vitro to determine if active ingredient and formulation of a range phosphite-based fungicides provided similar levels of Pythium blight suppression on perennial ryegrass and creeping bentgrass when applied at equivalent rates of phosphorous acid. They concluded that phosphorous acid EC_{50} values for isolates of 7 *Pythium* spp. were between 35.6 $\mu g \bullet ml^{-1}$ and 220.8 $\mu g \bullet ml^{-1}$, and that all phosphite treatments provided significant suppression of Pythium blight symptoms on perennial ryegrass and creeping bentgrass relative to the untreated controls. They also concluded no difference between potassium phosphite and fosetyl-Al with respect to suppression of Pythium blight when applied prior to symptom development at similar rates of phosphorous acid. However, the degree of Pythium blight suppression with phosphite-based fungicides may be greater on creeping bentgrass than on perennial ryegrass.

Several phosphite-based fungicides also were tested on a creeping bentgrass putting green to control dollar spot (*Clareedia* spp.), a common and troublesome foliar disease (Dernoeden, 2002). Overall, Vincelli and Dixon (2005) observed that phosphite products provided good control of dollar spot along with improved turf quality compared to untreated turfgrass.

Another major disease of highly managed turfgrass is anthracnose (*Colletotrichum cereale* Manns). On a creeping bentgrass/annual bluegrass putting green in Pennsylvania (USA), fosetyl-Al plus green pigment provided significant anthracnose suppression, but fosetyl-Al alone or potassium phosphite alone did not (Cook et al., 2006). In North Carolina (USA), fosetyl-Al plus green pigment or potassium phosphite alone provided excellent anthracnose control on an annual bluegrass putting green, but on a creeping bentgrass green, fosetyl-Al plus pigment was only moderately effective and potassium phosphite alone provided no significant anthracnose suppression on a creeping bentgrass green (Tredway and Wong, 2012). It was proposed that this inconsistency was due to differences in phosphite sensitivity of *Colletotrichum cereale* isolates. Others observed that disease severity, environmental conditions, presence of disease at the time of application, and number of applications before disease development were related to anthracnose control in published reports (Murphy et al., 2008). Phosphite tended to provide significant anthracnose suppression in the absence of weather extremes or excessive turfgrass stress, and when at least two applications were made before the onset of disease development (Murphy et al., 2008; Tredway and Wong, 2012).

6.1 Phosphite suppression of Microdochium nivale *infection in turfgrass*

Phosphite is very effective at controlling *Pythium* spp. in turfgrass, and some additional evidence suggests that phosphite is useful for the suppression of dollar spot and anthracnose (Dernoeden, 2002). In the early 2000s, phosphite was increasingly recommended to suppress or control disease caused by *Microdochium nivale* (Fr.) Samuels and Hallet in cool-season turfgrass, however, no published data was available to support that directive (Dempsey et al., 2012).

M. nivale is an ascomycete pathogen and causal agent for many disease complexes in numerous graminacea species (Smiley et al., 1992; Tronsmo et al., 2001). *M. nivale* produces conidia in large numbers that are readily dispersed by wind and rain splash and, along with soilborne mycelium, are the main source of inoculum (Tronsmo et al., 2001). In turfgrasses, *M. nivale* is regarded as the most damaging pathogen of temperate climates, thus infecting and causing disease in most cool-season species (Smiley et al., 1992). Under snow cover, *M. Nivale* causes pink snow mold disease, and during cold and wet conditions it causes Microdochium patch disease (Vargas, 2005).

The first report of the ability of phosphite to suppress *M. nivale* infection was by Dempsey and Owen (2010), and they observed significant reduction in *M. nivale*-caused disease incidence on creeping bentgrass from sequential applications of phosphite, although not to an acceptable level. Further research was conducted on velvet bentgrass (*Agrostis canina* L.) and annual bluegrass to evaluate Microdochium patch suppression from 14-day sequential applications of potassium phosphite, applied alone and in combinations with a fungicide (active ingredient = iprodione) or a biostimulant (containing N, K_2O, amino acids, micronutrients, kelp) from September through March in 2010–11 and again in 2011–12 (Dempsey et al., 2012). In both years and on both turfgrass species, Microdochium patch was most severe in untreated plots, and plots treated with the fungicide exhibited the best disease control. Sequential application of phosphite reduced disease incidence on average by more than 50%. The combination of biostimulant applied with the phosphite did not cause any significant reduction in disease incidence compared with the phosphite treatment alone. Of note, even though the fungicide applications were applied repeatedly on a 2-week interval, Microdochium patch was not fully controlled. The treatment most effective at reducing Microdochium patch incidence were the combination of fungicide and phosphite (Dempsey et al., 2012).

Further studies carried out at Oregon State University (Corvalis, OR, USA) between 2013 and 2019 examined several alternative means to control Microdochium patch in annual bluegrass putting greens (Mattox et al., 2014). Among these were the use of phosphite, applied alone or in combinations with iron sulfate heptahydrate, elemental sulfur, and the mineral oil Civitas Turf Defense (Intelligro; Mississauga, Ontario, Canada). The research primarily focused on using combinations of these different products in the absence of traditional fungicides. Phosphite applied alone every 2 weeks significantly suppressed Microdochium patch, and phosphite added as a tank-mix partner consistently improved Microdochium patch suppression in combination with iron sulfate heptahydrate, Civitas Turf Defense, or elemental sulfur. Mattox et al. (2020a,b) concluded the most consistent method of suppressing Microdochium patch on annual bluegrass putting greens and maintaining acceptable turfgrass quality without using traditional fungicides is to start in September with an application of Civitas Turf Defense tank-mixed with phosphite, followed 2 weeks later with a tank-mix of sulfur and phosphite. These tank-mixes are rotated every 2 weeks during the fall and continuing until early spring (Mattox et al., 2020a,b).

Because of these aforementioned published data and numerous anecdotal evidence, the use of phosphite to control not only *M. nivale,* but other fungal pathogens became widespread. The means by which phosphite achieved this inhibitory effect, however, needed to be determined.

6.2 Phosphite's mode of suppression on plant pathogens

While it was clear the phosphite has the ability to suppress some plant pathogenic infections, the mode of suppression was inconclusive and at best speculative. Most studies on phosphite-mediated inhibition of plant pathogens have been on its effects on oomycetes. Suppression of *Pythium* spp. by phosphite under field conditions was reported by Sanders (1983), but when no in vitro inhibition was demonstrated it was concluded that control resulted from enhanced host defenses. However, Fenn and Coffey (1984, 1987) demonstrated that phosphite inhibited 4 *Pythium* spp. and *Phytophthora cinnamomi* in vitro, and *P. cinnamomi* exhibited sensitivity to phosphite with EC_{50} values ranging from 4 μg•mL^{-1} to 148 μg•mL^{-1} (Wilkinson et al., 2001). In a later study, *Pythium* spp. were inhibited with EC_{50} values between 38.7 μg•mL^{-1} and 220.8 μg•mL^{-1} (Cook et al., 2009). This direct mode of inhibition seems to involve disruption of the pathogen's metabolism. For example, a study with three *Phytophthora* species showed that phosphite interfered with phosphate metabolism in pathogen cells by causing an accumulation of polyphosphate and pyrophosphate, diverting ATP from other metabolic pathways, resulting in reduced growth (Niere et al., 1994). Other studies determined that phosphite inhibited enzymes of the glycolytic and phosphogluconate pathways, disrupting P metabolism in *Phytophthora palmivora* by competing with phosphate as an allosteric regulator on several enzymes (Stehmann and Grant, 2000).

Less has been published on the in vitro effects of phosphite on fungal pathogens. Reuveni et al. (2003) showed inhibition of *Alternaria alternata* mycelial growth and conidial germination, while Burpee (2005) reported suppression of in vitro growth of *Colletotrichum cereale* (*Colletotrichum graminicola*). Mills et al. (2004) demonstrated that phosphite not only reduced mycelial growth but caused complete inhibition of sporulation of *A. alternata, Botrytis cinerea*, and *Fusarium solani*. Growth of *F. culmorum* and *F. graminearum* was reduced on phosphite-amended potato dextrose agar (Hofgaard et al., 2010). Also, mycelial growth of *M. majus* was reduced by more than 90% at the lowest phosphite concentration used of 10 μg•mL^{-1}, with full inhibition at concentrations of 100 μg μg•mL^{-1} (Hofgaard et al., 2010). Other studies determined that phosphite mediated significant reductions of *M. majus* infection in winter wheat (*Triticum aestivum*) (Hofgaard et al., 2010) and also *M. nivale* in creeping bentgrass (Dempsey and Owen, 2010; Dempsey et al., 2012) as previously mentioned. None of these studies, however, conclusively determined whether disease reduction was caused by either direct (i.e. fungistatic) means or stimulation of plant defenses.

6.3 In vitro suppression of Microdochium nivale *by phosphite*

The use of in vitro studies is an established method to assess a compound's ability either to reduce or inhibit the growth of, or to kill plant pathogenic organisms (Glynn et al., 2005; Hofgaard et al., 2010; Mann, 2002). When compiling a turfgrass disease protection program, an important factor is determining whether a compound is fungicidal or fungistatic (Smiley et al., 1992). It is possible that at sufficient concentrations, fungistatic compounds will prevent or reduce fungal growth and sporulation fully but, upon removal, the effects are reversed, and growth will recommence. This would have a significant influence on turfgrass disease management product application rates and application intervals.

Until 2018 there were no published data on the in vitro effect of phosphite on *M. nivale*. Dempsey et al. (2018) assessed the effect of phosphorous acid (H_3PO_3), phosphoric acid (H_3PO_4), dihydrogen potassium phosphite (KH_2PO_3), dihydrogen potassium phosphate (KH_2PO_4), and potassium hydroxide (KOH) on the in vitro mycelial growth and development of *M. nivale*. The *M. nivale* isolates were incubated on potato dextrose agar amended with 0 (unamended control), 10, 50, 100, and 250 µg•mL^{-1} of H_3PO_3, H_2PO_4, KH_2PO_3, KH_2PO_4, and KOH, respectively. Mycelial growth rate was determined by measuring colony radius at four points on each plate, from the edge of the initial inoculum to the extreme outer margin area of fungal mycelial development and growth rates (mm•day$^{-1)}$ calculated. Radial growth measurements were taken 1, 2, 3, 4, 5, 6, 7, 8, 9, and 10 days post-incubation (dpi). Mean values of each of 6 replicates were used to determine percent relative growth (PRG), which was calculated as: (radial growth on amended PDA/radial growth on unamended control PDA) × 100. Percent inhibition was calculated as 100 PRG. The effective concentrations that reduced mycelial growth by 50% (EC_{50}) and 90% (EC_{90}) were determined by probit transformation of PRG and regressing against $\log_{10}$ of amendment concentrations. Both H_3PO_3 and KH_2PO_3 caused significant inhibition of mycelial growth compared to all other compounds. The EC_{50} and EC_{90} values, as calculated at 5 dpi, were 40.99 µg•mL^{-1} and 80.90 µg•mL^{-1} for H_3PO_3, respectively, and 35.95 µg•mL^{-1} and 77.68 µg•mL^{-1} for the KH_2PO_3, respectively. In contrast, there were no significant growth inhibition with H_3PO_4, KH_2PO_4, and KOH amendments. The KH_2PO_3 PRG values were significantly lower than H_3PO_3. Mycelial growth of *M. nivale* was suppressed by PO_3^{3-} (i.e. phosphite) when compared to plates amended with H_3PO_4, KH_2PO_4, and KOH (Fig. 8).

To determination if phosphite had fungistatic rather than fungicidal properties, mycelial plugs of *M. nivale* were placed into 10 mL sterile distilled water (SDW) containing 0 (control), 10, 50, 100, and 250 µg•mL^{-1} of H_3PO_3, H_2PO_4, KH_2PO_3, KH_2PO_4, and KOH with 6 replications, and incubated in darkness in a growth chamber maintained at 18–20°C for 10 days (Dempsey et al., 2018). The plugs were retrieved, rinsed twice in SDW, and transferred onto fresh unamended PDA and grown in darkness at 18–20°C for a further 10 days. Growth responses were measured, and the presence or absence of growth determined if the concentrations were fungicidal or fungistatic. While there were significant differences in growth suppression following immersion in the 50, 100, 250, and 500 µg•mL^{-1} concentrations, there was no complete inhibition. Further evidence of the fungistatic rather than fungicidal properties of phosphite was determined by measurement of colony diameters growing on H_3PO_3 and KH_2PO_3 amended PDA at 10 dpi. It was concluded that phosphite reduces rather than fully inhibits *M. nivale* growth as colonies continued to

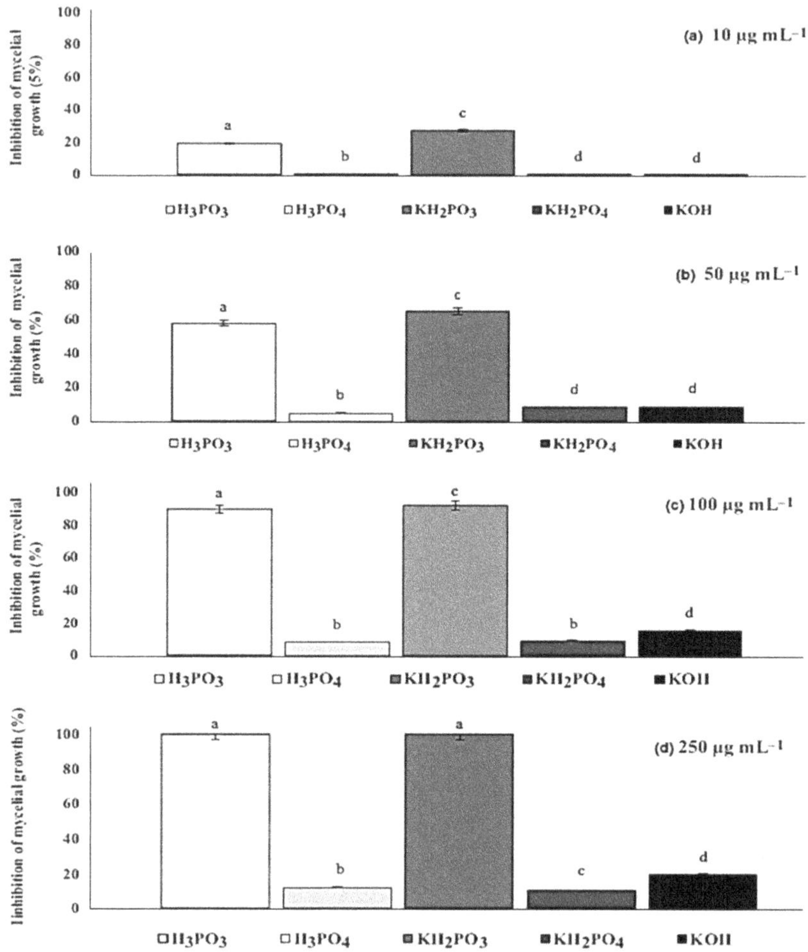

Figure 8 Inhibition of *M. nivale* mycelial growth on phosphorous acid (H_3PO_3), phosphoric acid (H_3PO_4), potassium dihydrogen phosphite (KH_2PO_3), potassium dihydrogen phosphate (KH_2PO_4), and potassium hydroxide (KOH) amended potato dextrose agar (PDA). Inhibition of *M. nivale* mycelial growth on PDA amended with (a) 10 µg•mL^{-1}; (b) 50 µg•mL^{-1}; (c) 100 µg•mL^{-1}; (d) 250 µg•mL^{-1} H_3PO_3, H_3PO_4, KH_2PO_3, KH_2PO_4, and KOH, presented as percent (%) inhibition of growth on unamended PDA. Growth rates calculated from pooled data of each of the four *M. nivale* isolates, n = 6, by measuring the colony radii at 4 points on each plate, 4 dpi. Bars are 95% confidence intervals. Letters indicate significant differences among compounds, as determined by Tukey Honest Significant Different Test at P= 0.05. Source: Dempsey et al. (2018).

grow, but at a slower rate in the phosphite-amended media at the end of the 10-day experimental period (Dempsey et al., 2018).

The effects of phosphite on *M. nivale* hyphal morphology was assessed using bright-field and fluorescence microscopy (Dempsey et al., 2018). The *M. nivale* mycelia, sampled from the outer margins of actively growing colonies, growing on PDA amended with 0 (unamended control), 10, 50, 100, and 250 μg•mL^{-1} of H_3PO_3, H_2PO_4, KH_2PO_3, KH_2PO_4, and KOH were examined. The fluorescent dye calcofluor white was used to visualize hyphae as described in Dubas et al. (2010). *The M. nivale* hyphae, viewed using bright-field microscopy in unamended control PDA, showed normal morphology as evidenced by smooth hyphal outlines. Hyphae grown on H_3PO_4 and KOH amended PDA appeared like those on unamended controls. The *M. nivale* hyphae grown on H_3PO_3 at concentrations of 75 μg•mL^{-1} and 100 μg•mL^{-1} amended PDA displayed an altered hyphal morphology, with the hyphae appearing swollen, short-branched, and stunted compared to hyphae grown on PO_4^{3-} and KOH amended plates (Fig. 9).

The effects of phosphite on conidial germination was determined by filtering *M. nivale* conidial suspensions through sterile cheesecloth and transferring 50 μL aliquots to 1.5 mL tubes and mixed with 1 mL solutions of 0 (control), 10, 50, 100, and 250 μg•mL^{-1} concentrations of H_3PO_3, H_2PO_4, KH_2PO_3, KH_2PO_4, and KOH (Dempsey et al., 2018). The 50 μL aliquots of the mixtures were pipetted onto depressions in cavity microscope slides and immediately placed on moist tissue paper in 9 cm diam. petri dishes and sealed, for a total of 6 replications. Following incubation in darkness in a growth chamber maintained at 18 ± 2°C for 48 h, the samples were agitated using an orbital shaker for 1 h then 20 μL pipetted onto fresh slides for assessment. The number of germinating conidia was counted, and percent germination calculated as: (conidia germinated/total conidia) ×100. Conidia were deemed to be germinated when the germ tube extended to at least twice the length of the conidium (Mills et al., 2004). The results showed that *M. nivale* conidia in all the unamended controls did not achieve 100% germination, with the highest rate of 85.6% determined in only one of 6 replicates. While there were only minor differences in germination rates in the 10 μg•mL^{-1} concentrations of all compounds, at the 50, 100. and 250 μg•mL^{-1} concentrations, germination rates in the H_3PO_3 and KH_2PO_3 amended plates were significantly less than germination rates in all other amendments (Dempsey et al., 2018).

The main conclusions of Dempsey et al. (2018) are that phosphite suppressed *M. nivale* mycelial growth, disrupted hyphal morphology, and reduced conidial germination. Both hyphae and conidia are infective propagules, providing inoculum for the turfgrass diseases caused by *M. nivale*. The incorporation of phosphite into growth media significantly suppresses the growth and development of *M. nivale* infective propagules in vitro and therefore supports research that demonstrates that phosphite can significantly reduce *M. nivale* infection of turfgrass in the field. Whether this direct fungistatic effect of phosphite on *M. nivale* was the sole means of disease suppression was not clear and the possible effect of phosphite being able to enhance turfgrass defenses also was examined (Dempsey et al., 2018).

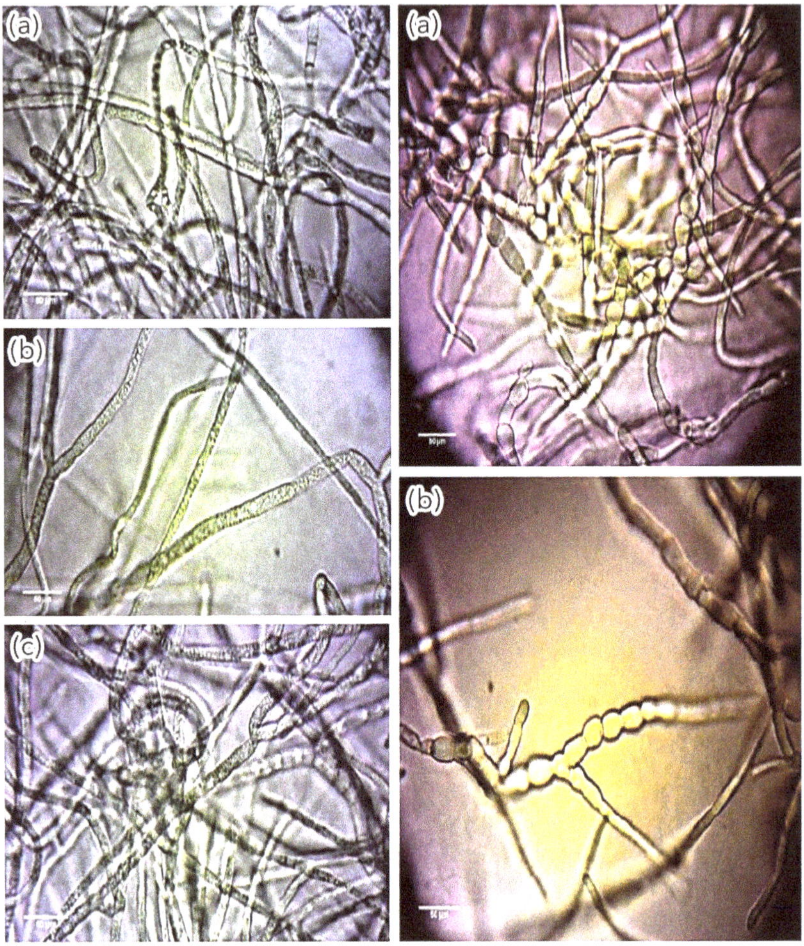

Figure 9 Bright-field micrographs of *M. nivale* hyphal growth in amended potato dextrose agar. Left: (a) Unamended control; (b) phosphoric acid (H_3PO_4) 100 μg•mL^{-1}; (c) potassium hydroxide (KOH) 100 μg•mL^{-1}. Right: (a) phosphorous acid (H_3PO_3) 75 μg•mL^{-1}; (b) phosphorous acid (H_3PO_3) 100 μg•mL^{-1}. Source: Dempsey et al. (2018).

6.4 Phosphite-mediated enhancement of defense responses in turfgrass infected by Microdochium nivale

Upon pathogen challenge, plants initiate an array of induced defenses involving numerous interconnected signaling pathways. Among these are the hypersensitive response (HR), generation of reactive oxygen species (ROS), programmed cell death (PCD), production of antimicrobial compounds such as pathogenesis-related (PR) proteins and phytoalexins, and salicylic acid (SA), which is key for induction of systemic resistance (Campbell and Reece, 2002).

Many of these defense compounds are classed as phenols, and synthesis and accumulation of total phenolic content (TPC) upon pathogen challenge has been shown to be an important defense response in plants. For example, the phenolic compound lignin is produced to strengthen cell walls, thus increasing resistance to pathogen ingression, and lignin levels have been shown to increase following pathogen infection in grasses (Campbell and Reece, 2002). Pociecha et al. (2009) concluded that increased levels of phenolic compounds gave rise to higher resistance to *M. nivale* in *Festulolium* species. Numerous other studies have identified defensive phenolic production in species of Graminaceae (Dubas et al., 2010). For example, phenolic compound accumulation was detected at the site of hyphal penetration, and it was concluded that this reaction was part of the *Triticale* sp. defense system against *M. nivale* (Dubas et al., 2010).

Hydrogen peroxide plays a major role in plant responses to both abiotic and biotic stress, regulating the expression of defense genes, as a key component of ROS and HR, and in strengthening cell walls (Campbell and Reece, 2002). Hydrogen peroxide also acts as a signal molecule initiating the transcription of resistance genes coding for PR proteins and for enzymes related to phytoalexin biosynthesis. Moreover, hydrogen peroxide stimulates the synthesis of SA, leading to induction of systemic acquired resistance (SAR; Campbell and Reece, 2002). The timely production of hydrogen peroxide and its accumulation in cells is a useful marker for determining the efficacy of a plant's response to pathogen challenge.

Priming and enhancement of plant defense responses by various means has been demonstrated in many plant systems and has led to a decrease in damage by pathogens (Heil and Bostock, 2002; Kuc, 1995). The consensus of published research is that while phosphite can act directly to inhibit many pathogens, the mode of disease suppression also includes a priming or enhancement of defenses in treated plants (Dempsey et al., 2022b). To determine if phosphite enhanced constitutive or inducible defenses, research into the responses of *M. nivale* infected turfgrasses was required. The aim of this research was to assess two important defense responses to pathogen challenge: synthesis and accumulation of TPC and generation of hydrogen peroxide (H_2O_2), and the effect of phosphite treatment on priming or enhancing of these responses (Dempsey et al., 2022b). Host defenses in creeping bentgrass and annual bluegrass in response to *M. nivale* infection were assessed by measuring TPC and H_2O_2 generation before and during the infection process, and to assess the effect of phosphite treatment on these induced responses and subsequent effects on *M. nivale* disease incidence.

To assess if *M. nivale* infection led to increased levels of TPC and increased H_2O_2 generation, leaf tissues of creeping bentgrass and annual bluegrass were collected from infected field trial plots and inoculated greenhouse plants, between September 2012 and March 2014 (Dempsey et al., 2022b). Tissue samples from the boundary zone between green, visually healthy, and dead desiccated parts of leaves were collected on four separate occasions during each of the 3 years. All field samples used were allowed to become infected under natural conditions, while greenhouse samples were infected using hyphal inoculum. Non-infected turfgrass served as control samples. The TPC was determined by modified extraction methods as described by Singleton et al. (1965), and TPC was calculated and reported as $mg{\bullet}g^{-1}$ dry weight of gallic acid equivalents (GAE). In field samples for both turfgrass species and during each year of assessment, TPC was significantly greater in infected tissues compared with uninfected tissues. The results were similar in greenhouse tissue samples with TPC significantly greater each year of assessment compared to the uninfected in both turfgrass species. Accumulation of TPC is a common response to pathogen challenge, and the speed of accumulation of phenolics prior to or following infection can influence the plant's level of susceptibility or resistance to a particular pathogen. It should be noted that TPC accumulation is also a response to abiotic stresses, for example, in response to mechanical injury, drought, UV radiation, and low temperature. It was therefore important to sample turfgrass tissues during a wide range of conditions and that treated turfgrasses and controls were harvested under identical situations. This ensured that TPC levels due to non-disease related pressures did not influence the assessments. The results from this study confirmed that increased levels of TPC are a response to *M. nivale* infection (Dempsey et al., 2022b).

A second defense response measured was H_2O_2 generation in response to infection and this was determined by analyses of infected greenhouse tissues only. Field samples were not used, as the generation of H2O2 upon infection is rapid and transitory and would have been difficult to assess under field conditions. Following hyphal inoculation of greenhouse grown creeping bentgrass and annual bluegrass, generation of H_2O_2 was determined over the course of 72 h post-inoculation (hpi) and compared with H_2O_2 in uninfected plants in the same greenhouses (Dempsey et al., 2022b). In both turfgrass species, there were significantly greater amounts of H_2O_2 generated in infected tissues than in uninfected tissues at 1 hpi and 6 hpi. At all other time periods there were no statistical differences between infected and uninfected tissues.

These data were important as they confirmed that in creeping bentgrass and annual bluegrass an increase in TPC and H_2O_2 generation are components of defense responses following biotic challenge (Dempsey et al., 2022b). Therefore, determination if phosphite treatment could stimulate or enhance these defense responses was a critical component of that research.

Whether phosphite pre-treatment can prime defense response was assessed by treating field trial plots and greenhouse samples of creeping bentgrass and annual bluegrass with phosphite and phosphate at $0.35g{\bullet}m^{-2}$, and SDW (control). Treatments were either sequential, as 6 treatments applied at 4-week intervals, or applied as a single treatment. Leaf tissues were harvested at 0, 1, 6, 24, 48, and 72 h post-application (hpa). The TPC and H_2O_2 generation were determined, and comparisons made between the phosphite, phosphate, and control levels. Results showed that a single phosphite treatment did influence TPC accumulations, leading to significantly higher levels, compared to controls in both field and greenhouse samples. Following phosphite treatment, TPC levels in field samples were significantly higher in creeping bentgrass tissues from 1 hpa to 72 hpa and in annual bluegrass from 12 hpa to 72 hpa. The increases in TPC were similar in treated greenhouse turfgrasses, with levels in creeping bentgrass significantly higher than controls from 12 hpa to 72 hpa, and in annual bluegrass at 1 hpa and from 12 hpa to 72 hpa (Dempsey et al., 2022b).

While these data indicate that a single phosphite treatment stimulated TPC accumulation, it can be argued that regarding induced defense responses in plants it was not different to the response elicited from the phosphate treatments (Dempsey et al., 2022b). Phosphate treatment, in both field and greenhouse turfgrasses, also led to significantly higher TPC levels compared to controls, from 12 hpa to 72 hpa, in a similar manner to phosphite treatment. However, while a single treatment of phosphite and phosphate elicited similar responses in TPC levels, sequentially applied treatments, over a period of 6 months, gave rise to significantly higher levels of TPC in phosphite-treated tissues compared with both phosphate and control tissues. Therefore, these data indicate that phosphite can indeed prime plants for stress responses, but also that several applications prior to infection are required for phosphite-mediated disease suppression to be successful (Dempsey et al., 2022b).

This conclusion, that sequential applications of phosphite are required to be effective in priming plant defenses and therefore suppressing plant disease, is supported by Koch and Hockemeyer (2021). One of the objectives of that research was to assess the efficacy of potassium phosphite to suppress snow mold (caused by *M. nivale*), when applied repeatedly throughout the fall. Potassium phosphite was applied 6 times, every 2 weeks, throughout the fall of 2019 at 3 locations in Wisconsin, USA, that typically produce high, medium, and low snow mold pressure. No reduction in snow mold disease relative to the nontreated control was observed at any location in response to the potassium phosphite treatments. They concluded that while phosphites have demonstrated excellent suppressive abilities against certain turfgrass diseases in past research, they demonstrated no suppressive abilities against three different snow mold diseases of Microdochium patch, gray snow mold (*Typhula* spp.), and speckled snow mold (*Typhula* spp.). Why the phosphite treatment failed to provide disease suppression was not entirely clear, but they surmised it was related to insufficient suppression of the pathogen during the fall and/or a prolonged period for pathogen recovery under the snow during the extended winter season in Wisconsin. In the midwestern USA, the long duration of snow cover prevents repeated applications throughout the winter, and any suppression of the fungus that may have occurred during the fall was inadequate to suppress disease development once snow cover arrived. This study therefore lends support to the data that several sequential applications of phosphite led to increased accumulations of TPC and subsequently increased resistance to pathogen challenge (Koch and Hockemeyer, 2021).

The question is: phosphite treatment primed the defense responses in healthy uninfected turfgrass, but did this also occur in *M. nivale* infected plants and lead to significantly reduced levels of disease? Following hyphal inoculation of greenhouse samples of creeping bentgrass and annual bluegrass, conditions were ideal for *M. nivale* infection, which led to rapid development of disease symptoms (Dempsey et al., 2022b). Phosphite, applied either as a single treatment or sequentially over six applications, led to significant reductions in patch diameters, compared to controls and phosphate treatments in both turfgrass species. Phosphite applied sequentially also gave rise to significant reductions in patch diameters compared to a single phosphite treatment in both turfgrass species. Analyses of infected creeping bentgrass and annual bluegrass tissues over the course of 10 dpi showed increasing accumulation of TPC in all samples with statistically significant differences between treatments each time. Both turfgrass species treated with either a single or 6 sequential treatments of phosphite accumulated significantly higher levels of TPC than phosphate-treated or control tissues, at all measurement periods from 2 dpi to 10 dpi. In both turfgrass species, phosphite from six sequential applications induced the largest TPC, which was significantly greater than all other treatments. The TPC in both turfgrass species treated with a single phosphite treatment was significantly greater than phosphate and control treatments at all time periods, except for at 0 dpi (Dempsey et al., 2022b).

There were also significant treatment effects on H_2O_2 generation in phosphite-treated turfgrass (Dempsey et al., 2022b). In creeping bentgrass, H_2O_2 generation in phosphite-treated plants was significantly greater at 1 hpi and 6 hpi than both the control and phosphate-treated plants. In annual bluegrass, there were no significant differences between treatments at any time except for the plants that received 6 sequential phosphite treatments where H_2O_2 generation was significantly greater than all other treatments at 1 hpi and 6 hpi.

Dempsey et al. (2022b) showed that phenolic compounds and H_2O_2 are components of responses to *M. nivale* infection in creeping bentgrass and annual bluegrass. Phosphite treatment led to enhanced TPC accumulations in both infected and uninfected plants and subsequently reduced severity of *M. nivale* infection. Results of H_2O_2 extractions indicated that while phosphite treatment increased H_2O_2 generation compared to controls, the effect was no different to the phosphate-treated plants. It was evident that phosphite treatment increased TPC, but how this occurs is not clear. The possibility that it is by induction of physiological stress or interference with the phosphate uptake mechanisms merits investigation. While this study has provided novel data on the process of *M. nivale* infection of turfgrass and subsequent plant responses, the mechanisms by which phosphite interacts with and influences these responses requires further research.

7 Conclusion and future trends

Phosphite use in turfgrass management primarily developed from a means to successfully control *Pythium* spp.-caused diseases in the 1980s, to its use as a biostimulant and plant protectant for several other fungal pathogens of turfgrasses. Along the way there has been much discussion, debate, critical review, and, subsequently, further research into the chemical properties of phosphite and its effects on plant physiology (Haylin and Schlegel, 2021).

Recent research has confirmed that when phosphite is applied as a foliar treatment to turfgrasses, it is rapidly taken up and translocated throughout the plant. Phosphite is stable within plant tissues and does not convert to phosphate. Once phosphite is in the soil, however, microorganisms can utilize it and convert it to plant-usable phosphate (Haylin and Schlegel, 2021).

With fungal pathogens, there are reports of phosphite-mediated suppression of dollar spot and anthracnose, but the bulk of research has been on its effect on *M. nivale*. Field trials in Europe and the USA determined significantly lower percentages of *M. nivale* incidence on phosphite-treated plots of turfgrass, when compared with untreated controls, with the addition of phosphite significantly enhancing fungicide efficacy (Dempsey et al., 2012; Tredway and Wong, 2012). Phosphite-mediated disease suppression appears to be two-fold: (a) phosphite has a direct fungistatic effect, being able to reduce *M. nivale* mycelial growth, disrupt hyphal morphology, and reduce conidial germination (Dempsey et al., 2022a) and (b) phosphite also has an indirect effect on disease infection by stimulating turfgrass defense responses (Dempsey et al., 2022b).

Turfgrass quality on phosphite-treated turf swards can be beneficial (Cook et al., 2006). Despite this aesthetic quality improvement, phosphite does not provide a source of P that can be metabolized by the turfgrass (Dempsey et al., 2022a). Long-term sequential phosphite applications can lead to cumulative increases in root meristematic tissues and subsequently contribute to increases in soil P levels. In P-sufficient root zones, foliar-applied phosphite can increase biomass in turfgrass but can also lead to a reduction in root-to-shoot ratios. In P-deficient root zones, foliar-applied phosphite can have a deleterious effect on growth and repress P-deficiency responses. Also, further research is needed to investigate phosphite applications and their effect on turfgrass defense mechanisms or pathways.

8 Where to look for further information

There are several specific areas in this subject that should be further explored.

The mode by which phosphite can stimulate turfgrass growth and development requires further study and how the ratio of phosphite to phosphate influence this beneficial effect.

The longevity of phosphite in the soil is of interest. Soil microorganisms convert phosphite to phosphate, what factors are involved in this and how do they effect the speed of conversion.

As most phosphite products used in the turfgrass industry are applied as foliar treatments with phosphite accumulating in the leaf, the *in planta* conversion of phosphite to phosphate requires study.

Regarding stimulation of turfgrass defense mechanisms, more detailed research is required to determine which specific pathways are affected, not only in regard to biotic challenge but also with abiotic challenges.

Research in the turfgrass industry especially regarding phosphite is quite limited. Oregon State University in the USA and the Scandinavian Turfgrass and Environmental Research Foundation have produced significant research data over the past few years and will hopefully continue in this area.

9 References

Albrigo, L. G. (1999). Effects of foliar applications of urea or nutriphite on flowering and yields of Valencia orange trees. *Proceedings of Florida State Horticultural Society* 112: 1-4.

Adams, F. and Conrad, J. P. (1953). Transition of phosphite to phosphate in soils. *Soil Science* 75(5): 361-371.

Bigelow, C., Fidanza, M., Ervin, E. and Zhang, X. (2021). Biostimulants and plant health products for turf systems: current and future trends. Available at: https://www.biostimulant.com/biostimulants-and-plant-health-products-for-turf-systems/.

Bigelow, C. A., Tudor, W. T. and Nemitz, J. R. (2012). *Facts about Phosphorus and Lawns*. Purdue University Cooperative Extension. AY-834-W.

Burpee, L. L. (2005). Sensitivity of *Colletotrichum graminicola* to phosphonate fungicides. *International Turfgrass Society Research Journal* 10: 163-169.

Butler, T., Bryan, J. and Frank, K. (2009). Turfgrass development as influenced by phosphate and phosphite nutrient applications. *International Turfgrass Research Journal* 11: 979-986.

Campbell, N. and Reece, J. (2002). *Biology San Francisco, Person Education*. San Francisco: Benjamin Cummings.

Carswell, C., Grant, B. R., Theodorou, M. E., Harris, J., Niere, J. O. and Plaxton, W. C. (1996). The fungicide phosphonate disrupts the phosphate-starvation response in Brassica nigra seedlings. *Plant Physiology* 110(1): 105-110.

Cook, P. J., Landschoot, P. J. and Schlossberg, M. J. (2006). Phosphonate products for disease control and putting green quality. *Golf Course Management*April: 93-96.

Cook, P. J., Landschoot, P. J. and Schlossberg, M. J. (2009). Inhibition of Pythium spp. and suppression of pythium blight of turfgrasses with phosphonate fungicides. *Plant Disease* 93(8): 809-814.

Danova-Alt, R., Dijkema, C. O. R., De Waard, P. and Köck, M. (2008). Transport and compartmentation of phosphite in higher plant cells - kinetic and 31P nuclear magnetic resonance studies. *Plant, Cell and Environment* 31(10): 1510-1521.

Dempsey, J. and Owen, A. G. (2010). The effect of phosphite treatments on the growth and disease susceptibility of *Agrostis stolonifera* L. 2nd European Turfgrass Society Conference, Angers, France. European Turfgrass Society.

Dempsey, J. J., Wilson, I. D., Spencer-Phillips, P. T. N. and Arnold, D. L. (2012). Suppression of *Microdochium nivale* by potassium phosphite in cool season turfgrasses. *Acta Agriculturae Scandinavica, Section B – Soil and Plant Science* 62(Suppl. 1): 70-78.

Dempsey, J. J., Wilson, I. D., Spencer-Phillips, P. T. N. and Arnold, D. L. (2018). Suppression of the *in vitro* growth and development of *Microdochium nivale* by phosphite. *Plant Pathology* 67(6): 1296-1306.

Dempsey, J. J., Wilson, I. D., Spencer-Phillips, P. T. N. and Arnold, D. L. (2022a). Uptake and translocation of foliar applied phosphite and its effect on growth and development in cool season turfgrass. *Journal of Plant Nutrition* 45(13): 2003-2022. https://doi.org/10.1080/01904167.2022.2044489.

Dempsey, J. J., Wilson, I. D., Spencer-Phillips, P. T. N. and Arnold, D. L. (2022b). Phosphite-mediated enhancement of defence responses in *Agrostis stolonifera* and *Poa annua* infected by *Microdochium nivale*. *Plant Pathology* 00: 1-10. https://doi.org/10.1111/ppa.13584.

Dernoeden, P. H. (2002). *Creeping Bentgrass Management: Summer Stresses, Weeds and Selected Maladies*. Hoboken, NJ: John Wiley & Sons Inc.

Dest, W. M. and Guillard, K. (1987). Nitrogen and phosphorous nutritional influence on bentgrass-annual bluegrass community. *Journal of the American Society for Horticultural Science* 112(5): 769-773.

Dubas, E., Golebiowska, G., Zur, I. and Wedzony, M. (2010). *Microdochium nivale* (Fr., Samuels & Hallett): cytological analysis of the infection process in triticale (×Triticosecale Wittm.). *Acta Physiologiae Plantarum* 33(2): 529-537.

Fabrıcio, W., Faquin, V., Da Silva Lobato, A., Baliza, D., Marques, D., Dos Passos, A., Bastos, C. and Guedes, E. (2012). Growth, phosphorus status, and nutritional aspect in common bean exposed to different soil phosphate levels and foliar-applied phosphorus forms. Scientific Research and Essays 725: 2195-2204.

Fenn, M. E. and Coffey, M. D. (1984). Studies on the *in vitro* and *in vivo* antifungal activity of Fosetyl-Al and phosphorus acid. *Phytopathology* 74(5): 606-611.

Fenn, M. and Coffey, M. D. (1987). Phosphonate fungicides for control of diseases caused by Phytophthora. *California Avocado Society* 71: 241-249.

Fidanza, M., Cisar, J. and Watschke, T. (2015). A perspective on biostimulants and sportsturf. SportsTurf 31(6): 8-10, 12.

Glynn, N. C., Hare, M. C., Parry, D. W. and Edwards, S. G. (2005). Phylogenetic analysis of EF-1 alpha gene sequences from isolates of *Microdochium nivale* leads to elevation of varieties *majus* and *nivale* to species status. *Mycological Research* 109(8): 872-880.

Gómez-Merino, F. C. and Trejo-Télle, L. I. (2015). Biostimulant activity of phosphite in horticulture. *Scientia Horticulturae* 196: 82-90.

Guest, D. and Grant, B. (1991). The complex action of phosphonates as antifungal agents. *Biological Reviews* 66(2): 159-187.

Haylin, J. L. and Schlegel, A. J. (2021). Review of phosphite as a plant nutrient and fungicide. *Soil Systems* 5(3): 52 https://doi.org/10.3390/soilsystems5030052.

Heil, M. and Bostock, R. M. (2002). Induced systemic resistance (ISR) Against pathogens in the context of induced plant defences. *Annals of Botany* 89(5): 503-512.

Hofgaard, I. S., Ergon, Å, Henriksen, B. and Tronsmo, A. M. (2010). The effect of potential resistance inducers on development of *Microdochium majus* and *Fusarium culmorum* in winter wheat. *European Journal of Plant Pathology* 128(2): 269-281.

Horvath, B. J., McCall, D. S., Ervin, E. H. and Zhang, X. (2007). Physiological effects of phosphite formulations on turfgrass challenged with pythium and heat stress. *Virginia Turfgrass Journal* (Jan./Feb): 14-15.

Jost, R., Pharmawati, M., Lapis-Gaza, H. R., Rossig, C., Berkowitz, O., Lambers, H.and Finnegan, P. M. (2015). Differentiating phosphate-dependent and phosphate-independent systemic phosphate-starvation response networks in *Arabidopsis thaliana* through the application of phosphite. *Journal of Experimental Botany* 66(9): 2501-2514.

Koch, P. L. and Hockemeyer, K. (2021). Iron sulfate and phosphite products fail to suppress snow mold on amenity turfgrass in Wisconsin. *Crop, Forage and Turfgrass Management* 7(2): e20138. https://doi.org/10.1002/cft2.20138.

Kuc, J. (1995). Phytoalexins, stress metabolism, and disease resistance in Plants. *Annual Review of Phytopathology* 33(1): 275-297.

Lobato, M. C., Olivieri, F. P., Daleo, G. R. and Andreu, A. B. (2010.). Antimicrobial activity of phosphites against different potato pathogens. *Journal of Plant Diseases and Protection* 117(3): 102-109.

Lovatt, C. J. (1990). *A Definitive Test to Determine Whether Phosphite Fertilization Can Replace Phosphate Fertilization to Supply P in the Metabolism of 'Hass' on 'Duke 7'* (vol. 74). California Avocado Society: 61-64.

Lovatt, C. J. (1999). Timing citrus and avocado foliar nutrient applications to increase fruit set and size. *hortTechnology* 9(4): 607-612.

Lovatt, C. J. and Mikkelsen, R. L. (2006). Phosphite fertilizers: what are they? Can you use them? What can they do? *Better Crops* 90: 11-13.

MacIntire, W. H., Winterberg, S. H., Hardin, L. J., Sterges, A. J. and Clements, L. B. (1950). Fertilizer evaluation of certain phosphorus, phosphorous, and phosphoric materials by means of pot cultures. *Agronomy Journal* 42(11): 543-549.

Mann, R. (2002). Disease survey 2001 - preliminary results. *International Turfgrass Society Research Journal* 215: 32-34.

Mattox, C. M., Kowalewski, A. R. and McDonald, B. W. (2014). *Evaluation of Fungicide Alternatives for the Control of Microdochium Patch on Annual Bluegrass in Western Oregon*. Plant Disease Management Report (vol. 2014).

Mattox, C. M., Kowalewski, A. R., McDonald, B. W., Lambrinos, J. G. and Pscheidt, J. W. (2020a) Combinations of rolling, mineral oil, sulfur, and phosphorous acid affect Microdochium patch severity. *Agronomy Journal* 112(5): 3383-3395. https://doi.org/10.1002/agj2.20191.

Mattox, C., Dumelle, M., McDonald, B., Gould, M., Olsen, C., Braithwaite, E. and Kowalewski, A. (2020b). Suppression of Microdochium patch using rotations of mineral oil, sulfur, and phosphorous acid. *Agronomy Journal* 113(5): 3819-3833. https://doi.org/10.1002/agj2.20557.

McDonald, A. E., Grant, B. R. and Plaxton, W. C. (2001). Phosphite (phosphorous acid): its relevance in the environment and agriculture and influence on plant phosphate starvation response. *Journal of Plant Nutrition* 24(10): 1505-1519.

Mills, A. A. S., Platt, H. W. and Hurta, R. A. R. (2004). Effect of salt compounds on mycelial growth, sporulation, and spore germination of various potato pathogens. *Postharvest Biology and Technology* 34(3): 341-350.

Murphy, J., Wong, F., Tredway, L., Crouch, J. A., Ingugiato, J., Clarke, B., Hsiang, T. and Rossi, F. (2008). Best management practices for anthracnose on annual bluegrass turf. *Golf Course Management* 76(8): 93-104.

Niere, J. O., Deangelis, G. and Grant, B. R. (1994). The effect of phosphonate on the acid-soluble phosphorus components in the genus Phytophthora. *Microbiology* 140(7): 1661-1670.

Ouimette, D. G. and Coffey, M. D. (1988). Quantitative analysis of organic phosphonates, phosphonate, and other inorganic anions in plants and soil by using high-performance ion chromatography. *Phytopathology* 78(9): 1150-1155.

Ouimette, D. G. and Coffey, M. D. (1990). Symplastic entry and phloem translocation of phosphonate. *Pesticide Biochemistry and Physiology* 38(1): 18-25.

Pociecha, E., Płażek, A., Janowiak, F., Waligórski, P. and Zwierzykowski, Z. (2009). Changes in abscisic acid, salicylic acid and phenylpropanoid concentrations during cold acclimation of androgenic forms of Festulolium (*Festuca pratensis Lolium multiflorum*) in relation to resistance to pink snow mould (*Microdochium nivale*). *Plant Breeding* 128(4): 397-403.

Raghothama, K. G. and Karthikeyan, A. S. (2005). Phosphate acquisition. *Plant and Soil* 274(1-2): 37-49.

Reuveni, M., Sheglov, D. and Cohen, Y. (2003). Control of moldy- core decay in apple fruits by b-aminobutyric acids and potassium phosphites. *Plant Disease* 87(8): 933-936.

Rickard, D. (2000). Review of phosphorous acid and its salts as fertilizer materials. *Journal of Plant Nutrition* 23(2): 16 -80.

Sanders, P. L. (1983). Control of *Pythium* spp. and Pythium blight of turfgrass with fosetyl aluminum. *Plant Disease* 67(12): 1382-1383.

Schroetter, S., Angeles-Wedler, D. and Kreuzig, R. (2006). Effects of phosphite on phosphorus supply in corn (Zea mays). Landbauforschung Volkenrode 56: 87-99.

Singleton, V. S. and Rossi, J. A. (1965). Colorimetry of total phenolics with phosphomolybdic-phosphotungstic acid reagent. *American Journal of Encology and Viticulture* 16: 144 -157.

Smiley, R., Dernoeden, P. and Clarke, B. (1992). *Compendium of Turfgrass Diseases* (2nd edn.). St Paul: APS Press.

Stehmann, C. and Grant, B. (2000). Inhibition of enzymes of the glycolytic pathway and hexose monophosphate bypass by phosphonate. *Pesticide Biochemistry and Physiology* 67: 1-12.

Thao, H. T. B. and Yamakawa, T. (2009). Phosphite (phosphorous acid): fungicide, fertilizer, or bio-stimulator? *Soil Science and Plant Nutrition* 55(2): 228-234.

Ticconi, C. A., Delatorre, C. A. and Abell, S. (2001). Attenuation of phosphate starvation responses by phosphite in *Arabidopsis*. *Plant Physiology* 127(3): 963-972.

Tredway, L. and Wong, F. (2012). Managing anthracnose with fungicides. *Golf Course Management* 80(6): 90–98.

Tronsmo, A. M., Hsiang, T., Okuyama, H. and Nakajima, T. (2001). Low temperature diseases caused by *Microdochium nivale*. In: *Low Temperature Plant Microbe Interactions Under Snow* Iriki, N., Gaudet, D. A., Tronsmo, A. M., Matsumoto, N., Yoshida, M. and Nishimune, A. (Eds). Sapporo, Japan: Hokkaido National Agricultural Experiment Station: 75.

Varadarajan, D. K., Karthikeyan, A. S., Matilda, P. D. and Raghothama, K. G. (2002). Phosphite, an analog of phosphate, suppresses the coordinated expression of genes under phosphate starvation. *Plant Physiology* 129(3): 1232–1240.

Vargas, J. (2005). *Management of Turfgrass Diseases*. Chichester: John Wiley and Sons.

Vincelli, P. and Dixon, E. (2005). Performance of selected phosphite fungicides on greens. *Golf Course Management* July: 77–81.

Wilkinson, C. J., Shearer, B. L., Jackson, T. J. and Hardy, G. E. St. J. (2001). Variation in sensitivity of Western Australian isolates of *Phytophthora cinnamomi* to phosphite *in vitro*. *Plant Pathology* 50(1): 83–89.

Zhang, J., Geng, J., Ren, H., Luo, J., Zhang, A. and Wang, X. (2011). Physiological and biochemical responses of *Microcystis aeruginosa* to phosphite. *Chemosphere* 85(8): 1325–1330.

Chapter 5

Advances in irrigation and water management of turfgrass

Marco Schiavon, Fort Lauderdale Research and Education Center, University of Florida, USA; and Matteo Serena, United States Golf Association, USA

1 Introduction

According to an estimate by Milesi et al. (2005), turfgrass occupies 163 812 km^2 of land in the continental United States. Turfgrass areas are one of the largest natural components within the urban landscape. Turf is the 'backbone' of residential lawns, parks, institutional and commercial landscapes, athletic fields and golf courses. Turf also plays a role outside of the cityscape, where large swathes of maintained grasses can resemble prairies and meadows.

Even though not all turfgrass areas in the United States are subjected to irrigation, Milesi et al. (2005) concluded irrigation was not necessary to maintain turf in northern states such as Michigan. By comparison, an estimated 125 cm of water•year^{1} are necessary to maintain turf in the desert climate of the southwest United States where evapotranspiration (ET) rates can exceed 1600 mm•year^{1}. For comparison, the average deficit between annual potential ET and rainfall can be higher than 1000 mm•year^{1} in inland California (Schiavon

http://dx.doi.org/10.19103/AS.2022.0110.05

et al., 2020), Phoenix, AZ, Las Vegas, NV (Schiavon et al., 2014c), and southern New Mexico (Serena et al., 2020a). Nevertheless, irrigation is needed even in hot humid climates where precipitation matches or even exceeds turf water needs, but soils are generally sandy in texture and low in organic matter and therefore root zones have low water holding capacity.

Water use rates vary greatly between turfgrass species, especially when taking environmental conditions (e.g. soil type and fertility status, water quality, etc.), and management practices (e.g. mowing, fertilization, irrigation) into account (Fig. 1). In general, turfgrass receives the majority of supplemental water necessary for its growth during the hot summer months when it is most needed (Schiavon et al., 2019). More than 50% of water used during the summer is for irrigation of outdoor landscape areas not only in the desert of the American southwest, but also in many other regions of the southern, midwestern and mid-Atlantic United States (Devitt and Morris, 2008; Haley et al., 2007; Kjelgren et al., 2000; Mayer et al., 1999) where turfgrass is the predominant groundcover.

Turfgrass species should be grown in environments in which they are well adapted (Christians and Engelke, 1994). Warm-season species water use typically ranges from 3 mm•day^{-1} to 9 mm•day^{-1}, while rates for cool-season turfgrasses are usually higher (4 mm•day^{-1} and 13 mm•day^{-1}) (Kenna, 2008). Not only has it been reported that switching from a cool-season to a warm-season species in a dry Mediterranean climate could result to a 20% water

Figure 1 Irrigation on putting green #16 while conducting an irrigation audit (i.e. catch-can test) on 10 September 2010 at the New Mexico State University Golf Course (Las Cruces, NM, USA).

savings through the year (Schiavon et al., 2013b), but also, the switch to warm-season turfgrass species could potentially eliminate the need for irrigation in humid transition zones altogether (Schiavon et al., 2016). Moreover, warm-season species not only use water more efficiently than cool-season grasses, but their growing season is shorter in the transition zones, reducing annual irrigation needs for growth during the warm summer months. Nevertheless, the loss of visual green color by these species during the winter months and the long dormancy period into the spring hinders their widespread use not only in the transition zone, but even in desert areas like inland California. Growing turfgrass species that are not well adapted to the environment results in intensified management practices to maintain quality and playability standards that may not be realistic, leading to excessive irrigation inputs.

Urban growth potential, potable water scarcity, limited annual precipitation, prolonged droughts and food crops prioritization sets limits to the availability of water that can be allocated to turfgrass areas, which are considered an unnecessary luxury by many communities. Moreover, policies that mandate restrictions on water allotment for outdoor irrigation, and monetary incentives for replacing turf with alternative groundcovers have been implemented by a number of municipalities in several US regions (Schiavon et al., 2013b). For these reasons, research has been focused extensively on water conservation efforts for turfgrasses in recent years. Aside from the use of warm-season turf species where possible, several potable water conservation strategies have been proposed and investigated in recent years. In this chapter, the focus is on estimating turfgrass water needs, sprinkler irrigation system advances, subsurface drip irrigation (SDI), alternatives to potable water, turfgrass management practices, use of plant growth regulators (PGRs) and soil surfactants.

2 Estimating turfgrass water needs

Since water limitations and restrictions are often implemented even outside arid and semi-arid regions, the old 'set and forget' irrigation scheduling that often provides water amounts that exceed those needed by turfgrass growth cannot be adopted by turfgrass managers (Bremer et al., 2012). The 'set and forget' schedule is often used to maintain turfgrass quality and functionality during the summer months when turfgrass water demands are at their peak, resulting in over-irrigation during the spring and fall. Best practice irrigation scheduling is based on turfgrass water requirements and turfgrass water use (Serena et al., 2020b). Turfgrass canopy water use is defined as the amount of water that is required for turfgrass growth and transpiration, together with water loss from the soil surface (also referred to as evaporation) (Beard, 1973). This amount is often reported as ET, it can be expressed in mm or inches per unit of time (day, week, month) and can be estimated directly by measuring soil moisture depletion by means of lysimeters or soil sensors, or indirectly through weather data. Climate parameters known to affect ET are radiation, air temperature, humidity and wind speed (Allen, 1989).

Irrigation practices have relied heavily on weather-based reference evapotranspiration (ET_o) calculation (Wherley et al., 2015). Estimation of ET_o is obtained using weather stations and the Penman–Monteith equation (Allen, 1989). More recently, a different standardized equation was proposed for short crop reference evapotranspiration (ET_{os}) (Allen et al., 2005; Romero and Dukes, 2016). However, replenishing full ET_o volumes may be cost prohibitive for the vast majority of turfgrass managers. Turfgrass ET (crop ET, ET_c) is usually lower than ET_o, hence a crop coefficient (K_c, where $K_c = ET_c/ET_o$) is usually applied to ET_o to estimate irrigation volumes (Colmer and Barton, 2017). Just as ET_c rates are extremely site specific, turfgrass K_c vary greatly among species, and even among cultivars within the same species (Wherley et al., 2015). Moreover, K_c could be affected by turfgrass management practices, hence, providing guidelines and K_c suggestions valid throughout the United States is extremely complicated. Comparing K_c of hybrid bermudagrass (*Cynodon dactylon* [L.] Pers. × *Cynodon transvaalensis* Burtt-Davy), zoysiagrass (*Zoysia japonica* Steud.), St. Augustinegrass (*Stenotaphrum secundatum* [Walter] Kuntze) and bahiagrass (*Paspalum notatum* Flugge) in Florida, Wherley et al. (2015) noted that K_c values fluctuate throughout the year, ranging from 0.8 during the growing season, to 0.3 in the fall, and therefore should be adjusted depending on growth potential and precipitation. However, since growing and dormancy periods vary between regions, these K_c values are hardly transferrable to other regions or climatic conditions.

Irrigating below ET_c level, referred to as deficit irrigation, has been proposed as a viable irrigation strategy for turfgrass swards since these areas are maintained for aesthetics and functionality rather than yield (Schiavon et al., 2019). Usually, K_c are developed under well-watered conditions (Romero and Dukes, 2016), however, there are certain degrees of drought stress that turfgrasses can tolerate. Deficit irrigation could be further implemented in areas with abundant precipitation. For example, bermudagrass could be maintained in Florida replacing as little as 15% ET (Wherley et al., 2014). Moreover, traffic may exacerbate detrimental deficit irrigation effects even in areas with abundant precipitation, where a K_c of 0.3 could be sufficient to maintain hybrid bermudagrass summer quality (Hejl et al., 2016). In more arid climates, other management practices such as use of soil surfactants or appropriate fertilization need to be implemented when deficit irrigation is used to irrigate turfgrass (Schiavon et al., 2019, 2021).

On the market, there are currently several irrigation controllers with two types of ET measurement functionality: one that connects to a network of weather stations and adjusts irrigation amount based on historic ET, and another with built-in sensors to measure weather parameters and correct irrigation time based on actual ET at the location (Grabow et al., 2013). Since the introduction of these ET-based irrigation controllers, several studies have documented a reduction in water consumption. Average irrigation reduction from using ET-based controllers across published studies in the last decade was 15%, however water savings are difficult to estimate as they are heavily depending on environmental conditions, turfgrass species, management intensity, amount of irrigation without ET sensors and human interaction with technology, as turfgrass managers could override the smart controllers and use irrigation on manual mode (Dukes, 2020). Moreover, when measuring the efficiency of ET-based controllers, the addition of rain sensors could further reduce irrigation, especially in environments where irrigation supplements rain (Dukes, 2020). Over a 15-month period, a 43% water savings was achieved in Florida landscapes consisting of a mix of 'Floratam' St. Augustinegrass and mixed ornamentals (Davis et al., 2009), as well as anywhere between 23% and 63% water savings in Florida lawns (McCready et al., 2009). Controversial results were presented by Grabow et al. (2013) in a study conducted in North Carolina, where ET-based controllers used 11% more water than time-based irrigation when irrigating tall fescue (*Schedonorus arundinaceus* [Schreb.] Dumort). Similarly, only two out of five ET-based controllers were able to reduce irrigation by 53% and 34% compared to a manual controller programmed to deliver 80% of historical ET_o when irrigating tall fescue turfgrass in a clay-loam soil in New Mexico (Al-Ajlouni et al., 2012); controversially, the other three sensors applied water in excess compared to the manually programmed controller. In Las Vegas, NV, Devitt et al. (2008) reported a 20% water savings in the same species. More recently, newer technologies were tested. Serena et al. (2020b) reported water savings from ET-based controllers versus constant irrigation runtime of 38% in tall fescue, and 42% in common bermudagrass (*Cynodon dactylon* [L.] Pers.). Cardenas et al. (2020) reported water savings from 51% to 63% when using ET-based technology, and from 28% to 63% when using smartphone app to control irrigation.

Another technology to control irrigation is based on soil moisture sensors (SMS), introduced about 40 years ago (Snyder et al., 1984). More recently, with the introduction of time domain reflectometry or time domain transmissometry, and frequency domain reflectometry or capacitance sensors, this technology resulted in a more accurate and inexpensive tool for landscape irrigation (Blonquist et al., 2006; Muños-Carpena et al., 2005). Similar to ET-based controllers, water savings from the use of SMS have been documented numerous times. For instance, in the humid subtropical environment of Florida, Cardenas-Lailhacar and Dukes (2010) documented water savings from 40% to 88% in bermudagrass swards during the rainy season, as well as savings of 16% to 83% during the dry or low rainfall season (Cardenas-Lailhacar and Dukes, 2010). Similar results were observed in Kansas, where SMS sensors were able to reduce irrigation compared to standard homeowner (3 days week[1]) frequency-based irrigation by 31–71% on tall fescue grown in a silty clay loam depending on yearly precipitations (Chabon et al., 2017). Similarly, the implementation of SMS was reported to reduce irrigation water usage in lawns in North Carolina (Grabow et al., 2013), Utah (Blonquist et al., 2006) and in Quebec, Canada (Fazackerley and Lawrence, 2010), as well as in drier regions of Western Australia (Pathan et al., 2007), New Mexico (Serena et al., 2020b), Nevada (Devitt et al., 2008), and California (Haghverdi et al., 2021). More recently, Sandor (2018) reported water savings from 61% to 72% when using SMS technology (e.g. Toro Precision Soil Sensor [Toro Company, Bloomington, MN, USA] and Rain Bird SMRT-Y [Rain Bird Corp., Azusa, CA, USA]) in a multiyear study conducted on bermudagrass in Fayetteville, AK. Additionally, Serena et al. (2020b) reported an average irrigation saving of 44% in tall fescue, and 39% in bermudagrass in a study conducted in the arid climate of New Mexico. The future of SMS would be the development of an adaptive irrigation system, capable of measuring soil root zone water content and salinity and automatically actuate water amount needed to keep turfgrass healthy and to remove completely human interaction from the irrigation operation (Weiss, 2020).

In general, it has been demonstrated numerous times and in under several climatic conditions that both ET-based and SMS technologies can provide significant water savings from turfgrass irrigation. Currently, the smartphone and apps (Fig. 2) have made irrigation more accessible and easier to use for the turfgrass practitioner and homeowner (Cardenas et al., 2020). A list of smartphone friendly controllers that can use smartphones as an interface can be found at: https://www.irrigation.org/SWAT/Smart-Systems/Smartphone-friendly_irrigation_controllers/SWAT/Smart-Systems/Smartphone-friendly_irrigation_controllers.aspx. However, in many situations, homeowners are not willing to implement the use of these technologies unless promoted by water conservation programs through rebates and discounts (Khachatryan et al., 2019). The use of irrigation sensors have been linked with customer satisfaction, however, studies showed that if users experience any challenges with the technology satisfaction decreased drastically (Dukes, 2020). Combining studies investigating ET-based sensors and SMS, smart controllers provided 51% water savings in experimental plots, and 30% water savings in studies conducted in homes and commercial landscape (Dukes, 2020). Research has shown that smart controllers could be a powerful tool to decrease irrigation amounts in a wide range of climates and soils, however, if compared to deficit irrigation water savings may be negligent (Dukes, 2020).

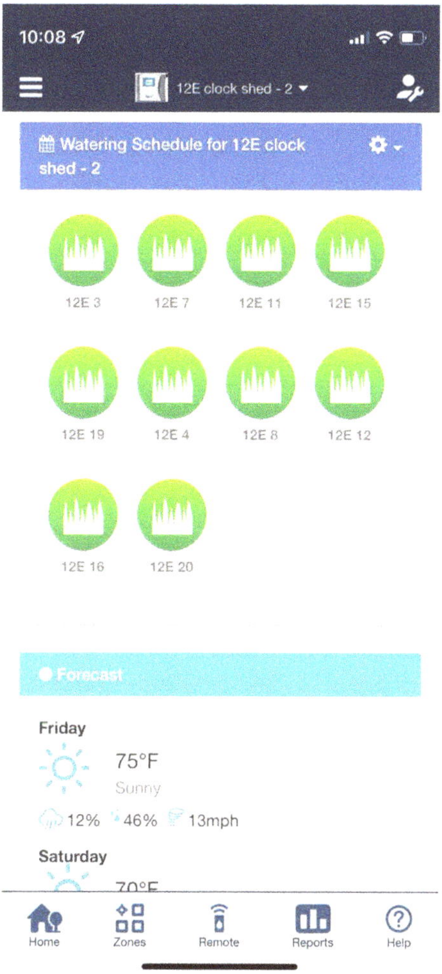

Figure 2 Example of a smartphone-friendly controller to remotely operate an irrigation system. Using the controller from the app, it is possible to schedule, or manually turn on, the irrigation system and also monitor weather conditions.

3 Advances with sprinkler devices and technologies

When it comes to turfgrass irrigation, many options are available to supplement water when needed. Some options for homeowners include flood irrigation, soaker hoses, subsurface drip, single sprinkler, and pop-up sprinklers (Leinauer and Smeal, 2012). In parks, sport fields and golf courses, the standard irrigation is the pop-up overhead sprinkler system, single head hoses and occasionally subsurface drip to supplement water to problematic areas (Leinauer and Smeal, 2012). Sprinkler irrigation has been used to supplement turfgrass with water since 1894 (Connolly, 2001). Since then, technology has evolved, in particular with the introduction of plastic materials. However, one of the more recent innovations was in 2002, with the introduction of MP rotators from Hunter Industries (San Marcos, CA). Unlike traditional sprinklers, MP rotators are composed of a multi-trajectory rotating streams delivering water at a steady rate. A study conducted in Egypt on seashore paspalum showed higher distribution uniformity (DU) coefficient for MP rotator sprinkler compared to spray nozzles which also lead to higher moisture content (Eid et al., 2019). Another study comparing four common types of sprinkler heads found that the (DU) varied greatly, depending on nozzle size and operating pressure (minimum, optimum and maximum) (Balcı and Orta, 2018). These results show the importance of a properly designed and maintained sprinkler irrigation system, with a frequent check on operating pressure. Recently (2019), California has introduced a government mandate on new installation requiring use of a pressure-regulated sprinkler body, to assure proper operating pressure at the head (https://www.nrdc.org/experts/ed-osann/ca-sets-standards-new-lawn-sprinklers-curb-waste). Some emerging technology is the use of low-pressure operating sprinklers, as they have demonstrated better water distribution and are less affected by wind (Li et al., 2019a). More recently, irrigation companies have introduced new sprinkler nozzles providing a streaming spray pattern (Fig. 3), with larger size droplets, higher precipitation rates and more uniformity (Molle et al., 2012). Very little research is available on different sprinkler irrigation technologies, and very little has changed in terms of sprinkler irrigation in the last 20 years. Most irrigation companies have focused on increasing sprinkler reliability,

Figure 3 Example of an advanced sprinkler nozzle providing a streaming spray pattern, for increasing precipitation rate and uniformity.

efficiency and ease of use and repair. Recent innovations have been in irrigation controllers, and communication and control devices associated with cell phone technology. The willingness of customers to pay the cost of upgrading irrigation technologies (Khachatryan et al., 2019), despite these new technologies having a payback time from zero to 12 months is unclear (Cardenas et al., 2020).

4 Subsurface drip irrigation

Sprinkler systems have been widely criticized due to their inefficiency and lack of consistent water DU (Schiavon et al., 2014b). Water applied to turfgrass stands by means of sprinklers is often subjected to overspray, runoff, wind drift and evaporation. For these reasons, several research groups across the United States have investigated SDI over the past 20 years as an alternative to irrigate turfgrass areas. Drip irrigation consists of plastic tubing equipped with emitters installed below the soil surface (Burt and Styles, 1999). With SDI, water is delivered directly into the root zone and the aforementioned issues that may occur with sprinkler irrigation is avoided. These SDI systems have been recommended for use in golf courses (Thompson, 2019), residential lawn and parks (Leinauer and Devitt, 2014) and occasionally mandated by public utilities due to their ability to conserve water in turfgrass systems (California Department of Water Resources, 2009).

Within the last decade, research has focused on establishing turfgrass from seed using subsurface irrigation. Seed germination and establishment delays may result from using SDI compared to sprinkler systems in sandy soils (Schiavon et al., 2012, 2013a). When seeded tall fescue and Kentucky bluegrass (*Poa pratensis* L.) were established under sprinkler irrigation or SDI in combination with either potable (0.6 $dS \bullet m^{1}$) or saline (2.9 $dS \bullet m^{1}$) water, sprinkler irrigation achieved faster and greater turfgrass cover by the end of the growing season compared to SDI in a NM sandy loam (Schiavon et al., 2013a). In a recent study conducted near Manhattan, KS, tall fescue was able to complete establishment faster with SDI compared to overhead sprinkler irrigation and aboveground drip irrigation in fine-textured soils (Bach et al., 2022). A similar study investigating the establishment of bermudagrass and seashore paspalum (*Paspalum vaginatum* O. Swartz) from seed irrigated with the same water qualities from SDI or sprinkler, resulted in slower establishment of SDI (Schiavon et al., 2012). Nevertheless, establishment of warm-season species was successful by the end of the growing season when dormant seeding was used in combination with SDI (Schiavon et al., 2012). Establishment of bermudagrass from sod in similar dormant conditions was also investigated, and similar results were obtained (Serena et al., 2014). Hence when establishing warm-season turf through SDI in the transition zone, anticipating seeding or sodding date is recommendable.

Other studies investigated the use of fertilizers on warm-season turf grasses irrigated from SDI (Serena et al., 2018a), or the use of soil surfactants and PGRs under deficit irrigation (Schiavon et al., 2014a). In a 3-year study, it was demonstrated that SDI can sustain acceptable warm-season turfgrass quality under deficit irrigation when compared to sprinkler irrigation (Schiavon et al., 2014a). Acceptable turfgrass quality of bermudagrass and seashore paspalum irrigated by means of SDI was also achieved with 90% less water than the ET_o demand over a 2-year period, however, unacceptable turfgrass quality was observed starting in the third year (Serena, 2019). The use of SDI also was investigated for other warm-season species, including zoysiagrass, manilagrass (*Zoysia matrella* [L.] Merr.) and buffalograss (*Buchloe dactyloides* [Nutt.] Engelm), comparing natural precipitation versus supplemental irrigation in Texas. Higher turfgrass quality was observed when supplemental irrigation was provided compared to natural rainfall only (Culpepper et al., 2020). An ongoing study on tee height creeping bentgrass (*Agrostis stolonifera* L.) has shown no difference in turfgrass quality compared to sprinkler irrigation (Leinauer, unpublished data). Limited research is available on the effectiveness of SDI to reduce disease occurrence. However, with SDI there is no contact between water and leaf surfaces resulting in decreased leaf wetness, and thus a less favorable environment for foliar pathogens (Bell et al., 1998). Preventing

Figure 4 Installed SDI (Netfim) in a tee box on 17 July 2017, at The Club at Las Campanas (Santa Fe, NM, USA). Source: Image courtesy of B. Leinauer.

contact between water and leaf surfaces could have potential benefits when wastewater is used as a source of irrigation because of the higher salt content that could burn the leaves (Taylor et al., 2006).

The use of SDI can easily lead to 30% water savings as compared to traditional sprinkler irrigation since the water is applied directly in the root zone (Serena et al., 2020a). Moreover, other advantages come from using SDI, including less disease pressure, fewer weeds, less soil compaction and fewer repairs needed compared to sprinklers. However, the higher installation costs limit its application on some turfgrass sites. In addition, cultivation management practices that may occur in golf courses, such as deep core aerification, could not be performed without damaging the drip system. However, some other disadvantages that were exposed in the past, such as clogging caused by root intrusions, have been addressed by the irrigation industry and do not pose a concern for SDI use today. Today, the Toro Company coats their drip line with a herbicide, and Rainbird includes a copper plate inside each emitter, to prevent root intrusion. Other companies such Netafim (Fig. 4) and Hunter created a physical barrier (i.e. check valve) in front of every emitter to prevent root intrusion and sediment deposit.

5 Alternatives to potable water

Water scarcity resulting in drought could impact the use of clean drinking water in arid and semi-arid regions for turfgrass landscape irrigation. As a result, utilization of low-quality recycled or reclaimed wastewater, gray water or even seawater for irrigation is becoming more common at turfgrass facilities such as SOD growers, golf courses, municipal parks and stadiums (Sevostianova and Leinauer, 2014). Salinization issues are an issue in arid and semi-arid regions, and also in cold weather climates where deicing salts with high sodium (Na) content are part of winter management practices on highways, roads and sidewalks, and in coastal areas where freshwater aquifers are being polluted by seawater intrusion (Marcum, 2006).

Salinity in irrigation water is the measurement of soluble salts content and is expressed as electrical conductivity (EC_w). Irrigation water is considered low in salinity if EC_w is lower than 0.25 $dS\bullet m^{1}$, medium if it ranges from 0.25 $dS\bullet m^{-1}$ to 0.75 $dS\bullet m^{1}$, high from 0.75 to 2.25, and high if EC_w is higher than 2.25 $dS\bullet m^{1}$ (US Regional Salinity Laboratory, 1954). Good practices to lower EC_w include blending saline water with fresh water, or alternating saline and fresh water irrigation events (Marcum, 2006). However, some turfgrass managers do not have fresh water sources available for blending or alternating and are forced to use saline water year-round. Soils are classified as saline if soil electrical conductivity (EC_e) is higher than 4.0 $dS\bullet m^{1}$ (US Regional Salinity Laboratory, 1954). Not only are salts (e.g. $NaCl$, Na_2SO_4, $CaCl_2$, $MgCl_2$, KCl) applied to soil through irrigation, but soil salinity is directly affected by soil water content as well. As water evaporates from the soil surface, capillarity forces water to rise from lower depths accumulating additional salts in the root zone (Marcum, 2006). High salinity levels are common in recycled and groundwater sources specially in arid areas and are responsible for lowering the osmotic potential of soil water, thereby reducing the amount of water available to the plant, and potentially inducing physiological drought stress with consequent loss of turfgrass quality and functionality (Sevostianova and Leinauer, 2014).

Applying a leaching fraction (LF), consisting of applying supplemental water above regular evapotranspirational needs, is the most common management practice to alleviate salinity stress (Schiavon and Baird, 2018). By depositing salt deeper in the soil below the turfgrass root zone, salinity issues can be alleviated and prevented (Schiavon et al., 2017). Currently LF for salinity affected areas has been calculated as (Rhoades, 1974):

$$LF = \frac{ECw}{(5 \times ECe) - ECw}$$

where EC_e is the turfgrass species salinity threshold as measured on a soil saturation extract, and EC_w is the salinity level of the irrigation water. Numerous studies have investigated the salinity threshold for cool- (Alshammary et al., 2004; Friell et al., 2013; Koch et al., 2011; Marcum, 2001) and warm-season (Chavarria et al., 2021; Marcum and Pessarakli, 2006; Qian et al., 2007) turfgrasses, all concluding that warm-season species are more salt tolerant than cool-season species. Choosing the best adapted turfgrass species and cultivars plays a fundamental role for successful abiotic stress management in general and plays a large role in coping with salinity stress. Halophytic species like *Puccinellia* spp. (alkali grasses) or inland saltgrass (*Distichlis spicata* L.) have been shown to possess the greatest salinity tolerance among turf-type grasses (Marcum, 2006), however, their potential to produce high quality turf surfaces is yet to be determined (Sevostianova et al., 2011a,b).

Most studies that investigated salinity thresholds among turfgrasses were conducted under controlled environmental and greenhouse conditions. Furthermore, using different criteria (e.g. shoot or root weight, weight reduction, visual injury, photosynthetic activity, etc.) to assess salinity tolerance results in different study outcomes. Therefore, salinity thresholds from the same cultivars can vary greatly depending on study conditions (Schiavon and Baird, 2018). For these reasons it has been suggested that absolute salinity tolerance values of different turfgrass genotypes cannot be established (Marcum, 2006).

Under field conditions, turfgrass is subjected to multiple abiotic stressors such as drought, cold, or heat, which may exacerbate the effects of salinity stress. Sevostianova et al. (2011b) demonstrated that warm-season species such as bermudagrass, inland saltgrass, seashore paspalum, and zoysiagrass can maintain acceptable summer quality if watered by means of overhead sprinkler irrigation or SDI with EC_w as high as 3.5 dS•m^1 for multiple years when irrigation is applied at 110% ET. Conversely, in a multi-year study conducted in inland California, Schiavon et al. (2017) demonstrated that the LF equation underestimates perennial ryegrass (*Lolium perenne* L.) requirements for leaching in a Mediterranean climate by 20%, and that summer heat stress contributed to loss of turfgrass quality just as much as salinity. In a three-year study conducted in New Mexico comparing tall fescue, alkaligrass (*Puccinellia distans* L. Parl), perennial ryegrass and fine fescue (*Festuca rubra* L.) watered at 120% ET with either 0.6 dS•m^{-1}, 2.0 dS•m^{-1}, or 3.5 dS•m^{-1}, only tall fescue maintained acceptable quality at the highest salinity level (Sevostianova et al., 2011a). Given the importance of leaching salts from the root zone, and the close relationship that soil moisture has with salinity, uniform water application is crucial for the controlled leaching of salts. Therefore, the need for sprinkler system maintenance and efficiency becomes even more essential to provide uniform leaching. Uneven distribution of root zone soil salinity as a result of poor sprinkler DU has been documented and it has been suggested to exacerbate salinity issues in urban areas (Ganjegunte et al., 2017).

Nevertheless, in areas where ET rates are extremely high during the summer months, applying a LF may be impossible for turf managers as irrigation run times may be too long to maintain playable and functional turf surfaces. In these situations, deficit irrigation may not provide sufficient leaching from the root zone, exacerbating salinity stress caused by increase of salts near the soil surface (Schiavon and Baird, 2018). Ganjegunte et al. (2017) demonstrated that even with relatively low EC_w (1.07 dS•m^1) if irrigation volumes do not match ET demand, soil salinity increases under arid conditions as the result of evaporative concentration of salts in the soil. If availability of saline water is insufficient to cover ET and LF needs, turf managers should choose salt and drought tolerant species like bermudagrass or seashore paspalum. Schiavon et al. (2014a,b) demonstrated that these species can sustain sufficient quality when irrigated at 50% ET with EC_w 2.3 dS•m^{-1}. The same results were observed in California when Tifway bermudagrass was watered at 75% ET with EC_w 4.4 dS•m^{-1} (Schiavon and Baird, 2018). Moreover, although in desert areas, natural rainfall during either the growing season or the dormancy period are limited, they are sufficient to leach salts from the root zone, even when turfgrass is irrigated with SDI that increase soil salinity levels near the soil surface (2014a,b). In areas where irrigation volumes are not sufficient to provide a LF, other practices that improve soil physical properties that enhance leaching (e.g. sand capping, topdressing, aerification) should be undertaken (Schiavon and Baird, 2018).

Not only can high salinity levels in irrigation water induce turfgrass physiological drought, but can also affect nutrient uptake. Chang et al. (2019)

demonstrated that fairway bermudagrass N uptake increased when EC_w was up to 5 dS•m^{-1} compared to reverse osmosis (0 dS•m^{-1}) and sodic potable (EC<1 dS•m^{-1}) water but decreased significantly when EC_w reached 10 dS•m^{-1}. Schiavon and Baird (2018) suggested that supplemental foliar N application above regular fertilization practices and the use of biostimulants could improve bermudagrass quality irrigated with EC_w of 4.4 dS•m^{1}. When comparing different N sources, ammonium sulfate produced lower quality of 'Champion' bermudagrass irrigated at 2.5 dS•m^{-1} compared to urea and potassium nitrate, even though N recovered in shoots were similar for the three fertilizers (Chang et al., 2020a).

High EC_w also hinders seedling germination along with establishment delays. Several greenhouse and growth chamber studies have reported lower seed germination rates for cool- and warm-season turfgrasses (Serena et al., 2012a; Zhang et al., 2011). Even germination of halophytes such as inland saltgrass could be inhibited by high EC_w (Shahba et al., 2008). Nevertheless, when experiments were conducted under field conditions using EC_w similar to those encountered by turfgrass managers, salinity did not appear to delay establishment of bermudagrass or seashore paspalum (Schiavon et al., 2012; Serena et al., 2012b). Lower percent area ground cover was observed after 200 days in tall fescue established with groundwater (EC = 2.8 dS•m^{-1}) compared with potable water irrigation (potable EC = 06 dS•m^{1}), even though days to reach 75% ground cover were similar with both water qualities. Comparing establishment of three cool-season (Kentucky bluegrass, tall fescue, perennial ryegrass) and two warm-season (bermudagrass and seashore paspalum) turfgrasses seeded in the spring and irrigated with potable (EC_w = 0.6 dS•m^{1}) and saline (EC_w = 2.3 dS•m^{1}) water, Serena et al. (2012b) found that turf irrigated with non-potable water established 26 days faster. The only other report of slower seed germination and establishment when watered with saline (EC_w = 2.3 dS•m^{1}) water instead of potable water is Kentucky bluegrass (Wilhelm et al., 2010).

Salinity in irrigation water is hardly caused by a single salt. Turfgrass stands have experienced negative impacts due to gradual accumulation of salts deposited, particularly because of excess sodium (Na) at the soil surface moving into the turfgrass root zone. The Na disrupts soil physical properties by displacing calcium (Ca) and magnesium (Mg) on the negatively charged sites of soil colloids, which leads to dispersed soil particles or deflocculation that contributes to loss of soil structure, an increase in soil compaction, a reduction of soil permeability, and anaerobiosis (Marcum, 2006). With soil compaction, which is exacerbated in trafficked areas, saturated hydraulic conductivity can decrease drastically, eliminating leaching potential and accelerating salinity built-up in the root zone. These problems are more frequent in fine texture soils that are more susceptible to deflocculation than sands (Marcum, 2006). Sodium status of water and soil is reported as sodium adsorption ratio (SAR) and it is calculated as:

$$SAR = \frac{[Na^+]}{\sqrt{\frac{[Ca^{2+}]+[Mg^{2+}]}{2}}}$$

where Na, Ca and Mg concentration are determined from saturated paste extract. When SAR ranges from 0 to 10 irrigation water is classified low for Na hazard; values from 10 SAR to 18 SAR are classified medium, and from 18 SAR to 24 SAR are classified high (US Regional Salinity Laboratory, 1954). According to Duncan et al. (2000), when SAR of irrigation water is > 6 soil structural deterioration problems begin to arise. Sodium permeability hazard starts, assuming medium hazard irrigation water quality, when soil SAR is > 2 and becomes severe when soil SAR > 12 (Carrow and Duncan, 2012). Soils that have EC_e < 4.0 and SAR > 12 are classified sodic, while soils with EC_e > 4.0 and SAR > 12 are classified saline-sodic (US Regional Salinity Laboratory, 1954). All sodic soils have permeability issues, except for very sandy soil. Absolute concentration of Na in water analysis has been given too much importance by turf managers, as it is not the presence of Na in irrigation water that is important, but its balance and interaction with Ca and Mg. In fact, Ca is the most important ion responsible for soil structure due to its double ionic charge that attracts contiguous soil colloids (Marcum, 2006). Magnesium possesses similar properties, although to a lesser extent than Ca.

While chemical amendment of saline soil is not needed, and gypsum-based products are shown not to lower EC_e in golf course fairway root zones (Li et al., 2019b; Schiavon and Baird, 2018), Ca applications are needed to displace Na in sodic and saline-sodic soils. Thanks to its low application cost and high accessibility, gypsum has been the preferred amendment to reduce Na permeability hazard by turfgrass managers (Schiavon and Baird, 2018). Light and frequent application of gypsum with typical rates of 22 kg•93 m^2 have been preferred to heavy single applications to remediate Na-affected soils (Marcum, 2006). In those situations, gypsum application should continue until exchangeable sodium percentage drops below 15% (Marcum, 2006).

Bicarbonate (HCO_3^-) and carbonate (CO_3^{2-}) that are often present in recycled, reclaimed water and well water, and can react with Ca and Mg to form insoluble calcium carbonate ($CaCO_3$) or magnesium carbonate ($MgCO_3$), hence precipitating Ca and Mg, via the reactions:

$$CO_3^{2-}\left(\text{or } H^+\_CO_3^{2-}\right)+Ca^{2+} \rightarrow CaCO_3, \text{ or less frequently}$$
$$CO_3^{2-}\left(\text{or } H^+\_CO_3^{2-}\right)+Mg^{2+} \rightarrow MgCO_3.$$

When Ca and Mg are bound, Na will attach to negatively charged sites of soil colloids inducing Na permeability hazard, resulting in lime deposition near the soil surface also called caliche. The caliche layer formation is more typical of arid zones, it is impermeable and prevents leaching of salts. Bicarbonate concentration is low at 0–120 mg•L[1], moderate at 120–180 mg•L[1], severe at 180–600 mg•L[1] or very severe when higher than 600 mg•L[1]. Moreover, high presence of bicarbonates dissolved in irrigation water could raise water pH causing alkalinity (pH > 8.0), and when present in extremely high concentrations could form deposits on turfgrass leaves. When high levels of bicarbonates are present in irrigation water, the result is an increase in Ca and Mg availability. To account for their influence in irrigation water when bicarbonate concentration is > 120 mg•L[1], adjusted SAR has been proposed (adj SAR). In this formula Ca concentration is substituted by Ca_x which is a factor that comes from a table of HCO_3^- versus EC_w (Ayers and Westcot, 1994).

Correction of high bicarbonate levels can be obtained by acidification of water, typically using sulfur or sulfuric acid (H_2SO_4) that react with bicarbonates forming carbon dioxide. Sulfuric acid also has the potential to react with soil calcium carbonates forming gypsum via the following reaction:

$$H_2SO_4 + CaCO_3 \rightarrow CaSO_4 + H_2O + CO_2.$$

Nevertheless, injection of sulfuric acid into water to obtain acidification is a specialized practice, and it should be done at a low rate to neutralize 80% ca. of bicarbonates to avoid a sudden drop of water pH (Marcum, 2006). Turf managers should consult with a specialized laboratory before deciding if acidification is required as well as the application rate.

6 Turfgrass management practices

Several management practices have been investigated to decrease water needs of turfgrass, including mowing, fertilization, soil surfactants, and PGRs. Mowing is the most frequently implemented cultural practice for turfgrasses. Higher mowing heights translate into higher turfgrass ET rates due to larger leaf areas, which expose more stomata (Fry and Huang, 2004). Previous research documented a 15% increase in water use for Kentucky bluegrass when mowed at 5.1 cm compared to 2 cm (Feldhake et al., 1984). Similarly, annual bluegrass and creeping bentgrass increased their ET rates when mowing height was increased from 6 mm to 12 mm (Fry and Butler, 1989). Nevertheless, since higher mowing height translates to a deeper root system in turfgrass, increasing the mowing height is linked to enhanced drought tolerance (Fry and Huang, 2004). Reducing mowing height combined with deficit irrigation resulted in lower root length and lower total root mass in seashore paspalum (Shahba et al., 2014). Low mowing height also reduce tolerance to abiotic stress such as drought and salinity by reducing turfgrass energy reserves (Shahba, 2010). Increasing mowing height has been shown to increase drought tolerance and salinity resistance of several seashore paspalum and bermudagrass cultivars by increasing leaf photochemical efficiency, total nonstructural carbohydrates, proline content, K and Ca under deficit irrigation and saline conditions (Elansary and Yessoufou, 2015; Shahba, 2010; Shahba et al. 2012, 2014). Higher mowing height has been reported to increase salinity tolerance in creeping bentgrass as well (Fu et al., 2005). In fact, creeping bentgrass mowed at 25.4 mm retained turfgrass quality at higher EC_w compared to 6.4 mm (Fu et al., 2005). Moreover, when grass was mowed at 6.4 mm root growth decreased significantly with increasing EC_w (Fu et al., 2005). Mowing height also influences drought-stressed turfgrass performance when subjected to traffic. In a study investigating two cool-season (Kentucky bluegrass and perennial ryegrass) and two warm-season turfgrasses (buffalograss and zoysiagrass) response to drought when subjected to golf cart traffic, Braun et al. (2021) observed that green cover and turfgrass quality loss was more pronounced at rough (5.6 cm) compared to fairway (1.6 cm) height of cut after 41 days without irrigation. Nevertheless, while warm-season species maintained sufficient quality during the drought period, cool-season species took 5 days to return to acceptable quality level after irrigation was restored when mowed at fairway height, while rough height took 7 days (Braun et al., 2021). The authors attributed better turfgrass performance at the lower mowing height to potential reduced water loss from the turfgrass canopy, since the effect of mowing height on root morphology and soil characteristics were negligible (Braun et al., 2021).

Nitrogen (N) fertilization is the single management practice with the greatest impact on turfgrass color, shoot density and growth (Rodriguez et al., 2002). As explained previously, enhanced turfgrass growth often translates to increased ET rates (Feldhake et al., 1984). Higher ET rates were observed in kikuyugrass [*Cenchrus clandestinus* (Hochst. ex Chiov.) Morrone] fertilized at 150 kg N•ha^{1} year^{1} compared to unfertilized treatments (Barton et al., 2009). Similar results were observed in bahiagrass and St. Augustinegrass although reducing N rates from 294 kg N•ha^{1} year^{1} to 98 kg N•ha^{1} year^{1} resulted in a maximum of 8% water use reduction likely due to reduced leaf area and reduced clipping yield (McGroary et al., 2011). Conversely, Ebdon et al. (1999) found that only extremely high N rates (e.g. 588 kg N•ha^{1} year^{1}) substantially increased ET of Kentucky bluegrass. Moreover, these N rates are usually too high, and their use would not be economically substantiated by turfgrass managers. The same study investigated the effect that macronutrients have on ET when applied alone or in combination. When N and phosphorous (P) were applied at high rates (294 kg N•ha^{1} year^{1} and 43 kg P•ha^{1} year^{1}), ET increased at increasing levels of applied potassium (K). However, applications of K were beneficial when N and P were applied at recommended rates for Kentucky bluegrass grown on a fine sandy loam (Ebdon et al., 1999). Impact of N sources on turfgrass water use have also been documented with slow-release fertilizers (i.e. polymer-coated sulfur-coated urea) reducing water consumption and increasing water use efficiency of St. Augustinegrass compared to ammonium sulfate (Saha et al., 2005).

Newer research suggests that N fertilization could mitigate the effect of water stress. In a study conducted in southern California on 'Princess 77' bermudagrass, Schiavon et al. (2019) documented that sufficient N fertilization during the growing season (25 g N•m^{-2} year^{-1}) applied through either organic N sources from seaweed extracts or other N sources containing humic acids can sustain bermudagrass quality and enhance its fall color retention when turf is watered at 40% ET_o throughout the season. Conversely, N sources with high burning potential such as calcium nitrate provided the least desirable results on deficit-irrigated bermudagrass (Schiavon et al., 2019). These results were attributed to high air temperatures during fertilizer application and a lack of sufficient water to dissolve the fertilizer (Cockerham and Minner, 2011). However, fertilizer source does not seem to have an effect on well-watered bermudagrass and seashore paspalum quality or their starch and total nonstructural carbohydrates content when turf is irrigated with non-potable groundwater (EC_w = 3.1 dS•m^{1}) as long as N rate is kept at least at 20 g N•m^{-2} year^{-1} (Serena et al., 2018a,b).

Biostimulants such as seaweed extract or humic acids also have been effective with mitigation of salinity stress in bermudagrass (Schiavon and Baird, 2018), and of drought stress as well as other abiotic stress such as heat by enhancing root biomass and root length and response to oxidative stress in other cool-season species of creeping bentgrass and tall fescue (Zhang and Schmidt, 2000; Zhang and Ervin 2004; Zhang et al., 2021). However, research involving biostimulants to alleviate abiotic stress is very limited in turfgrass settings, and it still unclear if the benefits are related to the additional N content or the physiological benefits of these products.

Nevertheless, excessive N fertilization increases maintenance costs for turfgrass managers and may have negative impacts on the environment causing surface and groundwater N contamination with subsequent water quality degradation (Petrovic, 1990; Galloway et al., 2008). These concerns are exacerbated in rapidly growing urban and suburban areas, especially coarse texture soils that have poor water holding capacity. Nevertheless, several studies demonstrated that mature, healthy and dense turf when managed properly is able to intercept N from fertilization practices (Fig. 5), hence serves as a filter against N leaching (Erickson et al., 2001; Telenko et al., 2015; Trenholm et al., 2012), and that concerns for nitrate leaching comes mostly from newly established sites (Geza et al., 2021; Trenholm et al., 2013), use of tailor reclaimed water for irrigation (Geza et al., 2021), or overirrigation (Sun et al., 2021). Conducting a watershed-scale assessment of landscaping practices, Bachman et al. (2016) found that moderately managed turf fertilized at 60 kg N•ha^{1} year^{1} reduced total N concentrations in the watershed compared to a meadow watershed that received no N fertilization, thus concluding that a turfgrass root system can reduce nutrient runoff and increase nutrient retention.

7 Use of plant growth regulators

Two types of PGRs are available to turf managers, and traditionally described as inhibitors of cell division (Type I), and inhibitors of gibberellic acid (GA) biosynthesis (Type II) (Fry and Huang, 2004). Refer to Chapter 12 (Advances in plant growth regulation in turfgrass) for a more detailed description and review of PGRs used in turfgrass. Before the introduction of Trinexapac-ethyl [(TE) 4-(cyclopropylhydroxymethylene)-3,5-dioxocyclohexanecarboxylic acid; Primo Maxx, (Syngenta Professional Products, Greensboro, NC)] in the market in the 1990s, PGR use was not as common in the golf industry as it is today. Prior to TE, growth suppression was sometimes inconsistent, and undesired effects of discoloration or phytotoxicity may have appeared on turfgrass stands (Fry and Huang, 2004). However, since its introduction in the market, PGRs have been included as a management practice in golf greens and fairways primarily to decrease mowing frequency, decrease seedheads or control weeds (Fry and Huang, 2004). However, new research suggests that TE could be a powerful tool to enhance turfgrass resistance to drought. As a Type II PGR, TE inhibits the formation of GA1 from GA20 and stops the conversion of GA1 to GA8 (Elansary and Yessoufou, 2015; Rademacher, 2000). Positive effects of TE application to perennial ryegrass subjected to dry down, and tall fescue ET rates, have been reported since the mid-1990s (Jiang and Fry, 1998; Marcum and Jiang, 1997), and physiological mechanism enhancing turfgrass drought tolerance have been studied extensively in recent years. McCann and Huang (2007) documented that TE application not only increased quality of creeping bentgrass during 21 days of combined heat and drought stress, but also increased leaf relative water content (RWC) and chlorophyll content. Meanwhile, untreated controls maintained constant canopy photosynthetic rate and photochemical efficiency during the stress period, suggesting that the positive effects of TE could have been linked to greater photosynthetic activity and osmotic regulation (McCann and Huang, 2007). Subsequent research indicated the benefits of TE application to increase osmotic adjustment due to Ca and K and soluble sugars accumulation in leaves (Bian et al., 2009). These results were confirmed in Kentucky bluegrass as well, with TE-treated plants maintain turfgrass quality, soil volumetric water content and leaf RWC for a longer period than untreated plants when subjected to 28 days of drought (McCann and Huang, 2008).

Figure 5 The botanical garden lawn at Huntington Library (San Marino, CA, USA) on 6 September 2021. Dense, healthy turfgrass has been shown to intercept nitrogen from fertilization and prevent nutrient leaching.

Authors suggest that TE applications lowered water use at the early stages of the drought.

Early increase of plant hormones (i.e. auxin and salicylic acid) associated with drought tolerance as well accumulation of various proteins and metabolites, including ferritin, catalase, glutathione-S-transferase, Rubisco, heat shock protein 70 and chaperonin 81, as well as fatty acids (palmitic acid, α-linolenic acid, linoleic acid and octadecanoic acid) was confirmed in drought-stressed Kentucky bluegrass (Krishnan and Merewitz, 2015; Xu and Huang, 2012). Similar results were confirmed in perennial ryegrass when water was withheld for 45 days (Mohammadi et al., 2017).

Positive effects of TE application on drought-stressed turfgrass have also been documented in warm-season species. Effects similar to those documented in Kentucky bluegrass and creeping bentgrass were detected on deficit-irrigated seashore paspalum (Elansary and Yessoufou, 2015). When comparing bermudagrass irrigated at 100% versus 30% actual evapotranspiration (ETa), TE mitigated the effect of drought stress under deficit irrigation for 6 weeks on 'Tifway' bermudagrass (Hejl et al., 2015).

Of note, all these aforementioned studies were conducted under controlled greenhouse or growth chamber conditions. Nevertheless, enhancement of drought stress in bermudagrass and seashore paspalum was confirmed only during field studies conducted in arid environments. Schiavon et al. (2014a) found increased normalized difference vegetation index of bermudagrass, and increased turfgrass quality and fall color retention of seashore paspalum when watered at 50% ET for three consecutive summers. Authors also reported increase leaf chlorophyll and carotenoids content and greatest superoxide dismutase (SOD) activity associated with TE applications.

Newer research has investigated the effects of TE in combination with soil surfactants. Serena et al. (2018c) documented that at 50% ET replacement applications of a commercially available soil surfactant, TE, and the combination of both had higher turfgrass quality than untreated bermudagrass in New Mexico, while at 65% ET replacement TE application had the highest impact on turfgrass quality. Serena et al. (2018a) suggested that using a PGR on bermudagrass fairways could result in a 30% reduction in water use compared to current irrigation practices. Conversely, in a similar study conducted in California, Schiavon et al. (2019) noted that TE did not help maintain turfgrass quality when bermudagrass was irrigated at 40% ET compared to 70% ET replacement. It still remains unclear whether TE has positive repercussions on turfgrass root architecture. While some studies reported increased root mass on TE-treated bermudagrass and seashore paspalum, even when deficit irrigated or irrigated with saline water (McCullough et al., 2005; Serena et al., 2020a), other studies did not support these findings (Baldwin et al., 2006; Fagerness and Yelverton, 2001; Schiavon et al., 2019), and documented decreased total root length following TE application (Marcum and Jiang, 1997). It is possible that TE effects on turf root architecture may vary depending on turfgrass species, and irrigation and agronomic management practice.

The use of TE has also been proposed as a powerful tool to overcome salinity stress. Baldwin et al. (2006) suggested that TE-treated 'TifEagle' and 'Champion' bermudagrass can tolerate salinity thresholds twice as high as untreated plants, increasing Ca and Mg content in shoot tissue by up to 77%. Similar results were reported also when bermudagrass was irrigated with different water quality sources. In a study investigating response of bermudagrass irrigated with three different water sources (reverse osmosis EC = 0 $dS \bullet m^{-1}$, sodic potable EC<1 $dS \bullet m^{-1}$ and saline EC = 7.5 $dS \bullet m^{-1}$), Hejl et al. (2015) found that quality of turfgrass treated was always similar or higher than untreated counterparts. Schiavon et al. (2014b) documented increased amounts of leaf chlorophyll and carotenoids content in bermudagrass and seashore paspalum irrigated with potable (EC = 0.6 $dS \bullet m^{-1}$) and saline (EC = 2.3 $dS \bullet m^{-1}$) water at 50% ET replacement. Nevertheless, research conducted in salt-sensitive species of Kentucky bluegrass suggested that the benefits of TE on salt tolerance varies depending on application rate and salinity levels, and that positive effects of TE applications may disappear when salinity levels increase (Arghavani et al., 2012; Pessarakli et al., 2006).

Use of other PGRs for water management purposes has not been investigated as widely as TE. Mohammadi et al. (2017) compared drought response of perennial ryegrass treated with TE, paclobutrazol [(PAC, Type II PGR; (2RS,3RS)-l-(4-chlorophenyl)-4,4-dimethyl-2-(1H-1,2,4-tri- azol-l-yl) pentan-3-ol], and abscisic acid (ABA, plant hormone) and concluded that PGRs increase antioxidant activity of drought-stressed plants, and all exogenous compounds enhanced RWC and decreased the electrolyte leakage and hydrogen peroxide content. The positive TE effects on drought-stressed plants were more pronounced compared to PAC and ABA (Mohammadi et al., 2017). Enhancement of drought tolerance following ABA application has also been reported in tall fescue (Jiang and Huang, 2002) and bermudagrass (Lu et al., 2009), nevertheless, no ABA-based product is currently available for turfgrass practitioners.

8 Soil surfactants

Hydrophobicity, also known as soil water repellency, is the result of the accumulation of hydrophobic coatings on the surfaces of soil particles (Dekker et al., 2001b). As a consequence, soils not only resist wettability, but also lose the ability to retain water (Dekker et al., 2001b). These coatings are usually organic compounds that range from root exudates (Dekker and Ritsema, 1996), to plant-derived waxes and organic acids (Kostka, 2000), bioproducts of microbes and fungi (Fidanza et al., 2007) or decaying organic matter (Fidanza et al., 2020). Soil water repellency in turfgrass systems has been extensively studied in the last 20 years, however there is no scientific consensus on what appears to be the primary cause of soil hydrophobicity, even if fungi were proposed to be its main drive (Hallett, 2007). While initially this problem was thought to be recurring in a diversity of soils, it was particularly evident in sand-based root zones due to low surface area per unit volume (Dekker et al., 2005). Turfgrass areas that are established on engineered sandy soils such as golf greens, are easily susceptible to soil water repellency (Cisar et al., 2000), and as water content of hydrophobic soils decreases, soil becomes progressively non-wettable (Dekker et al., 2001a). Prolonged hydrophobic conditions result in decreased infiltration, also called 'water resistance', and retention, with creation of soil preferential flow patterns (defined as the movement of water through specific pathways localized only in a portion of soil rather than the whole profile), uneven water distribution and finally with the development of localized dry spots (LDS) in turfgrass stands with eventual loss of turfgrass quality and functionality (Dekker et al., 2019). Normally, lower water DU from rainfall and irrigation is observed in soil hydrophobic soils, as well as higher runoff which increase the risk of non-point source pollution, and increased need of irrigation requirements due to reduced access to water for the turfgrass (Moore et al., 2010).

Soil surfactants, also called wetting agents by the turfgrass industry, have been used as a viable strategy to manage soil hydrophobicity (Cisar et al., 2000; Kostka et al., 1997). Soil surfactant molecules typically are amphihilic consisting of both a nonpolar/lipophilic group and a polar/hydrophilic group (Lehrsch et al., 2011). When applied to water repellent soils, the hydrophilic group bonds to water molecules, disrupting their cohesive forces and decreasing interfacial tension between the hydrophobic and hydrophilic surface of soil particles. Soil surfactants used in the turfgrass industry are normally nonionic (neutrally charged). It is estimated that out of the almost 200 soil surfactants in the US marketplace for the turfgrass industry 74% are nonionic, 9% are anionic, and the remaining are undisclosed, with no information on composition publicly available (Fidanza et al., 2020). The first soil surfactant (AquaGro) was introduced to the turfgrass market in the 1950s, and since then many advances on formulations have been made by the industry (Moore et al., 2010) with the block copolymer formulation being the most present (49% of block copolymer only, 58% of contained block copolymers or structurally modified block copolymers as formulation components) in the US market for the turfgrass industry, followed by polyoxyethylene and polyalkylenes blends, alkypolyglucosides, alcohol ethoxylates, a botanical and an organosilicate (Fidanza et al., 2020). Similar proportions among soil surfactant formulations have also been observed in the United Kingdom and the Republic of Ireland, even though more organosilicone soil surfactant products are available in these countries compared to the United States (Fidanza et al., 2020).

Two main soil surfactant 'types' are currently marketed to the turfgrass industry: 'retainers' and 'penetrants' (McCarty, 2016; Shaddox and Unruh, 2020; Xiang et al., 2021). The 'retainer' type contains a propylene oxide terminal functional group that is hydrophobic and repels water molecules, and is used where 'retaining' or 'holding' water in the soil (e.g. sandy soils, low organic matter, prolonged drought conditions) is needed. Conversely, the 'penetrant' type possesses an ethylene oxide terminal functional groups, which are hydrophilic and attract water molecules, and is marketed to provide uniform soil moisture distribution and create 'fast and firm' surface playing conditions. Nevertheless, it still remains unclear whether this distinction is solely a marketing claim, as scientific data to back up this distinction are still missing (Kostka and Fidanza, 2019; Xiang et al., 2021).

In a survey conducted by Gelernter et al. (2015), 94% of the golf course superintendents surveyed said they are using wetting agents as a mean to improve water conservation in 2013, compared to 88% of superintendents using them in 2005. The ability of a wetting agent to alleviate soil water repellency and control LDS or dry patch has been known for over two decades (Barton and Colmer, 2011; Cisar et al., 2000; Dekker et al., 2019; Kostka, 2000; Leinauer et al, 2007; Oostindie et al., 2008; Park et al., 2004), and newer research effort has investigated the potential for these products to support water conservation. In a study comparing three commercial wetting agents against an untreated control applied to a creeping bentgrass putting green and watered at either 100% or 30% ET_p, while no differences among treatments were visible when turf was well watered, all wetting agent treatments maintained higher quality than the control at 30% ET_p (Soldat et al., 2010). Authors also concluded that all wetting agents-treated plots had better soil moisture uniformity and lower water repellency than untreated plots. Even though during the second year of the study drought symptoms were visible in all plots, creeping bentgrass treated with surfactants showed a significant delay in drought symptoms development (Soldat et al., 2010). Similar results were found in Norway where a commercial soil surfactant lowered soil water repellency in creeping bentgrass putting greens, but increased turfgrass quality only under dry conditions (Aamlid and Pettersen, 2021). Other studies conducted on both cool- and warm-season golf greens documented that wetting agents did not improve surface firmness, nor turfgrass performance when turfgrass is grown under conditions not promoting the insurgence of soil water repellency (Bauer et al., 2017; Shaddox and Unruh, 2020). Despite several studies reported wetting agents may decrease soil volumetric water content when greens are irrigated with sufficient or excessive amount of water for turfgrass growth (Bauer et al., 2017; Laskowski et al., 2018; Shaddox and Unruh, 2020), this effect appears to reverse when greens are maintained under severe deficit irrigation (Laskowski et al., 2018). In a study conducted in Michigan, when the irrigation system was set to maintain soil water content at 8%, annual bluegrass and creeping bentgrass subjected to traffic stress, and treated with a modified methyl-capped block copolymers were able to maintain quality through the summer compared to untreated control (Laskowski et al., 2018).

Even though application of some wetting agents on fairways may be cost prohibitive for some golf course superintendents and turfgrass managers in general (Xiang et al., 2021); in particular when applications need to start before drought conditions develop, and are necessary at high application rates over long periods of time (e.g. for multiple months in the southern United States where growing season could be 6–8 months long) (Schiavon et al., 2019), soil surfactant research has focused also on their water conservation potential in large acreage fairway areas. Initial research focused on reduced application rates in combination with higher frequency. It was demonstrated that weekly applications at lower rates could improve bermudagrass quality (Park et al., 2004); however, total monthly amount of surfactant applied was the same or even more than the label rate sprayed at 28-day application intervals, providing no economical relief to turfgrass managers. Recently, a commercial soil surfactant improved turfgrass quality of seashore paspalum managed at fairway height and irrigated at 50% ET_o in NM, but no increase in turfgrass quality was detected when turf was treated with another commercially available soil surfactant, that nevertheless improved soil moisture uniformity (Schiavon et al., 2014a). Moreover, a commercially available soil surfactant improved root length density of bermudagrass and seashore paspalum irrigated at 50% ET_o by sprinkler irrigation (Serena et al., 2020a). Similarly, an experimental soil surfactant alone or in combination with topdressing sand improved quality and percent green cover of bermudagrass and lowered soil water hydrophobicity compared to topdressing sand alone when turf was irrigated at 50% ET_o (Alvarez et al., 2016). When comparing ten commercially available soil surfactants for their ability to conserve water in bermudagrass fairways in California, Xiang et al. (2021) found that all soil surfactants were able to maintain acceptable bermudagrass quality when irrigated at 55% ET_o during summer, suggesting a 20–30% water savings to current irrigation practices if a soil surfactant is applied to the turf (Serena et al., 2018c). Moreover, up to 16% volumetric water content with higher soil moisture uniformity was observed when any soil surfactant was applied (Xiang et al., 2021). Only minor differences were detected among soil surfactants treatments, hence there were not clear differences between 'retainers', which are marketed to help the soil retain water, or 'penetrants', which are marketed to help water infiltrate into soil (Kostka and Fidanza, 2019).

New research has investigated the effects of combining a soil surfactant with TE on deficit-irrigated bermudagrass managed at fairway height of cut. The soil surfactant alone or in combination with TE had a positive impact on bermudagrass quality irrigated at 50% ET_{os} in New Mexico (Serena et al., 2018c). However, in CA where summer precipitation is even more scarce, the soil surfactant had the greatest positive effect on turf irrigated at 40% ET_{os} whereas TE did not improve visual quality on the same plots (Schiavon et al., 2019). Use of soil surfactants have also been linked with increased N content in the soil thanks to better moisture distribution that improves microbial mediated N mineralization (Dekker et al., 2019), higher N uptake in plants and decreased N leaching (Abagandura et al., 2021). When two surfactants (10% alkoxylated polyols combined with 7% of glucoethers, and 10% oleic acid esters of block copolymer) were tested for their ability to enhance N recovery in bermudagrass grown either on sand, sandy loam or sandy clay loam, the application of either surfactant increased N uptake compared to the control. This increase in N efficiency was linked to an increase in soil water retention, therefore increasing irrigation water use efficiency, and a decrease in N leaching, particularly in sandy soil (Abagandura et al., 2021).

Even though initially soil surfactants were used almost exclusively by the golf industry (Fig. 6), new formulations and products are available for the landscape industry. Another commercially available soil surfactant has been shown to increase St. Augustinegrass quality that may be linked to enhanced nutrient uptake due to the soil surfactant product application (Chang et al., 2020b). Despite numerous studies documenting enhanced turfgrass quality when soil surfactants are applied, soil surfactants do not seem to facilitate a reduction of runoff volumes from non-hydrophobic soils (Chang et al., 2020b; Lehrsch et al., 2011). Moreover, soil surfactant applications do not seem to help with managing soil salinity (Li et al., 2019b; Schiavon et al., 2014a; Schiavon and Baird, 2018), although more research is needed particularly with the use of soil surfactants and irrigating with reclaimed water (Cisar et al., 2006).

9 Conclusion and future trends

The benefits of turfgrass to the community are well known and include: mitigation of the heat island effect in urban areas, controlling soil erosion, acting as a sink for carbon sequestration, fire hazard reduction and providing cheap, cool and safe surfaces for physical and athletic activities (Beard, 1973). Moreover, the landscape, golf, sports turf and parks and recreation represent multi-billion dollar (US$) industries worldwide. Despite these environmental, recreational and economic advantages, turfgrass is considered an unnecessary luxury by many communities especially those who are drought stricken (Schiavon et al., 2014a). Since turfgrass is used for its functional recreational and aesthetic purposes, and is a non-food and fiber producing crop, water allocation to turfgrass areas is often questioned and many times opposed. For these reasons, management strategies to conserve potable water are being continuously developed and will drive the development of novel water management projects.

The demands of the evolving industry surrounding turfgrass management has changed the way researchers tackle evaluating certain water demands. Water requirements are deeply affected by turfgrass species and cultivars, soil types, turfgrass management practices, irrigation type, irrigation water source and overall geographical region and location. When investigating drought resistance of different turfgrass species, early research focused on acute drought, assessing the response of turfgrass when irrigation was withheld for short periods of time (Schiavon et al., 2014c). These approaches, although providing important information since most turfgrass can survive short period of times without supplemental irrigation, were not representative of arid and semi-arid areas where precipitation events are scarce and concentrated outside turfgrass growing season. Furthermore, the overall majority of these research

Figure 6 Example of a 'split fairway' demonstration on a mixed species of cool-season turfgrasses on a well-drained sand root zone at Ludlow Golf Club (Ludlow, UK). Right side untreated versus left side treated with ProWet Evolve (TurfCare Group Limited, County Kildare, Ireland) soil surfactant to this unirrigated fairway, applied at 12.5 L•ha^{-1} product delivered in 800 L water•ha^{-1} on 7 July, 4 August and 1 September 2021; photo from 16 September 2021. Treated side of the fairway outperformed the untreated side with better turfgrass quality, uniformity, playability, resilience and improved and consistent root zone moisture content. Source: Image courtesy of S. Nightingale.

investigations were conducted under greenhouse conditions that are not representative of field conditions where turfgrass is normally grown. New water conservation research is examining longer term deficit irrigation where only a portion of ET is being replaced over multiple months per year, which is much more emulative of a practical commercial setting. In addition, early studies only investigated the effects of a single water management practice at the time. Newer trends follow real world approaches, where more than one management strategy is evaluated in tandem (Schiavon et al., 2019). For example, evaluating how turf behaves under traffic in combination with drought conditions will help refine how management strategies should be implemented (Braun et al., 2020; Laskowski et al., 2018).

Novel irrigation technologies, ET-based irrigation controllers, SMS and remote sensing technologies mounted on unmanned aerial vehicles are gaining popularity among turfgrass managers. Subsurface drip irrigation has been shown to be a viable alternative to sprinkler irrigation, but its application is still limited, and needs to be tested among a wide range of soil types and management practices. While the need for deep core aerification may hinder its widespread application especially in athletic fields, or golf greens, SDI could find its best fit in low management landscape areas and home lawns. Questions about appropriate spacing in different soil types and viability of long-term saline irrigation in combination with SDI still remain outstanding. Similar concerns have arisen with smart irrigation controllers and the accuracy of SMS when water sources alternative to potable water are adopted for turfgrass irrigation. Smart irrigation controllers and SMS can provide huge benefits for water conservation purposes compared to the old 'set and forget' irrigation schedule mentality. Newer concepts such as utilizing advanced sensors, precision agriculture technologies and remote sensing is gaining popularity to further pinpoint turfgrass watering needs and thus decrease irrigation volumes. While these newer technologies continue to emerge and become available, researchers will need to be diligent in ensuring experiments are conducted in appropriate settings where commercially relevant conclusions can be drawn from the results.

The chemical industry will continue to develop products that can lower turfgrass water use and help turf withstand prolonged periods with limited irrigation, or perhaps no irrigation at all. It is crucial to test new compounds that may become available in the market under unbiased conditions that clearly exhibit the benefits of application against hydrophobicity or drought stress. Identifying PGRs other than TE that can lower turf water requirements and evaluating their interactions with other chemical compounds like soil surfactants remain a priority for the golf, amenity turf and sports turf industries. Likewise, investigating whether wetting agent applications at rates below those specified on their labels can still have a positive effect on water conservation would help turfgrass managers decrease application costs while preventing LDS and lowering irrigation volumes for managed turfgrass.

10 Where to look for further information

10.1 Further readings

- A comprehensive review on turfgrass irrigation can be found in:

  Leinauer, B. and Devitt, D. A. (2014). Irrigation science and technology, In Stier, J. C. Horgan, B. P. and Bonos, S. A. (Eds), *Turfgrass: Biology, Use, and Management*, AgronomyMonograph (Vol. 56, pp. 1075–1131). ASA, CSSA, and SSSA, Madison, WI. doi:10.2134/agronmonogr56.c28.
- Good reviews about non-potable sources of water for irrigation:

  Marcum, K. B. (2006). Use of saline and non-potable water in the turfgrass industry: Constraints and developments, *Agric. Water Manag.* 80(1), 132–146. doi: 10.1016/j.agwat.2005.07.009.

  Sevostianova, E. and Leinauer, B. (2014). Subsurface-applied tailored water: combining nutrient benefits with efficient turfgrass irrigation, *Crop Sci.* 54(5), 1926–1938. doi: https://doi.org/10.2135/cropsci2014.01.0014.

10.2 Journals and conferences

- *Crop Science*, *Agronomy Journal*, *HortScience* and *HortTechnology* journals feature regularly turfgrass science articles.
- The ASA-CSSA-SSSA International Annual Meeting (www.agromony.org; www.crops.org; www.soil.org) is the biggest annual conference for turfgrass scientists.
- The annual American Society of Horticulture Science (www.ashs.org) conference often features a turfgrass science component.
- The European Turfgrass Society (www.turfgrasssociety.eu) hosts scientific conference every two years. Those conferences are attended not only by European but also by a good number of American scientists.
- The International Turfgrass Society (www.turfsociety.com) conducts a conference every four years that are attended by scientists from all over the world.

10.3 Major research projects

- The Specialty Crop Research Initiative (SCRI; https://nifa.usda.gov/funding-opportunity/specialty-crop-research-initiative-scri) funded by US Department of Agriculture (Beltsville, MD, USA) have been funding turfgrass research with the purpose of new cultivar development and water conservation.
- Research funding and support focusing on turfgrass water resources conservation also are provided by the United States Golf Association (www.usga.org) and The Lawn Institute (www.thelawninstitute.org).

11 References

Aamlid, T. S. and Pettersen, T. (2021). Benefits of a soil surfactant on putting greens under dry and wet conditions, *Intl. Turfgrass Soc. Res. J.* 14(1), 157–168. doi:10.1002/its2.75.

Abagandura, G. O., Park, D., Bridges Jr., W. C. and Brown, K. (2021). Soil surfactants applied with 15N labeled urea increases Bermudagrass uptake of nitrogen and reduces nitrogen leaching, *J. Plant Nutr. Soil Sci.* 184(3), 378–387. doi:10.1002/jpln.201900162.

Allen, R. G. (1989). A penman for all seasons, *J. Irrig. Drain. Engrg. ASCE* 112(4), 349-368.

Allen, R. G., Walter, I. A., Elliot, R. L., Howell, T. A., Itenfisu, D. and Jensen, M. E. (2005). *The ASCE Standardized Reference Evapotranspiration Equation*. Reston, VA: American Society of Civil Engineers. Available at: https://www.mesonet.org/images/site/ASCE_Evapotranspiration_Formula.pdf.

Al-Ajlouni, M. G., Vanleeuwen, D. M. and Hilaire, R. S. (2012). Performance of weather-based residential irrigation controllers in a desert environment, *J. AWWA* 104(12), E608–E621. doi:10.5942/jawwa.2012.104.0155.

Alshammary, S. F., Qian, Y. L. and Wallner, S. J. (2004). Growth response of four turfgrass species to salinity, *Agric. Water Manag.* 66(2), 97-111. doi:10.1016/j.agwat.2003.11.002.

Alvarez, G., Sevostianova, E., Serena, M., Sallenave, R. and Leinauer, B. (2016). Surfactant and polymer-coated sand effects on deficit irrigated Bermudagrass turf, *Agron. J.* 108(6), 2245-2255. doi:10.2134/agronj2016.06.0329.

Arghavani, M., Kafi, M., Babalar, M., Naderi, R., Hoque, M. A. and Murata, Y. (2012). Improvement of salt tolerance in Kentucky bluegrass by Trinexapac-ethyl, *Hort.Sci.* 47(8), 1163-1170. doi:10.21273/HORTSCI.47.8.1163.

Ayers, R. S. and Westcot, D. W. (1994). Water quality for agriculture, *FAO Irrigation and Drainage Paper* (vol. 29) Rev.1. Rome: United Nation Food and Agricultural Organization. http://www.fao.org/3/t0234e/t0234e00.htm.

Bach, A. P., Bremer, D. J., Lavis, C. C. and Keeley, S. J. (2022). Effects of drip irrigation and cultivation methods on establishment of seeded tall fescue, *Crop Forage Turf Manag.* 8(1), e20154. doi:10.1002/cft2.20154.

Bachman, M., Inamdar, S., Barton, S., Duke, J. M., Tallamy, D. and Bruck, J. (2016). A comparative assessment of runoff nitrogen from turf, forest, meadow, and mixed landuse watersheds, *J. Am. Water Resour. Assoc.* 52(2), 397-408. doi:10.1111/1752-1688.12395.

Balcı, F. and Orta, A. H. (2018). The comparison of water distribution patterns of pop-up style sprinklers used in landscape irrigation, *J. Tekirdag Agricul Fac.* 15(1), 26-37. Available at: https://www.cabdirect.org/cabdirect/abstract/20183291001.

Baldwin, C. M., Liu, H., McCarty, L. B., Bauerle, W. L. and Toler, J. E. (2006). Effects of Trinexapac-ethyl on the salinity tolerance of two ultradwarf Bermudagrass cultivars, *HortSci.* 41(3), 808-814. doi:10.21273/HORTSCI.41.3.808.

Barton, L., Wan, G. G. Y., Buck, R. P. and Colmer, T. D. (2009). Nitrogen increases evapotranspiration and growth of a warm-season turfgrass, *Agron. J.* 101(1), 17-24.

Barton, L. and Colmer, T. D. (2011). Granular wetting agents ameliorate water repellency in turfgrass of contrasting soil organic matter content, *Plant Soil* 348(1-2), 411-424. doi:10.1007/s11104-011-0765-3.

Bauer, S. J., Cavanaugh, M. J. and Horgan, B. P. (2017). Wetting agent influence on putting green surface Firmness, *Int. Turf. Soc. Res. J.* 13(1), 624-628. doi:10.2134/itsrj2016.06.0490.

Beard, J. B. (1973). *Turfgrass Science and Culture*. Prentice-Hall Int. Inc., London.

Bell, A. A., Liu, L., Reidy, B., Davis, R. M. and Subbarao, K. V. (1998). Mechanisms of subsurface drip irrigation-mediated suppression of lettuce drop caused by Sclerotinia minor, *Phytopathology* 88(3), 252-259. doi:10.1094/PHYTO.1998.88.3.252.

Bian, X., Merewitz, E. and Huang, B. (2009). Effects of Trinexapac-ethyl on drought responses in creeping bentgrass associated with water use and osmotic adjustment, *J. Am. Soc. Hort. Sci.* 134(5), 505-510. doi:10.21273/JASHS.134.5.505.

Blonquist Jr., J. M., Jones, S. B. and Robinson, D. A. (2006). Precise irrigation scheduling for turfgrass using a subsurface electromagnetic soil moisture sensor, *Agric. Water Manag.* 84(1-2), 153-165.

Braun, R. C., Bremer, D. J. and Hoyle, J. A. (2020). Simulated traffic on turfgrasses during drought stress: I. Performance and recovery of turf canopies, *Crop Sci.* 61(5), 2926-2938 doi:10.1002/csc2.20324.

Braun, R. C., Bremer, D. J. and Hoyle, J. A. (2021). Simulated traffic on turfgrasses during drought stress: II. Soil moisture, soil compaction, and rooting, *Intl. Turfgrass Soc. Res. J.* 14(1), 516-527 doi:10.1002/its2.62.

Bremer, D. J., Keeley, S. J., Jager, A., Fry, J. D. and Lavis, C. (2012). In-ground irrigation systems affect lawn-watering behaviors of residential homeowners, *HortTech* 22(5), 651-658.

Burt, C. M. and Styles, S. W. (1999). *Drip and Micro Irrigation for Trees, Vines, and Row Crops (p. 292).* Irrigation Training and Research Center, California Polytechnic State University, San Luis Obispo, CL.

California Department of Water Resources (2009). California Code of Regulations, Title 23. Waters, Division 2. Ch. 2.7. Model water efficient landscape ordinance. California Dep. Water Resour. Available at: http://www.water.ca.gov/ (accessed June 10th, 2021).

Cardenas, B., Migliaccio, K. W., Dukes, M. D., Hahus, I. and Kruse, J. K. (2020). Irrigation savings from smart irrigation technologies and a smartphone app on turfgrass, *Trans. ASABE* 63(6), 1697-1709.

Cardenas-Lailhacar, B. and Dukes, M. D. (2010). Precision of soil moisture sensor irrigation controllers under field conditions, *Agric. Water Manag.* 97(5), 666-672. doi:10.1016/j.agwat.2009.12.009.

Carrow, R. N. and Duncan, R. R. (2012). Sodic, saline-sodic, and alkaline soils, In: Carro, R. N. and Duncan, R. R. (Eds), *Best Management Practices for Saline and Sodic Turfgrass Soils. Assessment and Reclamation*. CRC Press Taylor and Francis Group, LLC, Boca Raton, FL.

Chabon, J., Bremer, D. J., Fry, J. D. and Lavis, C. (2017). Effects of soil moisture-based irrigation controllers, mowing height, and trinexapac-ethyl on tall fescue irrigation amounts and mowing requirements, *Int. Turf. Soc. Res. J.* 13(1), 755-760. doi:10.2134/itsrj2016.04.0242.

Chang, B., Wherley, B. G., Aitkenhead-Peterson, J. A. and West, J. B. (2019). Irrigation salinity effects on Tifway Bermudagrass growth and nitrogen uptake, *Crop Sci.* 59(6), 2820-2828. doi:10.2135/cropsci2019.01.0065.

Chang, B., Wherley, B. G., Aitkenhead-Peterson, J. A. and West, J. (2020a). Water chemistry and nitrogen source affect foliar uptake efficiency in 'champion' Bermudagrass, *J. Plant Nutr.* 43(16), 2480-2492. doi:10.1080/01904167.2020.1783310.

Chang, B., Wherley, B. G., Aitkenhead-Peterson, J. A., Ojeda, N., Fontanier, C. and Dwyer, P. (2020b). Effect of wetting agent on nutrient and water retention and runoff from simulated urban lawns, *Hort.Sci.* 55(7), 1005-1013. doi:10.21273/HORTSCI14982-20.

Chavarria, M., Wherley, B., Jessup, R. and Chandra, A. (2021). Physiological responses to salinity among warm-season turfgrasses of contrasting salinity tolerance, *J. Agron. Crop Sci.* 207(4), 669-678. doi:10.1111/jac.12501.

Christians, N. E. and Engelke, M. C. (1994). Choosing the right grass to fit the environment, In: Leslie, A. R. (Ed.), *Handbook of Integrated Pest Management for Turf and Ornamentals*. CRC Press, Boca Raton, FL.

Cisar, J. L., Williams, K. E., Vivas, H. E. and Haydu, J. J. (2000). The occurrence and alleviation by surfactants of soil-water repellency on sand-based turfgrass systems, *J. Hydrol.* 231-232, 352-358. doi:10.1016/S0022-1694(00)00207-9.

Cisar, J. L., Reuter, M., Snyder, G. H. and Fidanza, M. A. (2006). The use of non-saline reclaimed water for golf course irrigation in Florida, *Appl. Turfgrass Sci.* 3(1) doi:10.1094/ATS-2006-0210-02-TT.

Cockerham, S. T. and Minner, D. D. (2011). Turfgrass nutrition and fertilizers, In: Adous, D. E. (Ed.), *International Turf Management*. Routledge, New York.

Colmer, T. D. and Barton, L. (2017). A review of warm-season turfgrass evapotranspiration, responses to deficit irrigation, and drought resistance, *Crop Sci.* 57(S1), 98–110. doi:10.2135/cropsci2016.10.0911.

Connolly, J. (2001). Turning irrigation upside down, *Turf West*, 6–15.

Culpepper, T., Young, J. and Wherley, B. (2020). Comparison of four warm-season turfgrass species to natural rainfall or supplemental irrigation in a semiarid climate, *Agrosyst. Geosci. Environ.* 3(1), e20011. doi:10.1002/agg2.20011.

Davis, S. L., Dukes, M. D. and Miller, G. L. (2009). Landscape irrigation by evapotranspiration-based irrigation controllers under dry conditions in Southwest Florida, *Agric. Water Manag.* 96(12), 1828–1836. doi:10.1016/j.agwat.2009.08.005.

Dekker, L. W. and Ritsema, C. J. (1996). Variation in water content and wetting patterns in Dutch water repellent peaty clay and clayey peat soils, *CATENA* 28(1–2), 89–105. doi:10.1016/S0341-8162(96)00047-1.

Dekker, L. W., Doerr, S. H., Oostindie, K., Ziogas, A. K. and Ritsema, C. J. (2001a). Water repellency and critical soil water content in a dune sand, *Soil Sci. Soc. Am. J.* 65(6), 1667–1674. doi:10.2136/sssaj2001.1667.

Dekker, L. W., Oostindie, K., Ziogas, A. K. and Ritsema, C. J. (2001b). The impact of water repellency on soil moisture variability and preferential flow, *Int. Turf. Soc. Res.* 9, 498–505.

Dekker, L. W., Oostindie, K., Ritsema, C. J., Dekker, L. W., Oostindie, K. and Ritsema, C. J. (2005). Exponential increase of publications related to soil water repellency, *Soil Res.* 43(3), 403–441. doi:10.1071/SR05007.

Dekker, L. W., Ritsema, C. J., Oostindie, K., Wesseling, J. G. and Geissen, V. (2019). Effects of a soil surfactant on grass performance and soil wetting of a fairway prone to water repellency, *Geoderma* 338, 481–492. doi:10.1016/j.geoderma.2018.09.016.

Devitt, D. A., Carstensen, K. and Morris, R. L. (2008). Residential water savings associated with satellite-based ET irrigation controllers, *J. Irrig. Drain. Eng.* 134(1), 74–82. doi:10.1061/(ASCE)0733-9437(2008)134:1(74).

Devitt, D. A. and Morris, R. L. (2008). Urban landscape water conservation and the species effect, In Beard, J. B. and Kenna, M. P. (Eds), Water *Quality* and *Quantity Issues* for *Turfgrasses* in *Urban Landscapes* (pp. 171–192). Ames, IA: Counc. Agric. Sci. Technol.

Dukes, M. D. (2020). Two decades of smart irrigation controllers in U.S. Landscape irrigation, *Trans. ASABE* 63(5), 1593–1601. doi:10.13031/trans.13930.

Duncan, R. R., Carrow, R. N. and Huck, M. (2000). Understanding water quality and guidelines to management: an overview of challenges for water usage on golf courses for the 21st century, *USGA Green Sec. Rec.* 38(5):14.

Ebdon, J. S., Petrovic, A. M. and White, R. A. (1999). Interaction of nitrogen, phosphorus, and potassium on evapotranspiration rate and growth of Kentucky bluegrass, *Crop Sci.* 39(1), 209–218.

Eid, A. R., El-Ansary, M. Y., Farag, A. A. and Sultan, W. M. (2019). Evaluation of some turf irrigation sprinklers under deficit irrigation levels, *Misr J. Agr. Eng.* 36(2), 445–456. doi:10.21608/mjae.2019.94650.

Elansary, H. O. and Yessoufou, K. (2015). Growth regulators and mowing heights enhance the morphological and physiological performance of Seaspray turfgrass during drought conditions, *Acta Physiol. Plant.* 37(11), 232. doi:10.1007/s11738-015-1986-5.

Erickson, J. E., Cisar, J. L., Volin, J. C. and Snyder, G. H. (2001). Comparing nitrogen runoff and leaching between newly established St. Augustinegrass turf and an alternative residential landscape, *Crop Sci.* 41(6), 1889–1895. doi:10.2135/cropsci2001.1889.

Fagerness, M. J. and Yelverton, F. H. (2001). Plant growth regulator and mowing height effects on seasonal root growth of Penncross creeping bentgrass, *Crop Sci.* 41(6), 1901–1905. doi:10.2135/cropsci2001.1901.

Fazackerley, S. and Lawrence, R. (2010). Reducing turfgrass water consumption using sensor nodes and an adaptive irrigation controller, In: IEEE Sens. Appl. Symposium IEEE (vol. 2010), 90–94.

Feldhake, C. M., Danielson, R. E. and Butler, J. D. (1984). Turfgrass evapotranspiration. responses to deficit irrigation, *Agron. J.* 76(1), 85–89:doi:10.2134/agronj1984.00021962007600010022x.

Fidanza, M. A., Cisar, J. L., Kostka, S. J., Gregos, J. S., Schlossberg, M. J. and Franklin, M. (2007). Preliminary investigation of soil chemical and physical properties associated with type-I fairy ring symptoms in turfgrass, *Hydrol. Proc.* 21(17), 2285–2290. doi:10.1002/hyp.6757.

Fidanza, M., Kostka, S. and Bigelow, C. (2020). Communication of soil water repellency causes, problems, and solutions of intensively managed amenity turf from 2000 to 2020, *J. Hydrol. Hydromech.* 68(4), 306–312. doi:10.2478/johh-2020-0032.

Friell, J., Watkins, E. and Horgan, B. (2013). Salt tolerance of 74 turfgrass cultivars in nutrient solution culture, *Crop Sci.* 53(4), 1743–1749. doi:10.2135/cropsci2012.08.0476.

Fry, J. D. and Butler, J. D. (1989). Annual bluegrass and creeping bentgrass evapotranspiration rates, *Hort.Sci.* 24(2), 269–271.

Fry, J. D. and Huang, B. (2004). *Applied Turfgrass Science and Physiology*. John Wiley & Sons, Inc., Hoboken, NJ.

Fu, J. M., Koski, A. J. and Qian, Y. L. (2005). Responses of creeping bentgrass to salinity and mowing management: growth and turf quality, *HortSci.* 40(2), 463–467. doi:10.21273/HORTSCI.40.2.463.

Galloway, J. N., Townsend, A. R., Erisman, J. W., Bekunda, M., Cai, Z., Freney, J. R., Martinelli, L. A., Seitzinger, S. P. and Sutton, M. A. (2008). Transformation of the nitrogen cycle: recent trends, questions, and potential solutions, *Science* 320(5878), 889–892. doi:10.1126/science.1136674.

Ganjegunte, G. K., Clark, J. A., Sallenave, R., Sevostianova, E., Serena, M., Alvarez, G. and Leinauer, B. (2017). Soil salinity of an urban park after long-term irrigation with saline ground water, *Agron. J.* 109(6), 3011–3018. doi:10.2134/agronj2017.06.0369.

Gelernter, W. D., Stowell, L. J., Johnson, M. E., Brown, C. D. and Beditz, J. F. (2015). Documenting trends in water use and conservation practices on U.S. golf courses, *Crop Forage Turf Manag.* 1(1), 1–10. doi:10.2134/cftm2015.0149.

Geza, M., Deb, S. K., Leinauer, B., Stanek, S., Sevostianova, E. and Serena, M. (2021). Modeling NO3-N leaching during establishment of turfgrasses irrigated with tailored reclaimed water, *Vadose Zone J.* 20(3) doi:10.1002/vzj2.20112.

Grabow, G. L., Ghali, I. E., Huffman, R. L., Miller, G. L., Bowman, D. and Vasanth, A. (2013). Water application efficiency and adequacy of ET-based and soil moisture-based irrigation controllers for turfgrass irrigation, *J. Irrig. Drain. Engine* 139(2), 113–123.

Haghverdi, A., Reiter, M., Sapkota, A. and Singh, A. (2021). Hybrid Bermudagrass and tall fescue turfgrass irrigation in central California: I. Assessment of visual quality, soil moisture and performance of an ET-based smart controller, *Agronomy* 11(8), 1666. doi:10.3390/agronomy11081666.

Haley, M. B., Dukes, M. D. and Miller, G. L. (2007). Residential irrigation water use in central Florida, *J. Irrig. Drain. Eng.* 133(5), 427–434. doi:10.1061/(ASCE)0733-9437(2007)133:5(427).

Hallett, P. (2007). An introduction to soil water repellency, In: Gaskin, R. E. (Ed.). *Proceedings of the 8th International Symposium on Adjuvants For Agrochemicals (ISAA2007)*, International Society for Agrochemical Adjuvants (ISAA). Columbus, OH, 6–9.

Hejl, R. W., Wherley, B. G., Thomas, J. C. and White, R. H. (2015). Irrigation water quality and Trinexapac-ethyl effects on bermudagrass response to deficit irrigation, *Hort. Sci.* 50(7), 1081–1087. doi:10.21273/HORTSCI.50.7.1081.

Hejl, R. W., Wherley, B. G., White, R. H., Thomas, J. C. and Fontanier, C. H. (2016). Deficit irrigation and simulated traffic on 'Tifway' bermudagrass summer performance and autumn recovery, *Crop Sci.* 56(2), 809–817. doi:10.2135/cropsci2015.03.0197.

Jiang, H. and Fry, J. (1998). Drought responses of perennial ryegrass treated with plant growth regulators, *Hort.Sci.* 33(2), 270–273.

Jiang, Y. and Huang, B. (2002). Protein alterations in tall fescue in response to drought stress and abscisic acid, *Crop Sci.* 42(1), 202–207. doi:10.2135/cropsci2002.2020.

Kenna, M. P. (2008). Turfgrass and the environment, In: Beard, J. B. and Kenna, M. P. (Eds). *Water Quality and Quantity Issues for Turfgrasses in Urban Landscapes*. Ames, IA: Council Agr. Sci. Technol.

Khachatryan, H., Suh, D. H., Xu, W., Useche, P. and Dukes, M. D. (2019). Towards sustainable water management: preferences and willingness to pay for smart landscape irrigation technologies, *Land Use Policy* 85, 33–41. doi:10.1016/j.landusepol.2019.03.014.

Kjelgren, R., Rupp, L. and Kilgren, D. (2000). Water conservation in urban landscapes, *HortSci.* 35(6), 1037–1040.

Krishnan, S. and Merewitz, E. B. (2015). Drought stress and Trinexapac-ethyl modify phytohormone content within Kentucky bluegrass leaves, *J. Plant Growth Regul.* 34(1), 1–12. doi:10.1007/s00344-014-9434-0.

Koch, M. J., Huang, B. and Bonos, S. A. (2011). Salinity tolerance of Kentucky bluegrass cultivars and selections using an overhead irrigated screening technique, *Crop Sci.* 51(6), 2846–2857. doi:10.2135/cropsci2011.03.0174.

Kostka, S., Cisar, J., Short, J. and Mane, S. (1997). Evaluation of soil surfactants for the management of soil water repellency in turfgrass, *Int. Turfgrass Soc.* Res. J. 8, 485–494.

Kostka, S. J. (2000). Amelioration of water repellency in highly managed soils and the enhancement of turfgrass performance through the systematic application of surfactants, *J. Hydrol.* 231-232, 359–368. doi:10.1016/S0022-1694(00)00208-0.

Kostka, S. and Fidanza, M. (2019). Soil surfactant usage based on solid science. Golf Course Industry. Available at: https://www.golfcourseindustry.com/article/soil-surfactant-research-guidance/.

Laskowski, K., Frank, K. and Merewitz, E. (2018). Surfactant effects on creeping bentgrass and annual bluegrass exposed to different irrigation and traffic stress treatments, *Agron. J.* 110(1), 193–199. doi:10.2134/agronj2017.05.0261.

Lehrsch, G. A., Sojka, R. E., Reed, J. L., Henderson, R. A. and Kostka, S. J. (2011). Surfactant and irrigation effects on wettable soils: runoff, erosion, and water retention responses, *Hydrol. Proc.* 25(5), 766–777. doi:10.1002/hyp.7866.

Leinauer, B. and Smeal, D. (2012). *Turfgrass Irrigation*. NM State University, Cooperative Extension Service.

Leinauer, B. and Devitt, D. A. (2014). Irrigation science and technology, In J. C. Stier et al. (Eds.), Turfgrass: Biology, use, and management, *AgronomyMonograph* (Vol. 56, pp. 1075–1131). Madison, WI: ASA, CSSA, and SSSA. doi:10.2134/agronmonogr56.

Leinauer, B., Karcher, D., Barrick, T., Ikemura, Y., Hubble, H. and Makk, J. (2007). Water repellency varies with depth and season in sandy rootzones treated With ten wetting agents, *Appl. Turf Sci.* 4(1), 1-9. doi:10.1094/ATS-2007-0221-01-RS.

Li, H., Issaka, Z., Jiang, Y., Tang, P. and Chen, C. (2019a). Overview of emerging technologies in sprinkler irrigation to optimize crop production, *Int. J. Agricul Biol. Eng.* 12(3), 1-9. doi:10.25165/ijabe.v12i3.4310.

Li, L., Young, J. and Deb, S. (2019b). Effects of cultivation practices and products on Bermudagrass fairways in a semiarid region, *Agron. J.* 111(6), 2899–2909. doi:10.2134/agronj2019.04.0262.

Lu, S., Su, W., Li, H. and Guo, Z. (2009). Abscisic acid improves drought tolerance of triploid Bermudagrass and involves H2O2- and NO-induced antioxidant enzyme activities, *Plant Physiol. Biochem.* 47(2), 132–138. doi:10.1016/j.plaphy.2008.10.006.

Marcum, K. B. (2001). Salinity tolerance of 35 bentgrass cultivars, *HortSci.* 36(2), 374–376. doi:10.21273/HORTSCI.36.2.374.

Marcum, K. B. (2006). Use of saline and non-potable water in the turfgrass industry: constraints and developments, *Agric. Water Manag.* 80(1–3), 132–146. doi:10.1016/j.agwat.2005.07.009.

Marcum, K. B. and Jiang, H. (1997). Effects of plant growth regulators on tall fescue rooting and water use, *J. Turfgrass Manag.* 2(2), 13-27. doi:10.1300/J099v02n02_02.

Marcum, K. B. and Pessarakli, M. (2006). Salinity tolerance and salt gland excretion Efficiency of Bermudagrass turf cultivars, *Crop Sci.* 46(6), 2571-2574. doi:10.2135/cropsci2006.01.0027.

Mayer, P. W., DeOreo, W. B., Opitz, E. M., Kiefer, J. C., Davis, W. Y., Dziegielewski, B. and Nelson, J. O. (1999). Residential end uses of water, *Amer. Water Works Assn.* 310 pp.

McCann, S. E. and Huang, B. (2007). Effects of Trinexapac-Ethyl foliar application on creeping bentgrass responses to combined drought and heat stress, *Crop Sci.* 47(5), 2121–2128. doi:10.2135/cropsci2006.09.0614.

McCann, S. E. and Huang, B. (2008). Drought responses of Kentucky bluegrass and creeping bentgrass as affected by abscisic acid and Trinexapac-ethyl, *J. Am. Soc. Hort. Sci.* 133(1), 20–26. doi:10.21273/JASHS.133.1.20.

McCarty, L. B. (2016). *Applied Soil Physical Properties, Drainage, and Irrigation Strategies*. Springer, New York, NY.

McCready, M. S., Dukes, M. D. and Miller, G. L. (2009). Water conservation potential of smart irrigation controllers on St. Augustinegrass, *Agric. Water Manag.* 96(11), 1623–1632. doi:10.1016/j.agwat.2009.06.007.

McCullough, P. E., Liu, H. and McCarty, L. B. (2005). Response of six dwarf-type bermudagrasses to Trinexapac-ethyl, *HortSci.* 40(2), 460–462. doi:10.21273/HORTSCI.40.2.460.

McGroary, P. C., Cisar, J. L., Snyder, G. H., Erickson, J. E., Daroub, S. H. and Sartain, J. B. (2011). Water use of St. Augustinegrass and bahiagrass under varying nitrogen rates, *Agron. J.* 103(1), 100–106. doi:10.2134/agronj2010.0237.

Milesi, C., Elvidge, C. D., Dietz, J. B., Tuttle, B. T., Nemani, R. R. and Running, S. W. (2005). A strategy for mapping and modeling the ecological effects of US lawns, *J. Turfgrass Manag.* 1(1), 83–97.

Mohammadi, M. H. S., Etemadi, N., Arab, M. M., Aalifar, M., Arab, M. and Pessarakli, M. (2017). Molecular and physiological responses of Iranian Perennial ryegrass as affected by Trinexapac ethyl, paclobutrazol and abscisic acid under drought stress, *Plant Physiol. Biochem.* 111, 129–143. doi:10.1016/j.plaphy.2016.11.014.

Molle, B., Tomas, S., Hendawi, M. and Granier, J. (2012). Evaporation and wind drift losses During sprinkler irrigation influenced by droplet size distribution, *Irrig. Drain.* 61(2), 240–250. doi:10.1002/ird.648.

Moore, D., Kostka, S., Boerth, T., Franklin, M., Ritsema, C., Dekker, L., Oostindie, K., Stoof, C. and Wesseling, J. (2010). The effect of soil surfactants on soil hydrological behavior, the plant growth environment, irrigation efficiency and water conservation, *J. Hydrol. Hydromech.* 58(3), 142–148. doi:10.2478/v10098-010-0013-1.

Muñoz-Carpena, R., Li, Y. C., Klassen, W. and Dukes, M. D. (2005). Field comparison of tensiometer and granular matrix sensor automatic drip irrigation on tomato, *HortTech* 15(3), 584–590. doi:10.21273/HORTTECH.15.3.0584.

Oostindie, K., Dekker, L. W., Wesseling, J. G. and Ritsema, C. J. (2008). Soil surfactant stops water repellency and preferential flow paths, *Soil Use Manag.* 24(4), 409–415. doi:10.1111/j.1475-2743.2008.00185.x.

Park, D. M., Cisar, J. L., Williams, K. E. and Snyder, G. H. (2004). Alleviation of soil water repellency in sand based Bermudagrass in South Florida, *Acta Hortic.* 661(661), 111–115. doi:10.17660/ActaHortic.2004.661.13.

Pathan, S. M., Barton, L. and Colmer, T. D. (2007). Evaluation of a soil moisture sensor to reduce water and nutrient leaching in turfgrass (Cynodon dactylon cv. Wintergreen), *Australian J Experim, Agriculturists* 47, 215–222.

Pessarakli, M., Marcum, K. B. and Kopec, D. M. (2006). Interaction effects of salinity and Primo on the growth of Kentucky bluegrass, *J. Food Agr. Environ.* 4, 325–327.

Petrovic, A. M. (1990). The fate of nitrogenous fertilizers applied to turfgrass, *J. Environ. Qual.* 19(1), 1–14. doi:10.2134/jeq1990.00472425001900010001x.

Qian, Y. L., Fu, J. M., Wilhelm, S. J., Christensen, D. and Koski, A. J. (2007). Relative salinity tolerance of turf-type saltgrass selections, *Hort.Sci.* 42(2), 205–209. doi:10.21273/HORTSCI.42.2.205.

Rademacher, W. (2000). Growth retardants: effect of gibberellin biosynthesis and other metabolic pathways, *Annu. Rev. Plant Physiol. Plant Mol. Biol.* 51(1), 501–531.

Rhoades, J. D. (1974). Drainage for salinity control, In: van Schilfgaarde, J. (Ed.), *Drainage for Agricture, Agronomy Monograph No.17*, Amer. Soc. Of Agron. Madison Book Company.

Rodriguez, I. R., Miller, G. L. and McCarty, L. B. (2002). Bermudagrass establishment on high sand-content soils using various NPK ratios, *HortSci.* 37(1), 208–209.

Romero, C. C. and Dukes, M. D. (2016). Review of turfgrass evapotranspiration and crop coefficients, *Trans. ASABE* 59(1), 207–223. doi:10.13031/trans.59.11180.

Saha, S. K., Trenholm, L. E. and Unruh, J. B. (2005). Effect of fertilizer source on water use of St. Augustinegrass and ornamental plants, *HortSci.* 40(7), 2164–2166. doi:10.21273/HORTSCI.40.7.2164.

Sandor, D. (2018). Water conservation practices for irrigation of turfgrass lawns. Theses and dissertation. 3064, *University of Arkansas*. Available at: https://scholarworks.uark.edu/etd/3064.

Schiavon, M., Leinauer, B., Serena, M., Sallenave, R. and Maier, B. (2012). Bermudagrass and seashore paspalum establishment from seed using differing irrigation methods and water qualities, *Agron. J.* 104(3), 706–714. doi:10.2134/agronj2011.0390.

Schiavon, M., Leinauer, B., Serena, M., Sallenave, R. and Maier, B. (2013a). Establishing tall fescue and Kentucky bluegrass using subsurface irrigation and saline water, *Agron. J.* 105(1), 183–190. doi:10.2134/agronj2012.0187.

Schiavon, M., Barnes, B. D., Shaw, D. A., Henry, J. M. and Baird, J. H. (2013b). Strategies for converting tall fescue to warm-season turf in a Mediterranean climate, *hortTechnology* 23(4), 442–448. doi:10.21273/HORTTECH.23.4.442.

Schiavon, M., Leinauer, B., Serena, M., Maier, B. and Sallenave, R. (2014a). 'Plant growth regulator and soil surfactants' effects on saline and deficit irrigated warm-season grasses: I, *Crop Sci.* 54(6), 2815–2826. doi:10.2135/cropsci2013.10.0707.

Schiavon, M., Leinauer, B., Serena, M., Maier, B. and Sallenave, R. (2014b). 'Plant growth regulator and soil surfactants' effects on saline and deficit irrigated warm-season grasses: II, *Crop Sci.* 54(6), 2827–2836. doi:10.2135/cropsci2013.10.0708.

Schiavon, M., Green, R. L. and Baird, J. H. (2014). Drought tolerance of cool-season turfgrasses in a Mediterranean climate, *Eur. J. Hortic. Sci.* 79, 175–182.

Schiavon, M., Macolino, S., Leinauer, B. and Ziliotto, U. (2016). Seasonal changes in carbohydrate and protein content of seeded bermudagrasses and their effect on spring green-up, *J. Agron. Crop Sci.* 202(2), 151–160. doi:10.1111/jac.12135.

Schiavon, M., Pedroza, A., Leinauer, B., Suarez, D. L. and Baird, J. H. (2017). Varying evapotranspiration and salinity level of irrigation water influence soil quality and performance of perennial ryegrass (Lolium perenne L.), *Urban For. Urban Greening* 26, 184–190. doi:10.1016/j.ufug.2017.01.006.

Schiavon, M. and Baird, J. H. (2018). Evaluation of products to alleviate irrigation salinity stress on Bermudagrass turf, *Agron. J.* 110(6), 2136–2141. doi:10.2134/agronj2018.01.0064.

Schiavon, M., Orlinski, P., Petelewicz, P., Pudzianowska, M. and Baird, J. H. (2019). Effects of Trinexapac-ethyl, surfactant, and nitrogen fertilization on Bermudagrass water use, *Agron. J.* 111(6), 3057–3066. doi:10.2134/agronj2019.03.0225.

Schiavon, M., Mock, T., Stowell, L. J. and Baird, J. H. (2020). Management practices for optimal kikuyugrass quality and playing conditions, *Agron. J.* 112(5), 3402-3410. doi:10.1002/agj2.20198.

Schiavon, M., Shaddox, T. W., Williams, K. E., Gallo, S., Boeri, P. A., Unruh, J. B., Kruse, J. and Kenworthy, K. Nitrogen requirements for deficit-irrigated Bermudagrass (Cynodon spp.) fairways in South Florida, *J. Agron. Crop Sci.*, doi:10.1111/jac.12558.

Serena, M., Leinauer, B., Sallenave, R., Schiavon, M. and Maier, B. (2012a). Media selection and seed coating influence germination of turfgrasses under salinity, *Hort.Sci.* 47(1), 116–120. doi:10.21273/HORTSCI.47.1.116.

Serena, M., Leinauer, B., Sallenave, R., Schiavon, M. and Maier, B. (2012b). Turfgrass establishment from polymer-coated seed under saline irrigation, *Hort.Sci.* 47(12), 1789–1794. doi:10.21273/HORTSCI.47.12.1789.

Serena, M., Leinauer, B., Schiavon, M., Maier, B. and Sallenave, R. (2014). Establishment and Rooting Response of Bermudagrass Propagated with Saline Water and Subsurface Irrigation. *Crop Science*, 54: 827–836. https://doi.org/10.2135/cropsci2013.07.0512.

Serena, M., Schiavon, M., Sallenave, R. and Leinauer, B. (2018a). Nitrogen fertilization of warm-season turfgrasses irrigated with saline water from varying irrigation systems. 1. Quality, spring green-up and fall colour retention, *J. Agron. Crop Sci.* 204(3), 252–264. doi:10.1111/jac.12254.

Serena, M., Schiavon, M., Sallenave, R. and Leinauer, B. (2018b). Nitrogen fertilization of warm-season turfgrasses irrigated from varying irrigation systems: 2. Carbohydrate and protein content, *J. Agron. Crop Sci.* 204(3), 265–273. doi:10.1111/jac.12256.

Serena, M., Sportelli, M., Sevostianova, E., Sallenave, R. and Leinauer, B. (2018c). Combining Trinexapac-ethyl with a soil surfactant reduces Bermudagrass irrigation requirements, *Agron. J.* 110(6), 2180–2188. doi:10.2134/agronj2018.03.0148.

Serena, M., Sevostianova, E., Leinauer, B. (2019). Subsurface drip irrigation and minimum irrigation requirements to maintain bermudagrass and seashore paspalum. *ASA, CSSA and SSSA International Annual Meetings (2019).*

Serena, M., Schiavon, M., Sallenave, R. and Leinauer, B. (2020a). Drought avoidance of warm-season turfgrasses affected by irrigation system, soil surfactant revolution, and plant growth regulator trinexapac-ethyl, *Crop Sci.* 60(1), 485–498. doi:10.1002/csc2.20063.

Serena, M., Velasco-Cruz, C., Friell, J., Schiavon, M., Sevostianova, E., Beck, L., Sallenave, R. and Leinauer, B. (2020b). Irrigation scheduling technologies reduce water use and maintain turfgrass quality, *Agron. J.* 112(5), 3456–3469. doi:10.1002/agj2.20246.

Sevostianova, E., Leinauer, B., Sallenave, R., Karcher, D. and Maier, B. (2011a). Soil salinity and quality of sprinkler and drip irrigated cool-season turfgrasses, *Agron. J.* 103(5), 1503–1513. doi:10.2134/agronj2011.0162.

Sevostianova, E., Leinauer, B., Sallenave, R., Karcher, D. and Maier, B. (2011b). Soil salinity and quality of sprinkler and drip irrigated warm-season turfgrasses, *Agron. J.* 103(6), 1773–1784. doi:10.2134/agronj2011.0163.

Sevostianova, E. and Leinauer, B. (2014). Subsurface-applied tailored water: combining nutrient benefits with efficient turfgrass irrigation, *Crop Sci.* 54(5), 1926–1938. doi:10.2135/cropsci2014.01.0014.

Shaddox, T. W. and Unruh, J. B. (2020). Do wetting agents influence golf ball roll distance?, *HortTech* 30(3), 404–410. doi:10.21273/HORTTECH04576-20.

Shahba, M. A. (2010). Interaction effects of salinity and mowing on performance and physiology of Bermudagrass cultivars, *Crop Sci.* 50(6), 2620–2631. doi:10.2135/cropsci2010.04.0192.

Shahba, M. A., Qian, Y. L. and Lair, K. D. (2008). Improving seed germination of saltgrass under saline conditions, *Crop Sci.* 48(2), 756–762. doi:10.2135/cropsci2007.07.0382.

Shahba, M. A., Alshammary, S. F. and Abbas, M. S. (2012). Effects of salinity on seashore paspalum cultivars at different mowing heights, *Crop Sci.* 52(3), 1358–1370 doi:10.2135/cropsci2011.06.0337.

Shahba, M. A., Abbas, M. S. and Alshammary, S. F. (2014). Drought resistance strategies of seashore paspalum cultivars at different mowing heights, *Hort.Sci.* 49(2), 221–229. doi:10.21273/HORTSCI.49.2.221.

Soldat, D. J., Lowery, B. and Kussow, W. R. (2010). Surfactants increase uniformity of soil water content and reduce water repellency on sand-based golf putting greens, *Soil Sci.* 175(3), 111–117. doi:10.1097/SS.0b013e3181d6fa02.

Snyder, G. H., Augustin, B. J. and Davidson, J. M. (1984). Moisture sensor-controlled irrigation for reducing N leaching in Bermudagrass turf 1, *Agron. J.* 76(6), 964–969. doi:10.2134/agronj1984.00021962007600060023x.

Sun, H., Kopp, K., Cardon, G. and Jones, S. (2021). Numerical simulation of water and nitrogen transport in three turfgrass systems, *Intl. Turfgrass Soc. Res. J.* 14(1), 90–109doi:10.1002/its2.52.

Taylor, K. S., Anda, M., Sturman, J., Mathew, K. and Ho, G. (2006). Subsurface dripline tubing – an experimental design for assessing the effectiveness of using dripline to apply treated wastewater for turf irrigation in Western Australia, *Desalination* 187(1-3), 375-385. doi:10.1016/j.desal.2005.04.095.

Telenko, D. E. P., Shaddox, T. W., Unruh, J. B. and Trenholm, L. E. (2015). Nitrate leaching, turf quality, and growth rate of 'Floratam' St. Augustinegrass and common centipedegrass, *Crop Sci.* 55(3), 1320-1328. doi:10.2135/cropsci2014.09.0639.

Thompson, C. (2019). Subsurface drip irrigation reduces water use on tees. United States Golf Association Green Section. Available at: http://archive.lib.msu.edu/tic/usgamisc/know/2019-08-02.pdf.

Trenholm, L. E., Unruh, J. B. and Sartain, J. B. (2012). Nitrate leaching and turf quality in established 'Floratam' St. Augustinegrass and 'Empire' zoysiagrass, *J. Environ. Qual.* 41(3), 793-799. doi:10.2134/jeq2011.0183.

Trenholm, L. E., Unruh, J. B. and Sartain, J. B. (2013). Nitrate leaching and turf quality in newly sodded St. Augustinegrass, *J. Plant Nutr.* 36(12), 1935-1943. doi:10.1080/01904167.2013.819893.

U.S. Salinity Laboratory Staff (1954). Diagnosis and *Improvement* of *Saline* and *Alkali Soils*, USDA Handb. 60. U.S. Gov. Print. Office, Washington, DC.

Weiss, J. M. (2020). Development of an adaptive irrigation system to optimize turfgrass irrigation using reclaimed water. Doctoral dissertation. Colorado School of Mines.

Wherley, B., Heitholt, J., Chandra, A. and Skulkaew, P. (2014). Supplemental irrigation requirements of zoysiagrass and Bermudagrass cultivars, *Crop Sci.* 54(4), 1823-1831. doi:10.2135/cropsci2013.11.0753.

Wherley, B., Dukes, M. D., Cathey, S., Miller, G. and Sinclair, T. (2015). Consumptive water use and crop coefficients for warm-season turfgrass species in the Southeastern United States, *Agrictural Water Managment* 156, 10-18. doi:10.1016/j.agwat.2015.03.020.

Wilhelm, S., Alshammary, S. and Qian, Y. (2010). Establishment, growth and irrigation requirements of Kentucky bluegrass and tall fescue as influenced by two irrigation water sources, *Res. J. Environ. Sci.* 4(5), 443-451.

Xiang, M., Schiavon, M., Orlinski, P., Forconi, A. and Baird, J. H. (2021). Identification of wetting agents for water conservation on deficit-irrigated hybrid Bermudagrass fairways, *Agron. J.* 113(5), 3846-3856 doi:10.1002/agj2.20450.

Xu, C. and Huang, B. (2012). Proteins and metabolites regulated by Trinexapac-ethyl in relation to drought tolerance in Kentucky bluegrass, *J. Plant Growth Regul.* 31(1), 25-37. doi:10.1007/s00344-011-9216-x.

Zhang, Q., Wang, S. and Rue, K. (2011). Salinity tolerance of 12 turfgrasses in three germination media, *Hort.Sci.* 46(4), 651-654. doi:10.21273/HORTSCI.46.4.651.

Zhang, X. and Schmidt, R. E. (2000). Hormone-containing products' impact on antioxidant status of tall fescue and creeping bentgrass subjected to drought, *Crop Sci.* 40(5), 1344-1349.

Zhang, X. and Ervin, E. H. (2004). Cytokinin-containing seaweed and humic acid extracts associated with creeping bentgrass leaf cytokinins and drought resistance, *Crop Sci.* 44(5), 1737-1745. doi:10.2135/cropsci2004.1737.

Zhang, X., Goatley, M., McCall, D., Kosiarski, K. and Reith, F. (2021). Humic acids-based biostimulants impact on root viability and hormone metabolism in creeping bentgrass putting greens, *Intl. Turfgrass Soc. Res. J.* 14(1), 288-294. doi:10.1002/its2.37.

Chapter 6

Advances in maintenance practices of turfgrass

Adam W. Thoms, Iowa State University, USA; and Alex J. Lindsey, University of Florida, USA

1 Introduction

Managed turfgrass stands are most often found in and around urban settings. In just 10 years, from 1982 to 1992, the United States experienced a 25% increase in urban land development (Pendall, 1999). This shift in land use offers both environmental benefits and consequences. Milesi et al. (2005) reported that intensively managed turfgrass stands need many energy-consuming inputs, which may become more expensive or harder to justify in the future. On a positive note, residential lawns have been estimated on average to sequester soil organic carbon between 170 years and 198 years, depending on amounts of inputs with fertilizers, frequency of mowing, and pesticide usage (Selhorts and Lal, 2013). As those inputs shift or are no longer economically or sustainably practical, other turfgrass management factors or options must be increased or improved. One area that can be improved is turfgrass maintenance practices; however, turfgrass managers are challenged to produce the desired level of turfgrass quality and function with fewer or more targeted inputs (Turgeon and Kaminski, 2019). Over US$45 billion was spent on turfgrass maintenance in the United States, according to a 1994 estimate (Beard and Green, 1994), thus contributing a significant amount to the economy back then as well as today.

http://dx.doi.org/10.19103/AS.2022.0110.06

Turfgrass management has greatly changed from the time when sheep were used for mowing turf areas such as lawns and golf courses (Fig. 1) to the intensively managed areas today that are mowed with electric robotic (i.e. autonomous) mowers (Fig. 2) (Christians et al., 2017). Advancements in technology have allowed turfgrass managers to improve the performance and playability of turf for their golf courses, lawns, park and recreation areas, and athletic fields or sports pitches, and also increase the sustainability of those amenity sites. The millennial generation, combined with additional government regulations, is also driving the adaption of technology in the turfgrass industry (Michael, 2019). Today's turfgrass management practices include technologies such as autonomous mowers, improved controlled released fertilizers, data-driven irrigation systems, and more (Michael, 2019). Some of these technological additions, however, are due to the consumers having a lesser understanding of basic lawn maintenance, and a lesser desire to maintain a lawn (Michael, 2019).

An actively growing and dense turfgrass stand is one of the best ways to maintain a healthy and desired turfgrass stand and also limit competition from unwanted plants or weeds (Turgeon and Kaminski, 2019; Watschke et al., 2013). In the last 30 years, suggested fertility rates for turf, especially with nitrogen, have decreased (Christians et al., 2017). At the same time as lowering fertility rates, other inputs have been limited, such as water and pesticides. To promote and increase sustainability efforts in the turfgrass industry, there may need to be 'trade-offs' between natural and synthetic products (Johnson et al., 2013). Sustainability

Figure 1 Sheep grazing on the White House lawn in 1919. Source: Photo credit by Library of Congress, located at: www.whitehousehistory.org.

Figure 2 Example of a battery-powered robotic or autonomous mower. Source: Photo credit by amazon.com, located at: https://www.bobvila.com/articles/best-robot-lawn-mower/.

in this chapter refers to having fewer or more strategic and targeted inputs, such as lower nitrogen and phosphorous rates, less frequent mowing, lower carbon dioxide output from mowers, etc. The following sections present information about recent advancements in mowing, putting green management, turfgrass renovation, and turf colorants that can allow turfgrass managers the ability to lower inputs while still producing and maintaining high-quality turfgrass.

2 Mowing

In order to be a turfgrass, a plant must withstand regular defoliation or mowing (Christians et al., 2017; Turgeon and Kaminski, 2019). While there is research that advocates for the reduction of mowed turf in many urban settings (Sivakoff et al., 2018), this potentially comes at a cost in the form of increased habitat for insects such as mosquitoes that carry diseases harmful to human health (Hamer et al., 2008). Additionally, unmaintained land areas often accumulate trash and can become an eyesore for the community. Therefore, maintained turf is still desired and very popular in many urban communities (Brosnan et al., 2020b). Turf stands also provide areas for exercise, recreation, and socialization (Turgeon and Kaminski, 2019).

Mowers used on turf have seen great advancements since those first grass-cutting devices were utilized or invented. Early mower machines were human-powered and operated like a scythe or a modern reel-type mower. Edwin Budding invented the first mechanical mower in 1830, which is similar to today's modern reel-mower (White, 1979). Early mowers were very heavy, but in the 1960s, mowers started to incorporate plastics into their design to lessen the mower's physical weight (Shukitis, 2012).

Mowers can be divided into three groups: rotary (Fig. 3), reel (Fig. 4), and flail mowers. Rotary mowers work well for higher heights-of-cut turfgrass maintained as lawns and low-maintenance pastures with a horizontal blade striking or impacting the grass and mulching it. Reel mowers work better for lower heights-of-cut turfgrass maintained as putting greens, fairways, tees, and some athletic fields or sports pitches, and clippings are often collected. Flail mowers are less frequently used in turfgrass maintenance but still serve the

Figure 3 Example of a walk-behind mower equipped with a rotary mulching blade useful to 'chop-up' leaves of lawn-height turfgrass.

Figure 4 Reel mowers are used to produce the highest quality of cut for turfgrass.

purpose of cutting down taller and often overgrown grass stands. Flail mowers work with a shaft that runs parallel to the surface; attached to that shaft are blades that rotate, and those blades will shred the vegetative material. Often the decks of flail mowers are larger and can cut the biomass into smaller pieces so the material will not leave clumps.

Gelernter et al. (2017) reported that there are many types of energy sources in the golf course industry to power facilities and equipment including gas, diesel, electric, natural gas, propane, and heating oil. Mowers are powered by many of these energy sources. In the late 1960s, General Electric (General Electric Company, Boston, MA, USA) released the Elec-Trak, the first commercially produced electric garden tractor with a mower deck and also a snow thrower attachment; however, this device was eventually discontinued due to high cost, low demand, and mechanical issues (Nesbitt, 2006). In a survey of golf course superintendents, 25.5% of respondents have replaced fossil fuel-burning equipment with electric equipment since 2011 (Gelernter et al., 2017). Today, there are also electric mowers that are recharged from solar panel-powered charging stations (Dutta et al., 2016).

Zero-turn mowers are often used because of their small turning radius and higher working efficiency than other mower types (Wang et al., 2016). These mowers have experienced recent advancements of larger decks and mowing widths to make-up for the labor shortage, to electronic fuel injection to save up to 30% fuel usage, and to operator comforts such as comfortable seats and easier access to service points (i.e. grease fittings) (Williams-Villano, 2020). Now both riding and stand-on zero-turn mowers offer a lithium-ion battery which will power 6 h of continuous mowing (Rathmell, 2018). Electric mowers will most likely increase in popularity as more communities continue to place restrictions on emissions and noise (Rathmell, 2018).

Modern turf maintenance recommendations include returning grass clippings and therefore plant nutrients back to the soil, thereby reducing fertilizer inputs (Christians et al., 2017). Grass clippings returned to the lawn will provide about half (97.6 kg•ha^{1} year^{1}) of the nitrogen needs for a turfgrass stand in 1 year compared to recommended nitrogen application rates (195.2 kg•ha^{1} year^{1}) (Heckman et al., 2000). Kopp and Guillard (2002) reported that returning grass clippings back into the stand reduced nitrogen fertilizer inputs by 50% and did not lower turfgrass quality. Heckman et al. (2000) also reported the use of slow-release fertilizer reduced clipping yield in the spring thus potentially make returning clippings a more acceptable practice.

Mowing frequency has often been determined by never removing more than one-third of the leaf tissue in a single mowing event (Christians et al., 2017; Fidanza et al., 2022). Removal of more than 40% of the leaf tissue in a single mowing can result in stopping root growth for over a week (Crider, 1955). A lower height of cut, however, would require more frequent mowing to avoid removing more than one-third of the leaf tissue (Turgeon and Kaminski, 2019). High-quality golf course putting greens have seen a shift with reel mowers from utilizing walk-behind to ride-on mowers and now back to walk-behind mower over the past 20 years (Breuninger et al., 2013). Nikolai (2005) reported that reductions in mowing heights will result in an increase in putting green speed, down as low as 4 mm height of cut. Double cut or mowing over the same area twice in a single mowing event is another way to increase green speed (Dickson et al., 2019). Both Nikolai (2005) and Rossi (2008) note that low mowing heights and double mowing will lower putting green quality over time.

2.1 Autonomous mowers

Most autonomous mowers are placed into the rotary mower category. Autonomous mowers are promoted for their ability to save time because there is no human operator and are battery powered which is presumably better for the environment (Pirchio et al., 2018). The fact that autonomous mowers utilize battery power as opposed to gas or diesel makes these new mowers 'friendlier' for the environment and they also produce less noise. Autonomous mowers are designed to operate every day or every other day, and they do not collect clippings but instead return them to the turf (Brede, 2000). In fact, many of these mowers are often activated at night when people are asleep. Pirchio et al. (2018) reported that autonomous mowers could improve turf shoot density and turf quality compared to rotary mowers but not reel mowers. No matter what type of mower is used, proper maintenance will help them operate in peak condition and optimize inputs such as fuel consumption or oil/lubricant usage.

2.2 Vertical mowing

Vertical mowing is a common practice employed to limit or manage thatch accumulation and increase turfgrass density (Gaussoin et al., 2013). Thatch is defined as a 'layer' above the soil and below the living green tissue and is composed of a combination of dead leaves, stems, and roots (Christians et al., 2017). Vertical mowing or verticutting is a practice where vertical blades mounted on a shaft spin (Fig. 5) and remove vegetative horizontal growth, which is common with golf course putting greens and with warm-season turfgrasses (Rowland et al., 2009). Landreth et al. (2007) reported that closer spaced verticutting blades at 1–3 mm width can remove more surface organic matter than core aeration. Vertical mowing can also reduce 'grain' (i.e. leaves and stems that are growing in the same horizontal direction) and improve turfgrass shoot development (Duble, 1996; Dunnivant, 2008). Turf sites with excessive thatch accumulation can also require more irrigation inputs than a turf stand with lower amounts of thatch (Beard, 1973). Verticutting has benefits such as reducing thatch and increasing stand density, but it is also very destructive and often is noticeable for several days after the practice has been carried out (Landreth et al., 2007).

2.3 Grooming

A less disruptive maintenance practice to help remove horizontal turf growth and excess organic matter in putting greens is grooming (Williams, 2016).

Figure 5 Example of a groomer attachment placed in front of the vertical reel blade assembly.

Grooming typically allows the vertical blades to penetrate into the turf canopy at ≤3.8 mm (Fig. 5), and this practice results in improved mowing quality and an increase in turf density from stimulated stolon growth (McCarty et al., 2005). Thoms et al. (2011) reported that grooming reduced bermudagrass (*Cynodon dactylon* [L.] Pers.) cover under simulated athletic field traffic, while Haselbauer et al. (2012) reported that grooming did not improve or decrease bermudagrass cover under simulated athletic field traffic. A growing number of golf course superintendents have elected to add vertically mounted brushes to mowers to help improve the vertical 'standing up' of plants immediately before mowing (Lulis et al., 2021). Both Whitlark (2011) and Thomas (2012) report that brushing can reduce grain from horizontal leaves, increase density, reduce leaf texture, and increase ball roll distance. While Lulis et al. (2021), however, reported a lower ball roll distance and lower turfgrass quality with bushing.

3 Putting green management

Advancements in turfgrass management practices are perhaps the most obvious on golf courses. For example, cultivation (i.e. physical removal of organic matter) continues to be a common practice employed on golf course, putting greens to limit thatch accumulation and reduce the negative effects of traffic (Turgeon, 1991; Turgeon and Fidanza, 2017). Lightweight rolling of putting greens also continues to be used to increase ball roll distance without lowering the height of cut (Nikolai et al., 2001). Recent research has indicated that plant growth regulators could be effectively applied on an air temperature-directed growing degree-day basis compared to a traditional calendar-date basis (Kreuser and Soldat, 2011; Reasor et al., 2018). Another recent practice that is being implemented for putting greens is to measure the clipping volume. This practice measures the volume of clippings from mowing over the area of a putting green (Tritabaugh, 2021). The daily observations from clipping volume data can help determine the mowing and rolling frequency, and when or if plant growth regulator or nitrogen applications are needed (Tritabaugh, 2021). For example, the goal for putting greens at Hazeltine National Golf Club (Chaska, MN, USA) is to produce between 10 mL•m^2 and 20 mL•m^2 clippings per putting green over a 7-day average (Tritabaugh, 2021). Another benefit of measuring the volume of clippings collected from a putting green is to identify any issues with the reel mower, such as poor quality of cut due to the mower or operator error (Tritabaugh, 2021).

3.1 Cultivation

Cultivation is often carried out along with sand topdressing to help manage thatch production and organic matter accumulation (Turgeon, 1991).

Hollow-tine aerification is most often used to reduce thatch and decrease root zone compaction (McCullough et al., 2005). For example, hollow-tine cultivation is typically employed to reduce the bulk density of root zones where creeping bentgrass (*Agrostis stolonifera* L.) is maintained (Atkinson et al., 2012) and reduce and manage the organic matter of ultradwarf bermudagrass (*Cynodon dactylon* [L.] Pers. × *Cynodon transvaalensis* Burt Davy]) sites (Rowland et al., 2009). Solid-tine aerification has also been reported to reduce soil bulk density and root zone compaction (Prämaßing et al., 2009). Even small tines and blades, however, have promoted the introduction of more air into the root zone (Cockerham et al., 1993). While cultivation practices are necessary to reduce soil compaction, they can be very disruptive to the turf playing surface (Fig. 6).

A newer device called the ShockWave (manufactured by Imants BV, Reusel, Netherlands; distributed in the United States by AquaAid Solutions, Rocky Mount, NC, USA) functions as a linear decompaction device with offset vertical knives that slice into the soil (Dalsgaard et al., 2020). One pass of this device could help lower surface hardness comparable to that of traditional core cultivation or aerification without turf damage (Dalsgaard et al., 2020). The use of the ShockWave could help reduce root zone compaction and thus improve turfgrass growth (Fig. 7).

3.1.1 Injection

A newer maintenance practice is injecting dry sand into the root zone with a high-pressure stream of water (Dickson et al., 2018). Sand injection has been growing in popularity on golf courses (Lowe, 2017). There are various-sized

Figure 6 Example of hollow-tine cultivation, which is commonly used to remove organic matter from turfgrass profiles and root zones.

Figure 7 Example of the ShockWave linear decompactor (manufactured by Imants BV, Reusel, Netherlands; distributed in the United States by AquaAid Solutions, Rocky Mount, NC, USA) for turfgrass management, useful to relieve soil compaction, improve root zone aeration and remove surface water with minimal surface disruption. Source: http://www.aquaaidsolutions.com.

injection ports, which will change the size of the water stream and thus alter the amount of dry sand being placed into the root zone (Fig. 8). For example, dry sand injection can save 30–40% sand compared to hollow-tine aeration and topdressing (Broadbelt and McCoy, 2018). Pure air injection machines are also useful to alleviate compaction. These air injection machines have very minimal surface disruption on putting greens and on minimal influence on their performance (i.e. ball roll distance, green turf cover, and surface hardness) compared to traditional cultivation practices (Dickson et al., 2018). Dry sand injection also produces smaller disruptions to the putting green surface compared to hollow- or solid-tine cultivation (Dickson et al., 2018). Both the air injection and dry sand injection practices are not as effective as hollow-tine cultivation for improving water infiltration, but when these injection practices are combined with hollow-tine cultivation, they can be more effective in improving soil physical properties (Amgain and Fontanier, 2019).

3.1.2 Cultivation and sand topdressing

Cultivation is important on golf course, putting greens and fairways to improve turfgrass growing conditions, improve the air-filled porosity in the root zone, and improve overall turfgrass health (Christians et al., 2017). Sand topdressing is commonly employed to produce a fast, level playing surface for consistent

Figure 8 Example of sand injection into a putting green utilizing the DryJect (DryJect Inc., Hatboro, PA, USA); note the vertical sand channel placed in the center of the turf root zone sample. Source: Photo courtesy of J. Broadbelt.

and true ball roll, to dilute organic matter in the root zone (Ervin and Nichols, 2011), and to reduce scalping (Turgeon and Kaminski, 2019). The addition of sand topdressing has been associated with lower anthracnose (*Colletotrichum cereale* sensu lato Crouch, Clarke, and Hillman) severity on annual bluegrass (*Poa annua* L.) putting greens (Inguagiato et al., 2013). Inguagiato et al. (2012) also reported frequent small amounts of sand topdressing associated with a reduction in anthracnose incidence and severity. Golf course creeping bentgrass fairways, and mixed creeping bentgrass with annual bluegrass fairways, have also seen significantly lower dollar (*Clarireedia* spp.) spot incidence and increased turf density from the addition of sand topdressing (Skorulski et al., 2010; Green et al., 2019b).

3.1.3 Fraise mowing as a method of cultivation

Fraise mowing is a cultivation method that removes the aboveground biomass and soil to a specific depth (Daily, 2016). Fraise mowing is a very aggressive form of vertical mowing and operates similar to an asphalt grinder (Fig. 9). Fraise mowers can include up to 120 blades on a rotor to remove up to a 5 cm depth of biomass (Brosnan et al., 2020a). Fraise mowing has multiple uses in turfgrass renovation such as removing existing turfgrass, aiding turfgrass transition during overseeding and possible weed control.

Figure 9 Example of fraise mowing commonly used in turfgrass renovation to quickly remove the surface layer.

Fraise mowing has been used for weed removal (Brosnan et al., 2020a) and for turfgrass establishment (McCauley et al., 2019). Richardson et al. (2021) found that partial control of bermudagrass was achieved with fraise mowing. Fraise mowing alone also reduced the smooth crabgrass (*Digitaria ischaemum* [Schreb.] Muhl.) seedbank in fine fescue (*Festuca* spp.); however, it did not provide acceptable crabgrass control and further treatments were recommended (Sitko, 2020). On cool-season athletic fields or sports pitches, fraise mowing has been used to remove annual bluegrass with the addition of overseeding Kentucky bluegrass (*Poa pratensis* L.) after fraise mowing (Hansen and Christians, 2015). Bearss et al. (2020) reported that fraise mowing could be useful for the removal of surface organic matter in cool-season turfgrass fairways, but it did not help annual bluegrass control in a creeping bentgrass fairway scenario. Fraise mowing is a useful tool to remove existing turf or plant debris following a nonselective herbicide application during a renovation, and it also has the potential to be used with herbicide applications for weed control and turfgrass eradication during a renovation (Bearss et al., 2020; Richardson et al., 2021).

3.2 Rolling

Rolling is a cultural practice that is effective at smoothing the surface of an area, improving turfgrass quality, and increasing putting green speed (i.e. ball roll distance) (Cockerham et al., 1993; Nikolai et al., 2001) (Fig. 10). Mowing plus rolling daily produced the longest ball roll distance on an annual bluegrass putting green compared to other regimes of mowing and rolling (McDonald et al., 2013). However, mowing 4 days per week with rolling daily, or mowing daily with rolling 3 days per week, achieved similar ball roll distance (McDonald et al., 2013). Additionally, Roberts et al. (2012) found that rolling a putting green every other day immediately after mowing increased ball roll distance and annual bluegrass quality. Giordano et al. (2013) reported that rolling a putting green two times per day improved creeping bentgrass quality and ball roll distance compared to one time per day and not rolling. Giordano et al. (2012) also observed an increase in root zone volumetric water content associated with rolling a creeping bentgrass putting green, which potentially could be useful to reduce irrigation inputs. A higher height of cut of a putting green generally results in a lower ball roll distance; however, rolling every other day and/or increased mowing frequency resulted in a similar or greater ball roll distance compared to a lower height of cut only (Inguagiato et al., 2009).

There have been research studies into fairway rolling focusing on disease control (Roberson and McCall, 2018; Genova et al., 2018; Green et al., 2019b). Lightweight rolling of fairway-height creeping bentgrass has been shown to

Figure 10 Example of a lightweight roller used for putting green surfaces; rear wheels are lowered for transport.

suppress dollar spots by 61% while not having any detrimental impact on turfgrass quality (Roberson and McCall, 2018). Genova et al. (2018) reported that rolling creeping bentgrass fairways in the morning 6 days a week reduced dollar spot compared to 3 days a week. Coen (2018) noted that rolling has replaced dragging for removing dew off fairways and that less staff and resources were needed for rolling compared to dragging. In a challenging labor market, along with increases in the cost of fungicides, rolling could be a viable option for dollar spot management on fairways.

Although rolling is a useful cultural practice that can improve ball roll distance and turfgrass quality, excessive rolling can be detrimental (Giordano et al., 2013; Hartwiger et al., 2001; Young et al., 2015, 2017). Hartwiger et al.

(2001) found that rolling four or seven times per week decreased creeping bentgrass quality compared to not rolling or rolling one time per week. Additionally, rolling more than twice per day reduced creeping bentgrass quality (Giordano et al., 2013). Daily rolling also resulted in an increased potential for golf ball mark injury and increased creeping bentgrass putting green wear during environmental stresses compared to not rolling (Young et al., 2015, 2017). Reducing rolling, conducting that practice on nonconsecutive days, or utilizing target rolling techniques can reduce overall wear on the putting green surface (Gilhuly, 2006; Young et al., 2017). For example, target rolling is useful to disperse or reduce wear in close proximity to the hole but not the entire putting green (Gilhuly, 2006; Young et al., 2015).

Plant growth regulators are often used on putting greens to achieve greater ball roll distance, but rolling in combination with plant growth regulators can potentially reduce application rates (Reasor and Brosnan, 2020). Lightweight rolling with plant growth regulator applications at a lower application rate resulted in greater ultradwarf bermudagrass quality and similar ball roll distance compared to higher rates of plant growth regulator applications applied without lightweight rolling (Reasor and Brosnan, 2020). Thus, lightweight rolling could be utilized to reduce the amount of plant growth regulators needed for warm-season grass putting greens management.

Rolling can also be used as a technique to assist with sand incorporation following topdressing. Brushing combined with vibratory rolling improved topdressing sand incorporation and increased surface hardness on an ultradwarf bermudagrass putting green (Kauffman et al., 2011). Additionally, brushing, vibratory rolling, and the combination of the two decreased total biomass in the upper 2.5 cm of the root zone compared to a non-topdressed control (Kauffman et al., 2011). Thus, rolling combined with brushing may be a useful technique to improve the efficiency of sand incorporation following sand topdressing.

There have been numerous studies that associated rolling with turfgrass disease control. Rolling of a creeping bentgrass putting greens three times per week was associated with decreased dollar spot incidence, localized dry spot, and broadleaf weeds compared to unrolled greens (Nikolai, 2002; Nikolai et al., 2001), which could potentially reducing fungicide inputs. Rolling once per day or twice in the morning at 5 days per week also reduced the dollar spot on a creeping bentgrass putting green compared to not rolling (Giordano et al., 2012). Furthermore, rolling two to four times per week also reduced dollar spot on putting greens with red fescue (*Festuca rubra* L.), colonial bentgrass (*Agrostis capillaris* L.), perennial ryegrass (*Lolium perenne* L.) and annual bluegrass (Espevig et al., 2020). Mattox et al. (2018, 2020) reported that double rolling 5 days per week has the potential to suppress or inhibit Microdochium patch (*Microdochium nivale* [Fr.] Samuels and Hallett) on an annual bluegrass putting green. Rolling every other day after mowing and utilizing vibratory rolling for sand topdressing incorporation did not increase the severity of anthracnose on an annual bluegrass putting green (Inguagiato et al., 2009, 2013; Roberts et al., 2012). While often noted as reducing dollar spot severity on putting greens, regular rolling on a mixed stand of creeping bentgrass and annual bluegrass fairway did not regularly reduce dollar spot (Green et al., 2019b). Although it is not fully understood, there are potential mechanisms theorized on how rolling reduces dollar spot and other disease incidence and severity. Potential mechanisms on why rolling reduces disease incidence and severity include removal of dew and guttation water, increasing soil water content, increasing soil microbial populations, and immune response activation from turfgrass damage (Delvalle et al., 2011; Giordano et al., 2012; Green et al., 2019b; Hou et al., 2019; Liu et al., 1995; Mattox et al., 2020; Nikolai et al., 2001).

3.3 Firmness measurement

Recent advances in technology have produced devices that are now being used to provide golf course superintendents and greenkeepers with a quantifiable method to manage putting green surfaces (O'Brien et al., 2021). In many cases, this takes the uncertainty out of putting green management and can reduce inputs such as overwatering. A putting green surface typically is managed to be as firm as possible to get the greatest ball roll possible. One way to measure surface firmness is to calculate the depth produced from a golf ball's impact on the putting surface (Fig. 11). With the FieldScout TruFirm (Spectrum Technology, Aurora, IL, USA), a weighted ball is dropped from a set height, the ball impacts the surface, and then the indentation or depth is measured (O'Brien et al., 2021). The Clegg Impact Surface Tester (Lafayette Instruments, Lafayette, IN, USA) is commonly used to test surface hardness with units in gravities, with the higher numbers correlated with firmer surfaces (Thoms et al., 2016). While various Clegg Impact Surface Testers exist, the 2.25 kg hammer is the most common one for turfgrass, especially sports turf (Thoms et al., 2016). Of note, previous research indicates that a lower soil moisture content will result in a harder or more firm surface (Brosnan and McNitt, 2008; Rogers and Waddington, 1992) (Fig. 12).

Soil moisture meters are employed to monitor the volumetric water content of turfgrass root zones and especially putting greens, which is used to help identify when irrigation water is needed (Karcher et al., 2019). These soil moisture meters produce a value (i.e. percent volumetric water content) that can be used as a guide or target threshold for deciding when to implement irrigation and other cultural practices (Karcher et al., 2019). Soil moisture meters provide a quantitative measurement of root zone water content versus visual inspection and the 'feel method' with using a pocket knife or soil profiler (Fig. 13).

Figure 11 The FieldScout TruFirm (Spectrum Technology, Aurora, IL, USA) is used to test the surface firmness of putting greens without causing large indentions into the surface.

4 Renovation

Sometimes a turfgrass stand cannot be easily improved with traditional maintenance or cultural management practices and therefore needs to be replaced. Turfgrass renovation is defined as the process of replacing the existing turf with a new grass without major soil tilling or grading (Christians et al., 2017). Renovation is needed when the turf site is excessively damaged beyond recovery or if a new cultivar or different grass species is desired and perhaps better suited for the intended purpose and/or the environment. Turfgrass renovation is typically done by killing or removing the existing grass or vegetation and then seeding into that area. The timing for renovation depends on the turfgrass species being considered for the new stand. For example, cool-season and warm-season grasses are best established in later summer through early fall, and at early through late spring, respectively (Christians et al., 2017). There are some exceptions, however, to the timing and method of renovation followed by turf establishment.

One exception is the reestablishment of cool-season turfgrass putting greens and fairways from winter damage. A concern with spring reestablishment is low soil temperature, which can delay and/or reduce seed germination. Ebdon and DaCosta (2021) investigated species and

Figure 12 The Clegg Impact Surface Tester (Lafayette Instrument Co., Lafayette, IN, USA) is commonly used to measure the surface hardness of athletic fields/sports pitches.

cultivars of bentgrasses seeded for early spring reestablishment and reported that '007', '13-M' and 'Memorial' creeping bentgrass, and 'Capri' and 'Tiger II' colonial bentgrass, were equal to perennial ryegrass with producing a faster emergence and turf cover in lower soil temperatures (≈10°C). The faster establishment from those creeping bentgrass cultivars in the earlier stages of establishment was attributed to a heavier (i.e. larger) bentgrass seed compared to smaller-sized bentgrass seed (Ebdon and DaCosta, 2021). It is possible that the faster establishment time could translate to reduced irrigation water needs since more watering typically is required for seed establishment periods, especially during the spring (Christians et al., 2017).

Besides the conventional turfgrass establishment practices (Christians et al., 2017), other alternative or sustainable establishment methods have been investigated. Slice seeding is the process of slicing or cutting into the soil with a circular disk which produces a very narrow furrow or trench, followed by seed 'dropping' into the furrow and further pressed into the soil to ensure optimum seed placement and seed-to-soil contact (Fig. 14). Carroll et al. (2021b) found that slicing seed into the test site alone in late summer resulted in greater or equal creeping bentgrass cover compared to slicing

Figure 13 Example of a soil moisture meter (POGO TurfPro, Stevens Water Monitoring Systems, Portland, OR, USA) to measure volumetric water content in a turfgrass root zone. Source: @POGOTurfPro.

plus hollow-tine cultivation in late summer. Additionally, surface dimpling with a high seeding rate (73 kg•ha^{1}) provided similar establishment and turf cover compared slicing (Carroll et al., 2021b). Slicing alone or surface dimpling with a high seeding rate has the potential to be a more sustainable renovation practice relative to core cultivation or the combination of slicing and cultivation. Another alternative technique for turfgrass establishment is late fall sod installation. Late fall sod installation, either covered or uncovered after placement, resulted in increased rooting compared to spring or early summer sod installation, which could allow the site to be utilized earlier in the year (Henderson et al., 2009).

Figure 14 Example of a Vredo Seeder (manufactured by Vredo Dodewaard BV, Dodewaard, Netherlands; distributed in the United States by AquaAid Solutions, Rocky Mount, NC, USA); turfgrass seed is sown directly into the soil between a 5- and 25-mm depth setting.

4.1 Weed control

Killing the existing turfgrass and the invasive weeds is a critical step in turf renovation (Christians et al., 2017). Typically, this requires a nonselective herbicide that will translocate throughout the target plants to effectively kill off the existing vegetative ground cover. The most common nonselective herbicide active ingredient is glyphosate and is applied alone or in combination with other herbicides or management techniques to provide control of both cool- and warm-season turfgrasses during a renovation. As a sustainable practice, the effective application of glyphosate has reduced the amount of fossil fuels used in agricultural crop production systems (Cerdeira and Duke, 2006).

There have been multiple studies looking at ways to completely control cool-season grasses during renovations. Williams and Burrus (2004) reported that glyphosate was the only option that provided an acceptable renovation of perennial ryegrass to a seeded bermudagrass. Other herbicides and plant growth regulators did not transition as well from perennial ryegrass to bermudagrass compared to glyphosate alone (Williams and Burrus, 2004). Glyphosate also provided the most consistent control of tall fescue (*Festuca arundinacea* Schreb.) and Kentucky bluegrass (McCullough et al., 2006; Park and Landschoot, 2005). Furthermore, glufosinate-ammonium and trifloxysulfuron also have the potential to be utilized in controlling Kentucky bluegrass if glyphosate-resistant Kentucky bluegrass is present at the site (McCullough et al., 2006). Patton et al. (2004) reported that renovating from perennial ryegrass to a seeded zoysiagrass (*Zoysia japonica* Steud.) can be achieved by glyphosate application before seeding, followed by appropriate herbicide applications to reduce weed competition. An alternative technique for transition from a Kentucky bluegrass/perennial ryegrass turfgrass stand to a bermudagrass stand while maintaining greater turf quality are applications of glyphosate in strips rather than broadcast applications (Mittlesteadt et al., 2009). It appears that consistent control of cool-season grasses during a renovation can be achieved with glyphosate.

The timing of cool-season grass control in warm-season grasses often occurs when the warm-season grass is dormant. This allows the use of nonselective herbicides to control the cool-season grasses without long-term injury to the desired warm-season grass. Hoyle and Reeves (2017) found that applications of glyphosate plus a colorant did not antagonize annual bluegrass or tall fescue control and it did not delay the spring green-up of zoysiagrass, which means utilizing a turfgrass colorant to retain color and the glyphosate for weed or cool-season grass control can occur at the same time in dormant warm-season grasses.

4.1.1 Weed control during turf establishment

Another common issue with turfgrass renovation is weed control during the establishment period. Specifically, successful annual bluegrass control during renovation can be a challenge. Applications of the herbicides ethofumesate or mesotrione provided annual bluegrass control and suppression, respectively, during a perennial ryegrass renovation (Park et al., 2019). Bispyribac-sodium is another herbicide that can provide excellent to moderate control of annual bluegrass during a creeping bentgrass renovation (Branham and Sharp, 2011).

Herbicides and fraise mowing have been investigated for grass weed control during renovations. Combining glyphosate and fluazifop-p-butyl with fraise mowing improved bermudagrass control, especially with herbicide application after fraise mowing (Richardson et al., 2021). Interval treatments of dazomet and fraise mowing with tillage-incorporated dazomet provided increased control of annual bluegrass; however, it did not provide complete control and it impeded creeping bentgrass establishment (Bearss, 2020; Bearss et al., 2020). Furthermore, tarping-, tillage-, and irrigation-incorporated dazomet treatments and removing the top 25–40 mm of soil reduced annual bluegrass populations (Bravo et al., 2018; Green et al., 2019a; Landschoot et al., 2004; Park and Landschoot, 2003). Creeping bentgrass planted one day after dazomet application also resulted in lower annual bluegrass levels compared to 5–9 days after application (Branham et al., 2004). Beyond annual bluegrass control, dazomet was also effective at controlling redroot pigweed (*Amaranthus retroflexus* L.) and large crabgrass (*Digitaria sanquinalis* [L.] Scop.) (Brecke et al., 2005).

4.1.2 Warm-season grass control

Controlling warm-season grasses during renovation is often thought to be more challenging compared to cool-season grasses; however, multiple studies have had good success. Glyphosate (applied once or twice) and glyphosate plus fluazifop-p-butyl applied before winter dormancy provided common bermudagrass control in a renovation to an improved seeded bermudagrass (Munshaw et al., 2009). Teuton et al. (2005) reported that three applications of glyphosate plus fluazifop-p-butyl 4 weeks apart provided complete control of hybrid bermudagrass. Another way to control common bermudagrass is to apply glyphosate plus fluazifop-p-butyl before dazomet application (Jeffries et al., 2017). Dazomet is a soil fumigant that has multiple uses in a turfgrass setting to reduce pest and weed populations. Furthermore, Jeffries et al. (2017) reported that tillage incorporation of dazomet provided more consistent control than irrigation incorporation and that tarping is not needed with tillage incorporation. Skipping the tarping step would save the turf practitioner time and resources. Additionally, surface applications of dazomet provided greater control of seashore paspalum (*Paspalum vaginatum* Swartz) compared to glyphosate applied with or without fluazifop-p-butyl (Brosnan and Breeden, 2009). Overall, glyphosate applied with or without fluazifop-p-butyl and dazomet during renovation has been shown to provide the most consistent control of warm-season grasses.

4.1.3 Light exclusion

A sustainable or nonchemical method of turfgrass renovation is light exclusion. This method is known as the 'lights out' technique, and it utilizes black, geo-textile mats of woven polypropylene plastic to shade out turfgrass or weeds during a renovation (Michalski and Cheng, 2018). The 'lights out' technique provided better weed suppression than glyphosate and successfully renovated St. Augustinegrass (*Stenotaphrum secundatum* [Walter] Kuntze) to a seeded bermudagrass (Michalski and Cheng, 2018). The steps in the 'lights out' technique include: cover existing turfgrass area for 6 weeks to successfully control turf and weeds; remove cover and then remove dead weeds and turf; fertilize and provide irrigation to stimulate weed seed germination for 4 weeks; place a cover over the area again to remove those newly germinated and emerged weeds; and then several weeks later remove cover and proceed with establishing new turfgrass (Michalski and Cheng, 2018). The 'lights out' technique may be a sustainable or nonchemical method to kill existing turfgrass and vegetation during a renovation, but limitations include the specific site requirements where the renovation is to be conducted and the time and resources needed.

4.2 Seedbed preparation following herbicide use

A common issue with turf renovation when utilizing herbicides is the timing of seedbed preparation following herbicide application. Thompson et al. (2019) reported that scalping tall fescue 1 day after glyphosate application did not reduce herbicide efficacy, which is 2 days sooner than indicated on the glyphosate product label (Anonymous, 2021). Additionally, the cultural practices of verticutting, core cultivation or both employed 1-7 days after glyphosate application on a creeping bentgrass fairway was not detrimental to long-term glyphosate efficacy (Miele et al., 2017). Both of those practices, as described in Thompson et al. (2019) and Miele et al. (2017), shorten the total time required for successful turf renovation.

4.3 Nonchemical control of annual bluegrass

Nonchemical control options for annual bluegrass are also being explored due to annual bluegrass developing resistance to certain herbicides (Cain, 2020). Fraise mowing at a deeper depth reduced annual bluegrass populations compared to fraise mowing at a shallow depth; however, bermudagrass recovery was slower from fraise mowing at the deeper depth relatively compared to the shallow depth (Carroll et al., 2021a). Although fraise mowing can provide mechanical suppression of annual bluegrass in zoysiagrass, additional herbicide applications would be needed for successful control (Brosnan et al., 2020a). In addition, fraise mowing has the potential to reduce spring dead spot (*Ophiosphaerella* spp.) disease incidence in bermudagrass (Miller et al., 2017).

Murphy et al. (2005) found that seeding creeping bentgrass early to mid-summer reduced annual bluegrass competition due to higher soil temperatures suppressing annual bluegrass germination. Additionally, increasing phosphorus fertilization at perennial ryegrass establishment was associated with a decrease in the annual bluegrass population (Cain, 2020). A possible alternative method for managing annual bluegrass, large crabgrass, and common dandelion weeds (*Taraxacum officinale* spp. *Vulgare* [Lam.] Schinz and R. Keller) (Cain, 2020) is by utilizing Kentucky bluegrass clipping extracts as an allelopathic material (Cain, 2020). Nonchemical control of weeds is an alternative to herbicide applications; however, complete control can be difficult, and more research is needed to further explore the practicality of these options.

4.4 Other plant and soil issues

If there are other issues within the turf such as soil compaction, excessive thatch, and/or poor drainage, then other cultural management techniques will be needed to successfully renovate the site (Thoms and Christians, 2021). Often the thatch layer and excessive organic matter accumulation need to be removed prior to successful turfgrass renovation and establishment. Removing the thatch layer is typically done via mechanical methods such as verticutting, utilizing a power rake, core cultivation or fraise mowing. Drainage and soil physical properties of turfgrass root zones are important topics that warrant a much more expanded review and discussion and have always been a critical component of successful turfgrass management (Musser, 1962).

4.4.1 Thatch

If the thatch layer is excessive, stripping the dead sod/thatch layer, tilling the top soil with a rotovator or rototiller and regrading or reshaping the area may be necessary for a successful renovation. This is the most time-consuming type of renovation, but it also provides an opportunity to incorporate soil additives and amendments to improve soil quality prior to turfgrass establishment. Soil additives are typically organic amendments such as compost, peat-based products, and humic substances. Improving soil quality or soil health can help alleviate problems later by increasing the root zone's infiltration rate and also water-holding capacity, improving soil structure, and enhancing nutrient cycling and availability (Turgeon and Kaminski, 2019).

4.4.2 Soil root zone properties

Compost and peat-based products can improve soil physical properties such as increased soil aggregation, enhanced soil aeration, and increased water-holding capacity (Bigelow and Soldat, 2013). Increasing the water-holding capacity allows for the potential to reduce irrigation water amounts or frequency (Radovanovic and Bean, 2020). Beyond improving physical properties, compost can increase soil organic matter and provide long-term benefits in maintaining turfgrass nutrient status in the soil (Evanylo et al., 2016). Additionally, compost and humic substances have the potential to increase soil microbial populations and activity and nitrogen mineralization (Azeem et al., 2020; Lindsey et al., 2021a,b).

Furthermore, humic fertilizers have the capability to release more plant-available nitrogen over the course of the year (Lindsey et al., 2021c). Humic substances and humic fertilizers have the potential to reduce the amount of fertilizer applications and/or rates, thus offering these products as a more sustainable plant nutrition option. Humic substances also have the potential to increase the availability of phosphorus and micronutrients along with enhancing turfgrass rooting, all of which are important during establishment (Gerke, 2018; Lindsey et al., 2020). Adding soil amendments to improve soil quality is another sustainable management technique that can be included in a turf renovation program.

5 Turf colorants

Turf colorants applied to warm-season turfgrass sites are an alternative to overseeding and are useful to extend the warm-season turfgrass' green color into their winter dormancy period (Liu et al., 2007; Shearman et al., 2005). Overseeding is where a cool-season grass is seeded into a dormant warm-season grass to provide a green turf cover during the winter months (Christians et al., 2017). Colorants are used to improve turf color on dormant or damaged turf, and often the colorant product amount is increased gradually with multiple applications to avoid streaking (Cockerham et al., 1993).

There have been studies on how to best apply turfgrass colorants and in what conditions to achieve the best turfgrass color and longevity (Fig. 15). Bermudagrass, buffalograss (*Buchloe dactyloides* [Nutt.] Engelm.), and zoysiagrass green color and the duration of acceptable green color increased with colorant application volume or with sequential applications (Braun et al., 2016, 2017; Briscoe et al., 2010; Nascimento et al., 2020). Colorant-treated zoysiagrass showed improved visual color compared to tall fescue (Braun et al., 2016). Environmental conditions such as air temperature and rainfall (i.e. leaf wetness) can affect colorant performance. Pinnix et al. (2018) found that colorant applications on dormant hybrid bermudagrass and zoysiagrass at a lower temperature (−3.9°C) resulted in a greater risk of colorant transfer, which was measured by dragging an absorbent cloth over the treated are, compared to applications at higher temperatures (7.2°C or 15.6°C). Additionally, preapplication irrigation reduced colorant streaking on dormant hybrid bermudagrass and zoysiagrass, which is often caused by the use of high-viscosity colorants or increased colorant product application rates (Pinnix et al., 2019). It is important to consider environmental conditions and application rates when utilizing turf colorants.

Turfgrass colorants have had limited success with consistently improving spring green-up and physiological processes of warm-season turfgrass. Applications of colorants increased turf canopy and soil temperatures and buffalograss spring green-up (Shearman et al., 2005). However, applications of colorants have not been successful in accelerating zoysiagrass and bermudagrass spring green-up (Braun et al., 2016; Hargey et al., 2016). Additionally, pigment-containing products did not enhance creeping bentgrass physiological processes that are associated with improved quality and health

Figure 15 Turfgrass paint and colorant pigments can discolor the root zone if applied in large quantities over time.

during summer/heat stress (McCarty et al., 2013, 2014). Furthermore, ultraviolet-absorbing colorant applications were not effective at reducing the daily light integral requirement of creeping bentgrass (Russell et al., 2019).

Applications of turf colorants can be a useful and cost-saving alternative to overseeding, depending on the needs of turf managers (Carson, 2004; Liu et al., 2007). Turf colorants can improve aesthetic value and playability or performance; however, it has limitations compared to overseeding, especially in high-traffic areas due to the lack of turf recovery (Christians et al., 2017; Hargey et al., 2016; Liu et al., 2007).

6 Conclusion and future trends in research

Turfgrass maintenance practices and programs have evolved greatly over recent decades and will continue to change in the coming years. Concerns over the lack of labor, the cost of inputs and resources, and environmental factors (i.e. local weather and regional climate) will continue to influence how amenity turfgrass is cared for and maintained. Mowing technology is experiencing trends with engineering and design with larger decks (i.e. larger mowing swath to help compensate for lack of available labor), more electric-powered equipment, and further development and implementation of autonomous mowers into the turfgrass industry.

Turfgrass organic matter management has become a priority with golf course putting green management, with the use of vertical mowing, grooming, and sand injection practices supplementing and/or replacing traditional hollow-tine cultivation and sand topdressing on some golf courses. Fraise mowing is an emerging practice utilized for the mechanical removal of weeds without the use of herbicides, as well as for site-specific turf renovation programs.

Rolling also has become a common and potentially more sustainable cultural practice for golf course putting greens, with the aim of achieving adequate or desired green speeds. Furthermore, rolling practices have the potential to facilitate an increase in root zone water content, a decrease in disease incidence and severity, and a lower application rate of plant growth regulators, all of which could help better target and reduce inputs for putting green management. Daily measurement of clipping volume on putting greens is a guiding decision on cultural management inputs, and plant growth regulator applications are guided by air temperature-based growing degree-day ranges. Measurements of surface hardness and soil moisture are now made accurately and quantitatively by portable devices, with that data collected, stored, and analyzed to further support decisions about cultural management practices and inputs.

Turfgrass renovations incorporate herbicide application for effective grass and vegetation control; however, mechanical and other alternative methods are being explored to aid in renovations which can reduce overall pesticide usage. Additionally, seedbed preparation, turfgrass establishment timing and methods, and products to improve root zone properties are being evaluated to ultimately improve soil quality, provide for a faster and successful establishment time, and thus potentially reduce the need for further inputs and resources. Lastly, in areas without high-traffic stress, turf colorants can be utilized instead of overseeding as a potential sustainable management practice to provide green color during periods of turfgrass dormancy.

Golf course superintendents and greenkeepers, and athletic field and sports pitch managers, always have been innovative and creative (Lucas, 2020) and will continue to develop new and improved turfgrass cultural maintenance practices that are economical, environmentally beneficial, and agronomically sustainable.

7 Where to look for further information

Education resources located on the following sites:

- Golf Course Superintendents Association of America (GSCAA). https://www.gcsaa.org/environment/golf-course-environmental-profile.
- Turfgrass Producers International (TPI). https://www.turfgrasssod.org/resources/.
- Sports Field Management Association (SFMA). https://www.sportsfieldmanagement.org/.
- The Lawn Institute. https://www.thelawninstitute.org/.

8 References

Amgain, N. and Fontanier, C. H. 2019. Effect of alternative aerification practice to maintain soil physical properties of putting green. *Agron. Abr. ASA, CSSA, and SSSA International Annual Meetings, San Antonio, TX. 119865*. Effect of Alternative Aerification Practice to Maintain Soil Physical Properties of Putting Green (confex.com).

Anonymous 2021. *Roundup® Weed and Grass Killer Super Concentrate*. Monsanto Technology LLC, Saint Louis, MO.

Atkinson, J. L., McCarty, L. B. and Bridges Jr., W. C. 2012. Effect of core aerification frequency, area impacted, and topdressing rate on turf quality and soil physical properties. *Agronomy Journal* 104(6), 1710–1715. https://doi.org/10.2134/agronj2012.0224.

Azeem, M., Hale, L., Montgomery, J., Crowley, D. and McGiffen Jr., M. E. 2020. Biochar and compost effects on soil microbial communities and nitrogen induced respiration in turfgrass soils. *PLoS ONE* 15(11), e0242209.

Beard, J. B. 1973. *Turfgrass: Science and Culture*. Prentice Hall, Englewood Cliffs, NJ.

Beard, J. B. and Green, R. L. 1994. The role of turfgrasses in environmental protection and their benefits to humans. *Journal of Environmental Quality* 23(3), 452–460. https://doi.org/10.2134/jeq1994.00472425002300030007x.

Bearss, R. C. 2020. *Annual Bluegrass Emergence from Fraze Mowed Fairway Systems Fumigated with Dazomet*. Michigan State University, East Lansing, MI.

Bearss, R. C., Rogers III, J. N., Crum, J. R. and Silcox, C. A. 2020. Fairway renovation with fraise mowing cultivation and dazomet fumigation. *Hortscience* 55(8), 1222–1227. https://doi.org/10.21273/HORTSCI15164-20.

Bigelow, C. A. and Soldat, D. J. 2013. Turfgrass root zones: management, construction methods, amendment characterization, and use. *Turfgrass: Biology, Use, and Management* 56, 383–423.

Branham, B. E., Hardebeck, G. A., Meyer, J. W. and Reicher, Z. J. 2004. Turfgrass renovation using dazomet to control the *Poa annua* L. soil seed bank. *HortScience* 39(7), 1763–1767.

Branham, B. E. and Sharp, W. 2011. Annual bluegrass control in seedling creeping bentgrass with bispyribac-sodium. *Ats* 8(1), 1–8.

Braun, R., Fry, J., Kennelly, M., Bremer, D. and Griffin, J. 2016. Colorant application volume and color persistence on a 'Chisholm' zoysiagrass lawn. *hortTechnology* 26(3), 314–319.

Braun, R. C., Fry, J. D., Kennelly, M. M., Bremer, D. J. and Griffin, J. J. 2017. Single and sequential colorant application effects on buffalograss and zoysiagrass color during dormancy. *hortTechnology* 27(3), 393–398.

Bravo, J. S., Green, T. O., Crum, J. R., Rogers, J. N., Kravchenko, S. and Silcox, C. A. 2018. Evaluating the efficacy of dazomet for the control of annual bluegrass seed germination in renovated turf surfaces. *hortTechnology* 28(1), 44–47.

Brecke, B. J., Unruh, J. B. and Stephenson IV, D. O. 2005. Dazomet blended with rootzone mix for fumigation. *International Turfgrass Society Research Journal* 10, 1176–1178.

Brede, D. 2000. *Turfgrass Maintenance Reduction Handbook: Sports, Lawns, and Golf.* Sleeping Bear Press, Chelsea, MI.

Breuninger, J. M., Welterlen, M. S., Augustin, B. J., Cline, V. and Morris, K. 2013. The turfgrass industry. in J. C. Stier, B. P. Horgan and S. A. Bonos (Eds.), *Turfgrass: Biology, Use and Management*, pp. 37–104. American Society of Agronomy, Madison, WI.

Briscoe, K., Miller, G. and Brinton, S. 2010. Evaluation of green turf colorant as an alternative to overseeding on putting greens. *Applied Turfgrass Science* 7(1), 1–8.

Broadbelt, J. and McCoy, E. 2018. Organic matter management on sand-based fields: extract and replace vs. injection. *SportsTurf* 34(5), 22–25.

Brosnan, J. T., Breeden, G. K., Zobel, J. M., Patton, A. J. and Law, Q. D. 2020a. Nonchemical annual bluegrass (*Poa annua*) management in zoysiagrass via fraise mowing. *Weed Technology* 34(4), 482–488, https://doi.org/10.1017/wet.2019.136.

Brosnan, J. T., Chandra, A., Gassoin, R. E., Kowalewski, A., Leinauer, B., Rossi, F. S., Soldat, D. J., Stier, J. C. and Unruh, B. 2020b. A justification for continued management of turfgrass during economic contraction. *Agricultural and Environmental Letters*. doi .org/10.1002/ael2.20033.

Brosnan, J. T. and Breeden, G. K. 2009. Surface applications of dazomet provide nonselective control of seashore paspalum (*Paspalum vaginatum*) turf. *Weed Technology* 23(2), 270–273.

Brosnan, J. T. and McNitt, A. S. 2008. Surface conditions of highly maintained baseball fields in the northeastern United States. Part 1 – non-turfed basepaths. *Applied Turfgrass Science* 5(1), 1–8. https://doi.org/10.1094/ATS-2008-0520-01-RS.

Cain, A. B. 2020. Non-chemical approaches to managing annual bluegrass (*Poa annua L.*). M.S. Thesis. Oregon State University, Corvallis, OR.

Carroll, D. E., Brosnan, J. T., Unruh, J. B., Stephens, C. A., McKeithen, C. and Boeri, P. A. 2021a. Non-chemical control of annual bluegrass (*Poa annua*) in Bermudagrass (*Cynodon* spp.) via fraise mowing: efficacy and barriers to adoption. *Sustainability* 13(15), 8124.

Carroll, D. E., Kaminski, J. E. and Landschoot, P. J. 2021b. Influence of cultivation method, seeding rate, and fertilizer rate on the establishment of creeping bentgrass for putting green renovation. *Crop, Forage & Turfgrass Management* 7(1), e20085.

Carson, T. 2004. Overseed or paint? *Golf Course Management* 72(12), 22.

Cerdeira, A. L. and Duke, S. O. 2006. The current status and environmental impacts of glyphosate-resistant crops: a review. *Journal of Environmental Quality* 35(5), 1633–1658. https://doi.org/10.2134/jeq2005.0378.

Christians, N. E., Patton, A. J. and Law, Q. D. 2017. *Fundamentals of Turfgrass Management* (5th edn.). John Wiley and Sons Inc., Hoboken, NJ.

Cockerham, S. T., Gibeault, V. A. and Khan, R. A. 1993. Alteration of sports field characteristics using management. *International Turfgrass Society Research Journal* 7, 182-191.

Coen, C. 2018. *Successful Daily Maintenance in 2018*. The Reporter [IGCSA]. Sept. 4-5.

Crider, F. J. 1955. Root-growth stoppage resulting from defoliation of grass. U.S. Dept. of Ag. Washington, DC, USDA Tech. Bull. No. 1102:3-23 .

Daily, D. 2016. Is Fraze mowing right for you? *SportsTurf Magazine* 32(2), 16-18.

Dalsgaard, T. O., Thoms, A. W., Christians, N. E., Mertz, I. and Horton, R. 2020. Comparison of shockwave aerification and conventional aerification methods on athletic fields. *Agronomy Journal* 2, 1-8. https://doi.org/10.1002/agj2.20255.

Delvalle, T. C., Landschoot, P. J. and Kaminski, J. E. 2011. Effects of dew removal and mowing frequency on fungicide performance for dollar spot control. *Plant Disease* 95(11), 1427-1432.

Dickson, K. H., Strunk, W. D., Sorochan, J. C., Nikolai, T. A. and Hathaway, A. 2019. Impact of clip of the reel on creeping bentgrass (*Agrostis stolonifera* L.) putting greens. *Crop, Forage and Turfgrass Management* 4(1), 1-3 https://doi.org/10.2134/cftm2018.06.0049.

Dickson, K., Sorochan, J., Munshaw, G. and Thoms, A. 2018. Comparison of cultivation methods impact on playability of *Agrostis stolonifera* greens. *Journal of Testing and Evaluation* 46(3), 1256-1260, https://doi.org/10.1520/JTE20160419.

Duble, R. L. 1996. *Turfgrasses: Their Management and Use in the Southern Zone* (2nd edn.). Texas A&M University Press, College Station, TX.

Dunnivant, W. E. 2008. Grooming frequency and spacing effects on a TifEagle Bermudagrass putting green, M.S. Thesis. Auburn University, Auburn, AL.

Dutta, P. P., Baruah, A., Konwar, A. and Kumar, V. 2016. A technical review of lawn mower technology. *ADBU Journal of Engineering Technology* 4, 1-5.

Ebdon, J. S. and DaCosta, M. 2021. Soil temperature mediated seedling emergence and field establishment in bentgrass species and cultivars during spring in the Northeastern United States. *hortTechnology* 31(1), 42-52.

Ervin, E. and Nichols, A. 2011. Organic matter dilution programs for sand-based putting greens. *USGA Turfgrass and Environmental Research Online* 10(8), 1-6.

Espevig, T., Usoltseva, M. and Norman, K. 2020. Effects of rolling and N-fertilization on dollar spot and Microdochium patch on golf greens in Scandinavia. In BIO Web of Conferences (Vol. 18, p. 00008). EDP Sciences.

Evanylo, G. K., Porta, S. N., Li, J., Shan, D., Goatley, J. M. and Maguire, R. 2016. Compost practices for improving soil properties and turfgrass establishment and quality on a disturbed urban soil. *Compost Science and Utilization* 24(2), 136-145.

Fidanza, M., Baldwin, C., Koch, M., Freshour, L. and Harriman, B. 2022. Required Mowing events of transgenic Kentucky bluegrass cultivars compared with conventional turfgrass cultivar blends and species mixtures. *Crop, Forage and Turfgrass Management*. DOI:10.1002/cft2.20146.

Gaussoin, R. E., Berndt, W. L., Dockrell, C. A. and Drijber, R. A. 2013. Characterization, development, and management of organic matter in turfgrass systems. in J. C. Stier, B. P. Horgan, and S. A. Bonos (Eds.), *Turfgrass: Biology, Use, and Management*, pp. 425-456. American Society of Agronomy, Madison, WI.

Gelernter, W. D., Stowell, L. J., Johnson, M. E. and Brown, C. D. 2017. Documenting trends in energy use and environmental practices on US golf courses. *Crop, Forage and Turfgrass Management* 3(1). https://doi.org/10.2134/cftm2017.07.0044.

Genova, K. M., Clarke, B. B. and Murphy, J. A. 2018. *Impact of Rolling and Dew Removal on Dollar Spot Disease of Bentgrass Fairway Turf*. ASA, CSSA, SSSA Intl. Ann. Mtg. 113520.

Gerke, J. 2018. Concepts and misconceptions of humic substances as the stable part of soil organic matter: a review. *Agronomy* 8(5), 76.

Gilhuly, L. 2006. Green speed: trick or treat. *USGA Green Section Record* 44(4), 17-19.

Giordano, P. R., Johnson, C., Richards, J. S., Richardson, M. D., Karcher, D. E., Vargas Jr., J. M. and Nikolai, T. A. 2013. Investigating lightweight rolling thresholds for putting green quality and performance. *International Turfgrass Society Research Journal* 12, 157-162.

Giordano, P. R., Nikolai, T. A., Hammerschmidt, R. and Vargas Jr., J. M. 2012. Timing and frequency effects of lightweight rolling on dollar spot disease in creeping bentgrass putting greens. *Crop Science* 52(3), 1371-1378. https://doi.org/10.2135/cropsci2011.07.0373.

Green, T. O., Kravchenko, A., Rogers, J. N. and Vargas, J. M. 2019a. Annual bluegrass: emergence of viable seed in various putting green sites and soil removal depths. *hortTechnology* 29(4), 438-442.

Green, T. O., Rogers, J. N., Crum, J. R., Vargas, J. M. and Nikolai, T. A. 2019b. Effects of rolling and sand topdressing on dollar spot severity in fairway turfgrass. *hortTechnology* 29(4), 394-401.

Hamer, G. L., Kitron, U. D., Brawn, J. D., Loss, S. R., Ruiz, M. O., Goldberg, T. L. and Walker, E. D. 2008. *Culex pipiens* (Diptera: Culicidae): a bridge vector of West Nile virus to humans. *Journal of Medical Entomology* 45(1), 125-128.

Hansen, K. and Christians, N. 2015. Establishing Kentucky bluegrass after Fraze mowing: time to recovery after Fraze mowing can be affected by seeding rates and the use of turf covers. *Golf Course Management* 83, 88-93.

Hargey, D., Wherley, B., Reynolds, C., White, R. and Parker, G. 2016. Performance of winter overseeded and colorant-treated 'tifway' Bermudagrass receiving weekly irrigation and simulated traffic. *Hortscience* 51(5), 580-587.

Hartwiger, C. E., Peacock, C. H., DiPaola, J. M. and Cassel, D. K. 2001. Impact of light-weight rolling on putting green performance. *Crop Science* 41(4), 1179-1184.

Haselbauer, W. D., Thoms, A. W., Sorochan, J. C., Brosnan, J. T., Schwartz, B. M. and Hanna, W. W. 2012. Evaluation of experimental bermudagrasses under simulated athletic field traffic with perennial ryegrass overseeding. *hortTechnology* 22(1), 94-98.

Heckman, J. R., Liu, H., Hill, W., DeMilia, M. and Anastasia, W. L. 2000. Kentucky bluegrass responses to mowing practice and nitrogen fertility management. *Journal of Sustainable Agriculture* 15(4), 25-33. https://doi.org/10.1300/J064v15n04_04.

Henderson, J., Miller, N., Guillard, K., Harel, O. and Raman, B. 2009. Late fall sod installation produces equivalent or greater rooting strength of Poa pratensis than typical spring installations during the subsequent growing season. *Applied Turfgrass Science* 6(1), 1-11.

Hou, S., Liu, Z., Shen, H. and Wu, D. 2019. Damage-associated molecular pattern-triggered immunity in plants. *Frontiers in Plant Science* 10, 646.

Hoyle, J. A. and Reeves, J. A. 2017. Effect of colorant and glyphosate application timing on annual bluegrass and tall fescue control in dormant 'Meyer' zoysiagrass. *Itsrj* 13(1), 727-730.

Inguagiato, J. C., Murphy, J. A. and Clarke, B. B. 2009. Anthracnose disease and annual bluegrass putting green performance affected by mowing practices and lightweight rolling. *Crop Science* 49(4), 1454–1462.

Inguagiato, J. C., Murphy, J. A. and Clarke, B. B. 2012. Sand topdressing rate and interval effects on anthracnose severity of an annual bluegrass putting green. *Crop Science* 52(3), 1406–1415. https://doi.org/10.2135/cropsci2011.01.0010.

Inguagiato, J. C., Murphy, J. A., Clarke, B. B. and Roberts, J. A. 2013. Topdressing sand particle shape and incorporation effects on anthracnose severity of an annual bluegrass putting green. *International Turfgrass Society Research Journal* 12, 127–133.

Jeffries, M. D., Gannon, T. W., Reynolds, W. C., Yelverton, F. H. and Silcox, C. A. 2017. Herbicide applications and incorporation methods affect dazomet efficacy on Bermudagrass. *hortTechnology* 27(1), 24–29.

Johnson, P. G., Rossi, F. S. and Horgan, B. P. 2013. Sustainable turfgrass management in an increasingly urbanized world. In: *Turfgrass: Biology, Use, and Management*. https://doi.org/10.2134/agronmonogr56.c26.

Karcher, D., Richardson, M. and O'Brien, D. 2019. Use moisture meters for greater efficiency and healthier turf: the initial expense of purchasing a moisture meter may seem high, but many superintendents find that the savings from reduced water use and healthier turfgrass leads them to buy an additional meter – or meters. *Golf Course Management* 87(7), 40, 42, 44, 46.

Kauffman, J. M., Sorochan, J. C. and Brosnan, J. T. 2011. Brushing plus vibratory rolling enhances topdressing incorporation on ultradwarf Bermudagrass putting greens. *Applied Turfgrass Science* 8(1), 1–8.

Kopp, K. L. and Guillard, K. 2002. Clipping management and nitrogen fertilization of turfgrass. *Crop Science* 42(4), 1225–1231. https://doi.org/10.2135/cropsci2002.1225.

Kreuser, W. C. and Soldat, D. J. 2011. A growing degree day model to schedule trinexapac-ethyl applications on *Agrostis stolonifera* golf putting greens. *Crop Science* 51(5), 2228–2236. https://doi.org/10.2135/cropsci2011.01.0034.

Landreth, J., Karcher, D. and Richardson, M. 2007. Cultivating to Manage Organic Matter in Sand-Based Putting Greens: University of Arkansas. *USGA Turfgrass and Environmental Research Online* 6(19), 1–7.

Landschoot, P. J., Park, B. S., McNitt, A. S. and Fidanza, M. A. 2004. Effect of dazomet on annual bluegrass emergence and creeping bentgrass establishment in putting green turf. *HortScience* 39(6), 1478–1482.

Lindsey, A. J., Thoms, A. W. and Christians, N. E. 2020. Kentucky bluegrass and Bermudagrass rooting response to humic fertilizers during greenhouse establishment. *Agronomy Journal* 112(5), 3396-3401.

Lindsey, A. J., Thoms, A. W., McDaniel, M. D. and Christians, N. E. 2021a. Evaluation of humic fertilizers applied at full and reduced nitrogen rates on Kentucky bluegrass quality and soil health. *Agronomy* 11(2), 395.

Lindsey, A. J., Thoms, A. W., McDaniel, M. D. and Christians, N. E. 2021b. Evaluation of humic fertilizers on a sand-based creeping bentgrass putting green. *Crop Science* 61(5), 3734–3745.

Lindsey, A. J., Thoms, A. W., McDaniel, M. D. and Christians, N. E. 2021c. Plant-available soil nitrogen fluxes and turfgrass quality of Kentucky bluegrass fertilized with humic substances. *Crop Science* 61(6), 4416–4424.

Liu, H., McCarty, L., Sarvis, W., Baldwin, C. and Long, S. 2007. Comparison of bermudagrass green winter painting and overseeding. *Agron. Abr. ASA, CSSA, and SSSA International Annual Meetings, New Orleans, LA. 35944*. Comparison of Bermudagrass Green Winter Painting and Overseeding. (confex.com).

Liu, L. X., Hsiang, T., Carey, K. and Eggens, J. L. 1995. Microbial populations and suppression of dollar spot disease in creeping bentgrass with inorganic and organic amendments. *Plant Disease* 79(2), 144-147.

Lowe, T. 2017. *Are There Alternatives to Core Aeration? USGA Course Care Regional Update: Southeast Region*. United States Golf Association. 2.

Lucas, M. 2020. *History of Greenkeeping: Golf's Cause and Effect*. Links Counsellor South, Dartmouth, MA.

Lulis, T., Russell, T. R. and Kaminski, J. E. 2021. Influence of brush type and brushing frequency on putting green ball roll distance and turfgrass quality. *Crop Science* 61, 2989-2996.

Mattox, C. M., Kowalewski, A. R., McDonald, B. W., Lambrinos, J. G. and Pscheidt, J. W. 2018. Rolling and biological control products affect Microdochium patch severity on a sand-based annual bluegrass putting green. *Agronomy Journal* 110(6), 2124-2129.

Mattox, C. M., Kowalewski, A. R., McDonald, B. W., Lambrinos, J. G. and Pscheidt, J. W. 2020. Combinations of rolling, mineral oil, sulfur, and phosphorous acid affect Microdochium patch severity. *Agronomy Journal* 112(5), 3383-3395.

McCarty, L. B., Gregg, M. F., Toler, J. E., Camberato, J. J. and Hill, H. S. 2005. Minimizing thatch and mat development in a newly seeded creeping bentgrass golf green. *Crop Science* 45(4), 1529-1535.

McCarty, L. B., Gann, J. R., Wells, C. E., Bruce, T. F. and Gerard, P. D. 2013. Physiological responses of creeping bentgrass to pigment-containing products. *Agronomy Journal* 105(6), 1797-1802.

McCarty, L. B., Gann, J. R., Wells, C. E. and Gerard, P. D. 2014. Creeping bentgrass field response to pigment-containing products. *Agronomy Journal* 106(4), 1533-1539.

McCauley, R. K., Pinnix, G. D. and Miller, G. L. 2019. Fraise mowing as a spring transition aid. *Crop, Forage & Turfgrass Management* 5(1), 1-5.

McCullough, P. E., Liu, H. and McCarty, L. B. 2005. Mowing operations influence creeping bentgrass putting green ball roll following plant growth regulator applications. *HortScience* 40(2), 471-474.

McCullough, P. E., Hart, S. E., Askew, S., Dernoeden, P. H., Reicher, Z. and Weisenberger, D. 2006. Kentucky bluegrass control with postemergence herbicides. *Hortscience* 41(1), 255-258.

McDonald, B. W., Golembiewski, R. C., Cook, T. W. and Blankenship, T. M. 2013. Effects of mowing and rolling frequency, Primo Maxx, and roller weight on annual bluegrass putting green speed. *Ats* 10(1), 1-9.

Michael, D. J. 2019. *Turfgrass Science: An Industry Perspective on the Future of Turfgrass Research*. [Abstract]. CSSA, SSA Intl, ASA. Ann. Mtg, San Antonio, TX, 118047.

Michalski, J. and Cheng, Z. 2018. Effects of "lights out" turfgrass renovation on plants, soil arthropod and nematode communities. *Applied Soil Ecology* 127, 144-154.

Miele, K. M., Henderson, J. J. and Inguagiato, J. C. 2017. Glyphosate efficacy as influenced by cultivation practices on a creeping bentgrass fairway turf. *Hortscience* 52(11), 1621-1626.

Milesi, C., Elvidge, C. D., Dieta, J. B., TuĴle, B. T., Nemani, R. R. and Running, S. W. 2005. A strategy for mapping and modeling the ecological effects of US lawns. *Journal of Turfgrass Management* 1, 83-97.

Miller, G. L., Earlywine, D. T. and Fresenburg, B. S. 2017. Effect of Fraze mowing on spring dead spot caused by *Ophiosphaerella herpotricha* of Bermudagrass. *International Turfgrass Society Research Journal* 13(1), 225-228.

Mittlesteadt, T. L., Goatley, J. M., Askew, S. D., Hensler, K. L. and Askew, W. L. 2009. Low-impact conversion of cool-season turf to 'patriot' Bermudagrass. *International Turfgrass Society Research Journal* 11, 1205-1212.

Munshaw, G. C., Philley, H. W. and Beasley, J. S. 2009. No-till strategies for converting common Bermudagrass (*Cynodon dactylon* L.) to improved seeded cultivars. *International Turfgrass Society Research Journal* 11, 427-436.

Murphy, J. A., Samaranayake, H., Lawson, T. J., Honig, J. A. and Hart, S. 2005. Seeding date and cultivar impact on establishment of bentgrass in soil containing annual bluegrass seed. *International Turfgrass Society Research Journal* 10(1), 410-415.

Musser, H. B. 1962. *Turf Management*. McGraw-Hill, New York.

Nascimento, M. V. Ld, Santos, P. L. Fd, Costa, J. V., Martins, J. T., Villas Bôas, R. L. and Godoy, L. J. Gd 2020. Durability and concentration of organic colorant in the visual quality of Discovery™ Bermudagrass. *Ornamental Horticulture* 26(4), 621-632.

Nesbitt, S. R. 2006. The buzz on electronics. *Golf Course Management March* 74(3), 58, 60, 62-64.

Nikolai, T. A. 2002. Effects of rolling and fertility on different root zones. Doctoral dissertation. Michigan State University, East Lancing, MI.

Nikolai, T. A. 2005. *The Superintendent's Guide to Controlling Putting Green Speed*. John Wiley & Sons, Hoboken, NJ.

Nikolai, T. A., Rieke, P. E., Rogers, J. N. and Vargas Jr., J. M. 2001. Turfgrass and soil responses to lightweight rolling on putting green root zone mixes. *International Turfgrass Society Research Journal* 9(2), 604-609.

O'Brien, D., Karcher, D. and Richardson, M. 2021. Firmness measurements: OK great… Now what do I do with them. *Golf Course Management*, 64-68.

Park, B. S., Elmore, M. T. and Murphy, J. A. 2019. Using herbicides and perennial ryegrass to renovate turf dominated by annual bluegrass. *Crop, Forage & Turfgrass Management* 5(1), 1-7.

Park, B. S. and Landschoot, P. J. 2003. Effect of dazomet on annual bluegrass emergence and creeping bentgrass establishment in turf maintained as a golf course fairway. *Crop Science* 43(4), 1387-1394.

Park, B. S. and Landschoot, P. J. 2005. Nonselective herbicides for control of Kentucky bluegrass and tall fescue turf. *International Turfgrass Society Research Journal* 10, 1198-1204.

Patton, A. J., Williams, D. W. and Reicher, Z. J. 2004. Renovating golf course fairways with zoysiagrass seed. *HortScience* 39(6), 1483-1486.

Pendall, R. 1999. Do land-use controls cause sprawl? *Environment and Planning B* 26(4), 555-571.

Pinnix, G. D., McCauley, R. K. and Miller, G. L. 2018. Air temperature effects on turfgrass colorant transfer. *Crop, Forage & Turfgrass Management* 4(1), 1-6.

Pinnix, G. D., McCauley, R. K. and Miller, G. L. 2019. Leaf wetness influences turf colorant application. *Crop, Forage & Turfgrass Management* 5(1), 1-6.

Pirchio, M., Fontanelli, M., Frasconi, C., Martelloni, L., Raffaelli, M., Peruzzi, A., Caturegli, L., Gaetani, M., Magni, S., Volterrani, M. and Grossi, N. 2018. Autonomous rotary mowers versus ordinary reel mower- effects of cutting height and nitrogen rate on manila grass turf quality. *hortTechnology* 28(4), 509-515. HTTPS://doi.org/10.21273/HORTTECH04064-18.

Prämaßing, W., Reinders, A., Franken, H. and Schulz, H. 2009. Physical properties of different turfgrass soils influenced by aeration treatments. *International Turfgrass Society Research Journal* 11(2), 1139-1152.

Radovanovic, J. and Bean, E. 2020. Evaluation of amending compacted residential soils with compost on nutrient leaching. In: *World Environmental and Water Resources Congress 2020: Water Resources Planning and Management and Irrigation and Drainage*. American Society of Civil Engineers, Reston, VA. 78-90.

Rathmell, L. 2018. A greener mow:Greenworks debuted new battery-powered zero-turn mowers. *Lawn and Landscape* 39(9), 10-12.

Reasor, E. H. and Brosnan, J. T. 2020. Trinexapac-ethyl applications and lightweight rolling on ultradwarf Bermudagrass putting greens. *Crop, Forage & Turfgrass Management* 6(1), e20036.

Reasor, E. H., Brosnan, J. T., Kerns, J. P., Hutchens, W. J., Taylor, D. R., McCurdy, J. D., Soldat, D. J. and Kreuser, W. C. 2018. Growing degree day models for plant growth regulator applications on ultradwarf hybrid Bermudagrass putting greens. *Crop Science* 58(4), 1801-1807. https://doi.org/10.2135/cropsci2018.01.0077.

Richardson, M. D., Brosnan, J. T., McCalla, J. H. and Breeden, G. K. 2021. Fraise mowing can improve herbicidal control of Bermudagrass. *Agronomy Journal* 113(5), 3789-3799.

Roberson, T. L. and McCall, D. S. 2018. *Influence of Lightweight Fairway Rolling on Dollar Spot Suppression with Ferrous Sulfate*. [Abstract]. CSSA, ASA, SSSA Intl. Ann. Mgt. Baltimore, MD, 112067.

Roberts, J. A., Murphy, J. A. and Clarke, B. B. 2012. Lightweight rolling effects on anthracnose of annual bluegrass putting greens. *Agronomy Journal* 104(4), 1176-1181.

Rogers III, J. N. and Waddington, D. V. 1992. Impact absorption characteristics on turf and soil surfaces. *Agronomy Journal* 84(2), 203-209. https://doi.org/10.2134/agronj1992.00021962008400020016x.

Rossi, F. S. 2008. Does mower type effect turf performance. *CUTT* 19(1)1, 6-8.

Rowland, J. H., Cisar, J. L., Snyder, G. H., Sartain, J. B. and Wright, A. L. 2009. USGA ultradwarf Bermudagrass putting green properties as affected by cultural practices. *Agronomy Journal* 101(6), 1565-1572. https://doi.org/10.2134/agronj2009.0154.

Russell, T. R., Karcher, D. E. and Richardson, M. D. 2019. Daily light integral requirement of a creeping bentgrass putting green as affected by shade, trinexapac-ethyl, and a plant colorant. *Crop Science* 59(4), 1768-1778.

Selhorts, A. and Lal, R. 2013. Net carbon sequestration potential and emissions in home lawn turfgrasses of the United States. *Environmental Management* 51(1), 198-208.

Shearman, R. C., Wit, L. A., Severmutlu, S., Budak, H. and Gaussoin, R. E. 2005. Colorant effects on dormant buffalograss turf performance. *hortTechnology* 15(2), 244-246.

Shukitis, A. 2012. History of Lawn Mowers. Honors Theses. University of South Florida.

Sitko, C. 2020. *Non-chemical Vegetation Management Using Fraise Mowing in Naturalized Golf Course Grasslands*. Cornell University.

Sivakoff, F. S., Prajzner, S. P. and Gardiner, M. M. 2018. Unique bee communities within vacant lots and urban farms result from variation in surrounding urbanization intensity. *Sustainability* 10(6), 1926-1943. https://doi.org/10.3390/su10061926.

Skorulski, J., Henderson, J. and Miller, N. 2010. Topdressing fairways: more is better. *USGA Green Section Record* 48(2), 15–17.

Teuton, T. C., Bryan Unruh, J., Brecke, B. J., Miller, G. L. and Mueller, T. C. 2005. Hybrid Bermudagrass (*Cynodon dactylon* (L.) Pers. × *C. transvaalensis* Burtt-Davy) control with glyphosate and fluazifop. *Applied Turfgrass Science* 2(1), 1–6.

Thomas, R. 2012. Comb over brushing greens prior to mowing provides multiple benefits without the surface disruption. *Golf Course Industry* 24, 34–37.

Thoms, A. W., Brosnan, J. T., Sorochan, J. C. and Saxton, A. M. 2016. Exploring relationships in surface hardness data collected with different instruments. *Journal of Testing and Evaluation* 44(2), 945–951.

Thoms, A. W. and Christians, N. E. 2021. *Turfgrass Renovation*. Iowa State Education and Outreach Publications, HORT (vol. 3111).

Thoms, A. W., Sorochan, J. C., Brosnan, J. T. and Samples, T. J. 2011. Perennial ryegrass (*Lolium perenne* L.) and grooming affect Bermudagrass traffic tolerance. *Crop Science* 51(5), 2204–2211.

Thompson, C. S., Hoyle, J. A. and Braun, R. C. 2019. Scalping tall fescue as soon as one day after treatment does not reduce glyphosate efficacy. *Crop, Forage & Turfgrass Management* 5(1), 1–5.

Tritabaugh, C. 2021. Cranking up the volume. *Golf Course Management, June*, 43–50.

Turgeon, A. J. 1991. *Turfgrass Management*. Prentice-Hall Inc., Englewood Cliffs, NJ.

Turgeon, A. J. and Fidanza, M. A. 2017. Perspectives on the history of turf cultivations. *International Turfgrass Society Research Journal* 13, 629–635.

Turgeon, A. J. and Kaminski, J. E. 2019. Turfgrass management. *Turfpath LLC*. State College, PA.

Wang, X., Gao, L., Ayers, P., Su, S. and Yuan, C. 2016. The influence of the lift angle on the center of gravity: measurements for zero turning radius mowers. *Applied Engineering in Agriculture* 32(2), 189–199.

Watschke, T. L., Dernoeden, P. H. and Shetlar, D. J. 2013. *Managing Turfgrass Pests*. CRC Press. Boca Raton, FL.

White, K. S. 1979. *Onward and Upward in the Garden*. Farrar, Straus, Giroux, New York, 159–163.

Whitlark, B. 2011. Bring back brushing! A simple practice that will improve your greens. *USGA Green Section Record* 49(12), 1–2.

Williams-Villano, M. E. 2020. Cutting-edge trends. *Irrigation and Green Industry* 24(1), 30–35.

Williams, A. L. 2016. Beyond cultivation: the evolution of grooming and brushing has had an ongoing impact on the highest quality turf. Golf Course Industry 28(5), 54, 56, 58.

Williams, D. W. and Burrus, P. B. 2004. Conversion of perennial ryegrass to Bermudagrass using seeded cultivars, herbicides, and plant growth regulators. *HortScience* 39(2), 398–402.

Young, J., Richardson, M. and Karcher, D. 2015. Creeping bentgrass putting green response to combined mowing, rolling, and foot traffic under environmental stress. *Agronomy Journal* 107(6), 1959–1966.

Young, J., Richardson, M. and Karcher, D. 2017. Golf ball mark severity and recovery as affected by mowing height, rolling frequency, foot traffic, and moisture. *Agronomy Journal* 109(4), 1758–1764.

Chapter 7

Advances in turfgrass for athletic fields and sports pitches

Gerald M. Henry, University of Georgia, USA

1 Introduction

Sports venues create unique challenges for turfgrass growth and survival. From relentless traffic imposed by athletes and sunlight obstruction from stadium structures, turfgrass selection, establishment, management, and performance testing are all extremely important to ensure high-quality playing surfaces. Recent innovations in sports turf have led to a reduction in management inputs and a decrease in environmental impacts that have allowed for more sustainable turfgrass production while maintaining high levels of performance and player safety (Fig. 1). This chapter will focus on improvements in turfgrass breeding and genetics, maintenance practices designed to increase plant and soil health, field assessment, and the implementation of site-specific management.

2 Turfgrass species and cultivar selection

Typically, tropical climates are dominated by the use of warm-season turfgrass species while more temperate regions often utilize cool-season turfgrass species. However, increased expectations from end users and changes with local weather and regional climates over the past few decades have challenged these

http://dx.doi.org/10.19103/AS.2022.0110.07

Figure 1 Example of a professional sports field in the USA; at this facility, the field is retracted inside the stadium for the sporting event. Source: Z. Longnecker.

perceptions. Industry and university plant-breeding programs have focused their efforts on improving turfgrass tolerance to adverse weather conditions and excessive wear while selecting new cultivars that require less agronomic management inputs. As a result, several new cultivars of currently propagated species have been released while further expansion of underutilized turfgrass species continues to be investigated.

2.1 Warm-season turfgrass improvements

Hybrid bermudagrass [*Cynodon dactylon* (L.) Pers. × *C. transvaalensis* Burtt-Davy] is commonly used as a warm-season turfgrass for sports fields located in tropical climates (Brosnan et al., 2014). Although 'Tifway 419' hybrid bermudagrass is still one of the most popular cultivars, extensive breeding efforts have led to the introduction of several new hybrid bermudagrass varieties with a range of improved traits. 'TifTuf' hybrid bermudagrass, tested as DT-1, was co-released from the University of Georgia (Tifton, GA, USA) and the United States Department of Agriculture (USDA) (Tifton, GA, USA) in 2014 (Hanna and Schwartz, 2016). It has shown improvements with respect to traffic, shade, and cold tolerance (Kowalewski et al., 2015). Most importantly, 'TifTuf' has exhibited enhancements in water conservation and drought tolerance (Jespersen et al., 2019; Katuwal et al., 2020; Pinnix and Miller, 2019). 'Tahoma 31' hybrid bermudagrass, tested as OKC 1131, released from Oklahoma State University (Stillwater, OK, USA) in 2019 (Wu et al., 2020), has also shown increased traffic and cold tolerance (Gopinath et al., 2021). 'IronCutter' hybrid bermudagrass (Fig. 2), tested as JSC 2-21-18, was released from Oklahoma State University in 2019 (Taliaferro, 2020). Improvements from this cultivar include increased drought, traffic, and cold tolerance. 'Latitude 36' hybrid bermudagrass, tested as OKC 1119, was released from Oklahoma State University in 2010 (Wu et al., 2012) with advancements in cold and traffic tolerance (Su et al., 2013). 'Northbridge' hybrid bermudagrass, tested as OKC 1134, was also released

Figure 2 Example of 'IronCutter' hybrid bermudagrass growing on a sod farm. Source: Heartland Turf Farms, Columbia, IL, USA.

from Oklahoma State University in 2010 (Wu et al., 2013) with improvements in cold tolerance (Xiang et al., 2021). 'TifGrand' hybrid bermudagrass, tested as ST-5, was co-released from the University of Georgia and the USDA in 2009 (Hanna and Braman, 2008). It has demonstrated better shade tolerance than 'Tifway 419' and 'TifSport' when grown under 70% continuous shade (Hanna et al., 2010) and is often selected as a surface for sports fields with low light conditions.

Reductions in water quantity and poor water quality as a result of migratory influxes into urban areas have increased the opportunity and demand for reclaimed water use on sports fields (Harivandi, 2010). Enhanced salinity tolerance of seashore paspalum (*Paspalum vaginatum* Swartz) led several investigators to examine its use as an athletic field surface. In fact, improvements in wear and shade tolerance observed in new cultivars like 'Platinum TE' led the Federation Internationale de Football Association (FIFA; Zürich, Switzerland) to name it the exclusive turfgrass of the World Cup set to take place in Qatar in 2022 (Anonymous, 2019). Shade tolerance of zoysiagrass (*Zoysia* spp.) has also led investigators to examine its response to excessive wear and traffic in an attempt to introduce this species for use in sports fields. Karcher et al. (2005) observed faster recovery in 'Crown', 'Palisades', and 'Zorro' Japanese lawngrass (*Zoysia japonica* Stued.) from simulated divot injury, but this has proven to be more difficult when zoysiagrass is trafficked while growing in shade (Trappe et al., 2011). Bluemuda is a relatively new phenomenon where improved Kentucky bluegrass (*Poa pratensis* L.) cultivars are seeded into bermudagrass (*Cynodon* spp.) sports fields (Winka, 2018). Predominantly geared toward the transition zone, this system attempts to improve traffic tolerance, speed up recovery, and increase canopy density. The presence of year-round color with bluemuda can eliminate the need for fall overseeding while removing safety concerns associated with herbicides required for spring perennial ryegrass (*Lolium perenne* L.) removal.

2.2 Cool-season turfgrass improvements

An environmental initiative was created in the last decade to foster the development of new cool-season turfgrass cultivars that exhibit exceptional performance while utilizing less management inputs. The Alliance for Low Input Sustainable Turf (A-LIST; Willamette Valley, OR, USA), a non-profit partnership between industry and university cooperators, has begun to identify Kentucky bluegrass, tall fescue (*Festuca arundinacea* Schreb.), perennial ryegrass, and even fine fescues (*Festuca* L. spp.) that benefit the environment through water conservation, increased tolerance to biotic and abiotic stress, and decreased need for fertility and pesticides (A-LIST, 2021) (Fig. 3).

Kentucky bluegrass is one of the most widely planted cool-season turfgrass species for sports fields in sub-tropical climates (Ervin et al., 2008; Puhalla

Figure 3 Example of evaluating tall fescue cultivars for drought tolerance, as maintained under a covered hoop-house rain-out shelter. Source: M. Fidanza.

et al., 2020). Recent advances in breeding have made considerable improvements in seed germination, pest tolerance, and recovery from traffic and wear. The National Turfgrass Evaluation Program (NTEP) is a non-profit organization associated with the USDA (USDA; Beltsville, MD, USA) that assists in the independent assessment of experimental and commercial turfgrasses for the turfgrass industry in collaboration with university scientists. The 2017-2021 NTEP Kentucky bluegrass test consisted of 89 commercial and experimental entries tested across the United States (National Turfgrass Evaluation Program, 2017). In the Massachusetts NTEP trial (Amherst, MA, USA) in 2020, 'Finish Line', 'After Midnight', 'Syrah', 'Blue Knight', 'Nurush', and 'Twilight' all maintained mean turfgrass quality (TQ) ratings ≥ 6 when subjected to simulated traffic stress. Visual ratings of TQ are recorded on a scale of 1–9 with a rating of 6 considered acceptable TQ (Morris and Shearman, 2007). In the Virginia NTEP trial (Blacksburg, VA, USA) in 2020, 'Starr', 'Skye', 'Barvette HGT', 'Barserati', and 'Syrah' had the highest mean TQ ratings post-traffic stress. Meanwhile, 'Barvette HGT' maintained the greatest canopy fullness (green turfgrass cover) following traffic applied in the fall in the New Jersey NTEP trial (New Brunswick, NJ, USA) in 2020. Numerous experimental cultivars also performed well in several NTEP trial locations during 2020 in response to traffic, ensuring that more improved Kentucky bluegrass cultivars could potentially enter the market in the near future. Several of these new cultivars, including 'Barvette HGT' and '365ss', have been successfully incorporated for use in the bluemuda system due to increased wear and traffic tolerance.

Improvements in tall fescue over the last decade include increased traffic tolerance and recovery from wear due to selection for denser/shorter canopies and extensive rhizome production. The 2018–2022 NTEP tall fescue test consisted of 132 commercial and experimental entries tested across the United States (National Turfgrass Evaluation Program, 2018). Tall fescue was subjected to traffic and wear stress at two locations: Iowa (Ames, IA, USA) and New Jersey (New Brunswick, NJ, USA). In the Iowa NTEP trial in 2019, 'Birmingham', 'AST8118LM', 'AST8218LM', and 'Fayette' maintained mean TQ ratings ≥ 7 when subjected to simulated traffic stress. In the New Jersey NTEP trial in 2019, 'Degas', 'O'Keefe', 'Dragster', and 'GLX ACED' all exhibited ≥ 70% canopy fullness (green turfgrass cover) following traffic applied in the fall. Several experimental tall fescue cultivars responded favorably to traffic stress at those same NTEP locations.

Increased drought, salt, and wear tolerance are just a few of the improved traits observed in new perennial ryegrass cultivars. The 2016-2020 NTEP perennial ryegrass test consisted of 114 commercial and experimental entries tested across the United States (National Turfgrass Evaluation Program, 2016). Perennial ryegrass was subjected to traffic and wear stress at three locations: Oregon (Corvallis, OR, USA), Virginia (Blacksburg, VA, USA), and Wisconsin

(Madison, WI, USA). In the Oregon NTEP trial in 2019, 'Furlong', 'Allstar Fore', 'Fastball 3GL', 'Overdrive 5G', and 'Stellar 4GL' all maintained mean TQ ratings ≥ 6.6 when subjected to simulated traffic stress. In the Virginia NTEP trial in 2019, 'Xcelerator', 'Belize 2', 'Allstar Fore', and 'Superstar GL' all maintained mean TQ ratings ≥ 6.7 when subjected to simulated traffic stress. 'Stellar 4GL' was the only commercially available cultivar that exhibited a mean TQ rating ≥ 5.0 when subjected to simulated traffic in the Wisconsin NTEP trial during 2019. Enhancements in wear tolerance have been achieved by selecting for perennial ryegrasses that exhibit aggressive, prostrate growth habits and lateral-spreading shoots that produce a dense canopy and allow for quick recovery from wear (Charbonneau and Brownbridge, 2013). A recent patent was filed for a new subspecies of perennial ryegrass, *Lolium perenne* subsp. *stoloniferous*, that produces true stolons (Wipff and Singh, 2015). 'Saguaro', 'Allante', 'Sienna', 'RPR', and 'Savant' are just a few cultivar examples that exhibit this creeping, stoloniferous growth habit.

3 Establishing sports fields

Several factors including species/cultivar selection, budget, and speed of establishment play a role in the method in which sports fields are installed. Although sodding may be the preferred method for quick establishment, anecdotal evidence suggests that seeding and sprigging often lead to the development of more robust root systems that better tolerate traffic and adverse weather conditions. Several improvements in sports field construction and establishment aids/techniques have led to improved playing surfaces that can hold up to more wear and traffic. Furthermore, enhanced establishment resulting in healthier, denser turfgrass stands will ultimately require less agronomic management inputs.

3.1 Alternative root zone amendments

Sports fields can be constructed out of native soil or sand-based systems designed to increase water movement and resist soil compaction (Puhalla et al., 2020). Soil amendments are often added to improve drainage, increase turfgrass shear strength, or even enhance athlete traction. Soil reinforcing materials may consist of carpet fibers, polypropylene mesh or netting, and even shredded athletic shoes. McNitt and Landschoot (2005) reported that similar reinforcing amendments (fibers, filaments, and mesh) reduced divot size under high traffic simulation when incorporated with sand-based root zones. Chong and Ok (2006) observed increases in macro-porosity, hydraulic conductivity, and air permeability of native soils amended with ≥ 0.15 $g \bullet g^{-1}$ of crumb rubber. Although early studies theorized a link between crumb rubber and human cancer risk, none successfully identified significant human carcinogenic risk from exposure (Bleyer, 2017). Furthermore, Pochron et al. (2017) examined crumb rubber contamination in the soil and determined that its presence did not inhibit microbial respiration or earthworm survival; however, it did reduce earthworm weight gain by 14% and led to the deposition of toxic levels of zinc. High levels of zinc in the soil solution can be toxic to invertebrates and plants (Hou et al., 2018). Golestani et al. (2014) observed an increase in perennial ryegrass germination and establishment uniformity in response to sand amended with rice hull, tea factory waste, and leaf compost. This may be attributed to the high soil moisture retention and aeration characteristics of rice hull.

Native soil sports fields are often more difficult to manage due to soil compaction and drainage concerns; therefore, field managers have begun to examine sand-capping the underlying soil in order to improve these issues as well as provide a better media for root growth and firmer playing surfaces. Dyer et al. (2020) observed that sand-capping up to 10 cm produced more roots than at sand depths of 20 cm. More roots at shallower sand-capping depths could increase water availability during adverse weather conditions and thus stabilize the playing surface for improved playability and performance.

3.2 Seed coating and seed germination covers

Establishing turfgrass from seed can be challenging due to issues with seedling diseases, reduced water availability/wind desiccation, saline soil, poor nutrient availability, and uneven seed placement/poor seed-to-soil contact (Dexin et al., 2012). Poor stand establishment can create long-term management challenges as well as decrease sports field performance and playability. Seeds can be coated with materials that include hydration polymers, fungicides, fertilizers, and biologicals. Advantages of seed coating may include quicker germination, better seed distribution, increased water and nutrient absorption, enhanced root production, and improved storage shelf life.

Madsen et al. (2016) observed increased seed germination and enhanced turfgrass establishment under deficit irrigation when tall fescue and perennial ryegrass seed were coated with a nonionic copolymer surfactant using a polyvinyl alcohol binder compared to non-coated seed. Similarly, Leinauer et al. (2010) reported improved establishment at both recommended and reduced seeding rates of bermudagrass and cool-season turfgrass mixtures (including Kentucky bluegrass and perennial ryegrass) under reduced irrigation when seeds were coated with a super absorbent starch-based polymer. However, seed coating benefits aren't always observed under adequate growing conditions. Richardson and Hignight (2010) observed minimal effects of a fungicide/biostimulant and a starch-based polymer coating on tall fescue and Kentucky bluegrass seed germination in three soil types.

Seed germination covers/blankets are often employed to enhance germination and stand establishment through soil moisture retention, reduced erosion, and soil temperature modification. Cover technology may consist of permanent organic materials of wheat straw, wood fiber, and jute fiber as well as temporary polyethylene and polypropylene materials. Patton et al. (2010) reported better turfgrass coverage 35–45 days after seeding of warm-season turfgrass species in response to cover technology made from wood fibers, jute fibers, or spun polypropylene fabric. High temperatures recorded under the nonwoven polyethylene and polypropylene covers may have contributed to seedling damage and poor turfgrass establishment when compared to other materials (Patton et al., 2010). These types of temporary covers may provide better performance if removed within a few weeks following seed germination.

3.3 Hybrid turfgrass systems

Increased field use often demands high levels of resiliency from natural turfgrass sports fields. Hybrid turfgrass systems utilize plastic fibers to reinforce the field while the natural grass still provides benefits like transpirational cooling (Anderson et al., 2018). Hybrid turfgrass systems are defined by the arrangement of their reinforcing materials: (a) vertical fibers attached to a backing, (b) individual fibers oriented vertically in the root zone, or (c) randomly oriented fibers throughout the root zone (Thoms et al., 2021). New mat systems, which consist of vertical fibers attached to a backing, exhibited higher shear strength in response to simulated traffic compared to natural turfgrass systems (Thoms et al., 2021). These systems also retained more turfgrass cover under high traffic while providing a stable surface for continued use. Lulli et al. (2011) observed similar summer temperatures and abrasiveness of a new mat system trafficked with rugby and soccer when compared to a natural turfgrass pitch, but higher wear resistance versus natural turfgrass.

3.4 Thick cut sod

Entire or partial field replacement may occur each year, depending on the sport, level of play, and annual budget. Although sodding a field provides immediate coverage, it may take several weeks for sod roots to 'peg down' and for the field to be ready for play. Small sod pieces create numerous seams that can create undulations in the field and must be filled with topdressing material to avoid desiccation. Thick cut sod has a profile thickness of up to 5 cm. Hybrid bermudagrass and Kentucky bluegrass are typically grown on plastic and continuously topdressed with sand or native soil in order to match the root zone of the renovated field/intended destination (Price, 2014). Thick cut sod rolls are typically 1.2 m wide and 9–12 m in length and weigh approximately 1000 kg. For this reason, it takes specialized equipment to harvest, roll, transport, and install thick cut sod. Field use can resume within days of installation since the weight of the sod keeps it in place and rooting is not required (Mascitti et al., 2017). However, further delays in rooting may occur since there's often enough soil moisture and nutrients available in the thick cut sod profile to discourage movement of roots into the underlying root zone already in place.

4 Sports field maintenance

Although a considerable amount of effort is placed on the alleviation of soil compaction and surface hardness in response to wear and traffic, field managers are also concerned with remediation of environmental stress and enhancement of roots. The focus of sports field management has deviated considerably from producing high-quality athletic fields using limitless resources to the utilization of new techniques and technology that reduce agronomic inputs while still maintaining safe, sustainable playing surfaces.

4.1 Cultivation

The most common cultivation practice performed on turfgrass to relieve soil compaction is aerification (Fig. 4). Hollow-tine cultivation, which involves the removal of a core from the soil profile, is the most commonly conducted form of aerification (Atkinson et al., 2012; Murphy et al., 1993; Turgeon and Fidanza, 2017). Benefits from core removal include a reduction in penetration resistance, improvement in water and nutrient uptake through root stimulation, and increased gas exchange between the soil and ambient air (Fry and Huang,

Figure 4 Example of solid-tine core cultivation on a sports field. Source: The Toro Company, Bloomington, MN, USA.

2004; Wiecko et al., 1993). However, cores must either be removed from the playing surface or reincorporated into the soil profile using a drag mat. This disturbance may require field closure for a short period of time. A survey conducted in collaboration with the Sports Field Management Association (Lawrence, KS, USA) revealed that 47% of field managers (total of 218 surveyed) in the United States stated that they utilized open-spoon aerification to alleviate soil compaction (C. M. Straw, unpublished nationwide survey data from 2013, University of Georgia, Athens, GA). Although open-spoon aerification is a form of hollow-tine cultivation, tine penetration is relatively shallow and core production/disruption is considerably less. In fact, Straw et al. (2016a) concluded that long-term effects of open-spoon aerification, up to 4 aerification events per growing season, were minimal and similar to the non-treated check after 2 years.

Several new cultivation devices have become commercialized in the last few decades that aim at alleviating soil compaction, increasing water infiltration/drainage, enhancing gas exchange, and stimulating roots. The Air2G2 (Foley Company LLC., Prescott, WI, USA) injects pressurized air horizontally through the soil profile to a depth of 30 cm using 3 probes (20–30 cm in length) and a self-contained air compressor in order to fracture compaction layers. Although moving somewhat slowly, this device operates with little disturbance to the turfgrass canopy/root system and does not require core cleanup following use, so play may resume almost immediately (Dickson et al., 2017). The SISIS AER-AID system (Howardson Ltd., Kirk Langley, UK) operates similarly to the Air2G2; however, it utilizes shorter air-injection tines that are only 12.7 cm in length. Research pertaining to air-injection cultivation has been limited, but preliminary data suggests that the use of this technology may alleviate soil compaction and increase water infiltration without disruption in play (Bigelow and Soldat, 2013; Dickson et al., 2017).

Sand and gravel injection cultivation is often employed to amend turfgrass systems that exhibit poor water infiltration and drainage. The WaterWick machine (WaterWick Inc., Southampton, NY, USA) injects a narrow gravel vein 1.6–2.5 cm in width to a depth of 20–30 cm within the soil profile using vibratory plow blades spaced 30–60 cm apart (Turgeon and Fidanza, 2017). The DryJect (DryJect Inc., Hatboro, PA, USA) utilizes high-pressure (27 579 kPa) water injection to insert sand into holes measuring 0.6–1.3 cm spaced 8–25 cm apart to a depth of 25 cm through a vacuum created by the streams of water (Craft et al., 2016). The DryJect may also provide an option to directly inject materials like soil surfactants, nematicides, fungicides, and other soil amendments into the root zone where they can be most effective (Munnecke, 1972; Van Gundy and McKenry, 2012). A similar machine, the Graden Sand-Injection System (Graden USA, Inc., Richmond, VA, USA) is equipped with vertical mower blades set 4 cm apart that remove soil to a depth of 4 cm and replace it with sand (de Havilland, 2011).

The ShockWave (manufactured by Imants BV, Reusel, Netherlands; distributed in the United States by AquaAid Solutions, Rocky Mount, NC, USA) is a linear decompaction unit that utilizes 18–24 offset blades (12–15 mm in width) that slice through the soil profile to a depth of 12.7–25.4 cm. The side-to-side movement of the blades through the soil profile is directly attributed to their offset mounting and ultimately allows them to fracture a large volume of soil as the machine traverses across a field (Dalsgaard et al., 2020). A majority of the soil agitation occurs below the surface, so surface disruption is kept to a minimum and play can resume almost immediately. Dalsgaard et al. (2020) observed similar soil physical responses when comparing the ShockWave to solid-tine aerification under simulated athletic field traffic.

4.2 Topdressing

Topdressing is the surface application of sand, soil, or other amendments to fill recently produced aerification vents, alleviate imperfections in the playing surface, or minimize thatch accumulation below the turfgrass canopy (Carrow et al., 1987). However, a variety of new materials have recently been identified for application to sports fields in an attempt to enhance wear tolerance and increase field playability. Even though crumb rubber and calcined clay may provide some benefit to sports fields, sand topdressing often results in better long-term turfgrass quality. Miller (2007) observed a reduction in cleat traction (16–31%) in response to high rates (up to 17 mm) of crumb rubber and calcined clay applications, and both materials were associated with reductions in turfgrass stand density when subjected to simulated wear and traffic. Surface hardness was reduced by 24% in response to crumb rubber, while sand and calcined clay applications caused slight increases (1–2%) (Miller, 2007). Goddard et al. (2008) also observed a decrease in surface hardness along with a reduction in soil bulk density and shear resistance of Kentucky bluegrass and hybrid bermudagrass topdressed with crumb rubber (to a depth of 5.1 cm) when subjected to simulated traffic. Furthermore, Kowalewski et al. (2011) noted that trafficked Kentucky bluegrass maintained the highest percent living ground cover (85– 90%) when topdressed with 2.5 cm of crumb rubber.

4.3 Irrigation

Soil moisture is often highly correlated to other soil and plant parameters measured on sports fields, but those relationships may be weaker on sand-based fields compared to those on fields constructed with native soil (Straw et al., 2017a, 2020a). For this reason, it is extremely important to evaluate irrigation system performance (Fig. 5). Irrigation audits are typically conducted using the catch-can method and are designed to assess the distribution

Figure 5 Example of irrigation on a sports field. Source: Hunter Industries, San Marcos, CA, USA.

uniformity of a system (Irrigation Association, 2009). Evenly spaced cups are set out in a specified area to capture water emitted from the irrigation system (i.e. sprinklers). Although this method accurately measures application uniformity, it does not measure the efficiency of the system. Distribution and application uniformity simply refer to how uniformly water is applied to the surface of an area that is being watered. Irrigation system efficiency can be determined using a global position satellite (GPS) technology-equipped soil moisture probe to create maps of volumetric water content (Krum et al., 2010; Straw et al., 2018a). This method tracks where water enters and occupies the soil profile. Maps in conjunction with irrigation schematics and/or catch-can audits may help identify agronomic and/or system deficiencies that can be corrected for the conservation of water and improved uniformity and playability of sports fields.

Along with inefficiency, irrigation systems are also subject to overwatering and improper scheduling. However, irrigation water savings have been well documented in response to the addition of smart irrigation controllers. Water savings of approximately 43% in the landscape were reported by Davis et al. (2009) when evapotranspiration (ET)-based irrigation controllers were utilized for 15 months. Cardenas-Lailhacar and Dukes (2010) and Cardenas-Lailhacar et al. (2010) observed up to 88% water savings in home lawns, regardless of weather conditions, when irrigation controllers were equipped with soil moisture sensors (SMS). Serena et al. (2020) reported up to 42% and 44% water savings on bermudagrass and tall fescue, respectively, when irrigation scheduling was based on SMS, ET Climate Logic controllers, or 60% of historic ET for bermudagrass and 80% of historic ET for tall fescue. Not only do smart irrigation controllers demonstrate water conservation, but they may also help strengthen sports field root systems in deficit irrigation schedules. Wherley et al. (2014) observed more bermudagrass and zoysiagrass roots at depths greater than 25 cm in response to deficit irrigation, while Fu et al. (2007) noted more root proliferation of tall fescue between 8.7 cm and 36.3 cm depths when subjected to 20% ET.

As stated previously, more and more sports venues have the access and ability to apply reclaimed water as an alternative source for irrigation. Reclaimed water is water that went through domestic use, was treated at a wastewater facility, and is now suitable for non-potable use. Although the use of reclaimed water alleviates stress placed on potable water sources, high levels of soluble salts and nitrogen (N) compounds present within the water may create management challenges (Harivandi, 2010). Accumulation of salt (Na^+, K^+, Ca^{2+}, Mg^{2+}, and Cl^-) in the soil profile may compromise its structure, while continuous application of N compounds may result in excessive leaching. However, Devitt et al. (2013) reported that N losses in response to reclaimed water application were on average 8 ± 14 kg•ha^{-1} yr^{-1} or 2 ± 3% of the total N applied in the second year of their study, suggesting that an established turfgrass root system could successfully act as a bio-filter. Similar results regarding N loss were observed by Evanylo et al. (2010), but bermudagrass growth and turfgrass quality were reduced, leading to an overall reduction in bermudagrass N utilization.

4.4 Grow lights and cooling fans

Sports stadiums often create unique challenges that require additional maintenance/equipment to produce high-quality turfgrass. Partial roofs and other obstructions may reduce the quantity and quality of light that reaches the field surface. Shade reduces the photosynthetic capability of turfgrass that may ultimately lead to reductions in canopy cover, density, and rooting (Dudeck and Peacock, 1992; Gardner and Wherley, 2005). Richardson et al. (2019) reported that non-overseeded bermudagrass subjected to simulated shade exhibited a decline in turf coverage when the daily light integral was reduced from 41 mol m^{-2} d^{-1} to 26.0 mol m^{-2} d^{-1}. Therefore, grow lights (high-pressure sodium or light-emitting diode) attached to a moveable framework placed directly above sports field surfaces may help counteract stadium shade and enhance turfgrass recovery from wear (Mann, 2016) (Fig. 6).

The same structures that create low light conditions often reduce air circulation and increase relative humidity, making turfgrass more susceptible to diseases like powdery mildew and snow mold (Butt, 1978; McBeath, 2003). Permanent or portable oscillating fans increase air flow in constricted areas,

Figure 6 Example of grow lights placed on a sports field. Source: SeeGrow; Surrey, United Kingdom.

remove excess dew/guttation on leaf surfaces, and may temporarily reduce canopy temperatures. Rodriguez et al. (2005) reported that the use of a fan plus misting reduced creeping bentgrass (*Agrostis stolonifera* L.) putting green canopy, soil surface, and soil temperatures by as much as 9°C, 7°C, and 6°C, respectively. However, the addition of misting caused an increase in relative humidity and soil moisture (Rodriguez et al., 2005).

4.5 Painting techniques

Sports field managers are required to paint field lines and logos as part of their routine maintenance practices; however, designs are dependent on the sport or event (Fig. 7). The pigments contained within paint can alter the spectral quality of light accessible by the same turfgrass plants they cover. Paint color, thickness, and dilution all influence plant performance in response to painting. Reynolds et al. (2012) reported that non-diluted red paint reduced total canopy photosynthesis by 75% over a 6-week period while 1:1 diluted paint only resulted in a 19% reduction. This is likely due to the fact that red paint absorbed 51% of incident photosynthetically active radiation (PAR) while it transmitted and reflected 6% and 43%, respectively. On the contrary, transmitted and reflected PAR was 5% and 95%, respectively, in response to white paint (Reynolds et al., 2012). Paint color can also have an influence on canopy temperature. Segars and Moss (2014) recorded higher canopy temperatures on turfgrass surfaces painted with green and black paint compared to that on surfaces painted with white. Sand-based root zones are designed to facilitate high water infiltration and reduce the occurrence of

Figure 7 Example of an autonomous painter during a demonstration of applying white paint in a straight line.

soil compaction; however, May (2019) observed an increase in sand particle size due to binding in response to paint moving into the soil profile. In an attempt to rectify the negative impacts that paint can have on soil and plant parameters, Ogle (2020) evaluated different paint application nozzle sizes and pressure settings on turfgrass systems. No differences were observed between treatments with respect to turfgrass canopy paint coverage or soil dynamics; however, painting with a nozzle and pressure setting to deliver a flow rate of 470 $ml \bullet m^{-2}$ was significantly less expensive to paint an entire field compared to the higher output settings (Ogle, 2020). In the future, paint manufacturers should attempt to further identify pigments that are less harmful to turfgrass plants, while sports field managers should continue to investigate alternative painting methods that are less injurious but still produce vibrant colors and markings.

5 Performance testing and site-specific management

Field construction, management practices, local weather, and field usage continuously influence the growth and performance of turfgrass surfaces used as sports fields. Natural turfgrass fields often exhibit considerable within-field variations with respect to soil and plant parameters including moisture, compaction, and rooting (Straw and Henry, 2018; Straw et al., 2020a). Variability not only affects field playability but can also determine the need for and response to management inputs. For this reason, sports turf managers need to obtain accurate, real-time information about their fields in order to guide management decisions.

5.1 Performance testing

Performance testing (i.e. site assessment) of natural turfgrass sports fields involves the use of sampling devices and sensor technology to obtain information about surface properties (Carrow et al., 2010; McAuliffe, 2008; Straw et al., 2017b). In the United States, the most widely adopted testing methodology is the American Society for Testing and Materials F1702-10 (ASTM International, West Conshohocken, PA, USA), which requires a lightweight handheld apparatus in the field to measure impact attenuation for several different sports (ASTM, 2018). The international organization FIFA also created testing procedures to evaluate the performance of soccer fields (FIFA, 2015). Current testing protocols rely on the use of handheld devices to obtain data at only 5–12 locations, which are often sport dependent. The creation of surface maps through spatial analysis of field properties provides a more detailed approach to performance testing that accurately detects field variability. Straw et al. (2017b) evaluated sampling grid size and determined that sampling on a 4.8 m × 9.6 m grid was required to accurately describe soil moisture and penetration resistance on sports fields. Capturing sports field variability through the creation of spatial maps has proven to be important, because within-field variability may affect an athlete's perception of playing quality and safety (McClements and Baker, 1994; Straw et al., 2019). Field variations may require athletes to make abrupt or frequent adjustments to normal movements during the game. Therefore, certain sporting leagues have devised standards for specific field properties that must be adhered to in order to accommodate participating athletes. Furthermore, sports field variability of soil and plant parameters may also increase athlete injury risk (Hales and Johnson, 2019; Straw et al., 2018b).

Surface hardness, an indication of the firmness of the playing surface, is the most frequently quantified field characteristic and is commonly measured with the handheld Clegg Impact Tester (Aldahir and McElroy, 2014; Clegg,

1976). A soil penetrometer may also be utilized to measure surface hardness, but typically this device provides information about penetration resistance related to subsurface compaction deeper in the soil profile. However, positive correlations have been observed between data generated from the Clegg Impact Tester and the soil penetrometer (Caple et al., 2012).

Soil moisture is one of the most fluctuating measurable variables and is typically determined using time domain reflectometry or capacitance sensors (Leib et al., 2003). Measurements are reported as percent volumetric water content (% VWC). Handheld options have become one of the most widely utilized devices for turfgrass managers (Fig. 8).

Linear, horizontal, vertical, and rotational frictional forces applied to a sports field surface are referred to as traction (Aldahir and McElroy, 2014). An athlete's ability to change direction and speed is greatly influenced by the traction that exists between their shoe and the playing surface (Thomson et al., 2019). A handheld shear strength tester, comprising a torque wrench atop a shaft with a studded disc or shear vane foot, is inserted through the turfgrass canopy and into the upper soil profile to measure rotational traction (Straw et al., 2020b). More advanced devices like the Pennfoot (McNitt et al., 1997), Tennessee Athletic Field Tester (Thoms et al., 2013), and the BioCORE Elite Shoe-Surface Tester (Kent et al., 2011) use hydraulic pressure to simultaneously calculate several of the frictional forces previously described.

Although most sports fields are professionally constructed with exact specifications, undulations and imperfections can occur over time. The levelness of a sports field is indicated by surface evenness, which can be quantified with

Figure 8 Example of a POGO TurfPro (Stevens Water Monitoring Systems, Inc., Portland, OR, USA) for measuring soil moisture content (percent volumetric water content), and a soil water content map generated from monitoring the site. Source: C. Magro.

either a straight edge or a profile gauge (Straw et al., 2020b). The straight edge records localized changes in evenness of the turfgrass canopy while it moves along the surface (Bartlett et al., 2009; McClements and Baker, 1994). The profile gauge measures surface evenness through independently moving rods affixed to a frame that become displaced when undulations are detected (Baker, 1999). Turfgrass quality was historically rated visually until the introduction of spectral reflectance techniques in the 1980s that were capable of distinguishing turfgrass from soil through the interpretation of reflected light in the visible and near infrared spectra (Haggar et al., 1983). The most commonly used vegetation index in turfgrass is normalized difference vegetation index (NDVI) (Carrow et al., 2010). Handheld NDVI sensors are popular in research, but their use in situ on sports fields has been minimal due to cost and labor concerns.

Mobile data acquisition devices equipped with GPS are necessary for intensive sampling of entire fields or large sports complexes with multiple fields. However, few mobile devices are currently available for use in turfgrass systems. The Toro Precision Sense 6000 (PS6000) (The Toro Company, Bloomington, MN, USA), developed in 2005, was the first and only mobile, multi-sensor sampling device engineered for use on turfgrass sites. This device simultaneously measures soil moisture (% VWC), soil compaction (MPa), soil salinity (electrical conductivity), and plant performance (NDVI) with all samples geo-referenced with the use of an onboard GPS unit (Carrow et al., 2010; Straw et al., 2017b). Small unmanned aerial systems (sUAS) have become popular in the last decade and are often utilized by turfgrass managers to acquire aerial photographs. However, recent advances in aerial platforms and sensor technology may expand their use for early detection of drought stress and plant health/performance (Hong et al., 2019; Marín et al., 2020; Zhang et al., 2019). Limitations may include cost, data processing and interpretation as well as flight restrictions and associated permits/licenses.

Handheld sampling may be the only option without the creation of additional and accessible mobile, multi-sensor devices in the near future. Sampling efficiency has increased with recent improvements to handheld devices. The addition of GPS compatibility to handheld devices allows for the creation of maps from geo-referenced data. Straw et al. (2016b) determined that it would take 1.5 h for a team of 3 people to collect soil moisture, soil compaction, and NDVI measurements using handheld devices on a 4.8 m × 9.6 m sampling grid. That same data was obtained with the mobile PS6000 in less than one hour and operated by one person. Additionally, Straw et al. (2016b) compared soil moisture (% VWC), soil compaction (penetration resistance), and plant performance (NDVI) data obtained on sports fields using both handheld and mobile platforms. Data for % VWC and NDVI had moderate to strong correlations between handheld and mobile sampling devices; however, correlations for penetration resistance were weak to moderate. Similarly, Friell and Straw (2021) observed significant, moderate to strong correlations between proximal (PS600) and aerial (sUAS) NDVI data, but had difficulty correlating proximal (PS600) % VWC and aerial (sUAS) NDVI data.

The sampling automation achieved with mobile devices provides consistent measurements and reduces user errors often associated with handheld sensors (Straw et al., 2016b; Twomey et al., 2011). For example, the rate of probe penetration of the penetrometer may influence the accuracy of the measurements (Vaz et al., 2001). The mobile PS6000 unit overcomes this issue through the use of a rotating arm and wheel-driven shaft that enters the soil at a more consistent, steady rate. In the future, sampling devices should be incorporated into daily maintenance practices with the potential for sensors to be affixed on equipment (i.e. mowers) for increased efficiency and rapid data collection.

5.2 Site-specific management

The concept of precision agriculture, which has been around for nearly a century, relies heavily on spatial analysis for implementation (Franzen and Mulla, 2015). Although recently adopted in turfgrass, precision turfgrass management (PTM) focuses on enhancing input efficiency and management decisions through the application of inputs (i.e. fertilizer, irrigation, and cultivation) where, when, and in the amount needed (Carrow et al., 2010; Krum et al., 2010; Straw and Henry, 2018). Detailed site information through intensive spatial sampling described previously is required since PTM was developed and based on the premise of site-specific management (Carrow et al., 2010). The creation of spatial maps from obtained data can be used to delineate site-specific management units (SSMUs), or management zones, that contain similar plant and soil parameters.

Most published researches involving the use of SSMUs in turfgrass involve golf course fairways. Soil moisture SSMUs based on % VWC and plant performance (NDVI) were developed by Krum et al. (2010) for site-specific irrigation. Site-specific aerification SSMUs were created from the comparison of % VWC and penetration resistance data obtained at field capacity and under drier conditions (Flitcroft et al., 2010). Grubbs et al. (2019) utilized a soil apparent electrical conductivity (ECa) device in conjunction with physical soil samples to map the spatial variability of several soil properties (clay content, cation exchange capacity, organic matter, and soil pH) to determine site-specific fertility SSMUs. Only a few studies have specifically examined spatial variability of sports field properties. Straw et al. (2018a) conducted catch-can irrigation audits in conjunction with % VWC in order to compare irrigation application uniformity and irrigation system efficiency, respectively, to correct both field agronomic and irrigation system issues for water conservation and improved field uniformity. Further research concluded that field dry-down can have a significant impact on short-term spatiotemporal variations of SSMUs (% VWC, NDVI, and penetration resistance) due to the inverse relationship between soil compaction and soil moisture/drought stress (Straw and Henry, 2018). This relationship may affect SSMU delineation and identification of site-specific deficient areas within a field, so considerations should be made prior to field mapping.

6 Conclusion

A variety of anthropogenic and environmental stresses imposed on sports fields create challenges for turfgrass managers to maintain safe and consistent playing surfaces while minimizing agronomic management inputs. Advances in turfgrass breeding and genetics offer field managers access to improved cultivars that tolerate stress and require less management. New soil/root zone amendments, seed coating materials, and planting techniques have helped decrease field establishment time while increasing playability and performance. Several novel cultivation devices, enhancements to irrigation systems, grow lights, and even changes to field painting methodology have increased turfgrass tolerance to stress and recovery from damage. Most recently, the integration of sensor technology and performance testing guidelines have improved athlete safety while promoting site-specific management and efficient use of agronomic inputs.

7 Where to look for further information

Future research will continue to create and evaluate new cultivars for improved traffic, wear, and shade tolerance. Performance testing and safety standards will focus on comparisons between natural turfgrass and synthetic turf systems. Sensor technology must continue to evolve and eventually be integrated into maintenance equipment in order to aid in plant/soil information acquisition. Quick and accurate analysis of large data sets will be critical for real-time management decisions and the implementation of site-specific management for the overall reduction of turfgrass inputs. The Sports Field Managers Association (https://www.sportsfieldmanagement.com) is the professional organization for the men and women who manage sports fields worldwide. Their mission is to advance the professionalism in sports field management and safety through education, awareness, and industry development.

8 References

Aldahir, P. C. F. and McElroy, J. S. 2014. A review of sports turf research techniques related to playability and safety standards. *Agron. J.* 106(4):1297–1308. https://doi.org/10.2134/agronj13.0489.

Alliance for Low Input Sustainable Turf 2021. Available at: http://www.a-listturf.org.

American Society for Testing and Materials (ASTM) 2018. *F1702-10, Standard Test Method for Measuring Impact-Attenuation Characteristics of Natural Playing Surface Systems Using Lightweight Portable Apparatus*. American Society for Testing and Materials, West Conshohocken, PA.

Anderson, F. D., Fleming, P. R., Sherratt, P. J. and Severn, K. 2018. Novel field equipment for assessing the stability of natural and hybrid turfs. *Sports Eng.* 21(4):321–331. https://doi.org/10.1007/s12283-018-0273-2.

Anonymous (Golf Course Industry Staff) 2014. Platinum TE Paspalum Selected as Exclusive Turfgrass for 2022 FIFA World Club. Golf Course Industry 18 March 2019.

Atkinson, J. L., McCarty, L. B. and Bridges, W. C. Jr. 2012. Effect of core aerification frequency, area impacted, and topdressing rate on turf quality and soil physical properties. *Agron. J.* 104(6):1710–1715. https://doi.org/10.2134/agronj2012.0224.

Baker, S. W. 1999. The playing quality of turfgrass sports surfaces. In: Aldous, D. E. (Ed.), *International Turf Management Handbook* pp. 231–244. Butterworth-Heinemann, Oxford, UK.

Bartlett, M. D., James, I. T., Ford, M. and Jennings-Temple, M. 2009. Testing natural turf sports surfaces: the value of performance quality standards. *Proc. Inst. Mech. Eng. P* 223:21–29.

Bigelow, C. A. and Soldat, D. J. 2013. Turfgrass root zones: management, construction methods, amendment characterization, and use. In: Stier, J. C., Horgan, B. P. and Bonos, S. A. (Eds). *Turfgrass: Biology, Use, and Management* pp. 383–424. American Soc. of Agron, Madison, WI.

Bleyer, A. 2017. Synthetic turf fields, crumb rubber, and alleged cancer risk. *Sports Med.* 47(12):2437–2441. https://doi.org/10.1007/s40279-017-0735-x.

Brosnan, J. T., Dickson, K. H., Sorochan, J. C., Thoms, A. W. and Stier, J. C. 2014. Large crabgrass, white clover, and hybrid Bermudagrass athletic field playing quality in response to simulated traffic. *Crop Sci.* 54(4):1838–1843. https://doi.org/10.2135/cropsci2013.11.0754.

Butt, D. J. 1978. Epidemiology of powdery mildews. In: Spencer, D. M. (Ed.), *The Powdery Mildews* pp. 51–82. Academic Press, London.

Caple, M., James, I. and Bartlett, M. 2012. Mechanical behaviour of natural turf sports pitches across a season. *Sports Eng.* 15(3):129–141. https://doi.org/10.1007/s12283-012-0096-5.

Cardenas-Lailhacar, B. and Dukes, M. D. 2010. Precision of soil moisture sensor irrigation controllers under field conditions. *Agric. Water Manag.* 97:666–672. https://doi.org/10.1016/j.agwat. 2009.12.009.

Cardenas-Lailhacar, B., Duke, M. D. and Miller, G. L. 2010. Sensorbased automation of irrigation on Bermudagrass during dry weather conditions. *J. Irrig. Drain. Eng.* 136(3):184–193. https://doi.org/10.1061/(ASCE)IR.1943-4774.0000153.

Carrow, R. N., Johnson, B. J. and Burns, R. E. 1987. Thatch and quality of Tifway Bermudagrass turf in relation to fertility and cultivation 1. *Agron. J.* 79(3):524–530.

Carrow, R. N., Krum, J. M., Flitcroft, I. and Cline, V. 2010. Precision turfgrass management: challenges and field applications for mapping turfgrass soil and stress. *Precision Agric.* 11(2):115–134. https://doi.org/10.1007/s11119-009-9136-y.

Charbonneau, P. and Brownbridge, M. 2013. Rhizomatous tall fescue and regenerative perennial ryegrass performance in Ontario: Year 1. *Sports Turf Manag.* 26:14–18.

Chong, S. K. and Ok, C. H. 2006. Effect of rootzone mixes amended with crumb rubber on the physical properties. *Asian J. of Turfgrass Sci.* 20(1):83–91.

Clegg, B. 1976. An impact testing device for in situ base course evaluation. *Aust. Road Res. Board Proc.* 8:1–6.

Craft, J. M., Baldwin, C. M., Philley, W. H., McCurdy, J. D., Stewart, B. R., Tomaso-Peterson, M. and Blythe, E. K. 2016. Impact of dry-injection cultivation to maintain soil physical properties for an ultradwarf Bermudagrass putting green. *Hortscience* 51(9):1171–1175. https://doi.org/10.21273/HORTSCI11018-16.

Dalsgaard, T. O., Thoms, A. W., Christians, N. E., Mertz, I. and Horton, R. 2020. Comparison of Shockwave aerification and conventional aerification methods on athletic fields. *Agron. J.* 112:1–24. Available at: https://doi/10.1002/agj2.20255.

Davis, S. L., Dukes, M. D. and Miller, G. L. 2009. Landscape irrigation by evapotranspiration-based controllers under dry conditions in southwest Florida. *Agric. Water Manag.* 96:1828–1836. https://doi.org/10.1016/j.agwat.2009.08.005.

de Havilland, J. 2011. The anatomy of Graden contour sand injection. *Greenkeeper Int.* 11(1):44–44.

Devitt, D. A., Wright, L. and Young, M. H. 2013. Water and salt status of bare soil and turfgrass systems irrigated with recycled water. *Agron. J.* 105(4):1051–1060. https://doi.org/10.2134/agronj2012.0126.

Dexin, S., Evanylo, G. and Goatley, J. M. 2012. Effects of compost sources and seeding treatments on germination and emergence of four turfgrass species. *Compost Sci. Util.* 20(3):165–170. https://doi.org/10.1080/1065657X.2012.10737042.

Dickson, K., Sorochan, J., Munshaw, G. and Thoms, A. 2017. Comparison of cultivation methods impact on playability of Agrostis stolonifera Greens. *J. Test. Eval.* 46(3):1256–1260. https://doi.org/10.1520/JTE20160419.

Dudeck, A. E. and Peacock, C. H. 1992. Shade and turfgrass culture. In: Waddington, D. V., Carrow, R. N. and Shearman, R. C. (Eds.), *Agronomy Monographs*, Wiley, Hoboken, NJ, pp. 269–284.

Dyer, D. W., Wherley, B. G., McInnes, K. J., Thomas, J. C., Hejl, R. and Reynolds, W. C. 2020. Sand-capping depth and subsoil influences on 'Tifway' Bermudagrass response to irrigation frequency and drought. *Agron. J.* 112(5):3491–3499. https://doi.org/10.1002/agj2.20319.

Ervin, E. H., Zhang, X. and Roberts, J. C. 2008. Improving root development with foliar humic acid applications during Kentucky bluegrass sod establishment on sand. *Int. Conf. Turfgrass Sci. Manag. Sports Fields* 783:317–322. https://doi.org/10.17660/ActaHortic.2008.783.33.

Evanylo, G., Ervin, E. and Zhang, X. 2010. Reclaimed water for turfgrass irrigation. *Water* 2(3):685–701. https://doi.org/10.3390/w2030685.

Federation Internationale de Football Association (FIFA) 2015. FIFA quality programme for football turf handbook of test methods. Available at: https://football-technology.fifa.com/media/1238/fqp-handbook-of-test-methods-2015-v31-w-cover.pdf.

Flitcroft, I., Krum, J., Carrow, R., Rice, K., Carson, T. and Cline, V. 2010. *Spatial Mapping of Penetrometer Resistance on Turfgrass Soils for Site-Specific Cultivation*. [CD]. International Society of Precision Agriculture, Monticello, IL.

Franzen, D. and Mulla, D. 2015. A history of precision agriculture. In: Q. Zhang (Ed.), *Precision Agriculture Technology for Crop Farming* pp. 1–20. CRC Press, New York, NY.

Friell, J. and Straw, C. 2021. Comparing ground-based and aerial data at field scale during dry down on golf course fairways. *Int. Turfgrass Soc. Res. J.* 14(1):377–384 https://doi.org/10.1002/its2.46.

Fry, J. and Huang, B. 2004. *Applied Turfgrass Science and Physiology*. John Wiley and Sons, Hoboken, NJ.
Fu, J., Fry, J. and Huang, B. 2007. Tall fescue rooting as affected by deficit irrigation. *Hortscience* 42(3):688–691. https://doi.org/10.21273/HORTSCI.42.3.688.
Gardner, D. S. and Wherley, B. G. 2005. Growth response of three turfgrass species to nitrogen and trinexapac-ethyl in shade. *HortSci.* 40(6):1911–1915.
Goddard, M. J. R., Sorochan, J. C., McElroy, J. S., Karcher, D. E. and Landreth, J. W. 2008. The effects of crumb rubber topdressing on hybrid Kentucky bluegrass and Bermudagrass athletic fields in the transition zone. *Crop Sci.* 48(5):2003–2009. https://doi.org/10.2135/cropsci2007.07.0405.
Golestani, M. A., Dolatkhahi, A., Vahdati, N. and Roudson, O. N. 2014. Utilization of rice hull as a new substrate for turf grass seed germination in sod production as a sustainable approach. *J. Ornamental Plants* 4:33–37.
Gopinath, L., Moss, J. Q. and Wu, Y. 2021. Evaluating the freeze tolerance of Bermudagrass genotypes. *Agrosyst. Geosci. Environ.* 4(2):e20170. https://doi.org/10.1002/agg2.20170.
Grubbs, R. A., Straw, C. M., Bowling, W. J., Radcliffe, D. E., Taylor, Z. and Henry, G. M. 2019. Predicting spatial structure of soil physical and chemical properties of golf course fairways using an apparent electrical conductivity sensor. *Precision Agric.* 20(3):496–519. https://doi.org/10.1007/s11119-018-9593-2.
Haggar, R. J., Stent, C. J. and Isaac, S. 1983. A prototype hand-held patch sprayer for killing weeds, activated by spectral differences in crop/weed canopies. *J. Agric. Eng. Res.* 28(4):349–358. https://doi.org/10.1016/0021-8634(83)90066-5.
Hales, M. E. and Johnson, J. D. 2019. The influence of sport-field properties on muscle-recruitment patterns and metabolic response. *Int. J. Sports Physiol. Perform.* 14(1):83–90. https://doi.org/10.1123/ijspp.2018-0004.
Hanna, W. W. and Braman, S. K. 2008. Bermudagrass 'ST-5'. U. S. Plant Patent PP21017. Date Issued: 8 Aug.
Hanna, W. W., Braman, S. K. and Schwartz, B. M. 2010. 'ST-5', a shade-tolerant turf Bermudagrass. *Hort. Sci.* 45(1):132–134. https://doi.org/10.21273/HORTSCI.45.1.132.
Hanna, W. W. and Schwartz, B. M. 2016. Bermudagrass named 'DT-1'. US Plant Patent 27. 392 P2. Date Issued: 15 Nov.
Harivandi, M. A. 2010. Irrigating turfgrasses with municipal reclaimed water. *Acta Hortic.* XXVIII(938):95–103. https://doi.org/10.17660/ActaHortic.2012.938.12.
Hong, M., Bremer, D. J. and Merwe, D. 2019. Thermal imaging detects early drought stress in turfgrass utilizing small unmanned aircraft systems. *Agrosystems Geosci. Environ.* 2(1):1–9. https://doi.org/10.2134/age2019.04.0028.
Hou, J., Wu, Y., Li, X., Wei, B., Li, S. and Wang, X. 2018. Toxic effects of different types of zinc oxide nanoparticles on algae, plants, invertebrates, vertebrates and microorganisms. *Chemosphere* 193:852–860. https://doi.org/10.1016/j.chemosphere.2017.11.077.
Irrigation Association 2009. Recommended audit guidelines. Irrigation Association, Falls Church, VA. Available at: http://www.irrigation.org/certification/pdf/AuditGuidelines_FINAL.pdf.
Jespersen, D., Leclerc, M., Zhang, G. and Raymer, P. 2019. Drought performance and physiological responses of Bermudagrass and seashore paspalum. *Crop Sci.* 59(2):778–786. https://doi.org/10.2135/cropsci2018.07.0434.
Karcher, D. E., Richardson, M. D., Landreth, J. W. and McCalla, J. H. Jr. 2005. Recovery of zoysiagrass varieties from divot injury. *Appl. Turfgrass Sci.* 2(1):1–8. http://doi.org/10.1094/ATS-2005-0728-01-RS.

Katuwal, K. B., Schwartz, B. and Jespersen, D. 2020. Desiccation avoidance and drought tolerance strategies in bermudagrasses. *Environ. Exp. Bot.* 171:103947. https://doi.org/10.1016/j.envexpbot.2019.103947.

Kent, R., Crandall, J., Forman, J., Lessley, D., Lau, A. and Garson, C. 2011. Development and assessment of a device and method for studying the mechanical interactions between shoes and playing surfaces in situ at loads and rates generated by elite athletes. *Sports Biomech.* 11(3):414-429.

Kowalewski, A. R., Dunne, J. C., Rogers, J. N. III and Crum, J. R. 2011. Heavy sand and crumb rubber topdressing improves Kentucky bluegrass wear tolerance. *Appl. Turfgrass Sci.* 8(1):1-9. https://doi.org/10.1094/ATS-2011-1223-01-RS.

Kowalewski, A. R., Schwartz, B. M., Grimshaw, A. L., Sullivan, D. G. and Peake, J. B. 2015. Correlations between hybrid Bermudagrass morphology and wear tolerance. *hortTechnology* 25(6):725-730. https://doi.org/10.21273/HORTTECH.25.6.725.

Krum, J. M., Carrow, R. N. and Karnok, K. 2010. Spatial mapping of complex turfgrass sites: site-specific management units and protocols. *Crop Sci.* 50(1):301-315. https://doi.org/10.2135/cropsci2009.04.0173.

Leib, B. G., Jabro, J. D. and Matthews, G. R. 2003. Field evaluation and performance comparison of soil moisture sensors. *Soil Sci.* 168(6):396-408. https://doi.org/10.1097/01.ss.0000075285.87447.86.

Leinauer, B., Serena, M. and Singh, D. 2010. Seed coating and seeding rate effects on turfgrass germination and establishment. *hortTechnology* 20(1):179-185. https://doi.org/10.21273/HORTTECH.20.1.179.

Lulli, F., Volterrani, M., Magni, S. and Armeni, R. 2011. An innovative hybrid natural-artificial sports pitch construction system. *Proc. Inst. Mech. Eng. P* 225(3):171-175.

Madsen, M. D., Fidanza, M. A., Barney, N. S., Kostka, S. J., Badrakh, T. and McMillan, M. F. 2016. Low-dose application of nonionic alkyl terminated block copolymer surfactant enhances turfgrass seed germination and plant growth. *hortTechnology* 26(4):379-385. https://doi.org/10.21273/HORTTECH.26.4.379.

Mann, R. L. 2016. Growing sports turf in shady environments. *Acta Hortic.* XXIX(1122):83-90. https://doi.org/10.17660/ActaHortic.2016.1122.11.

Marín, J., Yousfi, S., Mauri, P. V., Parra, L., Lloret, J. and Masaguer, A. 2020. RGB vegetation indices, NDVI, and biomass as indicators to evaluate C3 and C4 turfgrass under different water conditions. *Sustainability* 12(6):2160. https://doi.org/10.3390/su12062160.

Mascitti, E. C., McNitt, A. S. and Serensits, T. J. 2017. Divot resistance of thick cut sod as influenced by preharvest nitrogen and sand topdressing. *Itsrj* 13(1):330-337. https://doi.org/10.2134/itsrj2017.01.0010.

May, R. C. 2019. Evaluation of the effects of repeated applications of athletic field paint to a sand-based rootzone. M.S. Thesis. Iowa State University, Ames, IA.

McAuliffe, K. W. 2008. The role of performance testing and standards in the sports turf industry: A case study approach. In: Stier, J. C., Han, L. and Li, D. (Eds). *Proceedings of the 2nd International Conference on Turfgrass Management and Sports Fields* pp. 391-398. International Society for Horticultural Science, Belgium.

McBeath, J. H. 2003. Snow mold: winter turfgrass nemesis. *Golf Course Manag.* 71:121-124.

McClements, I. and Baker, S. W. 1994. The playing quality of natural turf hockey pitches. *J. Sports Turf Res. Inst.* 70:13-28.

McNitt, A. S. and Landschoot, P. J. 2005. The effects of soil reinforcing materials on the traction and divot resistance of a sand root zone. *Int. Turfgrass Soc. Res. J.* 10:1115-1122.

McNitt, A. S., Middour, R. O. and Waddington, D. V. 1997. Development and evaluation of a method to measure traction on turfgrass surfaces. *J. Testing. and Eval.* 25(1):99--107.

Miller, G. L. 2007. An evaluation of crumb rubber and calcined clay for topdressing sports fields. *II Int. Conf. Turfgrass Sci. Manag. Sports Fields* 783:381-390.

Morris, K. N. and Shearman, R. C. 2007. NTEP turfgrass evaluation guidelines. National Turfgrass Evaluation Program, Beltsville MD. Available at: http://www.ntep.org/pdf/rat-ings.pdf.

Munnecke, D. E. 1972. Factors affecting the efficacy of fungicides in soil. *Annu. Rev. Phytopathol.* 10(1):375-398.

Murphy, J. A., Rieke, P. E. and Erickson, A. E. 1993. Core cultivation of a putting green with hollow and solid tines. *Agron. J.* 85(1):1-9.

National Turfgrass Evaluation Program 2016. 2016 national perennial ryegrass test. Available at: http://www.ntep.org/reports/pr16/pr16_20-5/pr16_20-5.htm. National Turfgrass Evaluation Program, Beltsville, MD.

National Turfgrass Evaluation Program 2017. 2017 national tall fescue test. National Turfgrass Evaluation Program, Beltsville, MD. Available at: http://www.ntep.org/reports/kb17/kb17_21-5/kb17_21-5.htm.

National Turfgrass Evaluation Program 2018. 2018 national Kentucky bluegrass test. National Turfgrass Evaluation Program, Beltsville, MD. Available at: http://www.ntep.org/reports/tf18/tf18_20-1/tf18_20-1.htm.

Ogle, B. M. 2020. Evaluating the effects of various painting techniques on athletic field surface and root zone characteristics. M.S. Thesis. University of Tennessee, Knoxville, TN.

Patton, A. J., Trappe, J. M. and Richardson, M. D. 2010. Cover technology influences warm-season grass establishment from seed. *hortTechnology* 20(1):153-159. https://doi.org/10.21273/HORTTECH.20.1.153.

Pinnix, G. D. and Miller, G. L. 2019. Comparing evapotranspiration rates of tall fescue and Bermudagrass in North Carolina. *Agric. Water Manag.* 223:105725. https://doi.org/10.1016/j.agwat.2019.105725.

Pochron, S. T., Fiorenza, A., Sperl, C., Ledda, B., Lawrence Patterson, C., Tucker, C. C., Tucker, W., Ho, Y. L. and Panico, N. 2017. The response of earthworms (*Eisenia fetida*) and soil microbes to the crumb rubber material used in artificial turf fields. *Chemosphere* 173:557-562. https://doi.org/10.1016/j.chemosphere.2017.01.091.

Price, C. 2014. Durable sod for in-season replacement. *SportsTurf* 30(2):28-30.

Puhalla, J. C., Krans, J. V. and Goatley, J. M., Jr. 2020. *Sports Fields: Design, Construction, and Maintenance* (3rd edn.). Wiley & Sons, Hoboken, NJ.

Reynolds, W. C., Miller, G. L. and Rufty, T. W. 2012. Athletic field paint impacts light spectral quality and turfgrass photosynthesis. *Crop Sci.* 52(5):2375-2384. https://doi.org/10.2135/cropsci2012.01.0059.

Richardson, M. D. and Hignight, K. W. 2010. Seedling emergence of tall fescue and Kentucky bluegrass, as affected by two seed coating techniques. *hortTechnology* 20(2):415-417. https://doi.org/10.21273/HORTTECH.20.2.415.

Richardson, M. D., Mattina, G., Sarno, M., McCalla, J. H. and Karcher, D. E. 2019. Shade effects on overseeded Bermudagrass athletic fields: I. Turfgrass coverage and growth rate. *Crop Sci.* 59(6):2845-2855. https://doi.org/10.2135/cropsci2019.05.0310.

Rodriguez, I. R., McCarty, L. B. and Toler, J. E. 2005. Effects of misting and subsurface air movement on bentgrass putting greens. *Agron. J.* 97(5):1438–1442. https://doi.org/10.2134/agronj2004.0302.

Segars, C. A. and Moss, J. Q. 2014. The effects of athletic-field paint color on net photosynthesis and canopy temperature of overseeded perennial ryegrass. *Appl. Turfgrass Sci.* 11(1). https://doi.org/ 10.2134/ATS-2014-0015-BR.

Serena, M., Velasco-Cruz, C., Friell, J., Schiavon, M., Sevostianova, E., Beck, L., Sallenave, R. and Leinauer, B. 2020. Irrigation scheduling technologies reduce water use and maintain turfgrass quality. *Agron. J.* 112(5):3456–3469. https://doi.org/10.1002/agj2.20246.

Straw, C. M., Bowling, W. J. and Henry, G. M. 2017a. Rainfall versus irrigation influences penetration resistance and surface hardness on a recreational sports field. *Int. Turfgrass Soc. Res. J.* 13(1):619–623. https://doi.org/10.2134/itsrj2016.10.0842.

Straw, C. M., Henry, G. M., Love, K., Carrow, R. N. and Cline, V. 2017b. Evaluation of several sampling procedures for spatial analysis of natural turfgrass sports field properties. *J. Test. Eval.* 46(2):714–729. https://doi.org/10.1520/JTE20160467.

Straw, C. M., Carrow, R. N., Bowling, W. J., Tucker, K. A. and Henry, G. M. 2018a. Uniformity and spatial variability of soil moisture and irrigation distribution on natural turfgrass sports fields. *J. Soil Water Conserv.* 73(5):577-586. https://doi.org/10.2489/jswc.73.5.577.

Straw, C. M., Samson, C. O., Henry, G. M. and Brown, C. N. 2018b. Does variability within natural turfgrass sports fields influence ground-derived injuries? *Eur. J. Sport Sci.* 18(6):893–902. https://doi.org/10.1080/17461391.2018.1457083.

Straw, C. M., Grubbs, R. A. and Henry, G. M. 2020a. Short-term spatiotemporal relationship between plant and soil properties on natural turfgrass sports fields. *Agrosyst. Geosci. Environ.* 3(1). https://doi.org/10.1002/agg2.20043.

Straw, C. M., Samson, C. O., Henry, G. M. and Brown, C. N. 2020b. A review of turfgrass sports field variability and its implications on athlete-surface interactions. *Agron. J.* 112(4):2401–2417. https://doi.org/10.1002/agj2.20193.

Straw, C. M., Grubbs, R. A., Henry, G. M., Carrow, R. N. and Cline, V. 2016a. Long-term effect of open-spoon aerification on plant and soil properties of community level sports fields. *Crop Forage Turfgrass Manag.* 2(1):1–7. https://doi.org/10.2134/cftm2016.04.0031.

Straw, C. M., Grubbs, R. A., Tucker, K. A. and Henry, G. M. 2016b. Handheld versus mobile data acquisitions for spatial analysis of natural turfgrass sports fields. *Hortscience* 51(9):1176–1183. https://doi.org/10.21273/HORTSCI11019-16.

Straw, C. M. and Henry, G. M. 2018. Spatiotemporal variation of site-specific management units on natural turfgrass sports fields during dry down. *Precision Agric.* 19(3):395-420. https://doi.org/10.1007/s11119-017-9526-5.

Straw, C. M., Henry, G. M., Shannon, J. and Thompson, J. J. 2019. Athletes' perceptions of within-field variability on natural turfgrass sports fields. *Precision Agric.* 20(1):118-137. https://doi.org/10.1007/s11119-018-9585-2.

Su, K., Moss, J. Q., Zhang, G., Martin, D. L. and Wu, Y. 2013. Bermudagrass drought tolerance associated with dehydrin protein expression during drought stress. *J. Am. Soc. Hort. Sci.* 138(4):277–282. https://doi.org/10.21273/JASHS.138.4.277.

Taliaferro, C. M. 2020. Turf Bermudagrass 'JSC 2-21-18'. US patent PP32130.

Thoms, A. W., Bearss, R., Rogers, J. N. and Sorochan, J. C. 2021. An evaluation of mat hybrid turfgrass systems under simulated traffic. *Intl Turfgrass Soc. Res. J.* 14(1):385-389 https://doi.org/10.1002/its2.58.

Thoms, A. W., Brosnan, J. T., Paquette, M. R., Zhang, S. and Sorochan, J. C. 2013. A new device for simulating athlete-to-surface interactions on natural and synthetic turf. *J. Test. Eval.* 41(3):497-503. https://doi.org/10.1520/JTE20120103.

Thomson, A., Whiteley, R., Wilson, M. and Bleakley, C. 2019. Six different football shoes, one playing surface and the weather; Assessing variation in shoe-surface traction over one season of elite football. *PLoS ONE* 14(4):e0216364. https://doi.org/10.1371/journal.pone.0216364.

Trappe, J. M., Karcher, D. E., Richardson, M. D. and Patton, A. J. 2011. Shade and traffic tolerance varies for Bermudagrass and zoysiagrass cultivars. *Crop Sci.* 51(2):870-877. Available at: https://di.org/10.2135/cropsci2010.05.0248.

Turgeon, A. J. and Fidanza, M. A. 2017. Perspective on the history of turf cultivation. *Int. Turfgrass Soc. Res. J.* 13(1):629-635. Available at: https://doi.org/itsrj2016.05.0394.

Twomey, D. M., Otago, L., Ullah, S. and Finch, C. F. 2011. Reliability of equipment for measuring the ground hardness and traction. *Proc. Instit. Mech. Eng., Part P: J. Sport Eng. Technol.* 225(3):131-137.

Van Gundy, S. and McKenry, M. V. 2012. Action of nematicides. In: Horsfall, J. G. (Ed.), *Plant Disease: an Advanced Treatise* (1st edn.) pp. 263-283. Elsevier, Amsterdam, Netherlands.

Vaz, C. M. P., Bassoi, L. H. and Hopmans, J. W. 2001. Contribution of water content and bulk density to field soil penetration resistance as measured by a combined cone penetrometer-TDR probe. *Soil Till. Res.* 60(1-2):35-42. https://doi.org/10.1016/S0167-1987(01)00173-8.

Wherley, B., Heitholt, J., Chandra, A. and Skulkaew, P. 2014. Supplemental irrigation requirements of zoysiagrass and Bermudagrass cultivars. *Crop Sci.* 54(4):1823-1831. https://doi.org/10.2135/cropsci2013.11.0753.

Wiecko, G., Carrow, R. N. and Karnok, K. J. 1993. Turfgrass cultivation methods: influence on soil physical, root/shoot, and water relationships. *Int. Turfgrass Soc. Res. J.* 7:451-457.

Winka, B. 2018. Best management practices on bluemuda athletic fields. *SportsTurf.* 34(7):22-24.

Wipff, J. K. and Singh, D. 2015. *Lolium perenne* subsp. *stoloniferum*; perennial ryegrass with determinate-stolons. United States Plant Patent. U.S. Patent no. 8927,804.

Wu, Y., Martin, D. L., Moss, J. Q., Walker, N. R. and Fontanier, C. 2020. Bermudagrass plant named 'OKC 1131'. US patent PP31695. 20 Jan. 2021.

Wu, Y. Q., Martin, D. L., Taliaferro, C. M., Anderson, J. A. and Moss, J. Q. 2012. Latitude 36 turf Bermudagrass. U.S. Patent US PP24. 271P1.

Wu, Y. Q., Martin, D. L., Taliaferro, C. M., Anderson, J. A. and Moss, J. Q. 2013. Northbridge turf Bermudagrass. U.S. Patent US PP24. 116P3.

Xiang, M., Fry, J. and Wu, Y. 2021. Winter survival of experimental bermudagrasses in the upper transition zone of the US. *Intl Turfgrass Soc. Res. J.* 14(1):708-712 https:/doi.org/10.1002/its2.3.

Zhang, J., Virk, S., Porter, W., Kenworthy, K., Sullivan, D. and Schwartz, B. 2019. Applications of unmanned aerial vehicle based imagery in turfgrass field trials. *Front. Plant Sci.* 10:279. https://doi.org/10.3389/fpls.2019.00279.

Chapter 8

Advancements in turfgrass for ornamental lawns

Rebecca Grubbs Bowling, Texas A&M University, USA; and Joseph Young, Texas Tech University, USA

1 Introduction

While many other turfgrass systems such as golf courses or athletic fields can be more easily defined by a clear function or aesthetic, the definition of an ornamental lawn is less clear (Fig. 1). In the literature, a lawn may refer to anything from residential lawns (Hamido et al., 2016; Yue et al., 2012) and park spaces (Barnes and Watkins, 2022; Ramer et al., 2019) to turfgrass managed in cemeteries (Clayden et al., 2018; Quinton et al., 2019) and along roadsides (Keilsohn et al., 2018; Friell and Watkins, 2021). The term 'lawn' may also be used to describe areas that have little to no turfgrass at all but are still used in a manner at least partially consistent with a conventional turfgrass lawn (Barnes and Watkins, 2022; Smith and Fellowes, 2014).

Like all turfgrass systems, well-managed lawns can offer a wide range of ecosystem services ranging from temperature regulation and runoff mitigation to recreation and support of mental, physical, and social well-being (Brosnan et al., 2020; Monteiro, 2017). Ecosystem service potential will vary in response to both management practices and intended use. Park visitors surveyed in one study identified several common uses for public lawns, including picnics, play with children and pets, organized and recreational sports, rest and relaxation,

http://dx.doi.org/10.19103/AS.2022.0110.08

Figure 1 Residence and cool-season turfgrass lawn in the background, golf course area in the foreground. Source: M. Fidanza.

individual and group exercise, wildlife viewing, and aesthetic enjoyment (Barnes and Watkins, 2022) (Fig. 2). Conversely, some lawns may rarely, if ever, be used for recreation. Instead, they may serve more utilitarian purposes such as slope stabilization (Ma et al., 2021) and roadside protection (Friell and Watkins, 2021).

Perhaps most critical is the abundance and diversity of lawns compared with golf courses, athletic fields, and sod production farms. While the individual lawn is typically smaller than these other turfgrass systems, aggregated lawns cover a much greater area and can therefore have a significant social and environmental impact (Fig. 3). Historically, estimates for US lawn cover have ranged from 58 000 km^2 (Vinlove and Torla, 1994) to 120 000 km^2 (Roberts and Roberts, 1989), with total turfgrass cover estimated at 163 800 km^2 or approximately 1.9% of total US land area (Fidanza, 2022; Milesi et al., 2005). Proportionately, lawns account for a much greater percentage of land area in urban and suburban settings where people are more heavily concentrated.

Figure 2 Example of natural turfgrass in a park. Source: M. Fidanza.

Figure 3 Example of a cool-season turfgrass lawn. Source: www.thelawninstitute.org.

One recent study estimates that lawns occupy around 70-75% of open green spaces in cities (Ignatieva et al., 2015). This suggests that total lawn and turfgrass area has likely only grown from past estimates in response to US urbanization trends. In 1987, when Roberts and Roberts published their lawn cover estimates, approximately 74.8% of the US population lived in an urban area (United Nations, 2018). This increased to 79.9% by 2005 (Milesi et al.) and finally to 83.1% today (https://bit.ly/3PGtdO2). The United Nations (2018) predicts that by 2050, 89.2% of the US population will reside in urban systems, underscoring the escalating importance of sustainable urban green spaces.

Ornamental lawns have come under scrutiny in recent years for their potential impact on environmental crises like water scarcity (Canales-Ide et al., 2019; Litvak and Pataki, 2016; Salvaggio et al., 2014), dwindling biodiversity (Aronson et al., 2017; Ignatieva and Hedblom, 2018; Smith et al., 2015), and global warming (Gu et al., 2015; Saidani and Kim, 2021; Zhang et al., 2013). This has led to a growing call for research to advance a sustainable science around lawn management, with some advocating for alternatives to conventional turfgrass lawns (Ignatieva and Hedblom, 2018; Smith et al., 2015). In response, turfgrass scientists have invested considerable time and resources into the development of more environmentally sustainable cultivars/varieties (Grimshaw et al, 2018; Katuwal et al, 2022), technologies (Jain et al., 2015; Serena et al., 2020), and best management practices (BMPs) (Hejl et al., 2021). Simultaneously, many studies have sought to better understand the ecosystem service potential of lawns (Brosnan et al., 2020; Hamido et al., 2016; Wang et al., 2016) and the potential trade-offs of lawn alternatives (Wang et al., 2016) in the hopes of answering persistent questions around the planning, development, and management of resilient green spaces.

Unfortunately, the abundance and diversity of ornamental lawns presents a complex challenge for the dissemination of BMPs to end users. Further, the history of the American lawn has far-reaching cultural implications for consumer expectations and willingness to adopt new practices or technologies, even when they offer proven environmental and economic benefits (Robbins, 2012). Many researchers have seen this as an opportunity to conduct novel socioenvironmental research exploring the relationship between lawn management and behavior, hoping to identify more effective pathways for engaging target audiences.

Although there is considerable variability across ornamental lawns, there are some distinct attributes differentiating lawns from other turfgrass systems, allowing researchers to identify research questions unique to this turfgrass group. This chapter will explore several of these themes and consider what they may mean toward future research.

2 Advancements in turfgrass breeding and selection for ornamental lawns

Development and widespread adoption of low-input turfgrasses has been a priority for many turfgrass breeding programs due to possible financial and environmental benefits (Yue et al., 2021).

2.1 Consumer lawn preferences and implications for contemporary breeding and production

To further inform breeding priorities, several recent studies have explored consumer preferences for turfgrass attributes, particularly as it relates to residential lawns (Barnes and Watkins, 2022; Ghimire et al., 2019; Yue et al., 2017a,b, 2021). Findings indicate there is considerable heterogeneity in end-user preferences at national and international scales (Ghimire et al, 2019; Yue et al., 2017a). Characteristics that have been important to consumers in varying degrees across studies include maintenance attributes, aesthetic characteristics (e.g. texture, color, and density), shade adaptation, origin (e.g. native or non-native), and price, with low maintenance identified as the most influential attribute for consumers evaluated in multiple studies (Ghimire et al., 2016; Hugie et al., 2012; Yue et al., 2012). Prioritization of specific maintenance practices varies; however, mowing, fertilization, and irrigation requirements are among those recognized as important to many consumers (Yue et al., 2012, 2017a). Although most studies did not evaluate consumer preferences related to traffic tolerance, one study identified the ability to withstand foot traffic as one of the most influential attributes for consumers in both the USA and Canada (Yue et al., 2017a). Other stress tolerances, particularly drought and salt tolerance, have also been identified as important attributes for warm-season turfgrass consumers in two studies (Ghimire et al., 2016, 2019).

Despite considerable heterogeneity in consumer preferences, at least three studies have identified specific consumer segments that correspond to more consistent preferences for both warm- (Ghimire et al., 2019) and cool-season turfgrasses (Yue et al., 2017a,b, 2021). Market segmentation may help plant breeders, plant producers, and similar stakeholders tailor outreach approaches to diverse end-user groups, thus increasing the likelihood of adoption. Those three studies have progressively drawn from each other, with the two most recent studies (Ghimire et al., 2019; Yue et al., 2021) each identifying two key consumer segments. Although they use a slightly different nomenclature, both studies identify a 'reluctant' segment, primarily consisting of more mature or established homeowners, and a 'willing' segment, primarily consisting of more enthusiastic consumers willing to adopt or invest in newer, low-input turfgrasses (Ghimire et al., 2019; Yue et al., 2021). Willing adopters, often correlated with younger, more enthusiastic 'hobby gardeners', appear inclined to adopt and invest in new, resource-efficient turfgrass cultivars, indicating that this should be a priority for many turfgrass breeding programs and producers. In the case of more reluctant consumers, the concern is often around aesthetics, with many consumers equating more sustainable practices with less attractive lawns and landscapes. In this case, turfgrass breeders are challenged to consider how they can enhance the resource-use efficiency and resiliency of grasses while preserving aesthetic characteristics that are desirable to these more mature or traditional and conventional consumers.

2.2 Contemporary breeding and selection trends for ornamental turfgrass lawns

Species or cultivars advanced for use in ornamental lawns often differ from those advanced for other use scenarios such as golf course putting greens and fairways or athletic fields. While some turfgrass species such as hybrid bermudagrass (*Cynodon dactylon* × *C. transvaalensis* Burtt Davy), perennial ryegrass (*Lolium perenne* L.), or zoysiagrass (*Zoysia* spp.) have more versatile cultivars that may be used for a wide range of functions, others, including St. Augustinegrass (*Stenotaphrum secundatum* [Walt.] Kuntze), centipedegrass (*Eremochloa ophiuroides* [Munro] Hack.), bahiagrass (*Paspalum notatum* Flueggé), and tall fescue (*Schedonorus arundinaceus* [Schreb.] Dumort.), are used almost exclusively in lawn and lawn-type environments and in golf course roughs. This is primarily due to phenotypic characteristics (e.g. height of growth and coarser texture) that limit the use of these species in systems where a fine texture or a shorter mowing height (<2.5 cm) is required (Patton et al., 2017).

Most modern turfgrass breeding programs generally prioritize traits related to abiotic stress resistance, growth characteristics, and appearance (Yue et al., 2017c). To strengthen selection efforts, some breeding programs have also begun working collaboratively to identify genotypes that perform well across diverse climates (Gouveia et al., 2021a; Katuwal et al., 2022; Pirtle et al., 2022). These efforts are further strengthened by the National Turfgrass Evaluation Program (Beltsville, MD, USA) and other multi-state research efforts that assess cultivar performance under diverse management regimes and stress variables (Gouveia et al., 2021b; Watkins et al., 2014).

In the case of lawns, growing emphasis has been placed on drought stress tolerance (Fig. 4) in response to climate change and tightening irrigation restrictions (Katuwal et al., 2022; Schwartz et al., 2018). This is evidenced by the release of several new drought-tolerant turfgrass cultivars that can be used in ornamental lawns such as 'TifTuf' hybrid bermudagrass [experimental name (EN): DT-1], 'Cobalt' St. Augustinegrass (EN: DALSA 1618), and 'Lobo' zoysiagrass (EN: XL14069). Although these new drought-tolerant turfgrass

Figure 4 Evaluation of tall fescue cultivars maintained under a rainout shelter for drought stress tolerance. Source: C. Bigelow.

cultivars are sold at a premium price, end-users are expected to have considerable water savings and long-term economic benefits (Minor et al., 2020). For example, across eleven major cities, consumers that adopt TifTuf are expected to save between 2.7 cm and 386.7 cm of water per 0.41 ha (equivalent to 1 acre) over 5 years. Based on city water rates in 2017, this would mean an estimated cost savings of between US$83 and US$55 910 over the same 5-year time frame. Multilocational testing has been successful in identifying dozens of additional genotypes with superior drought performance, with top-performing St. Augustinegrass and zoysiagrass selections maintaining >50% green cover longer than standard cultivars under drought stress (Katuwal et al., 2022). This suggests that drought-tolerant options will only become more abundant and more readily available to consumers over time.

In addition to drought tolerance, shade (Amundsen et al., 2017; Meeks and Chandra, 2021; Wherley et al., 2013), traffic tolerance (Grimshaw et al., 2018), cold tolerance (Moseley et al., 2021), and maintenance requirements (Buhlman et al., 2022) have also been priorities for turfgrass selection and breeding efforts in lawns (Fig. 5). New cultivars such as 'CitraBlue' St. Augustinegrass (EN: FSA1602), 'Innovation' zoysiagrass (EN: KSUZ 802), and 'ProVista' St. Augustine and Kentucky bluegrasses (*Poa pratensis* L.) represent a growing number of new turfgrass options for home lawns that boast benefits like improved shade

Figure 5 Evaluation of Kentucky bluegrass cultivars (left) and bermudagrass cultivars at spring green-up (right), for use in lawns. Source: C. Bigelow.

tolerance, reduced mowing requirements, or greater pest resistance (Fidanza et al., 2022).

Other species, perhaps historically regarded as limited in their potential use, are also being reevaluated as options for low-maintenance landscapes. Buffalograss (*Buchloe dactyloides* [Nutt.] Engelm. syn. *Bouteloua dactyloides* [Nutt.] Columbus), an important native option for low-input amenity grass stands, has been generally regarded as intolerant of stress, traffic, and sandy soils. However, Amundsen etal. (2017) found that this may not be the case with modern germplasm selections, instead observing that buffalograss performance was largely unaffected by soil texture and some genotypes performed well in moderate traffic and up to 60% shade.

Fine fescues (*Festuca* spp.) represent another group of grasses receiving increased attention by turfgrass breeders due to their potential use in low-input lawns (Braun et al., 2020; Grimshaw et al., 2018). Fine fescues are well established as highly versatile, requiring little to no water, fertilization or pesticide inputs compared with other popular cool-season turfgrass lawn options such as Kentucky bluegrass and perennial ryegrass (Braun et al., 2020). However, one weakness of fine fescues is their ability to tolerate traffic. Breeders are now exploring heritability of traffic and wear tolerance in hundreds of fine fescue genotypes, anticipating that the improvement of this attribute is possible through recurrent selection of replicated clonally propagated genotypes (Grimshaw et al., 2018).

3 Advancements in sustainable ornamental lawn management

Many lawn-centric BMPs have emerged over the past decade. Several of the cultural practices common to ornamental lawn management will most likely parallel those discussed in other chapters of this textbook, including developments in establishment, mowing and cultivation, nutrient management, and irrigation.

3.1 Soil preparation, planting and establishment

Soil testing for physical and chemical properties prior to establishment of turfgrass is crucial to effectively establish turfgrass lawns (Turgeon and Kaminski, 2019). These tests can quickly demonstrate concerns with soil compaction through bulk density, organic matter content relating to plant nutrient and water-holding potential, soil pH, and essential plant nutrient availability. Remediation of soil properties will be easiest and more beneficial prior to establishing home lawns. The perennial nature of the turfgrass lawn makes it challenging to implement disruptive and large-scale and changes post-establishment.

Beyond basic soil testing, amending soil properties is more achievable prior to establishment of a home lawn. The opportunity to incorporate materials into the upper soil profile at higher amounts is unique to this pre-establishment phase. For example, coarse sand fractions can be included to increase water infiltration and drainage in finer textured soils. Likewise, compost or fine-textured topsoil materials may be added or incorporated into landscapes to enhance nutrient and water availability during turfgrass establishment and long-term storage and availability for the life of the landscape (Evanylo et al., 2016; Kranz et al., 2020).

Compost materials that have been evaluated for use in lawns and landscapes range from composted sewage sludge (Cheng et al., 2007) and biosolid materials (Stacey et al., 2019) to yard waste (Cannavo et al., 2014). Compost incorporation has been found to generally improve nutrient availability and positively impact the soil microbiome (Azeem et al., 2020; Stacey et al., 2019). In addition, a recent concise review of the effects of compost incorporation on soil physical properties revealed that compost incorporation often has a positive effect on bulk density, infiltration, hydraulic conductivity, and plant-available water (Kranz et al., 2020, 2023). Despite these studies, there are persistent knowledge gaps related to appropriate compost rates and depths, as well as longevity of potential effects when used as a pre-plant amendment (Kranz et al., 2020).

In addition to compost materials, studies have explored the use of non-organic amendments to improve soil physical properties. For example, expanded shale, a lightweight, porous aggregate, has been evaluated in several studies as a rootzone amendment for landscapes, green roofs, and stormwater control features, due to its documented ability to improve overall plant performance in clay soils (Sloan et al., 2002, 2012). This has been largely attributed to its capacity to decrease bulk density in amended soils, improving permeability and drainage properties with potential benefits for vegetative growth (Mechleb et al., 2014). Expanded shale also was found to increase hydraulic conductivity of an amended Texas clay soil by more than three orders of magnitude with 40% expanded shale (Mechleb et al., 2014).

Environmental concerns related to off target movement of soil nutrients following fertility applications led to research efforts that altered nutrient recommendations for turfgrass seed germination or vegetative establishment (Carey et al., 2012). Much of the adjusted practices involve phosphorus application to promote greater seed germination and root development. High phosphorus-containing fertilizers were recommended when establishing any turfgrass propagules, but soil test results should indicate inadequate phosphorus levels for selected turfgrass species before applying high phosphorus-containing fertilizers (Chang et al., 2014; Liu and Landschoot, 2018). Likewise, nitrogen (N) source and rate selection should be based on soil texture analysis considerations to mitigate nutrient losses (Erickson et al., 2001; Wu et al., 2010). Thus, obtaining nutrient analysis of soil prior to establishing turfgrass in a lawn is critical to best prescribe pre-plant fertility optimization.

After optimizing soil conditions for seed germination and root development, seed coating technologies and post-plant management practices can limit pest competition during establishment. Seed coatings have shown greatest benefits to establishment of turfgrass seed when conditions may not be optimum for seed germination such as salinity, reduced irrigation, or lower seeding rates (Madsen et al., 2016; Pedrini et al., 2017; Richardson et al., 2010; Leinauer et al., 2010). Managing weed populations, especially during the establishment of turfgrass from seed, can pose significant challenges and delays in full turfgrass coverage. However, research has documented safe herbicide applications that can minimize weed populations without limiting germination and growth of seeded turfgrasses (Pease and Stier, 2020).

3.2 Fertilization and nutrient management

Research to advance more sustainable ornamental lawns has included efforts to decrease fertilizer inputs. This can be a challenge, as urban soils are notoriously nutrient deficient due to development and land management practices (Carey et al., 2012). However, some studies estimate that N inputs exceed demands in urban systems by as much as 51%, with much of this attributed to household fertilizer use (Fissore et al., 2011). This indicates that many consumers are overestimating supplemental N needs in residential lawns and landscapes and often applying more fertilizers than are needed. Excess nutrient inputs have been linked to nonpoint source pollution, impacting water quality and overall watershed health (Hobbie et al., 2017; Toor et al., 2017). These effects can be exacerbated by environmental conditions and cultural practices that promote surface runoff, leaching, or volatilization, emphasizing the importance of a holistic lawn management approach. In addition to minimizing harmful environmental effects, reducing fertilizer inputs, as appropriate, will also offer cost savings benefits to some consumers.

Extension and outreach education efforts increasingly prioritize soil sampling to facilitate a customized nutrient management approach. The intention behind this is to encourage consumers to only apply what is needed based on their soil's physical and chemical properties. There is limited data, however, to support a relationship between soil testing and fertilizer input reductions. While soil testing has established benefits for various agricultural applications (Pandey et al., 2018), research is needed to further explore the impact of soil testing on fertilizer use behaviors in turfgrass lawns.

When supplemental N is needed to support ornamental lawn growth, emphasis should be placed on appropriate application timing to minimize environmental losses and prevent additional lawn stressors, including weeds, disease, and abiotic stress-induced injury. Slow- or controlled-release fertilizers are recognized as one possible option to improve nutrient-use efficiency in lawn systems (Wesolowska et al., 2021). However, the performance of these products can vary significantly based on environmental conditions and product type. Predictive models suggest that lawn fertilization should be gradually decreased with stand age to further improve N-use efficiency and reduce long-term nitrous oxide emissions (Zhang et al., 2013).

Turfgrass clipping management has been linked to N-use efficiency in several studies, with most studies identifying lawn clipping return as a BMP capable of reducing supplement N requirements (Grubbs et al., 2021; Law et al., 2017). However, specific N benefits are difficult to consistently estimate, as N cycling is dependent on a wide range of environmental factors that will vary not only across sites but within sites due to the spatial heterogeneity of soil properties, including microbial ecology (Grubbs et al., 2021).

Other cultural practices discussed previously in this chapter and in other chapters of this textbook may improve nutrient-use efficiency include appropriate turfgrass selection and utilization of compost and other soil amendments to improve soil physical and chemical properties. Mowing and cultivation and irrigation and water management practices discussed in the following sections will also contribute to the function and process of nutrient cycling in lawns.

3.3 Mowing and cultivation

Contrary to golf courses or athletic fields, ornamental lawns are generally not managed to support a particular type of ball or foot movement. Rather, lawns are more often designed and maintained to support activities that do not require an especially low height-of-cut (HOC). When a low HOC is not required, taller or higher mowing heights are generally adopted based on associated economic and environmental benefits, including deeper root growth, reduced mowing requirements, and greater resistance to abiotic stresses such as drought, cold, or traffic.

Recent studies have evaluated the effect of lawn height and/or mowing frequency on turfgrass resilience and ecosystem service potential. Less frequent mowing offers some environmental benefits, including the support of pollinator abundance and diversity (Lerman et al., 2018; Wastian et al., 2016;) and reductions in carbon dioxide emissions (Lerman and Contosta, 2019). However, less frequent mowing may have adverse effects on turfgrass health and N cycling (Grubbs et al., 2021). Selection of turfgrass species or cultivars that have a reduced mowing requirement coupled with following the 'one third rule' may promote the benefits of reduced mowing without jeopardizing turfgrass performance (Law et al., 2016).

Additional advancements in lawn mowing include the introduction of several autonomous mower technologies, including those that are solar powered and self-sufficient (Daniyan et al., 2020; Jain et al., 2015). Autonomous mowers have been shown to save energy and potentially improve turfgrass quality in some lawn environments (Grossi et al., 2016); however, additional research is needed to further improve mower performance (Saidani et al., 2021).

3.4 Irrigation and water management

Municipal watering restrictions and concerns of water scarcity have led to research seeking to advance water-use efficiency (WUE), with particular emphasis on improvements in turfgrass breeding and selection, irrigation scheduling, and smart watering technology. Optimization of WUE is also contingent on cultural practices, with particular emphasis on soil management practices that improve soil structural properties to optimize plant-available water without compromising drainage and infiltration.

In a 3-year study evaluating turfgrass cultivar performance under common lawn irrigation frequency regimes (0, 1, 2, 4, and 8x monthly), researchers found that when irrigation was restricted to less than once per week, turfgrass selection becomes particularly important (Hejl et al., 2021). In that study, conducted in Texas (USA), bermudagrass cultivars generally outperformed all other species evaluated, including St. Augustinegrass, seashore paspalum (*P. vaginatum* Swartz), and zoysiagrass. However, these species may become increasingly competitive with bermudagrass as new drought-tolerant cultivars are developed. In an Indiana (USA) study evaluating Kentucky bluegrass and tall fescue performance under deficit irrigation, researchers asserted that inclusion of drought-tolerant cultivars in situations where turfgrass blends are desired may support stand resilience and reduce overall irrigation needs (Powlen et al., 2021).

Many studies have presented irrigation scheduling technologies, or smart controllers, to aid in the implementation of evapotranspiration-based irrigation scheduling (Bartlett et al., 2015; Murthy et al., 2019; Serena et al., 2020). New irrigation technologies may also facilitate greater awareness and adoption of other water-efficient practices and technologies, including rainwater harvesting (Wanjiru and Xia, 2015), subsurface drip irrigation (Canales-Ide et al., 2019), and cycle and soak scheduling (Murthy et al., 2019). Such practices will also have implications for improved nutrient-use efficiency, as they can potentially reduce surface runoff potential and contaminate loading in stormwater. Homeowners appear to prefer smart controllers over traditional automated irrigation systems, suggesting a willingness to invest in this technology, but additional research to monitor evolving consumer preferences has been recommended (Khachatryan et al., 2019).

While considerable emphasis has been placed on water savings and drought, climate change predictions in many parts of the USA anticipate increased intermittent flooding and stormwater management challenges (https://tinyurl.com/2p8c47d8). Researchers must also consider the role that ornamental lawns will play in flood mitigation and the increased utilization of green stormwater infrastructure and similar nature-based solutions to stormwater management.

4 Non-traditional or alternative lawns

With growing public concerns around ecological resilience and the environmental impact of traditional lawn management, there has been increasing interest in non-traditional or alternative lawns that deviate from conventional turfgrass monocultures (Ignatieva and Ahrné, 2013; Pooya et al., 2013; Barnes and Watkins, 2022). These alternatives are generally evaluated for their ability to achieve an environmental or economic goal, with particular emphasis on enhanced biodiversity (Filazzola et al., 2019; Ignatieva and Ahrné, 2013; Smith and Fellowes, 2014; Wolfin et al., 2021) and the reduction of inputs including water, nutrients, and carbon (Ignatieva and Hedblom, 2018; McCurdy et al., 2014; Smetana and Crittenden, 2014). Heightened interest in lawn alternatives has also been driven by a growing number of rebate or 'cash for grass' programs that seek to incentivize turfgrass removal and replacement in urban landscapes (Chang et al., 2022; Grant et al., 2020; Zhang and Khachatryan, 2021).

Alternative or non-traditional lawn research can encompass anything from novel grass mixtures (Pinkney et al., 2021; Whitman et al., 2022) and flowering 'bee lawns' (Ramer et al., 2019; Wolfin et al., 2021) to artificial turf (Francis, 2018; Barnes and Watkins, 2022), xeriscapes (Yang and Wang, 2017; Chang et al., 2021), or 'grass-free' lawns that generally consist of native and adaptive plant mixtures (Smith and Fellowes, 2014; Smith et al., 2015). Despite growing interest in alternative lawns, there are persistent knowledge gaps related to ecosystem service trade-offs and suitability across diverse climates and geographies. While an alternative lawn may effectively reduce water use (Yang and Wang, 2017), it may also increase water runoff volume (Chang et al., 2021) or net warming when compared to turfgrass (Chow and Brazel, 2012). As the push for alternative lawns becomes stronger, researchers must be prepared to answer questions about the benefits of one landscape mesocosm over another based on an increasingly complex set of social, environmental, and economic variables.

4.1 Non-traditional or alternative living lawns

Biodiversity has been widely recognized as a critical component of urban ecosystem resilience (Aronson et al., 2017; Filazzola et al., 2019; Ignatieva and Ahrné, 2013; Oke et al., 2021). As a result, research exploring living lawn alternatives often prioritizes mixed species landscapes that may or may not include turfgrass. In some cases, simply mixing multiple turfgrass species or cultivars within the same grass stand has been considered a revolutionary concept, particularly in warm-season lawns where turfgrass blends are less common. A Florida (USA) study evaluating intraspecific cultivar blends of St. Augustinegrass found that mixing as many as four cultivars offered environmental and economic benefits, improving turfgrass coverage and longevity compared with conventional monocultures (Whitman et al., 2022). A similar study found that mixed-cultivar plantings may also ultimately reduce weed pressure, thereby reducing the need for some pesticides (Pinkney et al., 2021).

Other studies have instead explored interspecific mixtures of turfgrasses with low-growing flowering species. These so-called 'bee lawns' or 'pollinator lawns' often include native or adapted plants, including clovers (*Trifolium* spp.) and thyme (*Thymus* spp.) interplanted with low-input turfgrass species (Wolfin et al., 2021). Such systems provide an important source of forage to native bees and other pollinators, while still tolerating regular mowing to maintain a desirable height. In a qualitative study evaluating public perceptions of bee lawns in Minnesota (USA) neighborhood parks, park visitors were generally very supportive of flowering lawns, though some visitors expressed trepidation at the prospect of attracting bees (Ramer et al., 2019). Although there is additional research on the benefits of clover inclusion in turfgrass lawns (McCurdy et al., 2014), research on bee lawns and similar environments remains geographically limited. Future research should explore appropriate mixtures across climates and their ecosystem service potential if adopted.

4.2 Non-traditional or alternative artificial lawns

There are many in-depth comparisons of natural turfgrass versus synthetic turf (Cheng et al., 2014; Liu and Jim, 2021; Russo et al., 2022; Schneider et al., 2014) (Fig. 6). Advocates for artificial or synthetic turf have identified several possible benefits, including reduced input requirements, comparable safety, a lower annual maintenance cost, superior traffic tolerance, and year-round green cover (Cheng et al., 2014). However, research also indicates that the suitability of artificial turf is less clear across diverse climates and use scenarios, with questions raised around heat fluxes (Carvalho et al., 2021; Jim, 2017), general safety (Cheng et al., 2014), end-user preferences (Barnes and Watkins, 2022; Conrad et al., 2019), and the short- and long-term environmental implications of using a synthetic surface (Chang et al., 2021, 2022; Ignatieva and Hedblom, 2018). Interest in this topic has only increased as advancements in artificial turf technology and growing interest in low-input landscapes have led to more widespread adoption of artificial turf in lawns and lawn-type environments (Barnes and Watkins, 2022; Cheng et al., 2014; Francis, 2018; Simpson and Francis, 2021).

In a study evaluating runoff volume and water retention using controlled rainfall experiments, artificial lawns exhibited decreased water retention and increased runoff when compared to living lawns, with longer artificial turf (~37 mm pile height) generating significantly greater runoff than shorter artificial turf (~22 mm pile height) (Simpson and Francis, 2021). These findings are similar to those from another study that found that artificial turf was one of two landscape mesocosms (the other being xeriscaping) that generated significantly greater cumulative runoff volumes (>400 L•m^{-2}) than a St. Augustinegrass lawn (~290

Figure 6 Example of a natural turfgrass (bermudagrass) baseball field (left) and an artificial (synthetic) football field (right). Source: M. Fidanza.

L•m^{-2}), sandcapped St. Augustinegrass lawn (~180 L•m^{-2}), or mulch-covered landscape (~280 L•m^{-2}) (Chang et al., 2021). In that same study, artificial turf also exhibited elevated runoff nitrate-N amounts (Chang et al., 2022) and significantly greater surface temperatures (along with mulch) compared with all other evaluated landscape mesocosms (Carvalho et al., 2021). These characteristics parallel those associated with similar devegetated artificial surfaces that lead to environmental crises like 'urban stream syndrome' or the ecological degradation of urban freshwater streams and rivers (Miguel-Chinchilla et al., 2019). These findings indicate that artificial lawns may be inappropriate in environments with more intense rainfall events and water quality challenges.

Researchers must also continue to explore the societal implications of widespread artificial lawn adoption, as the removal and replacement of living vegetation and other natural elements can limit the well-being benefits of green spaces and can degrade the connection between humans and the natural environment over time (Francis, 2018).

5 Socioenvironmental research to advance sustainable lawn management

The abundance and diversity of ornamental lawns presents a challenge for widespread dissemination of BMPs. This parallels broader challenges faced by urban cooperative extension or outreach programs, as conventional frameworks for agricultural outreach are not always well aligned with the demands of cities and surrounding areas (Fox et al., 2017; Gaolach et al., 2017). Professional organizations that support the sod production, sports turf, and golf turf industries offer clear communication pathways to stakeholders. However, no comparable mechanism currently exists to connect the vast majority of ornamental lawn managers with researchers and educators. Within a single neighborhood or community, there may be hundreds or thousands of lawns, each managed uniquely based on individual preferences, budgets, and maintenance capabilities creating a kind of 'mosaic effect'.

In response, there is an expanding body of socioenvironmental research exploring the unique relationship between lawns and behavior (Eisenhauer et al., 2016; Robbins, 2012; Warner et al., 2019). Many of these studies hope to identify more effective methods to shift behavior by characterizing end-user priorities and preferences as it relates to the lawn. While some studies have found that environmental stewardship can be effective at shifting behavior (Eisenhauer et al., 2016), others find that social acceptance (Blaine et al., 2012; Carrico et al., 2013), aesthetic preferences (Carrico et al., 2013), and financial concerns (Carrico et al., 2013) may be more influential with some audiences.

Martini and Nelson (2015) found that many homeowners had limited knowledge of the environmental implications of lawn management or the way their lawn connects with a larger ecosystem. This environmental knowledge is important in influencing behavior and thus highlighting the value of dynamic outreach education strategies to enhance environmental literacy around appropriate lawn management practices.

6 Conclusion and future trends

Ornamental lawns remain a dominant fixture of urban green spaces, offering an array of ecosystem services when managed appropriately and attracting considerable criticism when managed inappropriately. In response to growing environmental problems, lawns will continue to be subjected to public scrutiny. As a result, government policy changes may seek to force dramatic landscape transformations with limited data on ecosystem service trade-offs or impact. It is increasingly important that turfgrass researchers not only understand the ecosystem service potential of turfgrasses but also the way turfgrasses compare and contrast with proposed alternatives. Lawn research must also continue to identify cost-effective, consumer-friendly technologies that simplify the adoption of BMPs for lawn managers, who may lack the foundational knowledge to otherwise change their behaviors while also developing innovative strategies to connect with this immense and diverse audience. Finally, research should continue to consider the impact of climate change on the role of turfgrass lawns in sustainable urban green spaces and identify opportunities for turfgrass to evolve and expand its environmental portfolio as a component of research in nature-based solutions to environmental challenges.

7 Where to look for further information

Several organizations are supporting activities to advance research and outreach related to more sustainable ornamental lawns and urban greenspaces. Researchers seeking to learn more about this topic may explore organizations such as The Lawn Institute, a 501(c)(3) charitable organization that seeks to fund research, promotion, and consumer education on the value of urban lawns. Other organizations that help advance messaging around sustainable lawn and landscape management include Turfgrass Producers International (https://www.turfgrasssod.org/), The Conservation Foundation (https://www.theconservationfoundation.org/conservation-home/earth-friendly-landscaping/), the Lady Bird Johnson wildflower.org Center and more. In addition, there are a number of statewide and regional turfgrass organizations intended to support turfgrass stakeholders working in the lawn and landscape industry. In Texas, for example, organizations such as the Texas Nursery and Landscape

Association (https://www.tnlaonline.org/), the Texas Turfgrass Association (https://texasturf.com/index.php/2021-winter-conference/), and the Turfgrass Producers Association (https://texasgrass.com/) all play instrumental roles in supporting lawn and landscape professionals through education, resource development, and the support of applied research. Federal entities such as the Environmental Protection Agency, the United States Department of Agriculture, and the National Science Foundation are increasingly supporting research and education that advances more sustainable urban greenspaces including ornamental lawns.

8 References

Amundsen, K., Li, L., Shearman, R. C. and Gaussoin, R. E. 2017. Addressing misperceptions regarding buffalograss tolerance to sandy soils, traffic, and shade. *Itsrj* 13(1), 358-363.

Aronson, M. F., Lepczyk, C. A., Evans, K. L., Goddard, M. A., Lerman, S. B., MacIvor, J. S., Nilon, C. H. and Vargo, T. 2017. Biodiversity in the city: key challenges for urban green space management. *Frontiers in Ecology and the Environment* 15(4), 189-196.

Azeem, M., Hale, L., Montgomery, J., Crowley, D. and McGiffen Jr., M. E. 2020. Biochar and compost effects on soil microbial communities and nitrogen induced respiration in turfgrass soils. *PLoS ONE* 15(11), e0242209.

Barnes, M. R. and Watkins, E. 2022. Differences in likelihood of use between artificial and natural turfgrass lawns. *Journal of Outdoor Recreation and Tourism* 37, 100480.

Bartlett, A. C., Andales, A. A., Arabi, M. and Bauder, T. A. 2015. A smartphone app to extend use of a cloud-based irrigation scheduling tool. *Computers and Electronics in Agriculture* 111, 127-130.

Blaine, T. W., Clayton, S., Robbins, P. and Grewal, P. S. 2012. Homeowner attitudes and practices towards residential landscape management in Ohio, USA. *Environmental Management* 50(2), 257-271.

Braun, R. C., Patton, A. J., Watkins, E., Koch, P. L., Anderson, N. P., Bonos, S. A. and Brilman, L. A. 2020. Fine fescues: a review of the species, their improvement, production, establishment, and management. *Crop Science* 60(3), 1142-1187.

Brosnan, J. T., Chandra, A., Gaussoin, R. E., Kowalewski, A., Leinauer, B., Rossi, F. S., Soldat, D. J., Stier, J. C. and Unruh, J. B. 2020. A justification for continued management of turfgrass during economic contraction. *Agricultural and Environmental Letters* 5(1), e20033.

Buhlman, J. L., Freshour, S. L., Harriman, R. W., Koch, M. J., Unruh, J. B., van Santen, E., Williams, K. E., Yurisic, C. A. and Kenworthy, K. E. 2022. Mowing frequency and performance of genetically modified St. Augustinegrass and standard commercial cultivars. *International Turfgrass Society Research Journal* 14(1), 652-662.

Canales-Ide, F., Zubelzu, S. and Rodríguez-Sinobas, L. 2019. Irrigation systems in smart cities coping with water scarcity: the case of Valdebebas, Madrid (Spain). *Journal of Environmental Management* 247, 187-195.

Cannavo, P., Vidal-Beaudet, L. and Grosbellet, C. 2014. Prediction of long-term sustainability of constructed urban soil: impact of high amounts of organic matter on soil physical properties and water transfer. *Soil Use and Management* 30(2), n/a-n/a.

Carey, R. O., Hochmuth, G. J., Martinez, C. J., Boyer, T. H., Nair, V. D., Dukes, M. D., Toor, G. S., Shober, A. L., Cisar, J. L., Trenholm, L. E. and Sartain, J. B. 2012. A review of turfgrass fertilizer management practices: implications for urban water quality. *hortTechnology* 22(3), 280–291.

Carrico, A. R., Fraser, J. and Bazuin, J. T. 2013. Green with envy: psychological and social predictors of lawn fertilizer application. *Environment and Behavior* 45(4), 427–454.

Carvalho, H. D. R., Chang, B., McInnes, K. J., Heilman, J. L., Wherley, B. and Aitkenhead-Peterson, J. A. 2021. Energy balance and temperature regime of different materials used in urban landscaping. *Urban Climate* 37, 100854.

Chang, B., Wherley, B., Aitkenhead-Peterson, J. A. and McInnes, K. J. 2021. Effects of urban residential landscape composition on surface runoff generation. *Science of the Total Environment* 783, 146977.

Chang, B., Wherley, B., Aitkenhead-Peterson, J., McInnes, K. and Dwyer, P. 2022. Short-term impacts of urban landscape conversion on surface runoff quality. *Urban Ecosystems* 25(5), 1561–1576.

Chang, Z., Jin, X. and Li, D. 2014. Phosphorus responses vary among cool-season turfgrasses during establishment from seed. *Agronomy Journal* 106(6), 1975–1980.

Cheng, H., Xu, W., Liu, J., Zhao, Q., He, Y. and Chen, G. 2007. Application of composted sewage sludge (CSS) as a soil amendment for turfgrass growth. *Ecological Engineering* 29(1), 96–104.

Cheng, H., Hu, Y. and Reinhard, M. 2014. Environmental and health impacts of artificial turf: a review. *Environmental Science and Technology* 48(4), 2114–2129.

Chow, W. T. L. and Brazel, A. J. 2012. Assessing xeriscaping as a sustainable heat island mitigation approach for a desert city. *Building and Environment* 47, 170–181.

Clayden, A., Green, T., Hockey, J. and Powell, M. 2018. Cutting the lawn– natural burial and its contribution to the delivery of ecosystem services in urban cemeteries. *Urban Forestry and Urban Greening* 33, 99–106.

Conrad, S. A., Pipher, J. and Haider, W. 2019. How current lawn attributes affect choices concerning water conserving lawn options: an individualized choice experiment in Kelowna, British Columbia. *Landscape and Urban Planning* 183, 147–156.

Daniyan, I., Balogun, V., Adeodu, A., Oladapo, B., Peter, J. K. and Mpofu, K. 2020. Development and performance evaluation of a robot for lawn mowing. *Procedia Manufacturing* 49, 42–48.

Eisenhauer, B. W., Brehm, J. M., Stevenson, N. and Peterson, J. 2016. Changing homeowners' lawn care behavior to reduce nutrient runoff. *Society and Natural Resources* 29(3), 329–344.

Erickson, J. E., Cisar, J. L., Volin, J. C. and Snyder, G. H. 2001. Comparing nitrogen runoff and leaching between newly established St. Augustinegrass turf and an alternative residential landscape. *Crop Science* 41(6), 1889–1895.

Evanylo, G. K., Porta, S. N., Li, J., Shan, D., Goatley, J. M. and Maguire, R. 2016. Compost practices for improving soil properties and turfgrass establishment and quality on a disturbed urban soil. *Compost Science and Utilization* 24(2), 136–145.

Fidanza, M. 2022. Turfgrass is not the largest irrigated crop. *Golf Course Management* 90(6), 62.

Fidanza, M., Baldwin, C., Koch, M., Freshour, L. and Harriman, B. 2022. Required mowing events of transgenic Kentucky bluegrass cultivars compared with conventional turfgrass cultivar blends and species mixtures. Crop, Forage & Turfgrass Management 8(1):e20146.

Filazzola, A., Shrestha, N. and MacIvor, J. S. 2019. The contribution of constructed green infrastructure to urban biodiversity: a synthesis and meta-analysis. *Journal of Applied Ecology* 56(9), 2131-2143.

Fissore, C., Baker, L. A., Hobbie, S. E., King, J. Y., McFadden, J. P., Nelson, K. C. and Jakobsdottir, I. 2011. Carbon, nitrogen, and phosphorus fluxes in household ecosystems in the Minneapolis-Saint Paul, Minnesota, urban region. *Ecological Applications* 21(3), 619-639.

Fox, J. M., Ruemenapp, M. A., Proden, P. and Gaolach, B. 2017. A national framework for urban Extension. *The Journal of Extension* 55(5), 21.

Francis, R. A. 2018. Artificial lawns: environmental and societal considerations of an ecological simulacrum. *Urban Forestry and Urban Greening* 30, 152-156.

Friell, J. and Watkins, E. 2021. Review of cool-season turfgrasses for salt-affected roadsides in cold climates. *Crop Science* 61(5), 2893-2915.

Gaolach, B., Kern, M. and Sanders, C. 2017. Urban Extension: aligning with the needs of urban audiences through subject-matter centers. *Journal of Human Sciences and Extension* 5(2), 10.

Ghimire, M., Boyer, T. A., Chung, C. and Moss, J. Q. 2016. Consumers' shares of preferences for turfgrass attributes using a discrete choice experiment and the best-worst method. *Hortscience* 51(7), 892-898.

Ghimire, M., Boyer, T. A. and Chung, C. 2019. Heterogeneity in urban consumer preferences for turfgrass attributes. *Urban Forestry and Urban Greening* 38, 183-192.

Gouveia, B. T., Rios, E. F., Nunes, J. A. R., Gezan, S. A., Munoz, P. R., Kenworthy, K. E., Unruh, J. B., Miller, G. L., Milla-Lewis, S. R., Schwartz, B. M., Raymer, P. L., Chandra, A., Wherley, B. G., Wu, Y., Martin, D. L. and Moss, J. Q. 2021a. Multispecies genotype× environment interaction for turfgrass quality in five turfgrass breeding programs in the southeastern United States. *Crop Science* 61(5), 3080-3096.

Gouveia, B. T., Raymer, P., Schwartz, B. M., Rios, E. F., Kenworthy, K. E., Unruh, J. B., Fontanier, C., Carbajal, E. M., Kruse, J., Restuccia, N. K. and Milla-Lewis, S. R. 2021b. Seashore paspalum (*Paspalum vaginatum* Sw.) performance under shade in multi-environment trials. *Grass Research* 1(1), 1-8.

Grant, S. B., Duong, K., Rippy, M. A., Pierce, G., Feldman, D., Zanetti, E. and McNulty, A. 2020. From yards to cities: a simple and generalizable probabilistic framework for upscaling outdoor water conservation behavior. *Environmental Research Letters* 15(5), 054010.

Grimshaw, A. L., Qu, Y., Meyer, W. A., Watkins, E. and Bonos, S. A. 2018. Heritability of simulated wear and traffic tolerance in three fine fescue species. *HortScience* 53(4), 416-420.

Grossi, N., Fontanelli, M., Garramone, E., Peruzzi, A., Raffaelli, M., Pirchio, M., Martelloni, L., Frasconi, C., Caturegli, L., Gaetani, M., Magni, S., McElroy, J. S. and Volterrani, M. 2016. Autonomous mower saves energy and improves quality of tall fescue lawn. *hortTechnology* 26(6), 825-830.

Grubbs, R. A., Henry, G. M. and Cabrera, M. L. 2021. Effects of mowing interval on turfgrass clipping tissue characteristics and soil nitrogen dynamics. *Soil Science Society of America Journal* 85(4), 1174-1184.

Gu, C., Crane, J., Hornberger, G. and Carrico, A. 2015. The effects of household management practices on the global warming potential of urban lawns. *Journal of Environmental Management* 151, 233-242.

Hamido, S. A., Guertal, E. A. and Wood, C. W. 2016. Carbon sequestration under warm season turfgrasses in home lawns. *Journal of Geoscience and Environment Protection* 4(9), 53-63.
Hejl, R. W., Wherley, B. G. and Fontanier, C. H. 2021. Long-term performance of warm-season turfgrass species under municipal irrigation frequency restrictions. *Hortscience* 56(10), 1221-1225.
Hobbie, S. E., Finlay, J. C., Janke, B. D., Nidzgorski, D. A., Millet, D. B. and Baker, L. A. 2017. Contrasting nitrogen and phosphorus budgets in urban watersheds and implications for managing urban water pollution. *Proceedings of the National Academy of Sciences of the United States of America* 114(16), 4177-4182.
Hugie, K., Yue, C. and Watkins, E. 2012. Consumer preferences for low-input turfgrasses: A conjoint analysis. *HortScience* 47(8), 1096-1101.
Ignatieva, M. and Ahrné, K. 2013. Biodiverse green infrastructure for the 21st century: from "green desert" of lawns to biophilic cities. *Journal of Architecture and Urbanism* 37(1), 1-9.
Ignatieva, M., Ahrné, K., Wissman, J., Eriksson, T., Tidåker, P., Hedblom, M., Kätterer, T., Marstorp, H., Berg, P., Eriksson, T. and Bengtsson, J. 2015. Lawn as a cultural and ecological phenomenon: a conceptual framework for transdisciplinary research. *Urban Forestry and Urban Greening* 14(2), 383-387.
Ignatieva, M. and Hedblom, M. 2018. An alternative urban green carpet. *Science* 362(6411), 148-149.
Jain, S., Khalore, A. and Patil, S. 2015. Self-efficient and sustainable solar powered robotic lawn mower. *Int. J. Trend Res. Dev.(IJTRD)* 2(6), 294-297.
Jim, C. Y. 2017. Intense summer heat fluxes in artificial turf harm people and environment. *Landscape and Urban Planning* 157, 561-576.
Katuwal, K. B., Jespersen, D., Bhattarai, U., Chandra, A., Kenworthy, K. E., Milla-Lewis, S. R., Schwartz, B. M., Wu, Y. and Raymer, P. 2022. Multilocational screening identifies new drought-tolerant, warm-season turfgrasses. *Crop Science* 62(4), 1614-1630.
Keilsohn, W., Narango, D. L. and Tallamy, D. W. 2018. Roadside habitat impacts insect traffic mortality. *Journal of Insect Conservation* 22(2), 183-188.
Khachatryan, H., Suh, D. H., Xu, W., Useche, P. and Dukes, M. D. 2019. Towards sustainable water management: preferences and willingness to pay for smart landscape irrigation technologies. *Land Use Policy* 85, 33-41.
Kranz, C. N., McLaughlin, R. A., Johnson, A., Miller, G. and Heitman, J. L. 2020. The effects of compost incorporation on soil physical properties in urban soils-a concise review. *Journal of Environmental Management* 261, 110209.
Law, Q. D., Bigelow, C. A. and Patton, A. J. 2016. Selecting turfgrasses and mowing practices that reduce mowing requirements. *Crop Science* 56(6), 3318-3327.
Law, Q. D., Trappe, J. M., Jiang, Y., Turco, R. F. and Patton, A. J. 2017. Turfgrass selection and grass clippings management influence soil carbon and nitrogen dynamics. *Agronomy Journal* 109(4), 1719-1725.
Leinauer, B., Serena, M. and Singh, D. 2010. Seed coating and seeding rate effects on turfgrass germination and establishment. *hortTechnology* 20(1), 179-185.
Lerman, S. B., Contosta, A. R., Milam, J. and Bang, C. 2018. To mow or to mow less: lawn mowing frequency affects bee abundance and diversity in suburban yards. *Biological Conservation* 221, 160-174.

Lerman, S. B. and Contosta, A. R. 2019. Lawn mowing frequency and its effects on biogenic and anthropogenic carbon dioxide emissions. *Landscape and Urban Planning* 182, 114–123.

Litvak, E. and Pataki, D. E. 2016. Evapotranspiration of urban lawns in a semi-arid environment: an in situ evaluation of microclimatic conditions and watering recommendations. *Journal of Arid Environments* 134, 87–96.

Liu, B. and Landschoot, P. J. 2018. Influence of phosphorus in starter fertilizer on the establishment of tall fescue. *HortScience* 53(12), 1897–1906.

Liu, Z. and Jim, C. Y. 2021. Playing on natural or artificial turf sports field? Assessing heat stress of children, young athletes, and adults in Hong Kong. *Sustainable Cities and Society* 75, 103271.

Ma, Q., Li, Z., Xiao, H., Hu, Z. and Pung, L. 2021. Mechanical properties of clay reinforced with Bermuda grass root under drying-wetting cycles. *Environmental Earth Sciences* 80(1), 1–10.

Madsen, M. D., Fidanza, M. A., Barney, N. S., Kostka, S. J., Badrakh, T. and McMillan, M. F. 2016. Low-dose application of nonionic alkyl terminated block copolymer surfactant enhances turfgrass seed germination and plant growth. *HortTechnology* 26(4):379–385.

Martini, N. F. and Nelson, K. C. 2015. The role of knowledge in residential lawn management. *Urban Ecosystems* 18(3), 1031–1047.

McCurdy, J. D., McElroy, J. S., Guertal, E. A. and Wood, C. W. 2014. White clover inclusion within a Bermudagrass lawn: effects of supplemental nitrogen on botanical composition and nitrogen cycling. *Crop Science* 54(4), 1796–1803.

Mechleb, G., Gilbert, R., Christman, M., Gupta, R. and Gross, B. 2014. Use of expanded shale amendment to enhance drainage properties of clays. In: *Geo-congress 2014: Geo-characterization and Modeling for Sustainability* (pp. 3444–3454).

Meeks, M. and Chandra, A. 2021. Performance of diploid and interploid hybrids of St. Augustinegrass under moderate shade. *Crop Science* 61(5), 3719–3733.

Miguel-Chinchilla, L., Heasley, E., Loiselle, S. and Thornhill, I. 2019. Local and landscape influences on turbidity in urban streams: a global approach using citizen scientists. *Freshwater Science* 38(2), 303–320.

Milesi, C., Running, S. W., Elvidge, C. D., Dietz, J. B., Tuttle, B. T. and Nemani, R. R. 2005. Mapping and modeling the biogeochemical cycling of turf grasses in the United States. *Environmental Management* 36(3), 426–438.

Minor, J., Campbell, B., Waltz, C. and Berning, J. 2020. Water savings and return on investment of a new drought resistant turfgrass. *Journal of Environmental Horticulture* 38(2), 56–62.

Monteiro, J. A. 2017. Ecosystem services from turfgrass landscapes. *Urban Forestry and Urban Greening* 26, 151–157.

Moseley, D. O., Trappe, J. M., Milla-Lewis, S. R., Chandra, A., Kenworthy, K. E., Liu, W. and Patton, A. J. 2021. Characterizing the growth and winter survival of commercially available and experimental genotypes of St. Augustinegrass. *Crop Science* 61(5), 3097–3109.

Murthy, A., Green, C., Stoleru, R., Bhunia, S., Swanson, C. and Chaspari, T. 2019. Machine learning-based irrigation control optimization. In: Proceedings of the 6th ACM International Conference on Systems for Energy-Efficient Buildings, Cities, and Transportation (pp. 213–222).

Oke, C., Bekessy, S. A., Frantzeskaki, N., Bush, J., Fitzsimons, J. A., Garrard, G. E., Grenfell, M., Harrison, L., Hartigan, M., Callow, D. and Cotter, B. 2021. Cities should respond to the biodiversity extinction crisis. *Npj Urban Sustainability* 1(1), 1-4.

Pandey, S., Bhatta, N. P., Paudel, P., Pariyar, R., Maskey, K. H., Khadka, J., Thapa, T. B., Rijal, B. and Panday, D. 2018. Improving fertilizer recommendations for Nepalese farmers with the help of soil-testing mobile van. *Journal of Crop Improvement* 32(1), 19-32.

Patton, A. J., Schwartz, B. M. and Kenworthy, K. E. 2017. Zoysiagrass (Zoysia spp.) history, utilization, and improvement in the United States: a review. *Crop Science* 57(S1), S-37.

Pease, B. W. and Stier, J. C. 2020. Hydroseeding with mesotrione in tankmix for establishment of perennial ryegrass. *Agronomy Journal* 112(5), 3429-3434.

Pedrini, S., Merritt, D. J., Stevens, J. and Dixon, K. 2017. Seed coating: science or marketing spin? *Trends in Plant Science* 22(2), 106-116.

Pinkney, J. L., Laplante, K. A. and Dale, A. G. 2021. Mixing warm-season turfgrass cultivars to reduce weed pressure and increase lawn quality. *International Turfgrass Society Research Journal* 14(1), 713-716.

Pirtle, T., Chavarria, M. R., Erickson, J. E., Kenworthy, K. E., Cox, K., Unruh, J. B., Kruse, J., Cardenas, B., Dukes, M., Milla-Lewis, S., Chandra, A., Wherley, B. and Moss, J. 2022. Effects of St. Augustinegrass genotype and irrigation frequency on turfgrass quality in a subtropical environment. *International Turfgrass Society Research Journal* 14(1), 683-693.

Pooya, E. S., Tehranifar, A., Shoor, M., Selahvarzi, Y. and Ansari, H. 2013. The use of native turf mixtures to approach sustainable lawn in urban landscapes. *Urban Forestry and Urban Greening* 12(4), 532-536.

Powlen, J. S., Bigelow, C. A., Patton, A. J., Jiang, Y. and Fraser, M. L. 2021. Minimal irrigation requirements of Kentucky bluegrass and tall fescue blends in the northern transition zone. *Crop Science* 61(5), 2939-2948.

Quinton, J. M., Duinker, P. N., Gallant, K. A., Steenberg, J. W. N. and Charles, J. D. 2019. To tree or not to tree: user and management perspectives of cemetery trees. *Urban Forestry and Urban Greening* 43, 126385.

Ramer, H., Nelson, K. C., Spivak, M., Watkins, E., Wolfin, J. and Pulscher, M. 2019. Exploring park visitor perceptions of 'flowering bee lawns' in neighborhood parks in Minneapolis, MN, US. *Landscape and Urban Planning* 189, 117-128.

Richardson, M., McCalla, J. and Hignight, K. 2010. Seedling emergence of tall fescue and Kentucky bluegrass as affected by two seed coating techniques. *Hort Technology* 20, 415-417.

Robbins, P. 2012. *Lawn People: How Grasses, Weeds, and Chemicals Make Us Who We Are*. Philadelphia, PA: Temple University Press.

Roberts, E. C. and Roberts, B. C. 1989. *Lawn and Sports Turf Benefits*. Lombard, IL: Lawn Institute.

Russo, C., Cappelletti, G. M. and Nicoletti, G. M. 2022. The product environmental footprint approach to compare the environmental performances of artificial and natural turf. *Environmental Impact Assessment Review* 95, 106800.

Saidani, M., Pan, Z., Kim, H., Wattonville, J., Greenlee, A., Shannon, T., Yannou, B., Leroy, Y. and Cluzel, F. 2021. Comparative life cycle assessment and costing of an autonomous lawn mowing system with human-operated alternatives: implication for sustainable design improvements. *International Journal of Sustainable Engineering* 14(4), 704-724.

Saidani, M. and Kim, H. 2021. Quantification of the environmental and economic benefits of the electrification of lawn mowers on the US residential market. *The International Journal of Life Cycle Assessment* 26(6), 1267-1284.

Salvaggio, M., Futrell, R., Batson, C. D. and Brents, B. G. 2014. Water scarcity in the desert metropolis: how environmental values, knowledge and concern affect Las Vegas residents' support for water conservation policy. *Journal of Environmental Planning and Management* 57(4), 588-611.

Schneider, D., Hypes, J. A. and Hypes, M. G. 2014. Synthetic Turf vs. natural Grass. *Journal of Facility Planning, Design, and Management* 2(2), 81-84.

Schwartz, B. M., Hanna, W. W., Baxter, L. L., Raymer, P. L., Waltz, F. C., Kowalewski, A. R., Chandra, A., Genovesi, A. D., Wherley, B. G., Miller, G. L., Milla-Lewis, S. R., Reynolds, W. C., Wu, Y., Martin, D. L., Moss, J. Q., Kenna, M. P., Unruh, J. B., Kenworthy, K. E., Zhang, J. and Munoz, P. R. 2018. 'DT-1', a drought-tolerant triploid turf Bermudagrass. *HortScience* 53(11), 1711-1714.

Serena, M., Velasco-Cruz, C., Friell, J., Schiavon, M., Sevostianova, E., Beck, L., Sallenave, R. and Leinauer, B. 2020. Irrigation scheduling technologies reduce water use and maintain turfgrass quality. *Agronomy Journal* 112(5), 3456-3469.

Simpson, T. J. and Francis, R. A. 2021. Artificial lawns exhibit increased runoff and decreased water retention compared to living lawns following controlled rainfall experiments. *Urban Forestry and Urban Greening* 63, 127232.

Sloan, J. J., Mackay, W. A., Colbaugh, P., George, S. W. and Feagley, S. 2002. The suitability of expanded shale as an amendment for clay soils. *hortTechnology* 12(4), 646-651.

Sloan, J. J., Ampim, P. A. Y., Basta, N. T. and Scott, R. 2012. Addressing the need for soil blends and amendments for the highly modified urban landscape. *Soil Science Society of America Journal* 76(4), 1133-1141.

Smetana, S. M. and Crittenden, J. C. 2014. Sustainable plants in urban parks: a life cycle analysis of traditional and alternative lawns in Georgia, USA. *Landscape and Urban Planning* 122, 140-151.

Smith, L. S. and Fellowes, M. D. E. 2014. The grass-free lawn: management and species choice for optimum ground cover and plant diversity. *Urban Forestry and Urban Greening* 13(3), 433-442.

Smith, L. S., Broyles, M. E. J., Larzleer, H. K. and Fellowes, M. D. E. 2015. Adding ecological value to the urban lawnscape. Insect abundance and diversity in grass-free lawns. *Biodiversity and Conservation* 24(1), 47-62.

Stacey, N. E., Lewis, R. W., Davenport, J. R. and Sullivan, T. S. 2019. Composted biosolids for golf course turfgrass management: impacts on the soil microbiome and nutrient cycling. *Applied Soil Ecology* 144, 31-41.

Toor, G. S., Occhipinti, M. L., Yang, Y. Y., Majcherek, T., Haver, D. and Oki, L. 2017. Managing urban runoff in residential neighborhoods: nitrogen and phosphorus in lawn irrigation driven runoff. *PLoS ONE* 12(6), e0179151.

Turgeon, A. J. and Kaminski, J. E. 2019. Turfgrass management. *Turfpath LLC*. State College, PA.

United Nations 2018. The World Urbanization Prospects (2018 revision). The United Nations, New York. Available at: https://population.un.org/wup/Publications/Files/WUP2018-Report.pdf

Vinlove, F. K. and Torla, R. F. 1994. Comparative estimations of US home lawn area. *Journal of Turfgrass Management* 1(1), 83-97.

Wanjiru, E. M. and Xia, X. 2015. Energy-water optimization model incorporating rooftop water harvesting for lawn irrigation. *Applied Energy* 160, 521-531.
Wang, Z. H., Zhao, X., Yang, J. and Song, J. 2016. Cooling and energy saving potentials of shade trees and urban lawns in a desert city. *Applied Energy* 161, 437-444.
Warner, L. A., Silvert, C. J. and Benge, M. 2019. Using adoption and perceived characteristics of fertilizer innovations to identify extension educational needs of Florida's residential audiences. *Journal of Agricultural Education* 60(3), 155-172.
Wastian, L., Unterweger, P. A. and Betz, O. 2016. Influence of the reduction of urban lawn mowing on wild bee diversity (Hymenoptera, Apoidea). *Journal of Hymenoptera Research* 49, 51-63.
Watkins, E., Gardner, D. S., Stier, J. C., Soldat, D. J., St. John, R. A., Christians, N. E., Hathaway, A. D., Diesburg, K. L., Poppe, S. R. and Gaussoin, R. E. 2014. Cultivar performance of low-input turfgrass species for the North Central United States. *Applied Turfgrass Science* 11(1), 1-7.
Wesolowska, M., Rymarczyk, J., Góra, R., Baranowski, P., Slawinski, C., Klimczyk, M., Supryn, G. and Schimmelpfennig, L. 2021. New slow-release fertilizers-economic, legal and practical aspects: a review. *International Agrophysics* 35(1), 11-24.
Wherley, B., Chandra, A., Genovesi, A., Kearns, M., Pepper, T. and Thomas, J. 2013. Developmental response of St. Augustinegrass cultivars and experimental lines in moderate and heavy shade. *Hortscience* 48(8), 1047-1051.
Whitman, B., Iannone, B. V., Kruse, J. K., Unruh, J. B. and Dale, A. G. 2022. Cultivar blends: a strategy for creating more resilient warm season turfgrass lawns. *Urban Ecosystems* 25(3), 797-810.
Wolfin, J., Watkins, E., Lane, I., Portman, Z. and Spivak, M. 2021. *Floral Enhancement of Turfgrass Lawns Benefits Wild Bees and Honey Bees (Apis mellifera).*
Wu, L., Green, R., Klein, G., Hartin, J. S. and Burger, D. W. 2010. Nitrogen source and rate influence on tall fescue quality and nitrate leaching in a southern California lawn. *Agronomy Journal* 102(1), 31-38.
Yang, J. and Wang, Z. H. 2017. Planning for a sustainable desert city: the potential water buffering capacity of urban green infrastructure. *Landscape and Urban Planning* 167, 339-347.
Yue, C., Hugie, K. and Watkins, E. 2012. Are consumers willing to pay more for low-input turfgrasses on residential lawns? Evidence from choice experiments. *Journal of Agricultural and Applied Economics* 44(4), 549-560.
Yue, C., Wang, J., Watkins, E., Bonos, S. A., Nelson, K. C., Murphy, J. A., Meyer, W. A. and Horgan, B. P. 2017a. Heterogeneous consumer preferences for turfgrass attributes in the United States and Canada. *Canadian Journal of Agricultural Economics/Revue Canadienne d'Agroeconomie* 65(3), 347-383.
Yue, C., Wang, J., Watkins, E., Bonos, S., Nelson, K., Murphy, J. A., Meyer, W. and Horgan, B. 2017b. Consumer preferences for information sources of turfgrass products and lawn care. *Agronomy Journal* 109(4), 1726-1733.
Yue, C., Wang, J., Watkins, E., Bonos, S. A., Nelson, K. C., Murphy, J. A., Meyer, W. A. and Horgan, B. P. 2017c. An investigation of trait prioritization in turfgrass breeding programs. *Hortscience* 52(11), 1544-1549.
Yue, C., Cui, M., Watkins, E. and Patton, A. 2021. Investigating factors influencing consumer adoption of low-input turfgrasses. *Hortscience* 56(10), 1213-1220.

Zhang, Y., Qian, Y., Bremer, D. J. and Kaye, J. P. 2013. Simulation of nitrous oxide emissions and estimation of global warming potential in turfgrass systems using the DAYCENT model. *Journal of Environmental Quality* 42(4), 1100–1108.

Zhang, X. and Khachatryan, H. 2021. Interactive effects of homeowners' environmental concerns and rebate incentives on preferences for low-input residential landscapes. *Urban Forestry and Urban Greening* 65, 127322.

Part 2

Biotic and abiotic stresses

Chapter 9

Advances in turfgrass disease management

James Kerns, North Carolina State University, USA

1 Introduction

Diseases of amenity turfgrasses have troubled turfgrass managers and practitioners once these plants were cultivated for sport and recreation. However, disease incidence and severity continue to increase as turfgrass managers face a changing climate that will present difficult environmental conditions. While this may sound foreboding, many advances in turfgrass disease management have occurred and continue to occur. The current technological improvements in molecular biology that aid in the detection of new and existing pathogens, identification of fungal strains and biotypes that are resistant to fungicides, and how soil and root zone microbial communities respond to turfgrass management practices have revolutionized turfgrass pathology. This trend will only continue as long as the numerous local, state, regional, and national organizations and government agencies continue to fund and support research. Yet, the most critical aspect that cannot be ignored is ensuring that academia will continue to fill turfgrass pathology faculty positions in the future.

http://dx.doi.org/10.19103/AS.2022.0110.09

A chapter on the advances of sustainable turfgrass disease management could not occur without the commitment from agencies, organization, and industry that provide funding support, and universities alike. This particular chapter will cover recent advances in turfgrass disease management that promote sustainable suppression of the pathogens that cause diseases. Although an entire book could be written on this particular topic (Couch, 1995; Smith et al., 1989; Vargas, 2005), this chapter will focus on only a few critically important turfgrass diseases.

Plant disease management is a complicated venture regardless of the crop (Agrios, 2005). The goal of plant disease management is to reduce any aesthetic damage and negative economic impact caused by plant pathogens. The term 'control' has traditionally been used to describe this goal, yet under current climatic conditions, environmental concerns, and economic issues, this term is too absolute and rigid. Therefore, most plant pathologists prefer to use the term 'management' when discussing options to reduce the growth or infective or disease-causing capabilities of plant pathogens. There are many strategies for managing plant diseases; however, they can succinctly be separated into preventative and curative strategies. Preventative refers to all efforts taken prior to the onset of disease symptoms, and curative refers to those actions taken after disease symptoms develop.

The specific approaches to managing plant diseases also is classified based on certain principles. H. H. Whetzel initially proposed principles of disease management in 1929 that included four unique strategies of exclusion, eradication, protection, and immunization (i.e. resistance) (Newhall, 1980). In 1968, however, the National Academy of Sciences (Washington, DC, USA) slightly amended Whetzel's proposal to six specific strategies or principles as follows:

1. Avoidance - presentation of disease by selection of a site devoid of inoculum or manipulation of the environment to ensure unfavourability to disease development;
2. Exclusion - prevention of inoculum introduction;
3. Eradication - elimination, destruction, or inactivation of inoculum;
4. Protection - prevention of infection through use of a pesticide or some other barrier;
5. Resistance - utilization of cultivars that prevent, delay, or reduce infection by a pathogen; and
6. Therapy - curative treatment of infected plants.

These six principles have guided turfgrass pathologists recently to examine management strategies that do not solely focus on the use of pesticides to suppress the pathogen. Moreover, numerous research programs have integrated multiple principles and have developed strong best management practices to manage important turfgrass diseases (Smiley et al., 2005).

Therefore, the overall goal of this chapter is to discuss research-based examples that sustainable disease management is possible with turfgrass diseases. A significant portion of turfgrass disease management is accurate and timely applications of fungicides along with cultivating a microenvironment that is well suited for turfgrass growth and health.

2 Anthracnose

One of the best examples of integrated turfgrass disease management is from the Northeast USA and in particular research on anthracnose conducted by faculty and graduate students at Rutgers, the State University of New Jersey (New Brunswick, NJ, USA). This work was pioneered by Drs. Bruce Clarke and James Murphy and sought to understand the biology of this destructive disease so areas within the disease cycle could be exploited to improve management (Murphy et al., 2008).

Anthracnose is a destructive disease of primarily cool-season grasses caused by the fungus *Colletotrichum cereale*, although related fungal species have been documented as pathogens of various warm-season grasses (Crouch et al., 2006; Smiley et al., 2005). Up until the mid-1990s, anthracnose was considered by some turfgrass pathologists as senectopathic (i.e. the pathogen could only incite disease in senescing tissue). While there was debate about the classification of this pathogen prior to the mid-1990s, currently there is little doubt that anthracnose is an extremely important pathogen of cool-season grasses especially annual bluegrass (*Poa annua* L.) and creeping bentgrass (*Agrostis stolonifera* L.) (Danneberger et al., 1984; Dernoeden, 2002; Landschoot and Hoyland, 1995; Mann and Newell, 2005; Murphy et al., 2008, 2018). The disease is most destructive on these two turfgrass species when they are cultivated for putting greens with increased ball roll and putting quality as the stress on the turf to produce these conditions can be severe. Additionally, increasing temperatures and erratic rainfall events predispose turfgrass to infection by the anthracnose-causing pathogen.

Despite the increase in incidence (number of plants that are infected within a defined population) and severity (percentage of the plant affected by the disease) of anthracnose on cool-season grass putting greens, progress in developing sound turfgrass management practices for anthracnose has occurred due to efforts of scientists within the NE-1025 and NE-1046 multistate projects (Murphy et al., 2008, 2018). Researchers within these USDA-funded regional projects have consistently demonstrated that anthracnose is most devastating on weakened turfgrass. Turfgrass cultural practices that limit fertility, focus on ultralow mowing heights, and do not focus on precision irrigation can predispose turfgrass plants to anthracnose (Hempfling et al., 2017; Murphy et al., 2018). Consequently, research focused on optimizing turfgrass fertility, sand topdressing, irrigation, and mowing height to limit or suppress anthracnose severity (Hempfling et al., 2020; Inguagiato et al., 2012; Roberts and Murphy, 2014).

The efforts by these researchers within these two multistate groups mentioned above accurately detail that anthracnose is a stress-induced disease. However, they have also definitively shown that wounding does not increase anthracnose severity. Therefore, a well-designed cultural management program that integrates sand topdressing and cultivation (i.e. verticutting, solid- and hollow-tine aerification) can be performed without concerns of increasing the incidence and/or severity of anthracnose on annual bluegrass and creeping bentgrass golf course putting greens.

2.1 Influence of plant nutrition on anthracnose severity

Anthracnose severity is compellingly linked with nitrogen (N) and potassium (K) deficiencies in annual bluegrass and soil pH can also affect the severity of anthracnose. Many studies demonstrate that turfgrass managers should be mindful of N fertility when challenged with anthracnose. Research investigating the influence of low N rates applied weekly (i.e. 'spoon-feeding') successfully reduced anthracnose severity in the absence of fungicides on annual bluegrass putting greens. Specifically, rates of N as low as 4.9 kg N•ha^{-1} per week in conjunction with sand topdressing and a higher mowing height sufficiently reduced anthracnose severity enough to not warrant fungicide applications (Murphy et al., 2018). While this only occurred in some years it clearly demonstrates the importance of integrating plant nutrition with other cultural practices to suppress anthracnose development. Indeed, using a specific N application rate, as previously mentioned, may not be practical in every environment where anthracnose is observed on putting greens, yet researchers confirmed that maintaining 3.35% N in turfgrass leaf tissue through the use of light-rate, frequent applications of soluble N optimized anthracnose suppression (Hempfling et al., 2015; Inguagiato et al., 2008; Schmid, 2016).

Managing ball roll could be a concern with additional N fertility especially given the current demands of golfers to consistently have ball roll distances exceed 3 m (i.e. 10 feet) (Latin, 2022). Research was able to illustrate, however, that plant growth regulators (PGRs; i.e. trinexapac-ethyl, mefluidide, and ethephon) commonly applied to reduce annual bluegrass seedhead development and plant growth did not increase anthracnose severity (Inguagiato et al., 2010). When trinexapac-ethyl was applied every 14 days occasionally minor reductions in anthracnose severity was observed, yet when applications were delivered based on 200 growing degree day accumulation then a more substantial reduction in anthracnose severity was observed

(Murphy et al., 2018). Furthermore, incorporating other techniques to ensure enhanced ball roll distances such as double cutting and lightweight rolling did not encourage anthracnose development (Murphy et al., 2018). Thus, these ball roll-facilitating practices could be used in conjunction with spoon-feeding N applications and well-timed PGR applications to promote acceptable putting green speeds that meet the current expectations of golfers without encouraging increased anthracnose severity (Inguagiato and Guillard, 2016; Murphy et al., 2018).

As the factors that favor anthracnose development continued to be dissected by this team of turfgrass researchers, they turned their attention to the best timing for granular N applications. Anthracnose suppression was most effective when granular N applications were applied in the spring compared to fall N applications. Therefore, turfgrass managers dealing with chronic or persistent anthracnose should favor spring N applications when applying granular sources of N (Chen, 2016; Schmid, 2016). Although N source did not affect anthracnose severity, these studies did indicate that K deficiencies could increase anthracnose severity (Hempfling et al., 2015).

Anthracnose can be observed on foliage (i.e. foliar anthracnose) and on turfgrass crowns (i.e. basil rot anthracnose) and is rarely if ever observed affecting turfgrass roots, stolons, or rhizomes (Smiley et al., 2005). Given that annual bluegrass is not a deeply rooted plant, much of the pathogen's survival and initial activity is likely within the top 5 cm of the root zone. Therefore, understanding the plant and soil health status within this critical root zone layer is important for anthracnose management. When soil K is below 50 $mg \bullet kg^{-1}$ anthracnose is more severe on annual bluegrass putting greens (Hempfling et al., 2015; Murphy et al., 2018). Yet, a soil sample collected to a depth of 10 to 15 cm could indicate soil potassium is well above this critical value. Consequently, for turfgrass managers that struggle with anthracnose, soil samples should be no deeper than 5 cm from the surface to ensure a more accurate quantification of soil potassium for anthracnose management. If K is found to be low, researchers confirmed that only 49 kg $K \bullet ha^{-1}$ fertilization was required to remediate the deficiency in that 5 cm root zone layer (Hempfling et al., 2015; Murphy et al., 2018). It is important to note that the researchers in these studies used a Mehlich-3 extractant method to estimate K availability (Carrow et al., 2001; Schimd, 2016). Of note, different extractant methods used by various soil test laboratories may not directly relate to the studies presented here. Finally, K concentration in turfgrass leaf tissue could indicate the plants will respond to K fertility inputs. In field experiments, turfgrass with tissues concentrations <1.9% K suffered from severe anthracnose and would likely respond well to applications of K (Hempfling et al., 2015; Murphy et al., 2018).

2.2 Influence of mowing, irrigation, and sand topdressing on anthracnose severity

To ensure excellent playability of putting green surfaces, mowing heights have been lowered since the mid-1990s and dry, firm putting surfaces has become desirable for the modern golfer (Murphy et al., 2018). Therefore, agronomic practices such as mowing height, rolling, irrigation, and sand topdressing are conducive to provide firm, fast putting surfaces (Murphy et al., 2018). Yet with pressure from diseases such as anthracnose, these practices could promote more stress that could lead to increase severity of diseases like anthracnose. It is not surprising that lower mowing height is associated with more severe anthracnose development since the disease is exacerbated by stress. However, practices that promote increased ball roll distance such as lightweight rolling and double cutting did not affect anthracnose severity (Inguagiato et al., 2009a,b). Interestingly, increased traffic on the perimeters of putting greens due to rolling or clean-up mowing also did not promote more anthracnose development (Inguagiato et al., 2009a,b; Roberts et al., 2012). Further validating that anthracnose is not primarily a turfgrass tissue wound-related disease. Therefore, an effective approach for minimizing anthracnose on putting greens while maximizing surface playability could involve a combination lightweight rolling and double cutting.

Anthracnose development is stimulated under overly dry and wet soil conditions, yet in order to achieve the desired firm, fast conditions of putting greens, some drying of the turf and root zone is necessary (Roberts et al., 2011; Smiley et al., 2005; Vargas and Turgeon, 2003). The key is to prevent plant wilt stress and attempt to irrigate at least above 40% evapotranspiration to limit anthracnose severity (Roberts et al., 2011). Moreover, turfgrass managers should employ methods that quantify soil moisture by using a soil moisture meter rather than relying on qualitative measurement practices. Repeated stress from excessive or deficient irrigation will promote anthracnose and maybe challenging to overcome if the stress becomes too severe.

Maintaining a firm, fast putting green surface is expected from modern golfers and to achieve these conditions, sand topdressing throughout the growing season can assist with achieving those desired conditions (Murphy et al., 2018). Although sand topdressing is associated with developing a better putting surface for golfers, does it affect disease severity and in particular anthracnose severity? Initially, researchers working on anthracnose hypothesized that even light, frequent applications of sand topdressing throughout the growing season could increase anthracnose severity due to increase stress from wounding (Hempfling et al., 2015; Roberts et al., 2011; Wang et al., 2018). Yet, increased anthracnose severity was not associated with light, frequent sand topdressing events. Light, frequent topdressing throughout the summer months on annual bluegrass putting greens actually reduced anthracnose severity consistently in several studies (Hempfling et al., 2015; Roberts et al., 2011; Wang et al., 2018).

To promote sustainable management of anthracnose, these turfgrass management practices detailed above can be combined to reduce the dependency on fungicides for the disease. At times, the combination of these practices could reduce fungicide inputs, but at the very least, it will allow turfgrass managers to manage anthracnose effectively and hopefully prevent costly use of fungicides needed at higher curative application rates (Latin, 2022). Based on the research completed within the two aforementioned regional research groups, anthracnose of turfgrass still remains problematic, yet the incidence and severity of the disease has steadily declined. Integrating all of the turfgrass management practices detailed above with a strong fungicide program will likely be most important in the future as the climate continues to change which will make it more challenging to manage cool-season turfgrass putting greens. Nonetheless, the comprehensive research on the biology and management of anthracnose by turfgrass scientists in the Northeast/Mid-Atlantic USA has proven extremely effective in assisting turfgrass managers with anthracnose management (Murphy et al., 2018). The recent research on anthracnose is an excellent example of methodically assembling research-based information to develop a sustainable disease management program in a turfgrass ecosystem.

3 Dollar spot

Dollar spot is considered the most common and widely distributed disease of all turfgrasses worldwide (Latin, 2022). Unfortunately for turfgrass managers, dollar spot has been observed in all areas where turfgrasses are grown including golf courses, home lawns, and athletic fields. Although the disease is documented in many countries worldwide, it has been extensively researched and characterized in North America, Europe, and Australasia. Visual symptoms of dollar spot typically are exemplified by straw-colored to white lesions that span the turfgrass leaf blade. If conditions remain conducive for disease development circular, sunken spots of varying diameters occur in the turfgrass stand. These infected circular, sunken spots can range in color from dark brown to white and can vary depending on the infected turfgrass species. For example, in cool-season grasses, dollar spot symptoms are readily identifiable in the field as the sward symptoms are unique and quite diagnostic. In warm-season grasses, however, symptoms may start as dark brown to brown areas that are much larger (i.e. up to 10 cm diameter) than the traditional silver dollar-sized circular spots (i.e. the diameter of USD silver dollar is about the same diameter of a golf ball) that commonly develop on cool-season grasses (Smiley et al., 2005).

A correct diagnosis of the disease is the first step in sustainable turfgrass disease management. Without a proper diagnosis, which unfortunately is common due to underutilization of university diagnostic services and/or overreliance on social media for diagnostic advice, plant diseases cannot be rapidly and effectively managed. Dollar spot is considered an easy disease to diagnose without the aid of a microscope, yet especially in warm-season turfgrasses, the sward symptoms of dollar spot could resemble other leaf spots (i.e. *Bipolaris*, *Curvularia*, and *Drechslera* spp.) or Pythium blight (*Pythium* spp.). Hopefully, the importance of dollar spot on warm-season turfgrasses such as zoysiagrasses (*Zoysia* spp.) and seashore paspalum (*Paspalum vaginatum* Sw.) is better appreciated since today's turfgrass managers look for plant symptoms characteristic of dollar spot or use a diagnostic service if they are unsure (Smiley et al., 2005).

In reference to disease diagnosis, the taxonomy of the dollar spot-causing fungus remained in flux until recent research explored this topic. The proper identification of the pathogen has been debated as many researchers showed it was not a member of the *Sclerotinia* genus. Research in 2018, however, placed the dollar spot fungus in the genus *Clarireedia* that belongs within the *Rutsroemiaceae* family (Salgado-Salazar et al., 2018). This work also identified the genus *Clarireedia* has four primary species: *C. homoeocarpa, C. bennettii, C. jacksonii*, and *C. monteithiana*. The first two species, *C. homoeocarpa* and *C. bennettii*, are typified by isolates collected in the UK on fine-leaf fescues (*Festuca* spp.) primarily. *Clarireedia jacksonii* is predominately collected from cool-season turfgrasses, and *C. monteithiana* is mostly collected from warm-season turfgrasses (Salgado-Salazar et al., 2018). Understandably, it may be challenging to understand why taxonomy could be related to disease management, which is a valid question. Dollar spot in cool-season turfgrasses is well documented, well researched, and well understood when compared to dollar spot in warm-season turfgrasses (Sapkota et al., 2022). Therefore, by simply acknowledging through molecular taxonomic methods that the species that primarily affect cool-season and warm-season turfgrasses are different is a step forward with sustainable disease management.

Similarly, to the regional groups developed for anthracnose management a recent regional cooperative research project was created to advance our knowledge of this critical disease. The overarching goal of the project is to foster sustainable management of dollar spot through cooperative research much like those efforts previously summarized for anthracnose. Much work has already been conducted to improve dollar spot management, in turfgrasses; however, most if not all of the research has been conducted on cool-season turfgrasses. This may seem obvious given the importance of the disease on cool-season grasses. Yet within the transition zone in the USA, dollar spot appears to be increasing in incidence and severity on warm-season grasses.

Although not formally documented, in my experience, dollar spot is actively managed in many warm-season home lawns and remains a major issue on golf courses and athletic fields that are planted with seashore paspalum. Thus, more research on *C. monteithiana* on warm-season turfgrasses is warranted to ensure the turfgrass community promotes sustainable dollar spot management on all turfgrasses.

Dollar spot management has traditionally focused on fungicide efficacy and perfecting the timing of fungicide applications, yet two other disease management strategies are quite important and continue to improve dollar spot management: cultural management of disease and deployment of host resistance. These strategies are effective when integrated into a fungicide program and may even reduce fungicide inputs if implemented consistently.

3.1 Cultural management of dollar spot

In order for any plant disease to occur, three items must be present at a given point-in-time for that disease to develop: a virulent pathogen, a susceptible host, and a conducive environment. This 'plant disease triangle' is converted to a 'plant disease pyramid' when time is added as a fourth item (Franci, 2001). Manipulation of the environment through the removal of dew/leaf wetness, or additions of N fertility have proven effective at reducing dollar spot severity. Numerous studies describe reductions in dollar spot severity when dew is removed using dew whips, mowing, or lightweight rollers, and simply mowing early in the morning reduced dollar spot severity by 81% or more (Ellram et al., 2007; Giordano et al., 2012; Sapkota et al., 2022). Thus, frequent mowing in the morning can be viewed as a somewhat effective cultural management strategy for dollar spot. This is interesting to consider as dollar spot typically is not problematic for golf course putting greens because they are mowed almost every morning and are frequently or routinely treated with fungicides. Larger land areas such as fairways may not be feasible to mow every day; hence, alternative methods could be employed to temporarily reduce leaf wetness. These alternate methods include dragging fairways to remove dew, application of dew suppressing products, or rolling of fairways. Implementing these methods on larger turf areas can improve the performance of fungicide programs and could lead to economic savings of fungicide inputs.

Dollar spot has long been considered a low N disease, meaning the disease is most severe under low N inputs. A few studies, however, indicate that the N rate required to reduce dollar spot severity is likely not practical. For example, N applied as low as 73.2 kg $N\bullet ha^{-1}$ when compared to 0 kg $N\bullet ha^{-1}$ decreased dollar spot severity by 57% or more (Williams et al., 1996). In a more recent study, Townsend et al. (2021) did not observe a rate response to varying rates of N and dollar spot only was significantly reduced when N rates were very high (Townsend, et al., 2021). Although the N rates evaluated in Townsend et al. (2021) might be impractical for the turfgrass practitioner, perhaps coupling more moderate N rates with dew removal practices and a strategic fungicide program based on an accurate disease forecasting model will most likely reduce overall fungicide inputs. Future research that integrates cultural management practices could reveal more effective and sustainable dollar spot management programs for amenity turfgrasses.

4 Microdochium patch and snow molds

Microdochium patch (*Monographella nivalis* [Schaffnit] E. Mull) is another extremely important turfgrass disease in cool, humid regions (Smiley et al., 2005). This disease is also referred to as pink snow mold, but does not require snow cover to develop (Mattox and Kowalewski, 2020). Other snow mold diseases such as gray and speckled snow mold (*Typhula* spp.) require 60 and 120 days of continual snow cover to develop (Koch and Kerns, 2012). Researchers in Canada, the Pacific Northwest USA, and the UK have conducted many studies on alternative management options for Microdochium patch. Similar to studies conducted on dollar spot suppression, lightweight rolling reduced Microdochium patch severity (Mattox and Kowalewski, 2020). The most significant reduction in Microdochium patch severity was observed when rolling was paired with a phosphite, sulfur, and/or mineral oil applications every other week. Based on results presented in the literature, combining these aforementioned methods produced disease suppression similar to conventional fungicides, thus in areas where fungicide use is limited there are viable alternatives to managing Microdochium patch (Mattox and Kowalewski, 2020). This is vital to know as populations of this pathogen can develop resistance to fungicides and many countries, and regional and local government municipalities, are restricting fungicide applications and use (Latin, 2022). Turfgrass managers, however, are still expected to produce high-quality playing surfaces regardless of regulatory constraints. So research such as the summary of the many studies presented above is crucial and warrants even further investigation as climate and political climates continue to change.

Turfgrass research at the University of Wisconsin (Madison, WI, USA) focuses on providing exceptional suppression of snow molds. The majority of research explores optimal fungicide timing for snow mold applications in areas that receive considerable, persistent snow cover. Although the research integrates the use of fungicides, it demonstrates that applications should be made as close to snowfall as possible (Gonzales et al., 2022). Following this timing provides outstanding suppression of snow molds and limits re-applications in late winter or early spring when fungicides are prone to run-off. Previous research demonstrated that fungicides will persist in winter environments even in the absence of snow cover as long as soil temperatures remain at or below freezing (Gonzales et al., 2022). Moreover, breaks in snow cover will also limit snow mold development, so this research promotes sustainable management of these important winter diseases. Another fascinating result of these research efforts is documenting that acropetal penetrant fungicides can enter creeping bentgrass plants when temperatures are ≤ 3.8°C. Having data like these demonstrates to turfgrass managers that fungicides do not have to be applied well before snow cover for preventative snow mould control. Surprisingly, plants are able to uptake fungicides at remarkably low temperatures (Gonzales et al., 2022).

5 Brown patch of cool-season turfgrasses and large patch of warm-season turfgrasses

Brown patch of cool-season grasses is yet another prominent disease that turfgrass managers' battle. A significant proportion of brown research centered on management prior to the release of the QoI fungicides in the late 1990s. Brown patch was managed with limited fertility during the summer months and fungicides that were likely reapplied every 14–21 days prior to the advent of the QoIs. However, the QoIs revolutionized brown patch management because they typically provide at least 21 days of suppression of brown patch (Latin, 2022). In landscapes in the transition zone, brown patch was the scourge of perennial ryegrass and tall fescue swards and still remains a major issue in lawns that are not treated with fungicides. An interesting research question developed in the Carolinas about the status of sensitivity to fungicides within populations of *Rhizoctonia solani*. Sampling was conducted throughout NC from multiple yards planted with tall fescue. Multiple pathogenic fungi were discovered at different sampling times, revealing that *R. solani* was only mainly isolated from diseased turf in June and July (Koehler and Shew, 2017). Other pathogens that are closely related to *R. solani* are collected in May and August. Fungicides are applied in lawns in the Southeast USA monthly so populations of *R. solani*, the traditional causal agent of brown patch, is only exposed to fungicides in mid-summer. This is critical as it indicates that the development of fungicide resistance would be minimal. As such, pathogens that induce brown patch are considered low risk for occurrence of fungicide resistance.

As mentioned above, the pressure to limit pesticide inputs is great; so current research has refocused on managing brown patch with alternative practices. Introduction of low application rates of N during the summer months did not increase the severity of brown patch in the Carolinas (USA) and preliminary research in the Midwest (USA) indicate that N during the summer months may limit disease severity (Butler et al., 2019). Other practices that have been examined are the use of soil surfactants and elite tall fescue cultivars that have superior resistance to brown patch can also dramatically limit brown patch severity. Future studies could integrate these management practices with fungicides only when needed to promote more ecological suppression of brown patch in lawns and landscapes.

Large patch is a problematic disease of warm-season grasses such as centipedegrass, St. Augustinegrass, zoysiagrass, and bermudagrass. Zoysiagrass is a popular alternative to cool-season home lawns in many regions and has also replaced bermudagrass in areas where winter survivability limits bermudagrass cultivation. Yet, large patch remains a challenging issue for numerous turfgrass managers. Timing of fungicides for preventative control is critical and varies depending on environmental and edaphic conditions. For example, in the central USA (Kansas, Arkansas, and Missouri) fall applications are necessary but fail unless fungicides are applied in early spring (Koehler and Miller, 2019). Fungicides applied in the spring need to be applied prior to significant green-up of zoysiagrasses in order to be effective. Convincing turfgrass managers that a fungicide is required when the turf appears dormant is a challenging discussion. Yet, the evidence is clear that spring applications in this region are critical. Contrast that with the Carolinas, where two fall applications are indispensable for suppressing large patch. Spring applications are occasionally needed in the Carolinas. Turfgrass managers should pay particular attention to research that is conducted in their specific region as it will improve disease management.

There are many fungicides that effectively suppress large patch, yet the disease can still be tricky to manage. Although fungicides are one of the only strategies for successful management of large patch, applications strategies for those fungicides can affect fungicide efficacy. For instance, an exceptional active ingredient for large patch, azoxystrobin, provided zero suppression of large patch when the fungicide was placed solely on the leaf blades (Benelli et al., 2018). The pathogen/s that induce large patch infect the leaf sheaths and crowns so if the fungicide deposit is not placed in that region, then the pathogen could invade the turfgrass plant unchecked. Indeed, researchers demonstrated that placement of fungicides were paramount for managing large patch. Based on research conducted in this area, university extension personnel can advise turfgrass managers that fungicides should be applied in at least 823 L water-carrier•ha^{-1} to ensure effective suppression of large patch with labeled, efficacious fungicides. Once again having a basic understanding of fungicide application strategies prevents unnecessary curative applications of fungicides as the disease can be successfully managed preventatively if appropriate steps are taken in terms of timing and application methods.

6 Root diseases of turfgrasses

Root diseases of turfgrasses continue to increase in incidence and severity as turfgrass health is pushed to the limits to enhance playability. Although these diseases are not researched as much as the diseases listed above, there are some interesting findings worth discussing and advance the idea of sustainable disease management in turfgrass stands. Fairy ring is a disease that is incited by numerous basidiomycete fungi and can be rather destructive on turfgrass grown in sand-based root zones (Fidanza et al., 2016). Studies examining optimal timing for preventative fungicide applications revealed that fairy ring can be adequately suppressed with two to four applications delivered when soil temperatures at 5 cm depth range between 12°C and 24°C (Miller at al., 2012). Numerous fungicides perform well for fairy ring suppression and now with more efficacy data published online (refer to Section 10: Where to look for further information), turfgrass managers can select fungicides or tank-mix fungicides that control fairy ring, take-all patch, Pythium root dysfunction, and summer patch. Similar to fairy ring, those aforementioned diseases start the infection process within the same environmental conditions needed for fairy ring growth, thus fungicides can be deployed to suppress many different diseases at a similar time-of-year (Miller et al., 2012).

Pathogens that invade below ground tissues of turfgrasses can be challenging to manage as damage accrues overtime, symptoms and signs are below the soil, and there are significant barriers to getting the fungicide in contact with the pathogen. Studying etiology, or the cause of disease, is an excellent way to gain more insights into the symptoms, signs, and how the pathogen damages the plant. For example, research clearly demonstrates that the summer patch pathogen is a major problem for creeping bentgrass cultivated for putting greens throughout the USA (Tredway, 2006). Summer patch is traditionally assumed to be an issue for bluegrasses and fine fescues, yet this disease is a huge issue for creeping bentgrass and is still commonly misdiagnosed as take-all patch by inexperience diagnosticians (Tredway, 2006). Simply documenting new hosts for existing pathogen is vital to promoting sustainable turf disease management as it allows turfgrass managers to select the most appropriate course of action for fighting the disease.

Pythium root diseases are challenging and destructive to almost all turfgrass species (Smiley et al., 2005). Yet they are more complex than many realize. Pythium root rot and Pythium root dysfunction are two separate diseases much like foliar anthracnose and basal anthracnose. Now Pythium root rot and root dysfunction could be interrelated, but the conditions that promote the two diseases are very different. Pythium root dysfunction is managed with preventative applications when soil temperatures are between 12°F and 24°F in the spring and fall (Kerns and Butler, 2018). The disease is most severe on newly constructed sand-based rootzones and has only been formally documented on creeping bentgrass. Other turfgrass species could be susceptible to the disease, but formal documentation of other hosts has not occurred (Kerns and Butler, 2018).

Pythium root rot develops on almost all turfgrasses but is most severe on golf course putting greens (Kerns and Butler, 2018). Symptoms occur when turf growth is limited and soil moisture remains at or above field capacity for 3 or 4 days or more. Preventative applications should begin when soil temperatures reach 18°C and continue throughout the period of limited growth. For example, for creeping bentgrass putting greens in the Southeastern USA applications typically start around mid-May (i.e. 15 May) and continue every 2–4 weeks throughout the summer months (Kerns and Butler, 2018). Using an experienced diagnostic lab based at a university can assist with distinguishing these two important diseases and further fosters sustainable management of turfgrass diseases.

Getting the fungicide to the target site for management of root diseases is challenging. Many studies demonstrate best management practices for ensuring that fungicides are delivered to the root surface to prevent infection by root pathogens (Hutchens et al., 2019; Latin, 2022; Stephens et al., 2021). However, the most important consideration when selecting a fungicide for managing a root disease is to pick the most efficacious product (Latin, 2022). Once the most efficacious product has been selected, the best practice for ensuring that the active ingredient comes in contact with the pathogen is to irrigate immediately after the fungicide application with 3.2 mm with water (Hutchens et al., 2019; Stephens et al., 2021). Even with immediate irrigation of 25.4 mm of water after fungicide application, 50% of the active ingredient remains in the top 2.54 cm of soil (Hutchens et al., 2019). So there is little risk of moving the fungicide past the target site or facilitating leaching of the fungicide. Incorporating a soil surfactant can aid in movement of the fungicide into the root zone and still presents little risk for moving the fungicide too far (Hutchens et al., 2020).

7 Host resistance

Host resistance is a disease management strategy that was largely ignored in turfgrass ecosystems until advances were made during the early 2000s (Sapkota et al., 2022). Most cultivated varieties of turfgrass species were historically selected based on agronomic qualities, and disease resistance was only documented after-the-fact. Today, university scientists (i.e. breeders and plant pathologists) and seed companies strongly emphasize disease resistance in their turfgrass breeding programs. As a result, turfgrass cultivars have been released that display good-to-excellent partial resistance to diseases such as dollar spot (Koch and Kerns, 2012; Sapkota, et al., 2022). Plant disease resistance is the plant's ability to reduce, delay, or prevent development of a pathogen or disease (Smiley et al., 2005). Resistance can be complete or partial, and at least within the dollar spot pathosystem, all resistance is partial as complete resistance remains elusive (Sapkota et al., 2022).

Partial dollar spot resistance in creeping bentgrass cultivars, specifically 'Declaration' or 'Memorial', were observed over a three-year period in the Upper Midwest USA, where these cultivars also exhibited partial resistance to gray snow mold (*Typhula* sp.) (Koch and Kerns, 2012). Disease resistance within creeping bentgrasses is consistent across geographic regions which is another desirable trait for disease management (Thompson et al., 2019). Consequently, partial resistance to a disease such as dollar spot may yield some resistance to other diseases. Numerous other cultivars in other turfgrass species have been characterized for relative disease resistance and many also have partial disease resistance phenotypes. For example, cultivars of seashore paspalum have shown partial resistance to dollar spot and are most likely used in the breeding programs for seashore paspalum improvement (Benda et al., 2017; Flor et al., 2013). While partial resistance does not sound as appealing as complete resistance, numerous studies in other crops demonstrate that deploying partial resistance can be equivalent to an application of a fungicide or two (Brock, 2019; Fry, 1982). Therefore, when turfgrass areas are renovated, it is strongly recommended to check what current cultivars are available and know if they possess resistance to dollar spot. Incorporating dollar spot resistant cultivars into new or existing turfgrass swards can further reduce turfgrass managers reliance on fungicides, and future research is likely to remain focused on plant breeding programs aimed at improving dollar spot resistance, and with other turfgrass diseases as well.

8 Fungicide management

Dollar spot is considered one of the most important turfgrass diseases in terms of economic impact or expenditures for fungicides. Numerous research projects and programs focus on management of dollar spot with fungicides and incorporation of new fungicides into management programs targeting dollar spot. This is incredibly important on fairway turf in temperate regions as the areas where dollar spot can incidence could be vast. Fungicide resistance remains a problematic issue with dollar spot management. The vast majority of dollar spot-causing pathogen isolates collected display varying degrees of resistance to benzimidazole and demethylation inhibitor (DMI) fungicide classes (Hsiang et al., 1997, 2007; Latin, 2022; Stephens and Kaminski, 2019). Research recently has shown that isolates of the dollar spot pathogen are showing resistance to succinate dehydrogenase inhibitor (SDHI) fungicides as well (Popko et al., 2018). As new chemistries are released in these fungicide classes, turfgrass managers should not immediately discount their usefulness for dollar spot suppression. Research also indicates that cross resistance within the DMI class may not be as prevalent as previously speculated or assumed (Latin, 2022). Newer DMI active ingredients have proven effective in dollar spot fungicide efficacy field trials where other DMIs failed due to resistance (Latin, 2022). The mechanism behind these observation remains elusive for DMI fungicides, but numerous turfgrass pathologists have documented this trend (Latin, 2022).

The SDHI fungicides have emerged as reliable products for dollar spot suppression as they are highly efficacious and provide suppression in many cases for longer than 14 days (Latin, 2022). This is important because research indicates SDHI fungicide residues are not detected or have limited concentration within the turfgrass foliage, thatch, or soil at 14 days after the initial application (Latin, 2022; Stephens et al., 2021). Thus, the active ingredients within the SDHI fungicide class must cause substantial damage to the dollar spot fungus, thus making it more challenging for the fungus to rebound after exposure to the chemical (Latin, 2022). Given that highly effective SDHI chemistry coupled with a strategic cultural management program, it is conceivable to extend reapplication intervals for dollar spot suppression.

Resistance to SDHI fungicides has been document and was inevitable with the dollar spot fungus as a result of the tendency for resistant pathogen populations to persist (Popko et al., 2018). Yet, current research indicates that cross resistance within the SDHI class within populations of the dollar spot population may be limited and governed by certain genetic markers (Lee et al., 2021). Using these markers could facilitate more targeted research of populations in areas outside of the Northeast USA and promote novel fungicide programs that minimize failures due to fungicide resistance. Moreover, fungicide programs that utilize new DMI and SDHI active ingredients such as mefentrifluconazole and pydiflumetofen, respectively, have upgraded dollar spot management substantially (Latin, 2022). Furthermore, incorporation of the multisite and contact fungicide active ingredient of fluazinam adds an excellent rotation partner to a dollar spot fungicide program and should assist with limiting development of resistance.

Turfgrass managers that incorporate a sustainable cultural management program with the most advanced fungicide chemistry can expect excellent and slightly longer lasting dollar spot suppression. If a turfgrass manager is fortunate to undergo a putting green or fairway renovation then selecting a cultivar with improved resistance to dollar spot will facilitate even better dollar spot suppression that could encourage longer fungicide reapplication intervals. Finally, an accurate dollar spot forecast model is available and can limit and strategically target fungicide inputs even further. The Smith-Kerns model accurately predicts the onset of dollar spot in numerous locations within the USA

(Smith et al., 2018). The model is interactive and allows the turfgrass manager to accept different risk thresholds for dollar spot. Thus, selecting a more aggressive disease threshold would reduce fungicide advisories but would come with more risk for dollar spot breakthrough. The latest research indicates that coupling the forecast model with host resistance and cultural management practices can indeed minimize risk of dollar spot development (Clarke et al., 2020).

Dollar spot is yet another excellent example that details how research has facilitated the movement toward sustainable turfgrass disease management. Many of the practices described in this chapter are currently employed for dollar spot, especially on golf course fairways (Clarke et al., 2020). Although dollar spot remains a chronic problem for turfgrass managers, research-based advances with dollar spot management have been made and continue to be developed. It is not a controversial statement that dollar spot management is easier or more reliable than it was 20–30 years ago, as a result of efforts from many turfgrass scientists throughout the globe.

9 Conclusion and future trends

Sustainable turfgrass disease management is achievable as documented by the research described in this chapter. Future research should remain focused on integrating various disease management strategies with a goal to limit or strategically utilize inputs. As technology continues to evolve, more research will likely center on investigations into the micro- and phytobiomes to further enhance turfgrass disease management. Researchers have just started to explore the influence of soil and root zone microbiology on plant disease development and management. It is quite likely that in the near future, turfgrass disease management programs will focus on integrating many of the topics discussed in this chapter with novel materials and strategies designed to facilitate a certain population of soil or root zone microbes that reduce pathogen growth or prevent the pathogen's ability to cause infection or promote the plant's ability to defend themselves against the pathogen. The opportunities are endless for future research in turfgrass disease management as the climate continues to change and turfgrass management practices evolve to meet stakeholder expectations for sustainability.

10 Where to look for further information

There are numerous resources available to help guide future research. Listed below are a number of resources commonly used for turfgrass research. If interested in turfgrass science or turfgrass pathology, consider joining the American Phytopathological Society (www.apsnet.org), the Crop Science Society of America (www.crops.org), and/or the International Turfgrass Society (www.turfsociety.com). Each of these professional societies has numerous resources, conferences, and publications that house tremendous information on turfgrass diseases.

- The Turfgrass Information File: https://tic.lib.msu.edu/.
  - Excellent repository for journal articles, abstracts, and trade publications, all focused on turfgrass science.
- Plant Disease Management Reports: http://www.plantmanagementnetwork.org/pub/trial/pdmr/.
  - Turfgrass fungicide efficacy field trial results are published here.
- Chemical Control of Turfgrass Diseases: http://www2.ca.uky.edu/agcomm/pubs/PPA/PPA1/PPA1.pdf
  - Excellent publication listing turfgrass diseases and fungicide information for each disease in the USA.
  - University of Wisconsin – Turfgrass Pathology: https://tdl.wisc.edu/.
  - More information for turfgrass disease control in the Upper Midwest region of the USA.
  - Turffiles: https://www.turffiles.ncsu.edu/.
  - More information for turfgrass disease control in the Southeast region of the USA.

11 References

Agrios, G. N. (2005), *Plant Pathology* (5th edn.), San Diego, CA: Academic Press.

Benda, N. D., Flor, N. C., Harmon, P. F. and Kenworthy, K. E. (2017), Response of seashore paspalum genotypes to two isolates of *Sclerotinia homeoecarpa*, *Int. Turfgrass Soc. Res. J.*, 13, 454–458.

Benelli, J. J., Horvath, B. J., Womac, A. R., Ownley, B. H., Windham, A. S. and Sorochan, J. C. (2018), Large patch severity on Japanese lawngrass influenced by fungicide and application target site, *Crop Protect.*, 106, 23–28.

Brock, I. (2019), *Principles of Plant Disease Management*, New York, NY: Syrawood Publishing House.

Butler, E. L., Galle, G. H. and Kerns, J. P. (2019), Influence of nitrogen rate and timing, fungicide application method, and simulated rainfall after fungicide application on brown patch severity in tall fescue, *Crop Forage Turfgrass Manage.*, 5(1). doi:10.2134/cftm2019.03.0018.

Carrow, R. N., Waddington, D. V. and Rieke, P. E. (2001), *Turfgrass Soil Fertility and Chemical Problems*, New York: John Wiley & Sons, 218.

Chen, X. (2016). Influence of seasonal N fertilization, plant growth regulators, and fungicide application timing on anthracnose severity of annual bluegrass putting greens, Master's thesis 996. University of Connecticut Available at: http://digitalcommons.uconn.edu/gs_theses/996) Accessed January 6, 2018.

Clarke, B. B., Vincelli, P., Koch, P. and Munshaw, G. (2020), *Chemical Control of Turfgrass Diseases 2020*, Lexington, KY: University of Kentucky. Available at: http://www2.ca.uky.edu/agcomm/pubs/ppa/ppa1/ppa1.pdf.

Crouch, J. A., Clarke, B. B. and Hillman, B. I. (2006), Unraveling evolutionary relationships among the divergent lineages of Colletotrichum causing anthracnose disease in turfgrass and corn, *Phytopathology*, 96(1), 46–60.

Couch, H. B. (1995), *Diseases of Turfgrasses* (3rd edn.), Malabar, FL: Krieger Publishing.

Danneberger, T. K., Vargas, J. M. Jr. and Jones, A. L. (1984), A model for weather-based forecasting of anthracnose on annual bluegrass, *Phytopathology*, 74(4), 448–451.

Dernoeden, P. H. (2002), *Creeping Bentgrass Management: Summer Stresses, Weeds, and Selected Maladies*, Hoboken, NJ: John Wiley & Sons.

Ellram, A., Horgan, B. and Hulke, B. (2007), Mowing strategies and dew removal to minimize dollar spot on creeping bentgrass, *Crop Sci.*, 47(5), 2129–2137.

Fidanza, M., Settle, D. and Wetzel, H. (2016), An index for evaluating fairy ring symptoms in turfgrass, *Hortscience*, 51(9), 1194–1196.

Flor, N. C., Munoz, P., Harmon, P. and Kenworthy, K. (2013), Response of seashore paspalum genotypes to dollar spot disease, *Int. Turfgrass Soc. Res. J.*, 12, 119–126.

Franci, L. J. (2001), The disease triangle: a plant pathological paradigm revisited, *The Plant Health Instructor*. doi:10.1094/PHI-T-2001-0517-01.

Fry, W. E. (1982), *Principles of Plant Disease Management*, Cambridge, MA: Academic Press.

Giordano, P. R., Nikolai, T. A., Hammerschmidt, R. and Vargas, J. M. (2012), Timing and frequency effects of lightweight rolling on dollar spot disease in creeping bentgrass putting greens, *Crop Sci.*, 52(3), 1371–1378.

Gonzales, A. E., Hockemeyer, K. R., McConville, M., Remucal, C. K. and Koch, P. L. (2022), Assessment of temperature and time following application as predictors of propiconazole translocation in *Agrostis stolonifera*, *Agric. Sci. Technol.*, 3, 592–602.

Hempfling, J. W., Murphy, J. A. and Clarke, B. B. (2020), Midseason cultivation effects on anthracnose of annual bluegrass turf, *Agron. J.*, 112(5), 3411–3417. doi:10.1002/agj2.20202.

Hempfling, J. W., Clarke, B. B. and Murphy, J. A. (2015), Anthracnose disease on annual bluegrass as influenced by spring and summer topdressing, *Crop Sci.*, 55(1), 437–443.

Hempfling, J. W., Schmid, C. J., Wang, R., Clarke, B. B. and Murphy, J. A. (2017), Best management practices effects on anthracnose disease of annual blue- grass, *Crop Sci.*, 55(1), 602–610. doi:10.2135/cropsci2016.06.0492.

Hsiang, T., Liao, A. and Benedetto, D. (2007), Sensitivity of *Sclerotinia homoeocarpa* to demethylation-inhibiting fungicides in Ontario, Canada, after a decade of use, *Plant Pathol.*, 56(3), 500–507.

Hutchens, W. J., Gannon, T. W., Shew, H. D., Ahmed, K. A. and Kerns, J. P. (2020), Soil surfactants influence fungicide movement in US Golf Association putting green soil, *J. Environ. Q.*, 49(2), 450–459.

Hutchens, W. J., Gannon, T. W., Shew, H. D. and Kerns, J. P. (2019), Effect of post-application irrigation on fungicide movement and efficacy against *Magnaporthiopsis poae*, *Crop Protect.*, 122, 106–111.

Hsiang, T., Yang, L. and Barton, W. (1997), Baseline sensitivity and cross-resistance to demethylation-inhibiting fungicides in Ontario isolates of *Sclerotinia homoeocarpa*, *Eur. J. Plant Pathol.*, 103(5), 409–416.

Inguagiato, J. C. and Guillard, K. (2016), Foliar N concentration and reflectance meters to guide N fertilization for anthracnose management of annual bluegrass putting green turf, *Crop Sci.*, 56(6), 3328–3337.

Inguagiato, J. C., Murphy, J. A. and Clarke, B. B. (2008), Anthracnose severity on annual bluegrass influenced by nitrogen fertilization, growth regulators, and verticutting, *Crop Sci.*, 48(4), 1595–1607.

Inguagiato, J. C., Murphy, J. A. and Clarke, B. B. (2009a), Anthracnose disease and annual bluegrass putting green performance affected by mowing practices and lightweight rolling, *Crop Sci.*, 49(4), 1454–1462.

Inguagiato, J. C., Murphy, J. A. and Clarke, B. B. (2009b), Anthracnose of annual bluegrass putting green turf influenced by trinexapac-ethyl application interval and rate, *Int. Turfgrass Soc. Res. J.*, 11, 207–218.

Inguagiato, J. C., Murphy, J. A. and Clarke, B. B. (2010), Anthracnose development on annual bluegrass affected by seedhead and vegetative growth regulators, *Appl. Turfgrass Sci.*, 7(1), 1–19.

Inguagiato, J. C., Murphy, J. A. and Clarke, B. B. (2012), Sand topdressing rate and interval effects on anthracnose severity of an annual bluegrass putting green, *Crop Sci.*, 52(3), 1406–1415. doi:10.2135/cropsci2011.01.0010.

Kerns, J. P. and Butler, E. L. (2018), Demystifying Pythium diseases of golf course turfgrasses, *Golfdom*, 74, 30–34.

Koch, P. L. and Kerns, J. P. (2012), Relative resistance of creeping bentgrass cultivars to *Sclerotinia homoeocarpa* and *Typhula incarnata*, *Appl. Turfgrass Sci.*, 9(1), 1–5.

Koehler, A. M. and Shew, H. D. (2017), Seasonal dynamics and fungicide sensitivity of organisms causing brown patch of tall fescue in North Carolina, *Mycologia*, 109(4), 667–675.

Koehler, J. F. and Miller, G. L. (2019), Impact of nitrogen application timing, nitrogen source, and a single application on large patch of zoyisagrass, *Crop Sci.*, 59(3), 1285–1293.

Landschoot, P. and Hoyland, B. (1995), Shedding some light on anthracnose basal rot, *Golf Course Manage.*, 11:52–55.

Latin, R. (2022), *A Practical Guide to Turfgrass Fungicides*, St. Paul, MN: American Phytopathological Society.

Lee, J., Elliot, M. R., Kim, M., Yamada, T. and Jung, G. (2021), A rapid molecular detection system for Sdhb and sdhc point mutations conferring differential succinate dehydrogenase inhibitor resistance in *Clarireedia* populations, *Plant Dis.*, 105(3), 660–666. https://doi.org/105:660-66610.1094/pdis-04-20-0724-re.

Mann, R. L. and Newell, A. J. (2005), A survey to determine the incidence and severity of pests and diseases on golf course putting greens in England, Ireland, Scotland, and Wales, *Int. Turfgrass Soc. Res. J.*, 10, 224–229.

Mattox, C. and Kowalewski, A. (2020), Microdochium patch control alternatives, *Golfdom*, 76, 45.

Miller, G. L., Soika, M. D. and Tredway, L. P. (2012), Evaluation of preventive fungicide applications for fairy ring control in golf putting greens and in vitro sensitivity of fairy ring species to fungicides, *Plant Dis.*, 96(7), 1001–1007.

Murphy, J. A., Clarke, B. B. and Inguagiato, J. A. (2018), Update on best management practices for anthracnose on annual bluegrass turf, *Golf Course Manage.*, 86(2), 76–85.

Murphy, J. A., Wong, F., Tredway, L., Crouch, J. A., Inguagiato, J., Clarke, B., Hsiang, T. and Rossi, F. (2008), Best management practices for anthracnose on annual bluegrass turf, *Golf Course Manage.*, 76(8), 93–104.

Popko, J. T., Jr., Sang, H., Lee, J., Yamada, T., Hoshino, Y. and Jung, G. (2018), Resistance of *Sclerotinia homoeocarpa* field isolates to succinate dehydrogenase inhibitor fungicides, *Plant Dis.*, 102(12), 2625-2631.

Newhall, A. G. (1980), Herbert Hice Whetzel: pioneer American plant pathologist, *Annu. Rev. Phytopathol.*, 18(1), 27-36.

Roberts, J. A., Inguagiato, J. C., Clarke, B. B. and Murphy, J. A. (2011), Irrigation quantity effects on anthracnose disease of annual bluegrass, *Crop Sci.*, 51(3), 1244-1252. doi:10.2135/cropsci2010.07.0444.

Roberts, J. A. and Murphy, J. A. (2014), Anthracnose disease on annual bluegrass as affected by foot traffic and sand topdressing, *Plant Dis.*, 98(10), 1321-1325. doi:10.1094/PDIS-08-13-0877-RE.

Roberts, J. A., Murphy, J. A. and Clarke, B. B. (2012), Lightweight rolling effects on anthracnose of annual bluegrass putting greens, *Agron. J.*, 104(4), 1176-1181. doi:10.2134/agronj2011.0269.

Salgado-Salazar, C., Beirn, L. A., Ismaiel, A., Boehm, M. J., Carbone, I., Putman, A. I., Tredway, L. P., Clarke, B. B. and Crouch, J. A. (2018), *Clarireedia*: A new fungal genus comprising four pathogenic species responsible for dollar spot disease of turfgrass, *Fungal Biol.*, 122(8), 761-773.

Sapkota, S., Catching, K. E., Raymer, P. L., Martinez-Espinoza, A. D. and Bahri, B. A. (2022), New approaches to an old problem: dollar spot of turfgrass, *Phytopathology*, 112(3), 469-480.

Schmid, C. J. (2016), Influence of nutrient management and soil pH on anthracnose severity of annual bluegrass putting green turf. Ph.D. dissertation. Rutgers University Available at: https://rucore.libraries.rutgers.edu/rutgers-lib/51457/. Accessed January 6, 2018.

Smiley, R. W., Dernoeden, P. H. and Clarke, B. B. (2005), *Compendium of Turfgrass Diseases* (3rd edn.), St. Paul, MN: APS Press.

Smith, D. L., Kerns, J. P., Walker, N. R., Payne, A. F., Horvath, B., Inguagiato, J. C., Kaminski, J. E., Tomaso-Peterson, M. and Koch, P. L. (2018), Development and validation of a weather-based warning system to advise fungicide applications to control dollar spot on turfgrass, *PLoS ONE*, 13(3), e0194216.

Smith, J. D., Jackson, N. and Woolhouse, A. R. (1989), *Fungal Diseases of Amenity Turf Grasses*, London: E&F. N. Spon.

Stephens, C. M., Kerns, J. P., Ahmed, K. A. and Gannon, T. W. (2021), Influence of post-application irrigation and mowing timing on fungicide fate on a United States Golf Association golf course putting green, *J. Environ. Qual.*, 50(4), 868-876.

Stephens, C. M. and Kaminski, J. E. (2019), In vitro fungicide-insensitive profiles of *Sclerotinia homoeocarpa* populations from Pennsylvania and the surrounding region, *Plant Dis.*, 103(2), 214-222.

Thompson, C., Zhang, Q., Kennelly, M., Stier, J., Blume, C., Christians, N., Fry, J., Martin, D., Ostrander, J., Rincker, K., Settle, D. and Soldat, D. (2019), The dollar spot susceptibility of 25 bentgrasses is consistent across five states in the central USA, *Crop Forage Turfgrass Manage.*, 5, 1-4.

Townsend, R., Millican, M. D., Smith, D., Nangle, E., Hockemeyer, K., Soldat, D. and Koch, P. L. (2021), Dollar spot suppression on creeping bentgrass in response to repeated foliar nitrogen applications, *Plant Dis.*, 105(2), 276-284.

Tredway, L. P. (2006), Genetic relationships among *Magnaporthe poae* isolates from turfgrass hosts and relative susceptibility of 'penncross' and 'penn A-4' creeping bentgrass, *Plant Dis.*, 90(12), 1531–1538.

Vargas, J. M. (2005), *Management of Turfgrass Diseases* (3rd edn.), Hoboken, NJ: John Wiley & Sons.

Vargas, J. M. and Turgeon, A. J. (2003), *Poa Annua: Physiology, Culture, and Control of Annual Bluegrass*, Hoboken, NJ: John Wiley & Sons.

Wang, R., Hempfling, J. W., Clarke, B. B. and Murphy, J. A. (2018), Seasonal and annual topdressing effects on anthracnose of annual bluegrass, *Agron. J.*, 110(6), 2130–2135. doi:10.2134/agronj2018.01.0052.

Williams, D. W., Powell, A. J., Jr., Vincelli, P. and Dougherty, C. T. (1996), Dollar spot on bentgrass influenced by displacement of leaf surface moisture, nitrogen, and clipping removal, *Crop Sci.*, 36(5), 1304–1309.

Chapter 10

Advances in turfgrass insect pest management

Benjamin A. McGraw, Audrey Simard and Garrett Y. Price, Pennsylvania State University, USA

1 Introduction

Managed turfgrasses, including those utilized for golf courses, athletic fields, parks, roadsides, and home lawns, are of great ecological and economic importance in the USA, covering an estimated 163 800 km^2 (Milesi et al., 2005) and generating tens of billions of dollars in economic impact (Haydu et al., 2006). The management of pests, including insect herbivores and nuisance pests, both directly and indirectly affects the profitability of turfgrass culture, though economic impact studies have rarely been conducted. Great variability exists in the intensity of turfgrass culture between systems and regions, and thus the need for management or tolerance for insect-caused turf damage also varies greatly. However, most turfgrass insect pests are managed primarily based on esthetics rather than financial or yield-based considerations, which creates situations where pest tolerance or damage thresholds are set arbitrarily low.

Over the last 20 years, a considerable amount of industry- and university-sponsored research has been generated to address these unique challenges to control insects, including studies investigating the biology and ecology of major insect pests, novel surveillance techniques, and the discovery of novel controls including biological and biorational controls. However, given the need to create pristine esthetic conditions, low pest thresholds, and the need for interventions

http://dx.doi.org/10.19103/AS.2022.0110.10

to work rapidly, chemical controls remain the dominant means of controlling turfgrass insect pests. The reliance on synthetic chemical insecticides, as well as the continued discovery and marketing of new active ingredients and classes, has created a culture of preventive management. Ultimately, the effectiveness of chemical controls in most situations has limited the need to understand pest biology and ecology, which, in turn, has limited the ability to make progress in discovering alternative and sustainable control measures. The increased public demand for reducing chemical inputs in turfgrass and urban environments and greater regulatory scrutiny on existing chemical controls (Sparks and Lorsbach, 2017) is likely to accelerate the need for discovery and delivery of both novel active ingredients with improved environmental profiles and/or alternative strategies for reducing insect pests.

This chapter, though largely based on work conducted in North America, serves as a review of the recent advances that have been made in turfgrass entomological science and changes to existing management practices and identifies areas where research or industry support is needed to make changes to how insect pests are managed in urban and turfgrass environments. Notable changes over the last decade include changes to insect pest species distribution due to movement through plant materials and other anthropogenic factors as well as climatic changes allowing for insects to survive in new areas. In some regions, extreme weather conditions have exacerbated insect feeding and created situations where minor insect pests become major ones. In a rare instance, a previously neutral insect has emerged as a pest of turfgrass (Van Dyke and McGraw, 2021). University and industry professionals have risen to meet these challenges by generating much needed information on turfgrass insect biology and the development of novel pest controls that maximize pest control and minimize unintended effects on organisms within the turfgrass and the surrounding environment. These efforts will need to be sustained by a new generation of scientists if the industry is to continue to develop tools that are effective against a diversity of pests, overcome emerging insecticide resistance issues, and protect natural resources.

2 Recent studies of major insect pests

2.1 Annual bluegrass weevil – Listronotus maculicollis

The annual bluegrass weevil, *Listronotus maculicollis* Kirby (ABW), is the single-most destructive insect pest of golf course turfgrass in eastern North America (Fig. 1). The insect is believed to be a native pest, though since its detection on Long Island in 1957 (Vittum, 2020), reports of damaging populations have continued to expand radially from the metropolitan New York City area each year since. ABW populations are typically greatest in areas where populations are well established. Larval feeding damage on the turfgrass can be so

Figure 1 Example of an adult annual bluegrass weevil. Source: Photo by B. McGraw.

severe that multiple insecticides applications are made throughout the year (McGraw and Koppenhöfer, 2017). Traditionally, overwintering adults were targeted during spring emergence from overwintering to reduce oviposition and the potential for larval feeding. Recent behavioral studies have shown that controlling adults is essential for reducing egg laying and synchronizing larval populations, as females are capable of ovipositing up to 14 weeks if left uncontrolled (McGraw et al., 2021a). However, the number of insecticide applications employed against ABW has steadily increased over the last 30 years to now include sequential insecticide applications against both adults and larvae from spring (April–May) until third-generation adults enter overwintering sites in fall (September–October). Overuse of adulticides, in particular the pyrethroids, has led to the development of resistant populations in many areas throughout the region (Ramoutar et al., 2009a). A recent survey of golf course superintendents in the region indicates that approximately 20% of courses are affected by pyrethroid resistance (McGraw and Koppenhöfer, 2017).

In the last decade, ABW has not only expanded its distribution to the north, west, and south but has more frequently reported damaging creeping bentgrass (*Agrostis stolonifera* L.) as it moves into areas where less annual bluegrass (*Poa annua* L.) is encountered. Host preference for annual bluegrass has been investigated by examining larval distribution on golf courses (McGraw and Koppenhöfer, 2010, 2015) and through chemical ecology studies (McGraw et al., 2011; Kostromytska et al., 2018a). Field studies tracking adult distribution and larval density with respect to turfgrass species in mixed creeping bentgrass-annual bluegrass fairways showed a gradient of larvae from edges of fairways into centers (McGraw and Koppenhöfer, 2010). Adults did not appear to be selective, as larvae were found in both species. Subsequent damage ratings showed that creeping bentgrass was much more tolerant to damage. This was later confirmed in laboratory and greenhouse studies (Kostromytska et al., 2018a).

Electrophysiological studies have shown that adults respond to volatiles emitted from both annual bluegrass (McGraw et al., 2011) and creeping bentgrass (Kostromytska et al., 2018a) and that females and males may respond differently. Higher volatile emissions are produced from mechanically damaged plants, though clipped annual bluegrass repelled females, whereas intact plants were attractive. Kostromytska et al. (2018a) demonstrated that volatiles emitted from annual bluegrass were more attractive than *Agrostis* species, with the latter having significantly greater amounts of terpenoids than annual bluegrass.

*2.2 Billbugs (*Sphenophorus *spp.)*

Nearly a dozen species of weevils referred to as 'billbugs' (Coleoptera: Dryophthoridae) are known to damage turfgrasses in North America (Duffy et al., 2018b) (Fig. 2). Of these, the bluegrass billbug (*Sphenophorus parvulus* Gyllenhal; BB) and the hunting billbug (*S. venatus* Say; HBB) are the most widespread and damaging. HBB has long been a pest of warm-season turfgrasses in the southeastern USA (Doskocil and Brandenberg, 2012), though it can be commonly found within cool-season turfgrass stands in the transition zone and north. More recently, HBB has become problematic in the Midwest and Intermountain West USA regions, which has recently led to several life history studies to better predict stage development and time interventions. Duffy et al. (2018b) determined that BB was the most dominant species in cool-season turfgrasses in Indiana, whereas HBB was most commonly associated with warm-season turfgrasses. Studies conducted in Missouri suggest that HBB is more strongly attracted to *Zoysia japonica* (Steud.) than tall fescue (*Schedonorus arundinacea* Schreb.) and Kentucky bluegrass (*P. pratensis* L.) (Barrett et al., 2018). Molecular analyses of populations revealed that HBB was capable of overwintering as both larvae and adults, which resulted in asynchronous larval populations, or two distinct cohorts, highlighting the challenges in managing billbugs where species co-occur. Further complicating matters is that species

Figure 2 Example of an adult 'billbug'. Source: Photo by B. McGraw.

phenology may differ regionally. The HBB has been documented to be univoltine in Utah (Dupuy and Ramirez, 2016), whereas surveys in Indiana (Duffy et al., 2018b) and North Carolina (Doskocil and Brandenberg, 2012) document two overlapping generations per year. These findings have led to the development of growing degree day (GDD) models to assist turfgrass managers in the Intermountain West and Indiana USA regions (Duffy et al., 2018b; Dupuy et al., 2017). It is possible that HBB will become a more significant pest in the northern transition zone, mid-Atlantic, and northeastern USA as more zoysiagrass is used for golf course fairways and bunker surrounds.

2.3 White grubs

White grubs, or the root-feeding larvae of scarab beetles, are arguably the most important insect pests of turfgrasses globally (Vittum, 2020) (Fig. 3). They can be found in all turfgrass systems and regions within North America, with the most problematic species possessing an annual life cycle. The most severe pests are all well-established invasive species such as the Japanese beetle (*Popillia japonica* Newman), Oriental beetle (*Anomala orientalis* Waterhouse), Asiatic garden beetle (*Maladera castanea* Arrow), and the European chafer (*Amphimallon majale* Razoumowsky). All but the last also feed on several other plants. Despite the ubiquitous nature and potential damage of white grubs, little research has been conducted on this group in the last decade. This is largely due to the number and effectiveness of chemical control options, including the neonicotinoids and the anthranilic diamides.

Figure 3 Example of a white grub. Source: Photo by B. McGraw.

The Japanese beetle is the most widely dispersed and damaging of the white grub species in eastern North America. In recent years, populations have become established west of the Mississippi where it has become a pest of economic importance. The beetle's ability to disperse, its polyphagous nature, and its ability to detoxify plant defenses (Adesanya et al., 2017) increase the potential for turfgrass loss in areas surrounded by ornamental plantings. Despite their presence in numerous turfgrass systems, Japanese beetles (and other annual white grubs) are uncommon within the rootzones of golf course putting surfaces, despite the presence of susceptible host plants. In choice assays, female beetles avoided digging into soils maintained slightly above wilting (5% volumetric water content) (Obear et al., 2014). In non-choice assays, recovery of larvae was low from all moisture treatments, indicating that factors other than moisture may influence larval survival in these areas. Fungicides are routinely applied to putting surfaces in areas where Japanese beetles are present. In their following studies, Obear et al. (2016) examined the survival of eggs, and larvae were treated with commonly applied fungicides chlorothalonil and propiconazole. Chlorothalonil reduced first instar survival, whereas propiconazole had wider effects, delaying egg hatch and reducing egg and first instar survival. Both fungicides were shown to suppress detoxification enzymes in larval guts, suggesting that applications around the time of oviposition may have sublethal effects.

2.4 Fall armyworm

Fall armyworm (*Spodoptera frugiperda* J.E. Smith, FAW) is a highly polyphagous herbivore causing significant damage annually to maize, rice, cotton, turfgrass, and several other economically valued crops and ornamentals (Potter and Braman, 1991) (Fig. 4). Adult FAW overwinters in the southern USA and migrates as far north as southern Canada during the summer months, assisted by wind currents and severe weather events such as tropical storms. Periodic outbreaks from the east coast to the Rockies have occurred since the 1970s, including as recently as 2021, which resulted in millions of acres of turf loss (Anders, 2021). Current climate predictions and modeling suggest that with increasing temperatures and longer development periods, FAW outbreaks witnessed in 2021 are more likely to occur more frequently over larger sectors in the USA and around the world (Díaz-Álvarez et al., 2020). Recent investigations of host suitability have shown that FAW development in creeping bentgrass is comparable to bermudagrass (Hong et al., 2015), indicating that increased FAW activity in northern climates could lead to significant changes to caterpillar management and insecticide budgeting for managed turfgrass systems.

Figure 4 Example of a fall armyworm caterpillar. Source: Photo by B. McGraw.

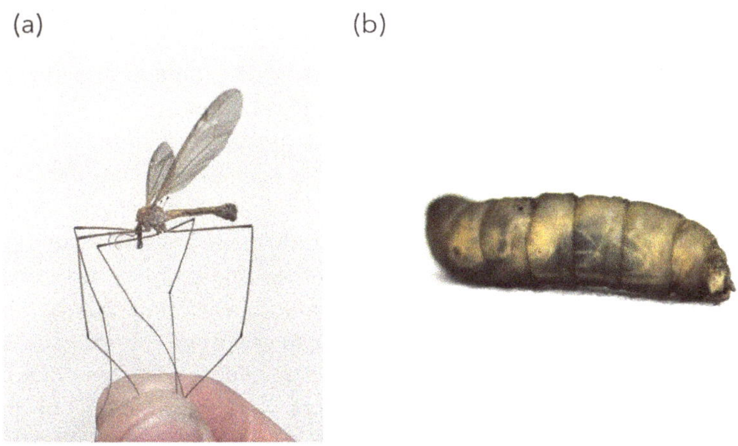

Figure 5 Example of an adult (a) and larvae (b) of the European crane fly. Source: Photo by B. McGraw.

3 Recent studies on emerging turf pests

3.1 European crane flies

Two European species of crane flies (ECFs) are known to damage turfgrass in North America where higher-than-average amounts of rainfall occur (Fig. 5). *Tipula paludosa* Meigen (the European crane fly) and *T. oleracea* L. (the marsh crane fly) have been introduced to North America separately on at least three different occasions since the 1950s. Their pest status is greatest in regions where annual rainfall is greater than average, including Atlantic Canada, the Great Lakes region, and the Pacific Northwest. The marsh crane fly, which is native to central and northern Europe and more tolerant of desiccating conditions, is a capable flier. It can be found damaging turfgrasses in a greater distribution throughout North America than the European crane fly, though both are suspected to be introduced to new areas through the transportation of sod (Petersen et al., 2011) or nursery stock such as populations that have recently been detected near Salt Lake City, Utah (Van Dyke et al., 2015).

The European crane fly tends to be more locally dispersed, though it has a greater potential for damage than the marsh crane fly. Factors important for *T. paludosa* generational survival and population persistence within an area have been shown to be linked to the water-holding capabilities (as determined by soil texture) and not turfgrass species (Petersen et al., 2013). Crane fly populations on soil with high sand (>47%) and low organic matter (<5%) content were either low or declined with age, while soils with high amounts of clay (>12%) and organic matter (>5%) generally supported greater populations through pupation. These findings may provide evidence for why ECFs are not typically found in putting surfaces in the USA, which may contain high percentages of sand either by design with rootzone construction or through modification with sand topdressing. However, in the UK, both insects are problematic within the interior of greens, suggesting a deviation from relict behaviors.

Ecological modeling of both species populations determined that niche shifts are likely to be occurring with these species at a larger scale in North America (Petersen, 2013). Distribution modeling used environmental parameters in native distribution with that in introduced regions of eastern and western North America to determine niche overlap and potential for spread into new regions. Petersen (2013) suggests that niche shifts are occurring with both species in North America and that the spread of the two species into new regions will occur in areas of greater or less annual deviation in temperature from native regions. However, these conclusions were based largely on the scale examined in the models. This plasticity in niche occurrence may suggest that these species, though different in their potential to disperse, may become greater pests in a wider range in North America.

3.2 Chaetocnema minuta

Recently, a novel pest of turfgrass has emerged, attacking both cool- and warm-season turfgrasses in the desert Southwest USA (Van Dyke and McGraw, 2021) (Fig. 6). *Chaetocnema minuta* Melsh., a flea beetle that is native to North America, was isolated from damaged perennial ryegrass (*Lolium perenne* L.) in southern Utah in 2016 and hybrid bermudagrass in Nevada in 2020.

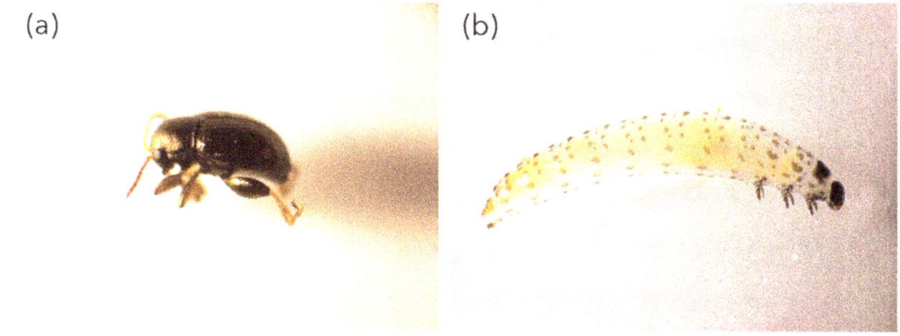

Figure 6 Example of an adult (a) and larvae (b) of the flea beetle (*Chaetocnema minuta*). Source: Photo by B. McGraw.

It is believed that *C. minuta* could have been causing mysterious turf losses as early as the late 1990s, though the small size and rapid arrival and then disappearance of the insects have led to great confusion in both the seasonal ecology and management of the beetle. Adults are often observed jumping above the turfgrass canopy of tee boxes and fairways or aggregating on white foam markers. Their presence precedes damage, which is often confused with drought stress. Very little is known about this insect, and surveys are much needed to address the geographic distribution of the pest, identify damaging life stages and susceptible hosts, as well as determine effective management tactics.

4 Advances in integrated pest management

4.1 Pest detection and monitoring

Detecting pest presence and accurately assessing population size is essential for managing turfgrass insects given low thresholds driven by esthetic concerns. Progress has been made in the last decade to better understand pest behavior to develop traps and to characterize seasonal phenology to build region-specific models to forecast life stage occurrence to better time controls.

Invasive mole crickets (*Neoscapteriscus* species) are severe pests of most turfgrass systems in the southeastern USA (Fig. 7). Most are capable fliers, invading sandy turfgrass sites in spring. Males may cause extensive damage while tunneling to create a burrow in which he will use as an amplifier for his mating call. Other males may be attracted to this call, indicating that the area is a suitable environment. The mating calls of two damaging species have been recorded and effectively used as a trap lure. Although mole cricket acoustic traps have been used for several decades, recent modifications to the traps have led to a trap that may be operated autonomously and produced for at low cost (Rohde et al., 2019).

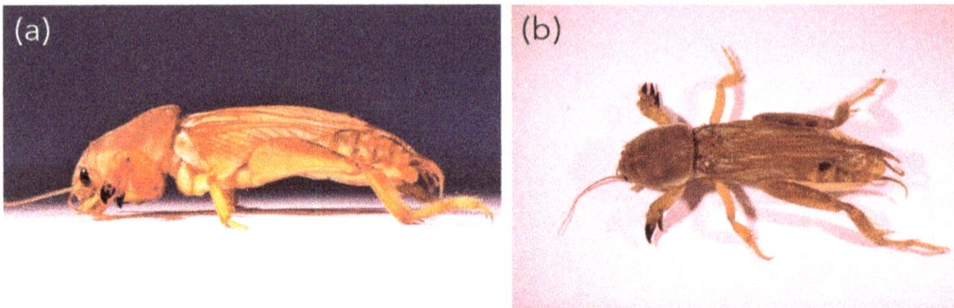

Figure 7 Example of an adult (a, b) mole cricket. Source: Photo by B. McGraw.

Pheromone-based monitoring programs have been limited to several scarab species (i.e. Japanese beetle and Oriental beetle) and lepidopteran pests (i.e. black cutworm and fall armyworm), though their use has been primarily for research purposes. The pheromone lures, which are either sex pheromones or sex pheromones combined with a floral/food scent, are highly effective in attracting adults from a great distance. Attraction to host plant volatiles has been examined in laboratory studies, with hopes of developing effective lures for traps for HBB and ABW (McGraw et al., 2011; Duffy et al., 2018a; Duffy and Richmond, 2020; Kostromytska et al., 2018a). Unfortunately, few studies have examined these effects in the field or have found success with volatile compounds or blends when deployed in the field (Kostromytska et al., 2018a). It is likely that the response to host plant-based attractants is less effective than sex pheromones if the signal is muted against the background volatile profile of the turfgrass system in which it is deployed.

A greater understanding of the effects of temperature on the physiological development, movement, and timing of life cycles has led to the development of predictive GDD models for insect monitoring. GDD models are designed to correlate a biological event with temperature accumulations from arbitrary start date (i.e. January 1 and March 1). These models have the potential to be integrated across many regions, should the event be driven largely by temperature and not influenced by multiple factors (i.e. photoperiod and precipitation). Occasionally, models may not align for the same pest in different regions, as observed with HBB in Indiana and Utah USA areas (Duffy et al., 2018b; Dupuy et al., 2017).

Unpublished GDD models are the basis for the Syngenta (Greensboro, NC) Weevil Trak(^{SM}), a regional ABW monitoring program which incorporates data from local weather stations, plant phenological indicators, and ground truthing by university and industry experts, to provide area-wide advice which is delivered to turfgrass managers through automated alerts. The success of the program has made it the most common monitoring tool used in ABW management (McGraw and Koppenhöfer, 2017).

4.2 Chemical control

Turfgrass insect pest management in the USA is primarily conducted with the use of synthetic, conventional insecticides to maintain public expectations of safety, playability, and esthetic quality. The need for discovering new classes and active ingredients is great, as new products are needed to provide efficacious control of new pests, mitigate insecticide resistance, and replace older chemistries removed through regular regulatory examination (Sparks, 2013). Recently, the latter has driven the greatest demand for novel discoveries for turfgrass, as adverse effects on non-target organisms have increased public pressure to remove these products for use in and around turfgrass sites.

The neonicotinoids have become the fastest growing insecticide class worldwide (Morrissey et al., 2015), given their low mammalian toxicity, residual activity, and broad and efficacious impact on a variety of pests. Their popularity in turfgrass insect control is largely attributed to their systemic nature, allowing for preventive control of most white grub species (Zenger and Gibb, 2001; Larson et al., 2017). However, the mobility and translocation of neonicotinoids from roots to flowers have contributed to the significant declines of natural enemies and pollinators within turfgrass and agricultural environments. In 2013, an application of dinotefuran to flowering shrubs in an Oregon shopping mall parking lot led to a widely publicized bee kill and drew media attention to the use of neonicotinoids in turfgrass and ornamental settings (Hatfield et al., 2021). Because of this, the future of the neonicotinoid class is uncertain in the USA. In Europe, major policy shifts have already begun, including in France with the banning of neonicotinoids along with significant investments in research into the discovery of non-chemical alternatives (Jactel et al., 2019).

The organophosphates, a class of acetyl choline esterase inhibitors, have been used in turfgrass insect control for several decades. The active ingredients trichlorfon, acephate, and chlorpyrifos are registered for use in turfgrass in the USA, all of which provide non-selective control of primarily surface-active insects, though trichlorfon is most commonly used as a curative white grub control. Acephate is frequently used against mole crickets in the southern USA, whereas chlorpyrifos is more broadly used against caterpillars, leatherjackets, and weevils. Chlorpyrifos, which has been registered since 1966, was banned from residential use in 2000 for its adverse neurotoxicological effects (Hites, 2021). However, in the same time period, chlorpyrifos became an important turfgrass insecticide as an alternative adulticide in managing pyrethroid-resistant ABW populations (McGraw and Koppenhöfer, 2017). However, recent petitions to the U.S. Environmental Protection Agency citing nervous system damage with exposure led to several court-ordered bans of the sale of chlorpyrifos. These decisions were later overturned with the changes in U.S. presidential administration in 2016, though state-level bans in California,

Hawaii, and New York were renewed. Ultimately, Corteva/Dow AgroSciences (Indianapolis, IN) - the producers of chlorpyrifos - announced that the active ingredient will no longer be produced after 2020.

4.3 Insecticide resistance

Insecticide resistance can be broadly defined as the ability of an insect population to tolerate levels of toxins that would otherwise be lethal to non-resistant members of the same species (Stenersen, 2004). With chemical management increasingly becoming a first resort against turfgrass pests, it should come as no surprise that resistance issues would emerge. Resistance development is accelerated when a pest population is routinely treated with a single chemical, which acts as strong selective pressure for resistance to develop. This is not an uncommon practice when controlling multivoltine insect pests in turfgrass. The southern chinch bug (*Blissus insularis* Barber) (SCB), FAW, and ABW are the most well-known examples of insecticide resistance development in turfgrass insect pests (Cherry and Nagata, 2005; Carvalho et al., 2013; Ramoutar et al., 2009a,b). FAW and SCB may be targeted with multiple insecticide applications per year, as both will complete several generations per year or may develop continuously throughout the year in semi-tropical areas of the USA. All three species have developed resistance to the pyrethroids, with ABW more recently being observed to be capable of developing cross-resistance to several pyrethroids (Kostromytska et al., 2018b). This resistance is achieved at least partially due to enhanced metabolic detoxification enzymes, though the exact mechanism is unknown (Ramoutar et al., 2009b). Worse yet, populations exhibiting high levels of resistance to pyrethroids tend to demonstrate 'multiple resistance' or show a greater tolerance for non-pyrethroid insecticides as well (Kostromytska et al., 2018b).

4.4 Non-target issues with chemical insecticides

Pesticides are often an essential and unavoidable tool in formulating an integrated pest management strategy against pathogenic diseases, insects, and undesirable weeds. However, declines in pollinator (i.e. honey bees and other bee species) abundance and health have led to fears of a 'pollination crisis' that threaten both agricultural productivity and biodiversity (Goulson et al., 2008). Previous studies have implicated the role of pesticides (primarily insecticides) in these declines, as exposure to these chemical agents has been associated with changes in bee behavior (Desneux et al., 2007) and reductions in colony queen production (Whitehorn et al., 2012). Larson et al. (2012, 2014) documented that clothianidin applications to turfgrass significantly reduced earthworms, earthworm casts, and soil microarthropods populations compared to the anthranilic diamide and the alternative white grub control of chlorantraniliprole. Clothianidin was also shown to affect bumble bees exposed to mixed white clover-turfgrass lawn turf, in which exposure led to delayed weight gain and reduced queen production compared to chlorantraniliprole (Larson et al., 2013). However, mowing weedy flowers prior to or after the application of clothianidin appeared to reduce exposure and any harmful effects of the application (Larson et al., 2015).

Pollinator insecticide exposure to insecticides may be through direct contact on plant foliar tissue, drift, and/or consumption of contaminated sources (i.e. pollen and water). In turfgrass, water-soluble systemic active ingredients, such as the neonicotinoid imidacloprid, have been detected within turfgrass guttation fluid post foliar and soil drench applications (Larson et al., 2015; McCurdy et al., 2017). Turfgrasses produce guttation fluid regularly and in high abundance compared to other agricultural grasses such as maize, wheat, and barley (Joachimsmeier et al., 2012; Girolami et al., 2009; Hoffmann and Castle, 2012). Guttation droplets may fall, evaporate, or be reabsorbed back into the plant (Chen and Chen, 2007). Arthropods that have been observed exploiting this resource include bees and other natural enemies (Shawki et al., 2006; Joachimsmeier et al., 2012). Negative effects have been observed with minute pirate bug (*Orius insidiosus* Say) in contact with guttation fluid exuded from Kentucky bluegrass (Larson et al., 2014). However, the magnitude of the effect of imidacloprid-guttation fluid on natural enemies and non-target insects may differ by the turfgrass community. McCurdy et al. (2017) found differences in concentrations of imidacloprid in dew between creeping bentgrass, St. Augustinegrass (*Stenotaphrum secundatum* (Walter) Kuntze)) , and bermudagrass (*Cynodon dactylon* (L) *Pers.* × *C. transvaalensis* Burtt-Davy) .

4.5 Biological control

Biological control, or the use of predators, parasites (including parasitoids), or pathogens to reduce pest populations, has had limited adoption in turfgrass insect management in the USA (Grewal, 1999), despite low probability of adverse environmental impacts and the potential for long-term suppression of target pests. Their expense relative to synthetic insecticides and survivability during shipment and storage put biological controls at a disadvantage to chemical controls and creates imbalance between supply and demand in the U.S. market.

Turfgrass managers are more likely to integrate biological controls in countries or regions where fewer chemical control options are available or are restricted altogether. Synthetic pesticides have been banned on school grounds to some degree in most states within the USA and entirely within Connecticut and New York (Grant, 2011). Feeding by subterranean pests such as white grubs and European crane flies creates challenges for maintaining not only the appearance but functional characteristics for a sports field in these areas. Entomopathogenic nematodes (EPNs) have been shown to be effective in managing both pests in these systems, though their impact is often variable and their persistence minimal given intense traffic. *Steinernema feltiae* Filipjev and *Heterorhabditis bacteriophora* Poinar have been shown to be effective in controlling third instar Japanese beetle larvae in athletic fields with sand rootzones (Helmberger et al., 2018), which may suggest that traffic impedes movement and survival in finer textured soils. In finer textured soils, the addition of a soil surfactant may improve the uniformity of soil moisture in the profile and potential dispersion of nematodes without affecting EPN survival (Hoctor et al., 2013; McGraw and Schlossberg, 2017).

Microbial based insecticides and related products represent less than 1% of the chemical control agents applied on golf courses, sports fields, and home lawns (Grewal, 1999). Of these, several fungal and bacterial products have shown promise in controlling key turfgrass insect pests while potentially affecting multiple life stages. *Metarhizium brunneum* (Petch) Strain F52 effectively controls a wide array of white grubs, including Japanese beetle, June beetles (*Phyllophaga* spp.), and masked chafers (*Cyclocephala* spp.) when applied when larvae are present, though adult beetles can also become infected as they enter the soil to lay eggs (Behle and Goett, 2016). Similarly, a commercial product containing *Bacillus thuringiensis galleriae* Berliner ('GrubGone' Phyllom BioProducts, Oakland, CA, USA) has also shown curative activity when applied against Japanese beetle larvae (Shetlar and Andon, 2015) as well as reduces adult feeding on hosts such as *Rosa* and *Tilia* spp. (Redmond et al., 2020). Though the product is primarily marketed for control of white grubs, the inconsistent field performance of the product (Hubble-Wirgler and Sutherland, 2017) and cost relative to preventive insecticides limit its current adoption.

Microbial insecticides may often provide narrow or species-specific control. This directed control may be advantageous from an environmental standpoint but a detriment when attempting to control mixed species complexes such as white grub larvae or caterpillar pests. Baculoviruses, such as nuclear polyhedrosis viruses (NPVs), have been shown to provide targeted control of caterpillar pests in turfgrass (Perelle and Harper, 1986) and other cropping systems with little potential for negatively affecting non-target insects. Additional benefits include the ability of virus particles to persist in the environment to infect future generations of pests. Infected caterpillars and larvae often suspend limp in an upside-down 'V' shape, with exoskeleton perforations allowing for viral polyhedral to cascade the pest host plant tissues to be consumed or contact with uninfected caterpillars and other pests. Black cutworms (*Agrotis ipsilon* Hufnagel) are naturally infected by a host-specific baculovirus *A. ipsilon* multicapsid nucleopolyhedrovirus (AgipMNPVs) (Prater et al., 2006). Cutworm mortality ranged from 75% to 90% among early instars 7 days postapplication on golf courses (Bixby-Brosi and Potter, 2010). Later instars may still cause feeding damage for a few days despite infection but ultimately succumb to the virus. Advantages of NPVs are compatible with other biological agents, alkaloid-producing endophytes, and chitin synthesis-inhibitors fungicides applications (Bixby-Brosi and Potter, 2012). Field applications of AgipMNPVs are subjected to ultraviolet degradation; however, the addition of adjuvants can prolong viral infectivity in the field.

4.6 Cultural control

Cultural controls, or management practices that directly or indirectly reduce pest development, survival, or dispersal, represent a non-toxic alternative to chemical controls and sustainable means of controlling insect pests (Kogan, 1998). Strategies can include manipulating management practices such as mowing, fertility, and irrigation or the selection of plant material that is either resistant or tolerant to pest attack.

Manipulating mowing practices has been examined for both its direct effects on pest removal and its indirect benefits to increased feeding tolerance (Potter et al., 1996; Williamson and Potter, 1997). Insects that inhabit close mown putting surfaces found on golf courses are likely to be frequently impacted by mowing. ABW larvae are disproportionately found in mowing heights less than 1.25 cm, though damage is rarely detected in putting surfaces maintained below 3 mm. Investigations into this phenomenon have shown that moderate percentages of ABW adults are removed through greens-height mowing, which might explain the uncommon occurrence of larval feeding damage (Czyzewski and McGraw, 2017). However, females could oviposit and larvae developed within turf maintained less than 3 mm, indicating that ABW may still damage low mown putting surfaces. Tracking ABW adults through fluorescent marks indicated that canopy activity was greatest between 15°C and 20°C (Czyzewski and McGraw, 2018), indicating that management practices in spring may need to be scheduled accordingly to improve removal from greens-height putting surfaces.

Multiple studies have attempted to reduce subterranean white grub populations through a multitude of cultural practices, including reducing irrigation during oviposition, applying inorganic fertilizers, and toughening or reducing the palatability of turfgrasses with silicon amendments (Potter et al., 1996). These efforts have been largely unsuccessful in providing meaningful suppression. Core cultivation is another such cultural effort that until recently had only been examined in non-replicated small plot trials and computer simulation. Computer modeling suggests that moderate-to-high grub reductions may be achieved through one-time mechanical disruption of the soil rootzone (Cranshaw and Zimmerman, 1989; Blanco-Montero and Hernandez, 1995). However, recent studies that examined the use of modern equipment with similar tine patterns would suggest otherwise (McGraw et al., 2021b). Only minimal (<10%) control was observed when single aerification events were employed over a range of white grub densities in field trials and computer simulations. Increasing the number of cultivation events improved the level and reliability of suppression but not to a level of control that meets or exceeds efficacy from chemical insecticides or expectations by managers of fine turf.

4.7 Host plant tolerance

Tolerance, or the ability of a plant to withstand herbivory, is viewed as a form of host plant resistance. Incorporating tolerant plants into turfgrass stands can be viewed as a non-chemical management practice that reduces damage, though it may not reduce pest densities.

Annual bluegrass is believed to be the preferred host plant of ABW, though the weevil is clearly capable of developing within *Agrostis* spp., including creeping bentgrasses (Kostromytska and Koppenhöfer, 2014). Field studies would suggest that creeping bentgrass is more tolerant and possibly recovers quicker than annual bluegrass (McGraw and Koppenhöfer, 2010). Additionally, ABW feeding damage within creeping bentgrass has been observed to be reduced by up to 42% compared to annual bluegrass (Kostromytska and Koppenhöfer, 2016). Female ABW oviposition injury and the number of eggs in addition to increase in larval mortality were also observed to be significantly less in *Agrostis* spp. and cultivars, suggesting that antixenosis and antibiosis are involved in host plant resistance to ABW within bentgrasses (Kostromytska and Koppenhöfer, 2014).

4.8 Host plant resistance

Developing plant stock that is resistant to insect herbivory is an effective management tool in most cropping systems, though it is the one that is rarely used in turfgrass management. The maintenance of high-performing turfgrass landscapes such as golf courses, athletic fields, and home lawns had led to the extensive use of fertilizers, increased irrigation requirements, and increased pest pressure. Reducing these inputs remains the focus of many university breeding programs, though selection for improved abiotic characteristics (i.e. heat and drought tolerance) and improved disease resistance is often of greater importance than selection for improved insect resistance. Investments in low-maintenance turfgrasses with high traffic and pest resilience are needed, though increasing the public demand for these sustainable turfgrasses across diverse environments and regions is a challenge for breeders and industry professionals.

Fine fescues (*Festuca* spp.) require low amounts of pesticides, fertilizer, and water inputs across various environmental conditions (i.e. low pH, shade, high salinity, and drought) and therefore are often a candidate for breeding improved pest tolerance (Bonos and Huff, 2013; Bertin et al., 2009). Enhanced resistance to abiotic and biotic stressor and low-put management are key characteristics currently being selected for across fescues (Braun et al., 2020). Chinch bugs (*Blissus* spp.) are arguably the biggest pests of fine fescues, though they can also be observed damaging turfgrasses such as bermudagrass, bentgrass, Kentucky bluegrass, tall fescue, perennial ryegrass, and zoysiagrass across the area from Southern Canada to Mexico (Vittum, 2020).

Warm-season turfgrasses have a wide range of levels of resistance to chinch bug feeding even within species. The bermudagrass cultivar'Mini Verde' displays high levels of tolerance compared to 'Tifway 419' and 'Jackpot' which have varying levels of susceptibility and resistance to chinch bug damage (Anderson et al., 2006a). St. Augustine grasses are generally known to provide control against southern and western chinch bugs by producing higher concentrations of oxidative enzymes in plant tissue after insect feeding, which reduces tissue damage (Rangasamy et al., 2009a). 'Captiva' St. Augustine grass resists southern chinch bug feeding because it has heavily sclerotized cells that the insects are unable to penetrate with their mouthparts (Anderson et al., 2006b; Rangasamy et al., 2009b). Bermudagrass (*Cynodon* spp.) cultivars such as 'Mini Verde' display levels of tolerance to chinch bug damage (Anderson et al., 2006a). Zoysiagrass (*Zoysia* spp.) have been a popular residential and commercial landscape for decades, with fine textured cultivars of *Zoysia matrella*(L.) Merr. increasing in demand in golf courses due to the slow growth and low fertilizer requirements. Zoysiagrass breeding and screening has been championed for its tolerance for hunting billbug, mole crickets, and sting nematode(*Belonolaimus longicaudatus* Steiner) (Reinert et al., 2011).

Endophytes (E+) are omnipresent across ecosystems regulating plant-microbe interactions under biotic stressors. Previous research on root E+ suggests that they may have the ability to both enhance tolerance (Barazani et al., 2005) or suppress inducible defenses against foliar-feeding herbivores (Zhou et al., 2016). Cultivars of ryegrass (*Lolium* spp.) and fescues (*Festuca* spp.) have mutualistic relationships with alkaloid-producing fungal E+ (Potter et al., 2008). Alkaloid by-products of E+ turfgrass cultivars are used as defense compound agents against primarily aboveground feeding insects (Koppenhöfer and Fuzy, 2003; Koppenhöfer et al., 2003). Tall fescue inoculated with E+ has been observed to decrease larval weights and reduce overall damage from the European crane fly (Petersen and Peck, 2013). Endophytes promote root regeneration and growth after herbivory stress of European chafer (Gan et al., 2017). Applications of pathogenic strains of E+ such as *Beauveria brongniartii* (Saccardo) Petch have been observed to suppress white grubs (Choo et al., 2002), and *B. bassiana* (Bals.-Criv.) Vuill. and *Epichloë (Neotyphodium) coenophialum* (Morgan-Jones & W.Gams) C.W. Bacon & Schardl have the same effect on black cutworms on tall fescue sports fields and golf courses (Baldauf et al., 2011; Gosselin et al., 2009).

Closely related insects may respond differently to feeding on E+ turfgrasses. Endophyte-enriched fine fescues are moderately resistant to western chinch bugs; however, regardless of endophyte presence they are highly susceptible to hairy chinch bugs. The highly variable nature of tolerance among turfgrasses highlights the importance of selecting resilient turf varieties to reduce injury without chemical control agents (Heng-Moss et al., 2003).

4.9 Plant growth-promoting rhizobacteria

Rhizobacteria are bacteria that form symbiotic relationships with plants that can be pathogenic, symbiotic, or neutral. Plant growth-promoting rhizobacteria (PGPRs) are beneficial bacterial symbionts that inhabit the rhizosphere (Backer et al., 2018). Turf roots with endophytic PGPRs provide exudate exchange for protection from plant pathogens and secure subterranean niches for growth and root expansion. Beneficial PGPR bacterial genera include *Azotobacter*, *Arthrobacter*, *Bacillus*, *Clostridium*, *Hydrogenophaga*, *Enterobacter*, *Serratia*, *Azospirillum*, *Paenibacillus*, *Pseudomonas*, and *Rhizobium*; however, some species within the aforementioned genera have been observed to be pathogenic (Benizri et al., 2001). Soil microbes such as PGPR's associates modulate turfgrass plant tissues to deter insect herbivores by eliciting inducible defense phytohormones such as jasmonic acid, salicylic acid, or ethylene, in addition to releasing organic volatile compounds that attract parasitoids and other nature enemies when herbivory or oviposition injury is incurred by the host plant (Ngumbi, 2011; Wilson and Faeth, 2001). FAW oviposition and larval development has been observed to be reduced and, in some cases, arrested within bermudagrass inoculated with PGPRs (Coy, 2014). The use of PGPR as biological control continues to be researched as a potential management strategy in efforts to reduce chemical control inputs within turfgrass systems.

5 Conclusion

The use of technology is likely to play a larger role in pest surveillance and delivery of controls in turfgrass management in the near future (Table 1). Pest scouting is labor intensive and often not performed due to relatively low cost and effectiveness of preventive insecticides. Pesticide restrictions or

Table 1 Summary of trends in insect management in turfgrass ecosystems

- Major pest groups continue to emerge in new regions across North America.
- Recent studies have improved our understanding of the life histories and ecology of major turfgrass insect pests.
- Turfgrass systems have traditionally been challenged by exotic insects. Invasive insect pests require careful examination following introduction to assess how environmental conditions impact behavior and population development.
- Recent studies in insect pest control emphasize the desire to reduce chemical insecticides use in perennial turfgrass systems.
- Despite numerous advances in pest control strategies, chemical management dominates pest control in the USA.
- Alternative strategies to control insects in turfgrass with chemicals (e.g. cultural and biological controls) in the USA must not only be as efficacious but be affordable and easily adoptable.

the removal of preventive insecticide classes in turf would create a greater need for determining when and where insect outbreaks are likely to occur. Precision agriculture is currently at the cusp of integrating 'big data' into pest management decision-making, but data collection currently outpaces our ability to interpret the meaning of the data and turn it into action. The use of drones and autonomous traps is becoming more commonplace in agriculture (Preti et al., 2021), though their adoption in turfgrass management lags agriculture. Given the esthetic demands on turfgrass pest management, autonomous pest surveillance will require the development of reliable technologies that deliver real-time alerts. For precision agriculture, such as drone-deployed controls, to become a reality in turfgrass management, these technologies must reduce labor compared to current efforts and require little technical expertise to operate.

There is a great need for the development of insect-resistant turfgrass cultivars, but the development and delivery of such innovations have not been realized in the twenty-first century. This may be in part due to the greater emphasis on abiotic traits (i.e. heat and drought tolerance) and disease tolerance. Current research into turfgrass defense mechanisms against insect herbivory may be a means around traditional plant selection and breeding to develop resilient turf. Plants have evolved constitutive or physical structures (i.e. trichomes) and inducible defenses such as secondary metabolite production (i.e. phytohormones) to deter or manage insect pests. The identification of inducible defense mechanisms elicited by biotic stressors, specifically insects within turfgrass ecosystems, is complex and not well understood. Phytohormones regulate essential physiology, including growth, defense against abiotic and biotic stresses, reproduction, and development. Mechanical damage elicited by insects induces phytohormone responses through herbivory or oviposition, primarily jasmonic acid and ethylene. Salicylic acid traditionally associated with pathogen defense is antagonistic to jasmonic acid; however, insects can activate the salicylic acid pathway via frass, saliva, and other extractions deposited in foliar wounds (Dafoe et al., 2013; Schwartzberg and Tumlinson, 2014). Insect feeding on a host plant activates transductive signals, resulting in a defense response in the form of jasmonic acid and/or salicylic acid (Mewis et al., 2005; Goggin, 2007; Louis and Shah, 2013). The characterization of secondary metabolites like phytohormones and volatile organic compounds within turfgrass cultivars have been hypothesized to enhance tolerance and/or promote resistance to a variety of turfgrass pests. The application of synthetic or upregulating phytohormones has the potential to enhance current pest management strategies and promote host plant resistance in addition to reducing applications and reliance on chemical control agents.

Insecticide resistance management is likely to impact the future management of insect pests in more regions across the country and likely in countries where few active ingredients are permitted. Understanding the mechanisms responsible for resistance development is essential for managing key pests that have already developed tolerance to insecticides. Ongoing research for both FAW and ABW has sought to characterize the contribution that the insect gut microbiome, in particular bacterial symbionts, can contribute to the development of resistance. Common relationships include the endosymbiont providing metabolic support, enabling access to an otherwise inaccessible resource or synthesizing essential nutrients for the insect host (Brune, 2014; Baumann, 2005; Wilson and Duncan, 2015), protection to its insect host against pathogens and parasitoids (Brownlie and Johnson, 2009; Bracewell and Six, 2015; Acevedo et al., 2017; Chu et al., 2013; Oliver et al., 2003), or those symbionts which protect the insect host by degrading toxins. Given the ability of some endosymbionts to enable their insect host to overcome the defensive toxins of plants, it is not surprising that some symbionts would do the same for synthetic insecticides especially when the insecticide will act as a strong selective pressure on both the pest insect and its symbionts. Several insecticide-resistant lines of FAW have been shown to degrade insecticides of several classes, with each resistant line of armyworm having a different dominant bacterial and microbial community (Almeida et al., 2017). While the commonly discussed mechanisms of resistance are ultimately based on the insect's genome, microbially mediated insecticide resistance comes from the endosymbiont's genome, with potentially no change in the host insect. Whereas standard resistance will require generations of insects exposed to the insecticide to become predominantly established, insecticide-degrading bacteria can potentially develop outside of the host and establish within a single generation through horizontal transfer, possibly even interspecies. However, this novel mechanism of resistance presents a unique opportunity to potentially target the symbionts as a way of managing insecticide resistance (Akami et al., 2019).

Turfgrass is a relatively young cropping system, and the diverse pests within these managed grassland systems continue to adapt to the changes in environmental conditions and management practices. Although turfgrass practitioners have many tactics to manage insect pests, chemical inputs continue to be the predominant means of control in the USA. The focus on chemical controls has hindered our understanding of pest ecology in these environments and emphasizes the need for continual discovery of new active ingredients by the agrochemical industry. However, fewer agricultural chemical companies are involved in the discovery of new or novel compounds, and the time to deliver new products to the market has become longer and increasingly more expensive. Furthermore, few university faculty positions devoted to studying turfgrass entomology remain, which limits the discovery of novel and innovative, non-chemical approaches to insect control. The advancement of turfgrass entomological science will require leveraging partnerships in the wider agricultural world to overcome environmental, regulatory, and economic challenges that are likely ahead.

6 Where to look for further information

- Vittum, P. J. (2020). *Turfgrass Insects of the United States and Canada*. Cornell University Press.

7 References

Acevedo, F. E., Peiffer, M., Tan, C. W., Stanley, B. A., Stanley, A., Wang, J., Jones, A. G., Hoover, K., Rosa, C., Luthe, D. and Felton, G. (2017). Fall armyworm-associated gut bacteria modulate plant defense responses. *Mol. Plant Microbe Interact.* 30(2), 127–137.

Adesanya, A. W., Held, D. W. and Liu, N. (2017). Geranium intoxication induces detoxification enzymes in the Japanese beetle, *Popillia japonica* Newman. *Pestic. Biochem. Physiol.* 143, 1–7.

Akami, M., Njintang, N. Y., Gbaye, O. A., Andongma, A. A., Rashid, M. A., Niu, C. Y. and Nukenine, E. N. (2019). Gut bacteria of the cowpea beetle mediate its resistance to dichlorvos and susceptibility to *Lippia adoensis* essential oil. *Sci. Rep.* 9(1), 6435.

Almeida, L. G. D., Moraes, L. A., Trigo, J. R., Omoto, C. and Consoli, F. L. (2017). The gut microbiota of insecticide-resistant insects houses insecticide-degrading bacteria: a potential source for biotechnological exploitation. *PLoS ONE* 12(3), e0174754.

Anders, C. (2021). Battalions of armyworms are chomping up fields across the nation - sometimes overnight. The Washington Post, WP company, 17 Sept. 2021, Available at: https://www.washingtonpost.com/nation/2021/09/17/armyworms-agriculture-outbreak-farming/.

Anderson, W. G., Heng-Moss, T. M., Baxendale, F. P., Baird, L. M., Sarath, G. and Higley, L. (2006a). Chinch bug (Hemiptera: Blissidae) mouthpart morphology, probing frequencies, and locations on resistant and susceptible germplasm. *J. Econ. Entomol.* 99(1), 212–221.

Anderson, W. G., Heng-Moss, T. M. and Baxendale, F. P. (2006b). Evaluation of cool-and warm-season grasses for resistance to multiple chinch bug (Hemiptera: Blissidae) species. *J. Econ. Entomol.* 99(1), 203-211.

Backer, R., Rokem, J. S., Ilangumaran, G., Lamont, J., Praslickova, D., Ricci, E., Subramanian, S. and Smith, D. L. (2018). Plant growth-promoting rhizobacteria: context, mechanisms of action, and roadmap to commercialization of biostimulants for sustainable agriculture. *Front. Plant Sci.* 9, 1473.

Baldauf, M. W., Mace, W. J. and Richmond, D. S. (2011). Endophyte-mediated resistance to black cutworm as a function of plant cultivar and endophyte strain in tall fescue. *Environ. Entomol.* 40(3), 639-647.

Barazani, O., Benderoth, M., Groten, K., Kuhlemeier, C. and Baldwin, I. T. (2005). *Piriformospora indica* and *Sebacina vermifera* increase growth performance at the expense of herbivore resistance in *Nicotiana attenuata*. *Oecologia* 146(2), 234-243.

Barrett, B. A., Patterson, M. E. and Xiong, X. (2018). Behavioral responses of Hunting Billbug (Coleoptera: Curculionidae) towards odors from different turfgrass species and adult conspecifics. *Environ. Entomol.* 47(6), 1541-1546.

Baumann, P. (2005). Biology of bacteriocyte-associated endosymbionts of plant sap-sucking insects. *Annu. Rev. Microbiol.* 59(1), 155-189.

Behle, R. W. and Goett, E. J. (2016). Dosage response mortality of Japanese beetle, masked chafer, and june beetle (Coleoptera: Scarabaeidae) adults when exposed to experimental and commercially available granules containing *Metarhizium brunneum*. *J. Econ. Entomol.* 109(3), 1109-1115.

Benizri, E., Baudoin, E. and Guckert, A. (2001). Root colonization by inoculated plant growth-promoting rhizobacteria. *Biocontrol Sci. Technol.* 11(5), 557-574.

Bertin, C., Senesac, A. F., Rossi, F. S., DiTommaso, A. and Weston, L. A. (2009). Evaluation of selected fine-leaf fescue cultivars for their turfgrass quality and weed suppressive ability in field settings. *hortTechnology* 19(3), 660-668.

Bixby-Brosi, A. J. and Potter, D. A. (2010). Evaluating a naturally occurring baculovirus for extended biological control of the black cutworm (Lepidoptera: Noctuidae) in golf course habitats. *J. Econ. Entomol.* 103(5), 1555-1563.

Bixby-Brosi, A. J. and Potter, D. A. (2012). Can a chitin-synthesis-inhibiting turfgrass fungicide enhance black cutworm susceptibility to a baculovirus? *Pest Manag. Sci.* 68(3), 324-329.

Blanco-Montero, C. A. and Hernandez, G. (1995). Mechanical control of white grubs (Coleoptera: Scarabidae) in turfgrass using aerators. *Environ. Entomol.* 24(2), 243-245.

Bonos, S. A. and Huff, D. R. (2013). Cool-season grasses: biology and breeding. *Turfgrass Biol. Use Manag.* 56, 591-660.

Bracewell, R. R. and Six, D. L. (2015). Experimental evidence of bark beetle adaptation to a fungal symbiont. *Ecol. Evol.* 5(21), 5109-5119.

Braun, R. C., Patton, A. J., Watkins, E., Koch, P. L., Anderson, N. P., Bonos, S. A. and Brilman, L. A. (2020). Fine fescues: a review of the species, their improvement, production, establishment, and management. *Crop Sci.* 60(3), 1142-1187.

Brownlie, J. C. and Johnson, K. N. (2009). Symbiont-mediated protection in insect hosts. *Trends Microbiol.* 17(8), 348-354.

Brune, A. (2014). Symbiotic digestion of lignocellulose in termite guts. *Nat. Rev. Microbiol.* 12(3), 168-180.

Carvalho, R. A., Omoto, C., Field, L. M., Williamson, M. S. and Bass, C. (2013). Investigating the molecular mechanisms of organophosphate and pyrethroid resistance in the fall armyworm *Spodoptera frugiperda*. *PLoS ONE* 8(4), e62268.

Chen, C. C. and Chen, Y. R. (2007). Study on laminar hydathodes of *Ficus formosana* (Moraceae). Salt injury of guttation on hydathodes. *Bot. Stud.* 48(2), 215-226.

Cherry, R. and Nagata, R. (2005). Development of resistance in southern chinch bugs (Hemiptera: Lygaeidae) to the insecticide bifenthrin. *Fla. Entomol.* 88(2), 219-221.

Choo, H. Y., Kaya, H. K., Huh, J., Lee, D. W., Kim, H. H., Lee, S. M. and Choo, Y. M. (2002). Entomopathogenic nematodes (*Steinernema* spp. and *Heterorhabditis bacteriophora*) and a fungus *Beauveria brongniartii* for biological control of the white grubs, *Ectinohoplia rufipes* and *Exomala orientalis*. *BioControl* 47(2), 177-192.

Chu, C. C., Spencer, J. L., Curzi, M. J., Zavala, J. A. and Seufferheld, M. J. (2013). Gut bacteria facilitate adaptation to crop rotation in the western corn rootworm. *Proc. Natl. Acad. Sci. U. S. A.* 110(29), 11917-11922.

Coy, R. M. (2014). Potential of plant growth-promoting rhizobacteria (PGPR) as a biological control agent against warm-season turfgrass pests (Doctoral dissertation).

Cranshaw, W. S. and Zimmerman, R. J. (1989). Biological, mechanical, and chemical control of turfgrass-infesting scarabs in Colorado. *Southwest. Entomol.* 114(4), 351-355.

Czyzewski, B. D. and McGraw, B. A. (2017). Mowing height influences *Listronotus maculicollis* (Coleoptera: Curculionidae) oviposition behavior and mechanical removal from golf course putting greens, but not larval development. *J. Econ. Entomol.* 110(5), 2165-2171.

Czyzewski, B. D. and McGraw, B. A. (2018). Detection of *Listronotus maculicollis* (Coleoptera: Curculionidae) turfgrass canopy activity with the use of a novel fluorescent marking system suggests opportunities for improved mechanical control. *Environ. Entomol.* 47(6), 1525-1531.

Dafoe, N. J., Thomas, J. D., Shirk, P. D., Legaspi, M. E., Vaughan, M. M., Huffaker, A., Teal, P. E. and Schmelz, E. A. (2013). European corn borer (*Ostrinia nubilalis*) induced responses enhance susceptibility in maize. *PLoS ONE* 8(9), e73394.

Desneux, N., Decourtye, A. and Delpuech, J. M. (2007). The sublethal effects of pesticides on beneficial arthropods. *Annu. Rev. Entomol.* 52, 81-106.

Díaz-Álvarez, E. A., Martínez-Zavaleta, J. P., López-Santiz, E. E., de la Barrera, E., Larsen, J. and del-Val, E. (2020). Climate change can trigger fall armyworm outbreaks: a developmental response experiment with two Mexican maize landraces. *Int. J. Pest Manag.*, 1-9.

Doskocil, J. P. and Brandenberg, R. L. (2012). Hunting billbug (Coleoptera: Curculionidae) life cycle and damaging life stage in North Carolina, with notes on other billbug species abundance. *J. Econ. Entomol.* 105(6), 2045-2051.

Duffy, A. G., Hughes, G. P., Ginzel, M. D. and Richmond, D. S. (2018a). Volatile and contact chemical cues associated with host and mate recognition behavior of *Sphenophorus venatus* and *Sphenophorus parvulus* (Coleoptera: Dryophthoridae). *J. Chem. Ecol.* 44(6), 556-564.

Duffy, A. G., Powell, G. S., Zaspel, J. M. and Richmond, D. S. (2018b). Billbug (Coleoptera: Dryophthoridae: Sphenophorus spp.) seasonal biology and DNA-based life stage association in Indiana turfgrass. *J. Econ. Entomol.* 111(1), 304-313.

Duffy, A. G. and Richmond, D. S. (2020). Hunting Billbug *Sphenophorus venatus* (Coleoptera: Dryophthoridae) adult feeding and attraction to warm-and cool-season turfgrasses. *Gr. Lakes Entomol.* 53(1), 8.

Dupuy, M. M. and Ramirez, R. A. (2016). Biology and management of billbugs (Coleoptera: Curculionidae) in Turfgrass. *J. Inte. Pest Manag.* 7(1), 6.

Dupuy, M. M., Powell, J. A. and Ramirez, R. A. (2017). Developing a degree-day model to predict billbug (Coleoptera: Curculionidae) seasonal activity in Utah and Idaho turfgrass. *J. Econ. Entomol.* 110(5), 2180–2189.

Gan, H., Churchill, A. C. L. and Wickings, K. (2017). Invisible but consequential: root endophytic fungi have variable effects on belowground plant-insect interactions. *Ecosphere* 8(3), e01710.

Girolami, V., Mazzon, L., Squartini, A., Mori, N., Marzaro, M., Di Bernardo, A., Greatti, M., Giorio, C. and Tapparo, A. (2009). Translocation of neonicotinoid insecticides from coated seeds to seedling guttation drops: a novel way of intoxication for bees. *J. Econ. Entomol.* 102(5), 1808–1815.

Goggin, F. L. (2007). Plant-aphid interactions: molecular and ecological perspectives. *Curr. Opin. Plant Biol.* 10(4), 399–408.

Gosselin, M. -E., Belair, G., Simard, L. and Brodeur, J. (2009). Toxicity of Spinosad and *Beauveria bassiana* to the black cutworm, and the additivity of sublethal doses. *Biocontrol Sci. Technol.* 19(2), 201–217.

Goulson, D., Lye, G. C. and Darvill, B. (2008). Decline and conservation of bumble bees. *Annu. Rev. Entomol.* 53, 191–208.

Grant, J. A. (2011). The child safe playing fields act: NY's ban on pesticide use on school and day care center grounds. *Cornell Univ. Turfgrass Times* 22, 1–4.

Grewal, P. S. (1999). Factors in the success and failure of microbial control in turfgrass. *Integr. Pest Manag. Rev.* 4(4), 287–294.

Hatfield, R. G., Strange, J. P., Koch, J. B., Jepsen, S. and Stapleton, I. (2021). Neonicotinoid pesticides cause mass fatalities of native bumble bees: a case study From Wilsonville, Oregon, United States. *Environ. Entomol.* 50(5), 1095–1104.

Haydu, J. J., Hodges, A. W. and Hall, C. R. (2006). Economic impacts of the turfgrass and lawncare industry in the United States. *EDIS* 2006(7), Available at: edis.ifas.ufl.edu/pdffiles/FE/FE63200.pdf.

Helmberger, M. S., Thaler, J. S., Shields, E. J. and Wickings, K. G. (2018). Entomopathogenic nematode performance against *Popillia japonica* (Coleoptera: Scarabaeidae) in school athletic turf: effects of traffic and soil properties. *Biol. Control* 126, 177–184.

Heng-Moss, T. M., Baxendale, F. P., Riordan, T. P., Young, L. and Lee, K. (2003). Chinch bug-resistant buffalograss: an investigation of tolerance, antixenosis, and antibiosis. *J. Econ. Entomol.* 96(6), 1942–1951.

Hites, R. A. (2021). The rise and fall of chlorpyrifos in the United States. *Environ. Sci. Technol.* 55(3), 1354–1358.

Hoctor, T. L., Gibb, T. J., Bigelow, C. A. and Richmond, D. S. (2013). Survival and infectivity of the insect-parasitic nematode *Heterorhabditis bacteriophora* Poinar in solutions containing four different turfgrass soil surfactants. *Insects* 4(1), 1–8.

Hoffmann, E. J. and Castle, S. J. (2012). Imidacloprid in melon guttation fluid: a potential mode of exposure for pest and beneficial organisms. *J. Econ. Entomol.* 105(1), 67–71.

Hong, S. C., Obear, G. R., Liesch, P. J., Held, D. W. and Williamson, R. C. (2015). Suitability of creeping bentgrass and Bermudagrass cultivars for black cutworms and fall armyworms (Lepidoptera: Noctuidae). *J. Econ. Entomol.* 108(4), 1954–1960.

Hubble-Wirgler, C. W. and Sutherland, A. M. (2017). Evaluation of a Granular Formulation of *Bacillus thuringiensis galleriae* for use against white grubs within municipal turf, 2016. *Arthropod Manag. Tests* 42(1), tsx119.

Joachimsmeier, I., Pistorius, J., Heimbach, U., Schenke, D., Kirchner, W. and Zwerger, P. (2012). Frequency and intensity of guttation events in different crops in Germany. *Julius Kühn-Archiv* 437, 87.

Jactel, H., Verheggen, F., Thiéry, D., Escobar-Gutiérrez, A. J., Gachet, E., Desneux, N. and Neonicotinoids Working Group (2019). Alternatives to neonicotinoids. *Environ. Int.* 129, 423–429.

Kogan, M. (1998). Integrated pest management: historical perspectives and contemporary developments. *Annu. Rev. Entomol.* 43, 243–270.

Koppenhöfer, A. M., Cowles, R. S. and Fuzy, E. M. (2003). Effects of turfgrass endophytes (Clavipitaceae: ascomycetes) on white grub (Coleoptera: Scarabaeidae) larval development and field populations. *Environ. Entomol.* 32(4), 895–906.

Koppenhöfer, A. M. and Fuzy, E. M. (2003). Effects of turfgrass endophytes (Clavicipitaceae: ascomycetes) on white grub (Coleoptera: Scarabaeidae) control by the entomopathogenic nematode *Heterorhabditis bacteriophora* (Rhabditida: Heterorhabditidae). *Environ. Entomol.* 32(2), 392–396.

Kostromytska, O. S. and Koppenhöfer, A. M. (2014). Ovipositional preferences and larval survival of annual bluegrass weevil, *Listronotus maculicollis*, on *Poa annua* and selected bentgrasses (Agrostis spp.). *Entomol. Exp. Appl.* 152(2), 108–119.

Kostromytska, O. S. and Koppenhöfer, A. M. (2016). Responses of *Poa annua* and three bentgrass species (Agrostis spp.) to adult and larval feeding of annual bluegrass weevil, *Listronotus maculicollis* (Coleoptera: Curculionidae). *Bull. Entomol. Res.* 106(6), 729–739.

Kostromytska, O. S., Rodriguez-Saona, C., Alborn, H. T. and Koppenhöfer, A. M. (2018a). Role of plant volatiles in host plant recognition by *Listronotus maculicollis* (Coleoptera: Curculionidae). *J. Chem. Ecol.* 44(6), 580–590.

Kostromytska, O. S., Wu, S. and Koppenhöfer, A. M. (2018b). Cross-resistance patterns to insecticides of several chemical classes among *Listronotus maculicollis* (Coleoptera: Curculionidae) populations with different levels of resistance to pyrethroids. *J. Econ. Entomol.* 111(1), 391–398.

Larson, J. L., Dale, A., Held, D., McGraw, B., Richmond, D. S., Wickings, K. and Williamson, R. C. (2017). Optimizing pest management practices to conserve pollinators in turf landscapes: current practices and future research needs. *J. Integr. Pest Manag.* 8(1), 1–10.

Larson, J. L., Redmond, C. T. and Potter, D. A. (2012). Comparative impact of an anthranilic diamide and other insecticidal chemistries on beneficial invertebrates and ecosystem services in turfgrass. *Pest Manag. Sci.* 68(5), 740–748.

Larson, J. L., Redmond, C. T. and Potter, D. A. (2013). Assessing insecticide hazard to bumble bees foraging on flowering weeds in treated lawns. *PLoS ONE* 8(6), e66375.

Larson, J. L., Redmond, C. T. and Potter, D. A. (2014). Impacts of a neonicotinoid, neonicotinoid-pyrethroid premix, and anthranilic diamide insecticide on four species of turf-inhabiting beneficial insects. *Ecotoxicology* 23(2), 252–259.

Larson, J. L., Redmond, C. T. and Potter, D. A. (2015). Mowing mitigates bioactivity of neonicotinoid insecticides in nectar of flowering lawn weeds and turfgrass guttation. *Environ. Toxicol. Chem.* 34(1), 127–132.

Louis, J. and Shah, J. (2013). Arabidopsis thaliana–*Myzus persicae* interaction: shaping the understanding of plant defense against phloem-feeding aphids. *Front. Plant Sci.* 4, 213.

Mewis, I., Appel, H. M., Hom, A., Raina, R. and Schultz, J. C. (2005). Major signaling pathways modulate Arabidopsis glucosinolate accumulation and response to both phloem-feeding and chewing insects. *Plant Physiol.* 138(2), 1149–1162.

McCurdy, J. D., Held, D. W., Gunn, J. M. and Barickman, T. C. (2017). Dew from warm-season turfgrasses as a possible route for pollinator exposure to lawn-applied Imidacloprid. *Crop Forage Turfgrass Manag.* 3(1), 1–6.

McGraw, B. A. and Koppenhöfer, A. M. (2010). Spatial distribution of colonizing *Listronotus maculicollis* populations: implications for targeted management and host preference. *J. Appl. Entomol.* 134(4), 275–284.

McGraw, B. A. and Koppenhöfer, A. M. (2015). Spatial analysis of *Listronotus maculicollis* immature stages demonstrates strong associations with conspecifics and turfgrass damage but not with optimal hosts on golf course fairways. *Entomol. Exp. Appl.* 157(3), 307–316.

McGraw, B. A. and Koppenhöfer, A. M. (2017). A survey of regional trends in annual bluegrass weevil (Coleoptera: Curculionidae) management on golf courses in Eastern North America. *J. Integr. Pest Manag.* 8(1), 1–11.

McGraw, B. A., Price, G., Simard, A. and Trotta, K. (2021b). Empirical and theoretical assessment of mechanical and injection cultivation methods to manage white grubs in turfgrass. *Intl. Turfgrass Soc. Res. J.* 14(1), 823–830.

McGraw, B. A., Price, G. Y., Simard, A. and Vittum, P. J. (2021a). Reproductive phenology and feeding patterns of *Listronotus maculicollis* during spring emergence: implications for spring management. *Crop Sci.* 61(5), 3197–3205.

McGraw, B. A., Rodriguez-Saona, C., Holdcraft, R., Szendrei, Z. and Koppenhöfer, A. M. (2011). Behavioral and electrophysiological responses of *Listronotus maculicollis* (Coleoptera: Curculionidae) to volatiles from intact and mechanically damaged annual bluegrass. *Environ. Entomol.* 40(2), 412–419.

McGraw, B. A. and Schlossberg, M. J. (2017). Fine-scale spatial analysis of soil moisture and entomopathogenic nematode distribution following release in wetting agent-treated turf. *Appl. Soil Ecol.* 114, 52–61.

Milesi, C., Running, S. W., Elvidge, C. D., Dietz, J. B., Tuttle, B. T. and Nemani, R. R. (2005). Mapping and modeling the biogeochemical cycling of turf grasses in the United States. *Environ. Manag.* 36(3), 426–438.

Morrissey, C. A., Mineau, P., Devries, J. H., Sanchez-Bayo, F., Liess, M., Cavallaro, M. C. and Liber, K. (2015). Neonicotinoid contamination of global surface waters and associated risk to aquatic invertebrates: a review. *Environ. Int.* 74, 291–303.

Ngumbi, E. N. (2011). *Mechanisms of Olfaction in Parasitic Wasps: Analytical and Behavioral Studies of Response of a Specialist (Microplitis croceipes) and a Generalist (Cotesia marginiventris) Parasitoid to Host-Related Odor*. Auburn University.

Obear, G. R., Adesanya, A. W., Liesch, P. J., Williamson, R. C. and Held, D. W. (2016). Fungicides affect Japanese beetle *Popillia japonica* (Coleoptera: Scarabaeidae) egg hatch, larval survival and detoxification enzymes. *Pest Manag. Sci.* 72(5), 966–973.

Obear, G. R., Williamson, R. C. and Liesch, P. J. (2014). Oviposition preference of the Japanese beetle (Coleoptera: Scarabaeidae) in golf putting greens under different soil moisture and fungicide regimes. *Appl. Turf Sci.* 11(1), 1–5.

Oliver, K. M., Russell, J. A., Moran, N. A. and Hunter, M. S. (2003). Facultative bacterial symbionts in aphids confer resistance to parasitic wasps. *Proc. Natl. Acad. Sci. U. S. A.* 100(4), 1803–1807.

Perelle, A. H. and Harper, J. D. (1986). An evaluation of the impact of sublethal dosages of nuclear polyhedrosis virus in larvae on pupae, adults, and adult progeny of the fall armyworm, *Spodoptera frugiperda*. *J. Invertebr. Pathol.* 47(1), 42–47.

Petersen, M. J., Olmstead, D. L. and Peck, D. C. (2011). Best management practices for invasive crane flies in Northeastern United States sod production. *J. Integr. Pest Manag.* 2(3), C1–C6.

Petersen, M. J. and Peck, D. C. (2013). Host plant resistance of cool-season (C3) turfgrasses to above- and belowground feeding by Tipula paludosa (Diptera: Tipulidae). *J. Econ. Entomol.* 106(3), 1463–1472.

Petersen, M. J., Seto, M. and Peck, D. (2013). Linking the spatio-temporal distribution of an edaphic crane fly to its heterogeneous soil environment. *Ecol. Entomol.* 38(6), 585–595.

Petersen, M. J. (2013). Evidence of a climatic niche shift following North American introductions of two crane flies (Diptera; genus Tipula). *Biol. Invas.* 15(4), 885–897.

Potter, D. A. and Braman, S. K. (1991). Ecology and management of turfgrass insects. *Annu. Rev. Entomol.* 36(1), 383–406.

Potter, D. A., Powell, A. J., Spicer, P. G. and Williams, D. W. (1996). Cultural practices affect root-feeding white grubs (Coleoptera: Scarabaeidae) in turfgrass. *J. Econ. Entomol.* 89(1), 156–164.

Potter, D. A., Tyler Stokes, J., Redmond, C. T., Schardl, C. L. and Panaccione, D. G. (2008). Contribution of ergot alkaloids to suppression of a grass-feeding caterpillar assessed with gene knockout endophytes in perennial ryegrass. *Entomol. Exp. Appl.* 126(2), 138–147.

Prater, C. A., Redmond, C. T., Barney, W., Bonning, B. C. and Potter, D. A. (2006). Microbial control of black cutworm (Lepidoptera: Noctuidae) in turfgrass using *Agrotis ipsilon* multiple nucleopolyhedrovirus. *J. Econ. Entomol.* 99(4), 1129–1137.

Preti, M., Verheggen, F. and Angeli, S. (2021). Insect pest monitoring with camera-equipped traps: strengths and limitations. *J. Pest Sci.* 94(2), 203–217.

Ramoutar, D., Alm, S. R. and Cowles, R. S. (2009a). Pyrethroid resistance in populations of *Listronotus maculicollis* (Coleoptera: Curculionidae) from southern New England golf courses. *J. Econ. Entomol.* 102(1), 388–392.

Ramoutar, D., Cowles, R. S. and Alm, S. R. (2009b). Pyrethroid resistance mediated by enzyme detoxification in *Listronotus maculicollis* (Coleoptera: Curculionidae) from Connecticut. *J. Econ. Entomol.* 102(3), 1203–1208.

Rangasamy, M., Rathinasabapathi, B., McAuslane, H. J., Cherry, R. H. and Nagata, R. T. (2009a). Oxidative responses of St. Augustinegrasses to feeding of southern chinch bug, *Blissus insularis* Barber. *J. Chem. Ecol.* 35(7), 796–805.

Rangasamy, M., Rathinasabapathi, B., McAuslane, H. J., Cherry, R. H. and Nagata, R. T. (2009b). Role of leaf sheath lignification and anatomy in resistance against southern chinch bug (Hemiptera: Blissidae) in St. Augustinegrass. *J. Econ. Entomol.* 102(1), 432–439.

Redmond, C. T., Wallis, L., Geis, M., Williamson, R. C. and Potter, D. A. (2020). Strengths and limitations of *Bacillus thuringiensis galleriae* for managing Japanese beetle (*Popillia japonica*) adults and grubs with caveats for cross-order activity to monarch butterfly (*Danaus plexippus*) larvae. *Pest Manag. Sci.* 76(2), 472–479.

Reinert, J. A., Engelke, M. C. and Heitholt, J. J. (2011). Hunting billbug (Coleoptera: Curculionidae) resistance among zoysiagrass (Zoysia spp) cultivars. *Fla. Entomol.* 94(3), 613-621.

Rohde, B. B., Allen, P. E., Benda, N., Brun, A., Mankin, R. W. and Dale, A. G. (2019). An acoustic trap to survey and capture two Neoscapteriscus species. *Fla. Entomol.* 102(3), 654-657.

Schwartzberg, E. G. and Tumlinson, J. H. (2014). Aphid honeydew alters plant defence responses. *Funct. Ecol.* 28(2), 386-394.

Shawki, M. A.-A., Kazda, T. D., Kohoutkova, J. J. and Taborsky, V. (2006). Toxicity to honeybees of water guttation and dew collected from winter rape treated with Nurelle D(R). *Plant Protect. Sci. Prague* 42, 9-14.

Shetlar, D. J. and Andon, J. (2015). Curative control of white grubs in turf, 2010. *Arthropod Manag. Tests* 40(1), 1-2.

Sparks, T. C. (2013). Insecticide discovery: an evaluation and analysis. *Pestic. Biochem. Physiol.* 107(1), 8-17.

Sparks, T. C. and Lorsbach, B. A. (2017). Perspectives on the agrochemical industry and agrochemical discovery. *Pest Manag. Sci.* 73(4), 672-677.

Stenersen, J. (2004). *Chemical Pesticides Mode of Action and Toxicology*. Boca Raton, FL: CRC Press..

Van Dyke, A. and McGraw, B. A. (2021). First report of *Chaetocnema minuta* (Coleoptera: Chrysomelidae) associated with turfgrass damage in the desert southwest United States. *J. Integr. Pest Manag.* 12(1), 19.

Van Dyke, A., Ramirez, R. A. and McGraw, B. A. (2015). First report of the invasive crane fly, *Tipula oleracea*, associated with turfgrass in Utah. *Crop Forage Turfgrass Mgmt.* 1(1), 1-3.

Vittum, P. J. (2020). *Turfgrass Insects of the United States and Canada*. Cornell University Press.

Whitehorn, P. R., O'connor, S., Wackers, F. L. and Goulson, D. (2012). Neonicotinoid pesticide reduces bumble bee colony growth and queen production. *Science* 336(6079), 351-352.

Williamson, C. R. and Potter, D. A. (1997). Nocturnal activity and movement of black cutworms (Lepidoptera: Noctuidae) and response to cultural manipulations on golf course putting greens. *J. Econ. Entomol.* 90(5), 1283-1289.

Wilson, A. C. and Duncan, R. P. (2015). Signatures of host/symbiont genome coevolution in insect nutritional endosymbioses. *Proc. Natl. Acad. Sci. U. S. A.* 112(33), 10255-10261.

Wilson, D. and Faeth, S. H. (2001). Do fungal endophytes result in selection for leafminer ovipositional preference? *Ecology* 82(4), 1097-1111.

Zenger, J. T. and Gibb, T. J. (2001). Impact of four insecticides on Japanese beetle (Coleoptera:Scarabaeidae) egg predators and white grubs in turfgrass. *J. Econ. Entomol.* 94(1), 145-149.

Zhou, W., Starr, J. L., Krumm, J. L. and Sword, G. A. (2016). The fungal endophyte *Chaetomium globosum* negatively affects both above-and belowground herbivores in cotton. *FEMS Microbiol. Ecol.* 92(10), fiw158.

Chapter 11

Advances in turfgrass weed management

Matthew T. Elmore, Rutgers University, USA; Aaron J. Patton, Purdue University, USA; Travis W. Gannon, North Carolina State University, USA; and James T. Brosnan, University of Tennessee, USA

1 Introduction

A weed can be simply defined as 'a plant growing where it is not wanted' (Harlan and deWet, 1965), which puts the onus on humans to determine which plants are weeds. For example, the dandelion (*Taraxacum officinale* F.H. Wigg.) is the epitome of a plant that can be considered both a weed and a desirable plant, depending on the location, purpose, and person. Brought from Europe to the New World intentionally by settlers as a medicinal herb, dandelions are cultivated for use in salad and to make wine, and they are admired for their beauty. Dandelion art and images are widespread in home décor, marketing, and even tattoos (Sanchez, 2006; Struwe, 2018). Contrast this idea with mass market consumer products for broadleaf weed control that contain the image of yellow dandelion flowers or puffball seedheads to attract the customer (Fig. 1). Clearly, there is a wide spectrum of dandelion appreciation.

Among professional turfgrass managers, there is more consensus about superior and inferior vegetation. Cultivated turfgrass species are superior to weeds and provide many environmental and economic benefits, as reviewed by Brosnan et al. (2020c). Weeds such as clover (*Trifolium* spp.) and crabgrass (*Digitaria* spp.) do not possess the full suite of functional and cultural benefits

http://dx.doi.org/10.19103/AS.2022.0110.11

Figure 1 A common consumer herbicide used to control dandelions in lawns.

provided by cultivated turfgrass species (Brosnan et al., 2014). As such, economical and effective weed control in sustainable turfgrass ecosystems is important to maintain both the functional and cultural benefits of turfgrass.

Rapid advances in turfgrass weed control occurred during the post-World War II era with advancements in selective herbicides. Led by the discovery that 2,4-D could selectively control dandelion and other broadleaf weeds without injury to Kentucky bluegrass (*Poa pratensis* L.), new herbicides were central to weed control innovation for many years (Marth and Mitchell, 1944). Highlighting this, L.W. Kephart of the United States Department of Agriculture (USDA; Washington, DC, USA) remarked of 'an intense preoccupation…with weed-killing chemicals and our seeming neglect of other avenues of approach to scientific weed control' during an address at the 1948 Northeastern Weed Control Conference (Kephart, 1948). The first turfgrass article published in the flagship Weed Science Society of America (https://wssa.net) journal *Weeds* was titled 'A comparison of five chemicals for crabgrass control in turf' (Engel et al., 1953). More recently, weed control research is including nonchemical practices more frequently but remains unbalanced toward chemical control research (Harker and O'Donovan, 2013).

While herbicides are important tools for weed management, weeds have developed resistance to herbicides, registration of new herbicide active ingredients has slowed, and government restriction of pesticide use continues to increase (Brosnan et al., 2020a; Duke, 2012). Economical and effective weed control will increasingly rely on strategies that integrate nonherbicide strategies as foundational elements of weed control programs. This will require practitioners to have more expertise on herbicide mode of action and specific cultural control strategies for each weed to provide the weed-free turfgrass desired by their customers. This chapter summarizes the current knowledge on weed control utilizing biological, physical, cultural, and organic control options as well as a look to the future on potential opportunities and challenges related to sustainable weed management in turfgrass ecosystems.

1.1 Herbicide development after 2,4-D

Prior to the development of synthetic herbicides, inorganic molecules were used. Salts of heavy metals including arsenicals were utilized in various systems in the early- to mid-twentieth century (Zimdahl, 2015); however, they were discontinued due to detrimental environmental and human health effects. The first synthetic organic herbicide was dinoseb and was introduced for selective grass and broadleaf weed control in large-seeded crops in 1932 (King, 1966). The first synthetic herbicide used in turfgrass systems was 2,4-D, a phenoxy herbicide, and was first synthesized in 1941 in the UK before obtaining a U.S. patent by F. D. Jones of American Chemical Paint Company (Philadelphia, PA, USA) in 1945 and was commercialized soon thereafter (King, 1966; Peterson et al., 2016). After 2,4-D was commercialized, other auxin mimic herbicides including additional phenoxy-carboxylic acids, benzoic acids, picolinic acids, and pyridinecarboxylic acids were introduced and continue to be commonly applied in cool- and warm-season turfgrass systems today (Table 1). Photosystem II (PSII)-inhibiting herbicides were developed soon thereafter, and atrazine was commercially available for maize in 1958 and later registered for warm-season turfgrass. According to United States Environmental Protection Agency (US EPA) in 2013, simazine (a PSII-inhibiting triazine) was the third most commonly used preemergence herbicide used in sod production and the second most commonly used preemergence herbicide used on golf courses (approximately, based on the amount of active ingredient applied) (US EPA, 2020). Triazine herbicides remain a very important herbicide family in the USA, although they are regulated and have additional use restrictions recently imposed due to groundwater contamination potential, among other reasons. Protoporphyrinogen oxidase (PPO)-inhibiting herbicides were initially discovered in the 1970s but were not registered in turfgrass until the late 1980s and are used both preemergence and postemergence. Common postemergence PPO herbicides include carfentrazone and sulfentrazone, while oxadiazon and flumioxazin are routinely used as preemergence herbicides although each has postemergence activity. The most unique attribute of oxadiazon is that it does not prohibit lateral spread and recovery of rhizomatous or stoloniferous turfgrasses and can be applied at sprigging of select turfgrass species.

Published by Burleigh Dodds Science Publishing Limited, 2023.

Table 1 Select herbicide mechanism of action in turfgrass ecosystems and date registered, as sorted by date

Mechanism of action	Examples of active ingredient(s)	Decade mechanism of action was described or synthesized (not necessarily in turf)	Product registered with US EPA	WSSA/HRAC group number[a]
Synthetic auxin mimics	2,4-D, MCPA, mecoprop, dicamba, clopyralid, fluroxypyr, triclopyr, quinclorac, and halauxifen-ethyl	1940s	Trimec (1970) Escalade 2 (2005)	4
Photosystem II inhibitor	Atrazine, simazine, and metribuzin	1950s	AAtrex (1969) Princep (1973)	5
Long-chain fatty acid synthesis inhibitor	Ethofumesate and bensulide	1960s	Prograss (1981)	15, unknown
Protoporphyrinogen oxidase inhibitor	Carfentrazone, flumioxazin, oxadiazon, and sulfentrazone	1960s	Dismiss (2005) Ronstar (1988) SureGuard (2003)	14
7,8-dihydropteroate synthetase inhibitor	Asulox	1970s	Asulam (1975)	18
Acetyl CoA carboxylase inhibitor	Fenoxaprop, fluazifop, sethoxydim, pinoxaden	1970s	Acclaim (1988) Fusilade II (1995) Manuscript (2018)	1
Cellulose biosynthesis inhibitor	Indaziflam, isoxaben	1970s	Gallery (1989) Specticle WP (2010)	29
Microtubule assembly inhibitor	Pendimethalin, prodiamine, and dithiopyr, pronamide	1970s	Barricade (1992) Pendulum (1990)	3

Acetolactate synthase inhibitor	Flazasulfuron, foramsulfuron, halosulfuron, imazosulfuron, metsulfuron, rimsulfuron, sulfosulfuron, trifloxysulfuron, imazapic, imazaquin, penoxsulam, pyrimisulfan, thiencarbazone	1980s	Image (1989) Katana (2009) Manage (1994) Manor (1999) Monument (2003) Revolver (2003) TranXit (2001)	2
Carotenoid biosynthesis inhibitor	Mesotrione, topramezone	1980s	Pylex (2012) Tenacity (2007)	27
Fatty acid thioesterase inhibitor	Methiozolin	1980s	PoaCure (2019)	30

[a]According to Weed Science Society of America Herbicide Site of Action Classification List (Weed Science Society of America, 2021).

Acetyl CoA carboxylase (ACCase)-inhibiting herbicides are commonly referred to as graminicides and were initially synthesized in the 1970s. ACCase herbicides, including fenoxaprop, fluazifop, clethodim, and sethoxydim, were registered in the 1980s and 1990s and possess selectivity in various turfgrass species, while pinoxaden was registered in 2018. Another agronomically important mechanism of action is microtubule assembly-inhibiting herbicides, including pendimethalin, prodiamine, dithiopyr, and pronamide, which were developed in the 1970s although not registered in turfgrass systems until the 1990s and remain important today for annual grass and small-seeded broadleaf weed control. Acetolactate synthase (ALS)-inhibiting herbicides were developed and registered in the 1980s and have multiple active ingredients used predominantly in warm-season turfgrasses for postemergence broadleaf and annual bluegrass (*Poa annua* L.) control and sedge (Cyperaceae) control in warm- and cool-season turfgrass. ALS-inhibiting sulfonylurea herbicides were known for 'revolutionizing' the herbicide industry as they possess high biological activity at low application use rates, which is a favorable trait among regulatory agencies (Waltz and Murphy, 2004) and has been an attribute of many herbicides registered recently. Although herbicidal activity at very low rates is viewed as a positive attribute, it can also be problematic due to off-target damage, as observed with the synthetic auxin herbicide aminocyclopyrachlor in 2011 (Patton et al., 2013a).

Other recently registered herbicides include mesotrione and topramezone (carotenoid biosynthesis inhibitors), indaziflam (cellulose biosynthesis inhibitor), and methiozolin (fatty acid thioesterase inhibitor) (Brabham et al., 2021). Mesotrione and topramezone were registered in 2007 and 2012, respectively, and constitute a major mechanism of action for grass and broadleaf weed control, primarily in cool-season turfgrasses. Indaziflam is the most recently registered preemergence herbicide (2010). Although it is not a new mechanism of action, it is critical for annual bluegrass resistance management in warm-season turfgrass systems. Similarly, methiozolin was registered in 2019 but was not a new mechanism of action.

1.2 Problems with relying too much on herbicides for weed control

Herbicides are effective tools for weed management, but there are consequences when turf managers rely on them as the primary control strategy. The primary consequence of relying entirely on herbicides for weed control is that the practice selects for plants that have evolved mechanisms to survive the application or repeated exposure to the herbicide active ingredient(s). Over 513 unique cases of herbicide resistance have been documented throughout agriculture, affecting over 267 different weed species and 21 of the 31 known physiological sites targeted by herbicides (Heap, 2022). The rapid onset of herbicide resistance in agricultural systems has revived interest in nonchemical practices to augment herbicide use including crop rotation, tillage, and the use of cover crops after harvest. In turfgrass, these nonchemical practices are often referred to as cultural controls.

Resistance to PSII-inhibiting herbicides was first identified in turf in Mississippi in 1996 (Kelly et al., 1999). Resistance to dinitroaniline herbicides has increased rapidly since 2010, especially in grassy weeds such as annual bluegrass and goosegrass (*Eleusine indica* [L.] Gaertn) particularly on golf courses (Brosnan et al., 2020a,d), and resistance to the pyridine herbicide dithiopyr was also recently identified in goosegrass (Elmore et al., 2022a; Russell et al., 2022). The development of herbicide-resistant weeds is problematic and needs to be addressed, but there are fortunately several factors unique to turfgrass systems that have reduced the evolution of herbicide resistance to date. Patton et al. (2018b) note nine factors that help to reduce herbicide resistance development in turf: (1) applying herbicide premixtures, (2) using grass species known to create dense swards to outcompete germinating weeds, (3) fertilizing to promote increased turf density and reduced weed competition, (4) applying herbicides in high spray volumes to optimize spray coverage, (5) removing tough-to-control weeds by hand-weeding, (6) mowing to decrease weed seed production, (7) harvesting grass clippings and inflorescences to reduce the weed seedbank, (8) reducing soil cultivation to minimize turf injury and recruitment of buried seed, and (9) using preemergence herbicides to prevent weed seed emergence.

For turfgrass managers, the consequence of developing resistant weed populations has been exacerbated in recent years by slowed herbicide discovery efforts amongst agrochemical manufacturers. This began with industry consolidation in the mid-1980s and further slowed by the adoption of transgenic cropping systems in the mid-1990s which reduced the need for additional active ingredients and mechanisms of action (Duke, 2012; Duke and Dayan, 2022). The cost to develop and register new herbicides has also increased. The last new herbicide mechanism of action was carotenoid biosynthesis inhibitors discovered in the 1980s (Beaudegnies et al., 2009). The onset of herbicide-resistant weeds has somewhat reinvigorated herbicide development, although innovations are primarily developed for major agronomic grass crops (e.g. maize, wheat, barley, and rice) and are typically commercialized in these crops for several years before they are registered for the turfgrass market. Exceptions include aminocyclopyrachlor, dithiopyr, ethofumesate, indaziflam, and methiozolin, among others, which were first registered in turfgrass and/or other specialty crop systems.

There may be novel mechanisms of action to aid turfgrass weed management in the future, but identifying new herbicide targets is quite challenging. An alternative approach, such as using natural products and pharmaceutical inhibitors to inspire new herbicides, may be useful, particularly when coupled with artificial intelligence to better understand the structural dynamics of new protein targets in plants (Duke and Dayan, 2022). Cyclopyrimorate (WSSA Group 33) and tetflupyrolimet (WSSA Group 28) represent two new herbicidal active ingredients that are now recognized by the Herbicide Resistance Action Committee as of 2021 (Duke and Dayan, 2022; HRAC, 2022). However, history suggests it will be several years until these products are tested for potential use in turfgrass.

2 Current knowledge and future outlook for integrated weed management strategies

Current methods, tactics, and tools for integrated turfgrass weed management include the use and optimization of synthetic herbicides, the use of organic and alternative herbicides including biologicals, physical weed management, cultural practices, improved turfgrass genetics, and the use of remote sensing and artificial intelligence in weed control programs.

2.1 Optimizing synthetic herbicide efficacy

For tough-to-control weeds, optimizing herbicide efficacy is important. This section will discuss strategies that optimize herbicide efficacy against the weeds directly and not consider their effect on the turfgrass sward (to be discussed later in the chapter) which can indirectly enhance herbicide efficacy.

Tank mixing herbicides (e.g. combining two more herbicide products in the same spray solution) to improve weed control or reduce injury to desirable turfgrass is a common strategy. For example, in cool-season turfgrass, annual bluegrass control is difficult to achieve with herbicides. Certain tank mixtures and programs can improve annual bluegrass control, but complete annual bluegrass control using herbicides alone is unlikely in cool-season turfgrass with current herbicide options. Combinations of preemergence and postemergence herbicides (Reicher et al., 2017), low-rate sequential application programs (Jeffries et al., 2013; Skelton et al., 2012), and tank mixtures of mesotrione + amicarbazone (Elmore et al., 2022b) can improve annual bluegrass control, but efficacy is consistently inconsistent across geographic locations and years. To better understand inconsistencies in annual bluegrass control in both cool- and warm-season turfgrass, future research should document management practices [e.g. irrigation, fungicides, and nitrogen (N) fertilizer] that promote or prevent summer annual bluegrass decline as well as the growth habit of annual bluegrass plants at the site using descriptions in Carroll et al. (2021b; i.e. creeping habit characteristic of *Poa annua* var. *reptans* [Hausskn.] T. Koyama, or upright habit characteristic of *Poa annua* subsp. *erecta*).

The most consistent chemical control of annual bluegrass involves paclobutrazol or flurprimidol application at 2- to 4-week intervals from annual bluegrass seedhead shatter until first frost (Diehl et al., 2021; Jeffries et al., 2013; McCullough et al., 2005; Patton et al., 2019a; Reicher et al., 2017; Woosley et al., 2003). The requirement for frequent applications limits practicality on many turf areas. Another drawback is frequent paclobutrazol or flurprimidol application can select for dollar spot (*Clarireedia* spp.) disease developing insensitivity to other demethylation inhibitors, including fungicides which are important tools for disease management (Allan-Perkins et al., 2017).

Methiozolin, recently registered as PoaCure® for annual bluegrass control, is a new tool for selective annual bluegrass control in cool-season turfgrass, particularly creeping bentgrass putting greens. Unlike other herbicides for annual bluegrass control that were briefly registered for use on putting greens before bentgrass injury issues prevailed, field research indicates methiozolin programs provide annual bluegrass control with little to no creeping bentgrass injury (Askew and McNulty, 2014; Brosnan et al., 2013; Koo et al., 2014; Patton et al., 2019a; McCullough et al., 2013; Xiong et al., 2015). Interestingly, the mechanism for selectivity of annual bluegrass in creeping bentgrass is not well understood, as absorption, translocation, and metabolism do not clearly explain differences in species tolerance (Flessner et al., 2013; McCullough et al., 2013; Yu and McCullough, 2014) and suggests that the deeper and more robust root system of creeping bentgrass that may play a role in selectivity (Brabham et al., 2021).

In warm-season turfgrass, use restrictions on the relatively broad-spectrum herbicide monosodium methanearsonate require the use of alternative, suboptimal herbicides for weed control. For example, the herbicide topramezone is by far the most effective herbicide for goosegrass control, but it also causes dramatic foliar bleaching of most warm-season turfgrass species. Strategies to reduce bleaching but maintain weed control efficacy include tank mixing with various herbicides (Brewer et al., 2022a,b; Carroll et al., 2021a; Cox et al., 2017; Gonçalves et al., 2021), chelated iron (Boyd et al., 2021), and irrigating immediately after application (Brewer et al., 2022b; Kerr et al., 2019). These strategies often reduce goosegrass control, and none eliminate bleaching of desirable turfgrass.

N fertilization at the time of herbicide application can improve weed control. N fertilizer rates as low as 12 kg N•ha^{-1} enhanced the efficacy of postemergence herbicides mesotrione and topramezone for both crabgrass (*Digitaria* spp.) and annual bluegrass (Beck et al., 2015; Elmore et al., 2012, 2013a). While ammonium sulfate is often tank mixed with herbicides to improve absorption, N fertilizer was applied to the soil in these experiments, which suggests that weed control was improved by mechanisms other than increased foliar herbicide absorption. Similar work using higher N rates (73 kg N•ha^{-1}) found that N applied just before flazasulfuron application improved annual bluegrass control, which was attributed to increased herbicide translocation (Brosnan et al., 2010). These studies found enhanced herbicide efficacy independent of competition from desirable turfgrass, and the effect may be greater in situations where turfgrass vigor is enhanced by N.

Seasonal application timing can improve herbicide efficacy for perennial weed control. While it is generally understood that immature annual weeds are easier to control, mature perennial weed control can be improved if herbicide applications are made at certain times of the year. Herbicides are most effective for warm-season weeds such as dallisgrass (*Paspalum dilatatum* Poir.) and common bermudagrass (*Cynodon dactylon* [L.] Pers.) at certain temperature-based growing degree-day (GDD) thresholds in the spring and cooling degree-day thresholds in the late summer or early autumn (Brosnan et al., 2011; Elmore et al., 2013b; Johnston and Henry, 2016). For perennial broadleaf weeds, fall applications after first frost are generally considered more effective than spring applications, but some of this is based on research that did not include spring applications for comparison (Kohler et al., 2004; Reicher and Weisenberger, 2007). Fall applications tend to be more effective, but for tough-to-control weeds such as ground ivy (*Glechoma hederacea* L.) herbicide selection is more important (Patton et al., 2017a). Proper spring application timing is also important as auxin mimic herbicides can be less effective against dandelion if made earlier than 135 GDD in the spring or during peak bloom (Raudenbush and Keeley, 2014; Schleicher et al., 1996). The ALS-inhibiting herbicide florasulam should be applied before 75 GDD in the spring for optimum dandelion control (Patton et al., 2018a). Early postemergence applications (200 to 250 GDD) are optimal for yellow nutsedge (*Cyperus esculentus* L.) control (Li et al., 2021).

Other factors important to optimize herbicide efficacy include ensuring the target weed is not drought stressed at the time of application (Shekoofa et al., 2020) and avoiding antagonism of ACCase and phenoxy herbicides (Dernoeden and Fidanza, 1994), or of weak acid herbicides by hard water (Schortgen and Patton, 2021). Failing to include adjuvants can dramatically reduce weed control from many herbicides including ALS inhibitors, carotenoid biosynthesis inhibitors, and quinclorac (Zawierucha and Penner, 2001). Anecdotally, a large proportion of practitioners do not fully comprehend when to include or exclude adjuvants with herbicides. Practitioners often indicate that the adjuvant was not available or not suggested by the salesperson at the point of purchase, or the adjuvant was included even though the product label indicates it should not be included (e.g. herbicides that contain PPO herbicides).

2.2 Synthetic herbicide safety and reducing dislodgeable residue

Regulatory agencies (e.g. US EPA) require registrants to conduct comprehensive testing to ensure herbicides do not adversely affect human or environmental health. There are many environmental and toxicological tests required for herbicide registration and reregistration, including tests to characterize environmental fate, residue, and potential off-target movement. In agronomic crops, tolerances are derived from comprehensive risk assessments establishing a maximum amount of a pesticide allowed in or on a food which is commonly referred to as maximum residue limit. These limits are based on multiple factors including the toxicity of the pesticide and metabolites, how much of the pesticide remains in or on food, as well as all potential exposure routes. Other priority areas for herbicide residue concerns revolve around environmental impacts, specifically off-target movement and subsequent damage to neighboring plant or animal species or natural resources. Herbicide registration and reregistration require extensive testing and are only allowed registration if certain criteria are met, ensuring they do not contaminate air, groundwater, and other water bodies or otherwise compromise environmental and/or human health.

In turfgrass ecosystems, herbicide residue concerns are much different compared to food and fiber crops and other cropping systems. In turfgrass ecosystems, concern about herbicide residue focuses on worker and nonworker exposure, residue in turfgrass clippings (e.g. which may be used in composting), and other environmental aspects including how long an herbicide persists as well as its mobility off-target at any time after the application. Specifically, human pesticide exposure for nonworkers in turfgrass systems may occur by multiple routes, including transfer from treated vegetation onto humans as well as multiple indirect routes. Similar to pesticide residue in agronomic crops, comprehensive human herbicide exposure risk assessments must be conducted prior to herbicide registration and reregistration to ensure adverse health effects do not occur following exposure to treated turfgrass areas. Within the occupational and residential exposure tests required by US EPA, foliar transferable residue dissipation studies are required to quantify herbicide residue on treated turfgrass surfaces that may be transferred via non dietary ingestion onto human skin or clothing as well as inhaled (e.g. dust or vapor). Otherwise stated, how much of an applied pesticide is dislodged or transferred to workers or nonworkers at various times after the application? These data are entered into risk models to ensure people are not exposed to adverse effect levels or concentrations. Select herbicides contain verbiage on product labels to address herbicide residue remaining in turfgrass clippings and if they can cause adverse effects. For example, clopyralid can persist in composted clippings, and the compost with clopyralid residue can damage sensitive plants. Dithiopyr and quinclorac have similar label verbiage, indicating that clippings should not be collected from treated areas soon after application to prevent off-target herbicide movement via clippings.

Various factors influence dislodgeable herbicide residue from treated turfgrass surfaces. Product formulation, irrigation, spray application parameters including application water carrier volume and adjuvant or surfactant inclusion, and turfgrass species among other factors influence dislodgeable herbicide residue in turfgrass ecosystems. Specifically, Thompson et al. (1984) reported that liquid 2,4-D, which is highly water-soluble and subject to dislodge, was 15-fold more dislodgeable than granular applications following application. When liquid 2,4-D was used, only <0.01% of applied 2,4-D was dislodgeable 1 day after a rainfall event. In the absence of rain, dislodgeable 2,4-D foliar residues consistently decreased after the application, as small amounts of 2,4-D were detected at 7 days after treatment but none detected beyond 12 days after treatment (Jeffries et al., 2016). Increasing application water carrier volume and spray droplet size can also decrease dislodgeable 2,4-D foliar residue from treated turfgrass surfaces. Specifically, dislodged 2,4-D decreased more than 25% when applied at 748 $L \bullet ha^{-1}$ compared to 187 $L \bullet ha^{-1}$ (Jeffries et al., 2017). Including a nonionic surfactant also reduced dislodgeable 2,4-D foliar residue (Maxwell et al., 2018). Time of day also influences dislodgeable foliar residue; specifically, 2,4-D dislodgeability consistently decreased from morning to afternoon, mostly due to canopy moisture (Jeffries et al., 2016). Therefore, it is possible that dislodgeable herbicide residues from turfgrass can be dramatically reduced with coordination between turf managers and scheduled events at the turf site.

2.3 Organic and alternative herbicides

Due to public concern over pesticide exposure, governments and other policymakers have severely restricted or banned pesticide use on school grounds and public areas in North America (Campbell and Wallace, 2020). Quebec, Canada, passed the Pest Control Products Act in 2003 that resulted in the ban on selling pesticides for use in lawns, including the herbicide 2,4-D (Conlin, 2008). The herbicide 2,4-D was included in the Quebec ban for precautionary reasons despite uncertain evidence about its effects on human health (Conlin, 2008). Thorough reviews of 2,4-D toxicology report 'no evidence that 2,4-D poses any health risk to humans' and 'no convincing or consistent evidence for any chronic adverse effect of 2,4-D in humans' (Burns and Swaen, 2012; Kennepohl et al., 2010). Ironically, in 2011, as part of a settlement with Dow AgroSciences (Indianapolis, IN, USA), the government of Quebec agreed that 'products containing 2,4-D do not pose an unacceptable risk to human health or the environment' (Dow AgroSciences LLC vs. Her Majesty the Queen in Right of Canada, 2011). Despite this declaration in 2011, Quebec did not remove 2,4-D from its list of banned pesticides. Later, Ontario, Canada, followed Quebec and outlawed the use of pesticides on lawns, gardens, hardscapes, cemeteries, parks, and schoolyards (Ontario Ministry of the Environment, 2009). Most recently, Montgomery County, MD, USA, passed legislation that bans the use of pesticides in public and private lawns (Janasie, 2019). Increasing regulation of pesticide use in turfgrass systems and an ever-increasing interest from the public on 'natural' or 'organic' products for weed control has resulted in a demand for organic and alternative herbicides. While these pesticide bans result in reductions in turfgrass and ground quality, especially in lower income areas, they are popular and will likely continue and be implemented by an increasing number of policymakers (Bartholomew et al., 2015).

Pesticide bans enacted by governments and policymakers provide a defined list of products permitted for use. The pesticide bans often allow exemptions for certain Organic Materials Research Institute (OMRI; https://www.omri.org)-approved horticultural oils and soaps as alternatives. Outside of these examples of legislation, defining what is permitted in a 'natural' and 'organic' program is difficult as the USDA National Organic Program is for agricultural products only. Currently, there are no federal standards for organic turfgrass management and no restrictions on using the word 'organic' by turfgrass practitioners. The Northeast Organic Farming Association of Connecticut (NOFA; https://ctnofa.org) established standards and an accreditation program for Organic Land Care in Connecticut, USA, which includes lawn management. Interestingly, the NOFA organic land care program allows for 'emergency non-organic rescue treatment' in extenuating circumstances and notification of the property owner. The use of a synthetic pesticide results in a property being designated as 'transitional', but the implications of the transitional designation nor the length of time it is designated transitional are not discussed (NOFA, 2017). This program also permits the use of conventional sod and seed.

There are multiple categories of products considered natural or organic. A category of products considered low risk is exempt from US EPA registration under the Federal Insecticide, Fungicide, and Rodenticide Act section 25(b). Another category includes products with a US EPA registration number and further certification status from organizations such as OMRI. This includes natural products and certain synthetic materials in Title 7 Code of Federal Regulations (CFR) 205.601. A third category is products marketed as natural or organic but do not have 25(b) exempt status and are not OMRI approved. Regardless of the category, nearly all the products are nonselective postemergence herbicides that destroy the cuticle layer and disrupt the cell membrane permeability but do not translocate. In cool-season turfgrass, these products cause severe injury to desirable turfgrass, and single applications do not provide weed control (Patton et al., 2019b). An initial investigation by Flessner et al. (2010) indicates these products have potential for weed control in dormant bermudagrass. Unlike many alternative products, chelated iron (Fe-HEDTA) is selective for broadleaf weeds in turfgrass (Patton et al., 2019b). While chelated iron products do not have 25(b) exempt status, they are often permitted where synthetic pesticides have been banned, including Ontario, Canada, and Montgomery County, MD, USA (Ontario Ministry of the Environment, Conservation and Parks: https://www.ontario.ca/page/using-pesticides-ontario; Kao-Kniffen, 2020).

Corn gluten meal is an organic preemergence herbicide widely sold for crabgrass control. Commonly used as an animal feed, corn gluten meal is the protein fraction of corn extracted during the wet milling process and contains 10% N by weight. A similar product, corn gluten hydrolysate, is derived from corn gluten meal and is slightly more effective in controlled environments (Liu et al., 1994). Despite widespread usage, demonstrated corn gluten meal efficacy in field studies at the recommended rate (980 $kg \bullet ha^{-1}$) with N fertilizer standards for comparison is limited to one study which found that two sequential applications at 730 $kg \bullet ha^{-1}$ provided some smooth crabgrass control (Dernoeden, 2001). Other work that included a fertilizer standard and corn gluten meal at registered use rates found that corn gluten meal and fertilizer standards provided similar weed control (Christians and Dant, 2005; St. John and DeMuro, 2013). The recommended corn gluten meal rate provides 98 $kg\ N \bullet ha^{-1}$, which exceeds the legal nutrient application limit in some states in the USA. Controlled environment research suggests that efficacy is dependent on a drought period shortly after weed emergence, and the most effective rates are higher than practical due to cost and excessive N fertilizer rate (Gardner et al., 1997; Liu and Christians, 1997). Efficacy may also depend on applying the product within 1 week prior to crabgrass germination (Christians, 1993). Interestingly, Gardner et al. (1997) found that corn gluten meal improved the efficacy of low rates of pendimethalin, but an N fertilizer standard was not included in the study. Corn gluten in combination with low rates of pendimethalin or other mitotic inhibiting herbicides would be an interesting area for further research since the primary mechanisms of action are similar (Unruh et al., 1997).

2.4 Biological herbicides

There are promising biological control agents in development for use in turfgrass, some even commercialized for a short period, but none are currently available. These efforts use the inundative or bioherbicide strategy where bacteria or fungi are applied at concentrations far exceeding those found naturally rather than the classical biological approach of releasing the organism to persist in the environment for a long period of time. While these agents often show promise in controlled environments and field research, they often fail commercially due to narrow weed control spectrum, high production costs, and specific environmental conditions required for efficacy (Watson and Bailey, 2013). The fungus *Sclerotinia minor* demonstrated moderate efficacy for dandelion control under specific environmental conditions and was sold as Sarritor in Canada (Abu-Dieyeh and Watson, 2007a,b). However, after just 1 year of commercialization, the introduction of Fe-HEDTA eliminated demand for Sarritor, and it is no longer commercialized (Watson and Bailey, 2013). More promising, a strain of *Phoma macrostoma* with broad-spectrum activity on dicots and safety to monocots was registered in Canada and the USA in 2011 and 2012, respectively, but the partnership of Agriculture and Agri-Food Canada (Ottawa, Ontario, Canada) with The O.M. Scotts Company (Marysville, OH, USA) did not result in a commercial product (Bailey and Falk, 2011; Bailey et al., 2011). Development of bacteria for biocontrol is a similar story of promising efficacy with no commercialized products. A *Pseudomonas fluorescens* isolate with proven cheatgrass (*Bromus tectorum* L.) control in large-scale multiyear field experiments was released commercially in agriculture for a short time (Kennedy et al., 1991). As part of this project, another *P. fluorescens* isolate was developed for annual bluegrass control in turfgrass (Kennedy, 2016) but never commercialized, despite field development efforts from the commercial sector. Similarly, a *Xanthomonas campestris* isolate for annual bluegrass control in creeping bentgrass (*Agrostis stolonifera* L.) was registered for a short time in Japan, but is no longer available (Imaizumi et al., 1997; Nishino and Tateno, 2000).

A biocontrol strategy that does not require introduction of a foreign agent uses the annual bluegrass weevil (*Listronotus maculicollis* Kirby) insect to control annual bluegrass. This insect is a common pest of golf course turfgrass in the northeastern USA and in Ontario, Canada. The annual bluegrass weevil larvae cause more damage to annual bluegrass than creeping bentgrass (Kostromytska and Koppenhöfer, 2014, 2016). Delaying insecticide applications until the weevil larvae cause substantial damage to the annual bluegrass can reduce annual bluegrass populations in golf course fairways. Using this biocontrol strategy in combination with the plant growth regulator paclobutrazol improves the annual bluegrass control compared to either strategy alone (Diehl et al., 2021, 2022).

Biopesticide sales increased from 2.4% of the crop protection market in 2009 to 5.0% in 2017 (Baker et al., 2020). If investment in developing biological options increases similarly, or legislation forces rapid change, effective options may be on the horizon. Future development should focus on finding bioherbicides that can reduce rates of synthetic pesticides or be combined with cultural management practices to provide commercially acceptable control instead of expecting the product to perform similarly to synthetic products alone. Given the price sensitivity toward synthetic pesticides, golf courses, athletic fields, and lawn care business models with less price sensitivity could be targeted for early adoption of these bioherbicide products. Golf courses and athletic fields with on-site managers are also better able to time biopesticide applications with proper environmental conditions often required for efficacy.

Lastly, fungal endophytes may confer indirect biological control of weeds. Several *Epichloë* spp. endophytic fungi exist in a mutualistic relationship with the host plant, providing important abiotic (e.g. drought) and biotic stress (e.g. disease and insect) tolerance (Funk et al., 1993). This host-plant resistance to stresses is conferred through the production of defensive alkaloids and antifungal proteins (Baldauf et al., 2011; Tian et al., 2017). While there is no direct evidence that the resulting changes in turfgrass physiology as a result of endophyte infection externally influence the germination or emergence of weeds, the host-plant resistance conferred by the endophytes does result in reduced stress to the turfgrass, which should lead to the maintenance of a dense sward and a subsequent reduction in weed recruitment. This concept merits further study.

2.5 Physical weed management

Optimal weed management programs are designed to reduce the overall abundance of existing weed seed in the soil profile, often termed the weed 'seedbank' as well as to limit the deposition of new weed seed into the soil (Norsworthy et al., 2018). In turfgrass management, practices such as fraise mowing and clipping collection after mowing can be implemented with this in mind. Fraise mowing is a turfgrass cultivation practice that removes aboveground biomass allowing swards to regrow from belowground vegetative tissue (Brosnan et al., 2020b). Research has shown that the debris removed during fraise mowing contains an abundance of weed seed. For example, debris generated after fraise mowing a zoysiagrass (*Zoysia japonica* Steud., 'Meyer') fairway contained seed from numerous winter and summer annual weed species, and 59% of those seeds were winter annual broadleaf species and 28% were annual bluegrass (Brosnan et al., 2020b). Summer fraise mowing bermudagrass (*Cynodon* spp.) resulted in 41% to 97% reductions in annual bluegrass the following spring; however, turfgrass managers identified numerous barriers to adopting this practice, including cost, labor, area closure, and debris removal (Carroll et al., 2021c).

These barriers may also prevent the adoption of flaming for weed control in turfgrass (Fig. 2). This practice can be selective as perennial turfgrasses can regrow from rhizomes or stolons after the aerial portions are killed by flaming. Using a flamer powered by liquefied petroleum gas in the spring provided moderate weed control in hybrid bermudagrass and seashore paspalum,

Figure 2 A flame treatment on cool-season turfgrass caused unacceptable cool-season turfgrass injury for more than 4 weeks and provided <50% weed control (Patton et al., 2019b).

but >70 $kg \bullet ha^{-1}$ of liquefied petroleum gas was required for weed control and 3 weeks was needed for the bermudagrass to recover from flaming (Martelloni et al., 2018, 2019). Flaming may have potential for weed control before cool-season turfgrass establishment from seed. Tillage, followed by flaming after weed seedling emergence but prior to tall fescue (*Festuca arundinacea* Schreb.; syn. *Schedonorus arundinaceus* [Schreb.] Dumort., nom. cons.) seeding, controlled several summer annual weed species as well as the fumigant dazomet (Hoyle et al., 2012), but was less effective than dazomet at removing perennial weeds (Patton et al., 2019b).

For selective mechanical broadleaf weed control without severe turfgrass injury, a device called the Weedbine is in development (Henderson, 2021). This device modifies the cutting unit on a triplex reel mower, removing the cutting reel and replacing it with a coarse brush. The bedknife is replaced with a grooved baseblade (Fig. 3). As the bristles pass through the grooved baseblade, the broader leaved plants are removed or severely injured while the narrower leaved monocots are less injured. Using the Weedbine once per week during the growing season provided dandelion and white clover control comparable to a synthetic herbicide program and similar or improved turfgrass quality compared to synthetic herbicide programs (Henderson, 2021).

On putting greens, clipping collection over a 2-year period has been shown to reduce viable annual bluegrass seed in the soil by 60%, which results in a more modest reduction in annual bluegrass establishment

Figure 3 Bristles of the coarse brush interacting with the grooved baseblade of the Weedbine. Source: Henderson (2021).

(Gaussoin and Branham, 1989). More commonly, clipping return decreases weed infestations, possibly a result of nutrient return from clippings and providing a light layer of mulch to prevent weed seed germination (Haley et al., 1985; Harivandi et al., 2001; Heckman et al., 2000). Perhaps timing clipping collection to coincide with weed flowering would be most effective and minimize barriers to adopting this practice over large acres of managed turfgrass.

2.6 Cultural weed management

Busey (2003) and Hahn et al. (2020) provide a comprehensive review on the effect of cultural practices such as mowing, N fertilization, irrigation, and cultivation on weed incidence in turfgrass ecosystems. Mowing at the highest height within the recommended range for a particular species and proper N fertilization to promote turfgrass canopy density usually decrease weed incidence. While the effect of mowing height on weed incidence is very powerful, Hahn et al. (2020) noted that most research examined large changes in cutting height instead of small changes in cutting height within practical limitations for usability. Another weakness of previous research is that mowing height treatments are not often accompanied by an increase in fertilizer input, mowing frequency, or other cultural practices that practitioners would employ to increase canopy density at lower mowing heights (Law et al., 2016). Investigations of small changes in cutting height that are not confounded by a reduction in canopy density or increase in turfgrass disease (as noted by Cutulle et al., 2014) would be useful to determine whether certain weeds are more or less adapted to certain mowing heights.

Interseeding turfgrass has utility for weed control, especially where effective herbicides are not available. Frequent interseeding that increases turfgrass cover can provide modest reductions of established perennial weeds such as dandelion and white clover (*Trifolium repens* L.) and annuals such as crabgrass (Elford et al., 2008; Miller and Henderson, 2012). Success of interseeding creeping bentgrass in fine turf where annual bluegrass is dominant is dependent on seeding during mid-summer when annual bluegrass is less competitive (Henry et al., 2005; Murphy et al., 2005).

The influence of cultural practices on annual bluegrass competitiveness has been evaluated more comprehensively than any other weed species, probably due to the ubiquity and cosmopolitan nature of annual bluegrass. Almost all the research evaluating cultural practices has been conducted in cool-season turfgrass where effective herbicide options for control are limited in the USA and nonexistent in Canada and Europe.

Annual bluegrass is more competitive under low mowing heights and greater irrigation. Specifically, greater irrigation frequency more than quantity promotes annual bluegrass infestation (Bell et al., 1999; Cain et al., 2021; Chen et al., 2018; Gaussoin and Branham, 1989; Youngner, 1959). High irrigation frequency is required to maintain annual bluegrass, particularly during the summer when the root system is most shallow and drought stress promotes anthracnose disease to which annual bluegrass is very susceptible (Roberts et al., 2011).

Low soil pH also reduces annual bluegrass competitiveness more so than desirable turfgrasses and has been recommended for annual bluegrass suppression (Goss et al., 1975; Varco and Sartain, 1986). The soil pH effect is thought to be a result of limited phosphorus (P) availability at low soil pH, which reduces annual bluegrass competitiveness more so than desirable turfgrass. This is corroborated by work that found that high elemental sulfur rates (>168 $kg \bullet ha^{-1}$ $year^{-1}$) reduced annual bluegrass on a putting green (Goss et al., 1975; Varco and Sartain, 1986). P requirements in turfgrass systems have been investigated for nearly a century, and annual bluegrass incidence often increases when phosphate fertilizer is applied (Sprague and Burton, 1937). This effect is usually only observed in sand root zones where P is low (<24 $mg \bullet kg^{-1}$ Mehlich-3-P), and in these experiments the lowest phosphate rate is associated with an increase in annual bluegrass compared to no phosphate application, but there is no further increase in annual bluegrass incidence with P rate (Lodge et al., 1990; Øgaard and Aamlid, 2020; Raley et al., 2013; Waddington et al., 1978). From this field research, it is not known whether P application enhances annual bluegrass establishment from seed or the competitiveness of mature plants. Comparing annual bluegrass and creeping bentgrass in monoculture, when no P was applied in a low P soil, creeping bentgrass yield and P uptake were greater than in annual bluegrass; however, as soil P increased (≥40 $mg \bullet kg^{-1}$), annual bluegrass yield and P uptake became superior (Kuo, 1993). However, it is not well understood how annual bluegrass and other turfgrass species compete under various soil P concentrations. Until the influence of P on annual bluegrass competitiveness is understood enough to provide a more defined set of soil test P values that will reduce the competitiveness of annual bluegrass without compromising desirable turfgrass quality, practitioner adoption of P management to reduce annual bluegrass encroachment is unlikely.

N also has significant effects on annual bluegrass encroachment, with higher N rates generally resulting in more annual bluegrass, with the exception being in a turf sward that is thin due to N deficiency. Especially for species like creeping bentgrass or velvet bentgrass (*Agrostis canina* L.) that require limited N, annual bluegrass encroachment is reduced when N is applied sparingly (Skogley, 1975; Tang et al., 2021). The time of year N is applied is also important, with late autumn and early spring applications as well as spoon-feeding N during the summer favoring annual bluegrass (Dernoeden, 2013). The N form is also important as acidifying N sources such as ammonium sulfate tend to favor desirable turfgrass species over annual bluegrass (O'Connor et al., 2018).

Annual bluegrass is more sensitive to high rates of iron (Fe) than creeping bentgrass (Xu and Mancino, 2001). In field studies, Fe has been investigated for annual bluegrass control in creeping bentgrass putting greens with mixed results. In a 2-year study, Han et al. (2017) found that ferrous sulfate applied at 12 $kg{\bullet}ha^{-1}$ or 49 $kg{\bullet}ha^{-1}$ nine times annually had a modest effect on annual bluegrass encroachment in a creeping bentgrass putting green. As that study continued 7 more years, the effect of ferrous sulfate was minimal (Tang et al., 2021). Similarly, ferrous sulfate applied six times annually at 50 $kg{\bullet}ha^{-1}$ in a 4-year, three-location study on creeping bentgrass putting greens provided little to no annual bluegrass control (Patton et al., 2019a). These results contrast with a 2-year study where biweekly applications of ferrous sulfate at 12 $kg{\bullet}ha^{-1}$ or 49 $kg{\bullet}ha^{-1}$ showed moderate reductions in annual bluegrass cover (Ervin et al., 2017).

Certain weed species, especially prostrate knotweed (*Polygonum aviculare* L.) and goosegrass, are very competitive in highly trafficked areas on golf courses, athletic fields, and parks. Whether wear tolerance, soil compaction, and lack of turfgrass cover are responsible for weed competitiveness in these areas is unknown. Goosegrass seedling and bermudagrass root and shoot growth were reduced similarly in compacted soils, suggesting that wear tolerance may be the mechanism for the competitive advantage of goosegrass in highly trafficked areas (Arietta et al., 2009). Nevertheless, core cultivation (Turgeon and Fidanza, 2017) is often recommended to reduce weed incidence in highly trafficked areas. Often, there is concern amongst practitioners that core cultivation will increase weed incidence or reduce efficacy of preemergence herbicides. Golf course superintendents try to schedule core cultivation to not coincide with annual bluegrass seedling emergence (Christians, 1996; Dernoeden, 2013). While this is logical, one long-term, multilocation study on the topic found that annual bluegrass incidence did not increase when creeping bentgrass putting greens were core cultivated in summer compared to early fall (Patton et al., 2019a). Information on the effect of cultivation practices (e.g. hollow-tine cultivation and vertical mowing) on weed infestations is often found in experiments where the primary objective was to investigate the efficacy of these practices on thatch removal or preemergence herbicides (Branham and Rieke, 1986; Elmore and Tuck, 2021; Johnson, 1979, 1982; Weston and Dunn, 1985). In the absence of preemergence herbicides, early spring vertical mowing and hollow-tine cultivation were found to increase large crabgrass (*Digitaria sanguinalis* [L.] Scop.), black medic (*Medicago lupulina* L.), and henbit (*Lamium amplexicaule* L.) in plots with poor turf quality that were severely thinned due to cultivation treatments (Johnson, 1979; Weston and Dunn, 1985). Where turfgrass quality was high throughout the experiment, aggressive core cultivation in the spring did not increase annual bluegrass, crabgrass, or goosegrass incidence in the absence or presence of preemergence herbicides (Branham and Rieke, 1986; Elmore and Tuck, 2021; Johnson, 1982; Patton et al., 2019a). Where turfgrass is dense and actively growing, evidence suggests that core cultivation should proceed without concern for annual weed encroachment. This is true even if a preemergence herbicide has been applied as core cultivation or vertical mowing do not reduce herbicide efficacy.

2.7 Increasing plant competition with weeds through genetics

Turfgrass species selection is a known way to suppress weeds either through a 'right-plant, right-place' approach to making sure the turfgrass species selected is well-adapted to the environment, pests, and management of the site or through enhanced competition with weeds conferred through high plant density or potential allelopathy. DeBels et al. (2012) compared the encroachment of weeds in Kentucky bluegrass (two cultivars), Chewings fescue (*Festuca rubra* L. ssp. *commutata* Gaudin), tall fescue, and perennial ryegrass (*Lolium perenne* L.) over the course of 3 years and while imposing three different mowing heights and three N fertilization programs in Wisconsin. DeBels et al. (2012) noted that dandelion was the most prevalent weed; tall fescue had the fewest weeds and was the only turf species with the same weed cover (3%) regardless of N fertilization rate. Unimproved 'Kenblue' Kentucky bluegrass had more weed cover (15% to 26%) which was not affected by N fertilizer rate (DeBels et al., 2012). Among the species, dandelion cover rankings were Kentucky bluegrass ≥ perennial ryegrass ≥ tall fescue = Chewings fescue. A similar study with multiple species and three mowing heights by Dernoeden et al. (1994) reported that smooth crabgrass and white clover cover were reduced in swards of sheep fescue (*Festuca ovina* L.) and hard fescue (*Festuca brevipila* Tracey) compared to tall fescue when irrigation and fertilizer inputs were withheld. Following 2 years of turf management without herbicides in Virginia (USA), hard fescue resisted weed invasion better than tall fescue (Askew et al., 2013). Overall, the fine fescue taxa are known to have similar or greater resistance to weed invasion than other cool-season turfgrass species (Braun et al., 2020). These experiments primarily highlight the 'right-plant, right-place' approach to reducing weed competition. Refer to Busey (2003) to review specific biotic and abiotic stresses for their effect on weed populations.

Few studies have directly compared warm-season turfgrass species for their ability to suppress encroaching weeds. Among warm-season turfgrass species, zoysiagrass (*Zoysia* spp.) is often noted for its ability to resist weed invasion (Patton et al., 2017b). Weeds are often most problematic amongst warm-season turfgrasses in winter in areas where the turfgrasses are dormant. Winter annual weeds such as annual bluegrass, common chickweed (*Stellaria media* L.), and henbit are often common in winter months. Although the high tiller density of zoysiagrass does not prevent all weeds as some have suggested (Forbes and Ferguson, 1947), zoysiagrass usually has less weed problems than bermudagrass (Patton et al., 2017b).

Plant breeders develop cultivars that have enhanced abiotic and biotic stress tolerance, which leads to improved turf performance, less plant death, and a resulting turf sward that has fewer weeds. Examples might include a bermudagrass cultivar with improved spring dead spot (*Ophiosphaerella herpotricha* [Fr.] Walker) resistance (Baird et al., 1998) that is accompanied by a decrease in weed cover (Fig. 4). Likewise, all turfgrass cultivars with improved abiotic and biotic stress tolerance achieved through plant breeding and improved genetics (or host-plant resistance conferred by endophytes as mentioned previously) should result in a reduction in weed competition. Cultivars can also resist weed invasion through enhanced tiller density. Beard et al. (2001) reported that significant cultural control of annual bluegrass could be gained by selecting cultivars of creeping bentgrass that have a high shoot density under close mowing. The increased density also increases competition and allows fewer opportunities for weed seed recruitment. While improved tiller density will likely ward off weeds in many turf species, a high tiller density may not necessarily prevent weeds, especially perennial weeds. Brede (1992) found that tall fescue cultivars with contrasting density had similar common bermudagrass cover. With tall fescue, the improved cultivar that provided higher turf quality and density did not provide additional perennial weed suppression over a 60+-year-old cultivar. While genetic (cultivar) advancements have been made, their value in weed control has probably been under-researched and thus is not well understood compared to the contribution of improved genetics for gains in drought, disease, or insect pest tolerance.

Figure 4 A common bermudagrass cultivar with high susceptibility to spring dead spot disease and the subsequent high weed encroachment.

The latest advancement in turfgrass cultivar development related to weed management is the use of biotechnology. The O.M. Scotts Company developed a transgenic Kentucky bluegrass and St. Augustinegrass (*Stenotaphrum secundatum* [Walter] Kuntze) cultivars with both a gene that produces a type of 5-enolpyruvylshikimate-3-phosphate synthase with reduced affinity for glyphosate (Blume et al., 2010; Buhlman et al., 2022; Harriman et al., 2019) and a gene responsible for the overexpression of GA2-oxidase to reduce functional gibberellic acid production, leading to improved sustainability through reduced lawn mowing and related emissions as well as increased shade tolerance (Fidanza et al., 2022; Reed et al., 2021). Several cultivars with these traits were developed and are branded under the names Scotts 'ProVista' (Fidanza et al., 2022). Because of the rapid development of glyphosate-resistant weeds in agricultural fields following the development of glyphosate-resistant crops (Shaner et al., 2012) and because of the current existence of glyphosate-resistant weeds in turfgrass ecosystems, such as annual bluegrass, there is great concern that the development of this technology will hasten the development of herbicide-resistant weed populations. Being mindful of this, the developers of ProVista™ grass varieties encourage the following best management practices to help steward the technology and prevent the development of herbicide-resistant weeds: (1) optimize turf health and quality to reduce the need to apply herbicides, (2) utilize preemergence herbicides for weed prevention, (3) do not rely solely on glyphosate for weed control, (4) rotate herbicide mode of action, and (5) tank mix other herbicides with glyphosate to improve the efficacy on target weeds (Pedersen, 2019). While transgenic genetic crops are now common, transgenic turfgrasses are novel and their value in weed control in turf ecosystems is under-researched and not well understood. Further, while the risk that this technology could lead to the evolution of herbicide-resistant weed populations is clear, it is unclear how quickly this may occur in turf ecosystems as there are fewer reports of herbicide-resistant weeds in turf than agronomic crops due to the previously mentioned factors that make turf ecosystems unique (Mithila et al., 2011; Patton et al., 2018b).

Allelopathy is the 'direct or indirect harmful effect by one plant on another through production of chemical compounds that escape into the environment' (Rice, 1984) (Fig. 5). In turfgrass ecosystems, annual bluegrass, colonial bentgrass (*Agrostis capillaris* L.), fine fescue taxa, Kentucky bluegrass, perennial ryegrass, and tall fescue have all been cited in works describing their ability to suppress weeds through allelopathy (Bertin et al., 2007, 2009; Braun et al., 2020; Breuillin-Sessoms et al., 2021; Buta and Spaulding, 1989; Fales and Wakefield, 1981; Lickfeldt et al., 2001; Lipinska and Sykut, 2012; Peters and Luu, 1985; Smith and Martin, 1994; Walters and Gilmore, 1976; Weston, 1996; Wu et al, 1998; Zuk and Fry, 2005). While there is much discussion about allelopathy, Breuillin-Sessoms et al. (2021) point out that 'this knowledge has not been translated into cultivar development' as no commercial cultivars are

Figure 5 The potential benefits of allelopathy are evident in this photo where a large patch of quackgrass (*Elymus repens* L.), which has a low tiller density, is excluding weeds compared to the Kentucky bluegrass and perennial ryegrass in this low maintenance lawn.

marketed for their value-added weed suppression. A challenge in perennial cropping systems, such as turfgrass, is separating the plant competition from allelochemical effects. In fact, Lickfeldt et al. (2001) concluded that weed suppression from these turfgrass species was likely due to competition for light, water, and nutrients and not from allelopathy. More recently, Breuillin-Sessoms et al. (2021) concluded that while the weed suppressive ability of some turfgrass species and cultivars can be confirmed, there is still a need to increase our knowledge of how this suppression works. Undoubtedly, turfgrass scientists will continue to study the mechanism of allelopathy with the potential to develop improved cultivars. However, weeds are just one of the pests of turf, and the importance of the 'right-plant, right-place' approach to selecting turfgrass species and cultivars, with both abiotic and biotic stress tolerance, should remain paramount.

2.8 Remote sensing and artificial intelligence for weed control

Spot spraying herbicides instead of broadcasting can dramatically reduce the amount of herbicide needed for weed control. This strategy reduces weed management costs if the spot application process is automated. In row crop agriculture, the See and Spray™ (Blue River Technology; Sunnyvale, CA, USA) system is commercially available for fallow field spraying. Other systems in development also use cameras and artificial intelligence to distinguish between crops and weeds. It is more challenging to detect grassy weeds in turfgrass than row crops as the grassy weeds and desirable turf have similar morphology and are intermixed, but recent work found that convolutional neural networks can detect grassy and broadleaf weeds from photographic images in both actively growing and dormant bermudagrass (Yu et al., 2019, 2020). With neural networks capable of detecting weeds in turfgrass, this technology could be coupled with a camera and sprayer system for field use. However, a commercially available system for spot spraying weeds in turfgrass is likely years away. It is unknown if any system exists in the late stages of development for weed control in turfgrass though there are considerable efforts underway in using remote sensing for early detection of abiotic stress and disease (Badzmierowski et al., 2019; Booth et al., 2021).

3 Current knowledge of integrated chemical and nonchemical management

Given the efficacy of chemical weed control, research that investigates herbicides or plant growth regulators to improve efficacy of nonchemical management practices will be the focus of this section. A summary of recent research integrating chemical and nonchemical practices for weed control can be found in Table 2. Compared to studies that examined chemical management alone, there have been surprisingly few studies that evaluate the effect of nonchemical methods in combination with chemical methods.

Several projects have investigated annual bluegrass control in creeping bentgrass putting greens. A 9-year study investigated how N rate, ferrous sulfate, and plant growth regulators affect annual bluegrass on a creeping bentgrass putting green. At the conclusion of year 2 of the study, flurprimidol reduced annual bluegrass cover from 43% to 8% with a low N program (24 kg N•ha^{-1} annually) but was less effective in the high N program (108–147 kg N•ha^{-1} annually) where annual bluegrass cover was 21% (Han et al., 2017). In years 3 through 9 of the study, flurprimidol was effective regardless of N program, although the high N program dramatically increased annual bluegrass cover when trinexapac-ethyl or no plant growth regulator was applied (Tang et al., 2021). Ferrous sulfate applied at 49 kg•ha^{-1} nine times annually resulted in a modest reduction in annual bluegrass cover with the low N program but was not effective in reducing annual bluegrass with the high N program in the first 2 years of the study (Han et al., 2017). In subsequent years, under the low N program, ferrous sulfate did not reduce annual bluegrass cover (Tang et al., 2021). Plant growth regulator programs muting effects of nonchemical management strategies were also reported by Ervin et al. (2017) and Diehl at el. (2021). Future annual bluegrass research should evaluate chemical management programs that would be moderately effective alone to determine whether they can be enhanced by nonchemical management. For example, Diehl et al. (2022) found that lower rates of paclobutrazol could be used for annual bluegrass control in a creeping bentgrass fairway when combined with the annual bluegrass weevil biocontrol.

Another biocontrol agent *S. minor* was moderately effective for dandelion control when applied alone. When cool-season turfgrass was seeded 10 days after *S. minor* application, dandelion cover was reduced by >90%, compared to 60% to 70% for *S. minor* alone. Seeding alone did not reduce dandelion cover (Abu-Dieyeh and Watson, 2007b).

Controlling perennial warm-season weeds such as dallisgrass and bermudagrass typically requires sequential herbicide applications. In research simulating lawn care scenarios, tall fescue interseeding in late summer or early autumn combined with herbicide programs increased control of warm-season perennials bermudagrass, dallisgrass, and false-green kyllinga (*Kyllinga gracillima* Miq.) compared to either practice alone. Three sequential applications of topramezone herbicide provided ≤50% bermudagrass control, which would be considered commercially unacceptable. When tall fescue was interseeded 3 weeks after the final topramezone application, bermudagrass was controlled 88% to 92% (Brosnan and Breeden, 2013). Moderately effective

Table 2 Recent research (within the last 20 years) integrating chemical and nonchemical strategies for weed management in turfgrass

Weed	Turfgrass species	Nonchemical practice	Herbicide, plant growth regulator, or biological agent	Findings	Reference
Annual bluegrass (*Poa annua*)	Creeping bentgrass (*Agrostis stolonifera*) (fairway)	Annual bluegrass weevil (*Listronotus maculicollis*) biocontrol	Plant growth regulator (paclobutrazol)	Allowing annual bluegrass weevil to cause damage combined with paclobutrazol controlled annual bluegrass more than either practice alone.	Diehl et al. (2021, 2022)
Annual bluegrass	Creeping bentgrass (putting green)	Low or high nitrogen (N) fertilizer regimen	Plant growth regulators (trinexapac-ethyl and flurprimidol)	Annual bluegrass cover was lowest at low N with flurprimidol. High N dramatically increased annual bluegrass cover in the absence of flurprimidol. A ferrous sulfate program was also evaluated but had little effect.	Han et al. (2017), Tang et al. (2021)
Annual bluegrass	Creeping bentgrass	Two core aerification timings	Eight treatments consisting of nutrients, herbicides, or plant growth regulators	Aerification timing did not influence annual bluegrass control.	Patton et al. (2019a)
Annual bluegrass	Creeping bentgrass	Biweekly ferrous sulfate application	Biweekly plant growth regulator paclobutrazol application	Ferrous sulfate reduced annual bluegrass alone but did not enhance paclobutrazol efficacy, likely because paclobutrazol was extremely effective alone.	Ervin et al. (2017)
Annual bluegrass	-	N fertilizer	Postemergence herbicide flazasulfuron	Flazasulfuron controlled annual bluegrass more when 73 kg N•ha^{-1} was applied prior to herbicide application and again 4 weeks later (150 kg N•ha^{-1} total).	Brosnan et al. (2010)

(*Continued*)

Table 2 *(Continued)*

Weed	Turfgrass species	Nonchemical practice	Herbicide, plant growth regulator, or biological agent	Findings	Reference
Common bermudagrass (*Cynodon dactylon*)	–	Fraise mowing	Postemergence herbicide glyphosate and glyphosate + fluazifop	Fraise mowing before or after herbicide application improved control compared to either practice alone.	Richardson et al. (2021)
Common bermudagrass	Tall fescue (*Schedonorus arundinaceus*)	Fall interseeding	Postemergence herbicide topramezone	Control increased from 27% to 50% with herbicide alone to >85% with seeding.	Brosnan and Breeden (2013)
Crabgrass, smooth (*Digitaria ischaemum*)	Tall fescue	Mowing height of 2.5–10 cm	Low rate of prodiamine	Prodiamine was less effective at 2.5 cm mowing height in one study year. Otherwise, mowing height had no effect.	Cropper et al. (2017)
Crabgrass (*Digitaria* spp.)	Common bermudagrass	Mowing	Preemergence herbicides	All herbicides provided greater crabgrass control at 3.8 cm mowing compared to 1.5 cm in one of two experiment years.	Gannon et al. (2015)
Crabgrass spp.	Tall fescue	N fertilizer	Postemergence herbicides mesotrione and topramezone	N fertilizer applied 3 days, 1 days, and 0 days before herbicide at ≥13 kg•ha^{-1} increases efficacy.	Beck et al. (2015), Elmore et al. (2012)
Crabgrass, smooth	Tall fescue	Fungicide to control brown patch disease	Preemergence herbicides pendimethalin, prodiamine, and oxadiazon	Azoxystrobin improved crabgrass control from pendimethalin by 15% to 20% but did not improve control from prodiamine and oxadiazon.	Ferrell et al. (2003)
Crabgrass, smooth	Kentucky bluegrass (*Poa pratensis*)	Interseeding with perennial ryegrass	Organic and conventional herbicide programs	Seeding reduced crabgrass infestations in organic programs but not the conventional program. The conventional program was extremely effective alone.	Miller and Henderson (2012)

Dandelion (*Taraxacum officinale*)	Cool-season turf mixture	Interseeding	Biological herbicide *Sclerotinia minor*	Seeding + *S. minor* reduced dandelion to 2% to 6% cover compared with 33% to 39% for *S. minor* alone. Interseeding alone had no effect.	Abu-Dieyeh and Watson (2007b)
Dallisgrass (*Paspalum dilatatum*)	Common bermudagrass	Vertical mowing	Postemergence herbicides	Vertical mowing 30 min prior to herbicide application improved dallisgrass control.	Brown et al. (2022)
Dallisgrass	Tall fescue	Fall interseeding	Postemergence herbicide fluazifop	Interseeding improved control from herbicide by 20% to 25%.	Elmore et al. (2013b)
False-green kyllinga (*Kyllinga gracillima*)	Tall fescue	Fall interseeding	Postemergence selective herbicides	Interseeding or herbicides alone resulted in <50% control. Herbicides + interseeding resulted in 90% to 98% control.	Elmore and Tuck (2022)
Ground ivy (*Glechoma hederacea*)	Kentucky bluegrass	Three mowing timings relative to herbicide application	Five postemergence selective herbicides	The timing of lawn mowing did not alter herbicide efficacy.	Beck et al. (2014)
Silvery thread moss (*Bryum argenteum*)	Creeping bentgrass	Hollow-tine cultivation and vertislicing	Carfentrazone herbicide	Hollow-tine cultivation and vertislicing slightly reduced moss cover, even in the absence of carfentrazone.	Raudenbush and Keeley (2017)
Silvery thread moss	Creeping bentgrass	Topdressing and N fertilizer	Carfentrazone herbicide	Carfentrazone followed by four biweekly applications of sand topdressing or 12 kg N•ha^{-1} controlled moss more than any practice alone.	Borst et al. (2010)

herbicide programs for dallisgrass and false-green kyllinga were similarly improved when tall fescue interseeding occurred following postemergence herbicide applications (Elmore et al., 2013b; Elmore and Tuck, 2022). These studies conducted seeding 1-4 weeks after herbicide programs concluded (based on herbicide label restrictions), but this may not be widely practical. Future research could evaluate whether seeding and herbicide applications can be effective if conducted at different times of year. These interseeding studies broadly suggest that combining herbicides with practices to enhance turfgrass vigor (e.g. N fertilizer and fungicide applications) is important to improve herbicide efficacy against perennial weeds, but this topic deserves more research.

While most research evaluating integrated strategies focused on weeds that cannot be controlled with herbicides alone, many weeds such as crabgrass can be controlled by high rates of preemergence herbicides alone and minimal regard for cultural management. But there may be potential to use lower herbicide rates if nonchemical management is integrated. Research evaluating various mowing regimens and heights found that a low rate of prodiamine (0.34 kg ha^{-1}) was less effective at the lowest mowing height of 1.3 cm than higher mowing heights, and this was not explained by a decrease in turfgrass density (Cropper et al., 2017). Other work in bermudagrass found that maximum labeled rates of several preemergence herbicides were more effective at a 3.8 cm mowing height compared to a 1.5 cm mowing height in 1 year, but there was no effect of mowing height the other year (Gannon et al., 2015). A better understanding of how to reduce herbicide rates or application frequency in combination with nonchemical approaches is needed. Could lower preemergence herbicide rates provide season-long crabgrass control if used in conjunction with an allelopathic cultivar or a 'right-plant right-place' approach? What is the potential to reduce herbicide rates for tall fescue treated to prevent brown patch disease? Ferrell et al. (2003) found that the preemergence herbicide pendimethalin controlled crabgrass more when tall fescue competitiveness was enhanced by azoxystrobin for brown patch disease control. Brosnan and Breeden (2013) attributed the aforementioned bermudagrass control enhanced by tall fescue interseeding partially to fungicide applications to prevent brown patch disease.

Future research projects designed to examine integrated management practices should include herbicide or plant growth regulator rates that do not completely control the weed on their own to avoid muting the effect of the nonchemical practice (Diehl et al., 2021). This would help practitioners rank the efficacy of various treatment combinations, better assess the value of nonchemical practices, and apply the knowledge to a wide range of scenarios they encounter in the field.

4 Barriers to integrated weed management among practitioners and how to address them

Herbicide restrictions are increasing worldwide. Legislation severely restricts herbicide options in many countries of Europe and the UK (Hahn et al., 2020). In the USA, restrictions are limited but increasing. An ideal situation would be a transition to a system that decreases herbicide use and increases other management techniques without forced legislation. However, integrated pest management is a >50-year-old concept with limited adoption to date (Stern et al., 1959). There are several barriers to the adoption of integrated weed management that are specific to the turfgrass industry (Ervin et al., 2022) and others that are common to other crops (Nowak et al., 1996). In each case, both the complexity of biological systems and the uniqueness of human individuals prevent greater adoption.

More research is needed to improve agronomic tools, methods, and strategies for integrated weed management. Harker and O'Donovan (2013) point out that the research on integrated weed management has increased in recent years, but most of the knowledge base is still focused on the use of herbicides. If integrated weed management is to increase, the research funding into biological, physical, cultural, and organic control options must also increase to provide a pathway to sustainability. A recent example of funding toward the development of integrated solutions is the project 'Research and Extension to Address Herbicide-Resistance Epidemic in Annual Bluegrass in Managed Turf Systems', funded by the USDA National Institute of Food and Agriculture Specialty Crops Research Initiative program (award no.: 2018-51181-28436) (http://resistpoa.org). Programs and projects like this must increase in the future if sustainable weed management is to be increased in turfgrass ecosystems.

Scientists and educators need to increase and improve relationships with community partners to both better meet their needs and collaborate to develop solutions (Gavazzi and Gee, 2018). Ervin et al. (2022) noted that universities play a key role in bringing together turf professionals for both formal and informal information sharing. Turf practitioners recognize university scientists as trusted information sources (Ervin et al., 2022), but the high cost of obtaining the information may prevent their adoption (Nowak et al., 1996). Practitioners may not always be able to afford to attend local or regional conferences and workshops in lieu of their already busy work schedule.

The challenges related to employee turnover and finding skilled, knowledgeable labor make the adoption of integrated weed management techniques difficult. The turf industry as well as the nursery and landscape industry is known to have high employee turnover (Mathers et al., 2010; Patton and Reicher, 2011). While new employees entering the workforce are trained on the safe use of pesticides prior to becoming a pesticide applicator, their supervisor is often the only staff member with training on integrated weed management. Businesses that invest in training new and experienced employees may reduce turnover and be better able to utilize integrated weed management. One example of this is a >69 ha Science, Employee Education, and Development campus being built by The Davey Tree Expert Company (Kent, OH, USA) to train and retain employees (https://www.davey.com/about/seed-campus).

Additionally, more information sharing is needed in the turf industry for non-English-speaking people or those who speak English as a second language (Ervin et al., 2022; Patton et al., 2013b). Reaching those that need training can be a challenge, although programs designed for entry-level workers in the turf industry are rated highly, with 95% of attendees feeling better equipped to do their jobs and 72% feeling better able to save money for their employer (Patton and Reicher, 2011). In 2022, The University of Florida (Gainesville, FL, USA) hosted a field day tour entirely in Spanish to improve information accessibility (https://tinyurl.com/y4c92mmc).

For the investment in these employee training efforts to be fully realized, it will be critical to find new ways to involve homeowners (mowing and irrigation) and landscape managers/contractors (mowing, irrigation, and planting decisions) in the management of the lawn to integrate their practices (Ervin et al., 2022). For example, yellow nutsedge is known to be more competitive in over-irrigated soils (Ransom et al., 2009). Lawn care operators trained to identify an overwatering or similar issue need to share this issue with the homeowner or person who controls the irrigation system.

Another challenge to integrated weed management related to training is the complexity of the system. While killing a weed may seem like a simple proposition, integrated weed management is more complex than choosing an herbicide and spraying a weed. Each weed responds differently to cultural, biological, physical, and chemical control tactics. Practitioners need to understand those many factors for each of the common weed species they encounter to then develop an effective integrated control strategy. For example, to control annual bluegrass in a golf course putting green, a practitioner must understand the influence of N, P, and Fe applications on both annual bluegrass and the desirable putting green species, the influence of mowing height and grass clippings collection on species dynamics, the influence of soil pH and root zone moisture on plant growth and development, the management of soil compaction and soil dynamics of weed seed recruitment from the seedbank, the effectiveness of interseeding techniques, timings, and choice to increase plant competition, the influence of shade/light management on turfgrass and weed species dynamics, the potential interactions between plant pathogens and insect pests on the population dynamics between the weed and the desirable turfgrass species, the influence of plant growth regulators with three differing plant physiological sites of action, and lastly the effectiveness of preemergence and postemergence herbicides from differing modes of action, and the subsequent government regulations surrounding their use on golf course putting greens. It is a challenge just to know this information, let alone to synthesize it into an effective weed management program specific to each practitioner's site, labor budget, and equipment and chemical inventory. Decision support tools that provide guidance on integrated solutions (e.g. chemical and cultural) can reduce the complexity of the system. Readily accessible tools for turfgrass weed management include the Purdue Turf Doctor App (https://www.entm.purdue.edu/turfdoctor), Wisconsin Turf Pest Control Tool (https://turfpests.wisc.edu), and University of Tennessee Mobile Weed Manual (https://www.mobileweedmanual.com/home). However, these only address basic needs such as herbicide selection and weed identification and do not provide information on how to integrate nonchemical practices or optimize herbicide efficacy. The 'Turfgrass Weed Control for Professionals' guide (Patton et al., 2022) discusses strategies to optimize herbicide efficacy but does not cover integrated weed management. Other tools such as the Michigan State University GDD Tracker (https://gddtracker.msu.edu) and Cornell University Fore Cast (http://turf.eas.cornell.edu) also help turfgrass managers optimize application timing for certain herbicides but do not address nonchemical techniques. Innovative, comprehensive, and accessible resources for turfgrass managers are needed but who will fund the development and maintenance of these resources?

Locally adapted information is important. Turf management and the challenging weed species found in turf are heterogeneous in each region and within each state. As such, there is no one-size-fits-all approach. Ervin et al. (2022) advocate for local efforts to improve integrated weed management, as regional and national policies and practices are likely to fail. Weed management guides certainly should be authored at the national and regional levels by scientists, specialists, and educators, but recommendations need to be locally adapted to provide the best possible information. Information that is from a neighboring state or from across the country may be viewed as irrelevant by stakeholders (Nowak et al., 1996). The ShortCUTT newsletter (https://turf.cals.cornell.edu/news/shortcutt-newsletter) released weekly during the growing season by Cornell University (Ithaca, NY, USA) is a great example of a local and timely resource to help turfgrass managers integrate nonchemical strategies. These local resources also require funding from university agricultural experiment stations, grants and other funding programs, and other noncommercial interests.

Another barrier to increased adoption of integrated practices is related to information sharing and organization schemes (Ervin et al., 2022). In some segments of the turf industry such as golf courses or athletic fields, the turf manager or superintendent has almost complete control of the management system. The supervisor works with their staff to control aspects of mowing, irrigation, fertilization, and traffic management and all aspects of pest management using physical, biological, and pesticide control strategies in addition to turfgrass species and cultivar selection and site preparation when planting new swards. Whereas in the professional lawn care industry, the homeowner or a separate landscape maintenance contractor often controls the initial turfgrass species and cultivar selection at establishment and the subsequent mowing, irrigation, and traffic management. The professional lawn care operator (LCO) only governs the fertilizer and pesticide applications. The mowing contractor may be inadvertently responsible for spreading weeds from one lawn to another (Ervin et al., 2022) or weeds may encroach from the neighboring properties. The lack of control of all the cultural management practices undermines the LCO ability to develop and implement a sustainable approach to lawn weed management (Ervin et al., 2022).

Cost is another important barrier to integrated weed management. Practitioners may be willing but unable to adopt integrated weed management strategies because of the economic expense (Nowak et al., 1996). Herbicides are the primary pesticide used in turf on an acreage basis (Clark and Kenna, 2010) because of their affordability. Compared to physical removal of weeds by laborers, the total cost (e.g. labor, herbicide, fuel, and equipment) of an herbicide application is much lower. If herbicides are affordable, safe, and effective, what is the incentive to adopt a new weed management technique? Additionally, there is a high labor cost associated with pest monitoring (Clark and Kenna, 2010). Often, the highly trained staff with higher salaries are entrusted with scouting to properly identify the pest and its growth stage. Researchers are also guilty of developing integrated pest management techniques without considering the economic feasibility of these recommendations on practitioners. Scientists and extension/outreach specialists need to better document the environmental and economic benefits of a holistic, long-term integrated pest management system in a real-world scenario, although a recent example of this is the work of Rossi and Grant (2009). Integrated weed management is perceived as more expensive, so research and extension/outreach efforts need to document the return on investment in both the short and long term (Moss, 2019). A return on investment could be aided by better understanding consumer preferences for weed management, similar to what has been done to understand consumer preferences for turfgrass attributes (Ghimire et al., 2019; Hugie et al., 2012; Yue et al., 2012), how these attributes influence adoption (Yue et al., 2021), and how integrated weed management can be facilitated by the turfgrass manager (Barnes et al., 2018). What premium would consumers pay for a lawn care service that reduces pesticide use through integrated weed management? Financial incentives for turfgrass, similar to the US EPA Pesticide Environmental

Stewardship Program Grants for the agricultural community could also hasten the adoption of integrated weed management strategies.

Another barrier to the adoption of integrated weed management techniques is an unwillingness to adopt (Nowak et al., 1996). This unwillingness could be for several reasons. The high cost of investment in new equipment, such as a GPS/GIS sprayer, may prevent practitioner adoption of this technology as part of an integrated strategy to control pests. High costs of water or fuel may prevent the proper mowing (e.g. frequency) or irrigation of a lawn. Integrated weed management practices also carry with them an increased risk (real or perceived) of a negative outcome (Nowak et al., 1996). Consumers (e.g. golfers, homeowners, and property managers) have a high expectation for a near perfect, esthetically beautiful turf product (Clark and Kenna, 2010) and practitioners may be unwilling to try new practices due to uncertainty and risk associated with the outcome (e.g. too many weeds) and the subsequent impact on their income and employment.

A final barrier to the adoption of integrated weed management techniques relates to legislation, regulation, and litigation to remove herbicides from the list of approaches available. Those drafting legislation often do not appreciate the complexity of the problem. They may assume that citizens have access to effective, alternative solutions when imposing herbicide restrictions and enact legislation without a careful study of alternatives to the products banned. In some cases, the alternatives proposed may have a higher safety risk to the applicator despite the perceived benefit to the environment that might accompany a change. For example, fearing potential litigation or human safety related to glyphosate use, businesses and municipalities banned the use of glyphosate and switched to the use of vinegar (containing acetic acid) as an alternative organic herbicide. These decisions were made without a careful review of the safety of the alternative product to applicators or its effectiveness to control weeds. Acetic acid herbicide solutions work as contact herbicides, offering short-term weed control, but they are known to burn skin and cause serious-to-severe eye injury, including blindness (Webber et al., 2018). In comparison, glyphosate is a systemic herbicide better able to kill weeds than acetic acid (Domenghini, 2020) without risk to the applicator (Kougias et al., 2021), and without risk of developing any form of cancer, including non-Hodgkin's lymphoma (Andreotti et al., 2018). Regulation of existing technology (e.g. herbicides) without careful study and funding to develop alternatives will limit the availability and access to the strategies (e.g. tools) that practitioners need to implement integrated weed management.

5 Conclusion and future trends

Weeds are major pests that reduce esthetics and functionality of turfgrass ecosystems. Knowledge required for effective and sustainable turfgrass weed management has evolved in recent years as weeds develop resistance to herbicides, government restriction of broad-spectrum herbicides has increased, and registration of new herbicide active ingredients has slowed. This chapter provided an in-depth discussion on relevant and recent developments in synthetic herbicides, alternatives to synthetic herbicides, and nonchemical strategies for integrated and sustainable turfgrass weed management, including cultural, physical, and genetic approaches. Advances in integrated turfgrass weed management strategies and barriers preventing adoption of those strategies were also presented in this chapter. Finally, this chapter also presented opportunities for research and development to improve an overall integrated and sustainable approach to turfgrass weed management.

6 Where to look for further information

Agronomy Monograph 56 titled *Turfgrass: Biology, Use, and Management*, edited by John Stier, Brian Horgan and Stacy Bonos contains chapters on both integrated pest management and weed management in turfgrass. (Stier, J. C., Horgan, B. P. and Bonos, S. A. (2013). *Turfgrass: Biology Use and Management. Agronomy monograph 56*. Madison, WI: Crop Sci Soc of America, 1307 p).

Current websites specific to various centers of research include the following:

- Resources from a collaboration of 16 university scientists funded by the USDA to limit the impact of annual bluegrass in turfgrass (Resistpoa.org).
- Annual research reports from the Rutgers Center for Turfgrass Science (https://turf.rutgers.edu/research/reports/).
- Resources for turfgrass practitioners from the University of Tennessee Turfgrass Weed Science Program (https://www.tnturfgrassweeds.org/resources).
- Resources for turfgrass practitioners from the Purdue University Turfgrass Program (https://turf.purdue.edu/turfgrass-weeds/).
- Resources for turfgrass practitioners from the North Carolina State University Turfgrass Program (https://www.turffiles.ncsu.edu/grasses/).
- Extension guidebook. *Turfgrass Weed Control for Professionals* from Purdue University (https://mdc.itap.purdue.edu/item.asp?Item_Number=TURF-100).

7 References

Abu-Dieyeh, M. H. and Watson, A. K. (2007a). Efficacy of *Sclerotinia minor* for dandelion control: effect of dandelion accession, age and grass competition. *Weed Res.* 47(1), 63–72. doi: 10.1111/j.1365-3180.2007.00542.x.

Abu-Dieyeh, M. H. and Watson, A. K. (2007b). Grass overseeding and a fungus combine to control *Taraxacum officinale*. *J. Appl. Ecol.* 44(1), 115-124. doi: 10.1111/j.1365-2664.2006.01247.x.

Allan-Perkins, E., Campbell-Nelson, K., Popko, J. T., Sang, H. and Jung, G. (2017). Investigating selection of demethylation inhibitor fungicide-insensitive *Sclerotinia homoeocarpa* isolates by boscalid, flurprimidol, and paclobutrazol. *Crop Sci.* 57(S1), S301-S309.

Andreotti, G., Koutros, S., Hofmann, J. N., Sandler, D. P., Lubin, J. H., Lynch, C. F., Lerro, C. C., De Roos, A. J., Parks, C. G., Alavanja, M. C., Silverman, D. T. and Beane Freeman, L. E. (2018). Glyphosate use and cancer incidence in the agricultural health study. *J. Natl. Cancer Inst.* 110(5), 509-516.

Arietta, C., Busey, P. and Daroub, S. H. (2009). Goosegrass and bermudagrass competition under compaction. *Agron. J.* 101(1), 11-16.

Askew, S. D. and McNulty, B. M. S. (2014). Methiozolin and cumyluron for preemergence annual bluegrass (*Poa annua*) control on creeping bentgrass (*Agrostis stolonifera*) putting greens. *Weed Technol.* 28(3), 535-542.

Askew, W. B., Goatley, J. M., Jr., Askew, S. D., Hensler, K. L. and McKissack, D. R. (2013). A comparison of turfgrasses for cemeteries and other low-input areas. *Int. Turfgrass Soc. Res. J.* 12, 245-250.

Badzmierowski, M. J., McCall, D. S. and Evanylo, G. (2019). Using hyperspectral and multispectral indices to detect water stress for an urban turfgrass system. *Agronomy* 9(8), 439. doi: 10.3390/agronomy9080439.

Bailey, K. L. and Falk, S. (2011). Turning research on microbial bioherbicides into commercial products - a *Phoma* story. *Pest Technol.* 5, 73-79.

Bailey, K. L., Pitt, W. M., Falk, S. and Derby, J. (2011). The effects of *Phoma macrostoma* on nonstarget plant and target weed species. *Biol. Control* 58(3), 379-386.

Baird, J. H., Martin, D. L., Taliaferro, C. M., Payton, M. E. and Tisserat, N. A. (1998). bermudagrass resistance to spring dead spot caused by *Ophiosphaerella herpotricha*. *Plant Dis.* 82(7), 771-774.

Baker, B. P., Green, T. A. and Loker, A. J. (2020). Biological control and integrated pest management in organic and conventional systems. *Biol. Control* 140, 104095. doi: 10.1016/j.biocontrol.2019.104095.

Baldauf, M. W., Mace, W. J. and Richmond, D. S. (2011). Endophyte-mediated resistance to black cutworm as a function of plant cultivar and endophyte strain in tall fescue. *Environ. Entomol.* 40(3), 639-647.

Barnes, M. R., Nelson, K. C., Meyer, A. J., Watkins, E., Bonos, S. A., Horgan, B. P., Meyer, W. A., Murphy, J. and Yue, C. (2018). Public land managers and sustainable urban vegetation: the case of low-input turfgrasses. *Urban For. Urban Greening* 29, 284-292.

Bartholomew, C., Campbell, B. L. and Wallace, V. (2015). Factors affecting school grounds and athletic field quality after pesticide bans: the case of Connecticut. *HortScience* 50(1), 99-103.

Beard, J. B., Croce, P., Mocioni, M., De Luca, A. and Volterrani, M. (2001). The comparative competitive ability of thirteen *Agrostis stolonifera* cultivars to *Poa annua*. *Int. Turfgrass Soc. Res. J.* 9, 828-831.

Beaudegnies, R., Edmunds, A. J. F., Fraser, T. E. M., Hall, R. G., Hawkes, T. R., Mitchell, G., Schaetzer, J., Wendeborn, S. and Wibley, J. (2009). Herbicidal

4-hydroxyphenylpyruvate dioxygenase inhibitors–a review of the triketone chemistry story from a Syngenta perspective. *Bioorg. Med. Chem.* 17(12), 4134–4152.

Beck, L. L., Patton, A. J., Law, Q. D., Weisenberger, D. V., Brosnan, J. T., Almodóvar, J. J. V., Breeden, G. K. and Kopsell, D. A. (2015). Mesotrione activity on crabgrass (*Digitaria* spp.) as influenced by nitrogen fertilization rate, source, and timing. *Weed Technol.* 29(2), 263-273.

Beck, L. L., Patton, A. J. and Weisenberger, D. V. (2014). Mowing before or after an herbicide application does not influence ground ivy (*Glechoma hederacea*) control. *Appl. Turfgrass Sci.* 11(1), 1-5. doi: 10.2134/ATS-2013-0017-RS.

Bell, G. E., Odorizzi, E. and Danneberger, T. K. (1999). Reducing populations of annual bluegrass and roughstalk bluegrass in creeping bentgrass fairways: a nutritional approach. *Weed Technol.* 13(4), 829–834.

Bertin, C., Senesac, A. F., Rossi, F. S., DiTommaso, A. and Weston, L. A. (2009). Evaluation of selected fine-leaf fescue cultivars for their turfgrass quality and weed suppressive ability in field settings. *HortTechnol.* 19(3), 660–668.

Bertin, C., Weston, L. A., Huang, T., Jander, G., Owens, T., Meinwald, J. and Schroeder, F. C. (2007). Grass roots chemistry: meta-tyrosine, an herbicidal non-protein amino acid. *Proc. Natl. Acad. Sci. U.S.A.* 43, 16964–16969.

Blume, C. J., Fei, S. Z. and Christians, N. E. (2010). Field evaluation of reduced-growth, glyphosate-resistant Kentucky bluegrass in a noncompetitive setting. *Crop Sci.* 50(3), 1048–1056.

Booth, J. C., Sullivan, D., Askew, S. A., Kochersberger, K. and McCall, D. S. (2021). Investigating targeted spring dead spot management via aerial mapping and precision-guided fungicide applications. *Crop Sci.* 61(5), 3134–3144.

Borst, S. M., McElroy, J. S. and Breeden, G. K. (2010). Silvery-thread moss control in creeping bentgrass putting greens with mancozeb plus copper hydroxide and carfentrazone applied in conjunction with cultural practices. *HortTechnol.* 20(3), 574–578.

Boyd, A. P., McElroy, J. S., McCurdy, J. D., McCullough, P. E., Han, D. Y. and Guertal, E. A. (2021). Reducing topramezone injury to bermudagrass using chelated iron and other additives. *Weed Technol.* 35(2), 289–296.

Brabham, C., Johnen, P., Hendriks, J., Betz, M., Zimmermann, A., Gollihue, J., Serson, W., Kempinski, C. and Barrett, M. (2021). Herbicide symptomology and the mechanism of action of methiozolin. *Weed Sci.* 69(1), 18–30.

Branham, B. E. and Rieke, P. E. (1986). Effects of turf cultivation practices on the efficacy of pre-emergence grass herbicides. *Agron. J.* 78(6), 1089-1091.

Braun, R. C., Patton, A. J., Watkins, E., Koch, P. L., Anderson, N. P., Bonos, S. A. and Brilman, L. A. (2020). Fine fescues: a review of the species, their improvement, production, establishment, and management. *Crop Sci.* 60(3), 1142–1187.

Brede, A. D. (1992). Cultural factors for minimizing bermudagrass invasion into tall fescue turf. *Agron. J.* 84(6), 919–922.

Breuillin-Sessoms, F., Petrella, D. P., Trappe, J. M., Mihelich, N. T., Patton, A. J. and Watkins, E. (2021). Field evaluation of weed suppression in fine fescue (*Festuca* spp.). *Crop Sci.* 61(4), 2812–2826.

Brewer, J. R., Askew, W. L. B. and Askew, S. D. (2022a). Differences in selectivity between bermudagrass and goosegrass (*Eleusine indica*) to low-rate topramezone and metribuzin combinations. *Weed Sci.* 70(1), 55–63.

Brewer, J. R., Craft, J. C. and Askew, S. D. (2022b). Influence of posttreatment irrigation timings and herbicide placement on bermudagrass and goosegrass (*Eleusine indica*) response to low-dose topramezone and metribuzin programs. *Weed Sci.* 70(2), 235-242.

Brosnan, J. T., Barrett, M. W. and Bhowmik, P. C. (2020a). Herbicide resistance in turfgrass: a chance to change the future? *Weed Technol.* 34(3), 431-436.

Brosnan, J. T., Breeden, G. K., Zobel, J. M., Patton, A. J. and Law, Q. D. (2020b). Nonchemical annual bluegrass (*Poa annua*) management in zoysiagrass via fraise mowing. *Weed Technol.* 34(4), 482-488.

Brosnan, J. T., Chandra, A., Gaussoin, R. E., Kowalewski, A., Leinauer, B., Rossi, F. S., Soldat, D. J., Stier, J. C. and Unruh, J. B. (2020c). A justification for continued management of turfgrass during economic contraction. *Agric. Environ. Lett.* 5(1), e20033. doi: 10.1002/ael2.20033.

Brosnan, J. T., Elmore, M. T. and Bagavathiannan, M. V. (2020d). Herbicide-resistant weeds in turfgrass: current status and emerging threats. *Weed Technol.* 34(3), 424-430.

Brosnan, J. T. and Breeden, G. K. (2013). Bermudagrass (*Cynodon dactylon*) control with topramezone and triclopyr. *Weed Technol.* 27(1), 138-142.

Brosnan, J. T., Breeden, G. K., Elmore, M. T. and Zidek, J. M. (2011). Application timing affects bermudagrass suppression with mixtures of fluazifop and triclopyr. *Weed Technol.* 25(4), 591-597.

Brosnan, J. T., Dickson, K. H., Sorochan, J. C., Thoms, A. W. and Stier, J. C. (2014). Large crabgrass, white clover, and hybrid bermudagrass athletic field playing quality in response to simulated traffic. *Crop Sci.* 54(4), 1838-1843.

Brosnan, J. T., Henry, G. M., Breeden, G. K., Cooper, T. and Serensits, T. J. (2013). Methiozolin efficacy for annual bluegrass (*Poa annua*) control on sand- and soil-based creeping bentgrass putting greens. *Weed Technol.* 27(2), 310-316.

Brosnan, J. T., Thoms, A. W., McCullough, P. E., Armel, G. R., Breeden, G. K., Sorochan, J. C. and Mueller, T. C. (2010). Efficacy of flazasulfuron for control of annual bluegrass (*Poa annua*) and perennial ryegrass (*Lolium perenne*) as influenced by nitrogen. *Weed Sci.* 58(4), 449-456.

Brown, A., Brosnan, J., Basinger, N., Porter, W. and Henry, G. (2022). Impact of verticutting on the efficacy and soil activity of ALS-inhibiting herbicides for dallisgrass control. *HortScience* 57(2), 290-296.

Buhlman, J. L., Freshour, S. L., Harriman, R. W., Koch, M. J., Unruh, J. B., van Santen, E., Williams, K. E., Yurisic, C. A. and Kenworthy, K. E. (2022). Mowing frequency and performance of genetically modified St. Augustinegrass and standard commercial cultivars. *Int. Turfgrass Soc. Res. J.* 14(1), 652-662. doi: 10.1002/its2.137.

Burns, C. J. and Swaen, G. M. (2012). Review of 2,4-dichlorophenoxyacetic acid (2,4-D) biomonitoring and epidemiology. *Crit. Rev. Toxicol.* 42(9), 768-786.

Busey, P. (2003). Cultural management of weeds in turfgrass: a review. *Crop Sci.* 43(6), 1899-1911.

Buta, J. G. and Spaulding, D. W. (1989). Allelochemicals in tall fescue - abscisic and phenolic acids. *J. Chem. Ecol.* 15(5), 1629-1636.

Cain, A., Braithwaite, E., McDonald, B., Kowalewski, A. and Elmore, M. (2021). Irrigation frequencies and reference evapotranspiration replacement affecting annual bluegrass encroachment in perennial ryegrass in western Oregon. *Int. Turfgrass Soc. Res. J.* 14(1), 555-559.

Campbell, J. H. and Wallace, V. H. (2020). Awareness, support, and perceived impact of the Connecticut pesticide ban. *HortTechnol.* 30(1), 96-101.

Carroll, D. E., Brosnan, J. T. and Breeden, G. K. (2021a). Applications of topramezone and SpeedZone® for POST goosegrass (*Eleusine indica*) control in hybrid bermudagrass. *Weed Technol.* 35(4), 598-603.

Carroll, D. E., Brosnan, J. T., Trigiano, R. N., Horvath, B. J., Shekoofa, A. and Mueller, T. C. (2021b). Current understanding of the *Poa annua* life cycle. *Crop Sci.* 61(3), 1527-1537.

Carroll, D. E., Brosnan, J. T., Unruh, J. B., Stephens, C. A., McKeithen, C. and Boeri, P. A. (2021c). Non-chemical control of annual bluegrass (*Poa annua*) in bermudagrass (*Cynodon* spp.) via fraise mowing: efficacy and barriers to adoption. *Sustainability* 13(15), 8124. doi: 10.3390/su13158124.

Chen, Y., Pettersen, T., Kvalbein, A. and Aamlid, T. S. (2018). Playing quality, growth rate, thatch accumulation and tolerance to moss and annual bluegrass invasion as influenced by irrigation strategies on red fescue putting greens. *J. Agron. Crop Sci.* 204(2), 185-195.

Christians, N. (1996). A historical perspective on annual bluegrass control. *Golf Course Management Magazine*, November, pp. 49-57.

Christians, N. E. (1993). The use of corn gluten meal as a natural preemergence weed control in turf. *Int. Turfgrass Soc. Res. J.* 7, 284-290.

Christians, N. E. and Dant, L. (2005/1999). Corn gluten meal/urea crabgrass control study - year, 6 - 2004. *Iowa State Turfgrass Res. Rep.* Ames, IA: Iowa State University.

Clark, J. M. and Kenna, M. P. (2010). Lawn and turf: management and environmental issues of turfgrass pesticides. In: Kreiger, R. (Ed.), *Hayes' Handbook of Pesticide Toxicology*. Elsevier Science & Technology, pp. 1047-1076.

Conlin, P. (2008). Notice of intent to submit a claim to arbitration. Available at: https://www.uncitral.org/res/transparency-registry/registry/data/can/dow_agrosciences_llc_html/dow-01.pdf.

Cox, M. C., Rana, S. S., Brewer, J. R. and Askew, S. D. (2017). Goosegrass and bermudagrass response to rates and tank mixtures of topramezone and triclopyr. *Crop Sci.* 57(S1), S310-S321. doi: 10.2135/cropsci2016.05.0439.

Cropper, K., Munshaw, G. and Barrett, M. (2017). Optimum seasonal mowing heights for smooth crabgrass reduction in tall fescue lawns. *HortTechnol.* 27(1), 73-77.

Cutulle, M., Derr, J., McCall, D., Nichols, A. and Horvath, B. (2014). Effect of mowing height and fertility on bermudagrass (*Cynodon dactylon*) encroachment and brown patch severity in tall fescue. *Weed Technol.* 28(1), 225-232.

DeBels, B. T., Griffith, S. E., Kreuser, W. C., Melby, E. S. and Soldat, D. J. (2012). Evaluation of mowing height and fertilizer application rate on quality and weed abundance of five home lawn grasses. *Weed Technol.* 26(4), 826-831.

Dernoeden, P. H. (2001). Reduced herbicide rates for smooth crabgrass control in the mid-Atlantic region. *Int. Turfgrass Soc. Res. J.* 9, 1005-1008.

Dernoeden, P. H. (2013). *Creeping Bentgrass Management*. Boca Raton, FL: CRC Press.

Dernoeden, P. H., Carroll, M. J. and Krouse, J. M. (1994). Mowing of three fescue species for low-maintenance turf sites. *Crop Sci.* 34(6), 1645-1649.

Dernoeden, P. H. and Fidanza, M. A. (1994). Fenoxaprop activity as influenced by auxin-like herbicide application timing. *HortScience* 29(12), 1518-1519.

Diehl, K. H., Elmore, M. T., Koppenhöfer, A. M., Murphy, J. A. and Kostromytska, O. S. (2021). Annual bluegrass weevil, paclobutrazol, and overseeding for annual bluegrass control in cool-season turfgrass. *Crop Sci.* 61(2), 1458-1467.

Diehl, K. H., Elmore, M. T., Koppenhöfer, A. M., Murphy, J. A. and Kostromytska, O. S. (2022). Annual bluegrass weevil and paclobutrazol for annual bluegrass control in cool-season turfgrass. *Weed Technol.* 36(1), 137–144.

Domenghini, J. C. (2020). Comparison of acetic acid to glyphosate for weed suppression in the garden. *HortTechnol.* 30(1), 82–87.

Dow AgroSciences LLC vs. Her. Majesty the queen in right of Canada. (2011). *Settlement Agreement*, April 13. Available at: https://www.international.gc.ca/trade-agreements-accords-commerciaux/assets/pdfs/disp-diff/dow-03.pdf (vol. 409-2011).

Duke, S. O. (2012). Why have no new herbicide modes of action appeared in recent years? *Pest Manag. Sci.* 68(4), 505–512.

Duke, S. O. and Dayan, F. E. (2022). The search for new herbicide mechanisms of action: is there a 'Holy Grail'? *Pest Manag. Sci.* 78(4), 1303–1313.

Elford, E. M. A., Tardif, F. J., Robinson, D. E. and Lyons, E. M. (2008). Effect of perennial ryegrass overseeding on weed suppression and sward composition. *Weed Technol.* 22(2), 231–239.

Elmore, M. T., Brosnan, J. T., Kopsell, D. A. and Breeden, G. K. (2012). Nitrogen-enhanced efficacy of mesotrione and topramezone for smooth crabgrass (*Digitaria ischaemum*) control. *Weed Sci.* 60(3), 480–485.

Elmore, M. T., Brosnan, J. T., Kopsell, D. A. and Breeden, G. K. (2013a). Effects of temperature and nitrogen fertilizer on mesotrione activity against annual bluegrass (*Poa annua* L.). *Int. Turfgrass Soc. Res. J.* 12, 657–662.

Elmore, M. T., Brosnan, J. T., Mueller, T. C., Horvath, B. J., Kopsell, D. A. and Breeden, G. K. (2013b). Seasonal application timings affect Dallisgrass (*Paspalum dilatatum*) control in tall fescue. *Weed Technol.* 27(3), 557–564.

Elmore, M. T., Diehl, K. H., Di, R., Chen, J., Patterson, E. L., Brosnan, J. T., Trigiano, R. N., Tuck, D. P., Boggess, S. L. and McDonald, S. (2022a). Identification of two *Eleusine indica* (goosegrass) biotypes of cool-season turfgrass resistant to dithiopyr. *Pest Manag. Sci.* 78(2), 499–505.

Elmore, M. T., Patton, A. J., Thoms, A. W. and Tuck, D. P. (2022b). Mesotrione and amicarbazone tank-mixtures for annual bluegrass (*Poa annua*) control in cool-season turfgrass. *HortScience* 57(1), 10–16.

Elmore, M. T. and Tuck, D. P. (2021). Effect of hollow-tine cultivation on crabgrass infestation in cool-season turf. *Int. Turfgrass Soc. Res. J.* 14(1), 770–775. doi: 10.1002/its2.5.

Elmore, M. T. and Tuck, D. P. (2022). Tall fescue interseeding and postemergence herbicides for false-green kyllinga (*Kyllinga gracillima*) control in turfgrass. *Weed Technol.* 36(4), 548–552. doi:10.1017/wet.2022.51.

Engel, R. E., Aldrich, R. J. and Ahlgren, G. H. (1953). A comparison of five chemicals for crabgrass control in turf. *Weeds* 2(1), 27–32.

Ervin, D. E., Dixon, L. M., Montry, A., Patton, A. J., Elmore, M. T., Gannon, T. W., Bowling, R. G., Kaminski, J. E., Kowalewski, A. R., McCurdy, J. D., McElroy, J. S., Unruh, J. B. and Bagavathiannan, M. (2022). Contemporary challenges and opportunities for improved lawn weed management: insights from U.S. lawn care operations. *Outlooks Pest Manag.* 33(3), 95–100.

Ervin, E. H., Reams, N., Zhang, X., Boyd, A. and Askew, S. (2017). An integrated nutritional and chemical approach to *Poa annua* suppression in creeping bentgrass putting greens. *Crop Sci.* 57(2), 567–572.

Fales, S. L. and Wakefield, R. C. (1981). Effects of turfgrass on the establishment of woody plants. *Agron. J.* 73(4), 605–610.

Ferrell, J. A., Murphy, T. R., Burpee, L. L. and Vencill, W. K. (2003). Effect of brown patch (caused by *Rhizoctonia solani*) control on preemergence herbicide efficacy in tall fescue (*Festuca arundinacea*). *Weed Technol.* 17(4), 747-750.

Fidanza, M., Baldwin, C., Koch, M., Freshour, L. and Harriman, B. (2022). Required mowing events of transgenic Kentucky bluegrass cultivars compared with conventional turfgrass cultivar blends and species mixtures. *Crop Forage & Turfgrass Manag.* 8(1), e20146. doi: 10.1002/cft2.20146.

Flessner, M. L., McElroy, J. S., Rose, J. J. and Guertal, E. A. (2010). Utilization of clove oil and middle-length fatty acids as natural herbicides for weed control in dormant bermudagrass. In: *Proceedings of the 2nd European Turfgrass Society Conference*, Angers, France.

Flessner, M. L., Wehtje, G. R. and McElroy, J. S. (2013). Methiozolin absorption and translocation in annual bluegrass (*Poa annua*). *Weed Sci.* 61(2), 201-208.

Forbes, I. and Ferguson, M. H. (1947). Observations on the Zoysia grasses. *Greenkeepers Rep.* 15, 7-9.

Funk, C. R., White, R. H. and Breen, J. P. (1993). Importance of Acremonium endophytes in turf-grass breeding and management. *Agric. Ecosyst. Environ.* 44(1-4), 215-232.

Gannon, T. W., Jeffries, M. D., Brosnan, J. T., Breeden, G. K., Tucker, K. A. and Henry, G. M. (2015). Preemergence herbicide efficacy for crabgrass (Digitaria spp.) control in common bermudagrass managed under different mowing heights. *HortScience* 50(4), 546-550.

Gardner, D. S., Christians, N. E. and Bingaman, B. R. (1997). Pendimethalin and corn gluten meal combinations to control turf weeds. *Crop Sci.* 37(6), 1875-1877.

Gaussoin, R. E. and Branham, B. E. (1989). Influence of cultural factors on species dominance in a mixed stand of annual bluegrass/creeping bentgrass. *Crop Sci.* 29(2), 480-484.

Gavazzi, S. M. and Gee, E. G. (2018). *Land-Grant Universities for the Future: Higher Education for the Public Good*. Baltimore, MD: Johns Hopkins University Press.

Ghimire, M., Boyer, T. A. and Chung, C. (2019). Heterogeneity in urban consumer preferences for turfgrass attributes. *Urban For. Urban Greening* 38, 183-192.

Gonçalves, C. G., Brown, A. M., Basak, S. and McElroy, J. S. (2021). Response of seashore paspalum and bermudagrass to topramezone and triclopyr mixtures. *Weed Technol.* 35(3), 501-508.

Goss, R. L., Brauen, S. E. and Orton, S. P. (1975). The effects of N, P, K, and S on *Poa annua* L. in bentgrass putting green turf. *J. Sports Turf Res. Inst.* 51, 74-82.

Hahn, D., Sallenave, R., Pornaro, C. and Leinauer, B. (2020). Managing cool-season turfgrass without herbicides: optimizing maintenance practices to control weeds. *Crop Sci.* 60(5), 2204-2220.

Haley, J. E., Wehner, D. J., Fermanian, T. W. and Turgeon, A. J. (1985). Comparison of conventional and mulching mowers for Kentucky bluegrass maintenance. *HortScience* 20(1), 105-107.

Han, K. M., Kaminski, J. E. and Lulis, T. T. (2017). Influence of nitrogen, plant growth regulators, and iron sulfate on annual bluegrass populations on a golf green. *Int. Turfgrass Soc. Res. J.* 13(1), 661-669.

Harivandi, M. A., Hagan, W. L. and Elmore, C. L. (2001). Recycling mower effect on biomass, nitrogen recycling, weed invasion, turf quality, and thatch. *Int. Turfgrass Soc. Res. J.* 9, 882-885.

Harker, K. N. and O'Donovan, J. T. (2013). Recent weed control, weed management, and integrated weed management. *Weed Technol.* 27(1), 1-11.

Harlan, J. R. and DeWet, J. M. J. (1965). Some thoughts about weeds. *Econ. Bot.* 19(1), 16-24.
Harriman, R., Lee, L., Stalker, D. and Torisky, R. (2019). *Plants Comprising Events. PP009-401, 009-415, PP009-469, Compositions, Sequences, and Methods for Detection Thereof.* Patent Number 10, vol. 501,753. U.S. Patent and Trademark Office.
Heap, I. (2022). The international herbicide-resistant weed database. Available at: www.weedscience.org.
Heckman, J. R., Liu, H., Hill, W., DeMilia, M. and Anastasia, W. L. (2000). Kentucky bluegrass responses to mowing practice and nitrogen fertility management. *J. Sustain. Agric.* 15(4), 25-33.
Henderson, J. J. (2021). A new device for selective mechanical broadleaf weed control in turfgrass. *Int. Turfgrass Soc. Res. J.* 14(1), 717-724. doi: 10.1002/its2.26.
Henry, G. M., Hart, S. E. and Murphy, J. A. (2005). Overseeding bentgrass species into existing stands of annual bluegrass. *HortScience* 40(2), 468-470.
Herbicide Resistance Action Committee (2022). HRAC mode of action classification 2022. Available at: https://hracglobal.com/tools/hrac-mode-of-action-classification-2022-map.
Hoyle, J. A., McElroy, J. S. and Rose, J. J. (2012). Weed control using an enclosed thermal heating apparatus. *Weed Technol.* 26(4), 699-707.
Hugie, K., Yue, C. and Watkins, E. (2012). Consumer preferences for low-input turfgrasses: a conjoint analysis. *HortScience* 47(8), 1096-1101.
Imaizumi, S., Nishino, T., Miyabe, K., Fujimori, T. and Yamada, M. (1997). Biological control of annual bluegrass (*Poa annua* L.) with a Japanese isolate of *Xanthomonas campestrispv.poae* (JT-P482). *Biol. Control* 8(1), 7-14.
Janasie, C. (2019). "Can local governments regulate pesticides under FIFRA?" A Maryland court decides. *The Sandbar* 18(4), 11-12.
Jeffries, M. D., Gannon, T. W., Brosnan, J. T. and Breeden, G. K. (2016). Comparing dislodgeable 2,4-D residues across athletic field turfgrass species and time. *PLoS ONE* 11(12), e0168086. doi: 10.1371/journal.pone.0168086.
Jeffries, M. D., Gannon, T. W., Brosnan, J. T. and Breeden, G. K. (2017). Sprayer setup affects dislodgeable 2,4-D foliar residue in hybrid bermudagrass athletic fields. *Weed Technol.* 31(2), 269-278.
Jeffries, M. D., Yelverton, F. H. and Gannon, T. W. (2013). Annual bluegrass (*Poa annua*) control in creeping bentgrass putting greens with amicarbazone and paclobutrazol. *Weed Technol.* 27(3), 520-526.
Johnson, B. J. (1979). Vertical mowing and herbicide treatments on thatch in bermudagrass (*Cynodon dactylon*) turf. *Weed Sci.* 27(1), 14-17.
Johnson, B. J. (1982). Oxadiazon treatments on overseeded putting-green turf. *Weed Sci.* 30(4), 335-338.
Johnston, C. R. and Henry, G. M. (2016). Dallisgrass (*Paspalum dilatatum*) control with thiencarbazone-methyl, foramsulfuron, and halosulfuron-methyl in bermudagrass turf. *HortScience* 51(6), 754-756.
Kao-Kniffen, J. (2020). Allowable herbicides for schools and day care centers. Available at: https://cpb-us-e1.wpmucdn.com/blogs.cornell.edu/dist/6/5969/files/2021/02/CornellTurfgrass-school-herbicide-list-2020-08.pdf.
Kelly, S. T., Coats, G. E. and Luthe, D. S. (1999). Mode of resistance of triazine-resistant annual bluegrass (*Poa annua*). *Weed Technol.* 13(4), 747-752.

Kennedy, A. C. (2016). Pseudomonas fluorescens strains selectively suppress annual bluegrass (*Poa annua* L.). *Biol. Control* 103, 210–217.

Kennedy, A. C., Young, F. L., Elliott, L. F. and Douglas, C. L. (1991). Rhizobacteria suppressive to the weed downy brome. *Soil Sci. Soc. Am. J.* 55(3), 722–727.

Kennepohl, E., Munro, I. C. and Bus, J. S. (2010). Phenoxy herbicides. In: Kreiger, R. (Ed.), *Hayes' Handbook of Pesticide Toxicology (vol. 2)*. San Diego, CA: Elsevier Science & Technology, pp. 1829–1847.

Kephart, L. W. (1948). New weed control developments in the United States. In: *Proceedings of the Second Annual Meeting of the Northeastern Weed Science Society*, New York City, pp. 20–26.

Kerr, R. A., McCarty, L. B., Cutulle, M. C., Bridges, W. and Saski, C. (2019). Goosegrass control and turfgrass injury following metribuzin and topramezone application with immediate irrigation. *HortScience* 54(9), 1621–1624.

King, L. (1966). *Weeds of the World: Biology and Control (vol. 3790)*. New York: Interscience Publishers Inc.

Kohler, E. A., Throssell, C. S. and Reicher, Z. J. (2004). Cultural and chemical control of ground ivy (*Glechoma hederacea*). *HortScience* 39(5), 1148–1152.

Koo, S. J., Hwang, K. H., Jeon, M. S., Kim, S. H., Lim, J., Lee, D. G. and Cho, N. G. (2014). Methiozolin[5-(2,6-difluorobenzyl)oxymethyl-5-methyl-3,3(3-methylthiophen-2-yl) -1,2-isoxazoline], a new annual bluegrass (*Poa annua* L.) herbicide for turfgrasses. *Pest Manag. Sci.* 70(1), 156–162.

Kostromytska, O. S. and Koppenhöfer, A. M. (2014). Ovipositional preferences and larval survival of annual bluegrass weevil, *Listronotus maculicollis*, on *Poa annua* and selected bentgrasses (*Agrostis* spp.). *Entomol. Exp. Appl.* 152(2), 108–119.

Kostromytska, O. S. and Koppenhöfer, A. M. (2016). Responses of *Poa annua* and three bentgrass species (*Agrostis* spp.) to adult and larval feeding of annual bluegrass weevil. *Listronotus maculicollis* (Coleoptera: Curculionidae). *Bull. Entomol. Res.* 29, 72–739.

Kougias, D. G., Miller, E., McEwen, A., Reamer, H., Kovochich, M. and Pierce, J. (2021). Risk assessment of glyphosate exposures from pilot study with simulated heavy residential consumer application of Roundup® using a margin of safety (MOS) approach. *Risk Anal.* 41(9), 1693–1715.

Kuo, S. (1993). Calcium and phosphorus influence creeping bentgrass and annual bluegrass growth in acidic soils. *HortScience* 28(7), 713–716.

Law, Q. D., Bigelow, C. A. and Patton, A. J. (2016). Selecting turfgrasses and mowing practices that reduce mowing requirements. *Crop Sci.* 56(6), 3318–3327.

Li, L., Sousek, M., Reicher, Z. and Gaussoin, R. (2021). Strategies for increased yellow nutsedge (*Cyperus esculentus*) control in turfgrass with halosulfuron, sulfentrazone, and physical removal. *Weed Technol.* 35(6), 894–900.

Lickfeldt, D. W., Voigt, T. B., Branham, B. E. and Fermanian, T. W. (2001). Evaluation of allelopathy in cool season turfgrass species. *Int. Turfgrass Soc. Res. J.* 9, 1013–1018.

Lipinska, H. and Sykut, M. (2012). The influence of cut *Festuca rubra* L. turf not removed from the lawn surface on the proportion of dicotyledons in the lawn. *Acta Agrobot.* 65(1), 123–128.

Liu, D. L. and Christians, N. E. (1997). Inhibitory activity of corn gluten hydrolysate on monocotyledonous and dicotyledonous species. *HortScience* 32(2), 243–245.

Liu, D. L., Christians, N. E. and Garbutt, J. T. (1994). Inhibitory activity of corn gluten hydrolysate on monocotyledonous and dicotyledonous species. *J. Plant Growth Regulat.* 13, 221-226.

Lodge, T. A., Colclough, T. W. and Canaway, P. M. (1990). Fertiliser nutrition of sand golf greens cover and botanical composition. *J. Sports Turf Res. Inst.* 66, 89-99.

Martelloni, L., Caturegli, L., Frasconi, C., Gaetani, M., Grossi, N., Magni, S., Peruzzi, A., Pirchio, M., Raffaelli, M., Volterrani, M. and Fontanelli, M. (2018). Use of flaming to control weeds in 'Patriot' hybrid bermudagrass. *HortTechnol.* 28(6), 843-850.

Martelloni, L., Fontanelli, M., Caturegli, L., Gaetani, M., Grossi, N., Magni, S., Peruzzi, A., Pirchio, M., Raffaelli, M., Volterrani, M. and Frasconi, C. (2019). Flaming to control weeds in seashore paspalum (*Paspalum vaginatum* Sw.) turfgrass. *J. Agric. Eng.* 50(3), 105-112.

Marth, P. C. and Mitchell, J. W. (1944). 2,4-Dichlorophenoxyacetic acid as a differential herbicide. *Bot. Gaz.* 106(2), 224-232.

Mathers, H. M., Acuna, A. A., Long, D. R., Behe, B. K., Hodges, A. W., Haydu, J. J., Schuch, U. K., Barton, S. S., Dennis, J. H., Maynard, B. K., Hall, C. R., McNeil, R. and Archer, T. (2010). Nursery worker turnover and language proficiency. *HortScience* 45(1), 71-77.

Maxwell, P. J., Gannon, T. W. and Cooper, R. J. (2018). Nonionic surfactant affects dislodgeable 2,4-D foliar residue from turfgrass. *Weed Technol.* 32(5), 557-563.

McCullough, P. E., De Barreda, D. G. and Yu, J. (2013). Selectivity of methiozolin for annual bluegrass (*Poa annua*) control in creeping bentgrass as influenced by temperature and application timing. *Weed Sci.* 61(2), 209-216.

McCullough, P. E., Hart, S. E. and Lycan, D. W. (2005). Plant growth regulator regimens reduce *Poa annua* populations in creeping bentgrass. *Appl. Turfgrass Sci.* 2, 1-5. doi: 10.1094/ATS-2005-0304-01-RS.

Miller, N. A. and Henderson, J. J. (2012). Organic management practices on athletic fields: part 1. The effects on color, quality, cover, and weed populations. *Crop Sci.* 52(2), 890-903.

Mithila, J., Hall, J. C., Johnson, W. G., Kelley, K. B. and Riechers, D. E. (2011). Evolution of resistance to auxinic herbicides: historical perspectives, mechanisms of resistance, and implications for broadleaf weed management in agronomic crops. *Weed Sci.* 59(4), 445-457.

Moss, S. (2019). Integrated weed management (IWM): why are farmers reluctant to adopt non-chemical alternatives to herbicides? *Pest Manag. Sci.* 75(5), 1205-1211.

Murphy, J. A., Samaranayake, H., Lawson, T. J., Honig, J. A. and Hart, S. (2005). Seeding date and cultivar impact on establishment of bentgrass in soil containing annual bluegrass seed. *Int. Turfgrass Soc. Res. J.* 10, 410-415.

Nishino, J. and Tateno, A. (2000). Camperico - a new bioherbicide for annual bluegrass in Japan. *Agrochem. Jpn.* 77, 13-14.

Norsworthy, J. K., Korres, N. E. and Bagavathiannan, M. V. (2018). Weed seedbank management: revisiting how herbicides are evaluated. *Weed Sci.* 66(4), 415-417.

Northeast Organic Farming Association of Connecticut, Inc. (2017). *The NOFA Standards for Organic Land Care* (6th edn.). Connecticut: Northeast Organic Farming Association of Connecticut, Inc. ISBN: 978-0-692-58435-4.

Nowak, P., Padgett, S. and Hoban, T. J. (1996). Practical considerations in assessing barriers to IPM adoption. In: Lynch, S., Greene, C. and Kramer-LeBlanc, C. (Eds.), *Proceedings of the Third National IPM Symposium: Broadening Support for 21st Century IPM*, Washington, DC, pp. 93-114.

O'Connor, K., Hébert, F., Powers, J. E., Jordan, K. S. and Lyons, E. M. (2018). Leaf morphology explains the disparity between annual bluegrass and creeping bentgrass growth under foliar fertilization. *J. Plant Nutr.* 41(5), 596-608.

Øgaard, A. F. and Aamlid, T. S. (2020). Temperature effects on phosphorus requirements for creeping bentgrass establishment and spring growth. *Agron. J.* 112(5), 3478-3490.

Ontario Ministry of the Environment (2009). Ontario's cosmetic pesticides ban. Available at: http://news.ontario.ca/ene/en/2009/03/ontarios-cosmetic-pesticides-ban.html.

Ontario Ministry of the Environment (2020). Conservation and parks. *List of Active Ingredients Authorized for Cosmetic Uses*. Available at: https://www.ontario.ca/page/using-pesticides-ontario.

Patton, A., Elmore, M., Kao-Kniffin, J., Branham, B., Voigt, T., Christians, N., Thoms, A., Nikolai, T., Watkins, E., Miller, L., Xiong, X., Gaussoin, R., Carroll, M., Li, D., Gardner, D., Li, D., Landschoot, P., Soldat, D. and Koch, P. (2022). *Turfgrass Weed Control for Professionals*. West Lafayette, IN: Purdue University Press.

Patton, A. J., Beck, L. L., Weisenberger, D., Brosnan, J. T. and Breeden, G. K. (2017a). Evaluation of herbicides for ground ivy (*Glechoma hederacea* L.) control at two application timings. *Int. Turfgrass Soc. Res. J.* 13, 692-697.

Patton, A. J., Schwartz, B. M. and Kenworthy, K. E. (2017b). Zoysiagrass (*Zoysia* spp.) history, utilization, and improvement in the United States: a review. *Crop Sci.* 57(S1), S37. doi: 10.2135/cropsci2017.02.0074.

Patton, A. J., Braun, R. C., Schortgen, G. P., Weisenberger, D. V., Branham, B. E., Sharp, B., Sousek, M. D., Gaussoin, R. E. and Reicher, Z. J. (2019a). Long-term efficacy of annual bluegrass control strategies on golf course putting greens. *Crop Forage Turfgrass Manag.* 5(1), 1-10. doi: 10.2134/cftm2018.09.0068.

Patton, A. J., Braun, R. C. and Weisenberger, D. V. (2019b). Single applications of natural postemergence weed control options do not provide effective ground ivy control. *Crop Forage Turfgrass Manag.* 5(1), 180101. doi: 10.2134/cftm2018.12.0101.

Patton, A. J. and Reicher, Z. J. (2011). Basic training: a 1-day education module for new clientele in the turf industry. *J. Extension* 49, 5TOT6. Available at: https://archives.joe.org/joe/2011october/pdf/JOE_v49_5tt6.pdf.

Patton, A. J., Ruhl, G. E., Creswell, T. C., Wan, P., Scott, D. E., Becovitz, J. D. and Weisenberger, D. V. (2013a). Potential damage to sensitive landscape plants from wood chips of aminocyclopyrachlor damaged trees. *Weed Technol.* 27(4), 803-809.

Patton, A. J., Trenholm, L. and Waltz, F. C. (2013b). Turfgrass extension and outreach programming. In: Stier, J. C., Horgan, B. P. and Bonos, S. A. (Eds.), *Agronomy Monograph 56. Turfgrass: Biology, Use, and Management*. Madison, WI: American Society of Agronomy, Crop Science Society of America, Soil Science Society of America, pp. 147-177.

Patton, A. J., Weisenberger, D. and Breuninger, J. M. (2018a). A growing degree day model is effective to schedule spring Defendor (florasulam) applications for suppressing dandelion flowers. *Crop Forage Turfgrass Manag.* 4, 170088. doi: 10.2134/cftm2017.12.0088.

Patton, A. J., Weisenberger, D. V. and Schortgen, G. P. (2018b). 2,4-D-resistant buckhorn plantain (*Plantago lanceolata*) in managed turf. *Weed Technol.* 32(2), 182-189.

Pedersen, J. (2019). Scotts ProVista: a quantum leap in turfgrass genetics. On Course *Midwest Assoc Golf Course Superintend* 73(3), 24-25.

Peters, E. J. and Luu, K. T. (1985). Allelopathy in tall fescue. In: Thompson, A. C. (Ed.), *The Chemistry of Allelopathy: Biochemical Interactions Among Plants*. Washington, DC: American Chemical Society, pp. 273-283.

Peterson, M. A., McMaster, S. A., Riechers, D. E., Skelton, J. and Stahlman, P. W. (2016). 2,4-D past, present, and future: a review. *Weed Technol.* 30(2), 303-345.

Raley, R. B., Landschoot, P. J. and Brosnan, J. T. (2013). Influence of phosphorus and nitrogen on annual bluegrass encroachment in a creeping bentgrass putting green. *Int. Turfgrass Soc. Res. J.* 12, 649-656.

Ransom, C. V., Rice, C. A. and Shock, C. C. (2009). Yellow nutsedge (*Cyperus esculentus*) growth and reproduction in response to nitrogen and irrigation. *Weed Sci.* 57(1), 21-25.

Raudenbush, Z. and Keeley, S. J. (2017). Cultivation reduces infestation of silvery-thread moss in a creeping bentgrass putting green. *Int. Turfgrass Soc. Res. J.* 13(1), 686-691.

Raudenbush, Z. J. and Keeley, S. J. (2014). Springtime dandelion (*Taraxacum officinale*) control with seven postemergence herbicides applied at three anthesis stages. *HortScience* 49(9), 1212-1216.

Reed, C. J., Gardner, D. S., Danneberger, T. K., Koch, M. J. and Baldwin, C. M. (2021). Overexpression of GA2-oxidase improves Kentucky bluegrass shade tolerance. In: *ASA, CSSA, and SSSA International Annual Meetings: A Creative Economy for Sustainable Development*. Salt Lake City, UT: Paper Publications, p. 118.

Reicher, Z. J., Sousek, M. and Giese, M. (2017). Herbicide programs for annual bluegrass (*Poa annua* L.) control in Nebraska. *Crop Forage Turfgrass Manag.* 3(1), 1-7. doi: 10.2134/cftm2015.0221.

Reicher, Z. J. and Weisenberger, D. V. (2007). Herbicide selection and application timing in the fall affects control of ground ivy. *Appl. Turfgrass Sci.* doi: 10.1094/ATS-2007-0831-01-RS.

Rice, E. L. (1984). *Allelopathy*. Orlando, FL: Academic Press, Inc.

Richardson, M. D., Brosnan, J. T., McCalla, J. H. and Breeden, G. K. (2021). Fraise mowing can improve herbicidal control of bermudagrass. *Agron. J.* 113(5), 3789-3799.

Roberts, J. A., Inguagiato, J. C., Clarke, B. B. and Murphy, J. A. (2011). Irrigation quantity effects on anthracnose disease of annual bluegrass. *Crop Sci.* 51(3), 1244-1252.

Rossi, F. S. and Grant, J. A. (2009). Long term evaluation of reduced chemical pesticide management of golf course putting turf. *Int. Turfgrass Soc. Res. J.* 11, 77-90.

Russell, E. C., Peppers, J. M., Rutland, C. A., Patel, J., Hall, N. D., Gamble, A. V. and McElroy, J. S. (2022). Mitotic-inhibiting herbicide response variation in goosegrass (*Eleusine indica*) with a Leu-136-Phe substitution in α-tubulin. *Weed Sci.* 70(1), 20-25.

Sanchez, A. (2006). *The Teeth of the Lion: The Story of the Beloved and Despised Dandelion*. Blacksburg VA: The McDonald and Woodward Publishing Company.

Schleicher, L., Throssell, C., Reicher, Z. and Weisenberger, D. (1996). Establishing growing degree-day thresholds for scheduling postemergence broadleaf herbicide applications in early spring In: *Turfgrass Research Summary*. Purdue University Publication No. B-732 1995, pp. 82-89.

Schortgen, G. P. and Patton, A. J. (2021). Influence of hard water on 2,4-D formulations for the control of dandelion. *Weed Technol.* 35(3), 371-379.

Shaner, D. L., Lindenmeyer, R. B. and Ostlie, M. H. (2012). What have the mechanisms of resistance to glyphosate taught us? *Pest Manag. Sci.* 68(1), 3-9.

Shekoofa, A., Brosnan, J. T., Vargas, J. J., Tuck, D. P. and Elmore, M. T. (2020). Environmental effects on efficacy of herbicides for postemergence goosegrass (*Eleusine indica*) control. *Sci. Rep.* 10(1), 20579. doi: 10.1038/s41598-020-77570-5.

Skelton, J. J., Sharp, W. and Branham, B. E. (2012). Postemergence control of annual bluegrass with mesotrione in Kentucky bluegrass. *HortScience* 47(4), 522-526.
Skogley, C. R. (1975). Velvet bentgrass putting greens - Fertilizer and topdressing management. *USGA Green Sec. Rec.* 13(5), 7-9.
Smith, A. E. and Martin, L. D. (1994). Allelopathic characteristics of three cool-season grass species in the forage ecosystem. *Agron. J.* 86(2), 243-246.
Sprague, H. B. and Burton, G. W. (1937). *Annual Bluegrass* (*Poa annua* L.) *and Its Requirements for Growth* (Bulletin 630). New Brunswick, NJ: New Jersey Agricultural Experiment Station.
St. John, R. and DeMuro, N. (2013). Efficacy of corn gluten meal for common dandelion and smooth crabgrass control compared to nitrogen fertilizers. *Appl. Turfgrass Sci.* 10, 1-8 doi: 10.1094/ATS-2013-0426-01-RS.
Stern, V. M., Smith, R. F., van den Bosch, R. and Hagen, K. S. (1959). The integration of chemical and biological control of the spotted alfalfa aphid: the integrated control concept. *Hilgardia* 29(2), 81-101.
Struwe, L. (2018). From seed to symbols: dandelion design in our lives. Available at: https://designobserver.com/feature/dandelion-design/39741.
Tang, K., Lulus, T. T., Russell, T. R. and Kaminski, J. E. (2021). Long-term influence of nitrogen, plant growth regulators, and iron sulfate on annual bluegrass populations and turfgrass quality on a golf green. *Crop Sci.* 61(5), 3145-3154.
Thompson, D. G., Stephenson, G. R. and Sears, M. K. (1984). Persistence, distribution and dislodgeable residues of 2,4-D following its application to turfgrass. *Pestic. Sci.* 15(4), 353-360.
Tian, Z., Wang, R., Ambrose, K. V., Clarke, B. B. and Belanger, F. C. (2017). The *Epichloë festucae* antifungal protein has activity against the plant pathogen *Sclerotinia homoeocarpa*, the causal agent of dollar spot disease. *Sci. Rep.* 7(1), 5643. doi: 10.1038/s41598-017-06068-4.
Turgeon, A. J. and Fidanza, M. A. (2017). Perspective on the history of turf cultivation. *Int. Turfgrass Soc. Res. J.* 13(1), 629-635.
United States Environmental Protection Agency. (2020). Simazine proposed interim registration review decision case, September 10. Available at: https://www.epa.gov/sites/default/files/2020-09/documents/simazine_id_signed.pdf (vol. 0070).
Unruh, J. B., Christian, N. E. and Horner, H. T. (1997). Herbicidal effects of the dipeptide alaninyl-alanine on perennial ryegrass (*Lolium perenne* L.) seedlings. *Crop Sci.* 37(1), 208-212.
Varco, J. J. and Sartain, J. B. (1986). Effects of phosphorus, sulfur, calcium hydroxide, and pH on growth of annual bluegrass. *Soil Sci. Soc. Amer. J.* 50(1), 128-132.
Waddington, D. V., Turner, T. R., Duich, J. M. and Moberg, E. L. (1978). Effect of fertilization on Penncross creeping bentgrass. *Agron. J.* 70(5), 713-718.
Walters, D. T. and Gilmore, A. R. (1976). Allelopathic effects of fescue on the growth of sweetgum. *J. Chem. Ecol.* 2(4), 469-479.
Waltz, C. and Murphy, T. R. (2004). Low use rates are the future for sulfonylurea herbicides: background and basics. *Turf Grass Trends* 60(1), 57-58, 60.
Watson, A. K. and Bailey, K. L. (2013). *Taraxacum officinale* F. H. Wigg, dandelion (*Asteraceae*). In: Mason, P. G. and Gillespie, D. R. (Eds.), *Biological Control Programmes in Canada 2001-2012*. Oxfordshire, UK: CAB International, pp. 383-391.

Webber, C. L. III, White, P. M., Jr., Shrefler, J. W. and Spaunhorst, D. J. (2018). Impact of acetic acid concentration, application volume, and adjuvants on weed control efficacy. *J. Agric. Sci.* 10(8), 1-6. doi: 10.5539/jas.v10n8p1.

Weed Science Society of America. (2021). Herbicide site of action classification list. Available at: https://wssa.net/wssa/weed/herbicides/#:~:text=WSSA%2DHerbicide%20Site%20of%20Action%20(SOA)%20Classification%20List) (accessed June 15, 2022).

Weston, J. B. and Dunn, J. H. (1985). Thatch and quality of Meyer zoysia in response to mechanical cultivation and nitrogen fertilization. In: Lemaire, F. (Ed.), *Proceedings of the Fifth International Turfgrass Research Conference*, Avignon, France, pp. 449-458.

Weston, L. A. (1996). Utilization of allelopathy for weed management in agroecosystems. *Agron. J.* 88(6), 860-866.

Woosley, P. B., Williams, D. W. and Powell, A. J. (2003). Postemergence control of annual bluegrass (*Poa annua* spp. *reptans*) in creeping bentgrass (*Agrostis stolonifera*) turf. *Weed Technol.* 17(4), 770-776.

Wu, L., Guo, X. and Harivandi, M. A. (1998). Allelopathic effects of phenolic acids detected in buffalograss (*Buchloe dactyloides*) clippings on growth of annual bluegrass (*Poa annua*) and buffalograss seedlings. *Environ. Exp. Bot.* 39(2), 159-167.

Xiong, X., Moss, J. Q., Haguewood, J. B. and Koh, K. (2015). Safety of sequential fall methiozolin applications on creeping bentgrass putting greens. *Appl. Turfgrass Sci.* 1(1), 1-8.

Xu, X. and Mancino, C. F. (2001). Annual bluegrass and creeping bentgrass response to varying levels of iron. *HortScience* 36(2), 371-373.

Youngner, V. B. (1959). Ecological studies on *Poa annua* in turfgrasses. *Grass Forage Sci.* 14(4), 233-237.

Yu, J. and McCullough, P. E. (2014). Methiozolin efficacy, absorption, and fate in six cool-season grasses. *Crop Sci.* 54(3), 1211-1219.

Yu, J., Sharpe, S. M., Schumann, A. W. and Boyd, N. S. (2019). Detection of broadleaf weeds growing in turfgrass with convolutional neural networks. *Pest Manag. Sci.* 75(8), 2211-2218.

Yu, J., Schumann, A. W., Sharpe, S. M., Li, X. and Boyd, N. S. (2020). Detection of grassy weeds in bermudagrass with deep convolutional neural networks. *Weed Sci.* 68(5), 545-552.

Yue, C., Cui, M., Watkins, E. and Patton, A. (2021). Investigating factors influencing consumer adoption of low-input turfgrasses. *HortScience* 56(10), 1213-1220.

Yue, C., Hugie, K. and Watkins, E. (2012). Are consumers willing to pay more for low-input turfgrasses on residential lawns? Evidence from choice experiments. *J. Agric. Appl. Econ.* 44(4), 549-560.

Zawierucha, J. E. and Penner, D. (2001). Adjuvant efficacy with quinclorac in canola (*Brassica napus*) and turfgrass. *Weed Technol.* 15(2), 220-223.

Zimdahl, R. L. (2015). *Six Chemicals That Changed Agriculture*. Boston: Academic Press.

Zuk, A. J. and Fry, J. D. (2005). Seeded zoysiagrass establishment in a perennial ryegrass sward. *Crop Sci.* 45(4), 1521-1528.

Chapter 12

Advances in plant growth regulation in turfgrass

David Gardner and Ed Nangle, The Ohio State University, USA

1 Introduction

Plant growth regulators (PGRs) are organic compounds that affect plant hormone production or action, thereby altering the growth and development of plants including those species utilized as turfgrass (DiPaola, 1988). There are several compounds marketed for this use in turfgrass management, and their current classification system is indicated in Table 1. An earlier system classified them as either type 1, which inhibits cell division or type 2, which inhibits cell elongation (Roberts et al., 2016; Ok et al., 2011).

The current classification system was developed in the 1990s and categorized PGRs into classes A–E based on their physiological mode of action and site of plant uptake (Watschke et al., 2013). Class A (late gibberellic acid (GA) synthesis inhibitor) includes Trinexapac-ethyl (TE) ((4-(cyclopropylhydroxyme

http://dx.doi.org/10.19103/AS.2022.0110.12

Table 1 Summary of plant growth regulator classification for turfgrass management

Class[a]	Type	Example plant growth-regulating compounds and trade names	Chemical mode of action
A	1	Trinexapac-ethyl 'Primo Maxx'	Late gibberellic acid synthesis inhibitor
		Prohexadione-Ca 'Anuew'	
B	1	Paclobutrazol 'Trimmit'	Early gibberellic acid synthesis inhibitor
		Flurprimidol 'Cutless'	
C	2	Mefluidide 'Embark'	Cell division inhibitor
D	N/A	Glyphosate 'Roundup'	Herbicide
		Methiozolin 'PoaCure'	
		2,4-D 'many trade names'	
E	N/A	Ethephon 'Proxy'	Phytohormone

[a]While not classified, some fungicides, such as the demethylase inhibitors (FRAC Code Group 3), display growth-regulating properties and caution should be taken when applying in warmer air temperature >29°C.

thylene)-3,5-dioxocyclohexanecarboxylic acid)), 'Primo Maxx' (Syngenta Crop Protection LLC, Greensboro, NC), and prohexadione-Ca (PC) ((calcium salt of 3,5-dioxo-4 propionylcyclohexane-carboxylic acid), 'Anuew', Cleary Chemicals, LLC, Alsip, IL). TE reduces the synthesis of GA by inhibiting hydroxylation of the 3β-position on the ent-gibberellin skeleton, which is a late step necessary for biological activity (Adams et al., 1992). PC has a similar mode of action. Both are foliar absorbed.

Class B (early GA synthesis inhibitor) includes paclobutrazol ((2*RS*,3*RS*)-1-(4-chlorophenyl)-4,4-dimethyl-2-(1,2,4-triazol-1-yl)pentan-3-ol), 'Trimmit', Syngenta Crop Protection LLC, Greensboro, NC) and flurprimidol ((2-methyl -1-pyrimidin-5-yl-1-[4-(trifluoromethoxy)phenyl]propan-1-ol), 'Cutless', SePRO Corporation, Carmel, IN). These are root-absorbed and inhibit GA synthesis at cytochrome P450 monooxygenase, which blocks the GA precursor ent-kaurenoic acid. This is an early step in GA synthesis.

Class C (cell division inhibitor) includes mefluidide (5'-(1,1,1-trifluoro-methanesulfonamido) acet-2',4'-xylidide), which is foliar absorbed and reduces cytokinin production, thereby reducing cell division and seedhead formation. Mefluidide's registration was voluntarily canceled by the manufacturer in 2019. Maleic hydrazide [l,2-dihydro-3,6-pyridazinedione] is an older, occasionally tested, but not widely used growth regulator for turfgrass. Class D (herbicide) includes glyphosate (N-(phosphonomethyl) glycine, methiozolin ([5-(2,6-difluorobenzyl)oxymethyl-5-methyl-3,3(3-methyl thiophen-2-yl)-1,2-isoxazoline]), and other herbicides used at very low rates to slow plant growth with or without causing plant death. These compounds are infrequently used as growth regulators, though their growth-regulating effects when applied at very low rates have been studied (Hoisington, 2013).

Class E (phytohormones) includes ethephon ((2-chloroethyl)phosphonic acid) 'Proxy', Bayer Environmental Science, Cary, NC), which is foliar absorbed and causes an increase in the production of ethylene which can be used as a short-term seedhead suppressant in turfgrass. There are other compounds that have growth-regulating effects including some of the fungicides that are registered for use in turfgrass management (Mitkowski and Chaves, 2013).

There are many ways in which PGRs are used in turfgrass. One of the most important uses is for regulation, usually decreasing vertical growth of the grass and thereby decreasing mowing requirements. However, there are several other uses for PGRs that have been investigated that pertain to the morphology of the plant: growth regulation, enhancing rooting and establishment, enhancing wear tolerance, and seedhead suppression. A major focus of PGR research in the last decade has also been on the management and control of *Poa annua* in golf course putting greens. PGRs have also been researched in order to determine their effectiveness in improving turfgrass tolerance to various biotic and abiotic stresses including disease, extremes of weather such as cold or drought, and shade. The goal of this chapter is to focus primarily on research that was reported during the period 2010–2020. Previous research on the use of PGRs has been summarized in review chapters that are referenced in 'Where to look for further information' at the end of this chapter.

2 Growth regulation

The original and most important use of growth regulators is for the reduction in vertical growth and therefore a decreased need for mowing (Fig. 1). It was not long after the release of TE in the 1990s that other potential uses for growth regulators began to be investigated, such as for enhancing sod strength (Bingaman et al., 2001). More recent research, published primarily from 2010 to 2020, that investigated other uses is reviewed in subsequent sections of the chapter.

Most of the use, and therefore research, of growth regulators is on highly maintained turfgrass on golf courses or, more recently, athletic fields. Relatively few studies have been published documenting the effects of PGRs in low-maintenance turfgrass. Pornaro et al. (2017) investigated different rates of TE on three different mixes of cool-season turfgrasses and found differences in response based on the dominant species in the mixture. It was also found that the greatest growth suppression occurred during times of most active growth and that as the turfgrass matured, the growth-regulating effect of TE decreased.

There are often changes in the suggested application rates or frequency of growth regulators compared to either the initial product label or the earlier research (Lickfeldt et al., 2001). For example, one of the issues with the use of scheduled application intervals for TE for season-long growth suppression is that

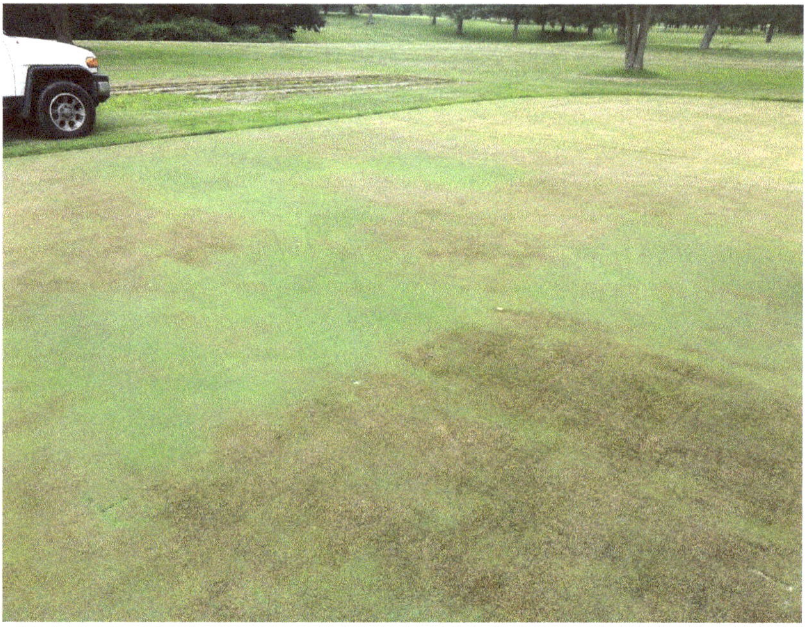

Figure 1 Suppression of growth resulted in less damage from scalping on putting green-height turf after mowing height was accidentally decreased.

it metabolizes more rapidly in warmer temperatures, resulting in a decrease in growth suppression during warmer months. Kreuser and Soldat (2011) developed a growing degree day (GDD) model for TE use on creeping bentgrass (*Agrostis stolonifera* L.) putting greens, where TE was applied at different intervals based on either 100, 200, 400, or 800 GDD (base = 0°C) compared to every 4 weeks. They found consistent 10–12% suppression with 100 GDD or 200 GDD intervals but alternating suppression and growth surge with the other intervals. In a follow-up field trial, they determined that TE applied every 200 GDD produced more uniform growth suppression compared to calendar-scheduled applications.

2.1 Cool-season turfgrass

Kreuser et al. (2018) published additional research on GDD model development using paclobutrazol, which is root-absorbed and applied either alone or in combination with TE on either a 6-week or 1000 GDD interval. The application rate of paclobutrazol did not influence the duration of growth regulation and suggested a reapplication interval of 269–302 GDD (base = 0°C). Both growth regulators were absorbed rapidly, and the observed rebound was thought to be the result of a buildup of metabolic precursors and total non-structural carbohydrates during the suppression phase.

The effect of TE on nitrogen (N) fertility requirements of creeping bentgrass was studied by Kreuser and Soldat (2012). The authors note that N and TE produce similar effects regarding turf color and density. Creeping bentgrass putting greens were fertilized with 3 rates of N either with or without TE applied every 200 GDD. Increasing N caused increases in clipping yield, tissue N content, and N removal, whereas TE enhanced color while causing decreased clipping yields and decreased N removal. Because of this, TE can reduce the N requirements of creeping bentgrass by 25–50%. The authors did not observe this level of suppression in previous experiments, and they speculate this was because the TE was applied every 3 weeks, which was not frequent enough to avoid cycles of post-suppression growth surge.

Whereas an 814 L•ha^{-1} water carrier rate was once considered standard, applications to turf using lower spray volumes have become more common in the last decade. The effect of TE applied in different spray volumes on the quality of 'Midnight' Kentucky bluegrass (*Poa pratensis* L.; KBG) was studied by Cho et al. (2019). Lowering the spray volume from 100 mL per meter to 1 mL per meter did not reduce quality in the greenhouse experiment and in the field study quality was improved on plots receiving the lower spray volume. Similarly, Kim et al. (2019) found that product dilution rate was not a significant factor in the growth regulation of creeping bentgrass from TE.

On a mixed creeping bentgrass/annual bluegrass (ABG) stand fertilized with soluble N, late autumn applications of TE were found to decrease shoot growth by as much as 20% during the following early spring (Rogan and Schlossberg, 2013). Significantly less N was removed by mowing from plots treated with TE. The authors suggest further study to investigate the effect of TE on N assimilation or removal following late autumn N applications.

Glab et al. (2020) used a portable spectrophotometer to study the visual quality of three perennial ryegrass (*Lolium perenne* L.) cultivars treated with five rates of either TE, paclobutrazol, flurprimidol, mefluidide, ethephon, or GA. GA application caused lower scores and more yellowing of the leaves. Paclobutrazol and TE improved color and quality by increasing the amount of red and blue hues and darkening leaves. However, flurprimidol and mefluidide did not improve quality.

The effect of TE on ball roll distance on ABG putting green was measured, along with different rolling and mowing schedules. The results were inconsistent with TE increasing ball roll in the second year of the study but not the first (McDonald et al., 2013). The authors suggest that the differences were due to sod being newly established in the first year of the study. One of the objectives was to see if TE could maintain ball roll speeds in the afternoon and this was observed in year 2; however, the 6.35-cm increase was not considered to be of practical significance.

A hydroponic study conducted with tall fescue (*Festuca arundinacea* Schreb.) showed that either TE or a decreased amount of N reduced shoot extension but did not alter the total mass of plant tissue produced (Sermons et al., 2017). The TE application increased verdure mass by 14–24% in the 3 temperatures that the study was conducted in (22/18°C, 26/22°C, and 30/26°C).

On glyphosate-tolerant perennial ryegrass, it was found that glyphosate applied at 0.57 kg•ha^{-1} had a growth-regulating effect like that of other PGRs (Baldwin et al., 2015). Glyphosate is considered a class D PGR but is not usually used for growth-regulating purposes. It was also shown that adding ammonium sulfate or any of the PGRs flurprimidol, paclobutrazol, or TE improved the perennial ryegrass color and increased control of ABG. The authors suggest future studies with combinations of different herbicides and PGRs to improve the safety of glyphosate-resistant perennial ryegrass and to decrease the chances of the development of herbicide-resistant biotypes of ABG.

2.2 Warm-season turfgrass

On 'Meyer' zoysiagrass (*Zoysia japonica* Steud.), sequential applications of bispyribac-sodium reduced visual quality, but this was not observed when PC or TE was applied (Marchi et al., 2013), and each PGR reduced clipping dry yield by 80%. The treatments reduced mowing requirements for up to 110 days. Marchi et al. (2015) tested different rates of TE, PC, and bispyribac-sodium on St. Augustinegrass (*Stenotaphrum secundataum* (Walt.) Kuntz). Each regulator produced some visual injury that dissipated by 28 days after treatment (DAT). One application of TE reduced turf height by 59.7% and dry matter production by 87%. The growth-regulating effects in that study were observed for up to 119 days.

Nineteen seeded cultivars of several warm-season turf species, including bermudagrass (*Cynodon dactylon* (L.) Pers.), buffalograss (*Bouteloua dactyloides* (Nutt.) Columbus), zoysiagrass, bahiagrass (*Paspalum notatum* Flüggé), seashore paspalum (*P. vaginatum* Sw.) and centipedegrass (*Eremochloa ophiuroides* (Munro)Hack.) were treated with either one or two applications of TE either in summer or autumn. Summer application of TE enhanced the turf color and quality of the tested species while the autumn application resulted in decreased quality, fall color retention, and delayed spring green-up (Severmutlu et al., 2012a). Clipping production was reduced by up to 72% (Severmutlu et al., 2012b). After the first TE application, the authors noted significantly lower turf quality for 1 week on treated bermudagrass and seashore paspalum, 4 weeks on buffalograss and bahiagrass, and 6 weeks on zoysiagrass. After 4 weeks when a sequential application was applied, plots with bermudagrass and centipedegrass treated with sequential applications of TE had higher quality compared to 1 application or the untreated. This improvement in quality and the duration of improvement varied by species. The authors speculate that a single late-season low-rate TE application may be able to provide color enhancement but not result in the negative effects noted in this experiment (Severmutlu et al., 2012a). One application of TE in the summer reduced vertical growth and clipping yields for up to 4 weeks and 2 applications reduced clipping yields for up to 8 weeks. One application in the autumn reduced clipping yields for up to 9 weeks. The authors suggest that TE application in autumn can aid in the transition to over-seeded cool-season turf by reducing growth and hastening the onset of winter dormancy.

TE decreased the quality of 'Tifway' bermudagrass 7 days after application but improved it at 14 and 21 DAT (McElroy et al., 2013). Clipping weights were decreased at 7 DAT but by 28 DAT clipping weights increased compared to the untreated. Because of the documented growth-regulating effect of ALS-inhibiting herbicides, this research also looked at combining TE with the ALS-inhibiting herbicide flucarbazone (3-methoxy-4-methyl-5-oxo-N-[2-(trifluoromethoxy)phenyl]sulfonyl-1,2,4-triazole-1-carboxamide), which improved quality and decreased clipping yields compared to TE alone.

The effect of TE alone or combined with flurprimidol on ball lie of 'TifSport' bermudagrass was studied by Kowalewski et al. (2012). Ball lie is defined as the proportion of the ball above the turfgrass canopy after a golf stoke. Different frequencies of PGR application alone or in combination with N were tested on different heights of cut (1.27 cm, 2.54 cm, and 3.81 cm). N alone correlated with worse ball lie while combining N with TE or TE plus flurprimidol improved ball lie and turf quality.

McCarty et al. (2011) evaluated TE, ethephon, flurprimidol, and the pairwise combinations of those PGRs to 'TifEagle' bermudagrass managed at three heights. All three PGRs reduced clipping yields, with flurprimidol being more effective than TE or ethephon. However, flurprimidol caused unacceptable quality (3.6–5.2 on a 1–9 scale) compared to TE (8.0–8.4) or ethephon (6.8–7.1). Flurprimidol increased ball roll more than TE did, and ethephon did not cause a difference. Lateral growth was impeded by flurprimidol but not TE or ethephon. Root length was improved with TE, unaffected by ethephon, and reduced with flurprimidol. Based on this work, the authors determined that TE was the best option for ultradwarf bermudagrass management and that no synergisms were detected when the PGRs evaluated were applied in combination.

Monthly applications of TE were made to 19 cultivars of 4 medium-to-coarse textured warm-season turfgrasses. Across all cultivars, the authors found that increasing TE from 60, 96, 180, or 240 $mg \bullet ha^{-1}$ reduced the number of required mowing by 15%, 17%, 24%, and 29%, respectively (Roche et al., 2013). The effect was more pronounced on bermudagrass, hybrid bermudagrass, and blue couch (*Digitaria didactyla*) compared to kikuyugrass (*Pennisetum clandestinum*).

Seashore paspalum is utilized for greens, tees, and fairways on the golf course and has several agronomic advantages. One issue is seedhead production during periods of active growth. On 'Sea Isle' seashore paspalum, TE applications reduced clipping yields 48–85% compared to the untreated. Seedhead cover was reduced on TE-treated plots by 13–52% for 1–6 weeks and 65–83% at 7–9 weeks after treatment (McCullough et al., 2014).

Reasor and Brosnan (2020) showed that lightweight rolling 5 days per week combined with a lower rate application of TE to 'MiniVerde' ultradwarf bermudagrass resulted in improved visual quality but did not change the ball roll distance appreciably compared to the higher rate of TE alone, which did reduce bermudagrass quality. In another study, the authors developed a GDD model that utilizes air temperatures and metabolism rates of PGRs to guide application decisions for bermudagrass as opposed to applying on a schedule (Reasor et al., 2018). These models had previously been developed to guide applications on creeping bentgrass. PC or TE was applied to three cultivars of bermudagrass at three separate field trial locations. Clipping yield suppression was greatest on TE-treated plots at 166–177 base 10°C GDDs and 92–97 base 10°C GDDs on PC-treated plots leading to recommended application timings of 216–230 GDD for TE and 120–126 GDD for PC. The authors recommend further research on the season-long implementation of GDD models for bermudagrass growth regulation programs.

Models based on GDD were also the subject of research by Brown et al. (2020a,b). Many GDD models are based solely on GDD, but in this research, the authors also looked at calendar date, global horizontal irradiance, soil temperature, and photosynthetically active radiation in order to predict application intervals for maximum suppression of 'MiniVerde' bermudagrass. Of the models tested, a base 0°C model was more useful and maximum suppression was observed at 262 GDD (Brown et al., 2020a). In a second field experiment, the authors tested applications of two rates of TE at intervals based on 100, 200, 400, and 600 GDD. Applications based on 100 GDD or 200 GDD produced uniform regulation and were dependent on the TE rate. In contrast, applications based on 400 GDD or 600 GDD resulted in less consistent suppression.

3 Enhancing rooting and establishment

Rooting in turfgrass stands is important for a variety of reasons. Stability in sports fields which is tied to rooting depth and quality is critical to player safety. On golf courses, rooting is considered indicative of overall plant health and shorter roots can cause concern for turfgrass managers due to reduced surface area impacting water and nutrient uptake. Basic agronomic practices that ensure sufficient levels of oxygen are a first step to maintaining root vigor and health, but other approaches may be used. This can be based on the turf use pattern of the facility and thus an interest in growth regulators which may suppress shoot growth in favor of root development is judged to be a valuable methodology for turfgrass managers.

Research has shown vertical growth reduction by TE is not at the expense of lateral shoot production. Less than 5% of applied TE was found in roots and rhizomes of KBG in a study by Fagerness and Penner (1998). It was hypothesized then that the diverted photosynthate that would be used for leaf tissue is instead used to produce stem or root tissue of the plant and thus well-timed applications of PGRs might improve rooting and establishment. Bingaman et al. (2001) found that applications of TE either 2 weeks prior to harvest increased root tensile strength by 22% or 2 weeks prior plus 2 weeks after transplant increased root tensile strength by 18%. However, Ervin and Koski (2001) found no change in root mass or TE-treated KBG. In warm-season turfgrass, TE application increased the oven-dry root mass of dwarf bermudagrass (McCullough et al., 2006) and zoysiagrass (Qian and Engleke, 1999).

Seeds of tall fescue 'Master' and perennial ryegrass 'Barrage' were soaked in different concentrations of paclobutrazol for 24 h prior to seeding. The height of the seedlings was significantly reduced 10 days after emergence, but after first mowing, few differences in seedling height, clipping dry weight, or leaf width were noted. The visual quality of perennial ryegrass was improved on the seeds treated with the highest rate of paclobutrazol but not the other rates (Shahrokhi et al., 2010).

Grasses are planted as a phytoremediation strategy on soils that are contaminated with petroleum hydrocarbons. Zhu et al. (2016) found that pretreating seeds with ethephon and the use of potting soil as an amendment could improve establishment of saltgrass (*Distichlis spicata* L. Greene) for phytoremediation. Saltgrass is more tolerant of saline conditions compared to other cool-season turfgrass species. However, the germinating seed of this species is more sensitive to excess salinity. Nickell and Qian (2013) found that ethephon could enhance the germination and establishment of saltgrass.

In a greenhouse trial, Volterrani et al. (2012) recorded the linear growth rate, production, and vitality of nodes and daughter plants of 'Tifway' bermudagrass. They found that TE applications did not affect stolon production, node vitality, or the number of nodes. The authors did find that stolon length was 24.6–52.9% greater versus untreated control at 2 weeks after treatment regardless of application rate. At 3 weeks and 4 weeks after treatment, the 2 highest rates tested (which were above label) still resulted in reductions in stolon length. Another observation the authors made was that with plants treated with higher than label rates; if nodes were excised at 4 weeks after treatment, the daughter plants showed the typical post-inhibition growth enhancement seen with TE applications. The authors suggest that if sprigs were harvested from previously treated turf there could be an improvement in the establishment rate.

A study was conducted to determine if TE could improve the root development of 'TamStar' and 'Floratam' St. Augustinegrass sod during a 35-day establishment. TE application did not affect the visual quality of 'TamStar' but decreased the visual quality of 'Floratam' (Hargey et al., 2016). While clipping yields were decreased due to TE, the authors did not observe any improvements to either root length, root development, or root extension rate.

A frequent practice in bermudagrass managed in the southern United States and transition zone is to overseed with perennial ryegrass to provide green cover during autumn and winter. Cox et al. (2013) studied the effects of different cultural and chemical applications, including the use of TE and found that in 1 year of the study TE increased ryegrass cover regardless of other cultural practices. However, in the second year of the field trial, the amount of perennial ryegrass did not differ from the untreated.

On sports-based fields that are grown on the sand, belowground biomass is important for stabilizing the system and providing traction for the athletes. As the biomass wears away, these structures can become less stable and divoting of the field can occur. For KBG fields that are only used for autumn sports, divot length can be reduced by up to 18% by applying TE from May to July (Serensits et al., 2017). The authors also found that TE increased the belowground root biomass by 8–9% compared to the untreated.

4 Enhancing wear tolerance

In addition to work that has investigated growth regulators for enhanced turf rooting and establishment, work has also been conducted to determine if the use of PGRs could potentially improve wear tolerance, which is of considerable interest to the sports turf industry. Several studies have documented the benefit of PGR application for improving tiller density (Beasley et al., 2005; Ervin and Koski, 2001). Ervin and Koski (2001) also cautioned against the use of TE on heavily trafficked turf because the reduced clipping yield could result in reduced wear tolerance. As of 2010, a limited amount of research that investigated the use of PGRs specifically for the enhancement of traffic tolerance had been published.

In a study with perennial ryegrass, TE was applied at either 0 kg ai•ha^{-1}, 0.25 kg ai•ha^{-1}, or 0.5 kg ai•ha^{-1} 3 times over a 6-week interval, after which traffic was applied using a cleated roller. The authors found that both traffic stress and applications of TE reduced the quality, vertical growth, and dry weight of clippings (Amiri-Khah et al., 2015). TE applications improved chlorophyll a, chlorophyll b, total chlorophyll, and total carbohydrates as well as the overall quality of the turf. However, applications of TE did not improve traffic tolerance.

On 'Riviera' bermudagrass subjected to simulated traffic, the turf color, quality, and percent cover were improved on plots treated with TE, ethephon and TE, or flurprimidol and TE, compared to plots treated with either ethephon or flurprimidol alone, suggesting that TE was the more important growth regulator for enhancing tolerance to traffic stress (Brosnan et al., 2010).

Deaton and Williams (2010) studied the effects of TE with or without overseeding of perennial ryegrass on four different cultivars of bermudagrass managed as a sand-based athletic field. They applied simulated traffic with a Brinkman traffic simulator. The authors found that TE did not improve traffic tolerance in either year of the study. The authors also utilized a Clegg shear tester to investigate the shear strength of bermudagrass cultivars 'Quickstand', 'Tifway 419', 'Riviera', and 'Yukon' that had been overseeded with perennial ryegrass at 0 kg PLS•ha^{-1}, 612 kg PLS•ha^{-1} or 1224 kg PLS•ha^{-1}. In this study, applications of TE did not have a significant effect.

In wheatgrass (*Agropyron desertorum* L.) and tall fescue, the application of TE can improve tolerance to traffic stress by altering the osmotic potential of the grass and increasing the activity of the antioxidants superoxide dismutase, ascorbate peroxidase, and peroxidase (Mohammadi et al., 2017b). In this study, traffic was simulated with a Brinkman traffic simulator. Traffic-stressed turf had lower quality and relative water content as well as lower total chlorophyll and antioxidant activity. The authors also found increases in electrolyte leakage and malondialdehyde in traffic-stressed turf. On turf treated with TE, the quality, relative water content, soluble sugar content, and total chlorophyll levels were improved and there were decreases in both electrolyte leakage and malondialdehyde.

In summation, the use of growth regulators on intensively used and managed sports playing surfaces provides turfgrass managers with options for enhanced wear tolerance. The expectation is that this pattern of use will continue to be adopted along with an increased understanding of the benefits of their use, particularly on highly used fields and pitches. Future focuses may include how combinations of wear and issues such as shade can be mitigated in stadiums using these products on a worldwide scale due to the needs and different microclimates where they could be utilized.

5 Seedhead suppression and seedhead production

The production of seedheads is a playability concern for many golf course superintendents as it results in bumpy inconsistent putting green surfaces (Fig. 2). Furthermore, the production of seedheads depletes carbohydrates ahead of the summer stress period of heat and humidity, and mitigating this loss can be

Figure 2 Seedhead suppression with plant growth regulators is an agronomic practice used regularly by turfgrass managers to reduce or eliminate the development of *Poa annua* seedheads on a yearly basis (left = untreated; right = treated) (Courtesy of D. Petrella).

critical to the success of summer playing conditions on those putting greens. A vast range of research has focused on either stopping seedhead production or limiting it using different growth regulators.

Some of the earliest work focused on ABG *Poa annua* which showed that multiple applications of maleic hydrazide (1,2-dihydro-3,6-pyridazinedione), chlorflurenol, and endothall at low rates could inhibit seedheads but that the most successful treatments led to objectionable discoloration of leaf tissue (Watschke et al., 1979). Other work surmised that applications of mefluidide and ethephon in a short period of time post spring greening of ABG led to desirable results regarding the control of seedheads (Gaussoin and Branham, 1989). Personal judgment and changes in application windows due to spring climate variation meant that getting this timing correct was a problem which required further work. Research on centipedegrass showed that 2 mefluidide applications at 0.56 kg ha^{-1} at 2-week intervals could suppress seedhead formation for 10 weeks while a combination application of flurprimidol and mefluidide led to damaged turf and limited regulation. Imazethapyr (5-ethyl-2-(4-isopropyl-4-methyl-5-oxo-4,5-dihydro-1H-imidazol-2-yl)nicotinic acid) applied twice at a rate of 0.30 kg•ha^{-1} and followed with 0.15 kg•ha^{-1} application suppressed seedhead on average for 8 weeks after application with no loss of turf density (Johnson, 1990). Carpetgrass (*Axonopus affinis* Chase.) seedhead formation was reduced by 88% and 86% following applications of sethoxydim (2-[(*E*)-*N*-ethoxy-*C*-propylcarbonimidoyl]-5-(2-ethylsulfanylpropyl)-3-hydroxycyclohex-2-en-1-one) and sulfometuron (methyl 2-{[(4,6-dimethylpyrimidin-2-yl)carbamoyl]sulfamoyl}benzoate), respectively, 21 days after application. Injury to the turf occurred with the sulfometuron methyl treatment while carpet grass seedhead production was restricted up to 6 weeks after applications of sethoxydim (0.17 kg•ha^{-1} and 0.22 kg•ha^{-1}) (Fry and Wells, 1990).

Warm-season turfgrasses are tolerant of many growth regulators and so their use for suppression of seedhead development has been used across many situations. Seashore paspalum (*Paspalum vaginatum*) is used in many areas where access to potable water is limited in warm temperate climates. During periods of growth, the species is prone to producing seedheads and applications of TE at a rate of 96 g•ai•ha^{-1} combined with flazasulfuron at a rate of either 16 g•ha^{-1} or 26 g•ha^{-1} suppressed clippings by 70–90% of untreated plots as well as consistently suppressing seedheads. There was some injury noted although this was never higher than 23% of treated areas (McCullough et al., 2011). Control of bermudagrass seedheads has also been investigated as they reduce the esthetics and functionality of the species. Further to this seedhead production is seemingly unaffected by day length (Beard, 1973). The suppression of 'Riviera' bermudagrass seedhead was limited to less than 25% 35 DAT using TE or mefluidide indicating a poor level of reduction (Brosnan et al., 2011). In the transition area of the United States, zoysiagrass is used for its improved cold and shade tolerance. During springtime, it also produces seedhead increasing maintenance requirements and reducing the visual quality of the surfaces due to issues with mowing quality. Applications of ethephon in late September at three sites across the Midwest (KY, KS, and IN) found that applications of 159 mL/100 m^2 reduced seedhead formation the following spring by greater than 90% with less than 10% injury reported (Patton et al., 2018).

Other work indicated that applications of bispyribac-sodium resulted in the visual injury of 'Meyer' zoysiagrass but that sequential applications of all growth regulators tested (TE, bispyribac-sodium, and PC) reduced both the height and development of rachis (Marchi et al., 2013). This result was not found when PC was applied to bahiagrass despite a reduction in clipping yield (Marchi et al., 2016). The visible injury was noted for a period of less than 28 days, however, on St. Augustinegrass (*Stenotaphrum secundatum*) when applications of TE, PC, and bispyribac-sodium were made. TE in the same piece of research reduced rachis emission (seedhead formation) by 96.4% (Marchi et al., 2015). The physiological responses can be harder to determine due to internal biological processes that turfgrass managers cannot see. Applications of radioactive labeled ^{14}C-ethephon to 5 different grasses resulted in foliar absorption increasing from 24 h to 96 h after treatment with no differences between species. The compound was distributed differently, however, within species. Bermudagrass and seashore paspalum translocated 25% more ^{14}C to the roots than the other 3 grasses which did result in higher injury levels for those 2 species compared to ABG, creeping bentgrass, or zoysiagrass (Sidhu et al., 2014).

A more recent focus has been refining the use of growth regulators to reduce the impact of seedheads from ABG. Since ABG is adapted to extremely low mowing heights on putting greens, the seedheads can have a direct effect on surface playability, thus causing balls to bounce and not hold the intended line of the putt. The application of growth regulators prior to seedhead formation has shown success and work has focused on autumn and winter applications due to inconsistent results being found with spring applications. Ethephon applications in January and February resulted in five times less seedheads compared to control plots without injury. Mefluidide did damage both creeping bentgrass and ABG when this methodology was applied (Askew, 2017). Other treatments in that research were more programmatic in nature and the early application of ethephon followed by variations in combinations of TE and ethephon and triadimefon resulted in greater seedhead suppression but also a slight increase in injury. Work with the addition of a pigmented oil-based product during autumn ethephon applications resulted in even greater reductions in seedhead development (Raudenbush et al. 2020). The ethephon uptake and translocation does influence seedhead suppression. Applications of ^{14}C to the flag leaf (last leaf to emerge) resulted in seven times more radioactivity in the inflorescence compared to applications made to the antepenultimate leaf. Further to this, ABG broke down ^{14}C-ethephon 1 day after being applied to the flag leaf, but a different compound was measured after 7 days. This indicates that applications and transport from the flag leaf are more important for seedhead suppression by ethephon than trying to get the product deeper into the canopy (McCullough and Sidhu, 2014).

5.1 Seed production

Growth regulators can be used to aid in seed production by reducing lodging and allowing for more efficient harvest. Chastain et al. (2014) found a 43% increase in seed yield on TE-treated perennial ryegrass. In addition to reducing lodging, TE application resulted in increased seeds per spikelet, mass of seed per spikelet, and seed set. The total number of spikes and total seed weight were not affected by TE.

A common practice in seed production is to make a single application of a growth regulator early in the season to reduce lodging and increase yield.

However, research has shown that PGR half-lives can be too short for this to be effective. A study was conducted on perennial ryegrass using PC and TE. Split applications of PC reduced lodging 70% and increased seed yield 36%. There were other benefits including reduced rates of and severity of stem rust (Koeritz et al. 2013). The authors suggest that the development of an application model based on growth stage, GDDs, and available soil N could improve growth regulator efficacy.

Sequential or split applications of growth regulators have also been used to try and increase seed yield from perennial ryegrass. Increases of 36% in seed yield were noted due to applications of PC along with plants that were 14 cm shorter resulting in 30% less lodging. The same trial did not see a similar result from applications of TE (Koeritz et al. 2013). Spring applications of TE to post-harvest fields of strong creeping red fescue (*Festuca rubra* L. subsp *rubra*) in the spring resulted in increases of up to 14% in floret spikelets and reduced lodging (Zapiola et al., 2014).

Growth regulators have been tested for weed suppression as well, with seed germination rates of dandelion (*Taraxacum officinale*) being reduced by treatments of ethephon at rates of 1.0 mM and 1.5 mM, respectively. Applications of salicylic acid also reduced seed germination rates compared to the control, but the effect was greater when combined with gibberellins (GA_3) (Kim and Lee, 2013).

The focus of seedhead suppression and production has different aims – however, in both situations, the potential arises for the use of growth regulators. Reductions in lodging and other issues tied to seed production are valuable to end users while golf course managers should consider the use of regulators to aid with playability issues tied to seedhead production. Caution is recommended if using any of these products for the first time and it is also strongly recommended that untreated control areas be used to compare the results.

6 Suppression and control of annual bluegrass

There has been much interest in trying to control ABG (*Poa annua* L.) due to the lack of abiotic and biotic stress tolerance it displays during the summer. ABG is strongly competitive in the 'shoulder' seasons, that is, early spring and late fall, on golf courses in the northern United States and Europe and can infest playing surfaces rapidly when allowed to germinate and produce inflorescence. The ability of growth regulators to specifically limit seedhead production as well as differentially impact ABG growth compared to creeping bentgrass is critical to the reason they have been used for this process. Consideration is given to the efficacy and longevity of treatments, potential non-target issues, and cost of the approach.

Early work on ABG control has vacillated between attempts to eradicate it chemically or best agronomic management practices to sustain it through stressful periods. Chemical eradication focused on the use of preemergence herbicides which prevent seed germination, particularly during the early autumn period in the northern hemisphere – a key time in the success of ABG development (Dernoeden, 1998). Active ingredients such as prodiamine, bensulide, dithiopyr, and pendimethalin are considered effective if applied in early autumn, but a second application may be needed ahead of spring seed germination. The main issues with these are the limitations on the amount of product that can be used annually thus requiring very precise application timings and only one product, bensulide, is labeled for use on putting greens due to non-target safety concerns, and only one application is allowed annually. Postemergence herbicides are equally limiting with both turf safety and lack of consistent ABG efficacy a problem (Dernoeden and Turner, 1988). Bispyribac-sodium has had success at ABG control but has concerns tied to it with non-target effects at cooler temperatures, and the timing of application initiation is also critical (McCullough and Hart, 2006). Previous growth regulator work did show efficacy against ABG development via applications of paclobutrazol and flurprimidol with beneficial responses from creeping bentgrass (Woosley et al., 2003); however, optimizing the timing of these applications through the season to include spring and autumn applications, and avoid frost, is needed.

Turf management approaches included work that showed TE was a valuable tool in aiding ABG quality during the summer period in New York and went as far as to state that TE should not be used in an ABG eradication program (Rossi, 2001). Interestingly, combining TE with flurprimidol resulted in control of ABG as well as a reduction in the phytotoxic response displayed post-treatment (Bigelow et al., 2007). Further work on this area has led to the development of a range of growth regulator combinations which focus on two areas: minimize negative impacts on desirable species while inhibiting ABG in a manner that does not reduce turf quality of the playing surface. Non-target effects have been detrimental to the success of many products marketed for control of ABG. A recent example is amicarbazone which can cause damage to desirable turfgrass species and may need to have limited seasonal applications in spring (McCullough et al., 2010). The addition of growth regulators, however, may enhance herbicide effects as well as allow for a reduction in the phytotoxic discoloration and damage associated with some of these products due to rate reductions and increased windows of application, and this is an ongoing area of interest. Paclobutrazol applications combined with amicarbazone were considered to provide control that was equal to or greater than individual applications of amicarbazone (Jeffries et al., 2013). Combinations of growth regulators can also be effective in the suppression of ABG, but concerns still arise with non-target species. Some discoloration was noted when weekly or bi-weekly bispyribac-sodium applications of 12.4 g•ha^{-1} and 24.8 g•ha^{-1} and paclobutrazol (224 g•ha^{-1} month^{-1}) were combined in spring treatments (Moss et al., 2012). Critically turf quality was not compromised below acceptable levels and this is an option for turfgrass managers.

Applications of paclobutrazol on a bi-weekly basis at a rate of 0.28 kg•ha^{-1} resulted in the greatest increases in creeping bentgrass coverage on putting greens in the Pacific northwestern United States over a 2-year period (Baldwin and Brede, 2011). The issue arises in this instance with climatic differences to other regions as well as the length of time the trial took. It has been shown that over an extended period (3–4 years) in three midwestern states (NE, IN, and MI) that applications of paclobutrazol did reduce ABG populations most effectively while flurprimidol was found to be almost as effective, but TE showed no impact on the ABG population (Reicher et al., 2015). It should be noted, however, that overall control levels were very low and that the applications of any growth regulator will need to coincide with cultural practice changes to allow for greater efficacy.

The simplistic approach of applying chemistry to the species has failed in many instances and it is now considered more appropriate to incorporate cultural control strategies to try and reduce ABG populations in putting greens in particular. The use of iron sulfate ($FeSO_4$) to reduce rootzone pH has received some interest in combination with flurprimidol and variations in N rates (Han et al., 2017). This 2-year trial, however, produced results that suggested further work was needed. The process of removing soil cores to improve air movement and drainage known as aeration, aerification, or core cultivation did not reduce the ABG population in three states in the midwestern United States on its own regardless of summer or autumn timing. Monthly applications of $FeSO_4$ also did not control ABG in the same trial over a 4-year period (Patton et al., 2019). The trial did incorporate newer chemistries, and of those, methiozolin, bispyribac-sodium, and paclobutrazol offered the greatest levels of reduction. At the time of this research publication, methiozolin was not available for use on greens in the United States and this has since changed. It has shown some excellent results without the need to add any other components such as ammonium sulfate or $FeSO_4$ to the applications (Flessner et al., 2017).

Growth characteristics of the bluegrasses also provide some interesting approaches for the use of growth regulators. Applications of ^{14}C radioactive labeled growth regulator PC were taken up in greater concentrations by ABG and KBG (*Poa pratensis* L.) when compared to perennial ryegrass and creeping bentgrass (*Agrostis stolonifera* L.) (Beam and Askew, 2007). There were also greater levels of root translocation which helps in part to explain why both KBG and ABG are suppressed to a greater extent by the application of this growth regulator. ABG also possesses a very diverse range of biotypes and it has been noted that differential responses between growth regulator applications and species composition can occur. An example indicates that light green colored ABG biotypes were controlled by PGRs to a much greater extent than dark green biotypes (Williams, 2014).

The use of growth regulators has been one of the most successful approaches to reducing or eradicating ABG in greens, but concerns still exist over turf safety, application timing, and cost, and much work continues to look at different modes of action, timings, and rates to aid with the problem of ABG development. The use of growth regulators also offers a more cost-effective approach to reducing ABG levels for larger areas of acreage such as fairways, and at the higher heights of cut, there is a reduction in concerns about safety. Turfgrass managers have begun to focus on the variations within the growth regulators and this precision will only increase as we learn more about these nuances.

7 Enhancing disease tolerance

Because some of the growth regulators are chemically similar to some of the fungicides used in turfgrass management, researchers have also investigated if applications of growth regulators could improve disease tolerance of turfgrasses. Earlier work in this area with TE is mixed. For example, Burpee et al. (1996) showed that when TE was applied in the absence of fungicides it had no effect on dollar spot. However, Golembiewski and Danneberger (1998) did observe suppression of dollar spot when TE was applied by itself. Research with flurprimidol and paclobutrazol indicates that these compounds are more likely to suppress dollar spot than TE (Burpee et al. (1996). Both flurprimidol and paclobutrazol have been shown to enhance dollar spot control when applied with fungicides (Fidanza et al., 2006; Stewart et al., 2007).

On fairway-height creeping bentgrass, Vincelli et al. (2011) did not observe a difference in dollar spot disease severity or the number of requisite fungicide applications on plots treated with TE. Huang et al. (2015) found that applications of TE to creeping bentgrass prior to the application of fungicides could reduce dollar spot severity, but this was not consistent and control with TE alone would not be considered acceptable on a high-end golf course. Also, dollar spot symptom expression and control were not affected. But, where fungicides cannot be applied, TE could provide some suppression.

Resistance of *Sclerotinia homeocarpa* (syn. *Clarireedia*) to fungicides and PGRs was studied. Some of the PGRs are chemically similar to the demethylation inhibitors (DMI) fungicides. On 'Putter' creeping bentgrass, Putman and Kaminski (2011) reduced dollar spot severity by 63–90% with paclobutrazol and 13–55% with TE. They also found that on plots treated with paclobutrazol the duration of control with fungicides was increased 28–84% whereas on plots treated with TE the duration of control was generally like that of plots receiving no PGR.

Gray leaf spot suppression was observed on perennial ryegrass treated with 0.5 mM ethephon prior to disease inoculation. However, higher concentrations increased susceptibility to the disease (Rahman et al., 2013).

Ok et al. (2011) found that repeated use of flurprimidol or paclobutrazol may cause decreased sensitivity of dollar spot to DMI fungicides and contribute to the selection of isolates that are resistant to DMI fungicides. Both PGRs caused overexpression of ShPDR1 which caused reduced sensitivity of dollar spot to DMI fungicides (Allan-Perkins et al., 2017). The authors state that PGR use should be considered when developing fungicide resistance plans and that PGRs that are not chemically related to the DMI fungicides should be used where DMI fungicides are applied.

Paclobutrazol applied alone did not reduce the severity of brown patch or dollar spot on creeping bentgrass. However, when paclobutrazol was combined with a DMI fungicide, Miller (2016) observed lower dollar spot severity compared to plots treated with fungicide alone.

Research has also been conducted to determine the growth-regulating effects of some of the demethylation inhibitor fungicides used in turfgrass management. A study examined the growth-regulating effect of 6 DMI fungicides compared to paclobutrazol on creeping bentgrass managed at 33 mm height-of-cut. In this study, none of the tested DMI fungicides reduced clipping yields but paclobutrazol did. Rooting depth was not significantly different among the treatments (Mitkowski and Chaves, 2013). The authors state that the potential negative growth-regulating impacts of DMI fungicides may be overstated.

Bacterial etiolation of creeping bentgrass caused by *Acidovorax avenae* results in elongated stems and leaves. While applications of TE can improve bentgrass quality (Bigelow, 2012) and was associated with a suppression of etiolation symptoms (Fidanza et al., 2013), it was also found to increase etiolation possibly due to the bacterium causing a phytohormone imbalance (Roberts et al., 2015). Roberts et al. (2016) found that switching to flurprimidol or paclobutrazol (early GA biosynthesis inhibitors) resulted in reduced etiolation in situations where bacterial etiolation could be an issue. In general, the use of growth regulators has shown benefits toward efforts to control fungal pathogens; however, there are areas of possible caution that require further research. Issues regarding the exacerbation of some diseases as well as caution over combining growth regulators with certain fungicide groups in the heat of the summer mean that turfgrass managers must still exercise caution when using these products.

8 Enhancing tolerance to environmental extremes

The use of growth regulators to enhance tolerances to the environment has been a recent development. Prior research was focused more so on the safety of products and the evaluation of their impact on agronomic practices. Evaluations of perennial ryegrass under drought conditions indicated that ethephon applications enhanced turfgrass quality during greenhouse trials (Jiang and Fry, 1997). Cold hardiness evaluations of 'Tifway' bermudagrass in the transition zone using TE showed no benefits (Fagerness and Yelverton, 1999). Zhang et al. (2003) indicated that the use of propiconazole resulted in reduced visual turf injury and improved photochemical efficiencies in heat-stressed sod. McCann and Huang (2007) found that TE aided creeping bentgrass to retain quality, relative water content, and chlorophyll content during 21 days of combined heat (35°C) and drought stress. This work was followed up with TE applications to both KBG and creeping bentgrass under drought conditions which observed that soil volumetric water content, leaf relative water content, and turf quality were all higher during 28 days of evaluation (McCann and Huang, 2008). In 2009, work from Bian et al. indicated that TE applications aided creeping bentgrass to retain visual quality and leaf photochemical efficiency under drought stress due to greater osmotic adjustment leading to a reduced rate of water depletion from the soil. The overwhelming majority of research papers published since 2010 report the use of growth regulators for the enhancement of drought tolerance, though some research investigated their use for ameliorating other stresses, such as cold stress.

8.1 Cold

Munshaw et al. (2010) investigated the effects of late-season applications of ethephon for the enhancement of cold tolerance in the bermudagrass cultivars 'Midiron', 'Tifway', 'Princess-77', and 'Riviera' in the transition zone. Their research found little benefit and some deleterious effects, and the authors do not recommend its use for this purpose. Similar short-term effects have been found in cool-season turf during times of frost development (Fig. 3).

In a study conducted in Scandinavia, the application of TE was tested to determine if it reduced the severity of snow mold caused by *Microdochium nivale*. No differences in disease severity were noted. However, in this trial, it was found that autumn applications of TE resulted in reduced leaf growth but also increased fructans both before and after winter. The authors suggest that autumn TE applications may be useful for improving survival through winter (Aamlid and Pettersen, 2013).

Applications of growth regulators to ABG in order to improve survival under ice cover were studied by Laskowski et al. (2018). Application of TE did

Figure 3 Applications of many plant growth regulators around periods of frost can lead to a non-lethal short-term phytotoxic response.

not improve survival but mineral oil (Civitas), propiconazole, and mefluidide did. The authors attributed the difference to alterations in the ratio of saturated and unsaturated fatty acids in the crowns of the plants.

8.2 Drought

Climate change and restrictions on water use for irrigating turfgrass have resulted in much research to enhance and improve the drought tolerance of turfgrass. The use of TE on tall fescue that was irrigated with soil moisture sensors as a method of reducing irrigation compared to a frequency-based irrigation schedule showed mixed results. TE reduced the growth of the tall fescue, but its effect on the quality of the turf was not consistent (Chabon et al., 2017). In a hydroponic study with tall fescue, Sermons et al. (2017) found few differences in mean water use, water use over time, or evapotranspiration due to the application of TE.

Some research has been done on the use of growth regulators at establishment in order to determine if there are benefits either in establishment rate or tolerance to environmental extremes. Shahrokhi et al. (2011) investigated the physiological response to drought stress of 'Barrage' perennial ryegrass and 'Master' tall fescue after establishing seeds that had been pre-treated (soaked) for 24 h with paclobutrazol at 20 mg, 30 mg, or 40 mg per liter. The authors found that plants from seeds soaked with a high concentration of paclobutrazol had improved drought tolerance due to decreased electrolyte leakage and increased total chlorophyll content. However, these results were dependent on rate and cultivar and were temporary.

Applications of TE, paclobutrazol, or abscisic acid increased the relative water content and decreased electrolyte leakage and resulted in the improved visual quality of drought-stressed perennial ryegrass (Mohammadi et al., 2017a). Levels of catalase and superoxide dismutase were greater in TE-treated plants though all three tested PGRs improved drought tolerance. The TE-treated plants had higher chlorophyll content, soluble sugar content, and proline compared to the abscisic acid or paclobutrazol-treated plants.

Applications of TE to tall fescue and wheatgrass increased drought stress resistance by increasing soluble sugar content as well as the activity of the antioxidants superoxide dismutase, ascorbate peroxidase, catalase, and peroxidase (Etamadi et al., 2015). The authors found decreased MDA content and electrolyte leakage in TE-treated turf. Soluble sugar content was also found to be at a higher concentration in TE-treated turf under drought stress. This was attributed to a decrease in demand for growth rather than an effect of TE on drought tolerance, but the increased SSC has been implicated in improving both drought stress and post-drought recovery. (Etamadi et al., 2015).

PC treatment increased the visual quality and relative water content of drought-stressed KBG (Fard et al., 2015). The treated plants also had higher levels of the enzymes superoxide dismutase, ascorbate peroxidase, and catalase compared to the untreated and these differences were larger on drought-stressed turf. Additionally, the authors observed lower levels of electrolyte leakage and malondialdehyde due to PC application.

Zhang et al. (2017) showed that the use of TE in combination with auxin could improve root viability and overall turf quality when managing creeping bentgrass under water-deficit conditions. Applications of TE alone or in combination with Daconil Action fungicide resulted in improved quality of creeping bentgrass managed as a golf course fairway that was subjected to drought stress (Jespersen and Huang, 2017). The authors evaluated turf quality, leaf membrane stability, and relative water content and canopy cover and found evidence of synergism between the two compounds. The authors also noted that the benefits were most significant during the most severe heat and drought stress. Jespersen and Huang suggested that further research could evaluate how the combination of treatments caused positive physiological effects. Sequential applications of TE prior to onset of drought followed by applications of the osmoregulant glycine betaine during drought improved creeping bentgrass tolerance to prolonged periods (up to 48 days) of drought as measured by turf quality and normalized difference vegetation index (NDVI) data (Burgess and Huang, 2016).

In a greenhouse experiment, Zhang et al. (2018) studied the use of ethephon as a seed treatment for KBG. The authors found higher relative water content and lower electrolyte leakage on seedlings treated with ethephon and then subjected to polyethylene glycol-induced drought conditions, suggesting that ethephon can enhance the drought tolerance of KBG seedlings.

Treatment of KBG with TE, which reduces GA biosynthesis, resulted in a 40% increase in the concentration of abscisic acid, auxin, jasmonic acid, and salicylic acid in a growth regulator trial with 'Kenblue' KBG (Krishnan and Merewitz, 2015). The authors propose that these alterations in hormone levels can have consequences for plant resistance to different biotic and abiotic stress and suggest that further study is warranted.

Drought-stressed KBG that had been treated with TE was found to have higher photosynthetic capacity and relative water content as well as decreased membrane leakage compared to untreated (Xu and Huang, 2012). They then analyzed the levels of different proteins and metabolites. TE treatment increased the accumulation of antioxidant proteins such as ferratin, glutathione-s-transferase, and catalase. Increased levels were also found of fatty acids such as palmitic acid, α-linolenic acid, linoleic acid, and octadecanoic acid. These help to maintain cell membrane stability and regulate energy dissipation in mitochondria. HSP 70 and chaperonin 81 levels were also increased, and the authors speculate that this is due to TE preventing or reversing the incorrect folding of these proteins in drought-stressed turf. Other proteins, such as Rubisco, were downregulated but not as much in TE-treated plants compared to the untreated.

In warm-season turfgrass, the research has focused mainly on either enhancing drought tolerance or improving tolerance to irrigation with non-potable water sources. Deficit irrigation is a method of reducing water used by irrigating with less than what the plant would maximally utilize. The use of growth regulators to improve quality in these conditions has been studied. Bermudagrass treated with either TE or a soil surfactant at 28-day intervals from May to October was subjected to 50%, 65%, or 80% of reference evapotranspiration. The authors found that either TE, the surfactant, or both improved soil water volumetric content, irrigation use efficiency, and visual color, and thus, bermudagrass could be maintained with 15–30% less irrigation (Serena et al., 2018).

Soil surfactants applied in combination with TE were evaluated by Schiavon et al. (2015a) to determine if they could improve the quality of 'Sea Spray' seashore paspalum and 'Princess 77' bermudagrass. The plots were deficit irrigated (50% of reference evapotranspiration) with either potable or saline water using sprinkler or subsurface drip irrigation. TE application and subsurface drip irrigation improved quality, NDVI, and fall color retention, particularly of seashore paspalum. In addition, the authors found the highest amounts of chlorophyll and carotenoids in the leaves of TE-treated 'Princess 77'. The effect of TE was less apparent on superoxide dismutases (SOD) activity, with a significant increase in activity noted in the first year of the study but not in the final 2 years (Schiavon et al. (2015b). After 3 years of drought stress, the visual quality of 'Princess 77' was enhanced by TE application and the authors attribute that to improvements in photochemical efficiency, pigment content, and SOD activity. These differences were not observed in TE-treated 'Sea Spray'. In a different field trial conducted by Serena et al. (2020), the application of TE to drought-stressed 'Sea Spray' seashore paspalum and 'Princess 77' bermudagrass did not result in increases in stolon or rhizome weight. The authors noted that applications of either TE or the soil surfactant Revolution (Aquatrols Corporation of America, Paulsboro, NJ) or the use of subsurface drip irrigation could improve turf quality and enhance drought avoidance. However, applying TE or Revolution did not provide further benefit to turf that was irrigated with subsurface drip irrigation. On plots irrigated with a sprinkler system, the application of TE improved root length and root weight. Further research was suggested by the authors to investigate the effects of TE on stolon and rhizome production because their results differed from previously published research that had documented a benefit. TE applied with abscisic acid improved the quality, photochemical efficiency, and increased the root length and dry weight while reducing evapotranspiration of drought-stressed 'Sea spray' seashore paspalum in a greenhouse trial (Elansary and Yessoufou, 2015).

Bermudagrass response to deficit irrigation with low-quality water has been studied by Hejl et al. (2015). In this study, 'Tifway' bermudagrass was irrigated with water treated by reverse osmosis, saline water, or sodic potable water. TE applications were able to offset some of the short-term visual quality impacts of deficit irrigation of drought-stressed bermudagrass. This effect was most beneficial during the spring studies, during which the bermudagrass was less stressed.

9 Enhancing shade tolerance

Shade alters the growth of turfgrass, and these differences are due both to changes in the quality and quantity of light under the shade canopy (Wherley et al., 2005). Decreasing the growth rate of the grass can result in significant improvements in shaded turfgrass quality. GA_3 affect plant responses to light stimuli. Overexpression of the gene GA2ox in creeping bentgrass caused the inactivation of bioactive GA_3 and improved the quality of the bentgrass when grown in shade (Studzinska et al., 2012).

Because of the way that shade alters turfgrass growth, GA_3-inhibiting growth regulators, such as TE, can improve the performance of shaded turfgrass by reducing some of the excessive vertical growth and thus growth regulators are widely used for this purpose. A large amount of research was published using PGRs for shaded turfgrass between 1995 and 2010 (Fig. 4). Many studies conducted under neutral density shade have demonstrated the benefit of repeated use of TE (Qian et al., 1998; Ervin et al., 2002; Goss et al., 2002; Steinke and Stier, 2003). However, TE use did not fully compensate for the alterations in the red:far red-light ratio when applied to turfgrass under tree shade (Gardner and Wherley, 2005; Bell et al., 2000). This may be because plant response to low red:far red-light is probably not mediated by GA (Zheng et al., 2001). There may be a range of factors at play under the tree line which includes the activity of the light receptor phytochrome which is mediated by the ratio of red to far red wavelengths (655–665 nm:725–735 nm) which is influenced by leaves in tree branches above. A normal fully sunlit site might expect to see a ratio of 1.15:1.2 while in shaded conditions depending on how deep the shade is 0.05:0.7 can be easily found (Franklin and Whitelam, 2005). More recent research has focused on better understanding the use of growth regulators with both traditional and potential alternative turfgrass species for golf course management.

The benefits of using PGRs for shaded turfgrass seem to be species-specific. When investigating the use of TE on shaded genotypes of hybrid

Figure 4 Plant growth regulators affect different species of grass in unique ways, particularly when the grass is already under stress. Some research has suggested that plant growth regulators can enhance tolerance of grass growing under shade, particularly artificial shade.

bluegrasses, leaf elongation was significantly reduced. However, this did not result in improved visual quality (Meeks et al., 2015). Similarly, Russel et al. (2019) studied daily light integral requirements for creeping bentgrass and investigated the effect of TE and turf colorants. The authors found that TE caused some improvement in quality during the summer months. However, neither TE nor the plant colorant was effective in reducing the DLI requirement of the bentgrass. In a study with 'Penncross' creeping bentgrass grown in either full sun, artificial shade, or deciduous tree shade, Nangle et al. (2012) found that TE in combination with urea would retain the quality of creeping bentgrass grown in shade. The authors also found that applications of TE increased flavonoid concentrations when grown under shade. This is a potential area of further study as flavonoids may play a role in improving disease resistance in shaded turfgrass.

In a study investigating TE use on shaded creeping bentgrass and velvet bentgrass, Pease and Stier (2018) found that neither species provided acceptable quality in 80% shade at the end of their study. The N rate was also more important than TE in determining final turfgrass quality. However, applications of TE along with annual rates of 49 kg of N per hectare improved the duration of acceptable quality and the authors recommended future work to look at the effect of TE when grown in natural versus artificial shade.

The use of growth regulators to ameliorate shade issues on close-mowed warm-season grasses for golf courses has been studied. In warm-season turfgrasses, some work has been done to use growth regulators to improve both traditionally used turfgrasses and in combination with species that may be more tolerant of low light conditions. Dunne et al. (2017) found that the use of TE resulted in increased cover when investigating shaded bermudagrass. Application of TE and paclobutrazol to shaded seashore paspalum increased the amount of chlorophyll, carotenoids, and total carbohydrates in the clippings (Hussein et al., 2012). The authors found that applications of either growth regulator improved the quality of seashore paspalum in levels of shade between 42% and 70%. The use of growth regulators with more shade-tolerant species has also been studied. Atkins et al. (2012) found that the use of TE could result in an acceptable quality of 'Diamond' zoysiagrass on putting greens in up to 60% shade.

10 Case study: how turf species and light quality determines effectiveness of using plant growth regulators for shaded turf

Shaded or light-restricted turf can be as much as 25% of the maintained grasses on a property (Beard, 1997). Shaded turfgrass has been associated with greater inputs and increased amounts of disease problems mainly due to changes in light wavelengths impacting the turfgrass leaf surfaces, changes in temperature, and limitations on air movement, thus extending periods of wetness on the leaf surfaces. The three driving factors can all be managed using growth regulators in the following way.

Variations in wavelengths under tree lines result in an alteration of the red:far red ratio (Bell et al., 2000). The resulting changes lead to morphological variations such as more elongated plants, loss of tillering, and an overall loss of density. One of the main responses is the increased concentrations of GA_3 in internodes of a range of plants not just turfgrasses (Yang and Li, 2017). This leads to increased cell elongation as plants try to grow toward the sunlight thus reducing tillering and turfgrass density. It has been shown that applications of TE prior to leaf development on deciduous trees and through the playing season will indeed help retain turfgrass quality on these surfaces. The reason for the benefit is that a reduction in elongation allows for the already limited carbohydrate to be utilized for recovery from wear and daily play.

The impact of shade whether artificial or natural generally results in reductions in air movement and research has shown this to be the case (Koh et al., 2003). The possibility that this reduction in air movement will lead to extended periods of leaf wetness has been a concern regarding disease development. Growth regulators have had variable responses to turfgrass diseases, but the use of mefluidide and ethephon has been shown to occasionally reduce the severity of anthracnose (Inguagiato et al., 2010) while applications of TE every 7 days have been tied to bacterial etiolation of creeping bentgrass and ABG (Roberts et al., 2016). Thus, when considering plant health options and turfgrass growth regulators in shaded conditions, caution should be taken for which disease is present most often and how much of an impact a regulator is known to have on it. Regulators are not standalone as a fungicide and all areas must be monitored for disease activity regardless of PGR applications.

In suppressing turfgrass growth, the number of inputs that are required will drop concurrently - N rates can be limited under shade for multiple reasons - it helps in reducing turfgrass elongation as well as reducing the amount of moisture in the plant. Further to this, growth regulators reduce the need for extra water and help maintain tree integrity on golf courses due to the retention of turfgrass quality. Water management is still a critical component, however, as tree roots are going to continue to play a dominant role in water uptake and pruning of roots may be a requirement regardless of the use of growth regulators. An important point to remember, however, is that just prior to tree leaf emergence is when the green surface will be optimal, and management of the playing surface must be aimed at retaining that quality over trying to improve it. Thus, this is where growth regulators are a valuable asset for many turfgrass managers.

It is suggested by the authors that a PGR program for shaded turfgrass should include the following characteristics:

1 First applications initiated past last frost at a given site;
2 Products that have shown the most promise for enhancing shade tolerance should be used;
3 Timing of applications should follow recommendations based on GDD models and continue through the season even as leaf emergence has occurred. Ensure fertility has been reduced, in particular, N levels to mitigate excessive production of leaf tissue. Ensure that tree roots are pruned every 2–3 years and where possible aim to remove branches that may aid in light penetration canopy; and
4 If frost is a concern at the facility, then PGR applications should be completed 3 weeks prior to the average annual first frost date to allow time for recovery; by this time, leaf fall should be beginning to occur and recovery due to increased light penetration will aid turfgrass quality.

11 Conclusion and future trends

Much of the research conducted between 2010 and 2020 focused not only on refining recommendations for use of PGRs for growth regulation but also for determining if, and defining recommendations for, whether the PGRs could be used for other purposes such as enhancing rooting and establishment, enhancing tolerance of turfgrass to disease, or enhancing tolerance to environmental extremes such as excess heat, drought, and shade. Much work was also done investigating the use of PGRs for suppression or control of ABG on golf course putting greens.

While conducting this review, the authors observed that much of the research involved the use of TE, even though it has been used for the longest period in the turfgrass industry among those compounds reviewed. In addition to the future research suggested by the authors, we believe there are opportunities to conduct more work with the other PGRs, either alone or in combination with TE.

12 Where to look for further information

12.1 Further reading

- Ervin, E. and Zhang, X. (2007). Applied physiology of natural and synthetic plant growth regulators on turfgrasses. in M. Pessarakli (Ed.), *Handbook of Turfgrass Management and Physiology*, pp. 171–200. St. Louis, MO: Taylor & Francis.

- Marchi, S. R., Martins, D. and McElroy, J. S. (2013). Growth inhibitors in turfgrass. *Planta Daninha*, Vicosa-MG, 31(3): 733–747.
- Reicher, Z. J., Dernoeden, P. H. and Richmond, D. S. (2013). Insecticides, fungicides, herbicides, and growth regulators used in turfgrass systems. in J.C. Stier, B.P. Horgan and S.A. Bonos (Eds), *Turfgrass: Biology, Use, and Management*. doi.org/10.2134/agronmonogr56.c24.
- Watschke, T. L., Prinster, M. G. and Brueninger, J. M. (1992). Plant growth regulators and turfgrass management. in. D. V. Waddington, R. N. Carrow and R. C. Shermon (Eds), *Agronomy Monograph*, pp. 557–588. John Wiley & Sons. DOI:10.2134/agronmonogr32.

12.2 Journals and conferences

These societies hold meetings that are well attended by researchers who study the use of growth regulators in turfgrass management:

- ASA (American Society of Agronomy), CSSA (Crop Science Society of America; Division C5 – Turfgrass Science), and SSSA (Soil Science Society of America) International Meetings are conducted annually. The societies publish *Agronomy Journal*, *Crop Science*, *Soil Science*, and *Crop, Forage and Turfgrass Management*.
- ITS (International Turfgrass Society) conducts a meeting every 4 years. Research is published every 4 years in the *International Turfgrass Society Research Journal*.
- ETS (European Turfgrass Society) conducts a bi-annual meeting that moves around sites in Europe with broad international participation. Research articles are published in the *European Turfgrass Society Research Journal*.
- ASHS (American Society of Horticultural Science) conducts a yearly meeting. The society publishes the journals *HortScience* and *HortTechnology*.

13 References

Aamlid, T. S. and Pettersen, T. (2013). Effect of plant growth regulator trinexapac-ethyl on turf quality, concentration of total nonstructural carbohydrates, and infection of *Microdochium nivale* in greens-type *Poa annua* in Scandinavia. *Int. Turf. Soc. Res. J.* 12:801–803.

Adams, R., Kerber, E., Pfister, K. and Weiler, E. W. (1992). Studies on the action of the new growth retardant CGA163935 (Primo). In: Karssen, C. M., van Loon, L. C. and Vreugdenhil, D. (Eds) *Progress in Plant Growth Regulation*. Kluwer Publishers, Dordrecht, the Netherlands, pp. 818-827.

Allan-Perkins, E., Campbell-Nelson, K., Popko, J. T., Sang, H. and Jung, G. (2017). Investigating selection of demethylation inhibitor fungicide-insensitive *Sclerotinia homoeocarpa* isolates by boscalid, flurprimidol, and paclobutrazol. *Crop Sci.* 57(S1):S301–S309.

Amiri-Khah, R., Eetemadi, N., Nikbakht, A. and Pessarakli, M. (2015). Effects of sequential trinexapac-ethyl applications and traffic on growth of perennial ryegrass (*Lolium perenne* L.). *Hort. Sci. Technol.* 33(3):340–348.

Askew, S. D. (2017). Plant growth regulators applied in winter improve annual bluegrass (*Poa annua*) seedhead suppression on golf greens. *Weed Technol.* 31(5):701–713.

Atkins, J. L., McCarty, L. B., Liu, H., Faust, J. and Toler, J. E. (2012). Diamond zoysiagrass golf green response to reduced light environments with the use of trinexapac-ethyl. *Agron. J.* 104(4):847–852.

Baldwin, C. M. and Brede, A. D. (2011). Plant growth regulator selection and application rate influence annual bluegrass control in creeping bentgrass putting greens. *Appl. Turf Sci.* 8(1). DOI: 10.1094/ATS-2011-0517-02-RS.

Baldwin, C. M., Brede, A. D. and Mayer, J. J. (2015). Growth regulation and tank mixing associated with a glyphosate tolerant perennial ryegrass cultivar. *HortTechnol* 25(2):214–220.

Beam, J. B. and Askew, S. D. (2007). Fate of prohexadione calcium in annual bluegrass (*Poa annua*) and three turfgrasses. *Weed Sci.* 55(6):541–545.

Beard, J. B. (1973). *Turfgrass: Science and Culture*. Prentice-Hall, Englewood Cliffs, NJ.

Beard, J. B. (1997). Shade stresses and adaptation mechanisms of turfgrasses. *Int. Turf. Soc. Res. J.* 8:1186–1195.

Beasley, J. S., Branham, B. E. and Ortiz-Ribbing, J. M. (2005). Trinexapac-ethyl effects on Kentucky bluegrass root architecture. *Hort. Sci.* 40:1549–1542.

Bell, G. E., Danneberger, T. K. and McMahon, M. J. (2000). Spectral irradiance available for turfgrass growth in sun and shade. *Crop Sci.* 40(1):189–195.

Bian, X., Merewitz, E. and Huang, B. (2009). Effects of trinexapac-ethyl on drought responses in creeping bentgrass associated with water and osmotic adjustment. *J. Am. Soc. Hort. Sci.* 134:505–510.

Bigelow, C. A. (2012). Plant growth regulators in bentgrass turf areas. *USGA Green Sect. Rec.* 50:1–4.

Bigelow, C. A., Hardebeck, G. A. and Bunnell, B. T. (2007). Monthly flurprimidol applications reduce annual bluegrass populations in a creeping bentgrass fairway. *App. Turf Sci.* DOI: 10.1094/ATS-2007-0508-02-RS.

Bingaman, B. R., Christians, N. E. and Gardner, D. S. (2001). Trinexapac-ethyl effects on rooting of Kentucky bluegrass (*Poa pratensis*) sod. *Int. Turf Res. J.* 9:832–834.

Brosnan, J. T., Breeden, G. K., Armel, G. R. and Vargas, J. J. (2011). Common bermudagrass seedhead suppression and growth regulation with fenoxaprop. *Weed Technol.* 25(3):404–410.

Brosnan, J. T., Thoms, A. W., Breeden, G. K. and Sorochan, J. C. (2010). Effects of various plant growth regulators on the traffic tolerance of 'Riviera' bermudagrass (*Cynodon dactylon* L.). *Hort. Sci.* 45(6):966–970.

Brown, A. M., Harris, J. R., Goncalves, C. G., Peppers, J. M., Magni, S., Volteranni, J. and McElroy, J. S. (2020a). Growing degree-days optimize trinexapac-ethyl reapplications on ultradwarf bermudagrass putting greens: I. Predicting the maximum suppression point. *Crop Sci.* 61(2):1446–1457.

Brown, A. M., Harris, J. R., Goncalves, C. G., Peppers, J. M., Magni, S., Volteranni, J. and McElroy, J. S. (2020b). Growing degree-days optimize trinexapac-ethyl reapplications on ultradwarf bermudagrass putting greens: II. Testing a reapplication schedule. *Crop Sci.* 61(2):1436–1445.

Burgess, P. and Huang, B. (2016). Effects of sequential application of plant growth regulators and osmoregulants on drought tolerance of creeping bentgrass (*Agrostis stolonifera*). *Crop Sci.* 54(2):837–844.

Burpee, L. L., Green, D. E. and Stephens, S. L. (1996). Interactive effects of plant growth regulators and fungicides on epidemics of dollar spot in creeping bentgrass. *Plant Dis.* 80(11):1245–1250.

Chabon, J., Bremer, D. J., Fry, J. D. and Lavis, C. (2017). Effects of soil moisture-based irrigation controllers, mowing height, and trinexapac-ethyl on tall fescue irrigation amounts and mowing requirements. *Soc. Res.* 13(1):755–760.

Chastain, T. G., Young III, W. C., Silberstein, T. B. and Garbacik, C. J. (2014). Performance of trinexapac-ethyl on *Lolium perenne* seed crops in diverse lodging environments. *Field Crops Res.* 157:65–70.

Cho, Y.-R., Cho, Y.-S. and Choi, J. (2019). Effects of foliar spray volumes and concentrations of trinexapac-ethyl on Kentucky bluegrass (*Poa pratensis* L.) growth. *Weed Turf Sci.* 8(3):279–287.

Cox, M. C., Mittlesteadt, T. L. and Askew, S. D. (2013). Improving perennial ryegrass overseeding in 'Patriot' bermudagrass using cultural and chemical treatments. *Int. Turf. Soc. Res. J.* 12:231–236.

Deaton, M. T. and Williams, D. W. (2010). Overseeding and trinexapac-ethyl effects on tolerance to simulated traffic of four bermudagrass cultivars grown as a sand-based athletic field. *HortTechnol* 20(4):724–729.

Dernoeden, P. H. (1998). Use of prodiamine as a preemergence herbicide to control annual bluegrass in Kentucky bluegrass. *Hort. Sci.* 33(5):845–846.

Dernoeden, P. H. and Turner, T. R. (1988). Annual bluegrass control and tolerance of Kentucky bluegrass and perennial ryegrass to ethofumesate. *Hort. Sci.* 23(3):565–567.

DiPaola, J. M. (1988). Turfgrass growth regulation. In: Kaufmann, J. E. and Westerdahl, H. E. (Eds) *Chemical Vegetation Management.* Plant Growth Regulation Society of America, San Antonio, TX, pp. 155–186.

Dunne, J. C., Miller, G. L., Arellano, C., Brandenburg, R. L., Schoeman, A. and Milla-Lewis, S. R. (2017). Shade response of bermudagrass accessions under different management practices. *Urban For. Urban Green.* 26:169–177.

Elansary, H. O. and Yessoufou, K. (2015). Growth regulators and mowing heights enhance the morphological and physiological performance of Seaspray turfgrass during drought conditions. *Acta Physiol. Plant.* 37(11):232.

Ervin, E. H. and Koski, A. J. (2001). Kentucky bluegrass growth responses to trinexapac-ethyl, traffic, and nitrogen. *Crop Sci.* 41(6):1871–1877.

Ervin, E. H., Ok, C. H., Fresenberg, B. S. and Dunn, J. H. (2002). Trinexapac-ethyl restricts shoot growth and prolongs stand density of 'Meyer' zoysiagrass fairway under shade. *HortSci.* 37(3):502–505.

Etamadi, N., Mohammadi, M. H. S., Nikbakt, A., Sabzalian, M. R. and Pessarakli, M. (2015). Influence of trinexapac-ethyl in improving drought resistance of wheatgrass and tall fescue. *Acta Physiol. Plant.* 37(3):53. DOI: 10.1007/s11738-015-1799-6.

Fagerness, M. J. and Penner, D. (1998). ^{14}C trinexapac-ethyl absorption and translocation in Kentucky bluegrass. *Crop Sci.* 38(4):1023–1027.

Fagerness, M. J. and Yelverton, F. H. (1999). Effects of trinexepac-ethyl on late-season development and cold hardiness of 'Tifway' bermudagrass. In: *Proceedings of the 53rd Annual Meeting Northeast Weed Science Society* 53:63.

Fard, J. R., Kafi, M. and Naderi, R. (2015). The enhancement of drought stress tolerance of Kentucky bluegrass by prohexadione-calcium treatment. J. Ornam Plants 5:197–204.

Fidanza, M., Gregos, J. and Brickley, D. (2013). Evaluation of fungicides for control of etiolation symptoms in perennial ryegrass, 2004, 2005 and 2006. *Plant Dis. Manag. Rep. (online)* Report 7:T014. DOI: 10.1094/PDMR07.

Fidanza, M. A., Wetzel, H. C. III, Agnew, M. L. and Kaminski, J. E. (2006). Evaluation of fungicides and plant growth regulator tank-mix programmes on dollar spot severity of creeping bentgrass. Crop Prot. 25(9):1032–1038.

Flessner, M. L., McElroy, J. S. and McCurdy, J. D. (2017). Annual bluegrass (Poa annua) control with methiozolin and nutrient tank-mixtures. Weed Technol. 31(5):761–768.

Franklin, K. A. and Whitelam, G. C. (2005). Phytochromes and shade-avoidance responses in plants. Ann. Bot. 96(2):169–175. DOI: 10.1093/aob/mci165.

Fry, J. D. and Wells, D. W. (1990). Carpetgrass seedhead suppression with plant growth regulators. HortSci. 25(10):1257–1259.

Gardner, D. S. and Wherley, B. G. (2005). Growth response of three turfgrass species to nitrogen and trinexapac-ethyl in shade. HortSci. 40(6):1911–1915.

Gaussoin, R. E. and Branham, B. E. (1989). Influence of cultural factors on species dominance in a mixed stand of annual bluegrass/creeping bentgrass. Crop Sci. 29(2):480–484.

Glab, T., Szewczyk, L., Gondek, K., Knaga, J., Tomasik, M. and Kowalik, K. (2020). Effect of growth regulators on visual quality of turfgrass Sci. Hortic. 267:109314.

Golembiewski, R. C. and Danneberger, T. K. (1998). Dollar spot severity as influenced by trinexapac-ethyl, creeping bentgrass cultivar, and nitrogen fertility. Agron. J. 90(4):466–470.

Goss, R. M., Baird, J. H., Kelm, S. L. and Calhoun, R. N. (2002). Trinexapac-ethyl and nitrogen effects on creeping bentgrass grown under reduced light conditions. Crop Sci. 42(2):472–479.

Han, K. M., Kaminski, J. E. and Lulis, T. T. (2017). Influence of nitrogen, plant growth regulators, and iron sulfate on annual bluegrass populations on a golf green. *Turfgrass Soc. Res. J.* 13(1):661–669. DOI: 10.2134/itsrj2016.05.0402.

Hargey, D., Wherley, B., Malis, A., Thomas, J. and Chandra, A. (2016). Effects of mowing and trinexapac-ethyl on St. Augustinegrass sod establishment during a simulated water variance period. HortTechnol 26(1):54–62.

Hejl, R. W., Wherley, B. G., Thomas, J. C. and White, R. H. (2015). Irrigation water quality and trinexapac-ethyl effects on bermudagrass response to deficit irrigation. Hort. Sci. 50(7):1081–1087.

Hoisington, N. R. (2013). Tolerance of bentgrass species and cultivars to methiozolin. M.S. Thesis. University of California, Riverside.

Huang, Y., Kaminski, J. E. and Landschoot, P. J. (2015). Regulation with trinexapac-ethyl and dew removal at the time of fungicide application did not influence dollar spot control. Hort.Sci. 50(3):496–500.

Hussein, M. M. M., Mansour, H. A. and Ashour, H. A. (2012). Response of Paspalum vaginatum turfgrass grown under shade conditions to paclobutrazol and trinexapac-ethyl as plant growth retardants. J. Hort. Sci. Ornam Plants 4:134–147.

Inguagiato, J. C., Murphy, J. A. and Clarke, B. B. (2010). Anthracnose development on annual bluegrass affected by seedhead and vegetative growth regulators. App. Turf Sci. 7(1):1-19. DOI: 10.1094/ATS-2010-0923-01-RS.

Jeffries, M. D., Yelverton, F. H. and Gannon, T. W. (2013). Annual bluegrass (Poa annua) control in creeping bentgrass putting greens with amicarbazone and paclobutrazol. Weed Technol. 27(3):520-526.

Jespersen, D. and Huang, B. (2017). Effects of trinexapac-ethyl and daconil action (acibenzolar-s-methyl and chlorothalonil) on heat and drought tolerance of creeping bentgrass. Crop Sci. 57(S1):138-146.

Jiang, H. and Fry, J. (1997). Drought responses of perennial ryegrass treated with plant growth regulators. Turfgrass Research [Kansas State University], May 1997:7-12.

Johnson, B. J. (1990). Influence of frequency and dates of plant growth regulator applications to centipedegrass on seedhead formation and turf quality. J. Am. Soc. Hortic. Sci. 115(3):412-416.

Kim, Y., Heo, H., Bae, E., Youn, J. and Lee, G. (2019). Inhibition responses of creeping bentgrass after applying trinexapac-ethyl as two spraying methods. *Weed Turf Sci.* 8:319-328.

Kim, Y. H. and Lee, I. J. (2013). Influence of plant growth regulator application on seed germination of dandelion (Taraxacum officinale). Weed Turf Sci. 2(2):152-158.

Koeritz, E. J., Watkins, E. and Ehlke, N. J. (2013). A split application approach to nitrogen and growth regulator management for perennial ryegrass seed production. Crop Sci. 53(4):1762-1777.

Koh, K. J., Bell, G. E., Martin, D. L. and Walker, N. R. (2003). Shade and airflow restriction effects on creeping bentgrass golf greens. Crop Sci. 43(6):2182-2188.

Kowalewski, A. R., Schwartz, B. M., Richardson, M. D., Karcher, D. E., McCalla, J. H., Patton, A. J. and Hanna, W. W. (2012). Effect of nitrogen, growth regulators, and mowing height on ball lie in TifSport bermudagrass. Appl. Turf Sci 9(1):1-10. DOI: 10.1094/ATS-2012-0625-01-RS.

Kreuser, W. C., Obear, G. R., Michael, D. J. and Soldat, D. J. (2018). Growing degree day models predict the performance of paclobutrazol on bentgrass putting greens. *Crop Sci.* 58(3):1402-1408.

Kreuser, W. C. and Soldat, D. J. (2011). A growing degree day model to schedule trinexapac-ethyl applications on Agrostis stolonifera golf putting greens. Crop Sci. 51(5):2228-2236.

Kreuser, W. C. and Soldat, D. J. (2012). Frequent trinexapac-ethyl applications reduce nitrogen requirements of creeping bentgrass putting greens. Crop Sci. 52(3): 1348-1357.

Krishnan, S. and Merewitz, E. B. (2015). Drought stress and trinexapac-ethyl modify phytohormone content within Kentucky bluegrass leaves. J. Plant Growth Regul. 34(1):1-12.

Laskowski, K., Frank, K. and Merewitz, E. (2018). Chemical plant protectants and plant growth regulator effects on annual bluegrass survival of ice cover. J. Agron. Crop Sci. 205(2):202-212.

Lickfeldt, D. W., Gardner, D. S., Branham, B. E. and Voigt, T. B. (2001). Implications of repeated trinexapac-ethyl applications in Kentucky bluegrass. Agron. J. 93(5):1164-1168.

Marchi, S. R., Martins, D., Costa, N. V. and Silva, J. R. V. (2013). Effect of plant growth regulators on growth and flowering of 'Meyer' zoysiagrass. Planta Daninha 31(3):695-703.

Marchi, S. Rd, Martins, D. and Costa, N. Vd (2015). Effects of plant growth regulators on the growth and flowering of Saint Augustine grass plants. Biosci. J. 31(3):785–793.

Marchi, S. Rd, Martins, D. and Costa, N. V. (2016). Growth and flowering inhibition of Paspalum notatum with application of trinexapac-ethyl and prohexadione-calcium. R. Bras. Eng. Agríc Ambient. 20(3):202–208.

McCann, S. E. and Huang, B. (2007). Effects of trinexapac-ethyl foliar application on creeping bentgrass responses to combined drought and heat stress. Crop Sci. 47(5):2121–2128.

McCann, S. E. and Huang, B. (2008). Drought responses of Kentucky bluegrass and creeping bentgrass as affected by abscisic acid and trinexapac-ethyl. J. Am. Soc. Hort. Sci. 133(1):20–26.

McCarty, L. B., Willis, T. G., Toler, J. E. and Whitwell, T. (2011). 'TifEagle' bermudagrass response to plant growth regulators and mowing height. Agron. J. 103(4):988–994.

McCullough, P. E. and Hart, S. E. (2006). Temperature influences creeping bentgrass (Agrostis stolonifera) and annual bluegrass (Poa annua) response to bispyribac-sodium. Weed Technol. 20(3):728–732.

McCullough, P. E., Hart, S. E., Weisenberger, D. and Reicher, Z. J. (2010). Amicarbazone efficacy on annual bluegrass and safety on cool-season turfgrasses. *Weed Technol.* 24(4):461–470.

McCullough, P. E., Liu, H., McCarty, L. B., Whitwell, T. and Toler, J. E. (2006). Nutrient allocation of 'Tifeagle' bermudagrass as influenced by trinexapac-ethyl. J. Plant Nutr. 29(2):273–282.

McCullough, P. E., Nutt, W., Murphy, T. R. and Raymer, P. (2011). Seashore paspalum seedhead control and growth regulation with flazasulfuron and trinexapac-ethyl. Weed Technol. 25(1):64–69.

McCullough, P. E. and Sidhu, S. S. (2014). Ethephon absorption and transport associated with annual bluegrass inflorescence suppression. *Crop Sci.* 54(2):845–850.

McCullough, P. E., Sidhu, S. S., Singh, R. and Yu, J. (2014). Seashore paspalum growth regulation with flucarbazone-sodium. Crop Sci. 54(3):1197–1204.

McDonald, B. W., Golembiewski, R. C., Cook, T. W. and Blankenship, T. M. (2013). Effects of mowing and rolling frequency, Primo Maxx, and roller weight on annual bluegrass putting green speed. *App. Turf Sci.* 10(1). DOI: 10.1094/ATS-2013-0529-01-RS.

McElroy, S. L., Flessner, M. L. and Murphy, T. (2013). Bermudagrass growth regulation with flucarbazone. Int. Turf. Soc. Res. J. 12:197–200.

Meeks, M., Chandra, A. and Wherley, B. G. (2015). Growth responses of hybrid bluegrass and tall fescue as influenced by light intensity and trinexapac-ethyl. Hort. Sci. 50(8):1241–1247.

Miller, G. L. (2016). Effect of watered-in demethylation-inhibitor fungicide and paclobutrazol applications on foliar disease severity and turfgrass quality of creeping bentgrass putting greens. Crop Prot. 79:64–69.

Mitkowski, N. A. and Chaves, A. (2013). Impact of demethylation inhibitor fungicides on clipping yield and rooting depth of creeping bentgrass. Hort. Sci. 48:1052–1055.

Mohammadi, M. H., Etemadi, N., Arab, M. M., Aalifar, M., Arab, M. and Pessarakli, M. (2017a). Molecular and physiological responses of Iranian perennial ryegrass as affected by trinexapac-ethyl, paclobutrazol and abscisic acid under drought stress. *Plant Physiol. Biochem.* 111:129–143.

Mohammadi, M. H. S., Etemadi, N., Nikbakht, A. and Pessarakli, M. (2017b). Physiological responses of two cool season grass species to trinexapac-ethyl under traffic stress. Hort.Sci. 52(1):99–109.

Moss, J. Q., Xiong, X., Su, K., Poudel, B. P. and Haguewood, J. B. (2012). Creeping bentgrass (Agrostis stolonifera) golf green tolerance to bispyribac-sodium tank-mixed with paclobutrazol. Weed Technol. 26(1):145-150.

Munshaw, G. C., Ervin, E. H., Beasley, J. S., Shang, C., Zhang, X. and Parrish, D. J. (2010). Effects of late-season ethephon applications on cold tolerance parameters of four bermudagrass cultivars. Crop Sci. 50(3):1022-1029.

Nangle, E. J., Gardner, D. S., Metzger, J. D., Street, J. R. and Danneberger, T. K. (2012). Impact of nitrogen source and trinexapac-ethyl application on creeping bentgrass (Agrostis stolonifera L.) physiology under neutral shade, deciduous tree shade and full sunlit conditions. Hort. Sci. 47(7):936-942.

Nickell, K. and Qian, Y. (2013). Saltgrass germination response to salinity with different seed treatments. Int. Turf. Soc. Res. J. 12:187-192.

Ok, C. H., Popko, J. T., Campbell-Nelson, K. and Jung, G. (2011). In vitro assessment of Sclerotinia homoeocarpa resistance to fungicides and plant growth regulators. Plant Dis. 95(1):51-56.

Patton, A. J., Braun, R. C., Schortgen, G. P., Weisenberger, D. V., Branham, B. E., Sharp, B., Sousek, M. D., Gaussoin, R. E. and Reicher, Z. J. (2019). Long-term efficacy of annual bluegrass control strategies on golf course putting greens. *App. Turf Sci.* 5(1). DOI: 10.2134/cftm2018.09.0068.

Patton, A. J., Schortgen, G. P., Hoyle, J. A., Harrell, M. S. and Reicher, Z. J. (2018). Fall applications of Proxy (ethephon) suppress spring seedheads of 'Meyer' zoysiagrass. Crop Forage Turf Mgmt 4(1):1-8. DOI: 10.2134/cftm2018.03.0012.

Pease, B. W. and Stier, J. C. (2018). Nitrogen rate and growth regulator effects on shaded velvet and creeping bentgrasses. Agron. J. 110(6):2151-2158.

Pornaro, C., Fiorio, S., Macolino, S. and Richardson, M. D. (2017). Growth and quality responses of low-maintenance turfgrasses to trinexapac-ethyl. Crop Prot. 98:236-242.

Putman, A. I. and Kaminski, J. E. (2011). Mowing frequency and plant growth regulator effects on dollar spot severity and on duration of dollar spot control by fungicides. Plant Dis. 95(11):1433-1442.

Qian, Y. L. and Engleke, M. C. (1999). Influence of trinexapac-ethyl on Diamond zoysiagrass in a shade environment. Crop Sci. 39(1):202-208.

Qian, Y. L., Engleke, M. C., Foster, M. J. V. and Reynolds, S. (1998). Trinexapac-ethyl restricts shoot growth and improves quality of 'Diamond' zoysiagrass under shade. Hort.Sci. 33:1019-1022.

Rahman, A., Kuldau, G. A. and Uddin, W. (2013). Induction of salicylic acid-mediated defense response in perennial ryegrass against infection by Magnaporthe oryzae. Phytopathol. 104(6):614-623.

Raudenbush, Z. R., Elmore, M. T., Nangle, E. J. and Tuck, D. P. (2020). December ethephon applications in combination with mineral oil improve annual bluegrass seedhead suppression. Crop Forage Turfgrass M*anag* 6(1). DOI: 10.1002/cft2.20016.

Reasor, E. H. and Brosnan, J. T. (2020). Trinexapac-ethyl applications and lightweight rolling on ultradwarf bermudagrass putting greens. Crop Forage Turfgrass M*ana*g. 6(1). DOI: 10.1002/cft2.20036.

Reasor, E. H., Brosnan, J. T., Kerns, J. P., Hutchens, W. J., Taylor, D. R., McCurdy, J. D., Soldat, D. J. and Kreuser, W. C. (2018). Growing degree day models for plant growth regulator applications on ultradwarf hybrid bermudagrass putting greens. Crop Sci. 58(4):1801-1807.

Reicher, Z., Sousek, M., Patton, A., Weisenberger, D., Hathaway, A. and Calhoun, R. (2015). Annual bluegrass control on putting greens from three or four years of season-long applications of herbicides or plant growth regulators in three states. *Crop Forage Turfgrass Manag.* 1(1): 1-7.

Roberts, J. A., Kerns, J. P. and Ritchie, D. F. (2015). Bacterial etiolation of creeping bentgrass as influenced by biostimulants and trinexapac-ethyl. Crop Prot. 72:119-126.

Roberts, J. A., Ritchie, D. F. and Kerns, J. P. (2016). Plant growth regulator effects on bacterial etiolation of creeping bentgrass putting green turf caused by Acidovorax avenae. Plant Dis. 100(3):577-582.

Roche, M. B., Penberthy, J. D. and O'Brien, L. (2013). Use of trinexapac-ethyl to reduce mowing requirements of warm season grasses in subtropical Australia. Int. Soc. Res. J. 12:829-830.

Rogan, C. M. and Schlossberg, M. J. (2013). Complimenting late-season nitrogen fertilization of cool-season turfgrass putting greens with trinexapac-ethyl. Agron. J. 103:1507-1514.

Rossi, F.S. (2001). Annual bluegrass population dynamics in response to growth regulators and herbicides. Intl. Turfgrass Soc. Res. J. 9: 906-909.

Russel, T. R., Karcher, D. E. and Richardson, M. D. (2019). Daily light integral requirement of a creeping bentgrass putting green as affected by shade, trinexapac-ethyl and a plant colorant. Crop Sci. 59(4):1768-1778.

Schiavon, M., Leinhauer, B. L., Serena, M., Maier, B. and Sallenave, R. (2015a). Plant growth regulator and soil surfactants' effects on saline and deficit irrigate warm-season grasses: I. Turf quality and soil moisture. Crop Sci. 54:2815-2826.

Schiavon, M., Leinhauer, B. L., Serena, M., Maier, B. and Sallenave, R. (2015b). Plant growth regulator and soil surfactants' effects on saline and deficit irrigate warm-season grasses: II. Pigment content and superoxide dismutase activity. Crop Sci. 54:2827-2836.

Serena, M., Schiavon, M., Sallenave, R. and Leinauer, B. (2020). Drought avoidance of warm-season turfgrasses affected by irrigation system, soil surfactant revolution, and plant growth regulator trinexapac-ethyl. *Crop Sci.* 60(1):485-498.

Serena, M., Sportelli, M., Sevostianova, E., Sallenave, R. and Leinauer, B. (2018). Combining trinexapac-ethyl with a soil surfactant reduces bermudagrass irrigation requirements. Agron. J. 110(6):2180-2188.

Serensits, T. J., McNitt, A. S. and Brosnan, J. T. (2017). Kentucky bluegrass divot resistance as affected by cultivar, trinexapac-ethyl, and soil cultivation. Crop Sci. 57(S1):218-226.

Sermons, S. M., Sinclair, T. R., Seversike, T. M. and Rufty, T. W. (2017). Assessing transpiration estimates in tall fescue: The relationship among transpiration, growth, and vapor pressure deficits. Environ. Exp. Bot. 137:119-127.

Severmutlu, S., Mutlu, N., Gurbuz, E., Shearman, R. C. and Riordan, T. P. (2012a). Effects of seasonal trinexapac-ethyl application on warm season turfgrass colour, quality and spring green-up under Mediterranean environment. Eur. J. Hort. Sci. 77:226-236.

Severmutlu, S., Mutlu, N., Gurbuz, E., Shearman, R. C. and Riordan, T. P. (2012b). Effects of seasonal trinexapac-ethyl application on warm season turfgrass growth under Mediterranean environment. Eur. J. Hort. Sci. 77:279-287.

Shahrokhi, M., Tehranifar, A., Hadizadeh, H. and Selahvarzi, Y. (2010). Turfgrass seedling height and quality in paclobutrazol- treated seeds of Lolium perenne L. Barrage and Festuca arundinacea L. Master. J. Biol. Environ. Sci. 4:109–115.

Shahrokhi, M., Tehranifar, A., Hadizadeh, H. and Selahvarzi, Y. (2011). Effect of drought stress and paclobutrazol-treated seeds on physiological response of Festuca arundinacea L. Master and Lolium perenne L. Barrage. J. Biol. Environ. Sci. 5:77–85.

Sidhu, S. S., Yu, J. and McCullough, P. E. (2014). Physiological behavior of ethephon in five turfgrasses. Crop Sci. 54(4):1816–1822.

Steinke, K. and Stier, J. C. (2003). Nitrogen and growth regulator applications for improving shaded turf performance. Crop Sci. 43(4):1399–1406.

Stewart, J. M., Latin, R., Reicher, Z. and Hallett, S. G. (2007). Influence of trinexapac-ethyl on the efficacy of chlorothalonil and propiconazole for control of dollar spot in creeping bentgrass. Appl. Turf Sci. DOI: 10.1094/ATS-2008-0319-01-RS.

Studzinska, A. K., Gardner, D. S., Metzger, J. D., Shetlar, D., Harriman, R. and Danneberger, T. K. (2012). The effect of PcGA2ox overexpression on creeping bentgrass (*Agrostis stolonifera* L.): Performance under various light environments. *Hort. Sci.* 47(2):280–284.

Vincelli, P., Williams, D. and Dixon, E. (2011). Early curative fungicide applications provide disease control on fairway-height creeping bentgrasss. *Ats* 8(1). DOI: 10.1094/ATS-2011-1025-01-RS.

Volterrani, M., Magni, S., Gactani, M. and Lulli, F. (2012). Trinexapac-ethyl effects on stolon activity and node vitality of 'Tifway' hybrid bermudagrass. *Hort. Technol* 22(4):479–483.

Watschke, T. L., Dernoeden, P. H. and Shetlar, D. J. (2013). *Managing Turfgrass Pests*. CRC Press, Boca Raton, FL.

Watschke, T. L., Long, F. W. and Duich, J. M. (1979). Control of *Poa annua* by suppression of seedheads with growth regulators. *Weed Sci.* 27(2):224–231.

Wherley, B. G., Gardner, D. S. and Metzger, J. D. (2005). Tall fescue photomorphogenesis as influenced by changes in the spectral composition and light intensity. *Crop Sci.* 45(2):562–568.

Williams, A. P. (2014). *Plant growth regulators and herbicides for management of Poa annua: Impact of biotypes and behavior of flurprimidol in turfgrass species*. Theses and Dissertations. Plant and Soil Sciences. https://uknowledge.uky.edu/pss_etds/45 (vol. 45).

Woosley, P. B., Williams, D. W. and Powell, A. J. Jr. (2003). Postemergence control of annual bluegrass (*Poa annua* spp. *reptans*) in creeping bentgrass (*Agrostis stolonifera*) turf 1. *Weed Technol.* 17(4):770–776.

Xu, C. and Huang, B. (2012). Proteins and metabolites regulated by trinexapac-ethyl in relation to drought tolerance and Kentucky bluegrass. *J. Plant Growth Regul.* 31(1):25–37.

Yang, C. and Li, L. (2017). Hormonal regulation in shade avoidance. *Front. Plant Sci.* 8:1527. DOI: 10.3389/fpls.2017.01527.

Zapiola, M. L., Chastain, T. G., Garbacik, C. J. and Young, W. C. (2014). Trinexapac-ethyl and burning effects on seed yield components in strong creeping red fescue. *Agron. J.* 106(4):1371–1378.

Zhang, N., Han, L., Xu, L. and Zhang, X. (2018). Ethephon seed treatment impacts on drought tolerance of Kentucky bluegrass seedlings. *HortTechnol* 28(3):319–326.

Zhang, X., Ervin, E. H. and Schmidt, R. E. (2003). Seaweed extract, humic acid, and propiconazole improve tall fescue sod heat tolerance and posttransplant quality. *Hort Sci.* 38(3):440–443.

Zhang, X., Ervin, E. H., Wu, W., Sharma, N. and Hamill, A. (2017). Auxin and trinexapac-ethyl impact on root viability and hormone metabolism in creeping bentgrass under water deficit. *Crop Sci.* 57:130–136.

Zheng, Z. L., Yang, Z., Jang, J. C. and Metzger, J. D. (2001). Modification of plant architecture in chrysanthemum by ectopic expression of the tobacco phytochrome *B1* gene. *J. Am. Soc. Hortic. Sci.* 126(1):19–26.

Zhu, H., Taylor, A. A., Astor, S. R. and Terry, N. (2016). Enhancing saltgrass germination and growth in a saline soil contaminated with petroleum hydrocarbons. *Plant Soil* 412(1-2):189–199.

Chapter 13

Advances in abiotic stress management in turfgrass

Charles Fontanier, Oklahoma State University, USA; and Chrissie A. Segars, Texas A&M University, USA

1 Introduction

Abiotic stresses of turfgrasses involve any environmental condition which limits plant growth, plant quality, and ecosystem functionality. Examples of commonly observed abiotic stresses in turfgrass ecosystems include drought, high temperature, low temperature, low irradiance, traffic, soil salinity, and soil chemical phytotoxicity (Fry and Huang, 2004). Regional climate and weather patterns, turfgrass species and cultivars, soils and soil biological, chemical, and physical properties, turfgrass maintenance practices, and societal pressures toward turfgrass can influence the relative importance of specific abiotic stresses to managed turfgrass ecosystems (Huang et al., 2014). These external factors have intensified in recent decades due to global climate change and population growth, resulting in diminishing water resources, increasing weather extremes, and increased demands for public green spaces (Adams et al., 1999).

1.1 Plant stress response

Plant physiologists have developed models for plant stress response with the simplest describing stress as any deviation from normal or optimal growing conditions (Salisbury and Ross, 1992). More commonly, the plant stress response is described as either elastic or plastic, which relates to the magnitude of the stress applied to the system (Salisbury and Ross, 1992). Plastic stress

http://dx.doi.org/10.19103/AS.2022.0110.13

responses are more severe and result in permanent changes to the plant's health or functionality. In contrast, elastic responses can be temporary and potentially useful in acclimating turfgrasses to subsequent stresses. Stress is often categorized as either being acute or chronic, and depending on the stressor can cause dysfunction within critical plant systems through mechanical injury to cell and plastid membranes, increased presence of reactive oxygen species within cells, or physical tearing and bruising of entire plant organs and tissues (Salisbury and Ross, 1992).

1.2 Goals in abiotic stress management

Managing abiotic stresses with reduced or strategically targeted inputs of irrigation, fertilizer, mowing and other cultural practices, and pesticides has become critical to the sustainability of the turfgrass industry (Miranda and Medlin, 2021). Historically, abiotic stress management once aimed to minimize or eliminate stress while maximizing the esthetic or functional traits of turfgrass ecosystems (Beard, 1973). Recent trends in the academic literature and with society's perception toward turfgrass suggest a more practical goal would be to achieve acceptable levels of stress mitigation that optimize resource inputs (Huang et al., 2014). This paradigm shift from maximization to optimization has been embraced by turfgrass science researchers and by turfgrass managers and practitioners who face criticisms toward their industry from stakeholders, society, and the government (Throssell et al., 2009). As a result, data-driven management tools have been developed to aid in the rapid detection of plant stress, predict plant input needs quantitatively, and ensure plant system growth, function, and purpose can be maintained with minimal inputs. Thus, the ability to monitor the environment, be resourceful with inputs, and detect plant stress response has become vital to progressive turfgrass managers around the world (Turgeon, 2005).

In this chapter, the goal is to describe the current literature related to abiotic stress management in turfgrasses with an emphasis on the sustainability of turfgrass ecosystems. The subject matter is organized into four approaches toward sustainable abiotic stress management: managing turfgrass growth, moderating the environment, enhancing or improving abiotic stress tolerance, and masking abiotic stress response, followed by an exploration of detecting abiotic stress and data-driven decision-making for managing abiotic stress.

2 Approaches to sustainable abiotic stress management

An overall goal of turfgrass management is to develop and implement sustainable strategies and practices for abiotic stress management. These would include managing turfgrass growth to support end-user expectations for quality and function, moderating or influencing the environmental conditions that contribute to abiotic stress, mitigation of drought, salinity, low temperature, and traffic stress conditions imposed on managed turfgrass ecosystems, and considerations with utilizing improved turfgrass species and cultivars, biostimulants, plant physiology mechanisms for responding to stress, and tactics that mask the appearance of abiotic stress.

2.1 Managing turfgrass growth

Several abiotic stresses can reduce the ability of the plant to maintain a positive carbohydrate balance or are worsened by poor carbohydrate status and function (Fry and Huang, 2004). Close, frequent mowing is necessary for acceptable performance in many turfgrass ecosystems but can interfere with plant photosynthetic capacity if residual leaf area is insufficient. Cultural turfgrass management practices which stimulate excessive growth can lead to increased removal of photosynthate at mowing and further compromise plant carbohydrate status. Conversely, turfgrass maintenance practices which minimize vertical growth without compromising plant density can create a healthier plant better adept at tolerating abiotic stresses (Fry and Huang, 2004; Law et al., 2016).

Plant growth regulators (PGRs) are commonly used in the turfgrass industry, particularly on golf courses, where they have been shown to improve turfgrass quality, reduce seedhead production, increase nutrient retention within the plant, and reduce clipping yield without sacrificing density (Haguewood et al., 2013; Kreuser and Soldat, 2012; McCullough et al., 2006, 2007). More recently, PGRs have been identified as contributing to enhanced response to abiotic stress, although the response has been variable depending on species and the particular stress involved. For example, combining the commonly used PGR of trinexapac-ethyl (TE) with soil surfactants reduced bermudagrass (*Cynodon dactylon* [L.] Pers. × *Cynodon transvaalensis* Burtt-Davy) irrigation water requirements in a desert climate (Serena et al., 2018). Under deficit irrigation, TE applications were associated with increased superoxide dismutase (e.g. an antioxidant enzyme) activity and increased turfgrass quality of bermudagrass (Schiavon et al., 2014a,b). Several studies have reported on TE enhancing drought or heat tolerance of cool-season turfgrasses (Jespersen and Huang, 2017; McCann and Huang, 2007; Xu and Huang, 2010). Others have reported on the benefits of using TE on warm- and cool-season turfgrasses managed under shade (Chen et al., 2021; Goss et al., 2002; Russell et al., 2019).

Overapplication of PGRs can create opportunities for compounding stress on the turfgrass plant. For example, warm-season turfgrasses maintained on putting green surrounds may be more prone to winterkill due to repeated growth regulation (Richardson et al., 2014). Optimal application intervals have been successfully defined using growing degree day (GDD) models, with reapplication every 216–230 GDD (10°C base air temperature) and every 200–220 GDD (0°C base air temperature) for ultradwarf bermudagrasses and creeping bentgrasses (*Agrostis* spp.), respectively (Carroll et al., 2021; Kreuser and Soldat, 2011; Kreuser et al., 2017; Reasor et al., 2018). The GDD models have largely been developed during the summer growing season, and less is known about their appropriateness for off-season periods. Late-fall applications of TE or prohexadione-Ca, particularly at the highest label application rates, contributed to delayed spring green-up and increased winterkill in annual bluegrass (*Poa annua* L.) (DaCosta et al., 2021). Currently, PGRs are commonly used tools for sustainable turfgrass management, but a broader understanding of mixing products, off-season application timings, and interactions between PGRs and other plant and soil health and protection products represents important areas of needed research.

2.2 Moderating the environmental conditions causing abiotic stress

Plant stress resistance is often described in terms of avoidance versus tolerance mechanisms (Salisbury and Ross, 1992). For example, drought resistance can be achieved through physiological tolerance to leaf dehydration as well as the ability of the plant to avoid cellular dehydration through traits such as deep rooting. In terms of managing abiotic stresses, the same terms of avoidance and tolerance can be applied. In this case, turfgrass managers can moderate the environment (i.e. avoidance) or enhance the stress tolerance of the plants through plant selection and their turfgrass management decisions and practices.

2.2.1 Drought stress mitigation

Drought stress is one of the primary environmental stresses impacting turfgrass health. Responses to drought stress vary greatly between cool-season (C3) and warm-season (C4) turfgrasses but can vary even further between species and cultivars within species (Fry and Huang, 2004; Huang et al., 2014). The primary tool used to mitigate drought stress in fine turf is supplemental irrigation through overhead sprinklers or sub-surface drip irrigation. Sustainable turfgrass management requires a working knowledge of plant evapotranspiration (ET) rates, soil or root zone water holding capacity, and irrigation system precipitation delivery rate and amount. Most commonly, irrigation is applied based on fractions of reference ET (ET_o), although soil moisture monitoring can also be an efficient approach to sustainable turfgrass irrigation practices.

Proper irrigation scheduling to meet plant water requirements has been extensively studied across many turfgrass species (e.g. for a more detailed discussion on turfgrass irrigation, refer to https://www.usga.org/course-care/digitalcollections/irrigation-management.html).

Hydrophobic soils and root zones can be a troublesome challenge that turfgrass managers often encounter (Cisar et al., 2000). Although all soil types can become hydrophobic, sand and sand-based soils and root zones are often more susceptible due to their lower surface area per unit volume than finer textured clay soils (Doerr et al., 2000; Franco et al., 1994; Wallis and Horne, 1992). Hydrophobic or water-repellent soils can be developed into localized dry spots as visually observed by severely drought-stressed turfgrass, and subsequent limited to no infiltration can contribute to surface water runoff during irrigation or rainfall events (Doerr et al., 2003; Mitra et al., 2006). Water lost as runoff can contribute to environmental contamination in the form of both nutrient and bacteria transport and deposition into water systems (Snyder et al., 1984). Soil surfactants, also referred to as wetting agents, typically will lower the surface tension of water, increase water infiltration and percolation of water through the soil profile, and decrease the risk or occurrence of surface runoff (Kostka et al., 1997; Zontek and Kostka, 2012). Research over the past few decades has consistently shown the value of wetting agents in turfgrass ecosystems for reducing soil hydrophobicity, increasing root zone soil moisture, treating localized dry spots, improving irrigation efficiency, and optimizing fungicide application and delivery to target soil-borne turfgrass diseases (Cisar et al., 2000; Fidanza et al., 2020; Leinauer et al., 2001; Mitra et al., 2004; Schiavon et al., 2015). Wetting agents represent tools that allow for more sustainable irrigation practices and have been shown to improve turfgrass quality under deficit irrigation programs in various climates and turfgrass systems (Cisar et al., 2000; Soldat, 2010; Schiavon et al., 2015). Recently, Chang et al. (2020) conducted a 2-year study with no drought stress and concluded that St. Augustinegrass (*Stenotaphrum secundatum* [Walt.] Kuntze) turfgrass quality and soil moisture were slightly improved by wetting agent and fertilizer treatments, but no significant treatment effects influenced water runoff volumes or fertilizer nutrient loss. Also recently, Abagandura et al. (2021) concluded from their study that applying wetting agents with urea (46N-0P-0K) can increase bermudagrass nitrogen uptake efficiency and reduce potential nitrogen leaching. The use of wetting agents and their possible influence on and within the turfgrass rhizosphere currently is being explored further (Bigelow, 2022; Fidanza and Kostka, 2022).

Wetting agents can have beneficial implications in sand root zones such as golf course putting greens and sand-based athletic fields (Kostka and Fidanza, 2019). In the southern region of the USA, low air temperature stress is often a concern for warm-season putting green turfgrass species. Recently, DeBoer et al. (2020) reported enhanced winter survival of ultradwarf bermudagrasses after a single late-fall application of a wetting agent that was timed prior to covering the turf at low-temperature thresholds.

2.2.2 Salinity stress mitigation

In many regions of the USA, water scarcity is the primary limitation for achieving sustainability within the turfgrass industry (Throssell et al., 2009). As a result, golf courses, athletic/sport facilities, and parks and recreation areas have increasingly utilized poor quality irrigation water, either from saline/sodic aquifers or reuse of treated effluent referred to as reclaimed or recycled water (Throssell et al., 2009). As demand increases for potable water, it is estimated that 35% of golf courses in the southwestern and southeastern USA use recycled water (Gelernter et al., 2015). Classically, alleviation of salinity or sodicity in turfgrass ecosystems has targeted improvement of the soil through the leaching of undesired cations (Gelernter et al., 2015; Turgeon, 2005). This typically has been coupled with gypsum applications to move sodium into solution and contribute to deflocculation of the soil. These approaches are still appropriate for the modern turfgrass ecosystem, although a number of novel products have been developed by the industry as salinity issues have become increasingly commonplace. Schiavon and Baird (2018) investigated seven products marketed for improving turfgrass performance under saline or sodic conditions and reported that products containing calcium in the form of calcium carbonate, calcium silicate, or calcium nitrate were effective at decreasing sodium in a bermudagrass root zone. Use of low salt-index fertilizers, plant health products like kelp extracts, and persistent efforts to reduce root zone salinity and sodicity are helpful to a quality salinity stress management program for turfgrass (Schiavon and Baird, 2018).

2.2.3 Low-temperature stress mitigation

Cold temperature stress can be put into two main categories: chilling and freezing stress. Particularly for warm-season turfgrass species, cold stress can limit the distribution and use of many species and cultivars. Warm-season turfgrasses are at a greater risk of injury from low temperatures than cool-season turfgrasses (Beard, 1973; Turgeon, 2005). To reduce the likelihood of low-temperature stress on turfgrasses, practices involving moisture management, protective winter covers, and the addition of wetting agents have been researched.

For example, on ultradwarf bermudagrass putting greens, a forecasted temperature threshold of −3.38°C was previously identified as the appropriate time to use protective covers (O'Brien and Hartwiger, 2013). A study in Fayetteville, AR, USA, investigated 4 different forecasted low-temperature thresholds at −3.38°C, −5.55°C, −7.77°C, and −9.44°C for 3 consecutive winters and the impacts on three ultradwarf bermudagrass cultivars: 'Champion', 'Mini Verde', and 'Tifeagle' (DeBoer et al., 2019). The protective cover was a black, woven polypropylene custom-made cover (Xton, Inc.; Florence, AL, USA), which is widely used in the region. This study indicated that the critical threshold for low-temperature injury can be lowered to −9.44°C without a significant change in winter survival (DeBoer et al., 2019). This change to −9.44°C compared to −3.38°C can significantly reduce the number of required covering events which allows for more rounds of golf to be played with less labor costs. Additional information on the effects of antecedent air and soil temperatures on winterkill risk may be required to generalize these recommendations across other regions of the USA and other global locations.

For cool-season species, low-temperature winterkill can also be a concern in higher latitudes where the ground freezes for an extended period of time. In these climates, winter covers also have been useful in mitigating winterkill risk. Prokopiuk et al. (2019) demonstrated winter covers improved turfgrass quality of cool-season turfgrasses in Poland. In contrast, Waalen et al. (2017) reported mixed results using plastic covers for six species in Norway. Specifically, covers only provided any observed benefit for annual bluegrass putting greens and, in some cases, increased disease severity of other turfgrass species (Waalen et al., 2017).

2.2.4 Traffic stress mitigation

Traffic stress is an important abiotic stress that negatively affects turfgrasses maintained on golf courses, athletic fields, residential and commercial lawns, and park and recreation areas (Turgeon, 2005). The expectation for year-round esthetically pleasing turfgrass quality, turfgrass function and playability, off-season activities, or concentrated in-season use can result in turfgrass failure (Carrow and Petrovic, 1992). Traffic stress directly impacts the health of the turfgrass plant through ripping and tearing of the above-the-ground structures and indirectly through physical changes in the soil such as compaction (Beard, 1973; Musser, 1950; Turgeon, 2005). Murphy and Ebdon (2013) have provided a thorough review of traffic stress mitigation for turfgrass management.

2.3 Improving abiotic stress tolerance

Recent advancements with methods and mechanisms in turfgrasses for their ability to tolerate abiotic stress have been in the areas of genetics, biostimulants, and plant physiology.

2.3.1 Genetic improvement of turfgrasses

Turfgrass breeding is a process that genetically improves the performance of cultivars of a species in the most efficient way possible (Fehr, 2011). With a changing climate, it is necessary to breed improved turfgrass cultivars with more tolerance to abiotic stress. To do this, turfgrass breeders must use a combination of breeding methods from conventional to new breeding techniques to not only improve genetics but maintain superior quality, function, and performance (Bushman and Warnke, 2013). Selection of improved turfgrasses with increased tolerances to abiotic and biotic stresses could ultimately lead to reduced inputs and the adoption of more environmentally responsible turfgrass management practices (Johnson et al., 2013; Reasor et al., 2018).

2.3.2 Development of Alliance for Low Input Sustainable Turf, Turfgrass Water Conservation Alliance and National Turfgrass Evaluation Program

Improvement of abiotic stress tolerance across major turfgrass species has seen an increase in formal interest from seed companies, sod producers, and university research programs over the past 40 years. Turfgrass seed and cultivar testing organizations such as the National Turfgrass Evaluation Program (NTEP; Beltsville, MD, USA), the Alliance for Low Input Sustainable Turf (A-LIST; Willamette Valley, OR, USA), and the Turfgrass Water Conservation Alliance (TWCA; Albany, OR, USA) have been established or adapted to meet these needs. The objectives of NTEP are to develop and coordinate uniform evaluation field trials of turfgrass cultivars and experimental selections in the USA and Canada. As a result, currently NTEP has collected data on tens of thousands of commercial and experimental turfgrasses and that information is available on their website (www.ntep.org). Historically, NTEP trials were focused on turfgrass quality, geographical adaptation, and to some extent biotic stress tolerance. More recent trials have been developed to specifically evaluate drought, traffic, and salinity tolerances. The A-LIST (www.a-listturf.org) seeks to test, identify, and promote cool-season turfgrasses that maintain excellent turfgrass quality under reduced inputs (e.g. water, chemicals, and fertilizers). Similarly, the TWCA (www.tgwca.org) has developed an objective protocol to evaluate turfgrass species and cultivars for drought response and recovery. Collectively, these three organizations illustrate the shift within the turfgrass industry toward developing sustainable turfgrasses that persist or are best suited within a variety of unfavorable environmental conditions.

2.3.3 Biostimulants and soil health

Biostimulants are traditionally defined as any substance or microorganism applied to plants to enhance nutrition efficiency, abiotic stress tolerance, and plant quality traits (du Jardin, 2015). Within both the turfgrass industry and the turfgrass research community, there has been an increased interest in biostimulants to promote plant and soil (e.g. root zone) health (Fidanza et al., 2015, 2019). An important component of sustainable abiotic stress management is the maintenance of a healthy soil. Conventional turfgrass cultural practices such as sand topdressing, aerification/core cultivation, and proper fertilizer programs can be combined with supplemental applications of soil health products. The most commonly used soil health products in turf typically include humic or fulvic acids as the primary active ingredient. Zhu and Li (2018) showed applications of humic acid improved the quality of golf course fairways irrigated with saline or recycled water. Lindsey et al. (2021) reported increased mineralizable carbon and nitrogen in soils after the application of humic-coated fertilizers, which allowed for sustained turfgrass quality at lower nitrogen rates than with urea alone. In contrast, Lindsey et al. (2020) reported no improvement in turfgrass growth or turfgrass quality with the addition of humic coating on fertilizers during establishment of Kentucky bluegrass (*Poa pratensis* L.) or bermudagrass.

Phytohormones, most notably cytokinin, are commonly used in turf management because of their counteracting effects on reactive oxygen species and the senescence-promoting hormones ethylene and abscisic acid (Fry and Huang, 2004). Hu et al. (2012) reported that post-stress applications of cytokinin can expedite recovery from drought stress in Kentucky bluegrass. In contrast, Munshaw et al. (2006) reported no effects of cytokinin-containing seaweed extract on bermudagrass freezing tolerance. Application of TE has also been shown to promote endogenous cytokinin concentrations, which suggests enhanced stress tolerance through the use of PGRs may be related not only to mitigating mower stress but also through the upregulation of anti-senescence phytohormones (Ervin and Zhang, 2007).

2.3.4 Designed stress response

The goals of preventing all stress from a turfgrass system are largely unrealistic and furthermore can be detrimental to plants that are suddenly exposed to a large stress event (Fry and Huang, 2004; Salisbury and Ross, 1992). That is, allowing for or inducing mild stress can stimulate plant stress responses, which enhance tolerance to future stresses (Huang et al., 2014). This process can be termed acclimation and is often critical for plant survival under extreme low temperature, salinity, or drought stress. Depending on the stress, plant responses to mild stress may include the reallocation of carbohydrates, increased production of antioxidants, or alterations to cellular and plastid membrane composition. Increases in membrane unsaturation have been reported as part of cold acclimation processes which promote freezing tolerance and post-dormancy regrowth (Fontanier et al., 2020; Hoffman et al., 2010).

Use of mild stress to acclimate turfgrass plants can be achieved through deficit irrigation strategies, timely usage of grow covers during winter, and application of stress-activating products. Nosalewicz et al. (2018) reported enhanced photosynthetic activity and maintenance of leaf hydration of tall fescue (*Schedonorus arundinaceus* [Schreb.] Dumort) and perennial ryegrass (*Lolium perenne* L.) experiencing water stress after having endured a previous drought. Mild stress of varying types can also interact to enhance plant overall stress tolerance (e.g. cross-stress interactions). For example, Munshaw et al. (2004) reported the use of saline irrigation water could subsequently contribute to enhanced cold hardiness of bermudagrass. Kong and Henry (2016) reported that freeze stress during late fall resulted in greater mortality in summer drought than the same freeze stress applied in spring. Although these treatments may not be intuitive management tools, a creative turfgrass manager can utilize knowledge of plant stress memory to 'harden-off' the plant.

Systemic acquired resistance (SAR) and induced systemic resistance (ISR) are terms commonly used to describe changes in the plant's physiological condition which are prime defenses against subsequent stress (Hsiang et al., 2011; Vallad and Goodman, 2004). Most often, these processes are applied to the prevention of biotic stresses, but the same pathways are thought to also contribute to abiotic stress tolerance mechanisms. With regard to turf management, chemical stimuli are used to promote these responses, hence the common label of plant health products serving as 'plant defense activators'. For SAR, products typically are applied foliarly to activate salicylic acid pathways, which are then translocated systemically throughout the plant. In contrast, ISR is derived from root activity, particularly due to the presence of plant growth-promoting rhizobacteria and fungi. Primary signal molecules for ISR include ethylene and jasmonic acid, which regulate genes associated with wounding. In some cases, SAR directly increases metabolites for stress tolerance, while in others, SAR and ISR are more useful in priming stress tolerance pathways for more timely response to subsequent stress.

Acibenzolar-*S*-methyl (ASM) is a component of some fungicides and has been associated with SAR in turfgrasses. Jespersen et al. (2017) and Jespersen and Huang (2017) reported ASM-enhanced heat and drought tolerance in creeping bentgrass (*Agrostis stolonifera* L.) through a multitude of physiological changes in protein synthesis and metabolite accumulation. Others have reported ASM to reduce the transpiration rate of creeping bentgrass under high vapor pressure deficits (Shekoofa et al., 2016), which presumably could have a benefit in mitigating plant water loss in geographic regions of high evaporative demand.

Exogenous applications of various plant protectants, antioxidants, phytohormones, and stress response promoters have been evaluated for use in moderating abiotic stress in turfgrasses. Laskowski et al. (2019) reported mefluidide, propiconazole, and a mineral oil product (Civitas™; Intelligro; Ontario, Canada) resulted in greater polyunsaturated fatty acids, particularly linolenic acid, in crown tissue of annual bluegrass which subsequently enhanced survival and regrowth after exposure to low temperature ice formation. Laskowski et al. (2019) speculate that mefluidide, through inhibition of 3-ketoacyl-CoA, influences long-chain fatty acid synthesis, which contributes to enhanced membrane stability during ice treatment. There is also evidence that mefluidide can increase abscisic acid content in plants, which has been associated with enhanced cold hardiness (Zhang et al., 1986; Laskowski et al., 2019). Civitas is a two-component product containing a mineral oil and a copper-based pigment. Civitas products have been associated with ISR activation in turfgrasses and subsequent increase in activity of the jasmonic pathway (Cortes-Barco et al., 2010). These findings were reinforced in a recent study that showed exogenous application of jasmonic acid or Civitas both improved recovery of annual bluegrass after ice encasement (Laskowski and Merewitz, 2020a). Interestingly, Civitas also has been associated with increased turfgrass clipping yields which suggest frequent usage of the product may negatively influence carbohydrate reserves in the absence of other growth regulators (Kreuser and Rossi, 2014).

Trifloxystrobin is a strobilurin fungicide which has shown evidence of promoting ISR in red pepper plants (Han et al., 2012). Trifloxystrobin fungicide products have been developed and marketed as plant-activating compounds for use in the turfgrass industry. The research-based confirmation of these responses in turfgrasses is lacking and therefore needs further study.

Silicon has in recent years been associated with a variety of improvements in stress tolerance, although many of these responses have been inconsistent or anecdotal and likely require additional research to confirm them. For example, foliar application of calcium silicate improved the wear tolerance of leaves of zoysiagrass (*Zoysia* spp.) (Saigusa et al., 2000), while potassium silicate did not improve wear tolerance of seashore paspalum (*Paspalum vaginatum* Sw.) (Trenholm et al., 2001). Soil applications of calcium magnesium silicate resulted in increased wear tolerance in perennial ryegrass after 3 years of repeated applications (Pruyne et al., 2019). These findings suggest benefits of silicon can be obtained through either foliar or soil-based applications, but only if the soil concentration reaches some minimal threshold. The most consistent responses associated with silicon have been reported for drought stress. Bae et al. (2017) reported sodium improved the growth of Kentucky bluegrass plants under drought stress, while similar results were discovered for tall fescue and creeping bentgrass (Eneji et al., 2008; Schmidt et al., 1999). The effects of silicon on salinity stress are less well-documented than for wear or drought, but there is evidence that silicon can inhibit the absorption of sodium in turfgrass seedlings (Chai et al., 2010). Similarly, foliar silicon applications have been reported to reduce salinity stress in both warm- and cool-season turfgrasses (Esmaeili et al., 2015). Schiavon and Baird (2018) also reported the benefits of using calcium silicate products to reduce sodium in the root zone, although this is more likely associated with the application of calcium than silicon. Refer to Guertal and Datnoff (2021) for a more extensive review of silicon effects on wear, salinity, and drought response in turf.

Other molecules showing potential as abiotic stress management tools include hydrogen peroxide, glycine betaine, melatonin, salicylic acid, jasmonic acid, and abscisic acid. These molecules are found endogenously in the plant as part of the normal stress response system. Exogenous application of these products has routinely shown an increase in endogenous concentrations and thus provides a boost to the normal stress response system. Melatonin has been shown to reduce heat and drought stress in tall fescue seedlings by promoting antioxidant enzyme activity, thus protecting plastid membrane function and pigment concentration in leaves (Alam et al., 2018, 2022). In another study, melatonin was reported to similarly enhance the chilling tolerance of bermudagrasses (Hu et al., 2016). Glycine betaine is widely known as a compatible solute important for osmotic balance, protein function, and membrane stability in plants. Foliar applications of glycine betaine have been reported to enhance drought tolerance in creeping bentgrass through the promotion of antioxidants and salinity tolerance in perennial ryegrass (Gan et al., 2018; Hu et al., 2012). Foliar applications of hydrogen peroxide have also been linked to increased cold stress in seashore paspalum (Soufi et al., 2016).

2.4 Masking abiotic stress

A growing component of abiotic stress management is masking injury by enhancing the greenness of turfgrass with colorants. Colorants are commonly used on golf courses and athletic fields to create a more esthetically pleasing appearance and to mask turfgrass problems (Gann, 2013; Kreuser and Rossi, 2014; McCarty et al., 2013). Products marketed as pigments have lower viscosity, metal content, minimal resin, and are designed for repeat applications, whereas paints have higher viscosity, higher metal content, are resin based, and are designed for infrequent use.

Colorants increase turfgrass quality in the short term, but the physiological response to colorants has been inconsistent (McCarty et al., 2013; Obear et al., 2017). Frequent use of colorants can alter the reflection, transmission, and absorption of light within the turfgrass canopy, which ultimately results in blocking the absorption of photosynthetically active radiation (PAR) and deterioration of turfgrass health over time (Reynolds et al., 2012, 2013, 2016; Segars and Moss, 2014). McCarty et al. (2017) screened multiple colorants for potential health benefits on creeping bentgrass heat stress and bermudagrass low-temperature stress. As a result, colorants increased heat stress and heavy metal accumulation in creeping bentgrass and had little effect on the health of bermudagrass (McCarty et al., 2017).

2.4.1 Summer and winter dormancy

Warm-season turfgrasses discolor after the first frost and can remain dormant until mid to late spring (Beard, 1973). This esthetically unappealing but important function of warm-season turfgrasses could be considered a form of abiotic stress, or at least an abiotic factor which reduces turfgrass quality and thereby creates goals of reducing the dormancy period or masking the discolored turf (e.g. with colorants) during the winter. Colorants have been reported to increase canopy temperatures, which in turn is thought to hasten spring green-up, but this has not always been reported to occur in the literature (Braun, et al., 2016; Hargey et al., 2016). Colorants can increase dormant turfgrass quality and varying products, rates, and timings have been reported (Braun, et al., 2016; Miller and Pinnix, 2018). Many of these products require multiple applications over the winter months but can be a more sustainable choice in comparison to winter overseeding systems using cool-season turfgrasses (Hargey et al., 2016).

3 Detecting abiotic stress and data-driven decision-making for managing abiotic stress

As previously mentioned, there has been a paradigm shift in abiotic stress management from maximizing turfgrass quality with minimal concern for resource inputs to optimizing those inputs for targeted responses. Although this would objectively be considered a positive shift, this new era creates added pressure for turf managers to be highly engaged in their facility and cognizant of all resource inputs. To do this, frequent and accurate diagnosis (e.g. scouting and monitoring) of plant and soil health is needed so the appropriate action can be performed if stress exceeds desired thresholds. During the past two decades, researchers have made advances in utilizing environmental monitoring devices and sensors to support and optimize resources needed for turfgrass management practices (Magro et al., 2022; Whitlark, 2020).

Collectively, these concepts contribute to what is commonly referred to as data-driven decision-making.

Shade is considered one of the more ubiquitous abiotic stresses affecting turfgrasses, with conventional estimates that 25% of turf endures some level of shade (Beard, 1973). Evaluation of genotypic variation in shade tolerance has been ongoing since the 1960s, but it has only been recently that scientists have made strides to quantify shade and standardize the development of minimum light requirements. Shade can be defined as any decrease in light within the PAR spectrum (400–700 nm) that results in a meaningful reduction in turfgrass growth, development, and performance. Measurement of PAR can be readily achieved using a quantum sensor and reported as photosynthetic photon flux density (PPFD) using units of $\mu mol\ m^{-2}\ s^{-1}$. The daily light integral (DLI) is defined as the accumulated PPFD reaching a surface over a 24-h period and more clearly describes the magnitude of light availability in biologically relevant terms (Faust and Logan, 2018; Korczynski et al., 2002).

Published minimum DLIs are available for at least one cultivar of bahiagrass (*Paspalum notatum* Flügge), centipedegrass (*Eremochloa ophiuroides* [Munro] Hack.), bermudagrass (*Cynodon* spp.), hybrid bermudagrass, seashore paspalum, St. Augustinegrass, and zoysiagrass (*Zoysia* spp.) with values ranging from 10 mol $m^{-2}\ d^{-1}$ to 12 mol $m^{-2}\ d^{-1}$ for zoysiagrass lawns up to 47 mol $m^{-2}\ d^{-1}$ for bermudagrass putting greens (Bunnell et al., 2005; Chen et al., 2021; Chhetri et al., 2019; Hodges et al., 2016; Miller et al., 2005; Zhang et al., 2017). Using published values, one can assess which combination of turfgrass and cultural management program would provide an acceptable turf, although direct application of these principles requires additional knowledge of edaphic properties of the site and the amount of root competition from nearby trees (Russell et al., 2020).

Managing turfgrasses by 'feel' and visual observations has been the cornerstone of many successful turf management careers. Modern turf managers should certainly desire these skills but can also utilize sensors to create objective data-driven assessments of plant health. Spectral reflectance or optical sensing was one of the first remote-sensing technologies to be widely implemented for turfgrass research purposes and is beginning to find use in industry applications. The process uses multi-band or hyperspectral sensors to detect specific wavelengths of light reflected from a surface and calculate an index representative of some plant trait. There are dozens of indices reported in the literature, but the most common approach involves quantifying the 'greenness' of the surface by detecting wavelengths typical of chlorophyll peak absorbance (e.g. 660–670 nm). By adjusting this response for background near-infrared radiation, sensors are quickly able to generate a reproducible number termed normalized difference vegetation index (NDVI). Reported values for NDVI of healthy turf have ranged from 0.73 to 0.95 and can vary by species and environmental conditions (Xiong et al., 2007; Bremer et al., 2011a,b). In research, handheld NDVI sensing has become a standard tool for objectively quantifying stress response and has been reported to be predictive of visual symptoms for several abiotic stressors including drought, salinity, low temperature, herbicide injury, wear, nitrogen stress, and shade (Brewer et al., 2017; Chhetri et al., 2019; Haghverdi et al., 2021; Hong et al., 2019; Kruse et al., 2006; Laskowski and Merewitz, 2020a,b; Trenholm et al., 1999, 2010; Xiong et al., 2007). Despite being able to discern differences in stressed and non-stressed turf, NDVI has been reported to be influenced by species or even environment of the turf (Bremer et al., 2011). Thus, use of NDVI for abiotic stress management requires some level of calibration for a particular site, and likely requires additional post-processing to normalize responses to baseline, non-stressed conditions. In terms of real-world application, NDVI sensors are readily available for handheld and aerial unmanned aircraft system (UAS) platforms, with the latter implemented by various commercial enterprises for rapid assessment of plant stress across large properties. Fenstermaker-Shaulis et al. (1997) suggested that 1 meter resolution could adequately assess turfgrass health for large area irrigation systems, but such programs are thus far limited to scouting, while development of algorithms that transfer real-time data into useful decision-making tools is lacking.

One drawback of using NDVI is that it saturates under high leaf area indices (e.g. turfgrass communities). Thus, NDVI has been less useful in discerning small differences in greenness of otherwise 100% green canopies, as might be important for detecting nutrient deficiencies (Kruse et al., 2006). The normalized difference red edge (NDRE) is possibly the next most widely used spectral reflectance index in turfgrass research and has good potential for precision turf management applications (Zhou and Soldat, 2022). Because NDRE does not saturate in the way the NDVI does, it has been effective in discerning between small differences in shades of green, as might be useful in targeted nitrogen fertilizer applications.

Although NDVI and NDRE are likely the most popular indices used for objective quantification of plant stress, many other indices have been developed for plant sciences. McCall et al. (2017) used a hyperspectral sensor with a narrow field of view to evaluate several published indices in creeping bentgrass subjected to soil moisture stress. The authors concluded the water band index (WBI) provided better prediction of soil water content than other indices including NDVI. In a separate paper, researchers showed WBI was highly correlated (−0.92) to visible wilt ratings and was able to predict wilt 8 h before NDVI and 4 h before the next best index, the green:red ratio index (Roberson et al., 2021). Caturegli et al. (2015) reported NDVI, WBI, and a third index (Structure Intensive Pigment Index) were highly correlated to drought stress on a linear gradient irrigation system. Additional research is needed to scale-up handheld spectral reflectance measurements commonplace in the literature to commercial applications such as ground or aerial-based platforms.

Thermal sensing has been used in turfgrass research since the 1980s, with specific applications in water stress management (Fitz-Rodríguez and Choi, 2002; Park et al., 2007). This approach uses measurements of surface temperature to predict plant water stress as a function of declining transpiration induced by soil moisture stress. Infrared thermometers initially were used by researchers, but more recent efforts have implemented thermal cameras either mounted to stationary posts or on unmanned aerial systems. When measured in conjunction with air temperature and vapor pressure deficit of the atmosphere, crop-specific indices can be developed to detect relative stress in the plant. Taghvaeian et al. (2013) developed the crop water stress index for various turfgrass species and coined the term grass water stress index. Crop water stress indices and thermal sensing procedures in general continue to be developed in academia but remain limited in application due to high variability in values over time (Haghverdi et al., 2021).

Remote sensing has tremendous potential for maintaining fine turf with minimal inputs (e.g. fertilizer, pesticides, irrigation). Continued research is needed to ground-truth sensors with turfgrass responses and most importantly create algorithms which directly relate to an action. Most of the remote sensing techniques require post-processing data to develop relative responses of stressed to non-stressed turf. To this end, absolute values for responses may be less important than variability of response and metrics such as variance in a particular turf system may be more indicative of stress than the mean. Stationary platforms can be deployed in high value spaces such as putting greens to create excellent spatial and temporal resolution datasets. Such platforms are at present cost-prohibitive for large acreage monitoring. Mobile platforms such as those on an UAS or ground-driven equipment (e.g. mowers) may have broader appeal for facility-wide stress detection. Although UASs have increasingly shown value for research activities, legal restrictions within urban and suburban communities where most turfgrass is maintained may limit their applicability for commercial use. Rather, ground equipment-mounted sensors may be more desirable as they can overcome the privacy concerns, wind issues, and resolution challenges associated with UAS platforms.

4 Conclusion and future trends

The sustainability of the turfgrass industry is predicated on adapting to changes in the environment (e.g. local weather and regional climate conditions), society's desires and expectations for managed green spaces and recreation areas, and government regulations and laws. Successful turfgrass managers should strive toward the ability to strategically utilize resources within environmental constraints and conditions to achieve the desired turfgrass quality and function required by turfgrass stakeholders. Further turfgrass research is needed to bridge the gap between academic theory and deliverable tools and tactics that allow practitioners to achieve a rapid assessment of abiotic stress and also implement sustainable turfgrass management practices.

5 Where to look for further information

The United States Golf Association Green Section Record maintains a collection of articles aimed at best management practices for dealing with various abiotic stresses including traffic, drought, and cold stresses (https://www.usga.org/course-care/green-section-record.html).

The Golf Course Superintendent Association of America similarly provides guidance on development of facilty-specific best management practices covering a wide range of sustaiabiltiy issues (https://www.gcsaa.org/environment/environmental-by-topic).

6 References

Abagandura, G. O., Park, D., Bridges, W. C. and Brown, K. 2021. Soil surfactants applied with ^{15}N labeled urea increases bermudagrass uptake of nitrogen and reduces nitrogen leaching. *Journal of Plant Nutrition and Soil Science* 184(3), 378–387. doi:10.1002/jpln.201900162.

Adams, R. M., Hurd, B. H. and Reilly, J. 1999. *Agriculture and Global Climate Change: A Review of Impacts to U.S. Agricultural Resources*. Pew Center for Global Climate Change, Arlington, VA.

Alam, M. N., Yang, L., Yi, X., Wang, Q. and Robin, A. H. K. 2022. Role of melatonin in inducing the physiological and biochemical processes associated with heat stress tolerance in tall fescue (*Festuca arundinaceous*). *Journal of Plant Growth Regulation*, 1–10.

Bae, E. J., Hong, A. R., Choi, S. M., Lee, K. S. and Park, Y. B. 2017. Silicon pretreatment alleviates drought stress and increases antioxidative activity in Kentucky bluegrass. *Itsrj* 13(1), 591–600.

Beard, J. B. 1973. *Turfgrass: Science and Culture*. Prentice Hall, Englewood Cliffs, NJ.

Bigelow, C. A. 2022. Plants as architects of soil. *TPI Turf News* 46(3), 54–55.

Braun, R., Fry, J., Kennelly, M., Bremer, D. and Griffin, J. 2016. Colorant application volume and color persistence on a 'Chisholm' zoysiagrass lawn. *Horttechnology* 26(3), 314–319.

Bremer, D. J., Lee, H., Su, K. and Keeley, S. J. 2011. Relationships between normalized difference vegetation index and visual quality in cool-season turfgrass: II. Factors affecting NDVI and its component reflectances. *Crop Science* 51(5), 2219–2227.

Brewer, J. R., Willis, J., Rana, S. S. and Askew, S. D. 2017. Response of six turfgrass species and four weeds to three HPPD-inhibiting herbicides. *Agronomy Journal* 109(4), 1777–1784.

Bunnell, B. T., McCarty, L. B., Faust, J. E., Bridges, W. C. and Rajapakse, N. C. 2005. Quantifying a daily light integral requirement of a 'TifEagle' bermudagrass golf green. *Crop Science* 45(2), 569–574.

Bushman, B. S. and Warnke, S. E. 2013. Genetic and genomic approaches for improving turfgrass. In: Stier, J. C., Horgan, B. P. and Bonos, S. A. (Eds) *Turfgrass: Biology, Use, and Management*. Agronomy Monograph 56. CSSA, ASA, SSSA, Madison, WI.https://doi.org/10.2134/agronmonogr56.c19.

Carroll, D. E., Brosnan, J. T., Reasor, E. H., Kerns, J. P., Stephens, C. M., Horvath, B. J., Dickson, K. H. and Tomaso-Peterson, M. 2021. Effects of growing degree day based trinexapac-ethyl application on ultradwarf hybrid bermudagrass putting green quality and ball roll. *International Turfgrass Society Research Journal*. doi:10.1002/its2.41.

Carrow, R. N. and Petrovic, A. M. 1992. Effect of traffic on turfgrasses. In: Waddington, D. V., Carrow, R. N. and Shearman, R. C. (Eds) *Turfgrass*. Agronomy Monograph 32. CSSA, ASA, SSSA, Madison, WI. doi:10.2134/agronmonogr32.c9.

Chang, B., Wherley, B., Aitkenhead-Peterson, J., Ojeda, N., Fontanier, C. and Dwyer, P. 2020. Effect of wetting agent on nutrient and water retention and runoff from simulated urban lawns. *HortScience* 55(7), 1005–1013.

Chen, Z., Wherley, B., Reynolds, C., Hejl, R. and Chang, B. 2021. Daily light integral requirements for bermudagrass and zoysiagrass cultivars: Influence of mowing height and growth regulators. *Crop Science* 61(4), 2837–2847. doi:10.1002/csc2.20515.

Chhetri, M., Fontanier, C., Koh, K., Wu, Y. and Moss, J. Q. 2019. Turf performance of seeded and clonal bermudagrasses under varying light environments. *Urban Forestry and Urban Greening* 43, 126355.

Cisar, J. L., Williams, K. E., Vivas, H. E. and Haydu, J. J. 2000. The occurrence and alleviation by surfactants of soil water repellency on sand-based turfgrass systems. *Journal of Hydrology* 231-232, 352–358.

Cortes-Barco, A. M., Hsiang, T. and Goodwin, P. H. 2010. Induced systemic resistance against three foliar diseases of *Agrostis stolonifera* by (2R, 3R)-butanediol or an isoparaffin mixture. *Annals of Applied Biology* 157(2), 179–189.

DaCosta, M., Ebdon, J. S., Miele, K., Bernstein, R. P. and Inguagiato, J. C. 2021. Plant growth regulator effects on winter hardiness of annual bluegrass putting green turf. *ITSRJ* 2021:1-11.

DeBoer, E. J., Karcher, D. E., McCalla, J. H. and Richardson, M. D. 2020. Effect of late-fall wetting agent application on winter survival of ultradwarf bermudagrass putting greens. *Crop, Forage and Turfgrass Management* 6(1), e20035.

DeBoer, E. J., Richardson, M. D., McCalla, J. H. and Karcher, D. E. 2019. Reducing ultradwarf bermudagrass putting green winter injury with covers and wetting agents. *Crop, Forage and Turfgrass Management* 5(1), 190019. doi:10.2134/cftm2019.03.0019.

Doerr, S. H., Ferreira, A. J. D., Walsh, R. P. D., Shakesby, R. A., Leighton-Boyce, G. and Coelho, C. O. A. 2003. Soil water repellency as a potential parameter in rainfall-runoff modelling: Experimental evidence at point to catchment scales from Portugal. *Hydrological Processes* 17(2), 363–377.

Doerr, S. H., Shakesby, R. A. and Walsh, R. P. D. 2000. Soil water repellency: Its causes, characteristics and hydro-geomorphological significance. *Earth-Science Reviews* 51(1-4), 33–65.

du Jardin, P. 2015. Plant biostimulants: Definition, concept, main categories and regulation. *Scientia Horticulturae* 196, 3–14.

Ervin, E. H. and Zhang, X. 2007. Influence of sequential trinexapac-ethyl applications on cytokinin content in creeping bentgrass, Kentucky bluegrass, and hybrid Bermudagrass. *Crop Science* 47(5), 2145–2151.

Faust, J. E. and Logan, J. 2018. Daily light integral: A research review and high-resolution maps of the United States. *HortScience* 53(9), 1250–1257.

Fehr, W. R. *Maximizing Genetic Improvement: Chapter Seventeen, Plant Breeding Class Notes* 2011.

Fenstermaker-Shaulis, L. K., Leskys, A. and Devitt, D. A. 1997. Utilization of remotely sensed data to map and evaluate turfgrass stress associated with drought. *Journal of Turfgrass Management* 2(1), 65–81.

Fidanza, M., Cisar, J. and Watschke, T. 2015. A perspective on biostimulants and sportsturf. *Sports Field Management* 31(6), 8–10, 12.

Fidanza, M. and Kostka, S. 2022. The turfgrass rhizosphere. *TurfNet*. https://www.turfnet.com/webinar_archives.html/mike-fidanza-the-turfgrass-rhizosphere-r263.

Fidanza, M., Kostka, S. and Bigelow, C. 2020. Communication of soil water repellency causes, problems, and solutions of intensively managed amenity turf from 2000 to 2020. *Journal of Hydrology and Hydromechanics* 68(4), 306–312.

Fidanza, M., Kostka, S., Ervin, E. and Bigelow, C. 2019. The European Union's view on biostimulants: What may be coming our way. *Golf Course Management* 87(9), 58–62.

Fitz-Rodríguez, E. and Choi, C. Y. 2002. Monitoring turfgrass quality using multispectral radiometry. *Transactions of the ASAE. American Society of Agricultural Engineers* 45, 865–871.

Fontanier, C., Moss, J. Q., Gopinath, L., Goad, C., Su, K. and Wu, Y. 2020. Lipid composition of three bermudagrasses in response to chilling stress. *Journal of the American Society for Horticultural Science* 145(2), 95–103.

Franco, C. M. M., Tate, M. E. and Oades, J. M. 1994. The development of water- repellency in sands: Studies on the physico-chemical and biological mechanisms. In: *Proceedings of the 2nd National Water Repellency Workshop*, Perth, Western Australia, pp. 18–30.

Fry, J. and Huang, B. 2004. *Applied Turfgrass Science and Physiology*. John Wiley & Sons, Hoboken, NJ.

Gan, L., Zhang, X., Liu, S. and Yin, S. 2018. Mitigating effect of glycinebetaine pretreatment on drought stress responses of creeping bentgrass. *HortScience* 53(12), 1842–1848.

Gelernter, W. D., Stowell, L. J., Johnson, M. E., Brown, C. D. and Beditz, J. F. 2015. Documenting trends in water use and conservation practices on U.S. golf courses. *Crop, Forage and Turfgrass Management* 1(1), 1–10. DOI: 10.2134/cftm2015.0149.

Goss, R. M., Baird, J. H., Kelm, S. L. and Calhoun, R. N. 2002. Trinexapac-ethyl and nitrogen effects on creeping bentgrass grown under reduced light conditions. *Crop Science* 42(2), 472–479.

Guertal, E. A. and Datnoff, L. E. 2021. Silicon in turfgrass: A review. *Crop Science* 61(6), 3861–3876.

Haghverdi, A., Reiter, M., Singh, A. and Sapkota, A. 2021. Hybrid bermudagrass and tall fescue turfgrass irrigation in central California: II. Assessment of NDVI, CWSI, and canopy temperature dynamics. *Agronomy* 11(9), 1733.

Haguewood, J. B., Song, E., Smeda, R. J., Moss, J. Q. and Xiong, X. 2013. Suppression of annual bluegrass seedheads with mefluidide, ethephon, and ethephon plus trinexapac-ethyl on creeping bentgrass greens. *Agronomy Journal* 105(6), 1832–1838.

Han, S. H., Kang, B. R., Lee, J. H., Lee, S. H., Kim, I. S., Kim, C. H. and Kim, Y. C. 2012. A trifloxystrobin fungicide induces systemic tolerance to abiotic stresses. *The Plant Pathology Journal* 28(1), 101–106.

Hargey, D., Wherley, B., Reynolds, C., White, R. and Parker, G. 2016. Performance of winter overseeded and colorant-treated 'Tifway' bermudagrass receiving weekly irrigation and simulated traffic. *HortScience* 51(5), 580–587.

Hodges, B. P., Baldwin, C. M., Stewart, B., Tomaso-Peterson, M., McCurdy, J. D., Blythe, E. K. and Philley, H. W. 2016. Quantifying a daily light integral for establishment of warm-season cultivars on putting greens. *Crop Science* 56(5), 2818–2826.

Hoffman, L., DaCosta, M., Ebdon, J. S. and Watkins, E. 2010. Physiological changes during cold acclimation of perennial ryegrass accessions differing in freeze tolerance. *Crop Science* 50(3), 1037–1047.

Hong, M., Bremer, D. J. and van der Merwe, D. 2019. Using small unmanned aircraft systems for early detection of drought stress in turfgrass. *Crop Science* 59(6), 2829–2844.

Hsiang, T., Goodwin, P. H. and Cortes, A. M. 2011. Plant defense activators and control of turfgrass diseases. *Outlooks on Pest Management* 22(4), 160–164.

Hu, L., Hu, T., Zhang, X., Pang, H. and Fu, J. 2012. Exogenous glycine betaine ameliorates the adverse effect of salt stress on perennial ryegrass. *Journal of the American Society for Horticultural Science* 137(1), 38–46.

Hu, Z., Fan, J., Xie, Y., Amombo, E., Liu, A., Gitau, M. M., Khaldun, A. B. M., Chen, L. and Fu, J. 2016. Comparative photosynthetic and metabolic analyses reveal mechanism of improved cold stress tolerance in bermudagrass by exogenous melatonin. *Plant Physiology and Biochemistry* 100, 94–104.

Huang, B., DaCosta, M. and Jiang, Y. 2014. Research advances in mechanisms of turfgrass tolerance to abiotic stress: From physiology to molecular biology. *Critical Reviews in Plant Sciences* 33(2–3), 141–189.

Jespersen, D. and Huang, B. 2017. Effects of trinexapac-ethyl and daconil Action (acibenzolar-S-methyl and chlorothalonil) on heat and drought tolerance of creeping bentgrass. *Crop Science* 57(S1) (Suppl.1), S-138.

Jespersen, D., Yu, J. and Huang, B. 2017. Metabolic effects of acibenzolar-s-methyl for improving heat or drought stress in creeping bentgrass. *Frontiers in Plant Science* 8, 1224.

Johnson, P. G., Rossi, F. S. and Horgan, B. P. 2013. Sustainable turfgrass management in an increasingly urbanized world. In: Stier, C., Horgan, B. P. and Bonos, S. A. (Eds) *Turfgrass: Biology, Use, and Management*. Agronomy Monograph 56. CSSA, ASA, SSSA, Madison, WI. doi:10.2134/agronmonogr56.c26.

Kong, R. S. and Henry, H. A. 2016. Prior exposure to freezing stress enhances the survival and recovery of *Poa pratensis* exposed to severe drought. *American Journal of Botany* 103(11), 1890–1896.

Korczynski, P. C., Logan, J. and Faust, J. E. 2002. Mapping monthly distribution of daily light integrals across the contiguous United States. *HortTechnology* 12(1), 12–16.

Kostka, S. and Fidanza, M. 2019. *Soil Surfactant Usage Based on Solid Science*. Golf Course Industry. https://www.golfcourseindustry.com/article/soil-surfactant-research-guidance/.

Kostka, S. J., Cisar, J. L., Short, J. R. and Mane, S. 1997. Evaluation of soil surfactants for the management of soil water repellency in turfgrass. *International Turfgrass Society Research Journal* 8, 485–494.

Kreuser, W. C. and Rossi, F. S. 2014. The horticultural spray oil, Civitas™, causes chronic phytotoxicity on cool-season golf turf. *HortScience* 49(9), 1217–1224.

Kreuser, W. C. and Soldat, D. J. 2011. A growing degree day model to schedule trinexapac-ethyl applications on *Agrostis stolonifera* golf putting greens. *Crop Science* 51(5), 2228–2236.

Kreuser, W. C. and Soldat, D. J. 2012. Frequent trinexapac-ethyl applications reduce nitrogen requirements of creeping bentgrass golf putting greens. *Crop Science* 52(3), 1348–1357.

Kreuser, W. C., Young, J. R. and Richardson, M. D. 2017. Modeling performance of plant growth regulators. *Agricultural and Environmental Letters* 2(1), 170001.

Kruse, J. K., Christians, N. E. and Chaplin, M. H. 2006. Remote sensing of nitrogen stress in creeping bentgrass. *Agronomy Journal* 98(6), 1640–1645.

Laskowski, K., Frank, K. and Merewitz, E. 2019. Chemical plant protectants and plant growth regulator effects on annual bluegrass survival of ice cover. *Journal of Agronomy and Crop Science* 205(2), 202–212.

Laskowski, K. and Merewitz, E. 2020a. Chemical priming to improve annual bluegrass survival of ice encasement. *Agronomy Journal* 112(6), 5002–5011.

Laskowski, K. and Merewitz, E. 2020b. Ethylene regulatory treatment effects on annual bluegrass survival of freezing temperature and ice cover. *Agronomy Journal* 112(2), 861–870.

Law, Q. D., Bigelow, C. A. and Patton, A. J. 2016. Selecting turfgrasses and mowing practices that reduce mowing requirements. *Crop Science* 56(6), 3318–3327.

Leinauer, B., Rieke, P. E., Van Leeuwen, D., Sallenave, R., Makk, J. and Johnson, E. 2001. Effects of soil surfactants on water retention in turfgrass rootzones. *International Turfgrass Society Research Journal* 9, 542–547.

Lindsey, A. J., Thoms, A. W. and Christians, N. E. 2020. Kentucky bluegrass and bermudagrass rooting response to humic fertilizers during greenhouse establishment. *Agronomy Journal* 112(5), 3396–3401.

Lindsey, A. J., Thomas, A. W., McDaniel, M. D. and Christians, N. E. 2021. Evaluation of humic fertilizers applied at full and reduced nitrogen rates on Kentucky bluegrass quality and soil health. *Agronomy* 11(2), 395.

Magro, C., Macolino, S., Pornaro, C., McMillan, M. and Fidanza, M. 2022. Considerations with determining the minimum number of volumetric water content measurements for turfgrass root zones. *Agronomy* 12(6), 1402. doi:10.3390/agronomy12061402.

McCann, S. E. and Huang, B. 2007. Turfgrass drought physiology and irrigation management. Pesserakli, M. (Ed.) In: *Handbook of Turfgrass Management and Physiology*, pp. 431-445. CRC Press. Boca Raton, FL.

McCarty, L. B., Gann, J. R., Wells, C. E., Bruce, T. F. and Gerard, P. D. 2013. Physiological responses of creeping bentgrass to pigment-containing products. *Agronomy Journal* 105(6), 1797–1802. doi:10.2134/agronj2013.0298.

McCarty, L. B., Gore, A. W., Brown, P. J., Martin, S. B. and Wells, C. E. 2017. Potential health benefits of pigment-containing products on creeping bentgrass and hybrid bermudagrass. *International Journal of Plant and Soil Science* 15(3), 1–13. Article No. IJPSS.32337. ISSN: 2320-7035.

McCullough, P. E., Liu, H., McCarty, L. B. and Toler, J. E. 2007. Trinexapac-ethyl application regimens influence growth, quality, and performance of Bermuda grass and creeping bentgrass putting greens. *Crop Science* 47(5), 2138–2144.

McCullough, P. E., Liu, H., McCarty, L. B., Whitwell, T. and Toler, J. E. 2006. Growth and nutrient partitioning of 'Tifeagle' Bermudagrass as influenced by nitrogen and trinexapac-ethyl. *Hortscience* 41(2), 453–458.

Miller, G. and Pinnix, D. 2018. Guide to using turf colorants. North Carolina State Extension. *Agriculturists* 843.

Miller, G. L., Edenfield, J. T. and Nagata, R. T. 2005. Growth parameters of Floradwarf and Tifdwarf bermudagrasses exposed to various light regimes. *The International Turfgrass Society Research Journal* 10, 879–884.

Miranda, S. and Medlin, J. 2021. Managing stress in turfgrass. *Golf Course Management*. https://www.gcmonline.com/voices/news/nachurs-turf-ornamental-turfgrass-stress.

Mitra, S., Chavez, A., Kostka, S. and Franklin, M. 2004. Conserving water on turfgrasses using systematic injection of wetting agents. *Abstractseries [ASA/CSSA/SSSA/CSSS]*. Annual Meeting (vol. 2004), p. 1.

Mitra, S., Vis, E., Kumar, R., Plumb, R. and Fam, M. 2006. Wetting agent and cultural practices increase infiltration and reduce runoff losses of irrigation water. *Biologia* 61(S19), S353–S357, doi:10.2478/s11756-006-0188-4.

Munshaw, G. C., Ervin, E. H., Shang, C., Askew, S. D., Zhang, X. and Lemus, R. W. 2006. Influence of late-season iron, nitrogen, and seaweed extract on fall color retention and cold tolerance of four bermudagrass cultivars. *Crop Science* 46(1), 273–283.

Munshaw, G. C., Zhang, X. and Ervin, E. H. 2004. Effect of salinity on bermudagrass cold hardiness. *HortScience* 39(2), 420–423.

Murphy, J. A. and Ebdon, J. S. 2013. Study and management of turfgrass traffic stress. In: Stier, J. C., Horgan, B. P. and Bonos, S. A. (Eds) *Turfgrass: Biology, Use, and Management*. Agronomy Monograph 56. CSSA, ASA, SSSA, Madison, WI. doi:10.2134/agronmonogr56.c27.

Musser, H. B. 1950. *Turf Management*. McGraw-Hill, New York.

Nosalewicz, A., Siecińska, J., Kondracka, K. and Nosalewicz, M. 2018. The functioning of *Festuca arundinacea* and *Lolium perenne* under drought is improved to a different extend by the previous exposure to water deficit. *Environmental and Experimental Botany* 156, 271–278.

Obear, G. R., Kreuser, W. C., Hubbard, K., De Bels, B. and Soldat, D. J. 2017. Plant colorants interfere with reflectance-based vegetation indices. *Crop Science* 57(2), 595–601. doi:10.2135/cropsci2016.05.0440.

O'Brien, P. and Hartwiger, C. 2013. *Covering Guidelines for Ultradwarf Bermudagrass Putting Greens*. United States Golf Association, Far Hills, NJ.

Park, D. M., Cisar, J. L., Williams, K. E., McDermitt, D. K., Miller, W. P. and Fidanza, M. A. 2007. Using spectral reflectance to document water stress in bermudagrass grown on water repellent sandy soils. *Hydrological Processes* 21(17), 2385–2389.

Prokopiuk, K., Żurek, G. and Rybka, K. 2019. Turf covering for sport season elongation cause no stress for grass species as detected by Chl a fluorescence. *Urban Forestry and Urban Greening* 41, 14–22.

Pruyne, D. T., Schlossberg, M. J. and Uddin, W. 2019. Perennial ryegrass wear resistance and soil amendment by Ca-and Mg-silicates. *Agronomy* 9(10), 578.

Reasor, E. H., Brosnan, J. T., Staton, M. E., Lane, T., Trigiano, R. N., Wadl, P. A., Conner, J. A. and Schwartz, B. M. 2018. Genotypic and phenotypic evaluation of off-type grasses in hybrid Bermudagrass [*Cynodon dactylon* (L.) Pers. x *C. transvaalensis* Burtt-Davy] putting greens using genotyping-by-sequencing and morphological characterization. *Hereditas* 155, 8. doi:10.1186/s41065-017-0043-3.

Richardson, M. D., Brosnan, J. T. and Karcher, D. E. 2014. Turfgrass winterkill observations from the transition zone. *Applied Turfgrass Science* 11(1), 1-4.
Roberson, T. L., Badzmierowski, M. J., Stewart, R. D., Ervin, E. H., Askew, S. D. and McCall, D. S. 2021. Improving soil moisture assessment of turfgrass systems utilizing field radiometry. *Agronomy* 11(10), 1960.
Russell, T. R., Karcher, D. E. and Richardson, M. D. 2019. Daily light integral requirement of a creeping bentgrass putting green as affected by shade, trinexapac-ethyl, and a plant colorant. *Crop Science* 59(4), 1768-1778.
Russell, T. R., Karcher, D. E. and Richardson, M. D. 2020. Daily light integral requirements of warm-season turfgrasses for golf course fairways and investigating in situ evaluation methodology. *Crop Science* 60(6), 3301-3313.
Saigusa, M., Onozawa, K., Watanabe, H. and Shibuya, K. 2000. Effects of porous hydrate calcium silicate on the wear resistance, insect resistance and disease tolerance of turf grass'Miyako. *Grassland Science* 45, 416-420.
Salisbury, F. and Ross, C. 1992. *Plant Physiology* (4th edn.). Wadsworth Publishing, Belmont, CA.
Schiavon, M. and Baird, J. H. 2018. Evaluation of products to alleviate irrigation salinity stress on bermudagrass turf. *Agronomy Journal* 110(6), 2136-2141.
Schiavon, M., Leinauer, B., Serena, M., Maier, B. and Sallenave, R. 2014a. Plant growth regulator and soil surfactants' effects on saline and deficit irrigated warm-season grasses: I. Turf quality and soil moisture. *Crop Science* 54(6), 2815-2826.
Schiavon, M., Leinauer, B., Serena, M., Maier, B. and Sallenave, R. 2014b. Plant growth regulator and soil surfactants' effects on saline and deficit irrigated warm-season grasses: II. Pigment content and superoxide dismutase activity. *Crop Science* 54(6), 2827-2836.
Schiavon, M., Petelewicz, P., Cremonese, G. and Baird, J. 2015. *Best Management Practices for Turf under Drought or Water Use Restrictions. 2015 Turfgrass and Landscape Research Field Day*, pp. 33-36. University of California-Riverside.
Segars, C. A. and Moss, J. Q. 2014. The effects of athletic-field paint color on net photosynthesis and canopy temperature of overseeded perennial ryegrass. *Applied Turfgrass Science* 11(1). doi:10.2134/ATS-2014-0015-BR.
Serena, M., Sportelli, M., Sevostianova, E., Sallenave, R. and Leinauer, B. 2018. Combining trinexapac-ethyl with a soil surfactant reduces bermudagrass irrigation requirements. *Agronomy Journal* 110(6), 2180-2188.
Shekoofa, A., Rosas-Anderson, P., Carley, D. S., Sinclair, T. R. and Rufty, T. W. 2016. Limited transpiration under high vapor pressure deficits of creeping bentgrass by application of Daconil Action. *Planta* 243(2), 421-427. doi:10.1007/s00425-015-2417-y.
Snyder, G. H., Augustin, B. J. and Davidson, J. M. 1984. Moisture sensor-controlled irrigation for reducing N leaching in bermudagrass turf. *Agronomy Journal* 76(6), 964-969.
Soldat, D. 2010. Wetting agent research update: wetting agents in wet conditions. *The Grass Roots* 39(5), 18-19.
Taghvaeian, S., Chávez, J. L., Hattendorf, M. J. and Crookston, M. A. 2013. Optical and thermal remote sensing of turfgrass quality, water stress, and water use under different soil and irrigation treatments. *Remote Sensing* 5(5), 2327-2347.
Throssell, C. S., Lyman, G. T., Johnson, M. E., Stacey, G. A. and Brown, C. D. 2009. Golf course environmental profile measures water use, source, cost, quality, management and conservation strategies. *Applied Turfgrass Science* 6(1), 1-20.

Trenholm, L. E., Duncan, R. R. and Carrow, R. N. 1999. Wear tolerance, shoot performance, and spectral reflectance of seashore paspalum and bermudagrass. *Crop Science* 39(4), 1147-1152.

Trenholm, L. E., Duncan, R. R., Carrow, R. N. and Snyder, G. H. 2001. Influence of silica on growth, quality, and wear tolerance of seashore paspalum. *Journal of Plant Nutrition* 24(2), 245-259.

Trenholm, L. E., Schlossberg, M. J., Lee, G., Parks, W. and Geer, S. A. 2000. An evaluation of multi-spectral responses on selected turfgrass species. *International Journal of Remote Sensing* 21(4), 709-721.

Turgeon, A. J. 2005. *Turfgrass Management* (7th edn.). Pearson Education, Upper Saddle River, NJ.

Vallad, G. E. and Goodman, R. M. 2004. Systemic acquired resistance and induced systemic resistance in conventional agriculture. *Crop Science* 44(6), 1920-1934.

Waalen, W. M., Espevig, T., Kvalbein, A. and Aamlid, T. S. 2017. The effect of ice encasement and protective covers on the winter survival of six turfgrass species on putting greens. *Itsrj* 13(1), 556-565.

Wallis, M. G. and Horne, D. J. 1992. Soil water repellency. *Advances in Soil Science* 20, 91-146.

Whitlark, B. 2020. The top five turf tools from 2020. *USGA Green Section Record*. https://tinyurl.com/y4f5kp5x.

Xiong, X., Bell, G. E., Solie, J. B., Smith, M. W. and Martin, B. 2007. Bermudagrass seasonal responses to nitrogen fertilization and irrigation detected using optical sensing. *Crop Science* 47(4), 1603-1610.

Xu, Y. and Huang, B. 2010. Responses of creeping bentgrass to trinexapac-ethyl and biostimulants under summer stress. *HortScience* 45(1), 125-131.

Zhang, C. L., Li, P. H. and Brenner, M. L. 1986. Relationship between mefluidide treatment and abscisic acid metabolism in chilled corn leaves. *Plant Physiology* 81(2), 699-701.

Zhang, J., Glenn, B., Unruh, J. B., Kruse, J., Kenworthy, K., Erickson, J., Rowland, D. and Trenholm, L. 2017. Comparative performance and daily light integral requirements of warm-season turfgrasses in different seasons. *Crop Science* 57(4), 2273-2282.

Zhou, Q. and Soldat, D. J. 2022. Evaluating decision support tools for precision nitrogen management on creeping bentgrass putting greens. *Frontiers in Plant Science* 13, 863211-863211.

Zhu, H. and Li, D. 2018. Using humus on golf course fairways to alleviate soil salinity problems. *HortTechnology* 28(3), 284-288.

Zontek, S. J. and Kostka, S. J. 2012. Understanding the different wetting agent chemistries. *USGA Green Section Record* 50, 1-6.

Chapter 14

Advances in managing organic matter in turfgrass ecosystems

Alec Kowalewski, Charles Schmid, Ruying Wang and Emily Braithwaite, Oregon State University, USA

1 Introduction

Beard (1973) defined thatch as a layer of dead and living stems and roots that accumulates between the green vegetation and soil surface (Fig. 1). Thatch has high lignin content and is composed of stems and nodes that do not readily decay (Ledeboer and Skogley, 1967). Thatch accumulation is a common problem in turfgrass ecosystems because such surface organic matter (OM) is resistant to decomposition. Many factors contribute to excessive thatch accumulation, including rapid and dense growth of the turfgrass species, excessive fertility and irrigation, and infrequent cultivation programs. Excessive thatch can be detrimental and cause desiccation and hydrophobicity during dry conditions, thus enhancing turfgrass drought stress. Water repellency was observed to be more severe with high OM soil in turf and more difficult to alleviate by using wetting agents (Barton and Colmer, 2011a,b). On the opposite extreme, thatch can also retain excess water, thereby restricting air exchange in the root zone during wet periods. Sampling of 24 golf course putting greens in New Zealand found surface infiltration rates declined as OM content increased (Glasgow et al., 2005). Early research demonstrated that thatch layer limits initial or early-time infiltration rate but not steady-state infiltration rate because thatch has a larger pore size than the underlying soil profile (Hurto et al., 1980; Taylor and Blake, 1982). Similarly, Liang

http://dx.doi.org/10.19103/AS.2022.0110.14

Figure 1 A putting green soil profile showing turfgrass, thatch, mat layer and the original rootzone.

et al. (2017) calculated early to intermediate infiltration rates (before approaching a constant infiltration rate) with a simulated rainfall method and reported that thatch held water, reduced runoff, and delayed infiltration. Furthermore, an examination of the physical properties of different sports turf root zone mixes indicated that saturated hydraulic conductivity (K_{sat}) was not affected by OM content but rather by pore size distribution (McCoy, 1992).

Although OM may not have a direct relationship with K_{sat}, OM can have an indirect effect by decreasing large pore spaces in the root zone. Large pores are an important factor influencing K_{sat}. As putting greens mature over the years OM will accumulate, which will decrease infiltration rates due to reductions in air-filled root zone porosity (Lewis et al., 2010; McClellan et al., 2009). Many researchers have shown evidence to support that decreases in air-filled porosity in sand root zones often result in decreased infiltration (Gibbs et al., 2000; Lewis et al., 2010; McCoy, 1992; Ok et al., 2003). Management practices such as sand topdressing and core cultivation introduce sand particles into the soil profile creating more pore spaces among OM. Measuring infiltration rates at five tensions from −5.5 to −0.5 cm, Wang et al. (2021) suggested that sand topdressing treatments increased K_{sat} by creating relatively larger pore spaces. Core cultivation physically removes thatch and plant tissue, and large amounts of sand are backfilled; therefore, a significant increase in infiltration was often observed, but the effects on OM reduction were inconsistent (McCarty et al., 2005, 2007; Schmid et al., 2014; Sidhu et al., 2014). This is likely because core cultivation is also very effective at promoting new growth/biomass.

2 Monitoring and measuring organic matter in turfgrass

Historically, researchers have quantified OM accumulation and/or reduction by measuring thatch thickness and OM content by combustion. However, with the increased frequency and quantity of topdressing applied to golf course putting greens and fairways, it has become very difficult to measure a thatch layer due to the mixing of sand or soil with thatch that produces a mat layer (Carrow et al., 1987). As a result, recent studies have focused on OM combustion, or loss on ignition (LOI), to quantify OM changes over time. Despite having numerous laboratory methods for determining OM in soils (Carrow et al., 1987; Nelson and Sommers, 1996), and putting green and sports turf root zone mixes (ASTM F1647-02, 2002), there is a lack of standardized sampling procedures for OM sampling in turf. Various sampling methods have been documented in the literature, including differing soil probe diameter size, soil sampling depth, whether the soil samples are stratified, and if the verdure is removed. Consequently, comparing results from OM studies is difficult due to nonuniformity in OM sampling/collection and measurement technique.

A summary of research papers related to OM management and their soil sampling methods is listed in Table 1. The diameter of soil cores used for OM analysis and the number of samples collected per plot vary greatly in the turfgrass literature. Most of these studies collected between two and five soil samples per plot (Barton et al., 2009; Carrow et al., 1987; Dunn et al., 1995; Espevig et al., 2012; Fu et al., 2009; McCarty et al., 2005; Murphy et al., 1993; Schmid et al., 2014; Stier and Hollman, 2003; Wang, 2015). Kauffman et al. (2013) conducted a field trial in which they investigated the effect of sample size (number per plot) on thatch/mat depth and OM content of four warm-season grasses. Results from this trial indicate that the number of samples required to detect a difference ($P = 0.05$) of 3 mm in thatch depth ranged from 10 to 13, depending on species/cultivar. Moreover, the number of samples required to detect a 0.5% change in OM content ranged from 3 to 17 samples. The discrepancy between the number of soil samples collected per plot in previous OM studies and the estimated number of samples required to detect a 0.5% OM content change might explain why few studies have been able to consistently detect differences in OM content. It should also be noted that Kauffman et al. (2013) used a 100-mm diameter soil core, which is larger than most of the previous studies investigating OM content. Although extensive research has been carried out on OM in turfgrass, no previous study has investigated the impact of soil core diameter on OM content, particularly with respect to cultivation treatments. It is important that soil core diameter is sufficiently large enough to capture variations in OM, while not being too disruptive to the turfgrass surface. Future research is needed to investigate the impact of soil core diameter samples on OM content measurement.

Table 1 A summary of research papers related to organic matter management and their soil sampling methods in the turfgrass science literature

Reference	Year	Sample diameter (mm)	Sample depth (mm)	Number of samples per plot
Carrow et al.	1987	54	Variable - thatch or mat layer	2
Murphy et al.	1993	50	Variable - thatch/mat layer	5
Dunn et al.	1995	100	Variable - mat layer	5
Stier and Hollman	2003	25	50	3
Glasgow et al.	2005	25 or 50	80	NS[a]
McCarty et al.	2005	19	51	2
Barton et al.	2009	70	0-50	2
Fu et al.	2009	25	80	2
McClellan et al.	2009	25	152	NS
Fontanier et al.	2011	NS	13-89	NS
Espevig et al.	2012	24	Variable - mat layer	2
Kauffman et al.	2013	100	25	25
Schmid et al.	2014	19	76	2
Fidanza et al.	2017	19	Variable - thatch and OM-'stained' layer	8

[a]NS, not stated.

Soil sampling depth for OM content in turfgrass research has varied greatly (Table 1). Several researchers have used a fixed soil core depth for OM samples and analyzed the entire soil sample minus the verdure (Barton et al., 2009; Fontanier et al., 2011; McCarty et al., 2005, 2007; Schmid et al., 2014; Stier and Hollman, 2003). While other researchers have also used a fixed soil core depth but stratified the soil core into specific depth increments prior to combusting the sample (Glasgow et al., 2005; McClellan et al., 2009; Windows and Bechelet, 2012). McClellan et al. (2009) divided the soil cores in their study into 0–76 mm and 76–152 mm increments, whereas Glasgow et al. (2005) and Windows and Bechelet (2012) both stratified their samples into 20-mm increments (0–20, 20–40, 40–60, and 60–80 mm). Other researchers have relied on variable depth sampling for OM based on the depth of the thatch/mat region (Carrow et al., 1987; Dunn et al., 1995; Espevig et al., 2012; Fidanza et al., 2017; Murphy et al., 1993; Wang, 2015). Early research by Carrow et al. (1987) looking at the effect of fertility and cultivation on Bermuda grass limited soil sampling to only the thatch region of the soil profile. The authors of this trial noted that it was difficult to make comparisons between treatments that received topdressing or core aeration (with cores returned) and all other treatments since those treatments form a mat layer, thus significantly reducing thatch. With increased frequencies and quantities of sand topdressing applied to putting greens and fairways, some studies have focused sampling depth on the mat or thatch/mat layer (Dunn et al., 1995; Espevig et al., 2012; Murphy et al., 1993; Wang, 2015). A recent study by Fidanza et al. (2017) stratified soil samples into three layers, verdure/thatch, OM-'stained' layer, and original root zone. The authors found that as the Bermuda grass putting greens aged, the depth of the OM-'stained' layer increased, but the OM level remained consistent and relatively low. In contrast, the depth of the thatch layer was relatively consistent, and the OM level fluctuated greatly throughout the season, which was attributed to the cultivation program.

Combustion temperature for measuring OM content using the LOI method also varies considerably in the turfgrass literature. In studies referenced previously, combustion temperature ranged from 440°C (Glasgow et al., 2005; Schmid et al., 2014) to 550°C (Espevig et al., 2012; Fidanza et al., 2017; Fu et al., 2009) to 600°C (Barton et al., 2009; Carrow et al., 1987; Dunn et al., 1995; Kauffman et al., 2013; Murphy et al., 1993; Stier and Hollman, 2003), and even up to 700°C (McCarty et al., 2005). To date, no published research has evaluated the effect of combustion temperature on OM content measurements in turfgrass. In general, researchers have used higher combustion temperatures for soils or thatch samples that have high OM to ensure complete combustion of OM (ASTM, 2020) or to reduce the time required to completely combust a sample. However, heating a soil sample to temperatures ≥1000°C could result in elevated OM measurements due to the combustion of carbonate C from $CaCO_3$, especially in calcareous soil. Rabenhorst (1988) determined that the appropriate combustion temperature for determining organic carbon (OM) content for a calcareous soil was 575°C, whereas carbonate C from $CaCO_3$ was combusted when samples were heated to 1000°C. Because of this, the authors recommend heating soil samples to a maximum of 575°C for the measurement of OM content using LOI in turfgrass.

To date, all previous studies investigating OM content in turfgrass have removed the verdure, or living green plant tissue above the soil surface, prior to laboratory analysis (Barton et al., 2009; Carrow et al., 1987; Dunn et al., 1995; Espevig et al., 2012; Fidanza et al., 2017; Fontanier et al., 2011; Fu et al., 2009; Glasgow et al., 2005; Kauffman et al., 2013; McCarty et al., 2005; McClellan et al., 2009; Murphy et al., 1993; Schmid et al., 2014; Stier and Hollman, 2003; Wang, 2015). However recently, researchers have suggested leaving the verdure intact for OM sampling (Woods, 2019). Subjective removal of verdure and loss of soil material during the verdure removal process may reduce the accuracy and precision of the measurement; however, to date, no study has addressed the implication of verdure removal.

As stated previously, the generalizability of much of the published research on OM is problematic due to the varied sampling procedures described above. It is critical for developing standardized sampling procedures for OM content, so it is more effective and easier to make comparisons between studies and provide guidelines for turfgrass managers. Further work is needed to establish the impact of soil core diameter and core depth and leave the verdure intact on OM measurements.

3 Organic matter and turfgrass diseases

The relationships between OM content and turfgrass diseases are not well understood, and what is known is largely based on empirical evidence. Conflicting reports of the influence of OM content on dollar spot (*Clarireedia jacksonii*) have been noted in the literature (Carrow et al., 1987; Fermanian et al., 1985; Moeller et al., 2008; Stier and Hollman, 2003). Several studies have shown that cultural management practices aimed at reducing OM content (cultivation, topdressing, and dethatching) can increase dollar spot severity (Carrow et al., 1987; Fermanian et al., 1985). Carrow et al. (1987) found that dollar spot was greater in plots that received core cultivation twice annually plus dethatching, which the authors speculated was due to weakened turf, rather than reduced OM content. Fermanian et al. (1985) observed that all plots receiving topdressing (sand or sand plus soil) had greater dollar spot severity compared to the non-treated control. In contrast, Moeller (2008) found consistent reductions in dollar spot severity in plots that received hollow tine cultivation, in combination with seasonal topdressing with medium-coarse sand, compared to the non-treated control. Finally, Stier and Hollman (2003) observed that cultivation and topdressing treatments applied to four creeping bentgrass (*Agrostis stolonifera* L.) cultivars had no effect on OM accumulation. On annual bluegrass putting greens, topdressing treatments were reported to decrease OM concentration in the mat layer by increasing the mat layer depth, and anthracnose disease was linearly reduced as the amount of topdressing increased up to 6.0 L m^{-2} yr^{-1} (Wang, 2015; Wang et al., 2018). Kerns et al. (2009) investigated the impact of OM content at the establishment on the severity of pythium root dysfunction (*Pythium volutum*) and found the disease was more severe in treatments containing 100% sand compared to treatments that contain a 70:30 (% v/v) sand to sphagnum peat moss ratio. The authors noted that this result was likely due to heat stress prior to inoculation that significantly reduced root depth in the 100% sand treatments. Recently, researchers have begun to investigate the effects of OM concentration on fungicide bioavailability. Stephens (2021) conducted an *in vitro* bioavailability assay using three fungicides (*pyraclostrobin*, *azoxystrobin*, and *propiconazole*) and four OM concentrations (0%, 0.5%, 1.0%, and 1.5% OM in potato dextrose agar) and found that more *pyraclostrobin* and *propiconazole* were required to inhibit take-all patch (*Gaeumannomyces graminis*) as OM concentration increased, whereas OM concentration did not influence *azoxystrobin* efficacy. This result suggests that some fungicides may have an affinity to bind to OM and this can reduce their efficacy. Further research is required to better understand the influence of OM on turfgrass diseases, as well as the impact of OM content on fungicide efficacy.

4 Organic matter and soil microbial populations

Putting green soils are often criticized for lacking microorganisms because they have high sand content and are intensively managed for OM reduction and control. Recent studies utilizing sequencing technology revealed that putting green turf in fact supports a diverse microbial community (Beirn et al., 2017; Allan-Perkins et al., 2019). In addition to OM management and control, putting green root zones are managed to increase porosity and avoid compaction, which also provides microorganisms with favorable habitats. Interestingly, golf course roughs, fairways, and putting greens vary in management intensity from low to high inputs but do not differ in microbial abundance and diversity (Allan-Perkins et al., 2019). One fairway study showed a high correlation between thatch microbial activities and thatch moisture (measured in the laboratory in $g\ g^{-1}$) but not thatch OM content (Mu and Carroll, 2013). Therefore, there is no evidence that management practices to limit OM accumulation of turfgrass could have a negative effect on the microbial community in the soil profile. Nevertheless, lacking soil OM is typically not a concern in managed turfgrass ecosystems. Ecological studies showed that turf used for golf courses accumulates soil OM at high rates, comparable to regenerating forests and fallowed cropland (López-Bellido et al., 2010; Qian and Follett, 2002; Qian et al., 2010).

5 Organic matter management practices

OM accumulation is often a problem when attempting to sustainably manage stoloniferous and rhizomatous turfgrass species such as creeping bentgrass, which is stoloniferous, Kentucky bluegrass, which is rhizomatous, zoysiagrass, kikuyugrass, and hybrid Bermud agrass which are both stoloniferous and rhizomatous. This problem, which is also often referred to as thatch accumulation, can be accelerated when maintaining newer, more aggressively spreading cultivars in sand-based systems like putting greens, tee boxes, and athletic fields or sports pitches (Fontainer et al., 2011; Stier and Hollman, 2003; White and Dickens, 1984). Excessive OM accumulation has been linked to a variety of problems, such as increased disease occurrence, reduced drainage, decline in turf appearance and playability, shallow rooting, localized dry spot or dry patch, and mower scalping (Carrow et al., 1987; Dunn et al., 1995; Fontainer et al., 2011; Murphy et al., 1993; Schmid et al., 2014; Waddington, 1992). Cultivation has long been accepted as one of the primary methods of OM removal in turfgrass (Beard, 1973). Hollow tine core cultivation and vertical mowing can remove OM (Hempfling et al., 2020; McCarty et al., 2007). However, research has determined that traditional cultivation methods (hollow tine, as well as solid tine, and vertical mowing) have little effect on OM levels in managed turfgrass systems when these practices are not combined with sand topdressing, or if the soil removed during cultivation is not reincorporated back into the root zone after the OM has been separated (Barton et al., 2009; Carrow et al., 1987; Smith, 1979; White and Dickens, 1984). In recent years, several novel cultivation methods have been developed such as air, water, and sand injection, as well as grooming and venting equipment, but again these cultivation practices have little effect on soil OM content (Fontainer et al., 2011; McCarty et al., 2007; Schmid et al., 2014). Considering this, the available research would suggest that sand topdressing is one of the most important cultural practices required when sustainably managing OM, particularly in the upper root zone (Barton et al., 2007; Carrow et al., 1987; Dunn et al., 1995; Stier and Hollman, 2003; Wang, 2015; White and Dickens, 1984). In more recent years, research has explored non-destructive OM control methods using the application of decomposition enhancing products with varying success (Berndt et al., 2014; Chamberlain and Crawford, 2000; Ledeboer and Skogley, 1967; McCarty et al., 2005; Sidhu et al., 2012, 2013, 2019; Weaver et al., 2021). This section will review the various cultural practices that have, and have not, been shown to effectively reduce OM accumulation.

The United States Golf Association recommendations for core cultivation or aerification are between 15% and 20% affected surface area per year (O'Brien and Hartwiger, 2001; O'Brien and Hartwiger, 2003). To achieve this affected surface area, researchers have explored the effects of core cultivation implemented from twice annually, to every other week, with varying tine diameters (6.4–15.8 mm) and spacings (50–76 mm), as well as varying tine depths (50–100 mm) (Barton et al., 2007; Carrow et al., 1987; Dunn et al., 1995; Fontainer et al., 2011; McCarty et al., 2007; Smith, 1979; White and Dickens, 1984).

However, hollow and solid tine cultivation (often referred to as aerification) have not been shown to effectively mitigate OM accumulation when soil cores are removed from hollow tine-cultivated turfgrass and sand topdressing is not applied to hollow tine- or solid tine-cultivated areas. Research conducted as early as 1979 by Smith found that hollow tine core cultivation every other week during the growing season did not adequately control thatch accumulation of a 'Tifdwarf' Bermuda grass green after 7 months, when frequent sand topdressing application or soil reincorporation was not implemented. Like this work, White and Dickens (1984) determined that monthly versus twice-yearly core cultivation without soil reincorporation or sand topdressing did not change the thatch accumulation of three different dwarf Bermuda grass hybrid greens ('Tifgreen', 'Dothan', and Tifdwarf) over a 2-year period. After this, Carrow et al. (1987) determined that core cultivation with the cores removed, and no sand topdressing, increased OM content in a 'Tifway' Bermuda grass stand maintained at fairway/athletic field height-of-cut. Dunn et al. (1995) determined that core cultivation without sand topdressing over a 3-year period had no effect on percent OM on 'Meyer' zoysiagrass maintained at fairway height-of-cut. Barton et al. (2007) determined that core cultivation, without sand topdressing, on a mature kikuyugrass provided minimal OM reductions when compared to treatments that incorporated sand topdressing. Fontainer et al. (2011) also found that hollow tine aeration and solid tine aeration (both ranging from 14% to 16.8% affected surface area) did not reduce OM on three different Bermuda grass putting greens (Tifdwarf, 'Tifeagle', and 'Miniverde') when soil cores were not reincorporated, or sand topdressing was not applied. While solid or hollow tine core cultivation can improve root zone drainage and gas exchange, without the addition of sand topdressing or soil core reincorporation, these cultural practices will not reduce OM accumulation.

In addition to hollow and solid tine core cultivation, a variety of other cultivation practices such as vertical mowing, grooming, venting, and injection (air, water, and sand) have been introduced. Vertical mowing utilizes vertically mounted blades that are mechanically driven that bring soil and OM up to the surface. Vertical mowing can be done with varying blade widths (2.0–3.6 mm wide), that are set to different spacings (10–38 mm apart) and depths (3–20 mm into the soil) (Carrow et al., 1987; Barton et al., 2007; Hempfling et al., 2020; McCarty et al., 2007). Grooming on the other hand is often done with mower attachments that are ground driven and cut into the thatch layer (Hempfling et al., 2020; Fontainer et al., 2011; McCarty et al., 2007). Grooming equipment typically does not bring OM to the soil surface. Venting equipment is typically a thin, solid tine or knife-like tine, and like grooming, it does not bring OM to the soil or turfgrass surface (Schmid et al., 2014). Venting equipment includes 'quad' needle tines, bayonet tines, and the PlanetAir (PlanetAir Turf Products, Naples, FL). Injection equipment inserts a tool or probe into the soil surface and then injects sand [DryJect (DryJect, Inc., Hatboro, PA) and Landpride (LandPride, Salina, KS)], water [HydroJect (Toro, Bloomington, MN), or air [Air2G2 (Foley Company, Prescott, WI)] into the soil (Amgain and Fontainer, 2021; Carrow, 2003). Like grooming and venting, injection equipment does not remove OM.

Like core cultivation, vertical mowing, grooming, or injection with air, water, or sand have not been shown to effectively reduce OM accumulation when sand topdressing is not combined with these practices. For instance, White and Dickens (1984) determined that various vertical mowing intensities, without soil reincorporation or topdressing, did not affect thatch accumulation in Bermuda grass hybrid greens. Carrow (2003) found that water injection (Hydroject) and sand injection (Landpride) did not decrease OM on a creeping bentgrass putting green. McCarty et al. (2007) found that walk mowing with a grooming attachment increased OM content on creeping bentgrass. Barton et al. (2007) determined that vertical mowing without sand topdressing, on a mature fairway-height kikuyugrass stand, provided minimal OM reductions. Fontainer et al. (2011) also found that venting aeration with the PlanetAir at various intensities (ranging from 7.4% to 37.8% affected surface area) did not decrease OM. Like these findings, Schmid et al. (2014) determined that the PlanetAir and Hydroject did not reduce OM content. Recent research by Amgain and Fontainer (2021) found that air and sand injection (Air2G2 and DryJect, respectively) into a 'Penncross' creeping bentgrass putting green had no effect on OM levels after 2 years of research. Though these cultural practices can provide advantages such as improved turfgrass appearance, drainage, and reduced disease occurrence, while producing minimal effects on playability (Amgain and Fontainer, 2021; Hempfling et al., 2020; Schmid et al., 2014), OM management practices, without the addition of sand topdressing, will not reduce OM content.

Although cultivation practices alone are not enough to mitigate OM accumulation, there seems to be some evidence to indicate that hollow or solid tine core cultivation practices, coupled with sand topdressing and/or soil core reincorporation, will provide the greatest amount of OM reductions. For example, Murray and Juska (1977) determined that 3 years of core cultivation with cores reincorporated back into common Kentucky bluegrass maintained at lawn height resulted in the greatest reduction in OM. Murphy et al. (1993) found that summer hollow tine core cultivation repeated over a 3-year period had the greatest reduction in the OM fraction within the thatch/mat layer of a 'Penneagle' creeping bentgrass green when soil cores were reincorporated back into the aerified plots. Later work completed by McCarty et al. (2007) observed the greatest reduction in OM content on an 'A-1' creeping bentgrass green over a 2-year period when cultivation cores were reincorporated back into the soil and combined with vertical mowing, grooming, and topdressing. Barton et al. (2007) determined that core cultivation combined with sand topdressing over a 2-year period provided the greatest OM reduction on a mature stand of fairway-height kikuyugrass. Research conducted by Schmid et al. (2014) found that quad needle tines, bayonet tines, hollow tines, and solid tines produced minor but significant OM reductions when repeated over a 2-year period on 'Providence' creeping bentgrass greens, when all treatments received relatively high annual sand topdressing treatments. Considering all of this evidence, it suggests that hollow tine core cultivation combined with soil core reincorporation or sand topdressing and solid tine cultivation combined with sand topdressing have consistently shown to be effective methods for reducing large percentages of OM in the soil profile.

If cultivation practices require the additional input of sand topdressing to effectively reduce OM levels, the next logical conclusion would be that topdressing alone is adequate for mitigating OM accumulation. Like the core cultivation work, a wide variety of sand topdressing frequencies (every other week to once a year) and rates [(0.01 cm (0.15 kg m^2) to 0.95 cm (14.65 kg m^2)] have been explored (Barton et al., 2009; Carrow et al., 1987; Dunn et al., 1995; Green et al., 2019; Kowalewski et al., 2010; McCarty et al., 2007; Obasa et al., 2013; Proctor et al., 2014; Smith, 1979; Schmid et al., 2014; Stier and Hollman, 2003; Wang et al., 2015). Together, these studies indicate the possibility of managing OM with sand topdressing alone. For example, in as early as 1984, White and Dickens determined that sand topdressing applied four times per year, compared to annual topdressing, and various cultivation practices (core cultivation and vertical mowing), had the greatest reduction on thatch accumulation on three different hybrid Bermuda grasses. Similar to these findings, Carrow et al. (1987) found the largest OM reduction in 'Tifway' Bermuda grass when sand topdressing was applied twice annually. This research also determined that sand topdressing applied once annually was better than dethatching or core cultivation with cores reincorporated in the soil. Dunn et al. (1995) determined that sand topdressing applied twice annually over a 3-year period produced the greatest reduction of OM in fairway-height 'Meyer' zoysiagrass. Stier and Hollman (2003) suggested that sand topdressing applied monthly, or every other week, was adequate enough to prevent the development of a visible thatch layer on three different bentgrass greens ('A-4', 'G-2', and 'Penncross') and a 'DW-182' annual bluegrass green over a 2-year period. Barton et al. (2007) determined that sand topdressing alone over a 2-year period provided a significant reduction in OM content of a mature kikuyugrass stand. This research also determined that sand topdressing was more effective at reducing OM than coring alone. Wang (2015) found that sand topdressing applied every 2 weeks reduced OM content on an annual bluegrass putting green. In addition to reduced OM, research has shown that sand topdressing can provide several benefits such as a decrease in disease occurrence, improved strength, increased shoot density, and faster water infiltration (Green et al., 2019; Kowalewski et al., 2010; Wang et al., 2018). However, there is a risk of developing a sand/thatch stratification layer when cultivation practices are not combined with sand topdressing. Considering this, the soil profile should be monitored regularly for the development of stratified layers if cultural practices are not combined with sand topdressing.

Ongoing research has been exploring non-destructive OM management methods with various degrees of success. Treatments of a liquid vermicompost extract (Worm Power; Aqua Aid Solutions, Rocky Mount, NC), a liquid humate derivative (EarthMAX; Harrell's LLC, Lakeland, FL), and blackstrap molasses (Plant Food Company, Inc., East Windsor, NJ) on zoysia golf greens did not affect thatch measured by weight (Weaver et al., 2021). Other commercial products Thatch-less (Novozymes Biologicals, Le Pecq, France) (Espevig et al., 2012) and Thatch-X (Ocean Organics/Emerald Isle Ltd., Ann Arbor, MI) (McCarty et al., 2005), as well as biological agents lignocellulolytic *Streptomyces* spp. (Chamberlain and Crawford, 2000) were not effective at reducing OM in putting green turf. However, another study showed some promising results with Thatch-X. Even though the thatch-mat depth was greater in the Thatch-X-treated plots compared with untreated control, the plots that received Thatch-X, as well as other cultural management practices, were able to maintain the OM at pre-study level, whereas the untreated control increased in OM concentration overtime (McCarty et al., 2007). Plant cell wall-degrading enzymes are hypothesized to accelerate the decomposition of thatch. Early research showed cellulase had no effect on OM (Ledeboer and Skogley, 1967), while a further investigation demonstrated that cellulase only accelerated short-term thatch-mat decomposition, but this effect was not sustained after 15 days (Berndt et al., 2014). Recent research has explored the potential of utilizing lignin-degrading enzymes such as fungal laccases because turfgrass thatch contains high lignin content and thereby resists microbial decomposition. Laccase applications showed promising effects on reducing OM accumulation in several turfgrass species and increasing K_{sat} (Sidhu et al., 2012, 2013). In contrast to cellulose, applying laccase for 6 months had a residual effect of OM reduction for another 6 months (Sidhu et al., 2019). The hypothesis is that laccase degrades lignin and loosens the cell wall structure allowing microorganisms to access and decompose the cell wall substances cellulose and hemicellulose.

6 Conclusion

OM accumulation will often become problematic when trying to sustainably manage turfgrass. While moderate OM can be beneficial and support a diverse microbial population, several detrimental effects and turfgrass diseases are linked to excess levels of OM and thatch in the turfgrass root zone. A variety of monitoring and measuring methods are available for evaluating changes in OM over time. There are also several cultural management practices for reducing accumulation, with the most consistently successful being a combination of cultivation and sand topdressing.

7 Where to look for further information

For basics on OM, management with cultivation and sand topdressing see textbooks and research by James Beard (Beard, 1973) and Bert McCarty

(McCarty, 2001; McCarty et al., 2005; McCarty et al., 2007). Future research needs on OM management in turfgrass ecosystems include the development of standardized sampling and analysis procedures. For potential OM sampling standardization protocol see American Standards for Testing and Materials (ASTM) F1647-02 (2002) and ASTM D2974-20 recommendations (2020). Another need is the exploration of more OM reduction methods that do not disrupt turfgrass playing surfaces. This includes further exploration into sand topdressing methods and thatch decomposition accelerators, such as cellulose and lignin-degrading enzymes. For research on these topics see recent works by Charles Fontanier at Oklahoma State University, Robert Carrow at the University of Georgia and Charles Schmid at Oregon State University. Other places to look for recent research on OM management include the Agronomy Monograph 'Turfgrass: Biology, Use, and Management' (Stier et al., 2013), and the annual Turfgrass (C-5) proceedings and abstracts of the American Society of Agronomy, Crop Science Society of America and Soil Science Society of America (ASA-CSSA-SSSA) International Annual Meeting. Funding organizations that have expressed a need for this research and often produce popular press articles on OM management include the United States Golf Association (USGA) and the Golf Course Superintendents Association of America (GCSAA).

8 References

Allan-Perkins, E., Manter, D. K. and Jung, G. (2019). Soil microbial communities on roughs, fairways, and putting greens of cool-season golf courses, *Crop Science* 59(4), 1753–1767. https://doi.org/10.2135/cropsci2018.04.0220.

Amgain and Fontainer (2021). Effects of air-injection or sand-injection on soil physical properties of a creeping bentgrass green, *Golf Course Management* 01(21), 113–119.

ASTM D2974-20 (2020). *Standard Test Methods for Determining the Water (Moisture) Content, Ash Content, and Organic Material of Peat and Other Organic Soils*, ASTM International, West Conshohocken, PA. Available at: www.astm.org.

ASTM F1647-02 (2002). *Standard Test Methods for Organic Matter Content of Putting Green and Sports Turf Root Zone Mixes*, ASTM International, West Conshohocken, PA. Available at: www.astm.org.

Barton, L. and Colmer, T. D. (2011a). Granular wetting agents ameliorate water repellency in turfgrass of contrasting soil organic matter content, *Plant and Soil* 348(1–2), 411–424. https://doi.org/10.1007/s11104-011-0765-3.

Barton, L. and Colmer, T. D. (2011b). Ameliorating water repellency under turfgrass of contrasting soil organic matter content: effect of wetting agent formulation and application frequency, *Agricultural Water Management* 99(1), 1–7.

Barton, L., Wan, G. G. Y., Buck, R. P. and Colmer, T. D. (2009). Effectiveness of cultural thatch-mat controls for young and mature kikuyu Turfgrass, *Agronomy Journal* 101(1), 67–74. https://doi.org/10.2134/agronj2008.0180.

Beard, J. B. (1973). *Turfgrass Science and Culture*, Prentice-Hall, Englewood Cliffs, NJ.

Beirn, L. A., Hempfling, J. W., Schmid, C. J., Murphy, J. A., Clarke, B. B. and Crouch, J. A. (2017). Differences among soil-inhabiting microbial communities in *Poa annua* turf throughout the growing season, *Crop Science* 57(S1), S-262-S-273. https://doi.org/10.2135/cropsci2016.06.0463.

Berndt, W. L., Gaussoin, R. E. and Vargas, J. M., Jr. (2014). Cellulase accelerates short-term decay of thatch-mat, *Agronomy Journal* 106(3), 781–788. https://doi.org/10.2134/agronj2013.0447.

Carrow, R. N. (2003). Surface organic matter in bentgrass greens, *United States Golf Association Turfgrass and Environmental Research Online* 2(17), 1–12.

Carrow, R. N., Johnson, B. J. and Burns, R. E. (1987). Thatch and quality of Tifway Bermudagrass turf in relation to fertility and cultivation 1, *Agronomy Journal* 79(3), 524–530.

Chamberlain, K. and Crawford, D. L. (2000). Thatch biodegradation and antifungal activities of two lignocellulolytic Streptomyces strains in laboratory cultures and in golf green turfgrass, *Canadian Journal of Microbiology* 46(6), 550–558. https://doi.org/10.1139/w00-025.

Dunn, J. H., Minner, D. D., Fresenburg, B. F., Bughrara, S. S. and Hohnstrater, C. H. (1995). Influence of core aerification, topdressing, and nitrogen on mat, roots, and quality of 'Meyer' Zoysiagrass, *Agronomy Journal* 87(5), 891–894.

Espevig, T., Molteberg, B., Tronsmo, A. M., Tronsmo, A. and Aamlid, T. S. (2012). Thatch control in newly established velvet bentgrass putting greens in Scandinavia, *Crop Science* 52(1), 371–382.

Fermanian, T. W., Haley, J. E. and Burns, R. E. (1985). The effects of sand topdressing on a heavily thatched creeping bentgrass turf, *International Turfgrass Society Research Journal*, 5, 439–448.

Fidanza, M. A., Snyder, G. H. and Cisar, J. L. (2017). Documenting Changes in USGA Specification Rootzone Properties in Ultradwarf Bermudagrass Greens, *International Turfgrass Research Society Journal*. https://doi.org/10.2134/itsrj2016.05.0327.

Fontanier, C. H., Steinke, K., Thomas, J. C. and White, R. H. (2011). Effects of venting aeration on neglected Bermudagrass putting greens, *Applied Turfgrass Science* [Online]. https://doi.org/10.1094/ats-2011-1201-01-rs.

Fu, J., Dernoeden, P. H. and Murphy, J. A. (2009). Creeping bentgrass color and quality, chlorophyll content, and thatch-mat accumulation responses to summer coring, *Crop Science* 49(3), 1079–1087.

Gibbs, R. J., Liu, C., Yang, M. H. and Wrigley, M. P. (2000). Effect of rootzone composition and cultivation/aeration treatment on surface characteristics of golf greens under New Zealand conditions, *Journal Turfgrass Science* 76, 37–52.

Glasgow, A., Gibbs, R., McAuliffe, K. W. and Liu, C. (2005). An investigation of organic matter levels in New Zealand golf greens, *International Turfgrass Society Research Journal* 10(2), 1078–1084.

Green, T. O., Rogers, J. N. III, Crum, J. R., Vargas, J. M. and Nikolai, T. A. (2019). Effects of rolling and sand topdressing on dollar spot severity in fairway Turfgrass, *hortTechnology* 29(4), 394–401.

Hempfling, J. W., Murphy, J. A. and Clarke, B. B. (2020). Midseason cultivation effects on anthracnose of annual bluegrass turf, *Agronomy Journal* 112(5), 3411–3417. https://doi.org/10.1002/agj2.20202.

Hurto, K. A., Turgeon, A. J. and Spomer, L. A. (1980). Physical characteristics of thatch as a turfgrass growing medium 1, *Agronomy Journal* 72(1), 165–167.

Kauffman, J. M., Sorochan, J. C. and Kopsell, D. A. (2013). Field sampling warm-season putting greens for thatch-mat depth and organic matter content, *hortTechnology* 23(3), 369–375.

Kerns, J. P., Shew, H. D., Benson, D. M. and Tredway, L. P. (2009). Impact of irrigation frequency, organic matter content and creeping bentgrass cultivar on the development of pythium root dysfunction, *International Turfgrass Society Research Journal* 11(1), 227–239.

Kowalewski, A. R., Rogers, J. N., Crum, J. R. and Dunne, J. C. (2010). Sand topdressing applications improve shear strength and Turfgrass density on trafficked athletic fields, *hortTechnology* 20(5), 867–872.

Ledeboer, F. B. and Skogley, C. R. (1967). Investigations into the nature of thatch and methods for its decomposition 1, *Agronomy Journal* 59(4), 320–323.

Lewis, J. D., Gaussoin, R. E., Shearman, R. C., Mamo, M. and Wortmann, C. S. (2010). Soil physical properties of aging golf course putting greens, *Crop Science* 50(5), 2084–2091.

Liang, X., Su, D., Wang, Z. and Qiao, X. (2017). Effects of turfgrass thatch on water infiltration, surface runoff, and evaporation, *Journal of Water Resource and Protection* 09(7), 799–810. https://doi.org/10.4236/jwarp.2017.97053.

López-Bellido, R. J., Lal, R., Danneberger, T. K. and Street, J. R. (2010). Plant growth regulator and nitrogen fertilizer effects on soil organic carbon sequestration in creeping bentgrass fairway turf, *Plant and Soil* 332(1–2), 247–255.

McCarty, L. B. (2001). Best Golf Course Management Practices. Prentice Hall, Englewood Cliffs, NJ.

McCarty, L. B., Gregg, M. F. and Toler, J. E. (2007). Thatch and mat management in an established creeping bentgrass golf green, *Agronomy Journal* 99(6), 1530–1537.

McCarty, L. B., Gregg, M. F., Toler, J. E., Camberato, J. J. and Hill, H. S. (2005). Minimizing thatch and mat development in a newly seeded creeping bentgrass golf green, *Crop Science* 45(4), 1529–1535.

McClellan, T. A., Gaussoin, R. E., Shearman, R. C., Wortmann, C. S., Mamo, M., Horst, G. L. and Marx, D. B. (2009). Nutrient and chemical properties of aging golf course putting greens as impacted by soil depth and mat development, *Hortscience* 44(2), 452–458.

McCoy, E. L. (1992). Quantitative physical assessment of organic materials used in sports turf rootzone mixes, *Agronomy Journal* 84(3), 375–381.

Moeller, A. C. (2008). Evaluation of putting green surface organic matter management programs. M.S. Thesis, Purdue University.

Mu, Y. and Carroll, M. J. (2013). Thatch and soil microbial activity in recently cultivated turfgrass, *International Turfgrass Society Research Journal* 12, 567–574.

Murphy, J. A., Rieke, P. E. and Erickson, A. E. (1993). Core cultivation of a putting green with hollow and solid tines, *Agronomy Journal* 85(1), 1–9.

Murray, J. J. and Juska, F. V. (1977). Effect of management practices on thatch accumulation, turf quality, and leaf spot damage in common Kentucky bluegrass 1, *Agronomy Journal* 69(3), 365–369.

Nelson, D. S. and Sommers, L. E. (1996). Total carbon, organic carbon, and organic matter. In: Sparks, D. L. (Ed.) (Series 5), *Methods of Soil Analysis, Part 3–Chemical Analysis, Soil Science Society of America (SSSA) Book*, Soil Science Society of America, Madison, WI, 961–1010.

Obasa, K., Fry, J., St. John, R. and Kennelly, M. (2013). Effect of cultivation and timing of nitrogen fertilization on large patch disease of zoysiagrass, *Plant Dis.* 97, 1075–1081.

O'Brien, P. and Hartwiger, C. (2001). Core aerification by the numbers, *USGA Green Section Record* 39(4), 8–9.

O'Brien, P. and Hartwiger, C. (2003). Aeration and topdressing for the 21st century, *USGA Green Section Record*, 41(2, March/April), 1–7.

Ok, C.-H., Anderson, S. H. and Ervin, E. H. (2003). Amendments and construction systems for improving the performance of sand-based putting greens, *Agronomy Journal* 95(6), 1583–1590.

Proctor, C. A., Johnston, W. J., Golob, C. T., Stahnke, G. K. and Williams, M. W. (2014). Topdressing Sand Color, Cultivation Timing, and Cultivation Method Effects on Disruption of a Creeping Bentgrass Golf Green in the Intermountain Pacific Northwest, https://doi.org/10.1094/ATS-2012-0166-RS.

Qian, Y. and Follett, R. F. (2002). Assessing soil carbon sequestration in turfgrass systems using long-term soil testing data, *Agronomy Journal* 94(4), 930–935.

Qian, Y., Follett, R. F. and Kimble, J. M. (2010). Soil organic carbon input from urban turfgrasses, *Soil Science Society of America Journal* 74(2), 366–371.

Rabenhorst, M. C. (1988). Determination of organic and carbonate carbon in calcareous soils using dry combustion, *Soil Science Society of America Journal* 66, 1878–1881.

Schmid, C. J., Gaussoin, R. E., Shearman, R. C., Mamo, M. and Wortmann, C. S. (2014). Cultivation effects on organic matter concentration and infiltration rates of two creeping bentgrass (*Agrostis stolonifera* L.) putting greens, *Applied Turfgrass Science* 11(1), ATS-2014-0032-RS. https://doi.org/10.2134/ATS-2014-0032-RS.

Sidhu, S. S., Huang, Q., Carrow, R. N. and Raymer, P. L. (2012). Use of fungal laccases to facilitate biodethatching: a new approach, *HortScience* 47(10), 1536–1542.

Sidhu, S. S., Huang, Q., Carrow, R. N. and Raymer, P. L. (2013). Efficacy of fungal laccase to facilitate biodethatching in Bermudagrass and zoysiagrass, *Agronomy Journal* 105(4), 1247–1252. https://doi.org/10.2134/agronj2012.0470.

Sidhu, S. S., Huang, Q., Carrow, R. N. and Raymer, P. L. (2014). Optimizing laccase application on creeping bentgrass (*Agrostis stolonifera* L.) to facilitate biodethatching, *Crop Science* 54(4), 1804–1815. https://doi.org/10.2135/cropsci2013.09.0612.

Sidhu, S. S., Huang, Q., Carrow, R. N. and Raymer, P. L. (2019). Short-term and residual effects of laccase application on creeping bentgrass thatch layer, *Hortscience* 54(9), 1610–1620.

Smith, G. S. (1979). Nitrogen and aerification on putting green thatch and soil, *Agronomy Journal* 71(4), 680–684.

Stephens, C. M. (2021). Etiology, epidemiology, and management of take-all root rot on golf course putting greens. Ph.D Dissertation, The North Carolina State University.

Stier, J. C., Horgan, B. P. and Bonos, S. A. (2014). *Turfgrass: Biology, Use, and Management (Agronomy Monograph)*. American Society of Agronomy, Madison, WI.

Stier, J. C. and Hollman, A. B. (2003). Cultivation and topdressing requirements for thatch management in A and G bentgrasses and creeping bluegrass, *Hortscience* 38(6), 1227–1231.

Taylor, D. H. and Blake, G. R. (1982). The effect of turfgrass thatch on water infiltration rates, *Soil Science Society of America Journal* 46(3), 616–619.

Wang, R. (2015). Impact of sand topdressing on turf quality, surface physical properties and disease severity of putting green turfs. Master's Dissertation, Rutgers, The State University of New Jersey, New Brunswick, NJ. 1–298.

Wang, R., Hempfling, J. W., Clarke, B. B. and Murphy, J. A. (2018). Seasonal and annual topdressing effects on anthracnose of annual bluegrass, *Agronomy Journal* 110(6), 2130–2135. https://doi.org/10.2134/agronj2018.01.0052.

Wang, R., Murphy, J. A. and Giménez, D. (2021). Velvet bentgrass putting green quality, water retention, and infiltration as affected by topdressing sand size and rate, *Agronomy Journal* 113(5), 3857–3870. https://doi.org/10.1002/agj2.20571.

Waddington, D. V. (1992). Soils, soil mixtures, and soil amendments. *Turfgrass. ASA, CSSA, and SSSA, Madison*, 32, 331–383.

Weaver, J. R., McCarty, L. B., Quisenberry, V. L., Hubbard, L. R., Bridges, W. G. and Brown, P. J. (2021). Evaluating biological thatch control on *Zoysia matrella* (L.) Merr. golf greens, *International Turfgrass Society Research Journal*, 1–7. https://doi.org/10.1002/its2.43.

White, R. H. and Dickens, R. (1984). Thatch accumulation in Bermudagrass as influenced by cultural practices, *Agronomy Journal* 70, 299–304.

Windows, R. and Bechelet, H. (2012). What's working?, *Bulletin for Sports Surface Management* 257, 14–19.

Woods, M. (2019). *Verdure Removal (Or Not) in Turfgrass Soil Profile Organic Matter Tests*, Bangkok, Thailand: Asian Turfgrass Center. Available at: https://www.asianturfgrass.com/post/verdure-organic-matter-testing/.

Chapter 15

Advances in biostimulants in turfgrass

Michael Fidanza, Pennsylvania State University, USA; Cale Bigelow, Purdue University, USA; Stanley Kostka, Pennsylvania State University, USA; Erik Ervin, University of Delaware, USA; Roch Gaussoin, University of Nebraska-Lincoln, USA; Frank Rossi, Cornell University, USA; John Cisar, Cisar Turfgrass Research Service, USA; F. Dan Dinelli, North Shore Country Club, USA; John Pope, Pope Soils Consulting and Counseling Services, USA; and James Steffel, Lehigh Agricultural and Biological Services, USA

1 Introduction

The term 'biostimulants' and biostimulant products have been misunderstood and often misused as potential 'miracle cures' in the turfgrass industry, and these products were often dismissed as 'snake oil', 'foo-foo juice' or 'foo-foo dust' (e.g. sarcastic reference to the mythical foo-foo tree). Some biostimulant products make performance claims substantiated by scientific research conducted in turfgrass ecosystems, while other products lack direct evidence of their benefit within a turfgrass management program (Fidanza et al., 2019).

The Merriam-Webster Dictionary (Springfield, MA, USA) lists 'biostimulant' as derived from 'bio' (- life) and 'stimulus' (a thing that evokes a specific functional reaction in an organ or tissue, or a thing that arouses activity or energy in something, or an interesting or exciting quality). Dr. Richard Schmidt (Emeritus Professor; Virginia Polytechnic Institute and State University, Blacksburg, VA, USA) is considered the pioneer of biostimulant research in turfgrass science. Dr. Schmidt along with Dr. Xunzhong Zhang (Virginia Polytechnic Institute and State University, Blacksburg, VA, USA) studied and evaluated various substances that promoted plant growth without the plant's response attributed to fertilizers, nutrients or pesticides. Drs. Schmidt and Zhang's concise definition

http://dx.doi.org/10.19103/AS.2022.0110.15

is 'Biostimulants are organic materials that when applied in small or minute quantities enhance plant growth and development' (Bigelow et al., 2021; Fidanza et al., 2015; Schmidt and Chalmers, 1993). The use of the word 'minute' in this definition was important and intended to differentiate the fact that these substances, compared to traditional nutrients and/or soil amendments, elicited a measurable and beneficial response at much lower application rates (Bigelow et al., 2021). Drs. Schmidt and Zhang explained the plant biostimulant effect as attributed to a hormonal response and the plant protection effect against abiotic stress as attributed to antioxidant production, and both of those effects are made possible by low concentrations of exogenous applications. Later, the term 'metabolic enhancers' was used, but the important distinction was that something positive was happening to the plant beyond what mineral nutrition supplied (Zhang and Schmidt, 1997).

More recently, du Jardin (2015) defined a plant biostimulant as 'any substance or microorganism applied to plants with the aim to enhance nutrition efficiency, abiotic stress tolerance, and/or crop quality traits, regardless of nutrient content'. Plant biostimulants also refer to commercial products containing mixtures of those aforementioned substances, compounds and microorganisms (du Jardin, 2015). The term 'plant biostimulant' often is used to describe the various categories of compounds and substances used in these products: plant growth hormones (e.g. abscisic acid, auxins, cytokinins, gibberellic acid), microorganisms (e.g. *Bacillus* spp., *Trichoderma* spp., mycorrhizae), amino acids, humic and fulvic acids, plant defense-activating substances, plant growth-promoting compounds, vitamins, pigments and oils, soil amendments and soil conditioners, composts and compost teas, and more (Bigelow et al., 2021; Crouch and Van Staden, 1993; Ervin, 2013a,b; Fidanza et al., 2015; Kostka and Fidanza, 2017).

The European Biostimulant Industry Council (EBIC; https://biostimulants.eu) defines biostimulants as follows: 'Agricultural biostimulants include diverse formulations of compounds, substances, and other products that are applied to plants or soils to regulate and enhance the crop's physiological processes, thus making them more efficient; biostimulants act on plant physiology through different pathways than nutrients to improve crop vigor, yields, quality and post harvest shelf life/conservation'. The EBIC also has a functional definition of plant biostimulants as follows: 'A material which contains substance(s) and/or microorganisms whose function, when applied to plants or the rhizosphere, is to stimulate natural processes to benefit nutrient uptake, nutrient efficiency, tolerance to abiotic stress, and/or crop quality, independently of its nutrient content'. Of note, the EBIC's functional definition expands beyond the 'plant' to also include the 'soil' (e.g. rhizosphere) (Fidanza et al., 2019). The Association of American Plant Food Control Officials (AAPFCO; https://aapfco.org) defines biostimulants as follows: 'any substance or compound other than primary (e.g.

N, P, and K), secondary (e.g. Ca, Mg, S), and microplant nutrients (e.g. Fe, Cu, etc.), that can be demonstrated by scientific research to be beneficial to one or more plant species when applied exogenously; ... a substance or material, with the exception of nutrients or pesticides, which has the capacity to beneficially modify plant growth'. Of note, the ASPFCO definition of biostimulants refers to the term 'beneficial substance'.

Biostimulants are often categorized by 'what they are' (e.g. how are these substances or compounds or component materials described chemically or physically?) and 'what they do' (e.g. how do these substances or compounds benefit the turfgrass plant or the turfgrass soil/root zone?) (Fidanza et al., 2019). Therefore, a classification strategy is needed to organize and clarify helpful information for biostimulants utilized in the turfgrass industry. Also, the turfgrass practitioner and stakeholder would benefit from knowing not only what a biostimulant is composed of, but how those commercially available biostimulant products may benefit managed turfgrass ecosystems.

2 Classification of biostimulants for turfgrass

A proposed classification method or strategy for listing biostimulants in turfgrass is presented in Table 1. Overall, biostimulants are listed as primarily targeting the plant or soil/rhizosphere and then further organized by category to describe their composition, followed by active or functional ingredients (e.g. compounds, substances, other descriptive terms) listed within each category. Examples of common names for biostimulant products are listed for each category.

2.1 Phytohormones

Salisbury and Ross (1991) define a plant hormone as follows: 'an organic compound synthesized in one part of a plant and translocated to another part, where in very low concentrations it causes a physiological response'. Plant hormones or phytohormones are considered chemical messengers in plants. Phytohormones are often referred to as 'signal molecules' that occur in very low concentrations and are vital to plant growth and development, and regulation and function of many physiological processes (Mauseth, 2019). Of note, each plant cell is capable of producing phytohormones, unlike animals which require specialized hormone-synthesizing glands. The most common phytohormones utilized as plant biostimulants are abscisic acid, auxins, cytokinins, ethylene and gibberellic acid.

2.1.1 Abscisic acid

Abscisic acid (ABA) is associated with water regulation in plants as indicated by an increase in ABA concentration in leaves under drought stress (Swamy and

Table 1 Proposed classification of biostimulants for turfgrass science and industry

	Plant		
	Category[1]	Examples of active or functional ingredients	Examples of biostimulant products[2]
I	Phytohormones	Abscisic acid	Algae, indoleacetic acid, benzyl-adenine, gibberellins, kelp, seaplant, seaweed extracts and more
		Auxins	
		Cytokinins	
		Ethylene	
		Gibberellic acid	
		Others	
II	Biopolymers, protein hydrolysates and other N-containing compounds	Amino acids (e.g. proline)	Amino acids, chitosan, glycine betaine, and more
		Antioxidants	
		Betaines	
		Chitin	
		Enzymes	
		Fatty acids	
		Non-protein amino acids	
		Peptides	
		Polyamines	
		Polysaccharides	
		Vitamins	
		Others	
III	Other botanical or synthetic bioactive compounds	Elicitor compounds	acibenzolar-*S*-methyl, jasmonic acid, salicylic acid, and more
		Induced systemic resistance compounds	
		Plant defense activator compounds	
		Others	

		Soil/Rhizosphere	
IV	Humic substances	Fulvic acid	Leondardite and more
		Humic acid	
V	Organics	Biochar	Biochar, composts, compost teas, kelp, seaplant, seaweed extracts, vermi/worm extracts and more
		Bioextracts	
		Biofertilizers	
		Composts/compost extracts	
		Soil amendments/supplements	
		Others	
VI	Inorganics/minerals	Al, Co, Na, Mo, Se, Si, etc.	Phosphite salts and more
		Phosphites	
		Others	
VII	Biologicals/microbials	Beneficial fungi -	Many
		Arbuscular *mycorrhizal* fungi	
		Trichoderma spp.	
		Others	
		Beneficial bacteria -	
		Bacillus spp. and other species	
		Plant growth-promoting rhizobacteria	
		Others	
		Other beneficial organisms	
VIII	Soil surfactants[3]	--?--	--?--
IX	Other naturally derived or synthetic bioactive compounds	Elicitor compounds	--?--
		Induced systemic resistance compounds	
		Plant defense activator compounds	
		Others	

[a]Category based on chemical and/or physical composition.
[b]Broad or general name of a biostimulant product listed; no product trade name provided. Should an additional column list the specific biostimulant function of those biostimulant products, and/or should the manufacturer include that information on their product label?
[c]Proposed location of soil surfactants as a category if some of those soil surfactants are to be considered or included as biostimulants.

Smith, 1999). ABA involvement with plant adaptation to drought stress includes regulation of stomatal aperture, transpiration, leaf senescence and cellular osmotic conditions (Chen et al., 2018). ABA also is associated with the plant's ability to mitigate abiotic stress from salinity and temperature, and possible biotic stress tolerance by signaling plant pathogen defense mechanisms (Mauch-Mani and Mauch, 2005; Wilkinson and Davies, 2002). Some examples of ABA research in turfgrasses include foliar application of ABA improved Kentucky bluegrass (*Poa pratensis* L.) growth under drought stress (Wang et al., 2003) and ABA's role in heat tolerance (Li et al., 2014), cold tolerance (Zhang et al., 2013a), salinity stress (Huang et al., 2021b) and photosynthesis (Chen et al., 2018).

2.1.2 Auxins

Auxins are produced in shoot and root tips and promote cell division and elongation and many aspects of plant growth and development (Salisbury and Ross, 1991). Auxin is responsible for phototropism (e.g. shoots growing upward, toward the light) and gravitropism (e.g. roots growing downward into the soil) (Mauseth, 2019). Indole-3-acetic acid (IAA) is the most common naturally occurring auxin and included as the auxin component of many biostimulant products (Sanderson et al., 1987). An example of auxin research in turfgrass is foliar applications of IAA with the shoot-absorbed plant growth regulator trinexapac-ethyl to creeping bentgrass (*Agrostis stolonifera* L.) maintained under water-deficit environmental conditions may promote root viability and drought tolerance resulting in improved turfgrass quality (Zhang et al., 2017).

2.1.3 Cytokinins

Cytokinins are involved with plant growth and development and stress-response processes, and in particular with cell division and delaying of leaf senescence (e.g. plant senescence is the process of aging in plants; plants have both stress-induced and age-related developmental aging) (Ervin, 2013b; Haberer and Keiber, 2002). This delay of leaf senescence or 'stay green' effect is a plant stress response in which cytokinins inhibit the action of senescence-inducing enzymes, slowing the degradation of chlorophyll, and maintaining photosynthetic rates and root viability (Ervin, 2013a; Haberer and Keiber, 2002). In plants, cytokinins are produced in healthy, actively growing root tips and translocated to shoots via the transpiration stream (Ervin, 2013a; Haberer and Keiber, 2002). Consequently, as summer root decline proceeds in cool-season turfgrasses, leaf tissue cytokinin levels also decline (Ervin, 2013a).

An example of a commonly used biostimulant product in this category is seaweed extract (SWE), also referred to as seaplant or kelp. SWE is derived from brown, green or red macroalgae and has been utilized in the turfgrass industry since the 1950s (du Jardin, 2015; Khan et al., 2009; Sanderson and Jameson, 1986; Sangha et al., 2014). SWE is widely used in turfgrass culture and management due to its positive effects on turfgrass growth and abiotic stress mitigation from drought, nutrient deficiency, soil salinity and extreme temperatures (Bigelow et al., 2010; Fidanza et al., 2015; Kostka and Fidanza, 2017; Wang et al., 2012, 2013, 2014; Zhang and Ervin, 2008). SWE from brown seaweed (*Ascophyllum nodosum* [L.] Le Jolis) has been shown to contain biologically active concentrations of natural cytokinins of trans-zeatin riboside and isopentenyl-adenine (Zhang and Schmidt, 1999). Although SWE is best known for containing cytokinins and cytokinin-like compounds, SWE potentially can contain many other beneficial and potentially biologically active compounds (e.g. polysaccharides, fatty acids, vitamins, mineral nutrients and more) (Craigie, 2011; Khan et al., 2009). Some examples of cytokinin research in turfgrass are as follows: benzyl-adenine increased gross CO_2 exchange rate, chlorophyll and nonstructural carbohydrate contents of Kentucky bluegrass, indicating antisenescence activity; foliar application of SWE has been shown to increase leaf cytokinin content and antioxidant activity while decreasing lipid peroxidation and delaying senescence in creeping bentgrass drought stress (Zhang and Schmidt, 1999, 2000), increase turfgrass tolerance to salinity stress (Nabati et al., 1994), to ultraviolet B irradiation (Ervin et al., 2004), and to heat stress (Liu et al., 2002; Zhang and Ervin, 2008).

2.1.4 Ethylene

While abscisic acid, auxin, cytokinin and gibberellic acid exist in the plant in liquid form, ethylene is a gaseous phytohormone that regulates plant growth (e.g. the development of leaves, flowers and fruits), senescence, response to environmental stresses (e.g. heat and freezing stresses), and often interacts with other phytohormones (Iqbal et al., 2017). Heat stress is a major contributor to leaf senescence and other physiological damage in cool-season grass species, and ethylene and cytokinins are two phytohormones associated with leaf senescence (Xu and Huang, 2009). Some examples of ethylene research in turfgrass are as follows: foliar application of an ethylene synthesis inhibitor or a synthetic cytokinin was associated with suppressed leaf senescence and improved heat tolerance of creeping bentgrass (Xu and Huang, 2009), and the inhibition of ethylene production in annual bluegrass (*Poa annua* L.) may improve winter survival during environmental conditions associated with ice encasement (Laskowski and Merewitz, 2020). Foliar application of ethephon, which is converted to ethylene in leaf tissue, has been shown to reduce rooting and hasten chlorosis in cool- and warm-season turfgrass species. Ethephon, however, has been shown to be a very effective annual bluegrass seedhead inhibitor, given proper application timing (Patton et al., 2018; Peppers et al., 2021).

2.1.5 Gibberellic acid

Gibberellic acid (GA) controls important plant growth functions such as cell elongation and stem growth, seed germination, flower development and flowering time (Schwechheimer, 2008). The number of known gibberellins exceeds 100 and they are classified as GA_1, GA_2, GA_3, etc. (Salisbury and Ross, 1991). GA_3 was the first commercially available gibberellin. The use of GA products is currently recommended to reverse the effects of an over-application of anti-GA plant growth regulators (Ervin, 2013b). In low-light or shade conditions, GA can accumulate in turfgrass, thus causing shoot elongation and leaf etiolation (Tan and Qian, 2003).

Some examples of gibberellic acid research in turfgrass are as follows: annual bluegrass seed germination is accelerated with scarification plus the exogenous application of GA (De Castro et al., 2020), increased stolon growth up to a certain concentration in bermudagrass (*Cynodon* sp.) and creeping bentgrass (Juska, 1958), and under shade, GA_3 reduced ultradwarf bermudagrass (*Cynodon dactylon* [L.] Pers. × *C. transvaalensis* Burtt-Davy) visual quality and chlorophyll levels but did not affect lateral regrowth of stolons (Bunnell et al., 2005). In addition, there are examples of research which explored the use of plant growth regulators in turfgrasses and their relationship to gibberellic acid, other phytohormones and plant physiological responses (Gaussoin and Branham, 1987; Gaussoin et al., 1997; Heckman et al., 2001a,b,c,d; Li et al., 2020). Of note, commercially available PGRs for turfgrasses (e.g. flurprimidol, paclobutrazol, prohexadione calcium, trinexapac-ethyl, etc.) are synthetic chemicals that interact with the gibberellin biosynthesis pathway, thus 'regulating' a phytohormone response versus 'eliciting' a biostimulant-induced phytohormonal response.

2.2 Biopolymers, protein hydrolysates and other nitrogen-containing compounds

Examples of compounds in this category include amino acids (e.g. betaine and proline) and chitin.

2.2.1 Amino acids

Amino acids are the 'building blocks' for proteins, enzymes, nucleic acids, antioxidants and other secondary compounds (Guo et al., 2021; Salisbury and Ross, 1991). The L-forms of amino acids are assimilated by plants, and these L-amino acids and short-chain peptides are reported to increase plant N uptake, increase root mass, activate natural defense mechanisms and enhance photosynthesis (Salisbury and Ross, 1991). For example, L-proline has been shown to improve water stress tolerance in plants (Ashraf and Fooland, 2007), and lysine metabolism is related to abiotic and biotic stress response in plants (Yang et al., 2020). These amino acids and short-chain peptide compounds are derived by chemical or enzymatic hydrolysis of animal or plant by-products, fermentation metabolites by microorganisms or chemical modification and synthesis (du Jardin, 2015). Glycine betaine (GB) is the amino acid betaine derived from glycine and is synthesized in the chloroplast, and is associated with the mitigation of abiotic stress from drought, salinity, extreme temperatures, ultraviolet radiation and heavy metals (Ashraf and Fooland, 2007). Commercial production of glycine betaine is from sugar beet (*Beta vulgaris* L.), which is purified from molasses during the sugar extraction process (Ashraf and Fooland, 2007).

Other responses from amino acid applications, for example phenylalanine, include increased production of antioxidants, the key components in dealing with abiotic stress (Kauffman et al., 2007). Although primarily plant-directed, when applied to soil, microflora (e.g. bacteria and fungi) can readily access and assimilate amino acids, resulting in increased soil microbial biomass, which could be an important component of soil health (Moody et al., 2007). Amino acids can be a readily available N source for rapid root and foliar update in turfgrass, suggesting that they also have potential as fertility management options or to maximize plant use of existing soil nutrients (McCoy et al., 2020; Radkowski et al., 2020).

Some examples of documented amino acid research in turfgrasses include: increase in creeping bentgrass visual quality, chlorophyll content and superoxide dismutase (antioxidant) content from amino acid applications, thus indicating a potential role for foliar amino acids in a creeping bentgrass fertilization program (Zhang et al., 2013b), foliar applications of branched-chain amino acids to creeping bentgrass increased shoot density and root weight (Mertz et al., 2019), GB applications to creeping bentgrass was shown to ameliorate drought stress and this effect was related to maintenance of the antioxidant enzyme system (Gan et al., 2018), perennial ryegrass under high-temperature stress and treated with hydrolyzed amino acids resulted in improved photosynthetic efficiency (Biotta, 2013), foliar application of GB to perennial ryegrass under low-temperature stress resulted in enhanced growth via osmotic adjustment (Mickelbart and Boine, 2020) and creeping bentgrass treated with tryptophan containing a fermentation by-product or purified tryptophan plus urea (46N-0P_2O_5-0K_2O) resulted in improved leaf and root IAA content, root biomass and overall turfgrass quality compared to applications of urea alone (Mertz et al., 2017).

2.2.2 Chitin

Chitin is the second most abundant natural biopolymer in the world, after cellulose (Shamshina et al., 2019). The main source of chitin is from shells of marine crustaceans, particularly crab and shrimp, but other living organisms biosynthesize chitin, including algae, insects and fungi (Yan and Chen, 2015). The chemical structure of chitin is a polysaccharide, and is reacted with an alkaline substance (e.g. sodium hydroxide) to produce chitosan. Therefore, chitosan is a linear polysaccharide composed of randomly distributed components of β(1→4)-linked D-glucosamine and *N*-acetyl-D-glucosamine (Shamshina et al., 2019).

Chitosan is used to stimulate plant growth and induce a biological response to abiotic and biotic stresses (Hou et al., 2021). The plant's response to chitin is determined by species and growth stage of development, and by chitin structure and concentration (Hou et al., 2021). Research on chitosan has focused on its use as an 'elicitor' (e.g. an elicitor is a molecule that triggers a hypersensitivity response in plants) of stress response signaling (e.g. stomatal closure to manage transpiration and water use) (du Jardin, 2015). Chitosan interacts with many cellular components (e.g. DNA, plasma membranes) to bind to specific receptor sites that activate defensive genes which results in increased protection from plant pathogens and insect pests, but more broadly to enhance tolerance to abiotic stress (e.g. drought, salinity, nutrient deficiency or toxicity) (Hou et al., 2021).

Current research is exploring the metabolic mode of action of chitin and chitosan polysaccharides in plant abiotic stress tolerance, and also the suppression of plant pathogens (du Jardin, 2015). Of note, complex polysaccharides in SWE may have similar effects (Sun et al., 1997). An example of chitin research in turfgrass is as follows: chitosan may be helpful to alleviate heat-induced leaf senescence in creeping bentgrass by regulating chlorophyll metabolism and influencing the production of antioxidants and heat shock proteins (Huang et al., 2021a).

2.2.3 Polysaccharides

Algae (e.g. seaweed) produce a diversity of biologically active compounds that can enhance plant growth and resistance to abiotic and biotic stresses (Stadnik and Freitas, 2014). These compounds include polysaccharides which are essentially branched or unbranched chains of monosaccharides (–carbohydrate molecules). Some examples of polysaccharides are cellulose, glycogen and starch. Similarly, microorganisms (e.g. bacteria, fungi, actinomycetes) also produce a diversity of polysaccharides that can influence how a microbe adsorbs onto surfaces, is protected from desiccation and interacts with plants to trigger plant's response to abiotic or biotic stresses. An example of a polysaccharide is ulvans. These water-soluble polysaccharides are obtained from the cell walls of green macroalgae (*Ulva* spp.) commonly known as 'sea lettuce' (Sangha et al., 2014). In plants, ulvans have been shown to signal a defense response to pathogen infection (Jaulneau et al., 2010). No specific research with ulvans or other seaweed polysaccharides and turfgrasses has been published, although this represents an opportunity to further explore biotic stress mitigation in amenity turfgrasses.

2.3 Other botanical or synthetic bioactive compounds

This is a 'place-holder' category for plant-directed compounds not yet described or fully understood, or for compounds that do not fit the description of the other categories. Other naturally derived plant compounds or synthetic materials may be listed in this biostimulant category (Table 1). An example of an organic compound in his category is acibenzolar-*S*-methyl (ASM), which is a synthetic analogue of salicylic acid (Kessman et al., 1996). ASM is referred to as a 'plant defense activator', because it is not directly toxic to fungal pathogens but produces an induced systemic resistance response and thus activates a plant's natural defense system (Cole, 1999). Of note, salicylic acid is involved in both abiotic and biotic stress defense due to its ability to signal for the activation of plant immune systems (Durner et al., 1997; Schmidt and Zhang, 2001). ASM is a component of commercial turfgrass fungicides that contain the active ingredients chlorothalonil and fluazinam (Clarke et al., 2020). Examples of ASM research in turfgrass are ASM associated with improved turfgrass disease management (Hsiang et al., 2011; Lee et al., 2003), and also ASM associated with the ability to mitigate abiotic heat and drought stress in turfgrass (Jespersen et al., 2017).

2.4 Humic substances

Humic substances (e.g. humic and fulvic acids) are natural decomposition constituents of soil organic matter, typically derived from leonardite (a natural form of humates), associated with the surface layers of 'brown coal' deposits. Leonardite describes the base material used in the production of humic acids and was named after A.G. Leonard, the first director of the North Dakota Geological Survey in recognition of his contributions in this field (Fidanza et al., 2015). Humic substances derived from leonardite are considered brown-black polymeric acids that exhibit both hydrophobic and hydrophilic characteristics and bind to soil mineral surfaces (Adani et al., 2006). Humic acid can be considered a natural soil conditioner because it contains high concentrations of trace minerals and has a greater effect on soil with alkaline pH and lower cation exchange capacity (Chen, 1996; Wang et al., 1995).

Benefits of three active humic compounds (e.g. fulvic acid, humic acid and humins) in humic substances include increased soil nutrient and water-holding capacity (e.g. increased cation exchange capacity), prevention and reduction in leaching of soil nutrients, chelators of organic molecules and minerals facilitating increased plant root absorption, enhanced soil enzyme and metabolic activity and more (Barrett, 2015; Liu et al., 1998; Schmidt and Zhang, 1998). Some examples of humics/humic acid research in turfgrasses include applications of SWE plus humic acid to tall fescue (*Festuca arundinacea* Schreb.; syn. *Schedonorus arundinaceus* [Schreb.] Dumort., nom. cons.) sod increased canopy photosynthesis efficiency and recovery from post-harvest heat injury (Zhang et al., 2003), applications of humic acid were beneficial to creeping bentgrass shoot and root growth (Hunter and Anders, 2004), foliar application of leonardite humic acid prior to dry-down resulted in increased quality, photochemical efficiency, root mass and leaf α-tocopherol levels during drought (Zhang and Ervin, 2004), and sequential foliar applications of leonardite- or peat-based humic acid during Kentucky bluegrass sod establishment resulted in greater root mass and root strength than the fertilizer control (Ervin and Roberts, 2007).

2.5 Organics

Many organic compounds, materials and products can be placed into this category (Table 1). Bigelow and Soldat (2013) and Waddington (1992) provided recent reviews and considerations with chemical and physical soil amendments for turfgrass ecosystems. Traditionally, organic amendments such as peat moss, manures, biosolids, composts and other materials have been added to sand-based turfgrass root zones to increase water and plant nutrient retention and availability, but these benefits can decline within a few years as those amendments decompose (Zhang et al., 2012; Waddington, 1992). Ideally, organic materials and substances applied to turfgrass soils should be sufficiently decayed and biologically stable and decompose very slowly so that their benefits or positive impact can be expressed over a long time (Bigelow and Soldat, 2013). Thus, due to their stability and low cost, peat mosses are the most frequently used organic amendment for synthetic sand-based root zones.

Another recent example of a stabilized organic material being utilized for turfgrass root zones is biochar (Brockhoff et al., 2010). Biochar represents a high carbon content, highly porous, and tremendously stable (e.g. extremely resistant to microbial degradation) substance produced from the pyrolysis (e.g. heating organic material >500 °C in the absence of oxygen) of various biomass feedstocks (e.g. wood chips, plant resides, animal manures or other agricultural waste products) (Spokas et al., 2012). Some examples of organic materials/products research in turfgrasses include the beneficial use of biochar for water and nutrient retention for sand-based root zones and turfgrass establishment (Vaughn et al., 2015), beneficial inclusion of compost and biochar on soil biology of turfgrass subjected to deficit irrigation (Hale et al., 2021), benefits of compost-amended soil for nutrient availability of turfgrass (Wright et al., 2007) and benefits of compost and compost tea for turfgrass (Bero and Soldat, 2021). Another example of an organic biostimulant product applied to turfgrass is Worm Power® (AquaAid Solutions; Rocky Mount, NC, USA), which is vermicompost extract utilized to improve the biological and physical health of the turfgrass root zone (for more information: https://tinyurl.com/2ch8ekmr).

2.6 Inorganics/minerals

Many inorganic/mineral compounds and products can be placed into this category (Table 1). Phosphite (PO_3^{3}) has become the most common inorganic compound incorporated into many turfgrass management programs, particularly with disease management and suppression (Havlin and Schlegel, 2021). In most products, phosphite is delivered as potassium phosphite in the form of either potassium dihydrogen phosphite (KH_2PO_3) or dipotassium hydrogen phosphite (K_2HPO_3) (Landschoot and Cook, 2005). Silicon in the form of silica (e.g. silicon dioxide [SiO_2]) is an example of an inorganic mineral with potential benefits for abiotic and biotic stress reduction of turfgrasses (Guertal and Datnoff, 2021; Schmidt et al., 1999; Zeller et al., 2021). Some examples of inorganic/mineral products research in turfgrasses include the use of potassium phosphite to suppress snow mold disease (*Microdochium nivale* [Fr.] Samuels and Hallett) in cool-season turfgrass (Dempsey et al., 2012), the use of silica to enhance disease suppression in warm-season turfgrass (Datnoff and Rutherford, 2003) and foliar applications of potassium silicate to bermudagrass, tall fescue and perennial ryegrass irrigated with saline water was helpful to alleviate sodium-induced stress (Esmaeili et al., 2015).

2.7 Biologicals/microbials

Numerous biological/microbial organisms can be placed into this category (Table 1). Arbuscular *mycorrhizal* fungi form a mutually symbiotic relationship with plant roots, in which roots provide carbohydrates for the fungi, and the fungi aid in access and transfer of nutrients and water to the plant roots, and also water balance, and abiotic and biotic stress tolerance or protection (du Jardin, 2015). *Trichoderma* spp. is another example of a fungal organism that has been studied for biopesticide (e.g. fungicide-like properties) and biocontrol (e.g. inducer of disease resistance) benefits and facilitating abiotic stress tolerance in plants (du Jardin, 2015). *Bacillus* spp. is the most common example of a bacterial organism utilized for biological control of plant pathogens, and this is achieved via direct suppression by the release of antipathogen compounds, or via an indirect mechanism such as outcompeting the pathogen for space or food, or activating or inducing plant defense systems (Lugtenberg and Kamilova, 2009; Schlatter et al., 2017). Current research is exploring plant growth-promoting rhizobacteria (PGPR) and their ability to confer beneficial effects on plant growth and development by increased nutrient uptake (e.g. nitrogen and phosphorus), synthesizing plant growth-promoting compounds, activating abiotic and biotic stress tolerance mechanisms and possibly more (Lugtenberg and Kamilova, 2009; Rosier et al., 2018).

Zuberer (2012) summarized 35 years of research in this area to conclude (i) microbial function and growth in turfgrass soils are similar to agricultural soils or native grassland soils; (ii) soil microbial populations are resilient and changes to the community are most likely achieved through altering the soil environment toward a more favorable habitat; and (iii) best practices for managing the microbiology in the turfgrass root zone are those that adhere to well-established agronomic principles and practices (e.g. healthy turfgrass will provide the necessary resources for soil microbes to thrive and function). Some examples of research on biologicals in turfgrasses include synthetic fungicides applied at reduced label rates in combination with biological control agents (e.g. *Bacillus* spp.) provided an opportunity for dollar spot disease management in creeping bentgrass (Marvin et al., 2020), environmental factors and an actively growing turfgrass root system may have a greater influence and impact on soil microbial activity (Bigelow et al., 2002), and SWE can impart positive effects on soil microbial populations, especially PGPR that facilitate nutrient uptake or provide a source of biopesticidal metabolites or plant defense activators such as ulvans (Mueller and Kussow, 2005; Sangha et al., 2014).

2.8 Soil surfactants

Agriculture, horticulture and turfgrass industry practitioners commonly refer to soil surfactant products as 'wetting agents' (Fidanza et al., 2019; Kostka and Fidanza, 2019). The term 'soil surfactant' is a more appropriate and descriptive label from a soil science perspective (Kostka and Fidanza, 2018). Soil surfactant chemistries used in turfgrass management have been reviewed in Zontek and Kostka (2012). With over 300 soil surfactants commercially available globally, Fidanza et al. (2020) proposed a classification strategy for soil surfactants in the turfgrass industry, the majority of which in the global marketplace are of non-ionic chemical composition. Worthy of note, the first patent of a non-ionic soil surfactant was US Patent #2,867,944, 'Method of treating soil by non-ionic surface active agents' by L.W. Fletcher (Havertown, PA, USA), which was first published in 1959 (https://patents.google.com/patent/US2867944). Should all or some specific soil surfactants be listed as a biostimulant? Can soil surfactants 'behave as biostimulants' or 'facilitate a biostimulant effect' when applied to turfgrass root zones? Scientists and stakeholders in academia, government and industry have begun to explore this question further (Fidanza et al., 2019; Kostka and Fidanza, 2019).

In managed amenity turfgrass ecosystems, soil surfactants are primarily and traditionally used for water conservation, improving irrigation use efficiency and ameliorating soil water repellency (Kostka et al., 2007; Soldat et al., 2010). The utilization of soil surfactants is considered the number one water conservation strategy among golf course superintendents in the United Sgtates (Gelernter et al., 2015). Some examples of research on soil surfactants in turfgrasses include the use of soil surfactants to alleviate soil water repellency in sand-based turfgrass root zones (Cisar et al., 2000; Kostka, 2000; Dekker et al., 2019), beneficial use of soil surfactants with irrigation programs (Mitra et al., 2006; Moore et al., 2010), use of soil surfactants to improve delivery and efficacy of soil-directed fungicides (Fidanza, 2015; Fidanza et al., 2007) and nematicides (Webb, 2022) and use of soil surfactants to increase nitrogen uptake of bermudagrass and reduce nitrate leaching (Abagandura et al., 2021a,b).

2.8.1 Soil surfactants and the turfgrass rhizosphere

In 1904, German agronomist and plant physiologist Lorenz Hiltner (1862-1923) first documented the term 'rhizosphere' to describe the plant-root interface (Hiltner, 1904). 'Rhiza' is a word of Greek origin that means 'root' or 'root-like', and Hiltner (1904) first described the rhizosphere as that specific area surrounding a plant root that is inhabited by microorganisms, and these microorganisms are influenced by chemicals released from the roots. The rhizosphere is a concept that pertains to the soil-root interface, and recently has been described as the zone of biological, chemical and physical influence generated by root growth and activity (Philippot et al., 2013; Pinton et al., 2007). The rhizosphere also has been defined as that zone of soil surrounding a plant root where the biology, chemistry and physical properties and activities of the soil are influenced by the root and compounds exuded by the root (e.g. mucilage), and these influences change both radially and longitudinally along the root (Fig. 1), and the rhizosphere also facilitates communication between the plant and the microbiome (Bias et al., 2004; McNear, 2013).

Hallett et al. (2022) provided a review of plant roots and their capacity to improve soil, and in particular the importance of root hairs: 'just as dabbing paint with a brush allows it to penetrate into nooks and crannies on surfaces, root hair can aid the influence of plant roots by penetrating into soil pores too small for roots and distributing rhizodeposits into a greater soil volume'. Root hairs are single-cell outgrowths from the root epidermis that increase root surface area and explore the bulk soil. Rhizodeposits are materials exchanged

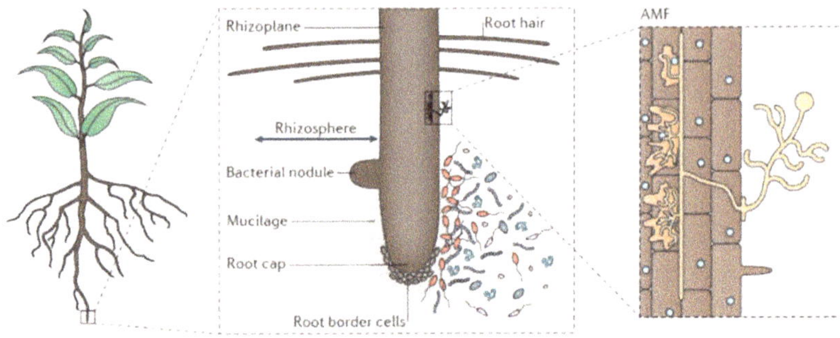

Figure 1 An example of a root section illustrating the rhizosphere (e.g. interface between plant roots and soil); AMF = arbuscular *mycorrhizal* fungi. Source: Philippot et al. (2013).

from plant to soil and dominated by root exudates, mucilage and sloughed cells. Root exudates are substances excreted by roots and composted of sugars, amino acids, organic acids and other carbon compounds. Mucilage typically composed of viscous polysaccharide (e.g. sugar) compounds secreted at the root tip and is needed to lubricate soil particles as the root pushes through that abrasive soil environment. Mucilage also represents a significant food source for soil microorganisms and can influence soil-water relations (Carminati et al., 2016). The rhizosheath is defined as the soil that adheres strongly to the root through the action of root hairs and rhizodeposits. Overall, root hairs not only help the plant access localized resources (e.g. air, nutrients and water), but they also sequester carbon (e.g. deposits of exudates, mucilage, rhizosheaths) and this ultimately improves the soil and root zone environment (e.g. binding soil particles, improving soil structure, etc.). (Bigelow, 2022; Hallett et al., 2022).

Plants invest in improving their root zone environment at the root-soil interface (e.g. the rhizosphere) (Gregory et al., 2022). By exuding materials such as mucilage, a high-molecular-weight biopolymer that absorbs and retains water, plants can remain hydrated for longer periods. Once dry-down cycles occur, however, water retention and rewetting are compromised due to the presence of small quantities of amphiphilic compounds that render the mucilage hydrophobic, thus impeding rehydration and access to water and nutrients in a diversity of species (Hallett et al., 2003; Carminati et al., 2010; Ahmed et al., 2018). Their work indicates that soil treated with a surfactant modifies mucilage behavior (e.g. reducing swelling, increasing mucilage viscosity and increasing the speed of wetting while resulting in a reduced rate of plant transpiration). Their work also suggests that in surfactant-treated soil, applied irrigation results in faster root rehydration and a larger proportion of that applied water is plant available. While the rate of plant transpiration was slowed, turgidity duration was markedly increased with concomitant maintenance of plant performance and tolerance to water stress. Soil surfactants may play an important role in modifying the turfgrass rhizosphere and thus influence the diversity of hydraulic processes and functions at the rhizosphere level (Figs. 2 and 3) (Bigelow, 2022; Carminati and Vetterlein, 2013; Carminati et al., 2011, 2016; Fidanza et al., 2019, 2020; Kostka and Fidanza, 2018, 2019; Singh et al., 2018).

In recent studies, Ahmadi et al. (2017) built on these earlier observations noting that not only did surfactant treatment increase and maintain access to water, but that surfactant modification of mucilage increased microbial activity in the rhizosphere, increased nutrient uptake efficiency and resulted in increased rooting and development of rhizosheaths. Moreover, the production and distribution of a range of plant and microbial enzymes in the rhizosphere were associated with soil health and suppression of soil-borne pests and pathogens (Ahmadi et al., 2018). As Hallett et al. (2022) so obviously state: *everything from the soil which passes through the plant passes through the rhizosphere* (Figs. 2 and 3). It is becoming apparent that strategies to 'engineer' the rhizosphere may have profound effects normally associated with biostimulants.

Roots are essential for the acquisition of water and nutrients/minerals from soils (Salisbury and Ross, 1991). Root exudates are important for the

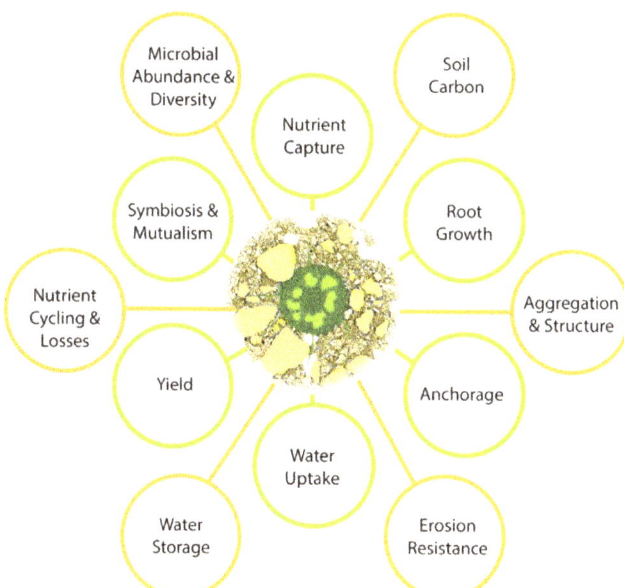

Figure 2 Illustration indicating processes and functions of the rhizosphere, and that *everything from the soil which passes through the plant passes through the rhizosphere.* Of note, 'Yield' pertains to crop production but could be substituted for 'Quality and Resilience' when pertaining to amenity turfgrass. Source: Hallett et al. (2022).

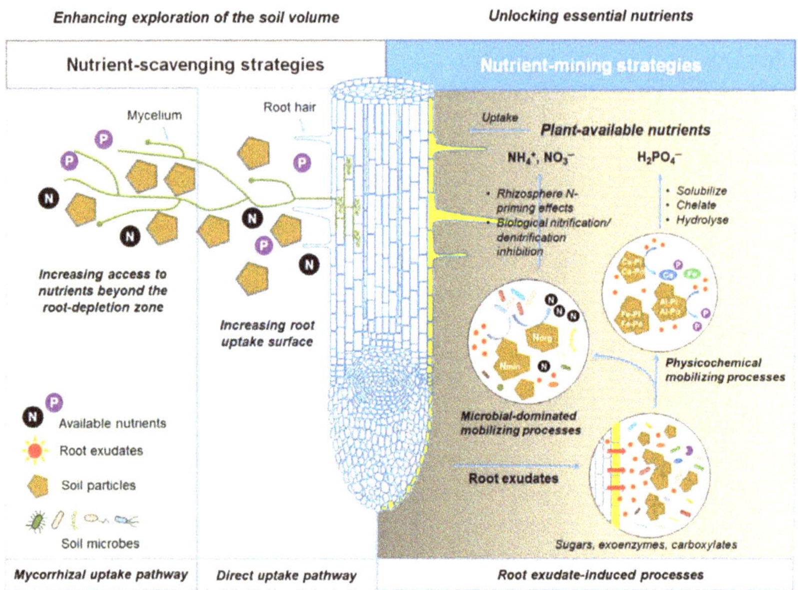

Figure 3 Illustration of root exudates functioning to enhance or influence nutrient acquisition with the rhizosphere. Soil surfactant applications in turfgrass root zones may play a beneficial role in this process. Source: Wen et al. (2022).

belowground, interspecies communication between plants and other plants and/or microorganisms, and thus modifications to roots and the rhizosphere can result in improved plant performance and increased stress tolerance (Ahmed et al., 2018; Bias et al., 2004; Rosier et al., 2018). Therefore, strategies to manipulate the rhizosphere to maintain or enhance the connectivity of roots and root hairs to soil particle surfaces can be expected to stimulate the rhizosphere microbiome to influence soil and plant health (Duddek et al., 2020, 2022). If root and root hair connectivity can be maintained, plant resilience should be optimized. Enhanced rhizosheath development has been documented in turfgrass grown in surfactant-treated soils (Ahmadi, personal communication). Preliminary, unpublished studies show that certain surfactants increase rooting in cool- and warm-season turfgrasses under greenhouse and field conditions (personal communications: T. Malehorn [iGin Research, Westminster, MD, USA], X. Zhang [Virginia Polytechnic Institute and State University, Blacksburg, VA, USA], S. B. Martin [S.B. Martin Turf Research, Florence, SC, USA], P. Kirby [Indigo Specialty Products, Rocklea, Australia]). Further, turfgrass remains turgid even as soil water deficits increase, suggesting that better root/root hair connectivity is being maintained (Fig. 4). A body of evidence is building that certain diverse root zone processes can be 'engineered' by surfactants to optimize rhizosphere and the biophysical, microbiological and chemical

Figure 4 An example of a beneficial application of a soil surfactant to improve the volumetric water content status of the root zone and possibly optimize the function of the rhizosphere of this creeping bentgrass fairway and putting green. (A) Left side of fairway was untreated. (B) Drought stress visible in the turfgrass sward, indicating root with compromised physiological function. (C) Illustration depicting poor root-to-soil contact. (D) Right side of fairway and putting green treated with soil surfactant (ProWet Evolve®; TurfCare; Kildare, Ireland). (E) Healthy and dense turfgrass visible on the right side, indicating that roots in the soil surfactant-treated turfgrass root zones maintain root-to-soil contact and access to water and nutrients. (F) Illustration depicting good or optimum root-to-soil contact. Photo by D. Jones. Parts (C) and (F) from Weil and Brady (2017). Parts (B) and (E) from Duddek et al. (2020, 2022).

properties of soil (Fig. 4) (Ahmadi et al., 2017, 2018; Bias et al., 2006; Duddek et al., 2020, 2022).

2.9 Other naturally derived or synthetic bioactive compounds

This category is a 'place-holder' for soil-directed compounds not yet described or fully understood, or for compounds that do not fit the description of the other categories. To date, no compounds are placed into this category.

3 Considerations with the use of biostimulants for sustainable turfgrass management

When asked to consider using a biostimulant product, a prudent and responsible turfgrass manager would ask '*What is in it?*' and '*What does it do?*', among other questions (Fidanza et al., 2019). Carrow (1993) published an inciteful list of eight questions to ask when evaluating a soil amendment or other product for turfgrass use and management (Table 2).

The purpose of asking those questions is to help guide the turfgrass manager toward making the best fact-based agronomic decision. Fidanza et al. (2019) further expanded on those important questions to ask when a turfgrass manager is considering a biostimulant product (Table 3). These questions further explore a biostimulant's intended use (e.g. abiotic or biotic stress, plant nutrient efficiency) and ability to produce or facilitate the desired turfgrass response.

Products that contain a biostimulant or combinations of biostimulants may provide turfgrass management options to maintain or improve turfgrass quality and function during abiotic and/or biotic stress conditions. Moderating and mitigating these stresses are an important strategy for establishing and maintaining a healthy, resilient and sustainable turfgrass sward. As these stresses increase, a number of plant physiological processes may be compromised: decline in the efficiency of light capture (photosynthesis) and carbon utilization (respiration); increase in the amount of destructive reactive oxygen species (free radicals) in leaves, shoots and roots; leaves and shoots stop allocating energy to roots and draw energy and resources from the roots; and root decline which precedes shoot decline and is often coupled with root infection from secondary plant pathogens (Yakhin et al., 2017). Therefore, should a biostimulant product or program become a valuable part of sustainable turfgrass management? The answer to that question may depend on what exactly the turfgrass practitioner

Table 2 Should I use this product on my turf?: Eight questions to ask when evaluating a soil amendment or other turf product to help you make sound, fact-guided agronomic decisions[a]

1. Is this product needed in my situation?
2. Are independent test results available?
3. What is the magnitude of response?
4. Does this product provide consistent results?
5. What is the duration of response?
6. Are there better alternatives?
7. Do benefits justify the costs?
8. Should I try this product in a trial area?

[a]Source: Carrow (1993).

Table 3 What are key questions golf course superintendents, greenkeepers, course care managers, sports field managers and lawn and landscape professionals should ask when considering a biostimulant product?[a]

- What's in it? What is the product's composition or active ingredient(s) or component(s)?
- What is its function? How does the product claim to benefit turfgrass?
- Will the product function in all climates, soil types, turfgrass species and turfgrass cultural practices and management programs?
- Does the product function best to help with abiotic (e.g. drought, heat, salt) or biotic (e.g. insects, pathogens, traffic) stresses?
- Where's the data? What does the research-based data show that the product does when the product is applied to turfgrass? Were the effects both qualitative and quantitative in the replicated field and/or controlled greenhouse research?
- Does the manufacturer have clear research-based data showing that all or most of the product's active ingredients are essential to its function? Does the data show how individual ingredients, when tested against the formulated product, no longer provide the functional benefit?

[a]Source: Adapted from Fidanza et al. (2019).

wants to accomplish (e.g. better rooting, better tolerance of heat or drought stress, improved recovery from heat or drought stress, traffic tolerance, turf recovery, disease prevention, turf recovery from disease, better color or visual quality, better playability).

Many biostimulant products are formulated from multiple components and/or are of undefined or vaguely defined composition (e.g. as components in fertilizer formulations or as soil amendments). While biostimulant activity may be measured from such formulations, is the observed plant or soil response a consequence of a single biostimulant component or from multiple components in the formulation or product? In studies conducted in growth chambers and in the greenhouse, the effects of single components can be measured. However, measuring such effects in the field is often more challenging. One or more complex formulations may need to be applied to observe a visual plant response. The challenge then becomes explaining which components were responsible for the observed effects.

Turfgrass practitioners should critically evaluate turf for the response desired. If a turf manager decides to include a biostimulant product as a component of their turf management program, it would be important to make many observations and measurements throughout the year and document what you are seeing so that you can make informed decisions going forward. Also be sure to review all available information and consider that the use of a biostimulant or any product should be predicated on results from independent and replicated third-party research.

4 Conclusion and future trends

Why should there be a functional definition of plant biostimulants? The European Union's view is that biostimulants should support ecology-based agricultural and food production systems. The EBIC's last part of their biostimulant definition: 'to stimulate natural processes ... independently of its nutrient content' indicates that although a biostimulant product may contain a fertilizer component (e.g. N or Fe), the observed demonstrated plant response is due to the biostimulant and not the fertilizer effect. Therefore, the European Union has proposed regulations to support and enforce a claims-based definition of plant biostimulants (e.g. biostimulant products should be defined by their function, not by the ingredients inside the product container) (Fidanza et al., 2019). The European Union also proposed that biostimulant products should have credibly demonstrated effects to support the product claims (e.g. documented research to support product label statements for the crops and markets targeted). For example, if a biostimulant product contains a microorganism that induces a plant defense response to a plant pathogen, then the manufacturer must provide data supporting that claim for that particular crop and use. Therefore, the manufacturer would have to prove the biostimulant inside the product performs the function for that particular crop or use as stated on the label (Fidanza et al., 2019; Ricci et al., 2019).

Currently, the EBIC is collaborating with the government, academia and industry to develop guidelines for compiling data and information obtained from consistent and reliable research results. The guidelines contain two main points: (i) data must be generated to support the biostimulant claim (e.g. biostimulant function), and (ii) data must be used to place biostimulant product into the European Union market and/or support commercial product claims. Of note, the data required will depend on the biostimulant's claim or declared function (Fidanza et al., 2019). Would biostimulant product manufacturers be required to include the specific biostimulant function or claim on the product label (Fig. 5)?

Regarding the required research-based data needed to support biostimulant claims, the European Union has proposed guidelines and criteria for laboratory, greenhouse and growth chamber studies that include proper data analysis and interpretation (e.g. the use of a 90% probability statistic) and field trials conducted under real-world conditions and under various environmental and geographic conditions. Of note, peer-reviewed scientific journal publications can be used to support a biostimulant claim if it is of acceptable quality under the European Union guidelines (Fidanza et al., 2019).

Should the US Department of Agriculture (USDA; Washington, DC, USA) adopt the European Union's claims-based definition of biostimulants for use in the United States? Recently, the US Environmental Protection Agency (USEPA;

Plant Root Bio-Extender[1]

A biostimulant for cool-season turfgrasses.

Composition	Biostimulant Function
Kelp .... 20% source: seaweed extract of brown macroalgae from Nova Scotia, Canada	heat stress tolerance
Bacillus subtilis .... 5% source: *Bacillus subtilis* strain XP175	activates plant defenses for brown patch (*Rhizoctonia solani*) disease
Water 75%	

Research to support claim of biostimulant function available at: www.manufacturerwebsite.com

[1]fictious name; no indication toward a commercial product is implied.

Figure 5 Example of an ideal biostimulant product label that would include information on both composition (e.g. active ingredients and sources) and biostimulant functions or claims.

Washington, DC, USA) has assembled a draft document of guidelines for clarifying which biostimulants, biological substances, mixtures and associated product label claims should be considered as plant growth regulators (source: https://www.epa.gov/pesticides/draft-guidance-plant-regulators-and-claims-including-plant-biostimulants). Those compounds identified as plant growth regulators will be subject to regulation under the Federal Insecticide, Fungicide and Rodenticide Act as pesticides. Of note, the USDA's proposed definition of biostimulant is as follows: 'Plant biostimulant is a substance(s), microorganism(s), or mixtures thereof, that, when applied to seeds, plants, the rhizosphere, soil or other growth media, act to support a plant's natural nutrition processes independently of the biostimulant's nutrient content. The plant biostimulant thereby improves nutrient availability, uptake or use efficiency, tolerance to abiotic stress, and consequent growth, development, quality or yield' (source: https://www.epa.gov/pesticides/draft-guidance-plant-regulators-and-claims-including-plant-biostimulants).

In the United States, however, a structured regulatory path to register biostimulants has not been established, thus precluding validation afforded by regulatory oversight and its value to developers and users of the technology. Therefore, the Biostimulant Industry Workgroup (BIW) was established in 2020 as a collaboration of the Biological Products Industry Alliance (BPIA; Oakton, VA, USA) and The Fertilizer Institute (TFI; Arlington, VA, USA). In response to US Department of Agriculture (USDA; Washington, DC, USA) and US Environmental

Protection Agency (USEPA; Washington, DC, USA), reports on biostimulants submitted to Congress in 2019, BWI prepared and released a draft document: 'United States Biostimulant Industry Recommended Guidelines to Support Efficacy, Composition, and Safety of Plant Biostimulant Products' (www.bpia .org). This draft document was developed by BIW representatives with input from a wide range of stakeholders and reviewers and released in 2022. The document offers clarification of terminology and a suggested framework for a regulatory process requiring scientifically sound data in support of product label claims and the safety of biostimulant products. The biostimulant industry awaits the outcome of those interactions among the USDA, USEPA, BPIA and TFI.

Regardless of the biostimulant product or strategy, it is important to note that biostimulants are not a substitute for essential mineral nutrients and a sound agronomic-based turfgrass management program. If the goal is to include or incorporate biostimulants as part of an overall plant and soil health program, then the research in turfgrass ecosystems has demonstrated that they must be applied in advance of those abiotic and biotic stresses to optimize their benefits. There are many exciting innovations on the nearby horizon and evidence-based efforts will lead the way toward a better understanding of how biostimulants will help maintain and improve plant and soil health (Yakhin et al., 2017). Today, much more scientific research is focused on the development, evaluation, use, function and benefits of biostimulants for sustainable agronomic practices in intensively managed amenity turfgrass ecosystems.

5 References

Abagandura, G. A., Park, D. and Bridges, Jr., W. C. 2021a. Surfactant and irrigation impacts on soil water content leachate of soils and greenhouse substrates. *Agroecosystems, Geosciences and Environment* 4:220153.

Abagandura, G. O., Park, D., Bridges, Jr., W. C. and Brown, K. 2021b. Soil surfactants applied with ^{15}N labeled urea increases Bermudagrass uptake of nitrogen and reduces nitrogen leaching. *Journal of Plant Nutrition and Soil Science* 184(3):378-387. doi: 10.1002/jpln.201900162.

Adani, F., Spagnol, M. and Genevini, P. 2006. Biochemical origin and refractory properties of humic acid extracted from the maize plant. *Biogeochemistry* 78(1):85–96.

Ahmadi, K., Razavi, B. S., Maharjan, M., Kuzyakov, Y., Kostka, S. J., Carminati, A. and Zarebanadkouki, M. 2018. Effects of rhizosphere wettability on microbial biomass, enzyme activities and localization. *Rhizosphere* 7:35–42. doi: 10.1016/j.rhisph.2018.06.010.

Ahmadi, K., Zarebanadkouki, M., Ahmed, M. A., Ferrarini, A., Kazyakov, Y., Kostka, S. J. and Carminati, A. 2017. Rhizosphere engineering: Innovative improvement of root environment. *Rhizosphere* 3:176–184. doi: 10.1016/j.rhisph.2017.04.015.

Ahmed, M. A., Zarenbanadkouki, M., Ahmadi, K., Kroener, E., Kostka, S., Kaestner, A. and Carminati, A. 2018. Engineering rhizosphere hydraulics: Pathways to improve plant adaptation to drought. *Vadose Zone Journal* 17(1):1-12. doi: 10.2136/vzj2016.09.0090.

Ashraf, M. and Fooland, M. R. 2007. Role of glycine betaine and proline in improving plant abiotic stress resistance. *Environmental and Experimental Botany* 59(2):206-216.

Barrett, K. D. 2015. *Fulvic and Humic Acids: Chemical Composition, Soil Applications and Ecological Effects*. Nova Science Publishers, Hauppauge, NY.

Bero, N. J. and Soldat, D. J. 2021. Effect of topdressing of compost, compost tea, and various fertilizers on soil and lawn characteristics. *International Turfgrass Society Research Journal* 14:110-120.

Bias, H. P., Park, S. W., Weir, T. L., Callaway, R. M. and Vivanco, J. M. 2004. How plants communicate using the underground information superhighway. *Trends in Plant Science* 9(1):26-32.

Bias, H. P., Weir, T. L., Perry, L. G., Gilroy, S. and Vivanco, J. M. 2006. The role of root exudates in rhizosphere interactions with plants and other organisms. *Annual Review of Plant Biology* 57:233-266.

Bigelow, C. 2022. Rooted in research: Plants as architects of soil. *TPI Turf News* 46(3):54-55.

Bigelow, C. A., Bowman, D. C. and Wollum, A. G. 2002. Characterization of soil microbial population dynamics in newly constructed sandbased rootzones. *Crop Science* 42(5):1611-1614.

Bigelow, C., Ervin, E. H., Zhang, X. and Nichols, A. 2010. Creeping bentgrass putting green cultivation recovery as affected by pre-stress conditioning with liquid fertilizer and biostimulant programs in the cool-humid region. *Applied Turfgrass Science*. doi: 10.1094/ATS-2010-0727-01-BR

Bigelow, C., Fidanza, M., Ervin, E. and Zhang, X. 2021. *Biostimulants and Plant Health Products for Turf Systems: Current and Future Trends*. https://www.biostimulant.com/biostimulants-and-plant-health-products-for-turf-systems/.

Bigelow, C. A. and Soldat, D. J. 2013. Turfgrass root zones: Management, construction methods, amendment characteristics, and use. In: Stier, J. C., Horgan, B. P. and Bonos, S. A. (Eds), *Turfgrass: Biology, Use, and Management. Agronomy Monograph* (vol. 56). CSSA, ASA-SSSA, Madison, WI.

Biotta, A. 2013. Enhancing plant tolerance to temperature stress with amino acids: An approach to their mode of action. *Acta Horticulturae* (1009):29-35. doi: 10.17660/ActaHortic.2013.1009.1.

Brockhoff, S. R., Christians, N. E., Killorn, R. J., Horton, R. and Davis, D. D. 2010. Physical and mineral-nutrition properties of sand-based turfgrass root zones amended with biochar. *Agronomy Journal* 102(6):1627-1631.

Bunnell, B. T., McCarty, L. B. and Bridges Jr., W. C. 2005. Tifeagle Bermudagrass response to growth factors and mowing height when grown at various hours of sunlight. *Crop Science* 42:472-479.

Carminati, A., Moradi, A., Vetterlein, D., Vontobel, P., Lehmann, E., Weller, U., Vogel, H.-J., and Oswald, S.E. 2010. Dynamics of soil water content in the rhizosphere. *Plant and Soil* 332:163-176.

Carminati, A., Schneider, C. L., Moradi, A. B., Zarebanadkouki, M., Vetterlein, D., Vogel, H.-J., Hildebrandt, A., Weller, U., Schüler, L. and Oswald, S. E. 2011. How the rhizosphere may favor water availability to roots. *Vadose Zone Journal* 10(3):988-998.

Carminati, A. and Vetterlein, D. 2013. Plasticity of rhizosphere hydraulic properties as a key for efficient utilization of scarce resources. *Annals of Botany* 112(2):277-290.

Carminati, A., Zarebanadkouki, M., Kroener, E., Ahmed, M. A. and Holz, M. 2016. Biophysical rhizosphere processes affecting root water uptake. *Annals of Botany* 118(4):561-571.

Carrow, R. N. 1993. Eight questions to ask: Evaluating soil and turf conditioners. *Golf Course Management* 61(10):56, 58, 60, 64, 70.

Chen, Z., Wang, Z., Yang, Y., Li, M. and Xu, B. 2018. Abscisic acid and brassinolide combined application synergistically enhances drought tolerance and photosynthesis of tall fescue under water stress. *Scientia Horticulturae* 228:1-9.

Chen, Y. 1996. Organic matter reactions involving micronutrients in soils and their effect on plants. In: Piccolo, A. (Ed.), *Humic Substances in Terrestrial Ecosystems*. Elsevier B.V., Amsterdam.

Cisar, J. L., Williams, K. E., Vivas, H. E. and Haydu, J. J. 2000. The occurrence and alleviation by surfactants of soil water repellency on sand-based turfgrass systems. *Journal of Hydrology* 231-232:352-358.

Clarke, B. B., Vincelli, P., Koch, P. and Munshaw, G. 2020. *Chemical Control of Turfgrass Diseases* (vol. 2020). http://www2.ca.uky.edu/agcomm/pubs/PPA/PPA1/PPA1.pdf.

Cole, D. L. 1999. The efficacy of acibenzolar-s-methyl, an inducer of system acquired resistance, against bacterial and fungal diseases of tobacco. *Crop Protection* 18(4):267-273.

Craigie, J. S. 2011. Seaweed extract stimuli in plant science and agriculture. *Journal of Applied Phycology* 23(3):371-393.

Crouch, I. J. and Van Staden, J. 1993. Evidence for the presence of plant growth regulators in commercial seaweed products. *Plant Growth Regulation* 13(1):21-29.

Datnoff, L. E. and Rutherford, B. A. 2003. Accumulation of silicon by Bermudagrass to enhance disease suppression of leaf spot and melting out. *USGA Turfgrass and Environmental Research Online* 2(18):1-6.

De Castro, E. B., McCurdy, J. D., Caputol, T. M., Brosnan, J., McCullough, P. E., Patton, A. J. and Carroll, D. E. 2020. Effects of temperature and gibberellic acid on the germination of Poa annua seeds in maintained turfgrass of Mississippi. *Proceedings of the Southern Weed Science Society* (vol. 73), p. 36.

Dekker, L. W., Ritsema, C. J., Oostindie, K., Wesseling, J. G. and Geissen, V. 2019. Effects of a soil surfactant on grass performance and soil wetting of a fairway prone to water repellency. *Geoderma* 338:481-492.

Dempsey, J. J., Wilson, I. D., Spencer-Phillips, P. T. N. and Arnold, D. L. 2012. Suppression of *Microdochium nivale* by potassium phosphite in cool-season turfgrasses. *Acta Agriculturae Scandinavica, Section B - Soil and Plant Science* 62(sup1)(Suppl.1):70-78.

du Jardin, P. 2015. Plant biostimulants: Definition, concept, main categories and regulation. *Scientia Horticulturae* 196:3-14. doi: 10.1016/j.scienta.2015.09.021.

Duddek, P., Ahmed, M., Zarebanadkouki, M., Koebernick, N., Ohmann, L., Lovric, G. and Carminati, A. 2020. *Root Hairs Bridge the Gap between Roots and Soil Water*. https://presentations.copernicus.org/EGU2020/EGU2020-11660_presentation.pdf.

Duddek, P., Carminati, A., Koebernick, N., Ohmann, L., Lovric, G., Delzon, S., Rodriguez-Dominguez, C. M., King, A. and Ahmed, M. A. 2022. The impact of drought-induced root and root hair shrinkage on root-soil contact. *Plant Physiology* 189(3):1232-1236.

Durner, J., Shah, J. and Klessig, D. F. 1997. Salicylic acid and disease resistance in plants. *Trends in Plant Science* 2(7):266-274.

Ervin, E. H. 2013a. Cytokinins and seaweed extracts for summer putting green health. *Golfdom* 69(4):36-39.

Ervin, E. H. 2013b. Plant health products: An overview of how they might boost summer stress tolerance. *The Commonwealth Crier*:9-11.

Ervin, E. H. and Roberts, J. C. 2007. Improving root development with foliar humic acid applications during Kentucky bluegrass sod establishment on sand. *Acta Horticulturae* 783:317-322. doi: 10.17660/ActaHortic.2008.783.33.

Ervin, E. H., Zhang, X. and Fike, J. H. 2004. Ultraviolet-B radiation damage on Kentucky bluegrass II: Hormone supplement effects. *HortScience* 39(6):1471-1474.

Esmaeili, S., Salehi, H. and Eshghi, S. 2015. Silicon ameliorates the adverse effects of salinity on turfgrass growth and development. *Journal of Plant Nutrition* 38(12):1885-1901.

Fidanza, M. 2015. Understanding and controlling fairy ring. *Pennsylvania Turfgrass* 4(1):10-16.

Fidanza, M., Cisar, J. L., Kostka, S. J., Gregos, J. S., Schlossberg, M. J. and Franklin, M. 2007. Preliminary investigation of soil chemical and physical properties associated with type-I fairy ring symptoms in turfgrass. *Hydrological Processes* 21(17):2285-2290.

Fidanza, M., Cisar, J. and Watschke, T. 2015. A perspective on biostimulants and sportsturf. *SportsTurf* 31(6):8-10.

Fidanza, M., Kostka, S. and Bigelow, C. 2020. Communication of soil water repellency causes, problems, and solutions of intensively managed amenity turf from 2000 to 2020. *Journal of Hydrology and Hydromechanics* 68(4):306-312.

Fidanza, M., Kostka, S., Ervin, E. and Bigelow, C. 2019. The European Union's view on biostimulants: What may be coming our way. *Golf Course Management* 87(9):58-62.

Gan, L., Zhang, X., Liu, S. and Yin, S. 2018. Mitigating effect of glycinebetaine pretreatment on drought stress responses of creeping bentgrass. *HortScience* 53(12):1842-1848.

Gaussoin, R. E. and Branham, B. E. 1987. Annual bluegrass and creeping bentgrass germination response to flurprimidol. *Horticulturalscience* 23:178-179.

Gaussoin, R. E., Branham, B. E. and Flore, J. A. 1997. Carbon dioxide exchange rate and chlorophyll content of turfgrasses treated with flurprimidol or mefluidide. *Journal of Plant Growth Regulation* 16(2):73-78.

Gelernter, W. D., Stowell, L. J., Johnson, M. E., Brown, C. D. and Beditz, J. F. 2015. Documenting trends in water use and conservation practices on U.S. golf courses. *Crop, Forage and Turfgrass Management* 1(1):1-10.

Gregory, P. J., George, T. S. and Paterson, E. 2022. New methods for new questions about rhizosphere/plant root interactions. *Plant and Soil* 476(1-2):699-712.

Guertal, E. A. and Datnoff, L. E. 2021. Silicon in turfgrass: A review. *Crop Science* 61(6):3861-3876.

Guo, N., Zhang, S., Gu, M. and Xu, G. 2021. Function, transport, and regulation of amino acides: What is missing in rice? *The Crop Journal* 9(3):530-542.

Haberer, G. and Keiber, J. J. 2002. Cytokinins: New insight into a classic phytohormone. *Plant Physiology* 128(2):354-362.

Hale, L., Curtis, D., Azeem, M., Montgomery, J., Crowley, D. E. and McGriffen, M. E. 2021. Influence of compost and biochar on soil biological properties under turfgrass supplied deficit irrigation. *Applied Soil Ecology* 168:104134. doi: 10.1016/j.apsoil.2021.104134.

Hallett, P. D., Gordon, D. C. and Bengouh, A. G. 2003. Plant influence on rhizosphere hydraulic properties: Direct measurements using a miniaturized infiltrometer. *The New Phytologist* 157(3):597–603.

Hallett, P. D., Marin, M., Bending, G. D., George, T. S., Collins, C. D. and Otten, W. 2022. Building soil sustainability from root-soil interface traits. *Trends in Plant Science* 27(7):688–698.

Havlin, J. L. and Schlegel, A. J. 2021. Review of phosphite as a plant nutrient and fungicide. *Soil Systems* 5(3):52. doi: 10.3390/soilsystems5030052.

Heckman, N. L., Gaussoin, R. E. and Horst, G. L. 2001a. Multiple trinexapac-ethyl applications reduce Kentucky bluegrass sod storage temperatures. *hortTechnology* 11(4):595–598.

Heckman, N. L., Horst, G. L. and Gaussoin, R. E. 2001b. Influence of trinexapacethyl on specific leaf weight and chlorophyll content of *Poa pratensis*. *Internatioal Turfgrass Society Research Journal* 9:287–290.

Heckman, N. L., Horst, G. L., Gaussoin, R. E. and Frank, K. W. 2001c. Storage and handling characteristics of trinexapac-ethyl treated Kentucky bluegrass sod. *Hortscience* 36(6):1127–1130.

Heckman, N. L., Horst, G. L., Gaussoin, R. E. and Young, L. J. 2001d. Heat tolerance of Kentucky bluegrass as affected by trinexapac-ethyl. *Hortscience* 36(2):365–367.

Hiltner, L. 1904. Ueber neuere Erfahrungen und probleme auf dem Gebiete der bodenbakteriologie und unter besonderer berucksichtigung der grundungung und brache. *Arb. Deut. Landw. Gesell.* 98:59–78.

Hou, J., Aydemir, B. E. and Dumanli, A. G. 2021. Understanding the structural diversity of chitins as a versatile biomaterial. *Philosphophical Transactions of the Royal Society a - Mathematical, Physical and Engineering Sciences* 379:20200331. doi: 10.1098/rsta.2020.0331.

Hsiang, T., Goodwin, P. H. and Cortes, A. M. 2011. Plant defense activators and control of turfgrass diseases. *Outlooks on Pest Management* 22(4):160–164.

Huang, C., Tian, Y., Zhang, B., Hassan, M. J., Li, Z. and Zhu, Y. 2021a. Chitosan (CTS) alleviates heat-induced leaf senescence in creeping bentgrass by regulating chlorophyll metabolism, antioxidant defense, and the heat shock pathway. *Molecules* 26(17):5337.

Huang, X., Amee, M. and Chen, L. 2021b. Bermudagrass *CdWRKY50* gene negatively regulates plants' response to salt stress. *Environmental and Experimental Botany* 188:1–11.

Hunter, A. and Anders, A. 2004. The influence of humic acid on turfgrass growth and development of creeping bentgrass. Nektarios, P. A. (Ed.) International Conference on Turfgrass Management and Science for Sports Fields. *ISHS Acta Horticulturae* 661:257–264. doi: 10.17660/ActaHortic.2004.661.32.

Iqbal, N., Khan, N. A., Ferrante, A., Trivellini, A., Francini, A. and Khan, M. I. R. 2017. Ethylene role in plant growth, development and senescence: Interaction with other phytohormones. *Frontiers in Plant Science* 8:475.

Jaulneau, V., Lafitte, C., Jacquet, C., Fournier, S., Salamagne, S., Briand, X., Esquerré-Tugayé, M. T. and Dumas, B. 2010. Ulvan, a sulfated polysaccharide from green algae, activates plant immunity through the jasmonic acid signaling pathway. *Journal of Biomedicine and Biotechnology* 2010:525291. doi: 10.1155/2010/525291.

Jespersen, D., Yu, J. and Huang, B. 2017. Metabolic effects of acibenzolar-s-methyl for improving heat or drought stress in creeping bentgrass. *Frontiers in Plant Science* 8:1224.

Juska, F. V. 1958. Some effects of gibberellic acid on turfgrasses. *USGA Journal and Turf Management* July issue:25–28.
Kauffman, G. L., Kneivel, D. P. and Watschke, T. L. 2007. Effects of a biostimulant on the heat tolerance associated with photosynthetic capacity, membrane thermostability, and polyphenol production of perennial ryegrass. *Crop Science* 47(1):261–267.
Kessman, H., Oostendorp, M., Ruess, W., Staub, T., Kunz, W. and Ryals, J. 1996. Systemic activated resistance - a new technology for plant disease control. *Pesticide Outlook* 7:10–13.
Khan, W., Rayirath, U. P., Subramanian, S., Jithesh, M. N., Rayorath, P., Hodges, D. M., Critchley, A. T., Craigie, J. S., Norrie, J. and Prithiviraj, B. 2009. Seaweed extracts as biostimulants of plant growth and development. *Journal of Plant Growth Regulation* 28(4):386–399.
Kostka, S. and Fidanza, M. 2017. Biostimulants, revisited. *SportsTurf* 33(9):8–13.
Kostka, S. and Fidanza, M. 2018. The quagmire that is soil surfactants in golf and sports turf management. ASA, CSSA and SSSA international annual meetings. Agronomy Abstracts, p. 113267.
Kostka, S. and Fidanza, M. 2019. *Soil Surfactant Usage based on Solid Science*. Golf Course Industry. https://www.golfcourseindustry.com/article/soil-surfactant-research-guidance/.
Kostka, S. J. 2000. Amelioration of water repellency in highly managed soils and the enhancement of turfgrass performance through the systematic application of surfactants. *Journal of Hydrology* 231–232:359–368.
Kostka, S. J., Cisar, J. L., Mitra, S., Park, D. M., Ritsema, C. J., Dekker, L. W. and Franklin, M. A. 2007. *Irrigation Efficiency: Soil Surfactants Can Save Water and Help Maintain Turfgrass Quality*. Golf Course Industry. https://www.golfcourseindustry.com/article/irrigation-efficiency--research-/.
Landschoot, P. and Cook, J. 2005. Sorting out the phosphonate products. *Golf Course Management* 73(11):73–77.
Laskowski, K. and Merewitz, E. 2020. Ethylene regulatory treatment effects on annual bluegrass survival of freezing temperature and ice cover. *Agronomy Journal* 112(2):861–870.
Lee, J., Fry, J. and Tisserat, N. 2003. Dollar Spot in Four Bentgrass Cultivars as Affected by Acibenzolar-s-methyl and organic fertilizers. *Plant Health Progress* 4(1):1–4. doi: 10.1094/PHP-2003-0626-01-RS
Li, F., Zhan, D., Xu, L., Han, L. and Zhang, X. 2014. Antioxidant and hormone responses to heat stress in two Kentucky bluegrass cultivars contrasting in heat tolerance. *Journal of the American Society for Horticultural Science* 139(5):587–596.
Li, L., Chestnut, E., Carlson, M., Kreuser, W. and Gaussoin, R. 2020. Field evaluation of preemergence activity of plant growth regulators on annual bluegrass. *Crop, Forage and Turfgrass Management* 6(1). doi: 10.1002/cft2.20073.
Liu, C., Cooper, R. J. and Fisher, D. S. 1998. Influence of humic substances on rooting and nutrient content of creeping bentgrass. *Crop Sci.ence* 38(6):1639--1644.
Liu, X., Huany, B. and Banowetz, G. 2002. Cytokinin effects on creeping bentgrass response to heat stress. I. Shoot and root growth. *Crop Science* 42(2):457–465.
Lugtenberg, B. and Kamilova, F. 2009. Plant-growth-promoting rhizobacteria. *Annual Review of Microbiology* 63:541–556.
Marvin, J. W., Kerr, R. A., McCarty, L. B., Bridges, W., Martin, S. B. and Wells, C. E. 2020. Curative evaluation of biological control agents and synthetic fungicides for *Clarireedia jacksonii*. *Hortscience* 55(10):1622–1625.

Mauch-Mani, B. and Mauch, F. 2005. The role of abscisic acid in plant-pathogen interactions. *Current Opinion in Plant Biology* 8(4):409–414.

Mauseth, J. D. 2019. *Botany: An Introduction to Plant Biology* (7th ed.). Jones and Bartlett Publishers Learning, Burlington, MA.

McCoy, R. M., Meyer, G. W., Rhodes, D., Murray, G. C., Sors, T. G. and Widhalm, J. R. 2020. Exploratory study on the foliar incorporation and stability of isotopically labeled amino acids applied to turfgrass. *Agronomy* 10(3):358. doi: 10.3390/agronomy10030358.

McNear, Jr., D. H. 2013. The rhizosphere - roots, soil and everything in between. *Nature Education Knowledge* 4(3):1.

Mertz, I., Christians, N., Ervin, E. H. and Zhang, X. 2017. Physiological responses of creeping bentgrass (Agrostis stolonifera L.) to a tryptophan-containing organic byproduct. *International Turfgrass Society Research Journal* 13:575–583.

Mertz, I. T., Christians, N. E. and Thoms, A. W. 2019. Utilizing branched-chain amino acids for increasing shoot density and establishment rate in creeping bentgrass. *hortTechnology* 30(1):102–106.

Mickelbart, M. V. and Boine, B. 2020. Glycinebetaine enhances osmostic adjustment of ryegrass under cold temperature. *Agronomy* 10(2):210. doi: 10.3390/agronomy10020210.

Mitra, S., Vis, E., Kumar, R., Plumb, R. and Mam, M. 2006. Wetting agent and cultural practices increase infiltration and reduce runoff losses of irigation water. *Biologia* 61(S19):S353–S357.

Moody, D., Schlossberg, M. J. and Fidanza, M. 2007. Does the systematic application of a molasses-based biostimulant influence organic matter decomposition? Abstracts of the International Annual Meetings of Asa. CSSA/SSSA. https://a-c-s.confex.com/a-c-s/2007am/techprogram/P37305.HTM.

Moore, D., Kostka, S. J., Boerth, T. J., Franklin, M., Ritsema, C. J., Dekker, L. W., Oostindie, K., Stoof, C. and Wesseling, J. 2010. The effect of soil surfactants on soil hydrological behavior, the plant growth environment, irrigation efficiency and water conservation. *Journal of Hydrology and Hydromechanics* 58(3):142–148.

Mueller, S. R. and Kussow, W. R. 2005. Biostimulant influences on turfgrass microbial communities and creeping bentgrass putting green quality. *HortScience* 40(6):1904–1910.

Nabati, D. A., Schmidt, R. E., Parish, D. J. 1994. Alleviation of salinity stress in Kentucky bluegrass by plant growth regulators and iron. *Crop Science* 43:198–202.

Patton, A. J., Schortgen, G. P., Hoyle, J. A., Harrell, M. S. and Reicher, Z. J. 2018. Fall applications of Proxy (ethephon) suppress spring seedheads of "Meyer" zoysiagrass. *Crop, Forage and Turfgrass Management* 4(1):1–8.

Peppers, J. M., Brewer, J. R. and Askew, S. D. 2021. Plant growth regulator and low-dose herbicide programs for annual bluegrass seedhead suppression in fairway and athletic-height turf. *Agronomy Journal* 113(5):3800–3807.

Philippot, L., Raaijmakers, J. M., Lemanceau, P. and van der Putten, W. H. 2013. Going back to the roots: The microbial ecology of the rhizosphere. *Nature Reviews. Microbiology* 11(11):789–799.

Pinton, R., Varanini, Z. and Paolo, N. 2007. *The Rhizosphere: Biochemistry and Organic Substances at the Soil-Root Interface*. CRC Press, Boca Raton, FL.

Radkowski, A., Rodkowska, I., Bocianowski, J., Sladkovska, T. and Wolski, K. 2020. The effect of foliar application of an amino acid-based biostimulant on lawn functional value. *Agronomy* 10(11):1656. doi: 10.3390/agronomy10111656.

Ricci, M., Tilbury, L., Daridon, B. and Sukalac, K. 2019. General principles to justify plant biostimulant claims. *Frontiers in Plant Science* 10:494.

Rosier, A., Medeiros, F. H. V. and Bias, H. P. 2018. Defining plant growth promoting rhizobacteria molecular and biochemical networks in beneficial plant-microbe interactions. *Plant and Soil* 428(1-2):35-55.

Salisbury, F. B. and Ross, C. W. 1991. *Plant Physiology* (4th ed.). Wadsworth Publishing Company, Belmont, CA.

Sanderson, K. J. and Jameson, P. E. 1986. The cytokinins in a liquid seaweed extract could they be the active ingredients? *Acta Horticulturae* 179:113-116.

Sanderson, K. J., Jameson, P. E. and Zabkiewicz, J. A. 1987. Auxin in a seaweed extract identification and quantitation of indole-3-acetic acid by gas chromatography-mass spectrometry. *Journal of Plant Physiology* 129:363-367.

Sangha, J. S., Kelloway, S., Critchley, A. T. and Prithiviraj, B. 2014. Seaweeds (macroalgae) and their extracts as contributors of plant productivity and quality. *Advances in Botanical Research* 71:189-219.

Schlatter, D., Kinkel, L., Thomashow, L., Weller, D. and Paulitz, T. 2017. Disease suppressive soils: New insights from the soil microbiome. *Phytopathology* 107(11):1284-1297.

Schmidt, R. E. and Chalmers, D. R. 1993. Late summer to early fall applications of fertilizer and biostimulants on bermudagrass. *International Turfgrass Society Research Journal* 7:715-721.

Schmidt, R. E. and Zhang, X. 1998. How humic substances help turfgrass grow. *Golf Course Management* 66(7):65-67.

Schmidt, R. E. and Zhang, X. 2001. Alleviation of photochemical activity decline of turfgrasses exposed to soil moisture stress or UV radiation. *International Turfgrass Society Research Journal* 9:340-346.

Schmidt, R. E., Zhang, X. and Chalmers, D. R. 1999. Response of photosynthesis and superoxidide dismutase to silica applied to creeping bentgrass grown under two fertility levels. *Journal of Plant Nutrition* 22(11):1763-1773.

Schwechheimer, C. 2008. Understanding gibberellic acid signaling - Are we there yet? *Current Opinion in Plant Biology* 11(1):9-15.

Shamshina, J. L., Berton, P. and Rogers, R. D. 2019. Advances in functional chitin materials: A review. *ACS Sustainable Chemistry and Engineering* 7(7):6444-6457.

Singh, R., Glick, B. R. and Rathore, D. 2018. Biosurfactants as a biological tool to increase micronutrient availability in soil: A review. *Pedosphere* 28(2):170-189.

Soldat, D. J., Lowery, B. and Kussow, W. R. 2010. Surfactants increase uniformity of soil water content and reduce water repellency on sand-based golf putting greens. *Soil Science* 175(3):111-117.

Spokas, K. A., Cantrell, K. B., Novak, J. M., Archer, D. W., Ippolito, J. A., Collins, H. P., Boateng, A. A., Lima, I. M., Lamb, M. C., McAloon, A. J., Lentz, R. D. and Nichols, K. A. 2012. Biochar: A synthesis of its agronomic impact beyond carbon sequestration. *Journal of Environmental Quality* 41(4):973-989.

Stadnik, M. J. and Freitas, M. Bd. 2014. Algal polysaccharides as a source of plant resistance inducers. *Tropical Plant Pathology* 39(2):111-118.

Sun, H., Schmidt, R. E. and Eisenback, J. D. 1997. The effect of seaweed concentrate on the growth of nematode-infected bent grown under low soil moisture. *International Turfgrass Society Research Journal* 8:1336–1342.

Swamy, P. M. and Smith, B. N. 1999. Role of abscisic acid in plant stress tolerance. *Current Science* 76:1220–1227.

Tan, Z. G. and Qian, Y. L. 2003. Light intensity affects gibberellic acid content in Kentucky bluegrass. *Hortscience* 38(1):113–116.

Vaughn, S. F., Dinelli, F. D., Tisserat, B., Joshee, N., Vaughan, M. M. and Peterson, S. C. 2015. Creeping bentgrass growth in sand-based root zones with or without biochar. *Scientia Horticulturae* 197:592–596.

Waddington, D. V. 1992. Soils, soil mixtures, and soil amendments. In: Waddington, D. V., Carrow, R. N. and Shearman, R. C. (Eds). *Turfgrass. Agronomy Monograph* (vol. 32). CSSA, ASA-SSSA, 331–383.

Wang, K., Zhang, X. and Ervin, E. 2012. Antioxidative responses in roots and shoots of creeping bentgrass under high temperature: Effects of nitrogen and cytokinin. *Journal of Plant Physiology* 169(5):492–500.

Wang, K., Zhang, X. and Ervin, E. 2013. Effects of nitrate and cytokinin on creeping bentgrass under supraoptimal temperatures. *Journal of Plant Nutrition* 36(10):1549–1564.

Wang, K., Zhang, X., Goatley, M. and Ervin, E. 2014. Heat shock proteins in relation to heat stress tolerance of creeping bentgrass at different N levels. *PLoS ONE* 9(7):e102914.

Wang, X. J., Wang, Z. Q. and Li, S. G. 1995. The effect of humic acids on the availability of phosphorus fertilizers in alkaline soils. *Soil Use and Management* 11(2):99–102.

Wang, Z., Huang, B. and Xu, Q. 2003. Effects of abscisic acid on drought responses of Kentucky bluegrass. *Journal of the American Society for Horticultural Science* 128(1):36–41.

Webb, S. 2022. *Experts Study How to Stop Nematodes in Their Tracks*. Golfdom [Online]. https://www.golfdom.com/experts-study-how-to-stop-nematodes-in-their-tracks/.

Weil, R. R. and Brady, N. C. 2017. *The Nature and Properties of Soils* (15th ed.). Pearson Education, New York.

Wen, Z., White, P. J., Shen, J. and Lambers, H. 2022. Linking root exudation to belowground economic traits for resource acquisition. *New Phytologist* 233(4):1620–1635.

Wilkinson, S. and Davies, W. J. 2002. ABA-based chemical signaling: The coordination of responses to stress in plants. *Plant, Cell and Environment* 25(2):195–210.

Wright, A. L., Provin, T. L., Hons, F. M., Zuberer, D. A. and White, R. H. 2007. Nutrient accumulation and availability in compost-amended turfgrass soil. *Hortscience* 42(6):1473–1477.

Xu, Y. and Huang, B. 2009. Effects of foliar-applied ethylene inhibitor and synthetic cytokinin on creeping bentgrass to enhance heat tolerance. *Crop Science* 49(5):1876–1884.

Yakhin, O. I., Lubyanov, A. A., Yakhin, I. A. and Brown, P. H. 2017. Biostimulants in plant science: A global perspective. *Frontiers in Plant Science* 7:2040.

Yan, N. and Chen, X. 2015. Sustainability: Don't waste seafood waste. *Nature* 524(7564):155–157.

Yang, Q., Zhao, D. and Liu, Q. 2020. Connections between amino acid metabolism in plants: Lysine as an example. *Frontiers in Plant Science* 11:928.

Zeller, W., Tubana, B., Rodrigues, F. A. and Datnoff, L. E. 2021. Silicon's role in plant stress reduction and why this element is not used routinely for managing plant health. *Plant Disease* 105(8):2033–2049.

Zhang, X. and Ervin, E. H. 2004. Cytokinin-containing seaweed and humic acid extracts associated with creeping bentgrass leaf cytokinins and drought resistance. *Crop Science* 44(5):1737–1745.

Zhang, X. and Ervin, E. H. 2008. Impact of seaweed extract-based cytokinins and zeatin riboside on creeping bentgrass heat tolerance. *Crop Science* 48(1):364–370.

Zhang, X., Ervin, E. H. and Schmidt, R. E. 2003. Seaweed extract, humic acid, and propiconazole improve tall fescue sod heat tolerance and posttransplant quality. *Hortscience* 38(3):440–443.

Zhang, X., Ervin, E. H. and Schmidt, R. E. 2013a. Antioxidant enzyme and photosynthesis responses to cold acclimation in two zoysiagrass cultivars. *International Turfgrass Society Research Journal* 12:437–443.

Zhang, X., Summer, P. and Ervin, E. H. 2013b. Foliar amino acid-based fertilizer impact on creeping bentgrass drought resistance. *International Turfgrass Society Research Journal* 12:429–436.

Zhang, X., Ervin, E. H., Wu, W., Sharma, N. and Hamill, A. 2017. Auxin and trinexapac-ethyl impact on root viability and hormone metabolism in creeping bentgrass under water deficit. *Crop Science* 57(S1):130–137.

Zhang, X. and Schmidt, R. E. 1997. The impact of growth regulators on the α-tocopherol status in water-stressed *Poa pratensis*. *International Turfgrass Society Research Journal* 8:1364–1371.

Zhang, X. and Schmidt, R. E. 1999. Antioxidant response to hormone-containing products in Kentucky bluegrass subjected to drought. *Crop Science* 39(2):545–551.

Zhang, X. and Schmidt, R. E. 2000. Hormone-containing products' impact on antioxidant status of tall fescue and creeping bentgrass subjected to drought. *Crop Science* 40(5):1344–1349.

Zhang, X., Zhou, D., Ervin, E. H., Evanylo, G. K., Cataldi, D. and Li, J. 2012. Biosolids impact antioxidant metabolism associated with drought tolerance in tall fescue. *HortScience* 47(10):1550–1555.

Zontek, S. J. and Kostka, S. J. 2012. Understanding the different wetting agent chemistries. *USGA Green Section Record* 50(15):1–6.

Zuberer, D. 2012. Soil microbes: Some practical perspectives for turfgrass systems. *USGA Green Section Record* 50(15):1–5.

Part 3

Case studies

Chapter 16

Considerations with using unmanned aircraft systems in turfgrass

Dale J. Bremer, Kansas State University, USA; Dana G. Sullivan, TurfScout, LLC, USA; Phillip L. Vines, University of Georgia, USA; David McCall, Virginia Polytechnic Institute and State University, USA; Jing Zhang, University of Georgia, USA; and Mu Hong, Colorado State University, USA

1 Introduction

Remote sensing is the practice of obtaining information about an object without coming into physical contact with that object. Remote sensing-based observations have been the subject of agricultural research for well over 50 years, accompanied by significant advancements in foundational research and accessible technology. During that time, sensors have been deployed on platforms based on the ground (i.e. handheld or vehicle-mounted) and in the air (i.e. manned aircraft or satellites). But in recent years, small unmanned aircraft systems (sUAS) have provided an alternative and more affordable means for remote data collection. Today, there are a wide variety of sUAS platforms ranging in size and operation (e.g. fixed-wing vs. multi-rotor). Along with the growing choice of platforms, more sensors (i.e. payloads) are becoming available, representing a range in spatial and spectral resolutions.

http://dx.doi.org/10.19103/AS.2022.0110.16

2 Brief overview of light reflectance by plants

Irradiant solar energy can be reflected, absorbed, or transmitted by plants (Campbell, 1996). Plants absorb light used for photosynthesis, but a majority of the remaining light is reflected. Sensors typically measure the amount of light reflected by the plant canopy, and this provides an assessment of how well a plant utilizes solar energy for photosynthesis and growth. A typical spectral reflectance curve for green vegetation exhibits a small peak in the green region (~550 nm) of the spectrum, followed by a sharp rise in the near-infrared (NIR) region (Figs. 1 and 2). The NIR is primarily influenced by biomass production, canopy geometry, and subsequent light scattering (Hatfield et al., 2008; Sullivan et al., 2004), while the visible spectra are influenced by energy absorption for photosynthesis (Campbell, 1996). Plants with greater photosynthesis absorb more energy (i.e. reflect less) in the visible and may produce more biomass (affecting NIR reflectance) than their counterparts with lower photosynthesis, which typically results in different spectral reflectance response curves. Examples include turfgrass mown at different heights (Fig. 1) and healthy vs. stressed turfgrass (Fig. 2). The visible spectra generally exhibit a much smaller range in reflectance compared to the NIR. The small range within the visible region of 'absorption' limits the visible region's sensitivity to plant stress. Examples of this limitation are often observed when applying

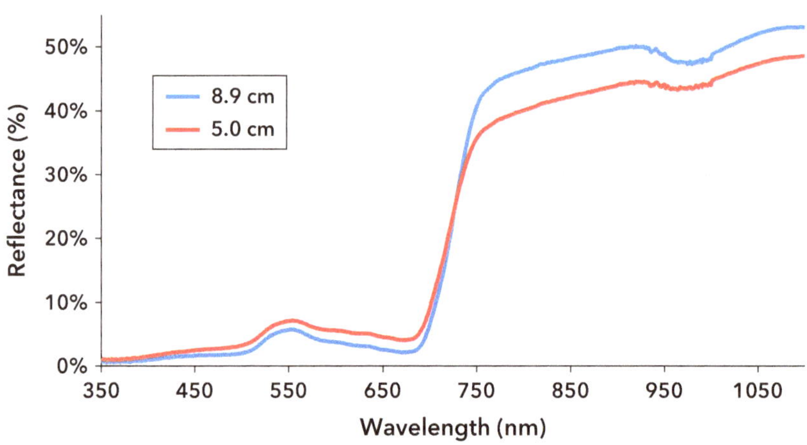

Figure 1 Spectral reflectance signatures of well-watered Kentucky bluegrass (*Poa pratensis* L.) turfgrass mown at two heights. Higher green leaf area index (LAI) at 8.9 cm (LAI = 2.13) than 5.0 cm (LAI = 1.28) resulted in lower reflectance in the visible (400–700 nm) and higher reflectance in the near-infrared (>780 nm) at 8.9 cm. The dramatic rise in reflectance between approximately 690 nm and 750 nm is the 'red edge'. Reflectance was measured on 28 September 2010 with a portable spectroradiometer (FieldSpec 3, ASD, Boulder, CO, USA) at the Rocky Ford Turfgrass Research Center (Manhattan, KS, USA) (An et al., 2015).

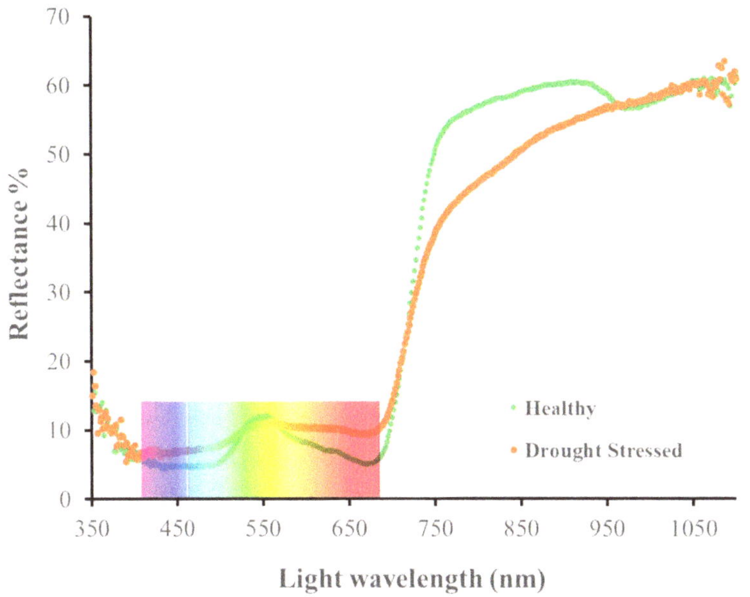

Figure 2 Spectral reflectance signatures of healthy and drought-stressed creeping bentgrass (*Agrostis stolonifera* L.) mowed at tee height (6.35 mm), measured with a portable spectroradiometer (PSR-1100F, Spectral Evolution, Haverhill, MA, USA) in a greenhouse study in Blacksburg, VA, USA. Reflectance is lower in the visible region of the spectrum (denoted by the colors corresponding to their respective wavelengths) because of absorption of energy by photosynthesis, compared with higher reflectance in the near-infrared (>~750 nm). Figure by Travis Roberson.

normalized vegetation indices (discussed in Section 2.1) that include visible spectra.

2.1 Benefits and limitations of vegetation indices

Images are the base product or layer from which additional information can be derived. The most common data layers generated from remote sensing are vegetation indices (VIs), which are mathematical combinations of two or more spectral wavebands and are also referred to as radiometric indices. VIs are designed to target specific plant characteristics such as canopy geometry, chlorophyll content, nutrient status, or water demand to name a few (Tucker, 1979; Sullivan et al., 2004, 2007; Gitelson et al., 2006). The earliest VIs were combinations of NIR and red band ratios with the specific intent to lessen the impact of variability in atmospheric conditions during the time of flight, as well as to separate the plant spectral response from the soil background (Hatfield et al., 2008). Specifically, VIs leveraged differences in soil and plant spectra to isolate plant spectral response prior to canopy closure in row crop agriculture.

This was needed because early measurements were typically obtained via satellite imagery (30 m^2 pixel resolution or greater) or sensors aboard manned aircraft (1 m^2 per pixel resolution), which often resulted in a mixture of plant and soil within any one pixel (Sullivan et al., 2007). In more recent years, high-resolution sUAS-based imaging has allowed practitioners to more easily segment images into plant and soil components for analysis.

One of the most well-known VIs, the normalized difference vegetation index (NDVI) (Rouse et al., 1974), is a normalized ratio of NIR and red spectral bands. The NDVI has been, in practice, a means to evaluate nitrogen and chlorophyll content, predict yields, and provide a measure of overall plant health status for more than five decades (Fenstermaker-Shaulis et al., 1997; Bell et al., 2004; Baghzouz et al., 2006, 2007; Sullivan and Holbrook, 2007; Caturegli et al., 2016, 2019). However, there are strengths and limitations associated with NDVI applications, particularly as applied to turfgrass. First, the NDVI is not linear with respect to plant response (Ritchie and Bednardz, 2005). The NDVI is a normalized measurement of NIR and red spectral response, but low ranges in red reflectance limit the sensitivity of this index over the dense canopy. Second, a basic assumption of the NDVI is that the spectral reflectance curve is that of a typical living plant, where reflectance in the NIR can be ≥70% higher than reflectance in the visible region of the light spectrum (Figs. 1 and 2). However, the topical application of some pigmented products (e.g. turf colorants, pesticides with pigmented additives) has been shown to greatly influence the spectral response patterns in turfgrass, which impacts the interpretation of NDVI when NIR wavelengths between 730 nm and 850 nm are used. In these cases, it is recommended to use alternative indices such as the green-to-red ratio index or photochemical reflectance index (PRI) that utilize regions of the light spectrum unaffected by the pigment but remain sensitive to plant status (McCall et al., 2021).

Other considerations include turfgrass cultural practices that may affect NDVI such as turfgrass species, mowing height, and sand topdressing, because these practices impact canopy geometry, shading, and surface features (Bremer et al., 2011a; Lee et al., 2011; An et al., 2015; Alvarez et al., 2016). Considering the impact of cultural practices on the response and interpretation of data, consistency in data collection, flight planning, and analysis can mitigate variability in spectral response associated with those cultural practices.

3 Critical mission planning

Important considerations must be made prior to initiating data collection protocols using sUAS in turfgrass. These considerations should clearly (i) identify the target of interest; (ii) specify resolution requirements to adequately assess the target of interest; (iii) determine the appropriate sensors to be used;

(iv) identify an aircraft suitable for mounting the desired sensors; (v) determine appropriate flying altitudes, image overlapping settings, and travel speeds to achieve the desired spatial resolution; and (vi) identify the frequency at which data should be collected. Once these parameters are defined, planning and implementation of flights may proceed.

3.1 Defining the target of interest

It is critically important to accurately identify the target of interest as the first step in planning for data collection, as this greatly influences other parameters. For many situations, turfgrass response to various abiotic and biotic stresses will be the target of interest. Examples include monitoring turfgrass performance during abiotic stresses such as drought, heat, cold, salinity, herbicide application, or traffic, as well as monitoring turfgrass performance during biotic stresses such as disease infection or insect feeding. However, the target of interest does not exclusively have to be turfgrass response. For certain missions, investigators may focus on quantifying disease severity or weed pressure, for example. In these situations, the targets of interest would be identifying specific disease symptoms or weed species present.

Consider a situation (scenario 1) where an investigator is interested in collecting data on small plot (0.9 m × 1.5 m) turfgrass research trials to monitor dollar spot (caused by *Clarireedia* spp.) disease activity. Alternatively, consider a second situation (scenario 2) where an investigator is interested in collecting data to identify large-scale drought stress patterns across a 40 ha golf course facility. The appropriate flight plans for these two situations will be extremely different from each other.

For scenario 1, there are at least two targets of interest that could be defined for the mission. The investigator could identify turfgrass performance as the target of interest and plan a mission to assess percent green cover (PGC) within a given area. However, the investigator could also identify dollar spot infection centers as the target of interest and plan a mission to quantify the number of infection centers within a given area. Both approaches are logical, and either could work for this scenario. Of greatest importance is that the investigator clearly defines the target of interest.

For scenario 2, there are also options available when identifying the target of interest. At this scale, the benefits of high-resolution imagery must be weighed against some logistical constraints, for instance, time to acquire data, number of batteries needed to complete the mission, and most importantly the size of the dataset to be processed (in gigabytes). Once the target resolution has been decided, imagery is most often used as a means to calculate PGC or measure relative differences in stress using common VIs.

3.2 Critical resolution requirements

In the earlier scenarios, the resolution requirement for identifying dollar spot infection centers in small plot research trials will be much higher than that needed to identify large-scale drought stress patterns across a golf course facility. In turn, this will directly influence subsequent decisions for sensor type, aircraft selection, flight altitude, image overlapping, and travel speed, all of which will impact total data collection time. Moreover, higher resolution data will increase the total size of resulting datasets; thus, ground resolution thresholds should be identified to adequately reflect the needs of the mission. For example, spatial resolution requirements should be determined based on the size of the target of interest. If the target of interest (i.e. disease expression, localized dry spot, nutrient stress) is most noticeable at 4 cm in diameter, a spatial resolution of 1–2 cm^2 per pixel may be necessary to accurately capture targets due to image overlapping and distortion concerns (discussed further in Sections 3.5 and 4.3).

3.3 Selection of appropriate sensors

There are various sUAS-mounted sensors available for monitoring turfgrass response, but there is often a single sensor that will be best suited for the task at hand. Many of the tools available at present are optical sensors that capture plant interactions with light. Examples include visible light sensors, spectral sensors, thermal sensors, and fluorescence sensors. There are important advantages and limitations to each of these, and they have been thoroughly reviewed previously (Deery et al., 2014; Li et al., 2014; Fahlgren et al., 2015; Zhang and Zhang, 2018; Chandel et al., 2020; Sangha et al., 2020; Tmušić et al., 2020; Feng et al., 2021).

Sensor options can be divided into four primary categories: visible light, multispectral, hyperspectral, and thermal. Visible light sensors, also called true color or red, green, blue (RGB) sensors, measure light in three wide bands (i.e. red, green, and blue) and are typically used for high-resolution data acquisitions necessary for fine feature detection (e.g. weed identification and disease onset). At the time of this publication, true color sensors are the least costly and provide numerous opportunities unique to turfgrass systems because of the relative uniformity across targeted fields of interest. While there are only three bands to explore, the combinations of digital pixel values within these three bands allow for differentiating approximately 16 million unique colors (Yucky et al., 2021). Many emerging uses of artificial intelligence and machine learning for pest identification utilize only true color images.

Multispectral sensors measure light spectra in discrete bands that are often 20–100 nm in width, while hyperspectral sensors measure light spectra in much smaller (<10 nm) increments that are usually continuous across the spectrum of measurement. For example, a hyperspectral sensor may record spectra in 2 nm increments from 450 nm to 1100 nm, while a multispectral instrument will record spectra in five or more select regions and each region will be 10 or more nanometers wide. These 'regions' of the light spectrum that are monitored are oftentimes referred to as 'bands'. Hyperspectral sensors are designed to collect discrete measurements of reflected energy in hundreds of short bandwidths across the spectrum; bandwidths are typically much narrower than those of multispectral sensors. Hyperspectral data allow scientists to isolate discrete regions of the plant reflectance spectrum that are related to very specific plant attributes (e.g. carotenoids, anthocyanins, lignins, water stress, and disease). Hyperspectral sensors produce more data, which have advantages and disadvantages. Analyzing data with narrow spectral bandwidth may provide opportunities for detecting subtle changes that a multispectral sensor may miss but comes at a cost in both equipment expense and time required for data processing.

Thermal bands are increasingly available as well and allow the user to assess how well a plant canopy can dissipate heat. When transpiration rates are compromised, plants will emit long wave energy as a means to dissipate heat (i.e. sensible heat flux) (Hatfield et al., 2008). It is less efficient than transpiration; thus long wave emittance is positively correlated with plant stress (Bremer and Ham, 1999; Bremer et al., 2001; Sullivan et al., 2007; Peterson et al., 2017).

3.4 Selection of suitable aircraft

Unmanned aircraft systems are typically categorized as either fixed-wing or multicopter aircrafts. These platforms are quite different in terms of payload, run time, maneuverability, initial costs, and maintenance costs (Li et al., 2014). Fixed-wing aircrafts are regarded as having higher payloads, longer run times, and faster travel speeds, meaning they can accommodate more onboard sensors and other data recording instruments, and they can cover more ground surface area in a given time, compared to multicopter aircrafts (Boon et al., 2017). However, operators must be conscious of potential image-blurring issues and ensure that onboard sensors will be compatible with the fast travel speeds of fixed-wing aircrafts. Many fixed-wing aircrafts also require large takeoff and landing areas and do not have the ability to hover in one location. Multicopter aircrafts have the capability to travel at slower speeds and maintain stable speeds at lower altitudes, often giving them an advantage for applications where high spatial resolution data are required (Thamm et al., 2015). However, multicopter aircrafts have lower payloads and shorter flight time capacities than fixed-wing aircrafts (Cai et al., 2014). Single-rotor and fixed-wing hybrid vertical takeoff and landing (VTOL) aircrafts are other options, though to date have been less used. Single-rotor helicopters are slower, with improved longevity and the capacity to carry heavier payloads but are also more expensive and potentially dangerous to operate. The fixed-wing hybrid has VTOL capacity and is an emerging option for improved flight endurance. Given the various differences in sUAS, much thought should be taken for the selection of the appropriate type of aircraft (Li et al., 2014).

3.5 Flight planning

Careful attention should be given to decisions concerning flight altitude, image overlapping, and travel speed. In general, higher flight altitudes will generate lower-resolution data. Nonetheless, more ground surface area can be covered in a given time with higher altitudes. Thus, the tradeoff in data collection time and data resolution must be considered. This underscores the previous statement that it is essential to identify appropriate resolution thresholds for a given target of interest, as there is no added benefit to acquiring higher resolution data than is needed to assess the defined target of interest. Image overlapping is the amount of redundancy between data taken from adjacent viewpoints. In general, the higher the overlapping rate, the higher the final data resolution will be. In most agricultural applications, a minimum frame front and side-lap (i.e. image overlap in the forward and lateral directions) of 70–80% is required for accurate stitching of individual frames to generate a single composite image, known as an orthomosaic. Again, the target resolution must be kept in mind when deciding on an appropriate image overlapping setting. Travel speed is another variable that can impact total data collection time. However, travel speed constraints will generally be determined by the type of aircraft that is being used and the shutter speeds of the onboard sensor equipment. Nonetheless, a higher travel speed will enable faster data collection across a larger ground surface area in a given time.

3.6 Data collection frequency

This is the time interval between flights required to adequately monitor the identified target of interest. The proper collection frequency will largely depend on the dynamics of the target of interest. For example, if the target of interest changes rapidly, such as foliar disease progression or drought stress development, a weekly or daily collection frequency may be required. However, if the target of interest is more stable, such as genetic color, a monthly collection frequency may be adequate. Also keep in mind that certain targets, such as some patch diseases or weed inflorescence, may have a finite period for data collection. Careful planning before this critical window of opportunity will lead to fewer mistakes and wasted efforts due to insufficient data collection. These types of decisions will require input from expert personnel who understand the various targets of interest.

4 Data processing considerations

4.1 Unique characteristics of small unmanned aircraft system-acquired remote sensing data

Unlike satellite and aircraft imagery, sUAS acquire imagery as a steady stream of frames, or images, which are 'stitched' together to produce a final image mosaic, as discussed previously (Section 3.5). The resulting mosaic is a product of platform (aircraft) choice, sensor (payload), and flight planning as well as time of day and atmospheric conditions during data capture. Time of day and atmospheric conditions are critical because, to date, sUAS-mounted sensors are typically passive (i.e. do not have their own light source, as opposed to active sensors) and, thus, are impacted by factors such as sun angle and cloud cover (Campbell, 1996).

4.2 Calibration

Although platform, sensor choice, and flight planning are critical first steps, image pre-processing and calibration significantly impact the digital integrity of the image product. Uncalibrated images represent an 'at-sensor' measurement of reflectance or more simply the amount of reflected light from the vegetation surface that was received by the sensor. At-sensor or uncalibrated measurements are subject to conditions during the instant of data capture such as sun angle (time of day), atmospheric conditions, surface conditions, and canopy geometry. Calibration techniques are designed to adjust uncalibrated values to 'at-target' values by correcting the at-sensor measurement for atmospheric conditions during the time of capture. When successful, calibrated measurements should provide a measurement of plant response that may be compared from flight-to-flight (time-series measurements).

Image calibration is typically accomplished using a combination of camera-specific settings or metadata, ground calibration targets, and paired measurements of downwelling (irradiant) and upwelling (reflected) energy. Downwelling light measurements are typically collected during flight by an upward-facing sensor on the aircraft, while calibration panel measurements are often collected manually pre- and post-flight with the onboard down-facing sensor. Although commonly used, reference panels can considerably impact the calibrated reflectance in both positive and negative ways. Considering that calibration panels are typically used pre- and post-flight, two new sources of error arise if calibration data are not acquired carefully: (i) human error during panel capture and (ii) potential variability in conditions during flight that is not representative of conditions during the time of panel capture. Examples that could negatively affect the representativeness of panel calibrations include (i) partly cloudy conditions and (ii) extended flight times when the angle of incident energy and/or cloud cover may vary significantly. For these reasons, the effective use of calibration panels and incorporation of downwelling irradiance at low altitudes is still the subject of ongoing research (Assmann et al., 2018; Delvapour et al, 2021).

4.3 Image processing packages: research grade versus edge-of-field

Image processing can be placed into two categories: pre-processing and post-processing. During the pre-processing phase, individual frames are mosaicked (stitched) into a single image composite (orthomosaic). It is during this phase that camera-specific settings, calibration data, and even the spatial resolution of the final image product are applied. The scope of the flight mission and the ultimate use of image products will impact image processing decisions. Considerations include overall area, need for real-time assessments, target size and spatial resolution requirements, and research-grade vs. field-grade requirements. For example, an edge-of-field solution, which is intended to provide near-real-time results, may be appropriate for a field-scale mission (e.g. >10 ha). Edge-of-field solutions are appropriate when image turn-around time needs to be fast and spatial resolution requirements are >10 cm^2 per pixel, such as with rapid assessments of drought stress. In this example, the number of key points selected during the stitching process may be greatly reduced and filters may be used to smooth the final image product.

Alternatively, a research-grade approach may be taken when the target area is smaller, spatial resolution requirements are higher (sub-centimeter to <10 cm^2 per pixel), and more rigorous image processing parameters are required to retain the digital integrity of the dataset (e.g. calibration, increased number of key points, reduced filtering/smoothing assumptions, bidirectional reflectance corrections, band to band alignment, geo-registration) (van der Merwe et al., 2020). Ultimately, a research-grade product increases the processing time required and can generate a much larger dataset, given its high spatial resolution requirements.

As a real-world example of how spatial resolution can impact a data product, consider a 38-ha location flown by one of the coauthors. When the as-flown image spatial resolution was 3 cm^2 per pixel, the generated image file size was 2.5 gigabytes. However, when the resolution was decreased to 6 cm^2 per pixel and 10 cm^2 per pixel, the generated image file sizes were only 615 megabytes and 222 megabytes, respectively. Therefore, as spatial resolution requirements of the target of interest increase, the raw file sizes used to generate these images increase by orders of magnitude.

Software service providers often estimate costs based on these parameters as well. Taking all aspects of a flight mission into consideration (elevation, speed, overlap, spatial resolution, number of bands, and temporal resolution) will define the amount of data in gigapixels that are acquired prior to any post-processing costs (time and expense). Gigapixels are determined based on the number of images multiplied by the image pixel width and height, number of bands, and number of flights. Since most cameras have a fixed image pixel count, the number of frames per mission increases/decreases with altitude, speed, and overlap. Although cloud-based image processing solutions are prevalent today, moving large amounts of data via the cloud is still inefficient.

5 Examples of small unmanned aircraft system applications in turfgrass

5.1 Drought assessment and detection in turfgrass

Few studies have investigated drought stress in turfgrass using sUAS. Those few studies have demonstrated that drought stress was successfully detected using spectral or thermal sensors mounted on sUAS (Table 1). In a 3-year study involving deficit-irrigated creeping bentgrass (*Agrostis stolonifera* L.) mowed at a golf course fairway height (15.9 mm), drought stress was detected with six of eight spectral VIs evaluated, which were derived from broadband reflectance in the NIR, green, and blue (e.g. Blue NDVI, NDVI Enhanced2, NIR BlueRatio, GreenBlue; Table 1) (Hong et al., 2019b). Correlations of those six VIs with visual turf quality (TQ) ratings and PGC from ground-based measurements with a digital camera, which were indicators of canopy drought stress, ranged from $r = 0.68$ to 0.87 (TQ) and $r = 0.71$ to 0.92 (PGC). In the same study, the NIR broadband also detected drought stress (TQ, $r = 0.65$ to 0.75; PGC, $r = 0.68$ to 0.84). Interestingly, early drought stress was detected with VIs before decreases in TQ and PGC (due to drought-induced leaf firing) were observed. Specifically, indices that detected early drought stress included Blue NDVI, NDVI Enhanced2, and NIR Blueratio in treatments irrigated at 15% and 30% reference evapotranspiration (ET_o) replacement ($P < 0.05$). However, the most consistently sensitive parameters of sUAS were the GreenBlue VI and the NIR broadband, which detected drought stress > 5 days before decreases in TQ over the 3-year study (Fig. 3). Thus, both reflectance from an individual broadband (NIR) and VIs derived from multiple broadbands demonstrated strong capabilities for detecting drought stress in turfgrass.

As discussed in Section 3.3, VIs calculated from wavebands strictly in the visible region of the spectrum may be produced from sUAS-based RBG images

Table 1 Spectral, red-green-blue (RGB), and thermal measurements and vegetation indices that detected drought stress from small unmanned aircraft systems (sUAS) and ground-based platforms in turfgrass field studies.

Sensor	Index	Description	Relationship with drought stress variables	References
sUAS-based				
Spectral	NDVI	Normalized difference vegetation index: (NIR – red)/(NIR + red)	Leaf relative water content, r = 0.96; Soil moisture r = 0.82 to 0.86.	Caturegli et al. (2020)
			Turf quality (TQ), R^2 = 0.13 to 0.68; Percent green cover (PGC), R^2 = 0.73 to 0.88; (2 species).	Zhang et al. (2019b)
	Blue NDVI	(NIR – blue)/(NIR + blue)	TQ, r = 0.77 to 0.87; PGC, r = 0.83 to 0.92.	Hong et al. (2019b)
	NDVI Enhanced2	(NIR + green – blue)/(NIR + green + blue)	TQ, r = 0.76 to 0.87; PGC, r = 0.82 to 0.92.	Hong et al. (2019b)
	NIR Blueratio	NIR - blue	TQ, r = 0.77 to 0.86; PGC, r = 0.83 to 0.90.	Hong et al. (2019b)
	WBI	Water band index: R_{900}/R_{970}	Soil moisture, r = 0.87 to 0.89; Leaf relative water content, r = 0.98.	Caturegli et al. (2020)
	NDRE	Normalized difference red edge: (NIR – red edge)/(NIR + red edge)	Spearman's rank correlation: TQ, r = 0.60; PGC, r = 0.68; (2 species).	Zhang et al. (2019b)
RGB	GreenBlue	(green – blue)/(green + blue)	TQ, r = 0.68 to 0.86; PGC, r = 0.71 to 0.91.	Hong et al. (2019b)
	VARI	Visible atmospherically resistant index: (green – red)/(green + red – blue)	TQ, R^2 = 0.05 to 0.63; PGC, R^2 = 0.69 to 0.89; (2 species).	Zhang et al. (2019b)
Thermal	Tc	Canopy temperature	TQ, r = −0.77; PGC, r = −0.78.	Hong et al. (2019a)
	Tc – Ta	Canopy and air temperature difference	TQ, r = −0.60; GC, r = −0.58.	Hong et al. (2019a)

Ground-based				
Spectral	NDVI	Normalized difference vegetation index: (NIR − red)/(NIR + red)	TQ, r = 0.87 to 0.90; PGC, r = 0.92 to 0.95.	Hong et al. (2019b)
			Soil volumetric moisture at 0% ET_o treatment, r = 0.54.	Badzmierowski et al. (2019)
	SAVI	Soil-adjusted vegetation index: $(1.0 + L)(R_{830} - R_{660})/(R_{830} + R_{660} + L)$; usually L = 0.5	Non-linear relationship with ET_o treatments (Lower SAVI at bottom ET_o treatments)	Taghvaeian et al. (2013)
RGB	PRI	Photochemical reflectance index: $(R_{531} - R_{570})/(R_{531} + R_{570})$	Color, $R^2 = 0.58$; Tissue moisture content, $R^2 = 0.73$.	Baghzouz et al. (2007)
	GRI	Green-to-red ratio index: R_{550} / R_{670}	Soil volumetric moisture at 0% ET_o treatment, r = 0.54.	Badzmierowski et al. (2019)
Thermal	Tc − Ta	Canopy and air temperature difference	Vapor pressure deficit of non-water-stressed turf, r = 0.64 to 0.88 (2 species).	Haghverdi et al. (2021)
	CWSI	Crop water stress index: $[(Tc - Ta)_{measured} - (Tc - Ta)_{lower}]/[(Tc - Ta)_{upper} - (Tc - Ta)_{lower}]$	Non-linear relationship with ET_o treatments (Lower CWSI at bottom ET_o treatments)	Taghvaeian et al. (2013)
			Seasonal CWSI average: A non-linear relationship with color; threshold for seasonally acceptable color = 0.1.	Emekli et al. (2007)

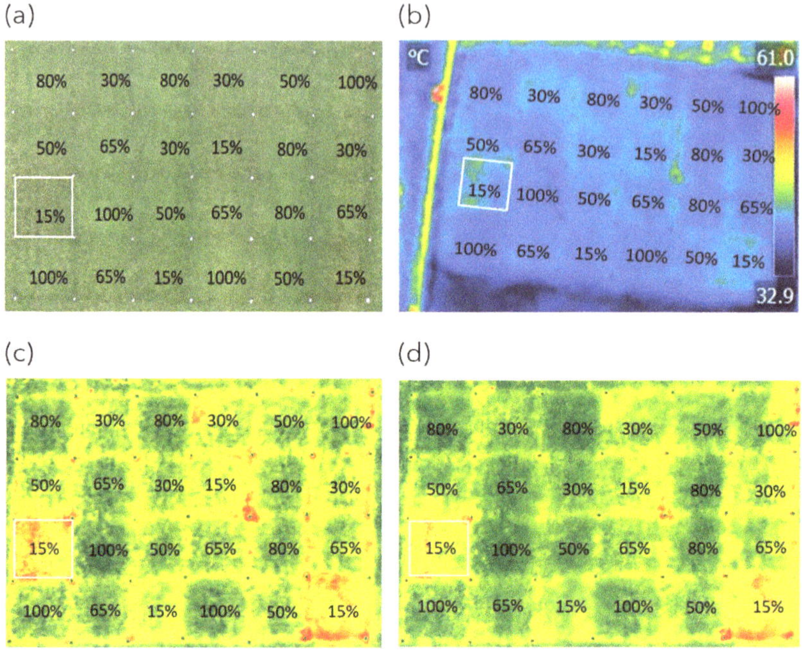

Figure 3 A red-green-blue image (a), thermal image (b), GreenBlue VI orthomosaic (c), and near-infrared reflectance orthomosaic (d) on 15 June 2017 of an sUAS study. Data were collected 6 days after deficit-irrigation treatments (15–100% ET_o) began (Hong et al., 2019a,b).

(i.e. without wavebands in the NIR or higher wavelengths) and used to evaluate drought stress in turfgrass (Caturegli et al., 2019; Hong et al., 2019b; Zhang et al., 2019b). For example, the GreenBlue VI mentioned earlier (in the 3-year sUAS study in deficit-irrigated creeping bentgrass plots) utilized green and blue wavebands in the visible region of the spectrum and was highly sensitive to drought stress in turfgrass, and it consistently detected drought stress earlier than seven other VIs (Table 1; Fig. 3) (Hong et al., 2019b). The visible atmospherically resistant index (Table 1), which uses green and red wavebands in the visible spectrum and was developed to reduce atmospheric effects (Gitelson et al., 2002), was strongly correlated with PGC of turfgrass under drought stress in an sUAS study (Zhang et al., 2019b). The dark green color index (DGCI) derived from sUAS RGB images was highly correlated with sUAS measurements of NDVI ($r = 0.85–0.96$; under different nitrogen fertilization regimes), which implies DGCI might be used in lieu of NDVI to detect drought-induced leaf firing when spectral sensors that measure NIR reflectance are not available (Caturegli et al., 2019); strong relationships between NDVI and PGC in turfgrass have been reported by others using sUAS-based measurements

(Bach et al., 2022a) and ground-based measurements (Bell et al., 2002; Bremer et al., 2011b).

Canopy temperature is another powerful indicator of early drought stress, including when measured from sUAS. For example, only 6 days after the initiation of deficit-irrigation treatments in small plots of creeping bentgrass, early drought stress was detected from a single image taken with an sUAS-mounted thermal camera (Hong et al., 2019a). Specifically, higher canopy temperature was detected in deficit-irrigated plots (15% ET_o and 30% ET_o; $P < 0.05$), before decreases in TQ and PGC were observed (Fig. 3). For larger areas such as golf courses, numerous thermal images would be required because of the limited footprint area per image and viewing angle effects among images, and thermal orthomosaics would need to be generated (van der Merwe et al., 2017, 2020). To evaluate changes over time, the timing of canopy temperature acquisition should be consistent and documented (e.g. data collected at approximately the same hour on each measurement day) (Mauri et al., 2021) and atmospheric conditions should be comparable among days (e.g. cloud-free) (Hong et al., 2019a). More research is needed to develop standard sampling and processing protocols for measuring canopy temperature with thermal cameras mounted on sUAS (Chandel et al., 2020; Sangha et al., 2020; Tmušić et al., 2020). There are tradeoffs regarding camera resolutions, sensitivities to viewing angle effects, and appropriate flying heights when selecting thermal sensors with different lenses that are beyond the scope of this discussion (Sangha et al., 2020).

5.1.1 Relationships between aerial and ground-based measurements

Most remote sensing research in turfgrass has been ground-based in regard to using spectral reflectance to detect drought stress. However, recent sUAS studies have indicated moderate to strong correlations between ground- and sUAS-based measurements of spectral reflectance in turfgrass. Correlations between ground- and sUAS-based NDVI ranged from r = 0.63 to 0.97 in a number of turfgrass studies (Caturegli et al., 2016; Hong et al., 2019b; Zhang et al., 2019b; Friell and Straw, 2021; Bach et al., 2022a, 2022b). Hong et al. (2019b) also reported strong correlations (r = 0.69 to 0.87) between sUAS-based broadband NIR (about 680–780 nm) and ground-based narrowband NIR (around 780 nm) in drought-stressed turfgrass. Therefore, ground-based measurements of drought-sensitive VIs, as well as individual wavebands would likely be applicable to their counterparts measured with sUAS-mounted sensors, although more research is required.

An example of a VI that is highly sensitive to drought but has only been evaluated using ground-based measurements in turfgrass is the PRI, which utilizes green and blue wavebands (Table 1). The PRI may have advantages over NDVI in detecting minor drought stress because of certain phytochemical changes that may occur before changes in biomass and color become evident (Zarco-Tejada et al., 2012; Gago et al., 2015; Barbedo, 2019). For example, stomatal conductance and water potential were more strongly correlated with PRI than with NDVI in an sUAS study conducted over an orange tree canopy (Zarco-Tejada et al., 2012). In turfgrass, initial results are promising from ground-based PRI applications in monitoring drought stress (Baghzouz et al., 2007; Table 1). Additional VIs obtained from ground-based RGB images have also been shown to be sensitive indicators of drought stress (Marín et al., 2020). The soil-adjusted VI (SAVI; Table 1), which uses the NIR broadband in its calculation, statistically detected drought stress in ground-based measurements of low vs. high irrigation treatments (Taghvaeian et al., 2013). In the future, drought-sensitive VIs that have only been evaluated in ground-based turfgrass studies need to be tested in sUAS research (Badzmierowski et al., 2019; see additional VIs summarized by Baghzouz et al., 2007).

Other ground-based research has identified specific narrowbands and broadbands in the visible to NIR spectrum that were indicators of drought stress in several turfgrass species and cultivars under field settings. For example, narrowbands (2–5 nm) that indicated drought stress ranged from 660 nm to 672 nm in several popular C3 turfgrasses and from 555 nm to 870 nm in C4 turfgrasses (Table 2). The water band index, centered around 970 nm and 900 nm, has proven effective for estimating drought stress in creeping bentgrass and hybrid bermudagrass [*Cynodon dactylon* (L.) Pers × *Cynodon transvaalensis* Burtt Davy], as well as in other crops (Peñuelas et al., 1993; McCall et al., 2017; Caturegli et al., 2020; Roberson et al., 2021). Also, individual broadbands centered at 830 nm (NIR) and 1650 nm (short-wave infrared) responded to drought stress, to a lesser extent than visible bands, in several turfgrass species (Taghvaeian et al., 2013). Field studies are warranted to use sUAS to leverage information obtained from ground-based remote sensing research regarding drought-sensitive wavebands that predict drought stress across different turfgrass species and cultivars.

Three to five broadbands, mostly from the absorption regions of chlorophyll (660–700 nm) and water (810–1480 nm), were used to develop optimum regression models for TQ (R^2 = 0.33–0.78) and leaf firing (R^2 = 0.16–0.83) under drought stress in 11 cultivars of five turfgrass species (Jiang and Carrow, 2007). Besides using multiple regressions of several bands (Baghzouz et al., 2006, 2007; Jiang and Carrow, 2007), the utilization of ground-based narrowbands across the entire visible and NIR spectrum and their derivatives has also shown advantages over VIs in the early detection of drought stress in pasture grass via the use of machine learning algorithms (Dao et al., 2021). Given the availability of sUAS-mounted hyperspectral sensors, selecting wavebands across wide regions, or

Table 2 Narrow spectral wavebands with high correlations to drought stress in C3 (top) and C4 (bottom) turfgrasses from ground-based field studies.

Species	Wavelength (nm)	Linear relationships with variables	References
C3 turfgrasses			
Tall fescue (*Festuca arundinacea* Schreb.)	671	Turf quality (TQ): r = −0.39 to −0.60; Leaf firing: r = 0.44 to 0.67 (3 cultivars)	Jiang and Carrow (2005)
Kentucky bluegrass (*Poa pratensis* L.)	672	Leaf water content: R^2 = 0.71 to 0.96	Suplick-Ploense et al. (2011)
Hybrid bluegrass (*Poa pratensis* L. × *Poa arachnifera* Torr.)	664, 668	Leaf water content: R^2 = 0.78, 0.80	Suplick-Ploense et al. (2011)
Perennial ryegrass (*Lolium perenne* L.)	660, 664	Leaf water content: R^2 = 0.88, 0.84	Suplick-Ploense et al. (2011)
Annual ryegrass (*Lolium multiflorum* Lam.)	693	Tissue moisture: R^2 = 0.71	Baghzouz et al. (2007)
C4 turfgrasses			
Zoysiagrass (*Zoysia japonica* Steud.)	687	TQ: r = −0.49; Leaf firing: r = 0.54	Jiang and Carrow (2005)
St. Augustinegrass [*Stenotaphrum secundatum* (Walt.) Kuntze]	687 693	TQ: r = −0.23 Leaf firing: r = 0.23	Jiang and Carrow (2005)
Hybrid bermudagrasses [*Cynodon dactylon* (L.) Pers. × *C. transvaalensis* Burtt Davy]	555	Tissue moisture, color, soil matric potential, leaf xylem water potential: R^2 = 0.66 (multiple regression predicting R_{555})	Baghzouz et al. (2007)
	667-693	TQ: r = −0.15 to −0.48; Leaf firing, r = 0.32 to 0.62; (4 cultivars)	Jiang and Carrow (2005)
Seashore paspalum (*Paspalum vaginatum* Swartz)	750, 775, 870	TQ, r = 0.15 to 0.48; Leaf firing, r = −0.40 to −0.47 (3 cultivars)	Jiang and Carrow (2005)

taking advantage of the full spectrum from visible to mid-infrared wavebands instead of merely a few wavebands for VIs may provide a more robust solution for detecting drought stress, and warrants further research in turfgrass field studies.

5.2 Strategic cultivar selection in turfgrass breeding programs

Turfgrass breeders are limited in their field phenotyping capability due to its high requirement of time and labor. Being able to collect comprehensive data during the early stages of selection and then later in advanced trials would benefit cultivar selection and improvement in turfgrass breeding programs. As technology advances, the use of sUAS in high-throughput phenotyping has increased rapidly in the past 6 years (Tattaris et al., 2016; Holman et al., 2016; Yang et al., 2017; Han et al., 2019; Li et al., 2020). Unlike many other agronomic crops, yield is not a breeding goal for turfgrass. Superior turfgrass lines are selected based on their phenotype, commonly known as phenotypic selection (Islam et al., 2014), which is ideal for evaluating with sUAS-based imagery. Zhang et al. (2019b) assessed the use of sUAS-based RGB and multispectral imageries on variety trials of two warm-season turfgrass species – bermudagrass (*Cynodon* spp.) and zoysiagrass (*Zoysia* spp.), using a low-cost sUAS platform. In the study, ground truth measurements were compared with sUAS-based measurements, and out of the top ten entries identified using ground measurements, 92% (bermudagrass) and 80% (zoysiagrass) overlapped with those using sUAS-based imagery. A collaborative project was initiated in late 2019 to equip turfgrass breeding programs in the southeastern United States with sUAS-based high-throughput phenotyping tools (Zhang et al., 2019a). This ongoing project aims to enhance the phenotyping capability of turfgrass breeders and further the application of sUAS in turfgrass breeding programs.

Despite the subjectivity of visual TQ ratings (Horst et al., 1984), it has been the primary evaluation method for many years (Karcher and Richardson, 2013). Digital image analysis (DIA) was implemented as a research tool starting around the 1980s, and turfgrass scientists adapted it into routine data collection with standardization in relevant equipment and image processing (Richardson et al., 2001). Equipment usually includes a 'light box', which is a metal box with light bulbs and wheels to be pushed across test plots. A camera is inserted on the top of the metal box to take pictures. Various turfgrass parameters are derived from DIA such as PGC, turf color, turfgrass establishment, drought stress, divot analysis, wear tolerance, and disease analysis (Patton et al., 2007; Steinke et al., 2010; Trappe et al., 2011a,b; Karcher and Richardson, 2013; Zhang et al., 2018; Bach et al., 2022a, 2022b). This method has worked well in turfgrass research, except that it remains time-consuming for turfgrass breeders to collect pictures on thousands of field plots in a timely manner. Based on previous validation work, an sUAS can collect images covering thousands of small plots (0.9 m × 1.5 m) in a few minutes vs. days of collecting data with a light box, during which time PGC could be changing among plots.

The general workflow of sUAS was described earlier but briefly includes image acquisition using sUAS, image processing and orthomosaic generation, and plot-level data extraction. In addition, consideration of data analytics is needed because larger amounts of data are generated from sUAS than with traditional phenotyping methods. Image acquisition involves sUAS and sensor selections as well as best practices in flight operations (described earlier).

Image processing is carried out by commercial software such as Pix4DMapper (Pix4D SA, Lausanne, Switzerland) and Agisoft metashape (Agisoft LLC, St. Petersburg, Russia) to generate orthomosaics. Plot-level data extraction can be done using ArcMap or QGIS (Wilber et al., 2021). But for large data extraction, faster workflows can be built in different programming languages such as R (R Core Team, 2021) and python (Van Rossum and Drake, 2011). Commercial software such as Phenix™ (Progeny Drone Inc., Lafayette, IN, USA) and TurfScout & AgSpect (TurfScout, LLC. Greensboro, NC, USA) is available for combining image processing and data extraction.

Similar to the aforementioned DIA using a light box, sUAS-based image analysis can extract traits such as turfgrass cover in both dormant and green vegetation and plot-based average vegetation and color indices. There are different methods to obtain PGC from sUAS-based imagery. For multispectral imagery, PGC can be calculated by thresholding NDVI. For RGB imagery, two different methods can be used: one is thresholding color index from RGB imagery, and the other is converting RGB imagery to Hue/Saturation/Brightness and using threshold values for Hue and Saturation to extract green pixels. Wang et al. (2022) evaluated a comprehensive combination of approaches for estimating turfgrass PGC using sUAS imagery. These approaches, varying in levels of complexity, were based on VIs, supervised and unsupervised machine learning classification, and image processing methods. They found that both RGB image-based PGC estimation methods, including the Hue-Saturation-Value method and the support vector machine agreed with ground measurements of PGC (R^2 = 0.86–0.96).

Turfgrass cultivar field trials can be monitored monthly for these traits with or without biotic and abiotic stress. Secondary traits such as rates of turfgrass establishment can be derived from repetitive measurements of turfgrass cover during establishment (Fig. 4). Each breeding program may vary in its process of compiling the information and making selections. For instance, one widely used mechanism is to rank the genotypes based on the number of times a given genotype enters the top statistical group over multiple traits of interest and/or multiple dates; this is known as the turf performance index (Wherley et al., 2011; Zhang et al., 2019b). As large amounts of comprehensive data are being collected using high-throughput phenotyping, more sophisticated algorithms for genotypic ranking and/or for identifying stress-tolerant genotypes are needed.

One of the challenges in analyzing sUAS-based imagery for field phenotyping in turfgrass variety trials is weed contamination. Weeds can be problematic in the early stages of breeding selection (for instance, single plant space nursery) when management practices are less proactive compared to advanced trials. Supervised classification can be incorporated to exclude weeds in the workflow (Rockstad et al., 2020), which slows down the process

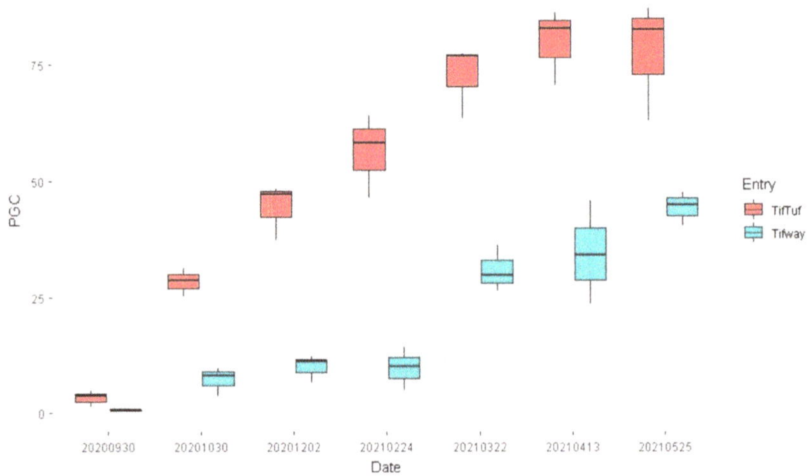

Figure 4 Boxplots of percent green cover (%) of two hybrid bermudagrass cultivars ('TifTuf' and 'Tifway') during establishment in Tifton, GA, USA.

due to the need for human input. In the future, it will likely be possible to implement machine learning models to mask weeds but that would require a large number of labeled images to be collected in order to train a model.

5.3 Predicting/managing turfgrass pests

Much of this chapter has provided a strong overview of the factors to consider when using sUAS and associated aerial imagery from a variety of sensors across turfgrass systems. The use of these tools provides a unique opportunity for monitoring and managing pest outbreaks across large surfaces that have been previously impractical because of time constraints and overlooked because of not being able to scout all areas effectively. While the use of sUAS to monitor and predict pest outbreaks is still in its infancy, there are a few examples that show promise as a shift toward precision turfgrass management.

5.3.1 Spatial distribution of pests

Most pest outbreaks are aggregated or occur in clusters rather than being uniformly distributed across larger surfaces (Campbell and Noe, 1985). This is particularly true on maintained turfgrasses, though documentation through peer-reviewed research is limited. A better understanding of the spatial variability and distribution of pest outbreaks provides the opportunity for targeted pesticide applications, providing both economic and environmental benefits.

Henry et al. (2009) concluded that two common paspalum species, dallisgrass (*Paspalum dilatatum* Poir.) and bahiagrass (*Paspalum notatum* Flügge), are not uniformly distributed across bermudagrass golf course fairways or roughs. Rather, they tended to cluster in areas with underlying issues, such as compacted soils. The authors reported that bahiagrass was impacted by mowing height, as it was more commonly found in roughs than in fairways, whereas dallisgrass grew in both shorter (fairway) and taller (rough) heights-of-cut. The understanding of underlying edaphic and environmental factors that drive paspalum outbreaks provides a unique opportunity for targeted cultural management strategies to alleviate these conditions.

Annual bluegrass weevil (ABW) is an economically important pest in the northeastern United States that causes damage in a predictable pattern. Damage from ABWs typically first appears closest to tree-lines and other out-of-play areas where the adults overwinter (Diaz and Peck, 2007). Similarly, another economically important turfgrass insect pest, the hunting billbug (*Sphenophorus venatus vestitus* Chittenden), was shown to also have an aggregate (clustered) distribution across sod farms in Georgia, USA (Gireesh et al., 2021). Spurlock (2009) reported that a large patch caused by *Rhizoctonia solani* (Kühn) re-occurred in the same location of golf courses from year to year. This reoccurrence suggests an opportunity for strategic fungicide applications through disease incidence mapping.

Horvath et al. (2007) showed a strong and stable spatial aggregation of dollar spot on creeping bentgrass and annual bluegrass (*Poa annua* L.) across seasons, even as overall disease pressure continued to increase. While the geospatial location of dollar spot aggregates changed from year to year, the clustering relationship remained the same across seasons. This suggests that historical disease incidence maps may not be useful for site-specific management of dollar spot, but aerial maps generated within season may provide valuable monitoring.

Most of the research to date that defines the spatial distribution of pests has been collected from intensive, time-consuming field sampling. While ground validation is critical for any confidence in pest estimations, much of the information could be collected rapidly and remotely using sUAS.

5.3.2 Pest mapping with aerial analysis

The most extensive use of sUAS to map a turfgrass pest for monitoring and management has been for assessing spring dead spot (caused by *Ophiosphaerella* spp.) of bermudagrass (Booth et al., 2021; Henderson, 2021). Booth et al. (2021) reported that spring dead spot developed in aggregates across golf course fairways in Virginia, USA. The authors demonstrated that spring dead spot frequently occurs in the same patches year after year, despite full bermudagrass recovery during the growing season. This understanding of the spatial and temporal dynamics of spring dead spot provided an opportunity for targeted fungicide applications. The authors reported using spring dead spot incidence maps to treat the disease site specifically using a high spatial resolution global positioning system sprayer without compromising efficacy. The result led to a decrease of 51% and 65% in fungicide use for the first and second years of the study, respectively.

The successful reduction of fungicide inputs using sUAS and targeted applications described by Booth et al. (2021) was not without challenges. Intensive manual selection of all spring dead spot within aerial imagery was not feasible for practical implementation by turfgrass professionals. Subsequent research provided a framework for automated detection without the need for manual selection or high-output computing (Henderson, 2021). The author also reported a detection accuracy with a 1 m buffer ranging from 53% to 93% compared to hand-validated pest maps. The author reported the diseased area with buffers covered between 4% and 21% of the fairways tested, providing the opportunity for substantial fungicide savings even greater than those reported by Booth et al. (2021). Follow-up research using spring dead spot mapping for targeted fungicide applications is ongoing.

The use of aerial pest mapping via sUAS provides a unique perspective on understanding the spatial and temporal dynamics of disease outbreaks or pest infestations. The disease incidence maps described earlier were used to help understand the underlying factors that drive spring dead spot epidemics. Hutchens et al. (2021) used the methods described by Henderson (2021) to compare zones of high, moderate, and low disease intensity. Results from this study suggested that the most influential factors in disease development included thatch accumulation and soil moisture, along with several key macro- and micronutrients. New research addressing the factors that drive dollar spot epidemics using the described strategies is ongoing (Henderson and McCall, 2021). Similar strategies using sUAS may be applied to numerous other diseases, weeds, and insect infestations to better understand their spread across turfgrass systems.

5.3.3 Remote pest detection using remote sensing and/or aerial imagery

The previous case studies have relied primarily on true color imagery, using only red, green, and blue bands. However, there are many opportunities for detecting diseases and other pests using wavelengths outside of the visible light spectrum. As has previously been discussed in this chapter, the spectral properties of both uniform, healthy turfgrass stands and those physiologically altered from various stressors can be used for rapid assessments. However, there are few documented reports in the literature where light outside of the visible spectrum has been used to detect either pathogen activity or subsequent damage to turfgrass canopies. Green et al. (1998) reported a significant correlation between Rhizoctonia blight (aka brown patch) and several visible or NIR wavelengths, with 810 nm being the most closely related. However, the authors also pointed out that there were numerous extraneous factors that contributed to the decline in canopy reflectance beyond simply disease degradation. Henderson (2021) reported that thermal image analysis could be a useful tool to detect brown patch in tall fescue and creeping bentgrass and could detect changes in pathogen load during pre-symptomatic pathogen infection and colonization.

5.3.4 Challenges with developing pest incidence maps

Developing any type of pest incidence map is merely a prediction of where the pests are located and will likely never be a perfect proxy for manual scouting. As with any type of modeling, the prediction is only as strong as the information provided when developing the model. Some challenges that may limit the accuracy of pest maps include the uniqueness of spectral properties of both the targeted pest and the intended turfgrass, the spectral or image resolution, the spatial resolution of both the aerial imagery and the ground validation dataset, and perhaps most importantly, the accuracy of proper ground validation of pests. Many pests or associated turfgrass damage may look very similar, particularly from dozens of meters above the surface. Predicted pest distributions from above will be rendered useless without proper and accurate verification from trained experts on the ground.

Another challenge with developing pest incidence maps for targeted management is that some expertise outside of the traditional fields of agronomy, plant pathology, entomology, and weed science is needed. This chapter includes excellent information on understanding the spectral properties of plant canopies, mission planning, and data processing. A basic understanding of these core subjects is critical for the proper adoption of remote pest mapping. Bock et al. (2010) discuss the benefits, drawbacks, and opportunities of using hyperspectral image analysis, among other things for plant disease detection. Wei et al. (2021) provide an additional overview of pathogen/disease detection of peanuts and other important agronomic crops with remote sensors. Many of the principles discussed apply to turfgrass pest detection. Additionally, a basic understanding of computer science will allow for a smooth transition into automated pest mapping. Useful strategies to automate pest mapping include simple computer coding to more complex types of image classification, simple machine learning, and more complex deep learning. Some of these strategies are discussed in a review by

Henderson (2021). Researchers have used a convolutional neural network to successfully identify and map various broadleaf and grassy weeds, spurges, and sedges in bermudagrass (*Cynodon dactylon* L.) sod fields (Zhang et al., 2021). Improvements in sUAS technology and sensor development, along with decreasing costs of equipment and increased computer processing outputs, provide a plethora of new opportunities to monitor, manage, predict, and better understand the behavior of various turfgrass pests.

6 Conclusion and future trends

The use of sUAS in turfgrass research and applied turfgrass management is emerging and offers many novel possibilities and unique challenges. Although remote sensing imagery acquired with sUAS has been the primary focus of this chapter, it is only part of the digital landscape. One promising strategy is to integrate sUAS remote sensing measurements with other data stream sources such as weather, soil moisture, as-applied fertilizer and pesticide maps, topography and basic soil descriptions, labor, and product inventory to enhance turfgrass management (Taghvaeian et al., 2013; Aboutalebi et al., 2019; Chávez et al., 2020; Chandel et al., 2021).

One example could include monitoring of drought stress by combining sUAS spectral reflectance or thermal measurements of the turf canopy with other critical data obtained from ground sensors or weather stations (e.g. soil moisture, forecasted reference ET_o). This information could be used to prevent drought stress while conserving water by improving irrigation recommendations, which could be further leveraged by using variable rate irrigation technology (Straw et al., 2019; Chávez et al., 2020; Dyer, 2022).

As we are able to acquire and create more data, 'big data' management becomes increasingly important. The successful measurement, collection, and analysis of big datasets could promote a greater understanding of how and why a plant responds in a given way and allow us to better model and even predict expected outcomes. Such efforts will likely require algorithm development and cloud computing services, possibly by using artificial intelligence through machine learning neural networks, to efficiently process large data streams (Zhang et al., 2021).

Harnessing and sharing digital data assets could be improved by the development of an integrated data management system (i.e. dashboard) that could offer a singular place where property managers, turfgrass practitioners, and agronomists can merge data from disparate sources and more effectively predict the impact of site-specific management choices (Fig. 5). To date, no singular software service provider has been able to universally address the demands of an integrated, client-facing dashboard management tool, although progress is being made by several organizations.

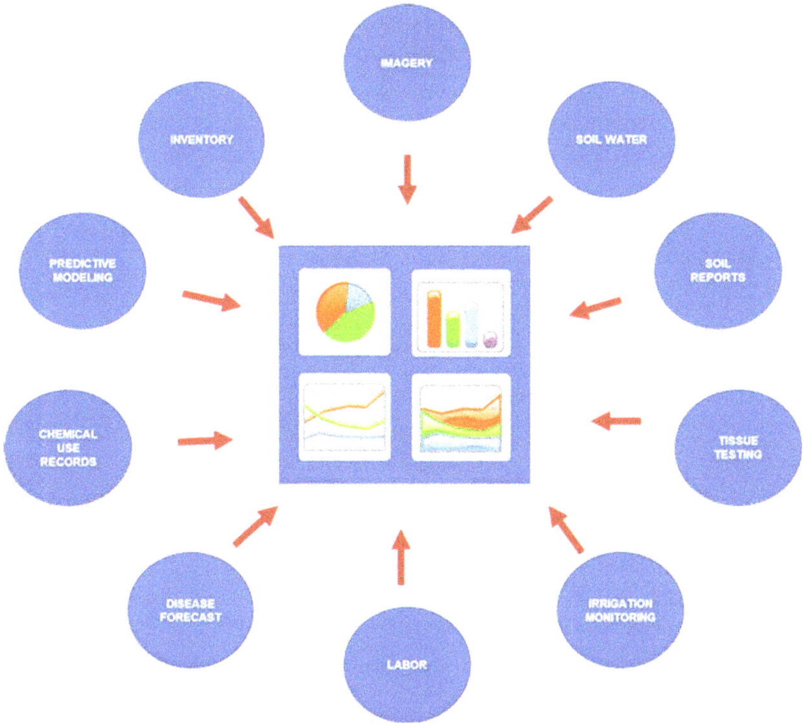

Figure 5 A potential vision of an integrated digital data hub, universally associating digital datasets within a singular, user-facing dashboard.

Determining thresholds for triggering various treatments such as irrigation or pesticide applications is critical to management in turfgrass. However, thresholds using spectral or thermal data are difficult to develop across species, soils, and environmental conditions (Barbedo, 2019). This is, in part, because the degree of detected stress is relative, and non-stressed reference areas may be needed for comparison purposes. Vegetation or thermal indices (e.g. NDVI, crop water stress index; Table 1) may be more robust than single spectral waveband or canopy temperature measurements alone in indicating stress (e.g. drought), with thresholds that could be calibrated for a given site and species (e.g. genetic color, texture) (Sullivan and Holbrook, 2007; Bremer et al., 2011a). Still, at present, ground truthing is needed to confirm the cause(s) of plant stress (e.g. drought vs. disease or multiple stressors). For example, NDVI has been related to nitrogen deficiency, salt stress, drought stress, and insect damage (Baghzouz et al., 2006; Bell et al., 2009; Johnsen et al., 2009; Xiang et al., 2017; Badzmierowski et al., 2019; Groover and Lawrence, 2020).

A rapidly growing field is crop spraying with sUAS, which can utilize remote sensing maps to target specific areas infested with weeds, disease, insects, etc.

This strategy may eventually provide effective control with reduced pesticide application amounts, but additional research is required in turfgrass (Koo et al., 2021). Autonomous flights are also becoming increasingly important and may become the standard, but attention to public safety and adherence to regulations are important considerations (van der Merwe et al., 2020). Detailed discussions of these and other topics are beyond the scope of this chapter, but it is expected that the use of sUAS will have novel and broad impacts on turfgrass science and management.

7 Where to look for further information

1. van der Merwe, D., Burchfield, D. R., Witt, T. D., Price, K. P., and Sharda, A. (2020). Chapter One - Drones in agriculture. In D. L. Sparks (Ed.), *Advances in Agronomy* (Vol. 162, pp. 1–30). Academic Press. https://doi.org/10.1016/bs.agron.2020.03.001.
2. Code of Federal Regulations - Part 107, Small Unmanned Aircraft Systems. https://www.ecfr.gov/current/title-14/chapter-I/subchapter-F/part-107.
3. Kansas State University - Unmanned Aircraft Systems Program. See "Resources" tab for latest free information for UAS pilots: https://www.salina.k-state.edu/research-training/applied-aviation-research-center/.

8 References

Aboutalebi, M. and Allen, N. Torres. (2019). Estimation of soil moisture at different soil levels using machine learning techniques and unmanned aerial vehicle (UAV) multispectral imagery. In: Rua, A. F., McKee, M., Coopmans, C., Thomasson, J. A., McKee, M. and Moorhead, R. J. (Eds) *Autonomous Air and Ground Sensing Systems for Agricultural Optimization and Phenotyping IV*, SPIE, Vol. 11008, p. 26. https://doi.org/10.1117/12.2519743.

Alvarez, G., Sevostianova, E., Serena, M., Sallenave, R. and Leinauer, B. (2016). Surfactant and polymer-coated sand effects on deficit irrigated bermudagrass turf, *Agronomy Journal* 108(6): 2245–2255. https://doi.org/10.2134/agronj2016.06.0329.

An, N., Goldsby, A. L., Price, K. P. and Bremer, D. J. (2015). Using hyperspectral radiometry to predict green leaf area index of turfgrass, *International Journal of Remote Sensing* 36(5): 1470–1483.

Assmann, J. J., Kerby, J. T., Cunliffe, A. M. and Myers-Smith, I. H. (2018). Vegetation monitoring using multispectral sensors - Best practices and lessons learned from high latitudes, *Journal of Unmanned Vehicle Systems* 7(1): 54-75.

Bach, A. P., Bremer, D. J., Lavis, C. C. and Keeley, S. J. (2022a). Effects of drip irrigation and cultivation methods on establishment of seeded tall fescue, *Crop, Forage and Turfgrass Management* 8(1): 1–10. https://doi.org/10.1002/cft2.20154.

Bach, A. P., Bremer, D. J., Lavis, C. C. and Keeley, S. J. (2022b). Establishing seeded tall fescue with covers and drip irrigation methods, *International Turfgrass Society Research Journal* 14(1): 441–449. https://doi.org/10.1002/its2.95.

Badzmierowski, M. J., McCall, D. S. and Evanylo, G. (2019). Using hyperspectral and multispectral indices to detect water stress for an urban turfgrass system, *Agronomy* 9(8): 439.

Baghzouz, M., Devitt, D. A. and Morris, R. L. (2006). Evaluating temporal variability in the spectral reflectance response of annual ryegrass to changes in nitrogen applications and leaching fractions, *International Journal of Remote Sensing* 27(19): 4137–4157. https://doi.org/10.1080/01431160600851843.

Baghzouz, M., Devitt, D. A. and Morris, R. L. (2007). Assessing canopy spectral reflectance of hybrid bermudagrass under various combinations of nitrogen and water treatments, *Applied Engineering in Agriculture* 23(6): 763–774. https://doi.org/10.13031/2013.24055.

Barbedo, J. (2019). A review on the use of unmanned aerial vehicles and imaging sensors for monitoring and assessing plant stresses, *Drones* 3(2): 40. https://doi.org/10.3390/drones3020040.

Bell, G. E., Howell, B. M., Johnson, G. V., Raun, W. R., Solie, J. B. and Stone, M. L. (2004). Optical sensing of turfgrass chlorophyll content and tissue nitrogen, *HortScience* 39(5): 1130–1132.

Bell, G. E., Martin, D. L., Koh, K. and Han, H. R. (2009). Comparison of turfgrass visual quality ratings with ratings determined using a handheld optical sensor, *HortTechnology* 19(2): 309–316.

Bell, G. E., Martin, D. L., Wiese, S. G., Dobson, D. D., Smith, M. W., Stone, M. L. and Solie, J. B. (2002). Vehicle-mounted optical sensing: An objective means for evaluating turf quality, *Crop Science* 42(1): 197–201.

Bock, C. H., Poole, G. H., Parker, P. E. and Gottwald, T. R. (2010). Plant disease severity estimated visually, by digital photography and image analysis, and by hyperspectral imaging, *Critical Reviews in Plant Sciences* 29(2): 59–107.

Boon, M. A., Drijfhout, A. P. and Tesfamichael, S. (2017). Comparison of a fixed-wing and multi-rotor UAV for environmental mapping applications: A case study, *The International Archives of the Photogrammetry, Remote Sensing and Spatial Information Sciences XLII-2/W6*: 47–54.

Booth, J. C., McCall, D. S., Sullivan, D., Askew, S. A. and Kochersberger, K. (2021). Investigating targeted spring dead spot management via aerial mapping and precision-guided fungicide applications. Crop Science 61: 3134–3144. https://doi-org.er.lib.k-state.edu/10.1002/csc2.20623.

Bremer, D. J., Auen, L. M., Ham, J. M. and Owensby, C. E. (2001). Evapotranspiration in a prairie ecosystem: Effect of grazing by cattle, *Agronomy Journal* 93(2): 338–348.

Bremer, D. J. and Ham, J. M. (1999). Effect of spring burning on the surface energy balance in a tallgrass prairie, *Agricultural and Forest Meteorology* 97(1): 43–54.

Bremer, D. J., Lee, H., Su, K. and Keeley, S. J. (2011a). Relationships between normalized difference vegetation index and visual quality in cool-season turfgrass: I. Variation among species and cultivars, *Crop Science* 51(5): 2212–2218. https://doi.org/10.2135/cropsci2010.12.0728.

Bremer, D. J., Lee, H., Su, K. and Keeley, S. J. (2011b). Relationships between normalized difference vegetation index and visual quality in cool-season turfgrass: II. Factors affecting NDVI and its component reflectances, *Crop Science* 51(5): 2219–2227.

Cai, G., Dias, J. and Seneviratne, L. (2014). A survey of small scale unmanned aerial vehicles: Recent advances and future development trends, *Unmanned Systems* 2(2): 175–199.

Campbell, C. L. and Noe, J. P. (1985). The spatial analysis of soilborne pathogens and root diseases, *Annual Review of Phytopathology* 23(1): 129–148.

Campbell, J. B. (1996). *Introduction to Remote Sensing* (2nd edn.), The Gilford Press, New York, NY.

Caturegli, L., Corniglia, M., Gaetani, M., Grossi, N., Magni, S., Migliazzi, M., Angelini, L., Mazzoncini, M., Silvestri, N., Fontanelli, M., Raffaelli, M., Peruzzi, A. and Volterrani, M. (2016). Unmanned aerial vehicle to estimate nitrogen status of turfgrasses, *PLoS One* 11(6): e0158268. https://doi.org/10.1371/journal.pone.0158268.

Caturegli, L., Gaetani, M., Volterrani, M., Magni, S., Minelli, A., Baldi, A., Brandani, G., Mancini, M., Lenzi, A., Orlandini, S., Lulli, F., de Bertoldi, C., Dubbini, M. and Grossi, N. (2019). Normalized difference vegetation index versus dark green colour index to estimate nitrogen status on bermudagrass hybrid and tall fescue, *International Journal of Remote Sensing* 41(2): 455–470. https://doi.org/10.1080/01431161.2019.1641762.

Caturegli, L., Matteoli, S., Gaetani, M., Grossi, N., Magni, S., Minelli, A., Corsini, G., Remorini, D. and Volterrani, M. (2020). Effects of water stress on spectral reflectance of bermudagrass, *Scientific Reports* 10(1): 15055. https://doi.org/10.1038/s41598-020-72006-6.

Chandel, A. K., Khot, L. R., Molaei, B., Peters, R. T., Stöckle, C. O. and Jacoby, P. W. (2021). High-resolution spatiotemporal water use mapping of surface and direct-root-zone drip-irrigated grapevines using UAS-based thermal and multispectral remote sensing, *Remote Sensing* 13(5): 954. https://doi.org/10.3390/rs13050954.

Chandel, A. K., Molaei, B., Khot, L. R., Peters, R. T. and Stöckle, C. O. (2020). High resolution geospatial evapotranspiration mapping of irrigated field crops using multispectral and thermal infrared imagery with METRIC energy balance model, *Drones* 4(3): 52. https://doi.org/10.3390/drones4030052.

Chávez, J. L., Torres-Rua, A. F., Woldt, W. E., Zhang, H., Robertson, C. C., Marek, G. W., Wang, D., Heeren, D. M., Taghvaeian, S., Neale, C. M. U. and Torres, W. (2020). A decade of unmanned aerial systems in irrigated agriculture in the western U.S., *Applied Engineering in Agriculture* 36(4): 423–436. https://doi.org/10.13031/aea.13941.

Dao, P. D., He, Y. and Proctor, C. (2021). Plant drought impact detection using ultra-high spatial resolution hyperspectral images and machine learning, *International Journal of Applied Earth Observation and Geoinformation* 102: 102364. https://doi.org/10.1016/j.jag.2021.102364.

Deery, D., Jimenez-Berni, J., Jones, H., Sirault, X. and Furbank, R. (2014). Proximal remote sensing buggies and potential applications for field-based phenotyping, *Agronomy* 4(3): 349–379 https://doi.org/10.3390/agronomy4030349.

Delvapour, N., Koparan, C., Nowatzki, J., Bajwa, S. and Sun, X. (2021). A technical study on UAV characteristics for precision agriculture applications and associated practical challenges, *Remote Sensing* 13(6): 1204.

Diaz, M. D. C. and Peck, D. C. (2007). Overwintering of annual bluegrass weevils, *Listronotus maculicollis*, in the golf course landscape, *Entomologia Experimentalis et Applicata* 125(3): 259–268.

Dyer, D. W. (2022). *Integrating canopy dynamics, soil moisture, and soil physical properties to improve irrigation scheduling in turfgrass systems*. PhD Dissertation, Kansas State University.

Emekli, Y., Bastug, R., Buyuktas, D. and Emekli, N. Y. (2007). Evaluation of a crop water stress index for irrigation scheduling of bermudagrass, *Agricultural Water Management* 90(3): 205–212.

Fahlgren, N., Gehan, M. A. and Baxter, I. (2015). Lights, camera, action: High-throughput plant phenotyping is ready for a close-up, *Current Opinion in Plant Biology* 24: 93-99 https://doi.org/10.1016/j.pbi.2015.02.006.

Feng, L., Chen, S., Zhang, C., Zhang, Y. and He, Y. (2021). A comprehensive review on recent applications of unmanned aerial vehicle remote sensing with various sensors for high-throughput plant phenotyping, *Computers and Electronics in Agriculture* 182: 106033 https://doi.org/10.1016/j.compag.2021.106033.

Fenstermaker-Shaulis, L. K., Leskys, A. and Devitt, D. A. (1997). Utilization of remotely sensed data to map and evaluate turfgrass stress associated with drought, *Journal of Turfgrass Management* 2(1): 65–81. Https://doi.org/10.1300/J099v02n01_06.

Friell, J. and Straw, C. (2021). Comparing ground-based and aerial data at field scale during dry down on golf course fairways, *International Turfgrass Society Research Journal* 14(1): 377–384. https://doi.org/10.1002/its2.46.

Gago, J., Douthe, C., Coopman, R. E., Gallego, P. P., Ribas-Carbo, M., Flexas, J., Escalona, J. and Medrano, H. (2015). UAVs challenge to assess water stress for sustainable agriculture, *Agricultural Water Management* 153: 9–19. https://doi.org/10.1016/J.AGWAT.2015.01.020.

Gireesh, M., Rijal, J. P. and Joseph, S. V. (2021). Spatial distribution of hunting billbugs (Coleoptera: Curculionidae) in sod farms, *Insects* 12(5): 402.

Gitelson, A. A., Kaufman, Y. J., Stark, R. and Rundquist, D. (2002). Novel algorithms for remote estimation of vegetation fraction, *Remote Sensing of Environment* 80(1): 76–87.

Gitelson, A. A., Keydan, G. P. and Merzlyak, M. N. (2006). Three-band model for noninvasive estimation of chlorophyll, carotenoids, and anthocyanin contents in higher plant leaves, *Geophysical Research Letters* 33(11): 111402 https://doi.org/10.1029/2006GL026457.

Green, D. E., Burpee, L. L. and Stevenson, K. L. (1998). Canopy reflectance as a measure of disease in tall fescue, *Crop Science* 38(6): 1603–1613.

Groover, W. L. and Lawrence, K. S. (2020). Plant health evaluations of *Belonolaimus longicaudatus* and *Meloidogyne incognita* colonized bermudagrass using remote sensing, *Journal of Nematology* 52: 2020–2109. https://doi.org/10.21307/JOFNEM-2020-109.

Haghverdi, A., Reiter, M., Singh, A. and Sapkota, A. (2021). Hybrid bermudagrass and tall fescue turfgrass irrigation in central California: II. Assessment of NDVI, CWSI, and canopy temperature dynamics, *Agronomy* 11(9): 1733.

Han, L., Yang, G., Dai, H., Yang, H., Xu, B., Li, H., Long, H., Li, Z., Yang, X. and Zhao, C. (2019). Combining self-organizing maps and biplot analysis to preselect maize phenotypic components based on UAV high-throughput phenotyping platform, *Plant Methods* 15(1): 57. https://doi.org/10.1186/s13007-019-0444-6.

Hatfield, J. L., Gitelson, A. A., Schepers, J. S. and Walthall, C. L. (2008). Application of spectral remote sensing for agronomic decisions, *Agronomy Journal* 100(S3): S-117-S-131.

Henderson, C. A. (2021). *Identification of disease stress in turfgrass canopies using thermal imagery and automated aerial image analysis*. M.S. thesis, Virginia Tech.

Henderson, C. A. and McCall, D. S. (2021). *Methods for Estimating Dollar Spot at Varying Altitudes of Aerial Imagery* [Abstract], CSSA, ASA, International, S. S. S. A. Annual Meeting, Salt Lake City, UT. Available at: https://scisoc.confex.com/scisoc/2021am/meetingapp.cgi/Paper/134083.

Henry, G. M., Burton, M. G. and Yelverton, F. H. (2009). Heterogeneous distribution of weedy Paspalum species and edaphic variables in turfgrass, *HortScience* 44(2): 447–451.

Holman, F., Riche, A., Michalski, A., Castle, M., Wooster, M. and Hawkesford, M. (2016). High throughput field phenotyping of wheat plant height and growth rate in field plot trials using UAV based remote sensing, *Remote Sensing* 8(12): 1031. https://doi.org/10.3390/rs8121031.

Hong, M., Bremer, D. J. and van der Merwe, D. (2019a). Thermal imaging detects early drought stress in turfgrass utilizing small unmanned aircraft systems, *Agrosystems, Geosciences and Environment* 2(1): 1-9. https://doi.org/10.2134/age2019.04.0028.

Hong, M., Bremer, D. J. and van der Merwe, D. (2019b). Using small unmanned aircraft systems for early detection of drought stress in turfgrass, *Crop Science* 59(6): 2829-2844. https://doi.org/10.2135/cropsci2019.04.0212.

Horst, G. L., Engelke, M. C. and Meyers, W. (1984). Assessment of visual evaluation techniques 1, *Agronomy Journal* 76(4): 619–622. https://doi.org/10.2134/agronj1984.00021962007600040027x.

Horvath, B. J., Kravchenko, A. N., Robertson, G. P. and Vargas, J. M., Jr. (2007). Geostatistical analysis of dollar spot epidemics occurring on a mixed sward of creeping bentgrass and annual bluegrass, *Crop Science* 47(3): 1206–1216.

Hutchens, W. J., Goatley, J. M., Kerns, J. P., Nita, M., Straw, C. M., Sullivan, D., Henderson, C. A. and McCall, D. S. (2021). *Environmental and Edaphic Factors That Influence Spring Dead Spot Epidemics*. CSSA, ASA, International, S. S. S. A. Annual Meeting, Salt Lake City, UT. Available at: https://scisoc.confex.com/scisoc/2021am/meetingapp.cgi/Paper/134068.

Islam, M. S., Studer, B., Møller, I. M. and Asp, T. (2014). Genetics and biology of cytoplasmic male sterility and its applications in forage and turf grass breeding, *Plant Breeding* 133(3): 299–312. https://doi.org/10.1111/pbr.12155.

Jiang, Y. and Carrow, R. N. (2005). Assessment of narrow-band canopy spectral reflectance and turfgrass performance under drought stress, *HortScience* 40(1): 242–245. https://doi.org/10.21273/HORTSCI.40.1.242.

Jiang, Y. and Carrow, R. N. (2007). Broadband spectral reflectance models of turfgrass species and cultivars to drought stress, *Crop Science* 47(4): 1611–1618.

Johnsen, A. R., Horgan, B. P., Hulke, B. S. and Cline, V. (2009). Evaluation of remote sensing to measure plant stress in creeping bentgrass (*Agrostis stolonifera* L.) Fairways, *Crop Science* 49(6): 2261–2274. https://doi.org/10.2135/cropsci2008.09.0544.

Karcher, D. E. and Richardson, M. D. (2013). Digital image analysis in turfgrass research. In: *Turfgrass: Biology, Use, and Management*, pp. 1133–1149.

Koo, D., Goncalves, C. G. and Askew, S. D. (2021). *Influence of Height and Speed on Spray Deposition Pattern of an Agricultural Spray Drone*. CSSA, ASA, International, S. S. S. A. Annual Meeting, Salt Lake City, UT. Available at: https://scisoc.confex.com/scisoc/2021am/meetingapp.cgi/Paper/139240.

Lee, H., Bremer, D. J., Su, K. and Keeley, S. J. (2011). Relationships between normalized difference vegetation index and visual quality in turfgrasses: Effects of mowing height, *Crop Science* 51(1): 323–332. https://doi.org/10.2135/cropsci2010.05.0296.

Li, F., Piasecki, C., Millwood, R. J., Wolfe, B., Mazarei, M. and Stewart, C. N. (2020). High-throughput switchgrass phenotyping and biomass modeling by UAV, *Frontiers in Plant Science* 11: 574073. https://doi.org/10.3389/fpls.2020.574073.

Li, L., Zhang, Q. and Huang, D. (2014). A review of imaging techniques for plant phenotyping, *Sensors* 14(11): 20078-20111.

Marín, J., Yousfi, S., Mauri, P. V., Parra, L., Lloret, J. and Masaguer, A. (2020). RGB vegetation indices, NDVI, and biomass as indicators to evaluate C3 and C4 turfgrass under different water conditions, *Sustainability* 12(6): 2160. https://doi.org/10.3390/su12062160.

Mauri, P. V., Parra, L., Yousfi, S., Lloret, J. and Marin, J. F. (2021). Evaluating the effects of environmental conditions on sensed parameters for green areas monitoring and smart irrigation systems, *Sensors* 21(6): 2255. https://doi.org/10.3390/s21062255.

McCall, D. S., Sullivan, D. G., Zhang, X., Martin, S. B., Wong, A. and Ervin, E. H. (2021). Influence of synthetic phthalocyanine pigments on light reflectance of creeping bentgrass, *Crop Science* 61(1): 804–813.

McCall, D. S., Zhang, X., Sullivan, D. G., Askew, S. D. and Ervin, E. H. (2017). Enhanced soil moisture assessment using narrowband reflectance vegetation indices in creeping bentgrass, *Crop Science* 57(S1): S161.

Patton, A. J., Volenec, J. J. and Reicher, Z. J. (2007). Stolon growth and dry matter partitioning explain differences in zoysiagrass establishment rates, *Crop Science* 47(3): 1237-1245.

Peñuelas, J., Filella, I., Biel, C., Serrano, L. and Save, R. (1993). The reflectance at the 950–970 nm region as an indicator of plant water status, *International Journal of Remote Sensing* 14(10): 1887–1905.

Peterson, K. W., Bremer, D. J. and Blonquist Jr., J. M. (2017). Estimating transpiration from turfgrass using stomatal conductance values derived from infrared thermometry, *ITSRJ* 13(1): 113–118. https://doi.org/10.2134/itsrj2016.09.0788.

R Core Team. (2021). *R: A Language and Environment for Statistical Computing*. Available at: https://www.R-project.org/. R Foundation for Statistical Computing, Vienna, Austria.

Richardson, M. D., Karcher, D. E. and Purcell, L. C. (2001). Quantifying turfgrass cover using digital image analysis, *Crop Science* 41(6): 1884–1888. https://doi.org/10.2135/cropsci2001.1884.

Ritchie, G. L. and Bednardz, C. W. (2005). Estimating defoliation of two distinct cotton types using reflectance data, *Journal of Cotton Science* 9: 182–188.

Roberson, T. L., Badzmierowski, M. J., Stewart, R. D., Ervin, E. H., Askew, S. D. and McCall, D. S. (2021). Improving soil moisture assessment of turfgrass systems utilizing field radiometry, *Agronomy* 11(10), 1960.

Rockstad, G., Austin, R., Yu, X., Melgar, E. M. C., Dunne, J. C., Miller, G. L., Jespersen, D. and Milla-Lewis, S. R. (2020). *Evaluation of UAV-Based Imagery for Drought Stress Traits in St. Augustinegrass, ASA, CSSA and SSSA Annual Meeting*.

Rouse, J. W., Haas Jr., R. H., Schell, J. A. and Deering, D. W. (1974). Monitoring vegetation systems in the Great Plains with ERTS. In: *Proceedings of the ERTS-1 Symposium, 3rd*, National Astronautics and Space Administration, Washington, DC, pp. 309-317.

Sangha, H. S., Sharda, A., Koch, L., Prabhakar, P. and Wang, G. (2020). Impact of camera focal length and sUAS flying altitude on spatial crop canopy temperature evaluation, *Computers and Electronics in Agriculture* 172: 105344. https://doi.org/10.1016/j.compag.2020.105344.

Spurlock, T. N. (2009). *Epidemiology and Etiology of Zoysiagrass Diseases in Northwest Arkansas*. University of Arkansas.

Steinke, K., Chalmers, D., Thomas, J., White, R. and Fipps, G. (2010). Drought response and recovery characteristics of St. Augustinegrass cultivars, *Crop Science* 50(5): 2076-2083.

Straw, C., Friell, J. and Horgan, B. (2019). Precision irrigation for golf courses using sensor and mapping technologies. In: *ASA-CSSA-SSSA Annual Meeting*. Available at: http://conservancy.umn.edu/handle/11299/214353.

Sullivan, D. G., Fulton, J. P., Shaw, J. N. and Bland, G. (2007). Evaluating the sensitivity of an unmanned thermal infrared aerial system to detect water stress in a cotton canopy, *Transactions of the ASABE* 50: 1955–1962.

Sullivan, D. G. and Holbrook, C. C. (2007). Using ground-based reflectance measurements as a selection criteria for drought- and aflatoxin-resistant peanut genotypes, *Crop Science* 47(3): 1040–1050.

Sullivan, D. G., Shaw, J. N., Mask, P. L., Rickman, D., Luvall, J. and Wersinger, J. M. (2004). Evaluating corn nitrogen variability via remote-sensed data, *Communications in Soil Science and Plant Analysis* 35(17–18): 2465–2483.

Suplick-Ploense, M. R., Alshammary, S. F. and Qian, Y. L. (2011). Spectral reflectance response of three turfgrasses to leaf dehydration, *Asian Journal of Plant Sciences* 10(1): 67–73. https://doi.org/10.3923/ajps.2011.67.73.

Taghvaeian, S., Chávez, J., Hattendorf, M. and Crookston, M. (2013). Optical and thermal remote sensing of turfgrass quality, water stress, and water use under different soil and irrigation treatments, *Remote Sensing* 5(5): 2327–2347. https://doi.org/10.3390/rs5052327.

Tattaris, M., Reynolds, M. P. and Chapman, S. C. (2016). A direct comparison of remote sensing approaches for high-throughput phenotyping in plant breeding, *Frontiers in Plant Science* 7: 1131. https://doi.org/10.3389/fpls.2016.01131.

Thamm, H. P., Brieger, N., Neitzke, K. P., Meyer, M., Jansen, R. and Mönninghof, M. (2015). SONGBIRD-an innovative UAS combining the advantages of fixed wing and multi rotor UAS, *International Archives of the Photogrammetry, Remote Sensing and Spatial Information Sciences* 40: 345–349.

Tmušić, G., Manfreda, S., Aasen, H., James, M. R., Gonçalves, G., Ben-Dor, E., Brook, A., Polinova, M., Arranz, J. J., Mészáros, J., Zhuang, R., Johansen, K., Malbeteau, Y., de Lima, I. P., Davids, C., Herban, S. and McCabe, M. F. (2020). Current practices in UAS-based environmental monitoring, *Remote Sensing* 12(6): 1001. https://doi.org/10.3390/rs12061001.

Trappe, J. M., Karcher, D. E., Richardson, M. D. and Patton, A. J. (2011a). Bermudagrass and zoysiagrass cultivar selection: Part 1, clipping yield, scalping tendency, and golf ball lie, *Applied Turfgrass Science* 8(1): 1–12. https://doi.org/10.1094/ATS-2011-0630-01-RS.

Trappe, J. M., Karcher, D. E., Richardson, M. D. and Patton, A. J. (2011b). Bermudagrass and zoysiagrass cultivar selection: Part 2, divot recovery, *Applied Turfgrass Science* 8(1): 1–12. https://doi.org/10.1094/ats-2011-0630-02-rs.

Tucker, C. J. (1979). Red and photographic infrared linear combinations for monitoring vegetation, *Remote Sensing of Environment* 8(2): 127–150.

van der Merwe, D., Burchfield, D. R., Witt, T. D., Price, K. P. and Sharda, A. (2020), Chapter One. Drones in agriculture. In: Sparks, D. L. (Ed) *Advances in Agronomy*. Academic Press 162: 1–30. https://doi.org/10.1016/bs.agron.2020.03.001.

van der Merwe, D., Skabelund, L. R., Sharda, A., Blackmore, P. and Bremer, D. (2017). Towards characterizing green roof vegetation using color-infrared and thermal sensors. In: *Cities Alive 15th Annual Green Roof and Wall Conference*, Seattle, WA, pp. 18–21.

Van Rossum, G. and Drake, F. L. (2011). *The Python Language Reference Manual*. Network Theory Ltd, Bristol, UK.

Wang, T., Chandra, A., Jung, J. and Chang, A. (2022). UAV remote sensing based estimation of green cover during turfgrass establishment, *Computers and Electronics in Agriculture* 194: 106721.

Wei, X., Aguilera, M., Walcheck, R., Tholl, D., Li, S., Langston Jr., D. B. and Mehl, H. L. (2021). Detection of soilborne disease utilizing sensor technologies: Lessons learned from studies on stem rot of peanut, *Plant Health Progress* 22(4): 436–444.

Wherley, B. G., Skulkaew, P., Chandra, A., Genovesi, A. D. and Engelke, M. C. (2011). Low-input performance of zoysiagrass (*Zoysia* spp.) cultivars maintained under dense tree shade, *HortScience* 46(7): 1033–1037.

Wilber, A. L., Czarnecki, J. M. P. and McCurdy, J. D. (2021). An ArcGIS Pro workflow to extract vegetation indices from aerial imagery of small plot turfgrass research, *Crop Science* 62(1): 503–511. https://doi.org/10.1002/csc2.20669.

Xiang, M., Moss, J. Q., Martin, D. L., Su, K., Dunn, B. L. and Wu, Y. (2017). Evaluating the salinity tolerance of clonal-type bermudagrass cultivars and an experimental selection, *HortScience* 52(1): 185–191. https://doi.org/10.21273/HORTSCI10773-16.

Yang, G., Liu, J., Zhao, C., Li, Z., Huang, Y., Yu, H., Xu, B., Yang, X., Zhu, D., Zhang, X., Zhang, R., Feng, H., Zhao, X., Li, Z., Li, H. and Yang, H. (2017). Unmanned aerial vehicle remote sensing for field-based crop phenotyping: Current status and perspectives, *Frontiers in Plant Science* 8: 1111. https://doi.org/10.3389/fpls.2017.01111.

Yucky, E. D. D., Putrada, A. G. and Abdurohman, M. (2021). IoT drone camera for a paddy crop health detector with RGB comparison (vol. 2021). In: *9th International Conference on Information and Communication Technology (ICoICT)*, pp. 155–159. https://doi.org/10.1109/ICoICT52021.2021.9527421.

Zarco-Tejada, P. J., González-Dugo, V. and Berni, J. A. J. (2012). Fluorescence, temperature and narrow-band indices acquired from a UAV platform for water stress detection using a micro-hyperspectral imager and a thermal camera, *Remote Sensing of Environment* 117: 322–337. https://doi.org/10.1016/j.rse.2011.10.007.

Zhang, J., Maleski, J., Jespersen, D., Waltz, F. C., Jr., Rains, G. and Schwartz, B. (2021). Unmanned aerial system-based weed mapping in sod production using a convolutional neural network, *Frontiers in Plant Science* 12: 702626. https://doi.org/10.3389/fpls.2021.702626.

Zhang, J., Poudel, B., Kenworthy, K. E., Unruh, J. B., Rowland, D., Erickson, J. E. and Kruse, J. (2018). Drought responses of above-ground and below-ground characteristics in warm-season turfgrass, *Journal of Agronomy and Crop Science* 205(1): 1–12. https://doi.org/10.1111/jac.12301.

Zhang, J., Schwartz, B. M., Maleski, J., et al. (2019a). *Application of Unmanned Serial Systems Based Imagery and Data Analytics in Turfgrass Field Trials, ASA, CSSA and SSSA Annual Meeting*.

Zhang, J., Virk, S., Porter, W., Kenworthy, K., Sullivan, D. and Schwartz, B. (2019b). Applications of unmanned aerial vehicle based imagery in turfgrass field trials, *Frontiers in Plant Science* 10: 279. https://doi.org/10.3389/fpls.2019.00279.

Zhang, Y. and Zhang, N. (2018). Imaging technologies for plant high-throughput phenotyping: A review, *Frontiers of Agricultural Science and Engineering* 5: 406–419 https://doi.org/10.15302/J-FASE-2018242.

Chapter 17

Considerations with selecting turfgrass varieties and cultivars

Kevin Morris, National Turfgrass Evaluation Program, USA; Yuanshuo Qu, Rutgers, The State University of New Jersey, USA; Len Kne, University of Minnesota, USA; and Steve Graham, University of Minnesota, USA

1 Introduction

The National Turfgrass Evaluation Program (NTEP) coordinates field research trials of turfgrass species and cultivars across the United States and Canada (Krans and Morris, 2007). NTEP is affiliated with the United States Department of Agriculture (USDA; Beltsville, MD, USA) and is one of the best-known and highly respected turfgrass research programs in the world (Morris and Kenna, 2017). Initiated in 1980, NTEP represents a mechanism to evaluate multiple turfgrass species and cultivars within the United States and Canada. At any one time, NTEP is coordinating the evaluation and testing of more than 600 grasses comprised of experimental and commercial cultivars, which encompasses about 20 species. The information collected and summarized by NTEP is currently requested by individuals, educators, turfgrass seed companies, other turfgrass industry companies and turfgrass organizations worldwide. Additionally, NTEP information is disseminated in trade publications, seed company advertisements and marketing publications, seed company fact sheets and university extension and outreach publications.

http://dx.doi.org/10.19103/AS.2022.0110.17

Seed companies rely heavily on NTEP data to support their turfgrass seed development, marketing and sales operations and initiatives. Plant breeders use NTEP data to gain further insight into the adaptability and utility of experimental turfgrass cultivars as evaluated in many different geographic locations, environmental conditions and management regimes. Turfgrass researchers and extension personnel use NTEP data to develop species and cultivar recommendations best suited for their geographical region and climate. Local and state government entities, such as parks and highway departments, use NTEP data for writing specifications to support turfgrass seed recommendations and purchasing. Most importantly, the golf course superintendent, athletic/sports field manager, parks and grounds manager, lawn care operator, sod grower and homeowner consult NTEP data during their decision-making process for purchasing seed or sod. It is the acceptance by the end-user or turfgrass practitioner that has made NTEP the standard for turfgrass species and cultivar evaluation and testing in the United States and many other countries worldwide.

NTEP data is published and freely available via the website: www.ntep.org. NTEP field trials are typically conducted for 4 to 5 consecutive years, with annual reports published and a final summary at the trial's conclusion. NTEP collects data on many characteristics such as turfgrass quality, genetic color, leaf texture, spring greenup, as well as response to various abiotic stresses (i.e. shade, drought, cold, traffic) and biotic stresses (i.e. diseases, insects) and more (https://ntep.org/reports/ratings.htm). This data is useful to identify top-performing turfgrasses (i.e. species and cultivars), particularly within a geographical region or maintained under a specific management regime (i.e. mowing height and/or frequency, irrigation, fertility, traffic, etc.).

NTEP data collected since 1981 is publicly available online. Data are collected on more than 100 parameters, including important factors related to sustainable turfgrass selection (Krans and Morris, 2007). The selection of turfgrasses with traits such as improved disease resistance, drought tolerance, heat tolerance, cold tolerance, persistence under poor soil conditions, persistence under irrigation with poor water quality and fast establishment is an important consideration for protecting the environment, reducing or optimizing management inputs, improving athlete safety and for achieving sustainability. To improve access to comprehensive and reliable turfgrass data, NTEP and the University of Minnesota (Minneapolis, MN, USA) have collaborated to develop a relational database accompanied with a user-friendly selection tool called the *Turfgrass Trial Explorer*. This tool is accessed via www.ntep.org/database.htm, and allows for additional selection, sorting, analyzing and downloading of NTEP data. Case studies are presented to demonstrate the current and future use of *Turfgrass Trial Explorer* for aid and guidance with making turfgrass species and cultivar selections that promote sustainability and environmental stewardship.

2 Factors that impact turfgrass sustainability

Sustainable turfgrass management impacts three familiar concerns: environmental, economic and societal, while also requiring the fourth concern of adequate usability or playability. Integrated Pest Management (IPM) is a key strategy to achieve or strive toward sustainable turfgrass management, with the prevention of pest damage (i.e. biotic stresses) as an essential step in that process (Beard, 2002). Selecting pest-resistant turfgrass species and cultivars is an important IPM strategy for pest prevention (USEPA, 2022). In addition, the same can be stated for turfgrass persistence during cold winters, warm summers, drought periods, low nutrient availability and/or use, poor water quality and other abiotic stresses. Choosing turfgrasses that persist and thrive with reduced management inputs, either by resisting or tolerating pests, surviving cold or heat or drought, maintaining adequate functional vegetative cover in poor soil conditions or tolerating damage from excessive play and traffic will ultimately result in sustainable turfgrass stands.

To choose a top-performing turfgrass species and cultivar, as well as manage that choice sustainably, understanding inputs and use requirements of a turfgrass site is essential. For example, questions to consider include: what are the soil characteristics at the site (i.e. pH, structure, organic matter content, salt content, nutrient and water holding deficiency or capacity); is the site in full sun or in shade, and if so, how many hours of sunlight does the site receive per day; is the site sloped such that runoff will increase and water infiltration will decrease?

Use characteristics to consider include wear and traffic. Does the site receive a significant foot or mechanical traffic (i.e. an athletic field or park)? If so, is this a seasonal pattern or year-round? Are fertilizer or pesticide restrictions inherent in the local jurisdiction or surrounding communities? Is water use curtailed, or is there a lack of potable water at the site? What, if any budget constraints limit the use of inputs, seed, sod, etc.? Any of these situations dictate a careful consideration of the species and cultivar planted as an improper choice can lead to increased input use, unsafe turfgrass cover and playability, increasing management costs and overall poor turfgrass performance.

Species vary in their ability to respond to situations such as shade impacts, reduced nutrient use, traffic and saline irrigation. An excellent chart in Brede (2000) (pp. 36–37) displays the performance of traditional turfgrass species under reduced maintenance. The species are rated as either favorable, moderate, fair or unfavorable in performance on 23 characteristics, with also a listing of their growth habit (i.e. bunch, rhizomonous or stoloniferous). For example, perennial ryegrass (*Lolium perenne* L.) is listed as favorable performance for establishment, wear recovery, mixing with other turfgrasses and initial seed costs, but as unfavorable for nitrogen requirements, mowing requirements and water use rate. In contrast, hard fescue (*Festuca trachyphylla* [Hack.] Krajinais) is listed with favorable performance for nitrogen and mowing requirements, but only fair for factors such as establishment and wear recovery. The good news is that there are cultivar differences within every species, hence the reason for NTEP evaluating each species and cultivars within species over time to document any changes and improvements in those cultivars for important or desired traits.

Geography is always an important consideration when choosing a sustainable turfgrass species or cultivar. Dodson (2005) outlines several key traits desired specific to sustainable golf course turf, including rapid recuperative ability, tolerance of low cutting heights and compacted soils, as well as adaptation to the use area. To aid in selecting an appropriate species, turfgrass climatic zones are outlined by Beard (2002) with delineations such as cool humid, cool semiarid, warm humid and humid transitional.

Obtaining performance information from locations similar to the target turfgrass site's climate is the first step in sustainable species and cultivar selection. Turfgrass extension information available in many states, such as Pennsylvania (Landschoot, 2016) and Utah (Kopp and Johnson, 2022), identifies species that are adapted to each climate. After a species is chosen based on geography or region, use and site considerations, cultivar information for that species can be obtained from NTEP trial sites that perform well in the region, as well as display the desired sustainable traits.

Establishment, including speed of germination or rate of spread, as well as the number of plants per unit area creates competition that reduces initial weed infestation (Mandel et al., 2013). After full establishment, other factors such as density, growth rate and regrowth from traffic or other stresses, as well as overall ground cover are other important traits to resist weed invasion. Choosing an appropriate species and cultivar can often resist the invasion of specific weeds, such as annual bluegrass (*Poa annua* L.), which is considered a common and invasive weed on golf courses particularly putting greens (Vargas and Turgeon, 2004). Annual bluegrass is not a sustainable species for golf courses or other high-use intensity sites, but often cannot be easily eliminated and therefore must be simply managed as a turf (Beard, 2002).

Sachs and Luff (2002) describe many other factors that are important in reducing pesticide use, a key factor in achieving sustainability. The authors discuss the choice of disease and insect-resistant cultivars, as well as grasses that survive well under poor soil conditions. Choosing a disease-resistant cultivar is the preferred method of disease prevention, reducing pesticide use, reseeding or re-establishment and overall enhancing sustainability (Beard, 1973; Christians et al., 2017).

Carrow and Duncan (1998) discuss a worldwide problem: salt-affected turfgrass sites due to poor quality irrigation water sources, salt spray in coastal areas and high salt soils. Seashore paspalum (*Paspalum vaginatum* Swartz) is a salt-tolerant species for subtropical and tropical regions, performing sustainably in low-quality, high-salt content soils or under saline irrigation water conditions (Duncan and Carrow, 2000). Unfortunately, the adaptability of seashore paspalum is limited to a small portion of the United States and the Caribbean. Therefore, throughout much of the United States and the world, other turfgrass species and cultivars must be utilized in salty soils or under low-quality irrigation water use to achieve sustainability. In addition, other soil factors such as low nutrient-holding capacity, suboptimal pH levels or soil compaction can also affect turfgrass sustainability (Brede, 2000).

Abiotic factors such as sustained drought, heat and cold affect a turfgrass' ability to persist and function (Beard, 2002). Drought damage can be seen during *acute* periods, which are short, intense droughts in regions of 700+ mm of annual precipitation, or *chronic* periods of sustained droughts throughout much or all of a growing season in regions of <700 mm of annual precipitation. Drought tolerance can be measured in several ways (refer to Case study 7: advanced selection of cultivars with improved drought resistance), and tolerance or resistance to drought and other abiotic factors are important factors to consider when selecting sustainable turfgrass cultivars.

Use factors such as intense traffic from athletic events, concerts, casual foot traffic or mechanized vehicles can result in unfavorable conditions. Soil compaction, reduced cover and weed invasion are often impacts of traffic and overuse (Puhalla et al., 1999). These impacts can lead to increased soil erosion, an increase in the use of pesticides or other inputs and a higher probability of athlete injury (Beard and Green, 1994). Selection of improved traffic and wear-tolerant cultivars can decrease the damage from these impacts.

A reduction in other inputs due to environmental sensitivity in the site or region, societal resistance, municipal regulations or budget constraints can affect sustainability. Nutrient use, whether nitrogen, phosphorus or potassium, pesticide use, irrigation available, the frequency of mowing and height of cut are just some of the inputs that affect the overall performance and persistence of turf stands.

Finally, seasonal adaptation of turfgrass cultivars is not widely discussed, although some variation in performance has been observed. If top performance in a particular season, for example summer, is essential for a turf's functionality, then choosing grasses that excel in summer may be useful for the best utility and sustainability. If, for instance, early seasonal growth is shown to reduce annual grassy weed populations in late spring, then the selection of cultivars with excellent spring greenup and spring seasonal performance may be justified.

3 Utilizing National Turfgrass Evaluation Program data to select appropriate cultivars

The NTEP website, organized by turfgrass species (Fig. 1), year of trial initiation and trial year (Fig. 2), can be utilized to choose top-performing, sustainable cultivars. In addition, a new database and selection tool – the *Turfgrass Trial*

Figure 1 Example of NTEP links to data for each species and trial (https://ntep.org/previous.htm).

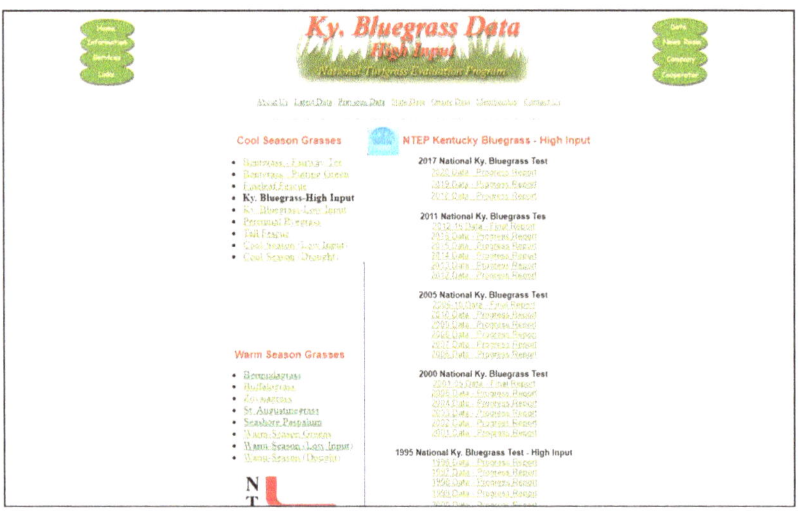

Figure 2 Example of links to select NTEP data by species, trial and year of data collection; Kentucky bluegrass – high input trial shown as a further example (https://ntep.org/kbh.htm).

Explorer - is available and improves the searching, sorting and analysis of NTEP data. In this chapter, case studies #1–6 use the *Turfgrass Trial Explorer* selection tool to evaluate NTEP data and identify sustainable cultivars. Case study #7 highlights potential future analysis and visualization of complex NTEP drought tolerance data using advanced tools.

Data on particular aesthetic and functional characteristics such as turfgrass quality, genetic color and various diseases are located on the left side of Fig. 3. As an example, maximizing summer density may be a good strategy to reduce weed infestation, thus improving sustainability of Kentucky bluegrass (*Poa pratensis* L.). Choosing 'Density' in Fig. 3 opens the links to seasonal density data in Fig. 4.

Figure 5 illustrates a typical NTEP table that displays summer density ratings from multiple locations, with a 'MEAN' (average of all ratings) to the right. Columns from left to right are NTEP entry 'NAME' (cultivar or experimental entry), trial locations (CT1 = Storrs, Connecticut, USA; IA1 = Ames, Iowa, USA, etc.) with trial location information found in the top left of Fig. 3 ('Locations/ Entry Info'). NTEP ratings are typically collected on a scale of 1–9, where 9 = ideal turfgrass quality, maximum density, dark green color, no disease, etc. Entry performance data is sorted from highest to lowest 'MEAN' in most NTEP tables. The statistical mean comparison procedure utilized in the Least Significant Difference Test (Williams and Abdi, 2010, 'LSD VALUE' - not shown in Fig. 5) is calculated and printed at the bottom of each NTEP table. In Fig. 5, the LSD = 0.6, which indicates all entries with mean (average) ratings within 0.6 are not statistically different.

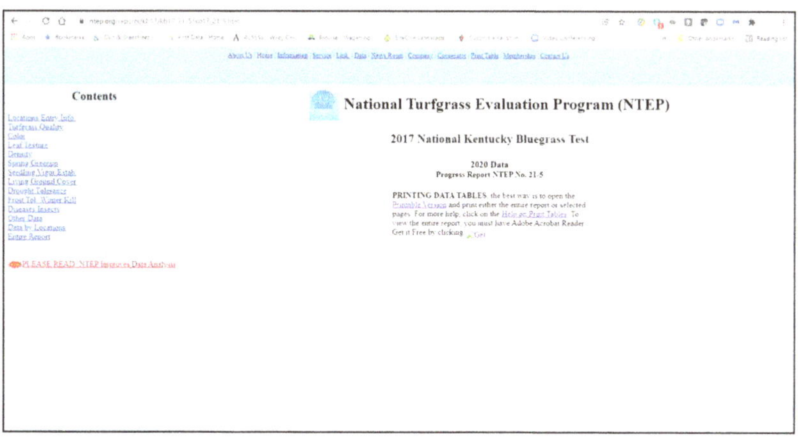

Figure 3 Example of the 2017 NTEP Kentucky bluegrass test (i.e. '2017' indicates trial was initiated in 2017), listing links to data from 2020.

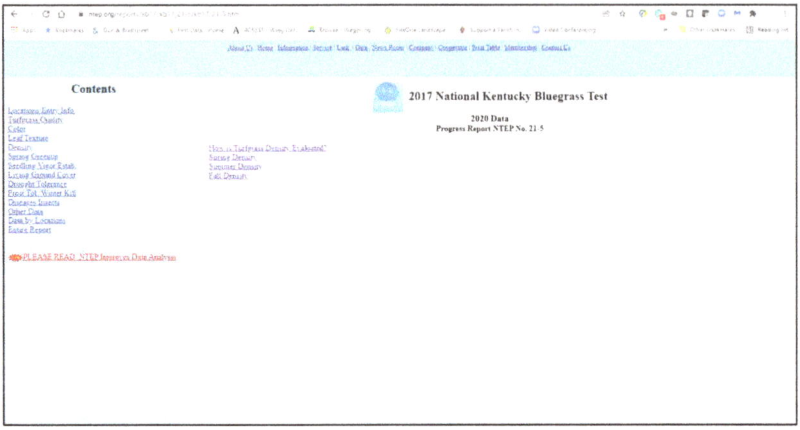

Figure 4 Example of density table links, to explore density data collected in 2020, from the 2017 NTEP Kentucky bluegrass test.

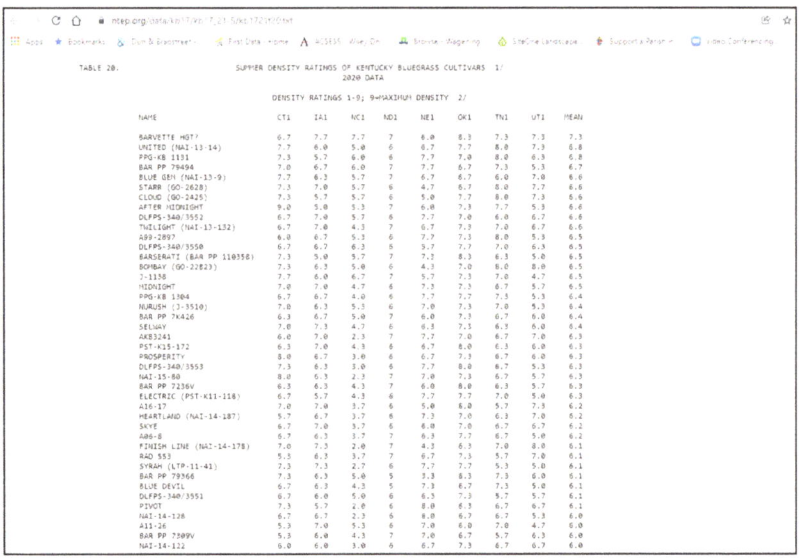

TABLE 20. SUMMER DENSITY RATINGS OF KENTUCKY BLUEGRASS CULTIVARS 1/
2020 DATA

DENSITY RATINGS 1-9; 9=MAXIMUM DENSITY 2/

NAME	CT1	IA1	NC1	ND1	NE1	OK1	TN1	UT1	MEAN
BARVETTE HGT?	6.7	7.7	7.7	7	6.0	8.3	7.3	7.3	7.3
UNITED (NAI-13-14)	7.7	6.0	5.0	6	6.7	7.7	8.0	7.3	6.8
PPG-KB 1131	7.3	5.7	6.0	6	7.7	7.0	8.0	6.3	6.8
BAR PP 79494	7.0	6.7	6.0	7	7.7	6.7	7.3	5.3	6.7
BLUE GEM (NAI-13-9)	7.7	6.3	5.7	7	6.7	6.7	6.0	7.0	6.6
STARR (GO-2628)	7.3	7.0	5.7	6	4.7	6.7	8.0	7.7	6.6
CLOUD (GO-2425)	7.3	5.7	5.7	6	5.0	7.7	8.0	7.3	6.6
AFTER MIDNIGHT	9.0	5.0	5.3	7	6.0	7.3	7.7	5.3	6.6
DLFPS-340/3552	6.7	7.0	5.7	6	7.7	7.0	6.0	6.7	6.6
TWILIGHT (NAI-13-132)	6.7	7.0	4.3	7	6.7	7.3	7.0	6.7	6.6
A99-2897	6.0	6.7	5.3	6	7.7	7.3	8.0	5.3	6.5
DLFPS-340/3550	6.7	6.7	6.3	5	5.7	7.7	7.0	6.3	6.5
BARSERATI (BAR PP 110358)	7.3	5.0	5.7	7	7.3	8.3	6.3	5.0	6.5
BOMBAY (GO-22823)	7.3	6.3	5.0	6	4.3	7.0	8.0	8.0	6.5
J-1138	7.7	6.0	6.7	7	5.7	7.3	7.0	4.7	6.5
MIDNIGHT	7.0	7.0	4.7	6	7.3	7.3	6.7	5.7	6.5
PPG-KB 1304	6.7	6.7	4.0	6	7.7	7.7	7.3	5.3	6.4
NURUSH (J-3510)	7.0	6.3	5.3	6	7.0	7.3	7.0	5.3	6.4
BAR PP 7K426	6.3	6.7	5.0	7	6.0	7.3	6.7	6.0	6.4
SELWAY	7.0	7.3	4.7	6	6.3	7.3	6.3	6.0	6.4
AKB3241	6.0	7.0	2.3	7	7.7	7.0	6.7	7.0	6.3
PST-K15-172	6.3	7.0	4.3	6	6.7	8.0	6.3	6.0	6.3
PROSPERITY	8.0	6.7	3.0	6	6.7	7.3	6.7	6.0	6.3
DLFPS-340/3553	7.3	6.3	3.0	6	7.7	8.0	6.7	5.3	6.3
NAI-15-80	8.0	6.3	2.3	7	7.0	7.3	6.7	5.7	6.3
BAR PP 7236V	6.3	6.3	4.3	7	6.0	8.0	6.3	5.7	6.3
ELECTRIC (PST-K11-118)	6.7	5.7	4.3	6	7.7	7.7	7.0	5.0	6.3
A16-17	7.0	7.0	3.7	6	5.0	6.0	5.7	7.3	6.2
HEARTLAND (NAI-14-187)	5.7	6.7	3.7	6	7.3	7.0	6.3	7.0	6.2
SKYE	6.7	7.0	3.7	6	6.0	7.0	6.7	6.7	6.2
A06-8	6.7	6.3	3.7	7	6.3	7.7	6.7	5.0	6.2
FINISH LINE (NAI-14-178)	7.0	7.3	2.0	7	4.3	6.3	7.0	8.0	6.1
RAD 553	5.3	6.3	3.7	7	6.7	7.3	5.7	7.0	6.1
SYRAH (LTP-11-41)	7.3	7.3	2.7	6	7.7	7.7	5.3	5.0	6.1
BAR PP 79366	7.3	6.3	5.0	5	3.3	8.3	7.3	6.0	6.1
BLUE DEVIL	6.7	6.3	4.3	5	7.3	6.7	7.3	5.0	6.1
DLFPS-340/3551	6.7	6.0	5.0	6	6.3	7.3	5.7	5.7	6.1
PIVOT	7.3	5.7	2.0	6	8.0	6.3	6.7	6.7	6.1
NAI-14-128	6.7	6.7	2.3	6	8.0	6.7	6.7	5.3	6.0
A11-26	5.3	7.0	5.3	6	7.0	6.0	7.0	4.7	6.0
BAR PP 7309V	5.3	6.0	4.3	7	7.0	6.7	5.7	6.3	6.0
NAI-14-122	6.0	6.0	3.0	6	6.7	7.3	6.7	6.7	6.0

Figure 5 Example of summary density table from data collected in 2020, from the 2017 NTEP Kentucky bluegrass test.

4 The updated National Turfgrass Evaluation Program database

NTEP data delivery via static pages on the current website is inefficient and greatly limits the querying power of the underlying data. For example, it takes a minimum of four mouse clicks on the current NTEP website to retrieve a report on a single trial (Xie et al., 2021). The report is a high-level summary of all results for individual entries, and does not allow for filtering of data for specific site characteristics or locations or performance trends. To address this issue, the University of Minnesota is collaborating with NTEP in a multi-year project to manage and interpret NTEP data. Surveys of turfgrass managers, seed company personnel, homeowners and other stakeholders determined that NTEP users fall into the following three categories, with differing needs and levels of expertise in turfgrass management:

1) BASIC - examples are homeowners, lawn hobbyists, water/conservation boards, homeowner associations, policymakers and many others.
2) PROFESSIONAL - examples are professional turfgrass managers, seed salespersons, marketers, brokers, sod producers and turfgrass extension specialists.
3) ADVANCED - plant breeders, turfgrass researchers and data scientists.

For the BASIC user, an online application was developed that allows homeowners to access appropriate regional data based on their postal code, and offers appropriate species and cultivars for their situation. For the ADVANCED user, NTEP and the University of Minnesota agreed to work on tools that further analyze and mine the database using advanced statistics and data science techniques. To address the needs of most NTEP users, focus was placed on developing a user-friendly interface targeting the needs of the PROFESSIONAL user.

Recently, as part of the 2017 USDA Specialty Crop Research Initiative, NTEP fineleaf fescue (*Festuca* spp.) data was ingested into the Postgres relational database (Xie et al., 2021). Before initiating the development of an interface for the PROFESSIONAL group, however, U-Spatial was utilized to import data from the remaining NTEP species into the Postgres relational database. These data files, currently more than one million records consisting over more than 100 possible variables, were collected from 1980 to the present. Additional tables were developed for trial-level information, site-specific information and participating cultivar information. Incorporating this data into a relational database allows the data to be efficiently explored and queried in one centralized place. Subsequently, the user interface, or selection tool, was developed and tested with the help of a group of about 40 volunteers, including golf course superintendents, seed company personnel, sod producers and researchers.

Therefore, the NTEP database and *Turfgrass Trial Explorer* selection tool (the PROFESSIONAL user interface) allows for further analysis of trial results and the components that impact overall species and cultivar performance, such as spring greenup, tolerance to drought and disease resistance. Conditions at trial sites can be explored in depth beyond geographical location, for instance, investigation of soil properties and turfgrass management parameters (i.e. mowing height, fertilizer practices, irrigation, etc.).

The *Turfgrass Trial Explorer* allows for customized analysis of the NTEP data, which the existing limited web-based static reports and tables cannot provide. Results can be sorted, subsetted and multiple variables can be compared in a single table. Conditions in the trials can be more closely correlated to conditions at a potential site to more closely match a cultivar's suitability and sustainability. Data can be easily downloaded for input into a spreadsheet for further sorting, development of fact sheets, electronic forwarding and printing. Because all of the data can be accessed via a single database, it can serve a much wider set of user groups, from the casual homeowner to the turfgrass manager to the academic researcher and seed company.

5 Case studies in selecting sustainable cultivars

The *Turfgrass Trial Explorer* is now accessible via the NTEP website at https://www.ntep.org/database.htm (Fig. 6). The *Turfgrass Trial Explorer* allows for the selection of individual cultivars, trials and evaluation parameters, as well as specific site characteristics. Figure 7 illustrates the selection of a single trial (2013 Bermudagrass - 2013–7 data) and evaluation parameter/variable (quality ratings). The table in Fig. 8 resulted from selections made in Fig. 7. In Fig. 8, trial locations providing data are listed across the top, with NTEP entries listed

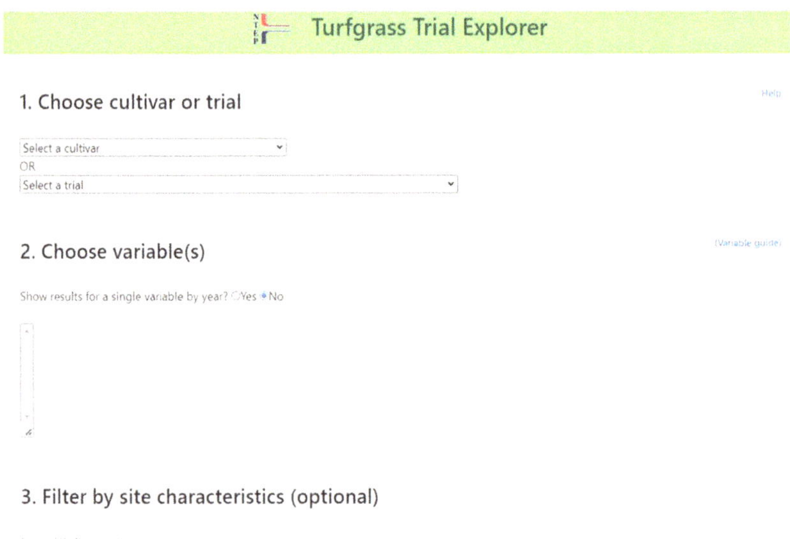

Figure 6 Opening screen of the *Turfgrass Trial Explorer*.

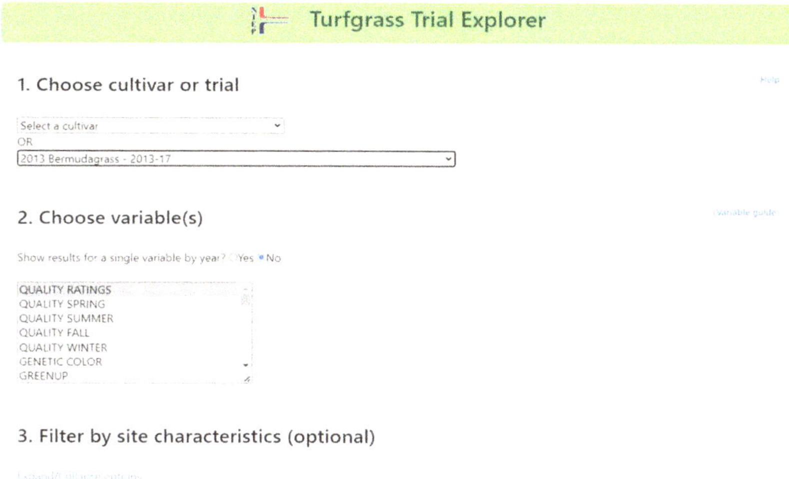

Figure 7 Selection of a single NTEP trial along with a single variable.

Search Reset

2013 Bermudagrass - 2013-17 - QUALITY RATINGS

* Top performing statistical group: green Export To CSV Link to your query:

NTEP Entry Name	♦Table avg♦	Auburn, AL	Fayetteville, AR	Tucson, AZ	Riverside, CA	Jay, FL	Griffin, GA	West Lafayette, IN	Wichita, KS	Lexington, KY	Lexington Traffic, KY	College Park, MD	Columbia, MO	Mississippi State, MS	Raleigh, NC	Stillwater, OK	Knoxville, TN
CV	1.5	9.4	6.5	7.2	4.1	4.2	12.4	4.7	5.2	5.0	7.2	8.9	11.0	2.7	6.4	2.4	2.8
LSD	0.1	0.7	0.7	0.7	0.4	0.4	1.1	0.5	0.4	0.5	0.8	0.8	0.8	0.1	0.5	0.2	0.1
TifTuf (DT-1)	6.9	7.1	7.2	7.4	6.7	6.3	6.3	7.0	5.5	7.3	7.3	6.7	6.6	7.0	6.5	6.1	7.5
Iron Cutter (JSC 2-21-18-v)	6.7	5.7	7.7	7.0	6.2	6.4	6.5	7.1	5.7	7.3	7.4	7.7	6.8	6.9	5.9	5.8	7.3
Tahoma 31 (OKC 1131)	6.7	5.7	8.0	6.7	5.8	5.7	6.0	7.7	5.5	7.3	7.7	7.6	6.9	6.8	5.7	6.2	7.4
11-T-510	6.6	6.6	7.2	7.0	6.4	6.0	5.6	6.5	5.5	7.0	7.3	6.6	6.4	6.8	6.1	6.0	7.4
Latitude 36	6.6	6.1	7.8	6.6	5.9	5.8	5.8	7.8	5.5	7.4	7.7	7.4	5.3	6.8	5.8	6.1	7.3
JSC 2-21-1-v	6.5	4.8	7.5	6.7	5.7	5.5	6.2	7.4	5.6	7.5	7.7	7.2	6.4	6.8	5.2	6.1	7.3
OKC 1163	6.5	5.1	8.6	7.1	6.2	4.7	6.1	7.0	4.9	7.1	6.9	6.2	6.6	6.7	4.9	6.0	6.9

Figure 8 Table resulting from selections made as indicated in Figs. 7 and 8.

by name and/or experimental designation in the far left side column. Note that an overall mean of all turfgrass quality ratings for each entry is listed in the 'Table Avg' column. Statistical values for LSD and coefficient of variation (LSD) are listed near the top of each column. Turfgrass quality ratings, like most NTEP ratings, are collected on a 1-9 scale, where 9 = ideal turf. In this instance, turfgrass quality results are sorted by the 'Table Avg' column (as noted by the ♦ designation). Results can be sorted from highest to lowest, or lowest to highest for any column by simply clicking on a column header. Also note that the green highlighted cells indicate those entries whose data are statistically equal to the top entry in that column. Finally, these results can be converted to a .csv file for further sorting and uploading into a spreadsheet by clicking the 'Export to CSV' button.

5.1 Case study 1: identify a geographical region

The table in Fig. 8 is useful in determining overall performance (i.e. turfgrass quality) for entries at each trial location. However, to identify the most or highly desired sustainable cultivars, top performance in multiple sites within a geographical region may be useful. For example, a site in Southern California is seeking data on sustainable cultivars. Figure 9 illustrates the selection of turfgrass quality ratings of the 2013 Bermudagrass trial in two southwestern US locations at Tucson, AZ, and Riverside, CA. The result is a turfgrass quality data table illustrated in Fig. 10, which was generated from selections in Fig. 9 of all locations in Fig. 8. For example, 'TifTuf' was the only entry in the top statistical group where all location data are averaged (Fig. 8), whereas in Fig. 10, five entries, including 'TifTuf', 'Tifway' and three experimental entries, were statistically equal in turfgrass quality at the two location average. Choosing the top-performing grasses in a geographical area can help achieve sustainability as those grasses often have less disease incidence and severity, reveal enhanced

Figure 9 Selection of turfgrass quality ratings in individual states.

2013 Bermudagrass - 2013-17 - QUALITY RATINGS

* *Top performing statistical group:* ***green*** Export To CSV Link to your query:

NTEP Entry Name	◆Table avg◆	Tucson, AZ	Riverside, CA
C.V.	4.1	7.2	4.1
LSD	0.4	0.7	0.4
TifTuf (DT-1)	7.1	7.4	6.7
Tifway	7.0	7.5	6.6
11-T-510	6.7	7.0	6.4
FAES 1325	6.7	7.2	6.1
OKC 1163	6.7	7.1	6.2
Iron Cutter (JSC 2-21-18-v)	6.6	7.0	6.2
11-T-251	6.3	6.6	5.9
FAES 1326	6.3	6.1	6.5
FAES 1327	6.3	6.8	5.8
Tahoma 31 (OKC 1131)	6.3	6.7	5.8
12-TSB-1	6.2	6.7	5.6
JSC 2-21-1-v	6.2	6.7	5.7

Figure 10 Table of turfgrass quality ratings from two trial locations and the mean (i.e. Table Avg).

performance relative to other entries and have an overall improved survivability profile.

5.2 Case study 2: use of seasonal analysis

Another turfgrass sustainability concept is understanding species and cultivars and their performance in different climatic seasons (Fig. 11). Seasonal quality is displayed in Fig. 12. For example, results sorted from highest to lowest 'Quality Summer' reveal entries 'Tahoma 31' and 'Iron Cutter' matched both spring and summer performance of 'TifTuf' (Fig. 12), which was not apparent in the information presented in Fig. 8.

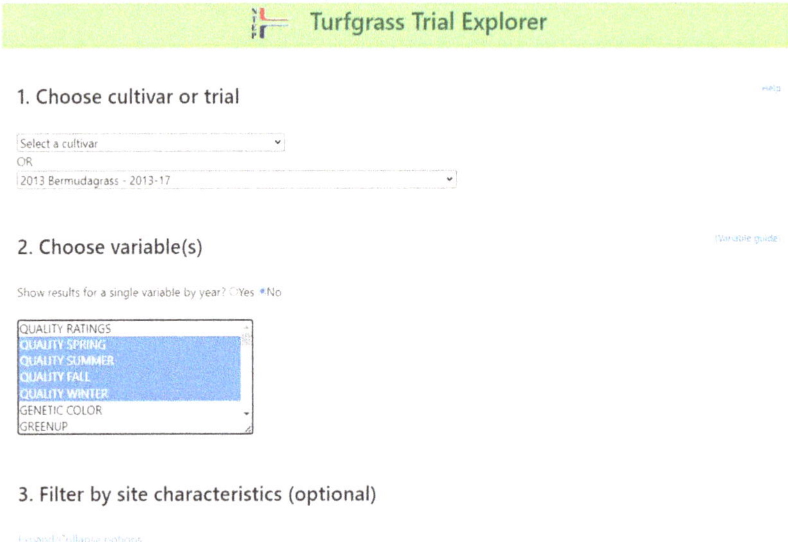

Figure 11 Choosing turfgrass quality ratings by season (i.e. spring, summer, fall, winter).

5.3 Case study 3: choose appropriate site characteristics

Evaluating soil and site characteristics as well as management level of turfgrass on the targeted site is important for choosing sustainable cultivars. Even though turfgrass soil properties can be modified, often turfgrass managers have limited funds and/or no opportunity to significantly impact site characteristics. Similarly, turfgrass management regimes are often determined by budgets and/or resources, limiting the options available to improve turfgrass quality, performance and sustainability. In this example, Fig. 13 illustrates the use of the *Turfgrass Trial Explorer* tool to select only trial locations with a soil pH of 4.6–5.5, which is considered below the optimum level for most turfgrass species and situations. The table in Fig. 14 contains data from 2013 NTEP Bermudagrass trial locations evaluated under low soil pH conditions. Multiple entries are statistically equal to 'TifTuf' (Fig. 14), in contrast to Fig. 8 where 'TifTuf' is the statistically superior entry in data averaged across all sites. The same analysis can be conducted for other soil characteristics, various mowing heights, nitrogen levels and other management parameters that may support sustainable cultivar choices. As seen in Figs. 13 and 14, sorting and analysis in this manner can lead to increased opportunities for sustainable cultivar selection.

2013 Bermudagrass - 2013-17 - QUALITY SPRING

* *Top performing statistical group:* ***green*** Export To CSV Link to your query: https://maps.umn.edu/nte

NTEP Entry Name	Quality Spring	◆Quality Summer◆	Quality Fall	Quality Winter
C.V.	2.2	1.5	1.6	3.9
LSD	0.2	0.1	0.2	0.3
Tahoma 31 (OKC 1131)	**6.4**	**7.0**	6.6	6.0
TifTuf (DT-1)	**6.4**	**7.0**	**7.0**	**6.8**
Iron Cutter (JSC 2-21-18-v)	**6.4**	**6.9**	6.6	6.1
Latitude 36	6.0	**6.9**	6.5	5.8
11-T-510	5.9	6.8	**6.8**	6.2
JSC 2-21-1-v	5.9	6.8	6.4	5.9
OKC 1163	**6.2**	6.7	6.3	5.6
OKC 1302	5.8	6.5	6.1	5.9
Patriot	5.8	6.5	6.1	5.2
Tifway	5.5	6.4	6.2	6.3
11-T-251	5.2	6.2	6.0	5.9

Figure 12 Display of turfgrass quality ratings by season (i.e. spring, summer, fall, winter).

5.4 Case study 4: identify key sustainability factors

To utilize warm-season species such as bermudagrass in the northern US sites where winter survivability is an issue, the *Turfgrass Trial Explorer* can create a table as illustrated in Fig. 15. Listing the entries by the lowest winter kill percentage (~the 2013–4 winter was particularly severe in the northern United States) highlights entries with the best chance to remain viable and sustainable in colder weather conditions. 'Greenup' (i.e. a measure of, or ability to achieve green color and growth early in spring) is an important usability trait in sites such as athletic fields and golf courses and enhances sustainability for those uses. The various drought tolerance measures (i.e. wilting, dormancy and recovery) in Fig. 15 document performance of these traits across the entries, and in this case, there were no statistical difference among the top entries.

5.5 Case study 5: locate and compare cultivars

Selecting a single creeping bentgrass cultivar ('777') in the *Turfgrass Trial Explorer* tool identifies the 2014 Bentgrass (Putting Green) trial with '777' as

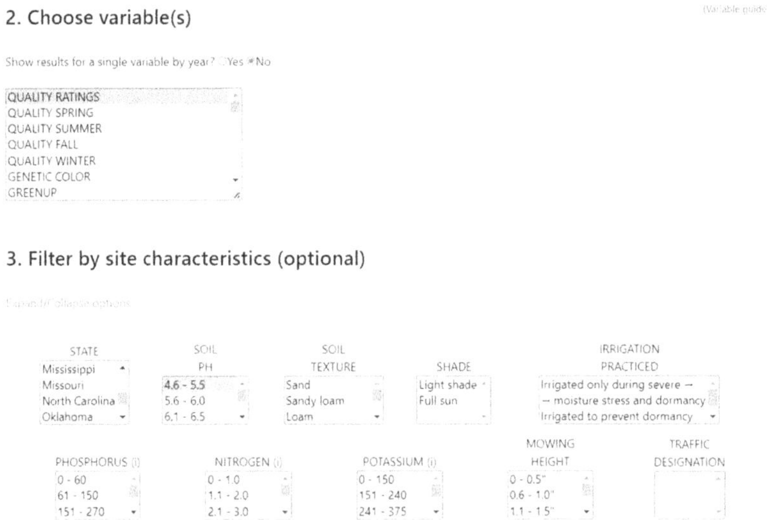

Figure 13 Selection of a soil characteristic to identify low soil pH locations.

2013 Bermudagrass - 2013-17 - QUALITY RATINGS

* *Top performing statistical group: green* Export To CSV Link to your query:

NTEP Entry Name	♦Table avg♦	Griffin, GA	Lexington, KY	Lexington Traffic, KY	Columbia, MO	Blacksburg, VA
C.V.	3.7	13.2	5.3	6.3	8.4	6.4
LSD	0.4	1.0	0.6	0.7	0.8	0.6
Iron Cutter (JSC 2-21-18-v)	6.7	6.6	7.4	7.0	5.5	6.4
JSC 2-21-1-v	6.7	6.3	7.4	7.3	5.7	6.4
Tahoma 31 (OKC 1131)	6.7	5.8	7.5	7.3	6.7	6.5
OKC 1163	6.6	6.1	6.9	7.0	5.3	6.9
Patriot	6.6	5.4	7.3	7.3	5.5	6.7
TifTuf (DT-1)	6.6	6.1	7.6	7.3	6.3	6.2
11-T-510	6.5	5.3	7.3	7.7	6.8	6.3
Latitude 36	6.5	5.6	7.4	7.3	5.7	6.3
Monaco (JSC 2007-13-s)	6.5	4.3	7.1	7.7	6.8	6.9
Riviera	6.5	4.4	7.1	7.7	6.5	6.7
JSC 2007-8-s	6.3	4.0	7.0	7.3	6.8	6.5

Figure 14 Low soil pH locations sorted by Table Avg from results across five locations.

2013 Bermudagrass - 2013-17 - GREENUP

** Top performing statistical group:* ***green*** Export To CSV Link to your query: https://maps.umn.

NTEP Entry Name	Greenup	◆Winter Kill◆	Wilting	Dormancy	Recovery
C.V.	3.0	22.4	52.5	25.1	41.1
LSD	0.2	21.6	3.7	1.9	3.4
Tahoma 31 (OKC 1131)	6.0	34.3	5.3	6.4	4.3
Yukon	5.0	37.8	3.5	5.4	5.0
JSC 2007-8-s	5.4	43.3	3.2	5.8	6.6
Riviera	5.5	44.5	5.0	5.4	4.3
Iron Cutter (JSC 2-21-18-v)	6.0	45.3	4.9	4.9	5.4
Rio (JSC 2009-6-s)	5.5	45.4	5.3	4.6	6.8
Patriot	5.1	47.0	6.4	4.6	5.3
OKC 1163	5.7	51.7	4.0	2.6	6.7
Monaco (JSC 2007-13-s)	5.3	53.0	2.6	4.5	6.0
JSC 2-21-1-v	5.4	55.1	2.3	6.1	3.0

Figure 15 Selecting data for greenup, winter kill and drought tolerance (i.e. wilting, dormancy, and recovery) parameters.

an entry (Fig. 16). This option of single cultivar choice helps locate a desired cultivar or experimental entry, and allows comparison of this choice with other tested entries for those traits selected. The resulting table in Fig. 17 compares '777' to other NTEP bentgrass entries for important or desired traits.

5.6 Case study 6: advanced selection of cultivars with improved drought resistance

The scarcity of water is one of the most significant environmental constraints to the survival and growth of turfgrasses. The selection of turfgrass species and cultivars with improved drought tolerance has been a major focus of NTEP. In 2016, NTEP initiated a multi-year and multi-location study on water use and drought tolerance of various cool-season cultivars (Table 1). Two approaches were proposed for the assessment of tested varieties. The first approach (Approach 1) focuses on estimating the minimal quantity of water required by each entry to maintain certain predefined quality attributes, while the second

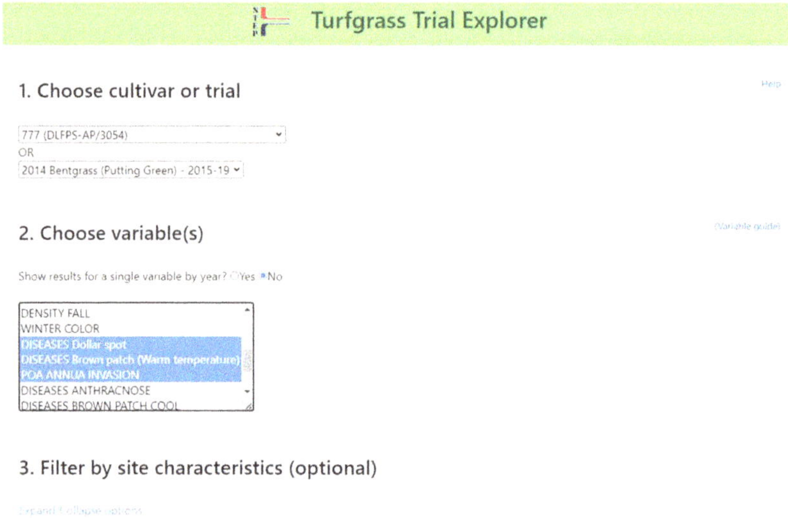

Figure 16 Choosing a single cultivar or experimental entry along with important traits with bentgrass sustainability for putting greens.

(Approach 2) examines the differences in turfgrass quality attributes when maintained under the same irrigation regime.

The following examples utilize Approach 1 to evaluate average quality during drought, and Approach 2 to evaluate quality-weighted water consumption.

5.6.1 *Evaluation of average quality during drought*

5.6.1.1 *Data processing and model fitting*

The listed entries (Table 1) were evaluated using Approach 2 at four locations from 2017 to 2020. These locations were Riverside, CA, Fort Collins, CO, Las Cruces, NM and Logan, UT. While Approach 1 measures the plot-level water consumption, Approach 2 focuses on the comparison of turfgrass quality when maintained under the same zone-level irrigation regime. During the period of drought treatment, all turfgrass entries were equally divided into three target irrigation potential evapotranspiration (ET_0) zones, which correspond to three levels of historical reference ET_0 at each testing location. Data collected via Approach 2 consist of green percent coverage data at regular intervals for 40%, 60% and 80% ET_0, respectively. The objective was to identify entries with the highest green percent coverage when subjected to the same irrigation levels. The target variable in the following analysis is average green percent coverage

2014 Bentgrass (Putting Green) - 2015-19 - QUALITY RATINGS

* Top performing statistical group: **green** [Export To CSV] Link to your query: https://maps.umn.edu/ntep/dev/?queryid=maeojon

◆NTEP Entry Name◆	Quality Value	Diseases Dollar Spot	Diseases Brown Patch Warm	Poa Annua Invasion
C.V.	1.2	2.3	5.6	8.4
LSD	0.1	0.3	0.6	1.0
777 (DLFPS-AP/3054)	6.3	7.0	7.2	7.7
Armor	5.7	4.5	7.0	7.5
Barracuda	6.2	7.7	7.2	7.0
Declaration	5.8	8.1	6.5	6.8
DLFPS-AP/3056	6.1	7.1	6.8	7.0
DLFPS-AP/3058	6.4	7.9	6.8	7.8
DLFPS-AP/3059	5.9	7.3	6.7	7.7
Kingdom	6.0	6.4	7.1	8.0
L-93 XD	6.3	8.0	6.4	6.8
Luminary	6.1	7.7	7.3	7.2
Macdonald (DLFPS-AP/3018)	6.4	7.4	6.9	8.0

Figure 17 Table generating results from selections made in Fig. 16.

(AVG_GPC) during drought treatment, which is defined by dividing the area under the curve of green percent coverage by the number of days in drought treatment. Specifically, for a given turfgrass plot with a set of observations $\{(x_i, y_i)\}$, where x_i is the number of days in drought and y_i is the green percent cover on that day,

1) a smooth curve, $f(x)$, was fitted using the local polynomial regression with a span parameter of 0.6, and a degree parameter of 2 as specified by package 'stats' in R (R Core Team, 2019);
2) calculate fitted values for $x = 0,1,,N$, where N is the maximum days in drought;
3) approximate area under the curve based on trapezoidal rule, i.e. $\int_0^N f(x)\mathrm{d}x \approx 0.5 \times \left[f(0) + 2f(1) + 2f(2) + \ldots + 2f(x_{N-1}) + f(x_N)\right]$;
4) divide area under the curve by the number of days in drought to obtain average green percent coverage.

Average green percent coverage data were subsequently subjected to analysis of variance (ANOVA), with the following formula:

$$AVG\_GPC = ENTRY + LOC + YEAR + ET_0 + ENTRY \times LOC + ENTRY \times YEAR$$

where AVG_GPC was partitioned into effects of entry, testing location, testing year, ET_0, the interaction between entry and location and interaction

Table 1 2016 National cool-season water use/drought tolerance test entries and sponsors

Entry No.	Name	Species	Sponsor
***1**	Barserati (BAR PP110358)	Kentucky bluegrass	Barenbrug USA
***2**	Barrari	Kentucky bluegrass	Barenbrug USA
***3**	Everest	Kentucky bluegrass	Jacklin Seed
***4**	Blue Note	Kentucky bluegrass	Mountain View Seeds
***5**	Babe	Kentucky bluegrass	Seeds, Inc.
6	NAI-13-132	Kentucky bluegrass	Columbia River Seed
7	NAI-13-14	Kentucky bluegrass	Columbia River Seed
***8**	Blue Devil	Kentucky bluegrass	Columbia River Seed
***9**	Dauntless	Kentucky bluegrass	Columbia River Seed
10	PST-K13-137	Kentucky bluegrass	Pure-Seed Testing, Inc.
***11**	Orion (PST-K13-143)	Kentucky bluegrass	Dye Seed Ranch, Inc.
12	PST-K15-169	Kentucky bluegrass	Pure-Seed Testing, Inc.
13	PST-K11-118	Kentucky bluegrass	Pure-Seed Testing, Inc.
14	PST-K13-141	Kentucky bluegrass	Pure-Seed Testing, Inc.
***15**	Midnight	Kentucky bluegrass	Standard entry
***16**	SR 4650	Perennial ryegrass	Standard entry
***17**	BarRobusto	Tall fescue	Barenbrug USA
18	BAR FA 121095	Tall fescue	Barenbrug USA
19	DLFPS 321/3677	Tall fescue	DLF Pickseed USA
20	DLFPS 321/3679	Tall fescue	DLF Pickseed USA
21	DLFPS 321/3678	Tall fescue	DLF Pickseed USA
***22**	Nonet	Tall fescue	Jacklin Seed
23	GO-AOMK	Tall fescue	Grassland Oregon
***24**	Supersonic	Tall fescue	Mountain View Seeds
***25**	Titanium 2LS	Tall fescue	Mountain View Seeds
***26**	Thor	Tall fescue	Columbia Seeds
***27**	Thunderstruck	Tall fescue	Columbia Seeds
28	RS4	Tall fescue	Landmark Turf & Native Seed
***29**	Kingdom	Tall fescue	Site One Landscape Supply
30	MRSL TF15	Tall fescue	Site One Landscape Supply
***31**	Catalyst	Tall fescue	Standard entry
***32**	Stetson II	Tall fescue	Site One Landscape Supply
33	PST-5SDS	Tall fescue	Pure-Seed Testing, Inc.
34	PST-R511	Tall fescue	Pure-Seed Testing, Inc.
***35**	Matisse (LTP-SYN-A3)	Tall fescue	Lebanon Seaboard Corp.

*Commercially available in the United States in 2021 or any other countries.

between entry and year. Given that only three ET_0 levels were implemented at each location, ET_0 was treated as a categorical variable. Estimated mean effects were compared using Fisher's Least Significant Difference Test with Bonferroni methods for correlation of multiple comparisons. Differences in average green percent coverage among species were also investigated by substitute entry with species in the above formula, which is done to avoid collinear variables in the model.

5.6.1.2 Results and discussion

The analysis of variance revealed significant effects of entry, location, year, ET_0 and interaction between entry and location on average green percent coverage during drought; however, no significant effect was found for the interaction between entry and year (Fig. 18). This suggested tested entries performed differently across various testing locations, but are consistent over years. Among all the entries, 'RS4', 'DLFPS 321/3678', 'Matisse (LTP-SYN-A3)', 'DLFPS 321/3679' and 'DLFPS 321/3677' demonstrated top average green percent coverage during the drought (Fig. 18). Of note, minimal segregation was found among all the entries. The listed 35 entries were divided into only 5 statistical groups. Since entry and location interaction was found statistically significant, average green percent coverage at each location was also investigated. A better segregation of entries was observed at Logan, UT, and Riverside, CA, compared to the other two locations. As far as differences among species, tall fescue demonstrated significantly higher average green percent coverage, with an estimated mean of 67.31 (±0.01), followed by perennial ryegrass (63.41±0.18) and Kentucky bluegrass (61.98±0.01). No statistical differences were detected in average green percent coverage between perennial ryegrass and Kentucky bluegrass.

5.6.2 Evaluation of quality-weighted water consumption

5.6.2.1 Data processing and model fitting

From 2017 to 2020, five other test locations adopted Approach 1 to evaluate listed entries. These locations include Amherst, Massachusetts; College Park, Maryland; Fayetteville, Arkansas; Griffin, Georgia; and West Lafayette, Indiana. Two trials were conducted at Amherst, Massachusetts, in 2020. Data collected via Approach 1 consist of two essential attributes for each entry (i.e. a time series of percent green coverage) captured by digital images (i.e. same as that in Approach 2) and total water consumption during the entire duration of drought (i.e. around 100 days). The objective was to identify entries with less water consumption while maintaining a relatively high percent green coverage

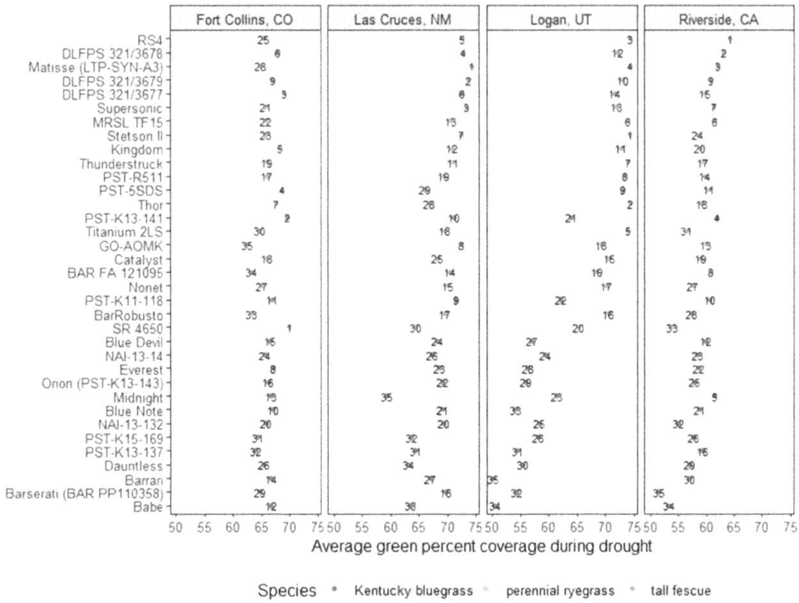

Figure 18 Mean and rank of entry effect on average green percent coverage estimated for each location from 2016 USGA-NTEP national cool-season water use/drought trial. The average green percent cover during drought (i.e. *y*-axis) was labeled by ranked entry effect estimated from all data collected from 2017–20 at four locations. Within each facet, entry effects at each location were estimated using the model: AVG_GPC = ENTRY + YEAR + ET_0 + ENTRY × YEAR. Numbers on the dots indicated rank of entry by average green percent coverage at each location.

during the period of drought. To accomplish this goal and to evaluate data from different locations, total water consumption during drought was weighted by the area under the curve of percent green coverage, which is hereinafter referred to as quality-weighted water consumption. For a given turfgrass plot, the first item to be calculated is the area under the curve of green percent coverage, as described in Section 5.7.1.1 and then divide observed water consumption by the area under the curve to obtain quality-weighted water consumption.

The transformed dependent variable (i.e. quality-weighted water consumption) can be interpreted as the amount of water required to maintain 1% of green coverage for one day. One limitation of this method is the failure to consider green percent coverage when there was no water consumption as the quotient will always be zero. To address this issue, 20 mm was added to the originally observed water consumption prior to being weighted. This will inflate the final estimates of quality-weighted water consumption, but allow better separation of entries with zero water consumption and preserve the relative rankings of the rest. Weighted water consumption data were subsequently subjected to analysis of variance, in which quality-weighted water consumption was partitioned into effects of entry (or species) and testing environment (i.e. location and year). Estimated mean effects were compared using Fisher's Least Significant Difference Tet with Bonferroni methods for correlation of multiple comparisons. Differences in water consumption among species were also investigated. Due to collinearity between entry and species, entry and species were not included in the same model as independent variables.

5.6.2.2 Results and discussion

When modeling data, four basic assumptions are made: (1) individual observations are mutually independent; (2) the data adhere to an additive model of variable effects and random errors; (3) random errors are normally distributed; and (4) within-group errors have the same variance across all treatment groups (Eisenhart, 1947). The first two assumptions are generally met in designed experiments. Both the original and transformed water consumption data in this study followed a heavy-tailed distribution, leading to non-normal error distributions. One potential method is to apply a regression model with *t*-distributed errors, which will be less sensitive to possible outliers. This method was not adopted for the following reasons: (1) Outliers are indications of inconsistency in entry performance and we want to penalize entries with outliers; (2) analysis of variance is reasonably robust against violations of the normality assumptions. When it comes to the assumption of equal variance, transformed data showed homogeneity of variance, while the original data did not (i.e. P-value $< 2.2^{e16}$) based on Levene's Test (Iachine et al., 2010). Histograms of both original and transformed data are presented in Fig. 19.

A two-way analysis of variance was performed to analyze the effects of entry and test environment on quality-weighted water consumption. The results revealed no statistically significant interaction between the effects of entry and environment, suggesting the consistent performance of different entries across various testing environments (Table 2). Simple main effects analysis showed both entry and environment have statistically significant effects on weighted water usage. Among all the entries evaluated, 'RS-4' topped the list with the latest amount of water consumption, followed by 'Titanium 2LS', 'DLFPS 321/3679', 'Kingdom' and 'DLFPS 321/3678' (Fig. 20), similar to the ranked list in Fig. 18 as estimated from Approach 2 data. Although a ranking of cultivars is provided, it is important to point out that many of them belong to the same statistical group (Fig. 20). Among different species, however, significant interactions are observed between the species and the environment. In four of the five testing locations, Kentucky bluegrass consumes more water than tall fescue to maintain the same level of green percent coverage for the same

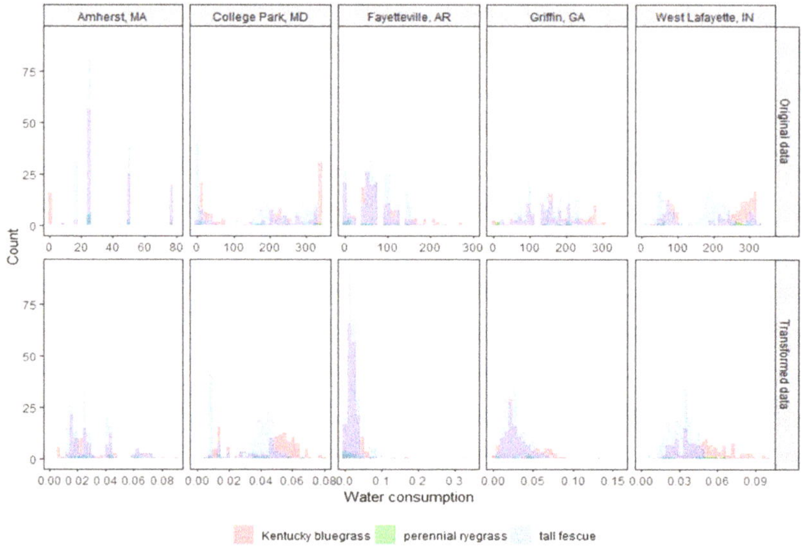

Figure 19 Distributions of original and transformed water consumption data.

Table 2 Analysis of variance evaluating effects of entry and test environment on quality-weighted water consumption in 2016 USGA-NTEP national cool-season water use/drought trial

	Degrees of freedom	Sum of squares	Mean squares	*F*-value	*P*-value
Entry	34	0.03094	0.000910	3.732	8e[12] ***
Environment	15	0.26759	0.017839	73.173	<2e[16] ***
Entry × environment	505	0.13146	0.000260	1.068	0.19
Residuals	1110	0.27061	0.000244		

Significance of *P*-values: '***' 0.001, '**' 0.01, '*' 0.05, '.' 0.1 and ' ' 1.

length of time in drought (Fig. 21). With all data pooled together, Kentucky bluegrass consumed significantly more water than tall fescue, and perennial ryegrass consumed significantly more water than tall fescue but not significantly different from Kentucky bluegrass (Fig. 21). This conclusion might not apply to other cultivars of perennial ryegrass, considering only one cultivar ('SR 4650') was evaluated.

6 Conclusion and future trends

NTEP data is valuable in identifying and selecting sustainable turfgrass species and cultivars. Locating that data is freely available through the NTEP website; however, static web pages limit the ability to easily identify and visualize

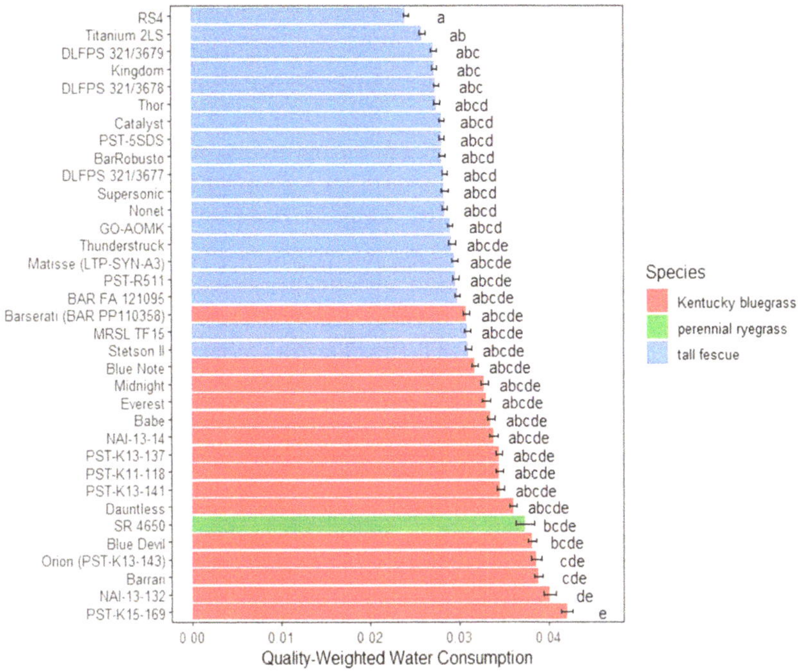

Figure 20 Mean and standard error of quality-weighted water consumptions for tested entries estimated from 2016 USGA-NTEP national cool-season water use/drought trial.

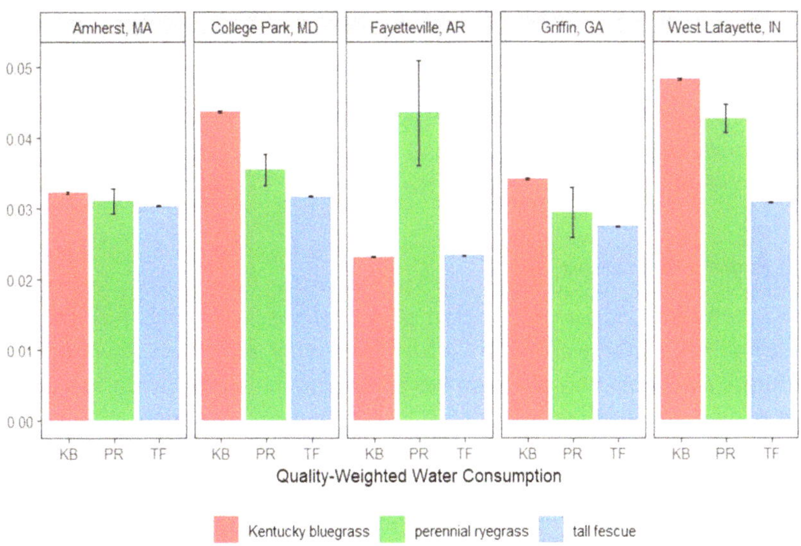

Figure 21 Mean and standard error of quality-weighted water consumptions at five testing locations in 2016 USGA-NTEP national cool-season water use/drought trial.

data on the various sustainable characteristics. A new database and tool, the *Turfgrass Trial Explorer*, simplifies and enhances sustainable turfgrass selection and will lead to reduced and optimum pesticide, water, nutrient and mowing requirements in turfgrass management. New developments and tools in data science, however, represent opportunities for improvements in the analysis and visualization of NTEP data as those appropriate technologies will most likely be incorporated into future versions of the *Turfgrass Trial Explorer*.

Different from agricultural commodities that are of the same grade, quality or type, each turfgrass cultivar has its own uniqueness. The aforementioned case studies focus on a specific trial; however, the same cultivars were also evaluated in other trials. Examining their performance in other trials for other variables will provide further information and insight into their sustainability across all tested locations or specific environments. Future work can be divided into two parts: knowing the past and facing the future.

6.1 Knowing the past

This part focuses on the analysis of historical data. Currently, summarized turfgrass ratings, either located on the NTEP website or via the *Turfgrass Trial Explorer* selection tool, would benefit from additional analysis and visualization options. For instance, robust and customized models are required to (1) compare cultivars across multiple locations, years and trials; (2) identify interactions between cultivars and environment; and (3) integrate information on the testing location (i.e. weather pattern, soil texture, management intensity, etc.). While data analysis turns facts, numbers and measurements into useful information, it is still essential to present the information in the most persuasive and understandable manner. Visual representations, such as graphs and charts, provide more accessible ways to understand trends, spot outliers and identify patterns in the results. This also helps to identify possible issues with current cultivar evaluation procedures.

6.2 Facing the future

The second part of the work is built on top of the first. Analysis of historical data helps identify sustainable cultivars and their adaptable environments. Turfgrass managers may utilize collected environment information to develop best management practices. Turfgrass breeders could investigate the pedigree record of these cultivars, thus making crosses to develop more sustainable varieties. Turfgrass researchers may be able to develop predictive models that correlate environment conditions, weather patterns and management levels to produce a recommendation for sustainable cultivars.

7 Where to look for further information

The following publications provide a good overview of the subject:

- Casler, M. D. and Duncan, R. R. (2003), Origins of the turfgrasses, In M. D. Casler and R. R. Duncan (Eds.), *Turfgrass Biology, Genetics and Breeding*, pp. 5–23, Madison, WI: American Society of Agronomy.
- Stier, J. C., Horgan, B. P. and Bonos, S. A. (2013), *Turfgrass Biology, Use and Management*, Agronomy Monograph No. 56, Madison, WI: American Society of Agronomy.

Key research in this area can be found at the following organizations:

- United States Department of Agriculture, Agricultural Research Service (www.ars.usda.gov).
- United States Golf Association (www.usga.org; www.usga.org/course-care/about-the-green-section.html).

8 References

Beard, J. B. (1973), *Turfgrass Science and Culture*, Englewood Cliffs, NJ: Prentice-Hall, Inc.

Beard, J. B. (2002), *Turf Management for Golf Courses*, Chelsea, MI: Ann Arbor Press.

Beard, J. B. and Green, R. L. (1994), The role of turfgrasses in environmental protection and their benefits to humans, *J. Environ. Qual.*, 23(3), 452–460.

Brede, D. (2000), *Turfgrass Maintenance Reduction Handbook: Sports, Lawn and Golf*, Chelsea, MI: Ann Arbor Press.

Carrow, R. N. and Duncan, R. R. (1998), *Salt-affected Turfgrass Sites: Assessment and Management*, Chelsea, MI: Ann Arbor Press.

Christians, N. C., Patton, A. J. and Law, Q. D. (2017), *Fundamentals of Turfgrass Management* (5th edn.), Hoboken, NJ: John Wiley and Sons, Inc.

Dodson, R. G. (2005), *Sustainable Golf Courses: A Guide to Environmental Stewardship*, Hoboken, NJ: John Wiley and Sons, Inc.

Duncan, R. R. and Carrow, R. N. (2000), *Seashore Paspalum: The Environmental Turfgrass*, Chelsea, MI: Ann Arbor Press.

Eisenhart, C. (1947), The assumptions underlying the analysis of variance, *Biometrics*, 3(1), 1-21.

Iachine, I., Petersen, H. C. and Kyvik, K. O. (2010), Robust tests for the equality of variances for clustered data, *J. Stat. Comput. Simul.*, 80(4), 365–377.

Krans, J. V. and Morris, K. (2007), Determining a profile of protocols and standards used in the visual field assessment of turfgrasses: a survey of national turfgrass evaluationprogram-sponsored university scientists, *Appl. Turfgrass Sci.*, 4(1), 1-6. https://doi.org/10.1094/ATS-2007-1130-01-TT.

Kopp, K. and Johnson, P. (2022), Turfgrass cultivars for Utah, Available at: https://extension.usu.edu/yardandgarden/research/turfgrass-cultivars-for-utah. Accessed July 7, 2022.

Landschoot, P. L. (2016), Turfgrass species for Pennsylvania, Available at: https://extension.psu.edu/turfgrass-species-for-pennsylvania. Accessed July 7, 2022.

Mandel, M., Chandran, R. and Balasko, J. (2013), Amending subsoil with composted poultry litter-II: effects on Kentucky bluegrass (*poa pratensis*) establishment, root growth, and weed populations, *Agronomy*, 3(4), 670–684. https://doi.org/10.3390/agronomy3040670.

Morris, K. N. and Kenna, M. P. (2017), A national cultivar evaluation of turfgrass water requirements and drought resistance, *Int. Turfgrass Soc. Res. J.*, 13(1), 119–122. https://doi.org/10.2134/itsrj2016.09.0831.

Puhalla, J., Krans, J. and Goatley, M. (1999), *Sports Fields: A Manual for Design, Construction and Maintenance of Sports Fields*, Chelsea, MI: Ann Arbor Press.

R Core Team (2019), R: A Language and Environment for Statistical Computing, Vienna, Austria: R Foundation for Statistical Computing, Available at: https://www.R-project.org/.

Sachs, P. D. and Luff, R. T. (2002), *Ecological Golf Course Management*, Chelsea, MI: Ann Arbor Press.

USEPA (2022), Integrated pest management (IPM) principles, Available at: https://www.epa.gov/safepestcontrol/integrated-pest-management-ipm-principles. Accessed July 5, 2022.

Vargas, J. M., Jr. and Turgeon, A. J. (2004), *Poa Annua: Physiology, Culture, and Control of Annual Bluegrass*, Hoboken, NJ: John Wiley and Sons, Inc.

Williams, L. J. and Abdi, H. (2010), Fisher's least significant difference (LSD) test, *Encyclopedia Res. Des.*, 218, 840–853.

Xie, Y., Farhadloo, M., Guo, N., Shekhar, S., Watkins, E., Kne, L., Bao, H., Patton, A. J. and Morris, K. (2021), NTEP-DB 1.0: a relational database for the national turfgrass evaluation program, *Int. Turfgrass Soc. Res. J.*, 14(1), 316–332. https://doi.org/10.1002/its2.76.

Chapter 18

Considerations with turfgrasses and pollinators

Michelle Wisdom and Michael Richardson, University of Arkansas, USA; and Paige Boyle, Utah State University, USA

1 Introduction

Ecosystem services are commonly defined as the benefits humans obtain from the surrounding network of organisms and the physical environment they interact with (Millennium Ecosystem Assessment, 2003). Ecosystem services can include a wide array of benefits such as production of food and fiber, soil conservation, water purification, and pollination. Many of these vital services are considered to be at risk (Daily, 1997; Palmer et al., 2004), especially in areas where the natural environment has become fragmented by an increasingly urbanized society. It is critical to understand the ecosystems' stressors (e.g. habitat fragmentation, pollution, and climate change) and methods for appropriate management (Cavender-Bares et al., 2018, 2020) in order to enhance and maintain ecosystem services, particularly in urban settings.

Pollinators are vectors for the transfer of pollen from the male part of a flower to the female part of a flower, facilitating seed set and fruit production. Pollination is not only a natural ecosystem service but also a managed production practice used extensively by farmers all over the world for crop

http://dx.doi.org/10.19103/AS.2022.0110.18

production (Gallai et al., 2009); its global value for agriculture has been estimated at US$235–577 billion (Intergovernmental Science-Policy Platform on Biodiversity and Ecosystem Services, 2016). Insect pollinators such as honey bees (Fig. 1) are necessary for the production of 84% of crop species cultivated in Europe (Williams, 1994). Globally, 87 of the top 124 food crops rely on pollination services (Klein et al., 2007).

The documented decline in pollinator populations has been attributed to a range of factors and has generated worldwide interest in the topic, and this chapter will discuss reasons behind this decline. While there are countless approaches being taken to address the issue, this chapter will propose and discuss some emerging thoughts on how to enhance currently underutilized habitats, such as lawns and other turf areas, to help expand and connect the currently fragmented and degraded pollinator habitats to support pollinating insects. Case studies highlight actions being taken in home lawns and golf course out-of-play areas to support pollinators. Potential drawbacks of mixed grass/forb (broadleaf flowering plants) plantings and future research needs are also addressed.

2 Land use change and pollinator decline

Over the past century, the human population has increased exponentially, growing from roughly 1.5 billion at the start of the twentieth century to a current worldwide population of almost 8 billion (United States Census Bureau,

Honey bee (*Apis mellifera*) foraging on crocus (*Crocus flavus* 'Golden Yellow') in a dormant bermudagrass (*Cynodon dactylon*) lawn in Arkansas, USA. (photo by M. Wisdom)	Honey bee (*Apis mellifera*) foraging on grape hyacinth (*Muscari* spp.) in planting bed in Arkansas, USA. (photo by M. Wisdom)	Bumble bee (*Bombus* spp.) foraging on self-heal (*Prunella vulgaris*) in a bermudagrass (*Cynodon dactylon*) lawn in Arkansas, USA. (photo by M. Richardson)

Figure 1 Pollinating insects found on several common forbs found in lawn sites. (a) Honey bee (*Apis mellifera*) foraging on crocus (*Crocus flavus* 'Golden Yellow') in a dormant bermudagrass (*Cynodon dactylon*) lawn in Arkansas, USA. Source: Photo by M. Wisdom. (b) Honey bee (*Apis mellifera*) foraging on grape hyacinth (*Muscari* spp.) in planting bed in Arkansas, USA. Source: Photo by M. Wisdom. (c) Bumble bee (*Bombus* spp.) foraging on self-heal (*Prunella vulgaris*) in a bermudagrass (*Cynodon dactylon*) lawn in Arkansas, USA. Source: Photo by M. Richardson.

n.d.). This unprecedented growth has resulted in significant changes to the landscape of where people live and how land is utilized. These changes have ultimately influenced and affected everything from climate, to food production, to energy demand, to natural ecosystems and the organisms found in those environments. In addition to the increase in the overall population, there is a growing trend in worldwide urbanization (Cohen, 2003), which is especially evident in the USA. Although approximately 97% of the U.S. landmass is considered to be rural, less than 20% of the population lives in those rural areas (US Census Bureau, 2017). Globally, it is estimated that by 2050, 70% of the world's population will be living in urban areas (UNDESA, 2018). As such, high-density urban landscapes are becoming more common across the planet, and those landscapes must first provide a range of functional services (e.g. housing, transportation, employment, and recreation) to the humans that occupy them. However, those landscapes should also be designed to promote valuable ecosystem services, by enhancing air and water quality, protecting soil and water resources, and providing habitat for other organisms that utilize those landscapes. The conundrum is how to blend functional services with ecosystem services, in order to benefit all, with as little detriment to each as possible (Grimm et al., 2000).

A further outcome of the increase in worldwide population is the demand that is now placed on agricultural systems to produce an adequate, affordable, and safe food supply. Over the last century, huge increases in crop productivity have been realized, but production has been forced to move outside the best agricultural lands to other natural and seminatural areas that previously were left uncultivated (Lark et al., 2020). This expansion of crop acreage has resulted in a significant loss of natural habitat and has created larger 'islands' of crop monocultures that can minimize species abundance and can significantly affect the organisms that once depended on native habitats (Raven and Wagner, 2021).

One group of organisms that has been especially affected by these shifts in human activity is the pollinating insect, and there have been well-documented declines in numbers of many native and exotic pollinators over the past decade. In Britain, in addition to pre-1980 losses of pollinators noted in previous studies (Carvalheiro et al., 2013), declines across a large portion of pollinator species occurred between 1980 and 2013. The declines are driven by a host of causes known to affect pollinators, including habitat loss, climate change, and the use of pesticides (Powney et al., 2019). In the USA, where the production of pollinator-dependent crops is valued at over US$50 billion per year (Bauer and Wing, 2016; Calderone, 2012), evidence suggests that pollinator decline has negatively impacted crop production (Reilly et al., 2020).

Land use changes and subsequent habitat fragmentation can lead to the loss of food and nesting resources for insect communities and can ultimately affect the behavior, dynamics, and composition of pollinator populations

(Cane et al., 2006; Hines and Hendrix, 2005; Potts et al., 2005; Tscharntke et al., 2005). Approximately 1 million hectares of U.S. farmland and natural habitat are converted annually to urban and suburban landscapes (USDA, 2009), and turfgrasses are a predominant land cover in these urbanized areas (Robbins et al., 2001), increasing by an estimated 155 000 ha across the USA annually (Robbins and Birkenholtz, 2003). A large expanse of landscape devoid of floral resources may act as a barrier for pollinator movement, while a landscape comprised of pockets of nearby flowering crops or other flowering plants can provide nectar and pollen as well as corridors for movement between additional flowering sites (Braaker et al., 2017; Chacoff and Aizen, 2006; Westphal et al., 2003).

It is worth noting that a growing body of research suggests urban settings may provide appropriate habitat and nutritional resources for pollinating insects, especially bees (Baldock, 2020; Theodorou et al., 2020). Urban green spaces comprise publicly and privately held lands such as parks, cemeteries, playing fields, churchyards, business, school, and industrial lawns, green roofs, utility rights-of-way, and roadsides. As urban expansion continues to displace the native species inhabiting these areas, there is still potential for these urban spaces to act as contributors to pollinator conservation (Wenzel et al., 2020), and it is possible for them to support a number of the affected pollinator groups. Incorporating floral resources into each of the abovementioned areas, with an emphasis on appropriate floral species and connectivity within the urban center, will encourage ecological harmony between humans and pollinating species.

3 Pollinators in the landscape

Pollinator abundance (number of individuals) and species richness (number of species) are positively associated with the abundance and species richness of flowering plants, and a decline in pollinating insect species can lead to a parallel decline of associated plant species that rely on them (Banaszak, 1996; Biesmeijer et al., 2006; Potts et al., 2003a; Steffan-Dewenter et al., 2001). Conversely, species richness of pollinators is equally affected by the diversity and availability of nectar sources, the ratio of pollen to nectar energy content, and floral morphology or structure (Bosch et al., 1997; Potts et al., 2003). Not all flowers support pollinators, so a variety of flowers with different bloom structures and nectar and pollen resources can benefit a wider range of pollinators (Fenster et al., 2004; Klein et al., 2007). Nectar is the main energy source for worker bees, while pollen provides proteins and lipids to both adults and larvae (Haydak, 1970; Keller et al., 2005; Requier et al., 2015). Diverse pollen resources are known to enhance the resistance of bees to disease, parasites, and pathogens and improve their tolerance to pesticides (Alaux et al., 2010; Di Pasquale et al., 2013). Seasonality is also an important aspect of floral resources, as a variety of season-long pollen and nectar resources are essential to support bee diversity (Williams et al., 2015).

Management and conservation of local agroecosystems and associated nearby natural, seminatural, and urban-managed pollinator habitats are receiving more attention, as appropriate pollinator habitat decreases (Klein et al., 2007). During monoculture mass-flowering events, honey bees will collect nectar from large expanses of crop flowers (Blitzer et al., 2012); however, they will still rely on the surrounding herbaceous weeds and woody plants for pollen resources (Requier et al., 2015). Additionally, between mass-flowering crop events, honey bees rely on other floral resources, with up to 40% of the diet during these gap periods consisting of weedy and woody species foraged from within adjacent seminatural habitats (Requier et al., 2015). Diverse floral resources must be available before, during, between, and after crop blooms to extend the foraging season for both native bees and honey bees (Menz et al., 2011; Requier et al., 2015), and surrounding urban landscapes can be thoughtfully planted to provide those additional resources. One way to incorporate supplemental floral resources is to add flowering plants into commercial and residential turfgrass ecosystems.

Healthy pollinator habitats should provide abundant forage resources, but these areas must also provide cover, access to water, and suitable nesting sites (Cane, 2001; Cane et al., 2007; Rands and Whitney, 2011). Manipulation of plant communities to conserve natural or seminatural habitats, including nesting sites, is currently the best management strategy for protecting wild bee abundance and diversity (Kremen et al., 2007; Kremen and Chaplin-Kramer, 2004; Morandin and Winston, 2006; Potts et al., 2005). Central-place foragers are pollinators that return to fixed nesting sites with food and water, which can then be stored and distributed (Mandelik et al., 2012; Ricketts et al., 2008). Pollinators that utilize fixed nesting sites will have a restricted foraging range around these nesting sites, which dictates that nesting sites must be located near appropriate floral resources. As such, this group of pollinators can be very sensitive to landscape changes that disrupt access to both of these critical resources (Kremen et al., 2007).

Multifunctionality of urban ecosystems is a key aspect of promoting land use as a pollinator habitat (Wratten et al., 2012), and it should be considered as an intentional step when establishing pollinator habitats, not a secondary consideration. Management techniques on farmland include planting and maintaining diverse flower-rich mixes in fallow fields, field margins, and conservation buffer strips (Burel, 1996; Haaland et al., 2011; Nichols and Altieri, 2013; Sidhu and Joshi, 2016; Wratten et al., 2012). Similar concepts could be implemented in urban landscapes to help support pollinators and connect fragmented habitats. Lawns, gardens, woody landscape plants, and fragments of natural and seminatural habitat may be established and managed to better support pollinator communities but doing so requires understanding the foraging and nesting requirements of urban pollinators and balancing the needs of pollinators with the needs and esthetic preferences of the public (Filibeck et al., 2016; Hernandez et al., 2009; Hoyle et al., 2017). If implemented in turf systems, the functional aspects of turf such as erosion control, water filtration, and safe recreational surfaces must also be maintained.

4 Urban grasslands as sources of pollinator habitat and forage

Choices regarding urban land use, land cover, and management will determine whether urban sites can provide ecosystem services and what those services are (Baker et al., 2007; Byrne et al., 2008; Cadennasso et al., 2007; Grimm et al., 2008). In the USA, the turfgrass lawn is a dominant part of the urban landscape (Bormann et al., 1993; Jenkins, 1994), and turfgrasses account for approximately 164 000 km^2, or almost 2%, of the continental USA (Milesi et al., 2005). While often viewed as a monoculture, most turfgrass sites are under minimal or no management (Augustin, 2007) and represent a diversity of monocots (grasses) and dicots (forbs) that can and do provide forage for pollinating insects. As urbanization increases, turfgrass areas provide opportunities to expand forage and habitat sites for pollinating species, especially in urban centers where habitat loss has been most prevalent.

Native pollinating insects can persist in urban settings with support from mindfully designed gardens, parks, and other managed green spaces (Ahrne et al., 2009; Baldock et al., 2019; Fetridge et al., 2008; Frankie et al., 2005; Hennig and Ghazoul, 2012; Hernandez et al., 2009; Matteson et al., 2008; McFrederick and LeBuhn, 2006; Owen and Owen, 1975; Tommasi et al., 2004; Tonietti et al., 2011). Affiliation programs like the Xerces Society's Bee City USA and Bee Campus USA encourage communities to preserve and supply pollinator habitat by planting native plants, providing nesting sites, reducing the use of pesticides, and educating the public about pollinators (The Xerces Society, 2021). Monarch City USA is a similar program dedicated to the preservation and promotion of monarch butterflies (Monarch City USA, 2020).

Although urban lawns are primarily used for esthetic, recreational, and economic purposes (such as increasing property values) (Grewal, 2012), turfgrasses also provide several important ecosystem services, including heat, noise, and glare reduction, water filtration, and runoff and erosion prevention (Beard and Green, 1994). Turfgrass systems may also be designed and managed to promote and support pollinators. Incorporating flowering plants into turfgrass and converting some of this land from turfgrass monocultures to mixtures of plants that provide pollinator forage are strategies that could support pollinators. In addition, management strategies that are implemented on existing urban land to encourage flowering forbs, such as raising mowing height, adjusting fertility levels, and limiting or practicing selective weed control, can also promote more biodiversity in turfgrass sites and support pollinators (Lerman et al., 2018; McCurdy et al., 2014; Silverton et al., 2006; Smith and Fellowes, 2014).

Golf courses can also be valuable urban sites for planting and managing foraging sites, nesting areas, and habitat corridors for pollinators (Colding, 2007; Shepherd, 2002). Although land use models predict that the conversion of golf courses to parkland or natural areas would increase pollinator abundance, golf courses are known to support a similar abundance of pollinators as suburban regions and more pollinators than residential and industrial areas (Lonsdorf et al., 2021). For example, golf courses may support a greater number of butterflies than nearby residential areas, office parks, and business districts (Blair and Launer, 1997), and golf courses in the UK were shown to harbor a greater diversity of birds and ground beetles and a greater abundance of beetles, butterflies, and bumble bees than the surrounding farmland (Tanner and Gange, 2005).

Roughs and out-of-play areas on golf course properties are especially valuable potential habitats and foraging areas for sensitive pollinating insect species (Dobbs and Potter, 2015; Wissman et al., 2016). Programs like the Audubon International and Environmental Defense Fund's Monarchs in the Rough program (Audubon International and Environmental Defense Fund, 2021), Monarch Watch's Monarch Waystation (Monarch Watch, 2021), and Syngenta's Operation Pollinator for Golf Courses (Syngenta, 2022) can support golf courses that desire to increase and improve pollinator habitat and forage availability. So far, 753 US golf courses (<5% of 16 165 total courses; National Golf Foundation, 2021) have participated in the Monarchs in the Rough program, and 104 courses (0.6% of total courses; National Golf Foundation, 2021) have registered Monarch Waystation (Monarch Watch, 2021). This indicates that although there is some existing interest in providing forage and habitat for pollinators on golf courses, there is still a great opportunity to expand these efforts to provide pollinator habitat on golf courses.

Utility turf areas such as roadside plantings can also serve as valuable habitat for pollinators. It is estimated that over 4 million hectares of managed roadsides exist in the USA (Forman et al., 2003), and these roadsides not only provide habitat for a broad range of wildlife and pollinators but also can serve as corridors to link fragmented habitat (Hopwood et al., 2016). Restoring roadsides to mixed grass/forb or native prairie plantings can provide excellent forage and nesting sites for native bees and butterflies, particularly when mowing and herbicide applications are carefully managed (Hopwood, 2008; Ries et al., 2001). State Departments of Transportation (DOTs) play a key role in this initiative, and the 2015 Fixing America's Surface Transportation (FAST) Act encourages the development of pollinator habitat on rights-of-way along highways (114th Congress, 2015).

Iowa's Integrated Roadside Vegetation Management program, for example, helps implement alternative roadside vegetation plantings (Iowa DOT, n.d.). To date, over 20 000 ha of federal, state, county, and city roadsides in Iowa have been converted to native grasses, wildflowers, and other alternative vegetation to help provide wildlife habitat, including pollinator forage (Iowa DOT, n.d.). Other states have also developed pollinator-friendly roadside programs, including Minnesota's I-35 Monarch Highway (MnDOT, 2021) and Washington State DOT's preservation of native habitat and restoration of disturbed sites along their 40 000 ha of non-paved roadsides (Carey, 2016). Regionally, the New England Transportation Consortium's ecoregional list of plants suitable for roadside plantings profiles 89 species of forbs and 37 species of grasses and grass-like native species which are appropriate for use on New England roadsides (Kuzovkina et al., 2016).

Residential and commercial lawns may also serve as valuable pollinator forage and habitat sites, if managed with pollinators in mind. In the USA, over 82 million households engage in lawn and garden activities (Grube et al., 2011). There are strong societal and, sometimes, legal, pressures to maintain dark-green, low-cut, turfgrass-only lawns (Blaine et al., 2012; Robbins and Sharp, 2003). Lawns, in general, are being more highly scrutinized by environmentalists and the general public, and there is a growing anti-lawn sentiment across the USA and Canada (Robbins and Sharp, 2003; Sandberg and Foster, 2005). In fact, several municipalities in Canada have banned the cosmetic use of lawn pesticides because of the potential negative health and environmental effects (Grewal, 2012). Highly managed monoculture lawns offer few foraging options for pollinators (Tonietto et al., 2011); however, most lawns could provide pollinator forage if flowering plants were allowed to persist in the lawn.

Weed conservation and management will be an important approach to promote pollinators in turf. Weeds are beneficial, and in many cases essential, for pollinators largely due to the species richness and continuous flowering season of many lawn weeds, which contributes to a diverse pollinator diet (Marshall et al., 2003; Nichols and Altieri, 2013; Requier et al., 2015). Additionally, purposeful incorporation of flowering plants into the lawn could also allow these lawns to act as pollinator forage sites and serve as corridors for pollinator movement between other habitats (Dearborn and Kark, 2010; Dover and Settele, 2009; Hennig and Ghazoul, 2012; Sidhu and Joshi, 2016). Promoting mixed-species lawns that support pollinators while still serving as functional turf may help counter anti-lawn sentiment by enhancing and expanding this important ecosystem service.

4.1 Case study: Ben Geren Park Golf Course

Location: Ben Geren Park Golf Course, Ft. Smith, Arkansas, USA.

Issue: A reduction in park budgets forced a reduction in the amount of maintained turf area around the golf course.

Action: In 2016, Sebastian County Parks Administrator and Certified Golf Course Superintendent, Jay Randolph, was facing reduced maintenance staff and operational budget cuts at Ben Geren Park in Fort Smith, AR. Ben Geren Park is a regional park that covers over 500 ha in the Arkansas River valley region and offers a broad diversity of facilities, including golf, biking and hiking trails, sports fields, and multi-use parkland areas. After looking for ways to reduce inputs and maintenance requirements across the course, staff and stakeholders ultimately decided to convert over 30 ha of bermudagrass (*Cynodon* sp.) and tall fescue (*Festuca* sp.) rough areas to native prairies that once existed on the property.

Result: Randolph began rehabilitating and reestablishing native tallgrass prairie sites in out-of-play areas of the golf course. Proposed renovation areas were mapped out, highlighted, and displayed in public places at the facility to avoid placing native prairies in areas that golfers would frequent. The areas selected for conversion were treated with herbicides to eradicate existing vegetation and grass, and forb seeds were purchased from native seed vendors; however, the cost of seeds limited the scope of the project. The course applied for native seed grants through the Arkansas Game and Fish Commission 'Acres for Wildlife' program, which provided seeds for an additional 8 ha. They also obtained permission to harvest native seeds from privately owned, small remnant prairies in the region. These efforts led to a broad-scale native prairie restoration in the park, eventually encompassing over 40 ha.

Converting these highly maintained areas to native prairies did have the primary effect of reducing labor, irrigation, fertilizer, and pesticide inputs. However, the restoration of the native prairies also provided desirable habitat for pollinators, rare invertebrates, and bird species. The converted area has become a native seed repository for additional restorations across the state of Arkansas and also provides educational and outreach opportunities and a prescribed burn training location for local fire departments (for more information: https://www.bengerengolf.com/massard-prairie/).

5 Plant selection

Refer to Table 1 for a complete list of plants discussed in the following section. Grasses themselves can be food sources for pollinators, though they are not a preferred or common pollen source. Poaceae pollen is often isolated in honey and suggests that some nutrition can be obtained from grass flowers (Jones, 2014; Puusepp and Koff, 2014; Salonen et al., 2009; Song et al., 2012; Stawiarz and Wroblewska, 2010). Hoverflies (*Melanostoma* spp.) have been observed foraging on false oat grass (*Arrhenatherum elatius*) and orchard grass (*Dactylis glomerata*) pollen (Gilbert, 1981), and solitary bees and honey bees have been observed gathering pollen from a variety of grasses (Immelman and Eardley, 2000). Although many of the grasses on which bees forage (Immelman and Eardley, 2000) are either grain crops or used for pasture, forage, fodder, or restoration, some grasses such as blue couch grass (*Digitaria didactyla*) can be used in a range of amenity situations (Turgeon and Kaminski, 2019). Turfgrasses, in particular, are not as widely studied regarding their ability to provide forage for pollinators, although hoverflies and sweat bees, bumble bees and honey bees have been observed foraging on centipedegrass (*Eremochloa ophiuroides*) inflorescences (Jones, 2014; Joseph et al., 2020). However, turfgrasses that are highly managed are also less likely to have pollen-producing seed heads due to frequent mowing and seed head removal.

A number of plants that are commonly considered weeds in manicured lawns can provide pollen and nectar to various insects. Historically, lawn seed included flowering legumes such as clovers (*Trifolium* spp.) to provide nitrogen for the associated turfgrass (Tyson and Grigsby, 1941). Though these plants are often viewed as undesirable weeds, mixed lawns are again being promoted for their ability to withstand drought, tolerate mowing, thrive in a range of soils, and supply nitrogen to other plants (Bigelow et al., 2020; McCurdy et al., 2013b; Sincik and Acikgoz, 2007). Leguminous plants such as white clover (*Trifolium repens*), strawberry clover (*Trifolium fragiferum*), bird's foot trefoil (*Lotus corniculatus*), ground plum (*Astragalus crassicarpus*), and black medic (*Medicago lupulina*) not only provide nitrogen to the lawn but can also be a potential food source for pollinators (Boyle et al., 2020; Larson et al., 2014a).

White clover is a common weed in both warm- and cool-season lawns and is well-known as a desirable plant for pollinators within turfgrass settings (Bigelow et al., 2020; Boyle et al., 2020; Cook, 2005; Lane et al., 2019b; McCurdy et al., 2013a, 2014). When incorporated into mixed lawns, white clover can provide foraging resources for several months during the spring and summer seasons and can attract honey bees, bumble bees, native bees, butterflies, and other pollinators (Boyle et al., 2020) (Fig. 2). White clover in lawns attracted a similar species richness of bees, though a greater abundance of *Apis mellifera* and fewer *Bombus* spp. were observed as the amount of hardscape increased (Larson et al., 2014a).

Strawberry clover can persist in mixtures with various turf species and provide spring and summer forage but may be slower to flower than other species (Boyle et al., 2020; Cook, 2005). Pollinators have been observed

Table 1 Potential flowering species to support pollinators in mixed grass/forb lawns

Species name	Common name	Height (cm)	Bloom season	Pollinators supported	Other notes	Reference(s)
Achillea millefolium	Yarrow	30-90	SP, SU	Special value to native bees, butterflies	Native; soft, fern-like appearance	Cook (2005); US Forest Service (n.d.)
Astragalus crassicarpus	Ground plum	30-90	SU	Butterflies, moths	Native; legume	Lady Bird Johnson Wildflower Center (2012)
Chamaemelum nobile	Chamomile	20-61	SP, SU, FA	Bees, other pollinators	Non-native, ground cover	North Carolina State University Extension (n.d.a)
Claytonia virginica	Spring beauty	10-30	SP	Native bees	Native; corm	Lady Bird Johnson Wildflower Center (2021)
Crocus spp.	Crocus	7.5-15	SP	Bees, flies	Non-native; bulb	Wisdom et al. (2019)
Dalea candida	White prairie clover	30-60	SU	Birds, butterflies	Native, legume	Lady Bird Johnson Wildflower Center (2013)
Lamium amplexicaule	Henbit	40	SP	Honey bees, bumble bees	Non-native, mint family	Brown (2016)
Lamium purpureum	Deadnettle	15-30	SP	Bees, giant bee fly	Non-native, mint family	Brown (2016)
Medicago lupulina	Black medic	15-20	SU	Bees, butterflies, flies	Non-native; legume, prostrate stems	Mahr (n.d.); Illinois Wildflowers (n.d.a)
Muscari spp.	Grape hyacinth	10-20	SP	Bees	Non-native; bulb	Wisdom et al. (2019)
Oxalis stricta	Wood sorrel	15-30	SP, SU, FA	Bees, butterflies, ants	Native; taproot, rhizomes, stolons	Lady Bird Johnson Wildflower Center (2020a)
Prunella vulgaris	Self-heal	30-90	SU	Bumble bees, butterflies	Native; mint, medicinal	Lady Bird Johnson Wildflower Center (2015); Boyle et al. (2020)

(Continued)

Table 1 (*Continued*)

Species name	Common name	Height (cm)	Bloom season	Pollinators supported	Other notes	Reference(s)
Stellaria media	Chickweed	10–30	WI, SP, SU	Small bees, butterflies, flies, parasitoid wasps, songbirds. Larval host butterflies, moths	Non-native annual; succulent, hairy stem	Illinois Wildflowers (n.d.b)
Taraxacum officinale	Dandelion	5–15	WI, SP	Bees	Non-native perennial; deep taproot, dense rosette	North Carolina State University Extension (n.d.b)
Thymus serpyllum	Creeping thyme	7.5	SU	Bees, butterflies	Non-native, dwarf, evergreen	Lane, et al. (2019b); North Carolina State University Extension (n.d.c)
Trifolium ambiguum	Kura clover	30	SU	Bees	Non-native perennial	Lane et al. (2019a); Speer and Allinson (1985)
Trifolium fragiferum	Strawberry clover	20–35	SP	Bees	Non-native; legume	Boyle et al. (2020); Oregon State University (2018)
Trifolium repens	White clover	10–15	SP, SU	Bees, butterflies, moths	Non-native; legume	Boyle et al. (2020); Larson et al. (2014a)
Viola sororia	Blue violet	7–20	SP, SU	Bees, fritillary butterfly larvae	Native; stemless, rhizomes	Lady Bird Johnson Wildflower Center (2020b)

SP, spring; SU, summer; FA, fall, WI, winter.

Native bee foraging on white clover (*Trifolium repens*) in Utah, USA. (photo by P. Boyle)	Bumble bee (*Bombus* spp.) foraging on white clover (*Trifolium repens*) in a bermudagrass (*Cynodon dactylon*) lawn in Arkansas, USA. (photo by M. Wisdom)	Bumble bee (*Bombus* spp.) foraging on white clover (*Trifolium repens*) in a bermudagrass (*Cynodon dactylon*) lawn in Arkansas, USA. (photo by M. Wisdom)

Figure 2 Pollinating insects foraging on white clover (*Trifolium repens*). (a) Native bee foraging on white clover (*Trifolium repens*) in Utah, USA. Source: Photo by P. Boyle. (b) Bumble bee (*Bombus* spp.) foraging on white clover (*Trifolium repens*) in a bermudagrass (*Cynodon dactylon*) lawn in Arkansas, USA. Source: Photo by M. Wisdom. (c) Bumble bee (*Bombus* spp.) foraging on white clover (*Trifolium repens*) in a bermudagrass (*Cynodon dactylon*) lawn in Arkansas, USA. Source: Photo by M. Wisdom.

foraging on strawberry clover in mixed turf/clover stands (Boyle et al., 2020). Kura clover (*Trifolium ambiguum*) can also establish well into Kentucky bluegrass (*Poa pratensis*) and hard fescue (*Festuca brevipila*) turf (Lane et al., 2019a) and is known to support honey bees (Speer and Allinson, 1985; Townsend, 1985).

There are a number of early-spring weeds that can be a valuable food source for pollinators during a period when other floral resources are limited. Dandelion (*Taraxacum officinale*) is a widely adapted lawn weed (Gray et al., 1973) that can provide early-spring nectar and pollen (Steinkraus, 2010), especially for small Syrphidae pollinators (Larson et al., 2014a) (Fig. 3). Henbit (*Lamium amplexicaule*) and purple deadnettle (*Lamium purpureum*) have also been recognized to provide early-spring forage resources for pollinators (Brown, 2016). Henbit has been shown to provide pollen and nectar for long-tongued honey bees and bumble bees, while purple deadnettle provides nectar to honey bees, bumble bees, digger bees (Halictid, Andrenid, and Colletid bees) and giant bee flies (*Bombylius major*) (Brown, 2016).

Other flowering perennials or self-seeding annuals have also been shown to persist in lawns and provide pollen for numerous insects. Self-heal (*Prunella vulgaris*) is known to persist in both warm-season bermudagrass and cool-season hard fescue lawns and attracts honey bees, bumble bees, native bees, butterflies, and other pollinators throughout the summer and early autumn (Boyle et al., 2020; Lady Bird Johnson Wildflower Center, 2015; Lane et al., 2019b) (Figs 1 and 4). Ground plum (*Astragalus crassicarpus*) is a native legume with showy purple or white flowers. Ground plum can thrive in an assortment of environmental conditions, including dry or moist areas, and full to part-sun. Butterflies utilize ground plum as a source of nutrition in the late spring and early summer months (Lady Bird Johnson Wildflower Center, 2012).

Grape hyacinth (*Muscari* spp.) in a mixed, cool-season turfgrass lawn in Arkansas, USA. (photo by M. Richardson)	Flowering dandelions (*Taraxacum officinale*) in a bermudagrass (*Cynodon* spp.) lawn in Arkansas, USA. (photo by M. Richardson)	Several cultivars of Crocus (*Crocus* spp.) flowering in a dormant buffalograss (*Buchloe dactyloides*) lawn in Arkansas, USA. (photo by M. Richardson

Figure 3 Examples of mixed grass and forb lawns. (a) Grape hyacinth (*Muscari* spp.) in a mixed, cool-season turfgrass lawn in Arkansas, USA. Source: Photo by M. Richardson. (b) Flowering dandelions (*Taraxacum officinale*) in a bermudagrass (*Cynodon* spp.) lawn in Arkansas, USA. Source: Photo by M. Richardson. (c) Several cultivars of Crocus (*Crocus* spp.) flowering in a dormant buffalograss (*Buchloe dactyloides*) lawn in Arkansas, USA. Source: Photo by M. Richardson.

Yarrow (*Achillea millefolium*) is another lawn weed with the potential for pollinator lawns. Yarrow's fine-textured leaves blend well with turf, can persist in mixed plantings with cool-season turfgrasses (Cook, 2005), and can attract butterflies, bees, and other insects with summer blooms (US Forest Service, n.d.). Creeping thyme (*Thymus serpyllum*) can be established in hard fescue (Lane et al., 2019b) and is known to attract and support a variety of butterflies during the summer months (North Carolina State University Extension, n.d.c).

Many spring-flowering bulbs are commonly observed to naturalize in grassy areas like meadows and pastures (Bryan, 2002; Leeds, 2000) but may also be more fully utilized in lawns. Bulbs such as *Crocus* spp. and *Muscari* spp. (Fig. 2) have been shown to persist in zoysiagrass (*Zoysia japonica*), bermudagrass (*Cynodon dactylon*), and buffalograss [*Buchloe dactyloides* (Nutt.) J.T. Columbus] lawns (Richardson et al., 2015; Wisdom et al., 2019) and can also provide late winter or early spring forage for pollinators when these warm-season grasses are dormant (Steinkraus, 2010; Wisdom et al., 2019) (Figs. 1 and 4). Spring beauty (*Claytonia virginica*) is another important early spring floral resource for pollinators (Marlin and LaBerge, 2001; Robertson, 1926, 1929a, 1929b) and can persist when corms are co-planted with turfgrass (Cook, 2005; Lane et al., 2019b).

In the development of mixed-species, pollinator-friendly turfgrass systems, plant selection depends on many factors. The flowering plants need to be able to persist and flower under whatever level of management is necessary for that particular turf system. The selection should be visually pleasing,

Honey bee (*Apis mellifera*) foraging on grape hyacinth (*Muscari* spp.) in a dormant bermudagrass (*Cynodon dactylon*) lawn in Arkansas (photo by M.Wisdom)	Self-heal (*Prunella vulgaris*) growing in a bermudagrass (*Cynodondactylon*) lawn in Arkansas (photo by M. Wisdom)

Figure 4 Forbs in bermudagrass lawns in Arkansas, USA. (a) Honey bee (*Apis mellifera*) foraging on grape hyacinth (*Muscari* spp.) in a dormant bermudagrass (*Cynodon dactylon*) lawn in Arkansas. Source: Photo by M. Wisdom. (b) Self-heal (*Prunella vulgaris*) growing in a bermudagrass (*Cynodon dactylon*) lawn in Arkansas. Source: Photo by M. Wisdom.

adding color and texture to the turfgrass without exceeding the height of the turfgrass canopy. Some species will certainly react differently when planted in specific environments or within different turf species. For example, self-heal planted into bermudagrass in Arkansas established quickly and eventually suppressed the warm-season turfgrass growth (Boyle et al., 2020), while self-heal co-planted with hard fescue in Minnesota achieved a balance with the cool-season turfgrass species (Lane et al., 2019b). Careful consideration should be taken when selecting species, particularly non-native species, as they may exhibit invasive tendencies (Minnesota Wildflowers, 2022). Unfortunately, there have not been enough studies carried out across environments which looked at a range of turfgrass species and flowering forbs to accurately predict what combinations of plants might or might not work. Careful observation – and trial and error – may be required to find forbs that fit well with the turfgrass species. Observing local turfgrass sites where forbs have persisted for many years can be a good indicator of what might be most suitable for a specific location.

5.1 Case study: Home lawn in Utah

Location: Home lawn in Logan, Utah, USA.

Issue: Buffalograss is on the edge of its habitat range in Logan, UT, and takes a long time to green up in the spring.

Figure 5 *Crocus* spp. bulbs flowering in a dormant buffalograss lawn in Logan, Utah, USA. Source: Photo by M. Wisdom.

Action: Dr. Paul Johnson (Professor of Turfgrass Management and Genetics and Department Head, Department of Plants, Soils and Climate, Utah State University, Logan, UT) started planting *Crocus* spp. bulbs to add spring color to his buffalograss lawn.

Result: The bulbs have multiplied naturally over the years (Fig. 5), and more bulbs have been planted to help supplement the blooms. *Muscari* spp. have also started emerging, after entering the lawn naturally from surrounding areas. In spring, the grass around the bulbs is mowed, and bulb leaves are mowed when they turn yellow in late spring. Honey bees, bumble bees, and other native bees have been observed foraging on the bulbs in the lawn.

6 Management of pollinator-friendly mixtures

If the ultimate goal of establishing a mixed-species lawn is to support pollinators, then other management practices such as mowing and pesticide use may need to be adjusted to accomplish that goal. General esthetic preferences have increased the demand for weed-free expanses of home and commercial lawns, resulting in greater use of chemical inputs and irrigation (Bormann et al., 1993). Treatment of flowering lawns with pesticides, especially insecticides, could have a detrimental effect on pollinators. Bermudagrass and St. Augustine grass (*Stenotaphrum secundatum*) dew and guttation can contain sublethal amounts of imidacloprid, and those amounts increase significantly when the pesticide is applied as a foliar spray, which is the most common method of application to lawns (McCurdy et al., 2017). Colonies of native bumble bees (*Bombus impatiens*) allowed to forage on white clover treated with clothianidin or a mixture of clothianidin and bifenthrin, common turfgrass insecticides, had reduced numbers of live workers, increased numbers of dead workers, and fewer honey pots (Larson et al., 2014b). Neonicotinoids, including clothianidin, should not be applied to blooming nectar-producing plants if pollinators frequent the area for foraging (Larson et al., 2013). Chlorantraniliprole residue had no adverse effects on bumble bees (Larson et al., 2013, 2014b) but demonstrates similar efficacy against turf-damaging caterpillars, scarab grubs, billbug and other weevil larvae, and crane flies (Held and Potter, 2012), indicating that chlorantraniliprole may be a more suitable alternative for treating these particular turf pests in mixed forb/turf stands (Larson et al., 2014b).

Alternatively, blooms may be removed with mowing prior to the application of pesticides to reduce the potential hazard to bumble bees and other pollinators (Larson et al., 2013, 2015). In addition to the obvious effects that applied insecticides can have on pollinators, potential toxicity from other pesticides, such as herbicides and fungicides, which may be applied to lawns should also be considered. Although poorly studied, recent evidence suggests that some herbicides and fungicides may also impact the long-term health of pollinators both directly, from exposure, and indirectly, from their impact on pollinating plants (Belsky and Joshi, 2020).

Cultural practices such as fertilization can influence interactions between turfgrass and flowering species. Increased nitrogen fertilization generally reduces plant diversity (Kidd et al., 2017; Silvertown et al., 2006), favors grass growth (Silvertown et al., 2006), and especially reduces the growth of flowering legumes (Bigelow et al., 2020; Silvertown et al., 2006). Nitrogen fertilization has been shown to reduce dandelion cover compared to stands receiving no nitrogen fertilizers (DeBels et al., 2012). Alternatively, potassium fertilization may increase dandelion and legume abundance (Silvertown et al., 2006; Tilman et al., 1999). Increased nitrogen fertilization rate may reduce desirable flowering plants in turf stands, but turf species is also an important factor, as not all species show a fertilizer effect on weed abundance (DeBels et al., 2012).

Mowing practices can also influence flowering and associated bee abundance and diversity in flowering lawns. Regardless of flowering species and turfgrass species, more intensively managed turf areas that are mowed several times each week may not be good candidates for incorporating pollinating plants. In a comparison of mowing regimes where suburban lawns in Massachusetts were mown either every 7 days, every 14 days, or every 21 days, the 21-day mowing treatment resulted in 2.5 times more dandelion and clover flowers; however, this infrequency of mowing resulted in an unkempt and unsightly lawn that was not favorable to homeowners and neighbors (Lerman et al., 2018). Mowing every 14 days resulted in a more favorable esthetic and greater bee abundance compared to a 7-day or 21-day mowing frequency (Lerman et al., 2018). Lower mowing heights (3.5 cm) can increase weed abundance (primarily dandelion) in tall fescue, fine fescue, and Kentucky bluegrass (DeBels et al., 2012) but will likely have a deleterious effect on flower abundance. The impact of mowing practices on pollinator forage will vary depending on the grass and flowering species used (Lane et al., 2019b) but would be a productive area for future research.

7 Additional considerations

7.1 Beneficial predators

Setting aside habitat to encourage natural enemies can also contribute to other pest control strategies and may reduce the need to apply more aggressive control measures such as synthetic insecticides (Gontijo et al., 2013; Landis et al., 2000; MacLeod et al., 2004; Walton and Isaacs, 2011). Lawns, golf courses, and native wildflower plantings can support natural enemy communities that may deter undesirable insects (Blaaw and Isaacs, 2015; Held and Potter, 2012; Lee and Hempel, 2005). For example, parasitoid wasps (Patt et al., 1997) and flower flies (Ramsden et al., 2015) benefit from floral resources and may also serve as pollinators themselves (Jauker and Wolters, 2008). The encyrtid wasp *Copidosoma bakeri* (Howard) is known to sporadically parasitize black cutworm eggs, braconid wasps and tachinid flies will sporadically parasitize turfgrass-feeding caterpillars (Bixby-Brosi and Potter, 2012), and root-feeding scarab grubs are parasitized by native tiphiid wasps (*Tiphia pygidialis* Allen) (Redmond et al., 2012).

Naturalized areas on golf courses may harbor beneficial predators, and although predation benefits may not extend to adjacent low-mown turf (Dobbs and Potter, 2014), these areas may also attract insectivorous birds, which may serve as a post-insecticide repopulation source of beneficial insects to surrounding turf and connect fragmented pollinator habitat (Colding, 2007). These areas may also provide opportunities to educate the public about insects, conservation, native plants, and ecology (Dobbs and Potter, 2015).

7.2 Lawn care options, conversions, and the need for public education

Maintained turfgrass lawns are well entrenched as a status symbol in North American cultural psychology (Blaine et al., 2012; Bormann et al., 1993; Jenkins, 1994), and a vast majority of Americans would not replace their home's turfgrass with anything that does not resemble a grass lawn (Grewal, 2012). In fact, weed-free, maintained lawns are often considered a status symbol in many communities and essential for maintaining property values (Blaine et al.,

2012). Homeowner associations, when present, may include bylaws specifically dictating acceptable heights of cut and enforcing weed control practices in lawns (Robbins et al., 2001). The desire and inability of homeowners to achieve these rigid standards promotes overuse and misuse of lawn care products (Grewal, 2012). As such, promoting the value of flowering lawns as pollinator habitat could encourage wider acceptance of flowering plants in turfgrass systems and provide educational opportunities related to pollinators and conservation (Bormann et al., 2001; Forbes and Kendle, 1997; Larson et al., 2014a).

Convincing the general public to adopt widespread conversion to pollinator-friendly lawns is likely going to be a slow process; however, it is important to note that the entire lawn need not be converted for it to be effective. Establishing pollinator plantings in small conservation strips surrounding or next to urban turf systems (for example, side- or backyards, fence lines, park surrounds, golf course out-of-play areas, parking strips, parking medians, etc.) can still help provide floral resources, habitat, and overwintering sites to pollinators and other invertebrates (Frank and Shrewsbury, 2004).

Promoting mixed-species lawns for pollinator support can also engage suburban residents in the process of 'reconciliation ecology', which encourages biodiversity in human-dominated ecosystems such as urban centers (Bormann et al., 2001; Rosenzweig, 2003). Flowering lawns can be managed to reduce inputs and support pollinators but must also remain functional for recreation and normal activities (Cook, 2005). Mixed pollinator-friendly turf may not be suitable for all situations, as it has been well documented that weed species such as white clover will not tolerate traffic as well as a desirable turfgrass such as bermudagrass (Brosnan et al., 2014). As such, a pollinator-friendly lawn or utility turf that receives heavy use may not persist as well as a traditional grass-only lawn. An increased presence of bees may also be an issue of concern in turf areas where children or others with bee allergies may recreate (Ramer et al., 2019). Strategic placement of pollinator-friendly lawns, especially in public spaces, will be critical to assure their long-term success and minimize risks to the general public.

8 Conclusion and future trends

Although pollinator lawn systems are drawing interest from a wide range of researchers and practitioners, there is still a significant lack of research evaluating flowering forbs in turfgrass systems. It is unclear how different plants will respond to mixed plantings with various turfgrass species or in different environments (i.e. climate and management). Although many forbs have been tested, location-specific management practices and environmental conditions vary, and more testing needs to be done to evaluate these mixed plantings in different locations and under different management conditions. There are countless common lawn weeds and flowering plants, such as white prairie clover (*Dalea candida*) (Lady Bird Johnson Wildflower Center, 2013), chickweed (*Stellaria media*) (Illinois Wildflowers, n.d.b), black medic (Illinois Wildflowers, n.d.a; Mahr, n.d.), blue violet (*Viola sororia*) (Lady Bird Johnson Wildflower Center, 2020b), chamomile (*Chamaemelum nobile*) (North Carolina State University, n.d.a), wood sorrel (*Oxalis stricta*) (Lady Bird Johnson Wildflower Center; 2020a), and many others, which can be valuable sources of forage for pollinators and may be potential candidates for intentional mixed-lawn plantings.

Turfgrass systems represent a large part of urban landscapes and provide a range of ecosystem services that will be desirable in the future. However, many of those turfgrass areas can also be managed to enhance and even connect pollinator habitats in urban ecosystems. Future efforts to better understand the ecology and management of mixed turfgrass/forb plantings to maximize those benefits will certainly be important to help move these ideas forward.

9 Where to look for further information

Literature investigating turfgrass (lawns) to support pollinators is sparse with few scientific publications addressing pollinator-lawns pre-2018. However, increased public interest in the topic has recently produced research opportunities and peer-reviewed results.

The University of Minnesota is a leader in the field of *Bee Lawns*, with fact sheets defining: what is a bee lawn, what plants to use in a bee lawn, what pollinators are supported in a bee lawn, when and how to establish a bee lawn, seeding rates, advantages of bee lawns, and how to renovate an existing lawn (https://extension.umn.edu/landscape-design/planting-and-maintaining-bee-lawn).

A number of governmental agencies, including Pennsylvania State University, Michigan State University, Iowa State University, Mississippi State University, and the Smithsonian Gardens, have resources specific to *Pollinator Lawns*, which include bees and other pollinating insects and mammals:

- https://pollinators.msu.edu/resources/pollinator-planting/pollinator-lawns/.
- https://gardens.si.edu/learn/for-educators/what-is-a-pollinator-lawn/.
- https://hortnews.extension.iastate.edu/how-create-pollinator-lawn.
- http://extension.msstate.edu/news/feature-story/2021/msu-extension-leads-pollinator-research-project.

A current project designed to address urgent issues affecting pollinator health and decline within the built environment is *The Refuge Lawn*, a

USDA-NIFA-funded project connecting people, plants, and pollinators (http://www.refugelawn.com/about-us). This project 'seeks to answer key questions regarding the value of common forbs and weeds within southeastern lawns. Stakeholder training and best management practices will be developed to help better manage turfgrass systems to benefit pollinators'.

Additional resources include:

- Pollinator-friendly native plants for your eco-region:
  - https://xerces.org/pollinator-conservation/pollinator-friendly-plant-lists.
- Eco-lawns:
  - https://extension.oregonstate.edu/gardening/techniques/ecolawn-less-work-water.
  - https://www.asianturfgrass.com/post/tom-cooks-extraordinary-lawn/.
  - https://twitter.com/osubeaverturf/status/1298649674329423873.

10 References

114th Congress. (2015). H.R.22 - FAST act. Available at: (https://www.congress.gov/bill/114th-congress/house-bill/22/text); accessed July 22, 2021.

Ahrne, K., Bengtsson, J. and Elmquist, T. (2009). Bumble bees (*Bombus* spp.) along a gradient of increasing urbanization, *PLoS ONE* 4(5), e5574.

Alaux, C., Ducloz, F., Crauser, D. and Le Conte, Y. (2010). Diet effects on honey bee immunocompetence, *Biol. Lett.* 6(4), 562–565.

Audubon International and Environmental Defense Fund. (2021). Monarchs in the rough. Available at: (https://monarchsintherough.org/); accessed February 4, 2022.

Augustin, B. (2007). Perception vs. reality: How much nitrogen do homeowners put on their lawn? In: *Abstracts. Asa, CSSA, and SSSA Annual Meeting*, Madison Book Company, WI.

Baker, L., Brazel, A., Byrne, L., Felson, A., Grove, M., Hill, K., Nelson, K. C., Walker, J. and Shandas, V. (2007). Effects of human choices on characteristics of urban ecosystems, *Bull. Ecol. Soc. Am.* 88, 404–409.

Baldock, K. C. R. (2020). Opportunities and threats for pollinator conservation in global towns and cities, *Curr. Opin. Insect Sci.* 38, 63–71.

Baldock, K. C. R., Goddard, M. A., Hicks, D. M., Kunin, W. E., Mitschunas, N., Morse, H., Osgathorpe, L. M., Potts, S. G., Robertson, K. M., Scott, A. V., Staniczenko, P. P. A., Stone, G. N., Vaughan, I. P. and Memmott, J. (2019). A systems approach reveals urban pollinator hotspots and conservation opportunities, *Nat. Ecol. Evol.* 3(3), 363–373.

Banaszak, J. (1996). Ecological bases of conservation of wild bees. In: Matheson, A., Buchman, S. L., O'Toole, C., Westrich, P. and Williams, I. H. (Eds) *The Conservation of Bees*, Academic Press, London, pp. 55–62.

Bauer, D. M. and Wing, I. (2016). The macroeconomic cost of catastrophic pollinator declines, *Ecol. Econ.* 126, 1–13.

Beard, J. B. and Green, R. L. (1994). The role of turfgrasses in environmental protection and their benefits to Humans, *J. Environ. Qual.* 23(3), 452–460.
Belsky, J. and Joshi, N. K. (2020). Effects of fungicide and herbicide chemical exposure on Apis and non-Apis bees in agricultural landscape, *Front. Environ. Sci.* 8, 81.
Biesmeijer, J. C., Roberts, S. P. M., Reemer, M., Ohlemüller, R., Edwards, M., Peeters, T., Schaffers, A. P., Potts, S. G., Kleukers, R., Thomas, C. D., Settele, J. and Kunin, W. E. (2006). Parallel declines in pollinators and insect-pollinated plants in Britain and the Netherlands, *Science* 313(5785), 351–354.
Bigelow, C. A., Macke, G. A., Johnson, K. and Richmond, D. S. (2020). Cool-season lawn performance as influenced by 'Microclover' inclusion and supplemental nitrogen, *Int. Turfgrass Soc. Res. J.* 2021, 1–12.
Bixby-Brosi, A. J. and Potter, D. (2012). Beneficial and innocuous invertebrates in turf. In: Brandenberg, R. L. and Freeman, C. P. (Eds) *Handbook of Turfgrass Insect Pests* (2nd edn.), Entomological Society of America, Latham, MD, pp. 87–93.
Blaaw, B. R. and Isaacs, R. (2015). Wildflower plantings enhance the abundance of natural enemies and their services in adjacent blueberry fields, *Biol. Control* 91, 94–103.
Blaine, T. W., Clayton, S., Robbins, P. and Grewal, P. S. (2012). Homeowner attitudes and practices towards residential landscape management in Ohio, USA, *Environ. Manage.* 50(2), 257–271.
Blair, R. B. and Launer, A. E. (1997). Butterfly diversity and human land use: Species assemblages along an urban gradient, *Biol. Conserv.* 80(1), 113–125.
Blitzer, E. J., Dormann, C. F., Holzschuh, A., Klein, A., Rand, T. A. and Tscharntke, T. (2012). Spillover of functionally important organisms between managed and natural habitats, *Agric. Ecosyst. Environ.* 146(1), 34–43.
Bormann, F. H., Balmori, D. and Geballe, G. T. (1993). *Redesigning the American Lawn*, Yale University Press, New Haven, CT.
Bormann, F. H., Balmori, D. and Geballe, G. T. (2001). *Redesigning the American Lawn: A Search for Environmental Harmony* (2nd edn.), Yale University Press, New Haven, CT.
Bosch, J., Retana, J. and Cerda, X. (1997). Flowering phenology, floral traits and pollinator composition in a herbaceous Mediterranean plant community, *Oecologia* 109(4), 583–591.
Boyle, P. E., Wisdom, M. M. and Richardson, M. D. (2020). Testing flowering perennial plants in a Bermudagrass (*Cynodon* spp.) lawn, *Hortscience* 55(10), 1642–1646.
Braaker, S., Obrist, M. K., Ghazoul, J. and Moretti, M. (2017). Habitat connectivity and local conditions shape taxonomic and functional diversity of arthropods on green roofs, *J. Anim. Ecol.* 86(3), 521–531.
Brosnan, J. T., Dickson, K. H., Sorochan, J. C., Thoms, A. W. and Stier, J. C. (2014). Large crabgrass, white clover, and hybrid Bermudagrass athletic field playing quality in response to simulated traffic, *Crop Sci.* 54(4), 1838–1843.
Brown, D. (2016). Purple deadnettle and henbit: Two common spring weeds, Mich. State. Univ. Ext. Pub. Available at: (http://msue.anr.msu.edu/news/purple_deadnettle_and_henbit_two_common_garden_spring_weeds); accessed February 6, 2021.
Bryan, J. E. (2002). *Bulbs*, Timber Press, Portland, OR.
Burel, F. (1996). Hedgerows and their role in agricultural landscapes, *Crit. Rev. Plant Sci.* 15(2), 169–190.
Byrne, L. B., Bruns, M. A. and Kim, K. C. (2008). Ecosystem properties of urban land covers at the aboveground-belowground interface, *Ecosystems* 11(7), 1065–1077.

Cadennasso, M. L., Pickett, S. T. A. and Schwarz, K. (2007). Spatial heterogeneity in urban ecosystems: Reconceptualizing land cover and a framework of classification, *Front. Ecol. Environ.* 5(2), 80–88.

Calderone, N. W. (2012). Insect pollinated crops, insect pollinators and US agriculture: Trend analysis of aggregate data for the period 1992–2009, *PLoS ONE* 7(5), e37235.

Cane, J. H. (2001). Habitat fragmentation and native bees: A premature verdict?, *Conserv. Ecol.* 5(1), 3.

Cane, J. H., Griswold, T. and Parker, F. D. (2007). Substrates and materials used for nesting by North American *Osmia* bees (Hymenoptera: Apiformes: Megachilidae), *Ann. Entomol. Soc. Am.* 100(3), 350–358.

Cane, J. H., Minckley, R. L., Kervin, L. J., Roulston, T. H. and Williams, N. M. (2006). Complex responses within a desert bee guild (Hymenoptera: Apiformes) to urban habitat fragmentation (Hymenoptera: Apiformes), *Ecol. Appl.* 16(2), 632–644.

Carey, M. (2016). Promoting the health of pollinators along WSDOT's roadsides Available at: (https://wsdot.wa.gov/publications/fulltext/Roadside/PollinatorsFactSheet.pdf); accessed February 4, 2022.

Carvalheiro, L. G., Kunin, W. E., Keil, P., Aguirre-Gutiérrez, J., Ellis, W. N., Fox, R., Groom, Q., Hennekens, S., Van Landuyt, W., Maes, D., Van de Meutter, F., Michez, D., Rasmont, P., Ode, B., Potts, S. G., Reemer, M., Roberts, S. P., Schaminée, J., WallisDeVries, M. F. and Biesmeijer, J. C. (2013). Species richness declines and biotic homogenisation have slowed down for NW-European pollinators and plants, *Ecol. Lett.* 16(7), 870–878.

Cavender- Bares, J., Arroyo, M. T. K., Abell, R., Ackerly, D., Ackerman, D., Arim, M., Belnap, J., Castañeda Moya, F., Dee, L., Estrada-Carmona N., Gobin, J., Isabell, F., Köhler, G., Koops, M., Kraft, N., Mcfarlane, N., Martínez-Garza, C., Metzger, J. P., Mora, A., Oatham, M., Paglia, A., Pedrana, J., Peri, P. L., Piñeiro, G., Randall, R., Robbins, W. W., Weis, J. and Ziller, S. R. (2018). Status and trends of biodiversity and ecosystem functions underpinning nature's benefit to people. In: Balvanera, P. and Cerezo, A. (Eds) *Regional and Subregional Assessments of Biodiversity and Ecosystem Services: Regional and Subregional Assessment for the Americas*, IPBES, Medellin, Columbia, pp. 207–362.

Cavender-Bares, J., Cubino, J. P., Pearse, W. D., Hobbie, S. E., Lange, A. J., Knapp, S. and Nelson, K. C. (2020). Horticultural availability and homeowner preferences drive plant diversity and composition in urban yards, *Ecol. Appl.* 30(4), 1–16.

Chacoff, N. P. and Aizen, M. A. (2006). Edge effects on flower-visiting insects in grapefruit plantations bordering premontane subtropical forest, *J. Appl. Ecol.* 43(1), 18–27.

Cohen, J. E. (2003). Human population: The next half century, *Science* 302(5648), 1172–1175.

Colding, J. (2007). Ecological land-use complementation for building resilience in urban ecosystems, *Landsc. Urban Plan.* 81(1–2), 46–55.

Cook, T. (2005). Low maintenance turf? Available at: (http://horticulture.oregonstate.edu/content/low-maintenance-turf); accessed February 4, 2022.

Daily, G. C. (1997). *Nature's Services: Societal Dependence on Natural Ecosystems*, Island Press, Washington, DC.

Dearborn, D. C. and Kark, S. (2010). Motivations for conserving urban biodiversity, *Conserv. Biol.* 24(2), 432–440.

DeBels, B. T., Griffith, S. E., Kreuser, W. C., Melby, E. S. and Soldat, D. J. (2012). Evaluation of mowing height and fertilizer application rate on quality and weed abundance of five home lawn grasses, *Weed Technol.* 26(4), 826–831.

Di Pasquale, G., Salignon, M., Le Conte, Y., Belzunces, L. P., Decourtye, A., Kretzschmar, A., Suchail, S., Brunet, J. L. and Alaux, C. (2013). Influence of pollen nutrition on honey bee health: Do pollen quality and diversity matter?, *PLoS ONE* 8(8), e72016.

Dobbs, E. K. and Potter, D. A. (2014). Conservation biological control and pest performance in lawn turf: Does mowing height matter?, *Environ. Manag.* 53(3), 648-659.

Dobbs, E. K. and Potter, D. A. (2015). Forging natural links with golf courses for pollinator-related conservation, outreach, teaching, and research, *Am. Entomol.* 61(2), 116-123.

Dover, J. and Settele, J. (2009). The influences of landscape structure on butterfly distribution and movement: A review, *J. Insect Conserv.* 13(1), 3-27.

Fenster, C. B., Armbruster, W. S., Wilson, P., Dudash, M. R. and Thomson, J. D. (2004). Pollination syndromes and floral specialization, *Annu. Rev. Ecol. Evol. Syst.* 35(1), 375-403.

Fetridge, E. D., Ascher, J. S. and Langellotto, G. A. (2008). The bee fauna of residential gardens in a suburb of New York City (Hymenoptera: Apoidea), *Ann. Entomol. Soc. Am.* 101(6), 1067-1077.

Filibeck, G., Petrella, P. and Cornelini, P. (2016). All ecosystems look messy, but some more so than others: A case-study on the management and acceptance of Mediterranean urban grasslands, *Urban For. Urban Green.* 15, 32-39.

Forbes, S. and Kendle, T. (1997). *Urban Nature Conservation: Landscape Management in the Urban Countryside* (1st edn.), Taylor & Francis, London, England.

Forman, R. T., Sperling, D., Bissonette, J. A., Clevenger, A. P., Cutshall, C. D., Dale, V. H., Fahrig, L., France, R. L., Goldman, C. R., Heanue, K., Jones, J., Swanson, F., Turrentine, T. and Winter, T. C. (2003). *Road Ecology*, Island Press, Washington, DC.

Frank, S. D. and Shrewsbury, P. M. (2004). Effect of conservation strips on the abundance and distribution of natural enemies and predation of *Agrotis ipsilon* (Lepidoptera: Noctuidae) on golf course fairways, *Environ. Entomol.* 33(6), 1662-1672.

Frankie, G. W., Thorp, R. W., Schindler, M., Hernandez, J., Ertter, B. and Rizzardi, M. (2005). Ecological patterns of bees and their host ornamental flowers in two northern California cities, *J. Kans. Entolomol Soc.* 78(3), 227-246.

Gallai, N., Salles, J. M., Settele, J. and Vaissière, B. E. (2009). Economic valuation of the vulnerability of world agriculture confronted with pollinator decline, *Ecol. Econ.* 68(3), 810-821.

Gilbert, F. S. (1981). Foraging ecology of hoverflies: Morphology of the mouthparts in relation to feeding on nectar and pollen in some common urban species, *Ecol. Entomol.* 6(3), 245-262.

Gontijo, L. M., Beers, E. H. and Snyder, W. E. (2013). Flowers promote aphid suppression in apple orchards, *Biol. Control* 66(1), 8-15.

Gray, E., McGehee, E. M. and Carlisle, D. F. (1973). Seasonal variation in flowering of common dandelion, *Weed Sci.* 21(3), 230-232.

Grewal, P. S. (2012). From integrated pest management to ecosystem management: The case for the urban lawn. In: Abrol, D. P. and Shankar, U. (Eds) *Integrated Pest Management: Principles and Practice*, CAB International, pp. 450-488. https://www.cabidigitallibrary.org/doi/10.1079/9781845938086.0450.

Grimm, N. B., Faeth, S. H., Golubiewski, N. E., Redman, C. L., Wu, J., Bai, X. and Briggs, J. M. (2008). Global change and the ecology of cities, *Science* 319(5864), 756-760.

Grimm, N. B., Morgan Grove, J., Pickett, S. T. A. and Redman, C. L. (2000). Integrated approaches to long-term studies of urban ecological systems, *BioScience* 50(7), 571-584.

Grube, A., Donaldson, D., Kiely, T. and Wu, L. (2011). Pesticides industry sales and usage, 2006 and 2007 market estimates. Available at: (https://www.epa.gov/pesticides/pesticides-industry-sales-and-usage-2006-and-2007-market-estimates); accessed April 27, 2022, United States Environmental Protection Agency, Washington, DC.

Haaland, C., Naisbit, R. E. and Bersier, L. G. (2011). Sown wildflower strips for insect conservation: A review, *Inst. Cons. Div.* 4, 60-80.

Haydak, M. H. (1970). Honey bee nutrition, *Annu. Rev. Entomol.* 15(1), 143-156.

Held, D. W. and Potter, D. A. (2012). Prospects for managing turfgrass pests with reduced chemical inputs, *Annu. Rev. Entomol.* 57, 329-354.

Hennig, E. I. and Ghazoul, J. (2012). Pollinating animals in the urban environment, *Urban Ecosyst.* 15(1), 149-166.

Hernandez, J. L., Frankie, G. W. and Thorp, R. W. (2009). Ecology of urban bees: A review of current knowledge and directions for future study, *Cities Environ.* 2(1), 1-15.

Hines, H. M. and Hendrix, S. D. (2005). Bumble bee (Hymenoptera: Apidae) diversity and abundance in tallgrass prairie patches: Effects of local and landscape floral resources, *Environ. Entomol.* 34(6), 1477-1484.

Hopwood, E., Hoffman Black, S. and Fleury, S. (2016). *Pollinators and Roadsides: Best Management Practices for Managers and Decision Makers.*

Hopwood, J. L. (2008). The contribution of roadside grassland restorations to native bee conservation, *Biol. Conserv.* 141(10), 2632-2640.

Hoyle, H., Jorgensen, A., Warren, P., Dunnett, N. and Evans, K. (2017). "Not in their front yard" The opportunities and challenges of introducing perennial urban meadows: A local authority stakeholder perspective. *Urban For. Urban Green.* 25, 139-149.

Illinois Wildflowers. (n.d.a). Black medic. Medicago Lupulina. Available at: (https://illinoiswildflowers.info/weeds/plants/black_medic.htm); accessed February 4, 2022.

Illinois Wildflowers. (n.d.b). Common chickweed stellaria media. Available at: (http://www.illinoiswildflowers.info/weeds/plants/cm_chickweed.htm); accessed February 4, 2022.

Immelman, K. and Eardley, C. (2000). Gathering of grass pollen by solitary bees (Halictidae, *Lipotriches*) in South Africa, *Zoosyst. Evol.* 76, 263-268.

Intergovernmental Science-Policy Platform on Biodiversity and Ecosystem Services. (2016). The assessment report of the intergovernmental science-policy platform on biodiversity and ecosystem services on pollinators, pollination and food production. In: Potts, S. G., Imperatriz-Fonseca, V. L. and Ngo, H. T. (Eds) *Secretariat of the Intergovernmental Science-Policy Platform on Biodiversity and Ecosystem Services*, Bonn, Germany.

Iowa Department of Transportation. (n.d.). Iowa living roadway trust fund Available at: (https://iowadot.gov/lrtf/); accessed February 3, 2022.

Jauker, F. and Wolters, V. (2008). Hover flies are efficient pollinators of oilseed rape, *Oecologia* 156(4), 819-823.

Jenkins, V. S. (1994). *The Lawn: A History of American Obsession*, Smithsonian Institution Press, Washington, DC.

Jones, T. M. (2014). Why is the lawn buzzing?, *Biodivers. Data J.* 2(2), e1101.

Joseph, S., Harris-Shultz, K. and Jespersen, D. (2020). Evidence of pollinators foraging on centipedegrass inflorescences, *Insects* 11(11), 795.

Keller, I., Fluri, P. and Imdorf, A. (2005). Pollen nutrition and colony development in honey bees - part II, *Bee World* 86(2), 27-34.

Kidd, J., Manning, P., Simkin, J., Peacock, S. and Stockdale, E. (2017). Impacts of 120 years of fertilizer addition on a temperate grassland ecosystem, *PLoS ONE* 12(3), e0174632.

Klein, A. M., Vaissiere, B. E., Cane, J. H., Steffan-Dewenter, I., Cunningham, S. A., Kremen, C. and Tscharntke, T. (2007). Importance of pollinators in changing landscapes for world crops, *Proc. Biol. Sci.* 274(1608), 303-313.

Kremen, C. and Chaplin-Kramer, R. (2004). Insects as providers of ecosystem services: Crop pollination and pest control. In: Steward, A. J. A., New, T. R. and Lewis, O. T. (Eds) Insect Conservation Biology. *Proceedings of the Royal Entomological Society's 23rd Symposium*, CABI Publishing, Oxfordshire, UK, pp. 349-382.

Kremen, C., Williams, N. M., Aizen, M. A., Gemmill-Herren, B., LeBuhn, G., Minckley, R., Packer, L., Potts, S. G., Roulston, T., Steffan-Dewenter, I., Vázquez, D. P., Winfree, R., Adams, L., Crone, E. E., Greenleaf, S. S., Keitt, T. H., Klein, A. M., Regetz, J. and Ricketts, T. H. (2007). Pollination and other ecosystems services produced by mobile organisms: A conceptual framework for the effects of land-use change, *Ecol. Lett.* 10(4), 299-314.

Kuzovkina, Y., Campanelli, J., Schulthess, C. P., et al. (2016). Effective establishment of native grasses on roadsides in new England. Available at: (http://nenativeplants.uconn.edu/references_10_2517683307.pdf); accessed February 10, 2022.

Lady Bird Johnson Wildflower Center. (2012). Astragalus crassicarpus. Available at: (https://www.wildflower.org/plants/result.php?id_plant=ASCR2); accessed January 11, 2022.

Lady Bird Johnson Wildflower Center. (2013). Dalea candida. Available at: (https://www.wildflower.org/plants/result.php?id_plant=DACA7); accessed February 4, 2022.

Lady Bird Johnson Wildflower Center. (2015). Prunella vulgaris. Available at: (https://www.wildflower.org/plants/result.php?id_plant=prvu); accessed July 22, 2021.

Lady Bird Johnson Wildflower Center. (2020a). Oxalis stricta. Available at: (https://www.wildflower.org/plants/result.php?id_plant=OXST); accessed February 4, 2022.

Lady Bird Johnson Wildflower Center. (2020b). Viola sororia. Available at: (https://www.wildflower.org/plants/result.php?id_plant=VISO); accessed February 4, 2022.

Lady Bird Johnson Wildflower Center. (2021). Claytonia virginica. Available at: (https://www.wildflower.org/plants/result.php?id_plant=clvi3); accessed February 4, 2022.

Landis, D. A., Wratten, S. D. and Gurr, G. M. (2000). Habitat management to conserve natural enemies of arthropod pests in agriculture, *Annu. Rev. Entomol.* 45, 175-201.

Lane, I., Watkins, E. and Spivak, M. (2019a). Turfgrass species affect the establishment and bloom of Kura clover (*Trifolium ambiguum*) in lawns, *Hortscience* 54(5), 824-828.

Lane, I. G., Wolfin, J., Watkins, E. and Spivak, M. (2019b). Testing the establishment of eight forbs in mowed lawns of hard fescue (*Festuca brevipila*) for use in pollinator conservation, *Hortscience* 54(12), 2150-2155.

Lark, T. J., Spawn, S. A., Bougie, M. and Gibbs, H. K. (2020). Cropland expansion in the United States produces marginal yields at high costs to wildlife, *Nat. Commun.* 11(1), 4295.
Larson, J. L., Redmond, C. T. and Potter, D. A. (2013). Assessing insecticide hazard to bumble bees foraging on flowering weeds in treated lawns, *PLoS ONE* 8(6), e66375.
Larson, J. L., Kesheimer, A. J. and Potter, D. A. (2014a). Pollinator assemblages on dandelion and white clover in urban and suburban lawns, *J. Insect Conserv.* 18(5), 863-873.
Larson, J. L., Redmond, C. T. and Potter, D. A. (2014b). Impacts of a neonicotinoid, neonicotinoid-pyrethroid premix, and anthranilic diamide insecticide on four species of turf-inhabiting beneficial insects, *Ecotoxicology* 23(2), 252–259.
Larson, J. L., Redmond, C. T. and Potter, D. A. (2015). Mowing mitigates bioactivity of neonicotinoid insecticides in nectar of flowering lawn weeds and turfgrass guttation, *Environ. Toxicol. Chem.* 34(1), 127–132.
Lee, J. C. and Hempel, G. E. (2005). Impact of flowering buckwheat on Lepidopteran cabbage pests and their parasitoids at two spatial scales, *Biol. Control* 34(3), 290–301.
Leeds, R. (2000). *The Plantfinder's Guide to Early Bulbs*, Timber Press, Portland, OR.
Lerman, S. B., Contosta, A. R., Milam, J. and Bang, C. (2018). To mow or to mow less: Lawn mowing frequency affects bee abundance and diversity in suburban yards, *Biol. Conserv.* 221, 160–174.
Lonsdorf, E. V., Nootenboom, C., Janke, B. and Horgan, B. P. (2021). Assessing urban ecosystem services provided by green infrastructure: Golf courses in the Minneapolis-St. Paul metro area, *Landsc. Urban Plan.* 208, 1–10.
MacLeod, A., Wratten, S. D., Sotherton, N. W. and Thomas, M. B. (2004). 'Beetle banks' as refuges for beneficial arthropods in farmland: Long-term changes in predator communities and habitat, *Agric. Forest Entomol.* 6(2), 147–154.
Mahr, S. (n.d.). Black medic, Medicago lupulina. Available at: (https://hort.extension.wisc.edu/articles/black-medic-medicago-lupulina/); accessed April 9, 2021.
Mandelik, Y., Winfree, R., Neeson, T. and Kremen, C. (2012). Complementary habitat use by wild bees in agro-natural landscapes, *Ecol. Appl.* 22(5), 1535–1546.
Marlin, J. C. and LaBerge, W. E. (2001). The native bee fauna of Carlinville, Illinois, revisited after 75 years: A case for persistence, *Cons. Ecol.* 5(1), 9.
Marshall, E. J. P., Brown, V. K., Boatman, N. D., Lutman, P. J. W., Squire, G. R. and Ward, L. K. (2003). The role of weeds in supporting biological diversity within crop fields, *Weed Res.* 43(2), 77–89.
Matteson, K. C., Ascher, J. S. and Langellotto, G. A. (2008). Bee richness and abundance in New York City urban gardens, *Ann. Entomol. Soc. Am.* 101(1), 140–150.
McCurdy, J. D., Held, D. W., Gunn, J. M. and Barickman, T. C. (2017). Dew from warm-season turfgrasses as a possible route for pollinator exposure to lawn-applied Imidacloprid, *Crop Forage Turfgrass Manag.* 3(1), 1–6.
McCurdy, J. D., McElroy, J. S. and Guertal, E. A. (2013a). White clover (*Trifolium repens*) establishment within dormant Bermudagrass turf: Cultural considerations, establishment timing, seeding rate, and cool-season companion grass species, *Hortscience* 48(12), 1556–1561.

McCurdy, J. D., McElroy, J. S., Guertal, E. A. and Wood, C. W. (2013b). Dynamics of white clover decomposition in a southeastern Bermudagrass lawn, *Agron. J.* 105(5), 1277–1282.

McCurdy, J. D., McElroy, J. S., Guertal, E. A. and Wood, C. W. (2014). White clover inclusion within a Bermudagrass lawn: Effects of supplemental nitrogen on botanical composition and nitrogen cycling, *Crop Sci.* 54(4), 1796–1803.

McFrederick, Q. S. and LeBuhn, G. (2006). Are urban parks refuges for bumble bees *Bombus* spp. (Hymenoptera: Apidae)?, *Biol. Conserv.* 129(3), 372–382.

Menz, M. H., Phillips, R. D., Winfree, R., Kremen, C., Aizen, M. A., Johnson, S. D. and Dixon, K. W. (2011). Reconnecting plants and pollinators: Challenges in the restoration of pollination mutualisms, *Trends Plant Sci.* 16(1), 4–12.

Milesi, C., Running, S. W., Elvidge, C. D., Dietz, J. B., Tuttle, B. T. and Nemani, R. R. (2005). Mapping and modeling the biogeochemical cycling of turf grasses in the United States, *Environ. Manag.* 36(3), 426–438.

Millennium Ecosystem Assessment. (2003). *Ecosystem and Human Well-Being: A Framework for Assessment*, Island Press, Washington, DC.

Minnesota Department of Transportation. (2021). Pollinators and MnDOT. Available at: (http://www.dot.state.mn.us/pollinators/); accessed February 4, 2022.

Minnesota Wildflowers. (2022). Lotus corniculatus: Birds-Foot Trefoil. Available at: (https://www.minnesotawildflowers.info/flower/birds-foot-trefoil); accessed February 4, 2022.

Monarch City USA. (2020). Monarch city U.S.A. Available at: (http://monarchcityusa.com/); accessed February 17, 2022.

Monarch Watch. (2021) Available at: (https://monarchwatch.org/waystations/); accessed February 3, 2022, Monarch Waystation Program.

Morandin, L. A. and Winston, M. L. (2006). Pollinators provide economic incentive to preserve natural land in agroecosystems, *Agric. Ecosyst. Environ.* 116(3–4), 289–292.

National Golf Foundation. (2021). 2020 industry facts. Available at: (https://www.ngf.org/); accessed September 7, 2021.

Nichols, C. I. and Altieri, M. A. (2013). Plant biodiversity enhances bees and other insect pollinators in agroecosystems. A review, *Agron. Sustain. Dev.* 33(2), 257–274.

North Carolina State University. (n.d.a). Extension, *Chamaemelum nobile*. Available at: (https://plants.ces.ncsu.edu/plants/chamaemelum-nobile/); accessed July 9, 2021.

North Carolina State University. (n.d.b). Extension, *Taraxacum officinale*. Available at: (https://plants.ces.ncsu.edu/plants/taraxacum-officinale/); accessed July 9, 2021.

North Carolina State University. (n.d.c). Extension, *Thymus serpyllum*. Available at: (https://plants.ces.ncsu.edu/plants/thymus-serpyllum/); accessed January 11, 2022.

Oregon State University. (2018). *Strawberry clover Trifolium fragiferum* L. Available at: (https://forages.oregonstate.edu/forages/strawberry-clover); accessed July 9, 2021.

Owen, J. and Owen, D. F. (1975). Suburban gardens: England's most important nature preserve?, *Environ. Conserv.* 2(1), 53–59.

Palmer, M., Bernhardt, E., Chornesky, E., Collins, S., Dobson, A., Duke, C., Gold, B., Jacobson, R., Kingsland, S., Kranz, R., Mappin, M., Martinez, M. L., Micheli, F., Morse, J., Pace, M., Pascual, M., Palumbi, S., Reichman, O. J., Simons, A., Townsend, A. and Turner, M. (2004). Ecology. Ecology for a crowded planet, *Science* 304(5675), 1251–1252.

Patt, J. M., Hamilton, G. C. and Lashomb, J. H. (1997). Foraging success of parasitoid wasps on flowers: Interplay of insect morphology, floral architecture and searching behavior, *Entomol. Exp. Appl.* 83(1), 21-30.
Potts, S. G., Vulliamy, B., Dafni, A., Ne'eman, G. and Willmer, P. (2003). Linking bees and flowers: How do floral communities structure pollinator communities?, *Ecology* 84(10), 2628-2642.
Potts, S. G., Vulliamy, B., Roberts, S., O'Toole, C., Dafni, A., Ne'eman, G. and Willmer, P. (2005). Role of nesting resources in organising diverse bee communities in a Mediterranean landscape, *Ecol. Entomol.* 30(1), 78-85.
Powney, G. D., Carvell, C., Edwards, M., Morris, R. K. A., Roy, H. E., Woodcock, B. A. and Isaac, N. J. B. (2019). Widespread losses of pollinating insects in Britain, *Nat. Commun.* 10(1), 1018.
Puusepp, L. and Koff, T. (2014). Pollen analysis of honey from the Baltic region, Estonia, *Grana* 53(1), 54-61.
Ramer, H., Nelson, K. C., Spivak, M., Watkins, E., Wolfin, J. and Pulscher, M. (2019). Exploring park visitor perceptions of 'flowering bee lawns' in neighborhood parks in Minneapolis, MN, US, *Landsc. Urban Plan.* 189, 117-128.
Ramsden, M. W., Menendez, R., Leather, S. R. and Wäckers, F. (2015). Optimizing field margins for biocontrol services: The relative role of aphid abundance, annual floral resources, and overwinter habitat in enhancing aphid natural enemies, *Agric. Ecosyst. Environ.* 199, 94-104.
Rands, S. A. and Whitney, H. M. (2011). Field margins, foraging distances and their impacts on nesting pollinator success, *PLoS ONE* 6(10), e25971.
Raven, P. H. and Wagner, D. L. (2021). Agricultural intensification and climate change are rapidly decreasing insect biodiversity, *Proc. Natl Acad. Sci. U. S. A.* 118(2), e2002548117.
Redmond, C. T., Williams, D. W. and Potter, D. A. (2012). Comparison of scarab grub population and associated pathogens and parasitoids in warm- or cool-season grasses used on transitional zone golf courses, *J. Econ. Entomol.* 105(4), 1320-1328.
Reilly, J. R., Artz, D. R., Biddinger, D., et al. (2020). Crop production in the USA is frequently limited by lack of pollinators, *Proc. R. Soc. B Bol. Sci.* 287, 2020922.
Requier, F., Odoux, J. F., Tamic, T., Moreau, N., Henry, M., Decourtye, A. and Bretagnolle, V. (2015). Honey bee diet in intensive farmland habitats reveals an unexpectedly high flower richness and a major role of weeds, *Ecol. Appl.* 25(4), 881-890.
Richardson, M. D., McCalla, J., Buxton, T. and Lulli, F. (2015). Incorporating early spring bulbs into dormant warm-season turfgrasses, *hortTechnology* 25(2), 228-232.
Ricketts, T. H., Regetz, J., Steffan-Dewenter, I., Cunningham, S. A., Kremen, C., Bogdanski, A., Gemmill-Herren, B., Greenleaf, S. S., Klein, A. M., Mayfield, M. M., Morandin, L. A., Ochieng', A., Potts, S. G. and Viana, B. F. (2008). Landscape effects on crop pollination services: Are there general patterns?, *Ecol. Lett.* 11(5), 499-515.
Ries, L., Debinski, D. M. and Wieland, M. L. (2001). Conservation value of roadside prairie restoration to butterfly communities, *Conserv. Biol.* 15(2), 401-411.
Robbins, P. and Birkenholtz, T. (2003). Turfgrass revolution: Measuring the expansion of the American lawn, *Land Use Policy* 20(2), 181-194.
Robbins, P., Polderman, A. and Birkenholtz, T. (2001). Lawns and toxins, *Cities* 18(6), 369-380.
Robbins, P. and Sharp, J. T. (2003). Producing and consuming chemicals: The moral economy of the American Lawn, *Econ. Geogr.* 79(4), 425-451.

Robertson, C. (1926). Phenology of inquiline and nest-making bees, *Psyche* 33(4-5), 116-120.
Robertson, C. (1929a). *Flowers and Insects: Lists of Visitors to Four Hundred and Fifty-Three Flowers*, Science Press Printing Company, Lancaster, PA.
Robertson, C. (1929b). Phenology of oligolectic bees and favorite flowers, *Psyche* 36, 112.
Rosenzweig, M. L. (2003). Reconciliation ecology and the future of species diversity, *Oryx* 37(2), 194-205.
Salonen, A., Ollikka, T., Grönlund, E., Ruottinen, L. and Julkunen-Tiitto, R. (2009). Pollen analyses of honey from Finland, *Grana* 48(4), 281-289.
Sandberg, L. A. and Foster, J. (2005). Challenging lawn and order: Environmental discourse and lawn care reform in Canada, *Environ. Polit.* 14(4), 478-494.
Shepherd, M. (2002). Making room for native pollinators: How to create habitat for pollinator insects on golf courses. Available at: (https://xerces.org/sites/default/files/2018-05/06-001_02_XercesSoc_Making-Room-for-Native-Pollinators.pdf.); accessed April 27, 2022.
Sidhu, C. S. and Joshi, N. K. (2016). Establishing pollinator wildflower habitats in agricultural farmland to provide multiple ecosystem services, *Front. Plant Sci.* 7, 1-5.
Silverton, J., Poulton, P., Johnston, E., Edwards, G., Heard, M. and Biss, P. M. (2006). The park grass experiment 1856-2006: Its contribution to ecology, *J. Ecol.* 94(4), 801-814.
Sincik, M. and Acikgoz, E. (2007). Effects of white clover inclusion on turf characteristics, nitrogen fixation, and nitrogen transfer from white clover to grass species in turf mixtures, *Commun. Soil Sci. Plant Anal.* 38(13-14), 1861-1877.
Smith, L. S. and Fellowes, M. D. E. (2014). The grass-free lawn: Management and species choice for optimum ground cover and plant diversity. *Urban For. Urban Green.* 13, 433-442.
Song, X. Y., Yao, Y. F. and Yang, W. D. (2012). Pollen analysis of natural honeys from the central region of Shanxi, North China, *PLoS ONE* 7(11), e49545.
Speer, G. S. and Allinson, D. W. (1985). Kura clover (*Trifolium ambiguum*): Legume for forage and soil conservation, *Econ. Bot.* 39(2), 165-176.
Stawiarz, E. and Wroblewska, A. (2010). Melissopalynological analysis of multifloral honeys from the Sandomierska upland area of Poland, *J. Apic. Sci.* 54, 65-75.
Steffan-Dewenter, I., Munzenber, U. and Tscharntke, T. (2001). Pollination, seed set and seed predation on a landscape scale, *Proc. Biol. Sci.* 268(1477), 1685-1690.
Steinkraus, D. (2010). Early spring flowers in Northwest Arkansas: The excellent, the good, and the poor, *Am. Bee. J.* 150, 351-354.
Syngenta. (2022). Operation pollinator. Available at: (https://www.syngenta-us.com/beehealth/operationpollinator.aspx); accessed February 3, 2022.
Tanner, R. A. and Gange, A. C. (2005). Effects of golf courses on local biodiversity,*Landsc. Urban Plan.* 71(2-4), 137-146.
Theodorou, P., Radzeviciute, R., Lentendu, G., Kahnt, B., Husemann, M., Bleidorn, C., Settele, J., Schweiger, O., Grosse, I., Wubet, T., Murray, T. E. and Paxton, R. J. (2020). Urban areas as hotspots for bees and pollination but not a panacea for all insects, *Nat. Commun.* 11(1), 576.
Tilman, E. A., Tilman, D., Crawley, M. J. and Johnston, A. E. (1999). Biological weed control via nutrient competition: Potassium limitation of dandelions, *Ecol. Appl.* 9(1), 103-111.

Tommasi, D., Miro, A., Higo, H. A. and Winston, M. L. (2004). Bee diversity and abundance in an urban setting, *Can. Entomol.* 136(6), 851–869.

Tonietto, D., Fant, J., Ascher, J., Ellis, K. and Larkin, D. (2011). A comparison of bee communities of Chicago green roofs, parks, and prairies. *Lands Urban Plan.* 103, 102–108.

Townsend, C. E. (1985). Miscellaneous perennial clovers. In: Taylor, N. L. (Ed) *Clover Science and Technology, American Society of Agronomy*, Crop Science Society of America and Soil Science Society of America, Madison, WI, pp. 563–578.

Tscharntke, T., Klein, A. M., Kruess, A., Steffan-Dewenter, I. and Thies, C. (2005). Landscape perspectives on agricultural intensification and biodiversity – Ecosystem service management, *Ecol. Lett.* 8(8), 857–874.

Turgeon, A. and Kaminski, J. (2019). Turfgrass species. In: Turgeon, A. and Kaminski, J. (Eds) *Turfgrass Management*, Turfpath, State College, PA, p. 81.

Tyson, J. and Grigsby, B. (1941). Growing beautiful lawns. *Mich. Agr. Expt, Sta. Res. Bull.* 224.

United Nations Department of Economic and Social Affairs (UNDESA). (2018). 2018 revision of world urbanization prospects. Available at: (https://www.un.org/development/desa/publications/2018-revision-of-world-urbanization-prospects.html); accessed February 5, 2022.

United States Census Bureau. (2017). One in five Americans live in rural areas. Available at: (https://www.census.gov/library/stories/2017/08/rural-america.html); accessed February 10, 2022.

United States Census Bureau. (n.d.). U.S. and world population clock. Available at: (https://www.census.gov/popclock/); accessed April 27, 2022.

United States Department of Agriculture. (2009). Summary report: 2007 national resources inventory. Available at: (https://www.nrcs.usda.gov/Internet/FSE_DOCUMENTS/stelprdb1041379.pdf); accessed February 3, 2022.

United States Forest Service. (n.d.). Plant of the week common yarrow (Achillea millefolium Available at: (https://www.fs.fed.us/wildflowers/plant-of-the-week/achillea_millefolium.shtml); accessed February 5, 2021.

Walton, N. J. and Isaacs, R. (2011). Influence of native flowering plant strips on natural enemies and herbivores in adjacent blueberry fields, *Environ. Entomol.* 40(3), 697–705.

Wenzel, A., Grass, I., Belavadi, V. V. and Tscharntke, T. (2020). How urbanization is driving pollinator diversity and pollination – A systematic review, *Biol. Conserv.* 241, 108321.

Westphal, C., Steffan-Dewenter, I. and Tscharntke, T. (2003). Mass flowering crops enhance pollinator densities at a landscape scale, *Ecol. Lett.* 6(11), 961–965.

Williams, I. H. (1994). The dependence of crop production within the European Union on pollination by honey bees, *Agric. Zool. Rev.* 6, 229–257.

Williams, N. M., Ward, K. L., Pope, N., Isaacs, R., Wilson, J., May, E. A., Ellis, J., Daniels, J., Pence, A., Ullmann, K. and Peters, J. (2015). Native wildflower plantings support wild bee abundance and diversity in agricultural landscapes across the United States, *Ecol. Appl.* 25(8), 2119–2131.

Wisdom, M. M., Richardson, M. D., Karcher, D. E., Steinkraus, D. C. and McDonald, G. V. (2019). Flowering persistence and pollinator attraction of early-spring bulbs in warm-season lawns, *Hortscience* 54(10), 1853–1859.

Wissman, J., Ahrné, K., Poeplau, C., Hedblom, M., Marstorp, H., Ignatieva, M. and Kätterer, T. (2016). Multifunctional golf courses, Scandinavian turfgrass and environment research foundation report. Available at: (http://www.sterf.org/Media/Get/2335/multifunctional-golf-courses-final-report-31march2016.pdf); accessed January 11, 2021.

Wratten, S. D., Gillespie, M., Decourtye, A., Mader, E. and Desneux, N. (2012). Pollinator habitat enhancement: Benefits to other ecosystem services, *Agric. Ecosys Environ.* 159, 112–122.

Xerces Society. (2021). Available at: (https://beecityusa.org/); accessed February 4, 2022, Bee City USA & Bee Campus USA.

Chapter 19

Considerations with water for turfgrass in arid environments

Brian Whitlark, United States Golf Association, USA; Kai Umeda, University of Arizona, USA; Bernhard R. Leinauer, New Mexico State University, USA; and Matteo Serena, University of California-Riverside, USA

1 Introduction

Golf courses in the western USA are almost entirely reliant on the irrigation system to meet turf water demand (B. Whitlark, personal observation). Precipitation for many golf courses is scant and when it does occur, only 50% of the rain may be useful to offset irrigation needs. Utilizing an underground irrigation system to meet the turf demands with sprinklers typically spaced 18–24 m apart in varying conditions is extremely challenging. Even the best irrigation systems operate with 20% error and that is in the absence of wind (Gross, 2015). Enter wind exceeding 8 $km{\bullet}hr^{-1}$ into the equation and that error increases substantially. In the absence of rain, irrigation error is increased simply because of normal and routine use of the irrigation system. Consequently, turf managers in the western USA have learned out of necessity to become excellent at maintaining irrigation systems to deliver water as efficiently as possible. In this chapter, we will discuss tried-and-true strategies that turf managers in the arid western USA use daily to deliver water efficiently to thereby maximize soil moisture consistency and optimize turf playability.

http://dx.doi.org/10.19103/AS.2022.0110.21

2 Soil moisture

Golf course superintendents and turfgrass managers are tasked with producing and maintaining a playing surface that offers firm conditions and optimizes turf health (B. Whitlark, personal observation). A key component of that goal is to produce consistent soil moisture across playing surfaces. Managing water resources and inputs represents an important task for golf courses in arid environments.

2.1 Soil moisture uniformity is more important than distribution uniformity

Irrigation distribution uniformity (DU) has been the long-standing evaluative consideration for an irrigation system's efficiency, but the DU has a limited impact on soil moisture (i.e. soil water status in the rootzone) consistency across irrigated areas of the golf course (Zoldoske, 2003). Improving soil moisture consistency is the fundamental goal for golf courses that are looking to maximize water use efficiency and optimize playability in those turf surfaces. Focusing on soil moisture uniformity (MU) is much more impactful for golf course superintendents who are working to provide firm playing surface conditions and healthy turf in conjunction with efficient water use.

An irrigation system that delivers high DU (i.e. >80%) does not guarantee high MU across a golf course. There are many reasons for this, but at the top of the list are soil spatial variability, followed by wind, slope, sunlight exposure, shade, and rootzone compaction. Preliminary results from a field study in Texas, USA (Young et al., 2019), indicated high variability among soil spatial characteristics such as bulk density, volumetric water content, and infiltration rate. With such high variability, it becomes clear why there is so much soil moisture inconsistency across large areas of fairways and roughs. While an irrigation system operating at very high DU may deliver high MU when evaluated indoors and on a flat surface, the DU does not correlate well with MU when evaluated outdoors on actual golf courses.

What is one of the first things golf course superintendents do following installation of a new irrigation system? They change station percentages and adjust cycle and soak times to produce consistent soil moisture and thereby deliver a firm playing surface! These adjustments are made daily, especially in areas of the western USA where turf water demand is met almost entirely through irrigation (B. Whitlark, personal observation). Optimizing MU is a function of careful irrigation management and utilizing turfgrass cultural practices to improve the soil's ability to optimally 'accept' water. The next subsection will provide five irrigation management strategies that are delivering improved MU at golf courses in the southwest USA where water is limited, water quality is often poor, and water cost is generally higher than anywhere in the country. The five irrigation management strategies include deep watering, soil moisture monitoring, supplemental irrigation, use of wetting agents, and modification of soil and soil drainage. If these strategies are working well under such challenging irrigation conditions of the southwest USA, they will almost certainly prove beneficial wherever a golf course is located.

2.2 Methods to improve moisture uniformity

Before exploring the five identified soil moisture improvement strategies, it is first worth mentioning the obvious 'low hanging fruit' method to improve soil moisture consistency. There are many examples demonstrating how golf course superintendents, greenkeepers, course care managers, sports turf and pitch managers, and turfgrass maintenance practitioners are saving water and optimizing water use efficiency, and here are a few common ideas related to the irrigation system:

- Raising and leveling sprinklers;
- Edging around sprinklers;
- Cleaning organic and inorganic debris from water lines;
- Adjusting sprinkler arcs;
- Adjusting system operating pressure;
- Conducting proactive leak repair/prevention;
- Adjusting the irrigation schedule based on evaporative loss and precipitation;
- Replacing worn sprinkler nozzles;
- Maintaining the on-site weather station.

These are all basic and legitimate methods that need to be employed within all turf maintenance operations (Whitlark, 2012). Turf practitioners are encouraged to address these 'low hanging fruit' items and continue to manage and maintain the irrigation system proactively.

2.2.1 Deep watering

The 'deep and infrequent' irrigation strategy is a commonly accepted practice for maintaining high-quality turf (Whitlark, 2021). However, many golf course superintendents do not use this strategy outside of putting greens. The practice of 'frequent and shallow' irrigation limits gas exchange in the upper rootzone, increases compaction, decreases rooting depth, and compromises soil moisture variability (Murphy, 2002). Consider that even a new irrigation system can only be expected to deliver water at 80% efficiency, and that is with no wind present (B. Whitlark, personal observation). Over time, the error in the water delivery is additive, similar to accumulating compounding interest in a retirement bank account. Watering at a shallow depth and often or more frequently will not improve chronically wet or dry rootzones but only leads to chronically wet surface conditions (B. Whitlark, personal observation). Applying water deeply at routine intervals will improve soil moisture consistency and uniformity across the entire golf course. Deep watering will help overcome irrigation system inefficiencies and soil spatial variability by improving the soil moisture status in chronically dry areas, thus consequently reducing the need for spot watering (Fig. 1).

Brian O'Laughlin, golf course superintendent at Annandale Golf Course in California, USA, schedules irrigation for 'Tifsport' bermudagrass [*C. dactylon* (L.) Pers. × *C. transvaalensis* Burtt-Davy] fairways at rootzone depths ranging from 13 to 38 mm nightly, depending on the time of year and evaporative demand. The hydraulic limitations of the irrigation pump system and pipe size limit irrigation to no more than about three fairways per night. In 2020, fairways typically were irrigated every sixth night. Surprisingly, recently irrigated fairways were not noticeably wetter than fairways irrigated several nights before. Examination of soil profiles revealed deep moisture and deep roots. Soil moisture was consistent across the fairways and roughs with very little evidence of overly wet or dry areas.

Annandale Golf Course is not alone in achieving success with deep watering, but every golf course could potentially employ a slightly different tactic. At Annandale Golf Course, full-circle sprinklers are programmed to run for at least 30 min and part-circle sprinklers run for 15 min, with individual run times adjusted for certain sprinklers based on site-specific conditions. For

Figure 1 Scheduled irrigation to apply water deep into the soil can improve root density and length. *Source:* Image provided by B. Whitlark.

example, sprinklers in low areas may only run 10–15 min, while sprinklers on south-facing slopes may run up to 2 h. Also, soak times typically are set at 15–20 min.

Golf courses with poorly drained soils may have to start with run times only 7–10 min with soak times 20–25 min to avoid runoff. Additionally, golf courses that have had success with deep watering all note that they started with a goal of applying only 6–12 mm water for one evening and gradually increase to higher water volumes. Golf course superintendents can gain confidence by experimenting with one hole (i.e. a par-3 fairway) and utilize deep watering over a 4- to 6-week period before expanding to the entire golf course.

As the day length and sun angles change through the year, the irrigation schedule must change as well. Turf that received full sunlight in the summer may be shaded in the spring and fall and the irrigation schedule must change accordingly to address distinctly different turf water requirements at different times of the year. Turf managers must adapt to these seasonal changes by extending irrigation scheduling intervals during periods of less evaporative demand and/or decreasing the depth of water applied during irrigation events.

The deep and infrequent watering strategy is one that has proven results and golf course superintendents report annual water use is either unchanged or slightly lower when compared to using a frequent and shallow watering schedule (B. Whitlark, personal observation). For example, O'Laughlin reported a savings of US$85 000 in water costs in 2020 using that deep and infrequent irrigation scheduling strategy despite a hot and dry year that would normally require more irrigation inputs to maintain the turf.

2.2.2 Soil moisture monitoring

A high percentage of golf courses use handheld soil moisture meters to help manage water on putting greens (Moeller, 2012). However, if the golf industry is going to significantly reduce water use and improve soil moisture consistency and uniformity across fairways and roughs, the adoption of in-ground moisture sensors is possibly the next option. There are a few early pioneers or adopters using in-ground soil moisture sensors to schedule irrigation in fairways and roughs. Tyler Truman, golf course superintendent at Sun City Palm Desert (Palm Desert, CA, USA), is one of those pioneers. Truman has been using the Toro TurfGuard (Toro Corp., Bloomington, MN, USA) sensors for the past 4 years and has installed four sensors in strategic areas across 16 ha of overseeded Tifway 419 bermudagrass fairways on their Santa Rosa Golf Course. Sensors were placed into representative 'dry,' 'moderately dry,' 'moderately wet,' and 'wet' areas determined by mapping the fairways using the Toro Precision Sense 6000 (Toro Corp., Bloomington, MN, USA) tool. The turf care team reviews the sensor data daily, in addition to field scouting and monitoring evapotranspiration

(ET) data to schedule irrigation. This strategy has yielded more consistent soil moisture conditions across fairways and roughs and has resulted in a 10–14% water savings which translates into an equivalent savings in electrical costs associated with operating the irrigation pumps.

Several United States Golf Association (USGA; Liberty Corner, NJ, USA) funded research studies have also demonstrated significant water savings when using in-ground soil moisture sensors to schedule irrigation compared to using ET-based schedule or a calendar-based schedule (Dyer et al., 2020; Saxena et al., 2020; Straw et al., 2020). A three-year study at the Rocky Ford Turfgrass Research Center (Manhattan, KS, USA) has shown significant water savings after 2 years of study. The total water applied when utilizing data from in-ground soil moisture sensors during the 2019 summer growing season on 'Meyer' zoysiagrass (*Zoysia japonica* L.) turf was 80% less than a calendar-based approach and 44% less than irrigating the zoysiagrass at 60% of reference ET (Dyer et al., 2020). The repeated study demonstrated similar results in the summer of 2020 with the sensor-based irrigation schedule delivering 84% water savings compared to a calendar-based approach and 72% savings compared to the 60% reference ET irrigation (Dyer et al., 2020). Specifically, irrigation totaled 330 mm for the traditional frequency-based water scheduling, 183 mm for the 60% of ET irrigation, and only 51 mm when irrigation was applied using soil moisture sensors. Of note, total precipitation for the duration of the 2020 study was 334 mm.

A field study at California State Polytechnic University (Pomona, CA, USA) has shown test plots on hybrid bermudagrass mowed at fairway height irrigated with the guidance of soil moisture sensor data resulted in less water applied versus plots irrigated based on ET. All sensor and ET data were collected from April 1 through 31 October in 2018 and again in 2019. All plots, regardless of irrigation water inputs, showed acceptable turf quality and turf cover (Saxena et al., 2020).

A collaborative study between the University of Minnesota (St. Paul, MN, USA), Texas A&M University (College Station, TX, USA), and the Toro Corp. evaluated irrigation water inputs as scheduled from in-ground soil moisture sensors compared to traditional (i.e. visual observation of the turf) and ET-based irrigation scheduling at 60% of reference ET. This study was conducted on a golf course, specifically on nine fairways (six par 4's and three par 5's) at Edina Country Club in Minneapolis, MN, on perennial ryegrass (*Lolium perenne* L.) turf. Soil moisture maps were produced using the Toro Precision Sense 6000. Low, medium, and high soil moisture class categories were established on each fairway and one soil sensor is placed in each moisture zone. In the soil moisture sensor treatments, irrigation was triggered when plant available water decreased by 50% in each moisture classification. The plant available water was determined by the difference in soil moisture content at field capacity (i.e. soil water content about 48 h after a rain or saturating irrigation event) and permanent wilting point, which was determined when wet wilt became visually apparent to the golf course superintendent. This field study had only 1 year of data collection in 2020, but results indicated 60% water savings in areas with soil moisture sensor treatments compared to areas irrigated using the deficit ET strategy and 45% savings when compared to irrigation applied based on traditional methods (observational). The mean irrigation applied per week was 6.98 mm for the ET-based treatments, 5.02 mm for the traditional irrigation methods, and only 2.76 mm per week for the soil moisture sensor-based treatments.

2.2.3 Supplemental irrigation

Hand watering, while necessary and marginally effective, does not facilitate optimum moisture uniformity and depth that can be achieved with low-precipitation portable sprinklers (B. Whitlark, personal observation). The Desert Mountain Club (Scottsdale, AZ, USA) uses low-precipitation sprinklers effectively on a variety of different turf species including perennial ryegrass, overseeded Tifway 419 bermudagrass, and creeping bentgrass (*Agrostis stolonifera* L.) fairways, despite a very efficiently operated irrigation system. A trailer is used to store and transport sets of portable sprinklers around the golf course for targeted irrigation as needed. Scott Bower, Director of Operations at The Martis Camp Club (Truckee, CA, USA), uses portable low-precipitation sprinklers daily during the short growing season (i.e. May/June through September). There is a total of 36 sets of portable sprinklers and two sets are placed adjacent to each fairway for easy deployment. The irrigation control network operates those portable sprinklers for a set amount of time ahead of play in the morning and also following fairway mowing. The portable sprinklers are often placed near cart paths and other strategic areas in the rough during play. The portable sprinklers typically run for 30 min at a time. While this may sound like a simplistic approach, the routine use of low-precipitation portable sprinklers at the Martis Camp Club has been a successful supplemental irrigation practice (Fig. 2).

2.2.4 Wetting agents

A wetting agent, the turf industry's commonly used term for a soil surfactant, is used to improve soil moisture consistency and uniformity, especially in sand-based rootzones (Fidanza et al., 2020; Zontek and Kostka, 2012). The USGA Green Section has published extensive resources on wetting agent use for golf course turf (https://www.usga.org/content/usga/home-page/course-care/digitalcollections/understanding-wetting-agents.html). Several studies have demonstrated that using wetting agents throughout the growing season

Figure 2 Example of low-precipitation portable sprinklers employed to address chronic dry areas by deeply wetting the soil. *Source:* Image provided by B. Whitlark.

can maintain turf quality while reducing water use (Fidanza et al., 2020). For example, a two-year study at the University of California (Riverside, CA, USA) evaluated the ability of mineral oil, fungicides, and wetting agents to improve turf quality and increase soil volumetric water content (Serena and Baird, 2021). The study revealed that 'all tested products could maintain acceptable quality for at least 12 weeks when irrigated at 55% reference ET replacement, better than the untreated control.' The best-performing treatments in this study were the mineral oil and a mixture of fungicides which demonstrated acceptable turf quality at only 34% and 33% ET replacement compared to the control which received 69% ET replacement (Serena and Baird, 2021). This research suggests that wetting agents, a fungicide combination, and mineral oil can improve rootzone moisture uniformity, increase rootzone moisture content, and produce acceptable turf quality under deficit irrigation (Serena and Baird, 2021). While the economics of applying these treatments may not be pragmatic for some golf facilities, this study demonstrates there is potential to reduce water inputs in conjunction with applying products according to the protocol outlined in this study.

2.2.5 Modify soil and improve drainage

There are many golf courses where soil physical limitations severely restrict water infiltration, resulting in poor soil moisture consistency and uniformity, poor turf surface playability, and compromised turf health (Whitlark, 2014). Soil infiltration refers to the ability of the soil to allow water to move into and through the soil profile. A proven strategy to improve rootzone conditions is sand topdressing. A USGA-funded field study demonstrated the value of an annual sand topdressing program to improve fairway soil conditions (Whitlark, 2014). In that study, all courses that had applied ≥7.6 cm topdressing sand over several years (i.e. 25 mm sand is equal to approximately 358.6–403.5 metric tons•ha^{-1}) observed a 2000% increase in infiltration rate (Whitlark, 2014). Also, plant available water increased by 254% from sand topdressing when compared to no soil modification (Whitlark, 2014). Furthermore, soil moisture consistency across golf course playing areas has improved for those golf courses that modified their poor soils through sand topdressing (Whitlark, 2014). Specifically, they observed less wet and dry areas and therefore more consistent playing surfaces, which they relate to as improved soil consistency (Whitlark, 2014). Of note, sand topdressing (Fig. 3) may not be advisable for fairways on every golf course, and it is strongly recommended to work with a physical soil testing laboratory before initiating a fairway sand topdressing program (B. Whitlark, personal observation).

Improving surface and subsurface drainage is another key strategy to improve soil moisture infiltration and uniformity. For example, a rootzone drainage network allows for increased site-specific watering on mounds and south-facing slopes without creating saturated conditions in low-lying areas (Whitlark, 2020). Several courses in southern Arizona have demonstrated the value of a subsurface drainage network in flat or low-lying areas (B. Whitlark, personal observation). The drainage allows golf course superintendents to apply enough water to leach salts from the system and supply adequate soil moisture to produce healthy turf on adjacent mounds and south-facing slopes without creating waterlogged and chronically wet conditions in areas where drainage has been installed.

Figure 3 Fairway sand topdressing can improve rootzone conditions for soil with limited water permeability. *Source:* Image provided by B. Whitlark.

3 Case studies

3.1 Turfgrass conversion can result in significant water savings

Golf courses in the western USA face increasingly stringent demands to reduce irrigation water inputs. In addition to imposed water restrictions, water costs continue to increase, some have seen costs skyrocket by over 500% in the past few years. Many courses in California have addressed these challenges by converting from cool-season turfgrass to warm-season turfgrass such as bermudagrass, seashore paspalum, or zoysiagrass (John Marman, West Coast Turf; Scottsdale, AZ, USA; personal communication). The vast majority of those golf courses have converted tees, fairways, and roughs to hybrid bermudagrass. The following case studies illustrate examples from golf courses that have been successful in converting to bermudagrass.

3.1.1 Birnam Wood Golf Course

Birnam Wood Golf Course (Santa Barbara, CA, USA) historically maintained a mixed stand of cool-season turfgrasses, sometimes referred to as the 'California Turf Surprise' (B. Whitlark, personal observation). This turf composition produces acceptable playing conditions but never delivers a premier golf experience and the water demand is higher than it would be for warm-season turf. Historically, the golf course budgeted for about 23 hectare meter of water use annually (228 193 839 L) with a cost of nearly US$200 000.

The golf course's leadership recognized an opportunity to improve the consistency of playing conditions and reduce resource inputs, most notably irrigation water by converting the cool-season turf to warm-season bermudagrass. In 2015/2016, the golf course converted 21 ha to 'Santa Ana' bermudagrass in fairways and Tifway 419 bermudagrass in roughs. In addition, 4.9 ha of irrigated turf was removed and replaced with low water use landscaping, although the golf course irrigation system has yet to be modified to accommodate independent water delivery to the golf course turf and the landscape areas that have different water demands.

Post renovation, the golf course budget is about 16.6 hectare meter of water use annually, which meant a 25% reduction in annual water inputs. Of note, the cost of water has increased by over 500% in the past decade. With that increased cost, the 25% water reduction saves the facility over US$100 000 annually. The conversion to bermudagrass has also saved approximately US$100 000 per year in fertilizer, seed, and plant protection products. The cost of the project was about US$1.25 million, and within 7 years, the golf course has already achieved a 100% return on investment.

3.1.2 Menlo Country Club

Menlo Country Club (Menlo, CA, USA) converted the sixth fairway from perennial ryegrass to Santa Ana bermudagrass in 2018. Golf course superintendent Chris Eckstrom has documented significant water savings on this fairway. He noted that 25% water savings is very realistic and feels confident that 30% is achievable. The bermudagrass fairway is irrigated once every three nights during the summer months and generally at the same depth (i.e. same irrigation runtime) as the perennial ryegrass fairways, which are irrigated nightly. For example, while the perennial ryegrass receives 30 min of runtime over a three-day period (i.e. 10 min per night), the bermudagrass only receives 10 min of runtime over the same three-day period.

In 2021, the golf course converted the remaining 22.3 ha of perennial ryegrass fairways to Santa Ana bermudagrass and replaced an additional 1.6 ha of turf with low water use naturalized grasses. The golf course realized a 20% water savings during that first-year grow-in period. With water budget of nearly US$1 million annually and water costs increasing 5–8% every year, the return on investment on the US$2.6 million project cost will be less than 8 years. Additional economic savings have been realized with reduced use of plant protectants, and golf course members are pleased with the improvement in the consistent quality of those and year-round playing conditions of those bermudagrass fairways.

3.1.3 The Santa Lucia Preserve Golf Course

The Preserve Golf Club (Carmel, CA, USA) is reliant on rainwater to meet its turf water demand for its 18-hole facility. In drought years, there was not enough water to meet the demand of the cool-season turf and, consequently, water was strategically shut off to areas of the golf course with lower priority such as roughs, tee surrounds, and even in fairways between the tees and landing areas. Golf course leadership realized that cool-season turf is not a sustainable playing surface. Therefore, in 2016, the golf course converted 11.3 ha of fairways to Santa Ana bermudagrass. A fraise mowing technique was used to remove the existing turf and to pulverize the soil and surface organic matter layer with a RotaDairon (Dairon SAS, Mulsanne, France). Since conversion of those fairways to Santa Ana bermudagrass, the club is using 13–14% less irrigation water annually. They were using about 283 905 884 L of water annually and now use about 246 051 766 L. The water saving is less, however, than what other golf courses have achieved simply due to the amount of land converted. At the Preserve Golf Club, there is 29 ha of irrigated turf, with only 11 ha of bermudagrass. The remaining 18 ha remain as cool-season turf on tees, roughs, green surrounds, and greens. At some point, it will be necessary to convert those areas to bermudagrass to finally eliminate the 'water thirsty' cool-season turfgrass.

A common theme among these California golf courses (i.e. Birnam Wood Golf Course, Menlo Country Club, and The Preserve Golf Club) that have made the conversion to warm-season turfgrass is a 25% or more annual irrigation water reduction or savings. Furthermore, in drought years and when facing local water use restrictions, these golf courses are much better equipped to continue to produce acceptable turf playing conditions during imposed water use reductions, as well as much better equipped to promote turf recovery once water restrictions are removed or until the rains arrive. Overall, on those California golf courses, bermudagrass delivers more consistent, firmer playing conditions year-round when compared to the 'hodge-podge' of cool-season turf grasses.

3.2 Native grasses and alternative plant materials following turfgrass removal

Water in the desert areas of southwest USA is a precious resource, and persistent drought conditions exacerbate water's limited availability and quality for turfgrass and landscape areas. Golf courses with more than 40 ha of intensely managed turf are regulated to reduce playable acreage for only greens, tees, and target fairways to conserve irrigation water. When turf is removed from the perimeters of those playable areas, the land must still be maintained aesthetically and also retain its functionality. A substitute for removed turf can be native plant materials that require lower maintenance inputs and irrigation water. The University of Arizona (Tucson, AZ, USA) has been evaluating and comparing the establishment and performance of several native grass species for potential adoption and installation when and where turfgrasses are removed (B. Whitlark, personal observation). Golf courses in Arizona have been implementing the use of alternative plant materials with varying success.

3.2.1 Briarwood Country Club, Camelback Golf Club, and Wigwam Golf Club

Three field experiments were conducted with 17 plant species including 14 native grasses, an introduced annual forage grass, a native forb, and an introduced landscape groundcover (Table 1). The small plot replicated experiments were established on three golf courses: Camelback Golf Club (Scottsdale, AZ, USA) in May 2016; Briarwood Country Club (Sun City West, AZ, USA) in June 2017; and Wigwam Golf Club (Litchfield Park, AZ, USA) in June 2019. The USDA Plant Material Center (Tucson, AZ, USA) provided seed for most of the native grasses and a native forb; the former University of Arizona

Table 1 Native grasses and plant species were evaluated for planting at those sites where turfgrasses were removed in Arizona, USA

Plant species	Common name and cultivar	Planting rate (kg ha^{-1})	Location[a]
Andropogon halii	Sand bluestem 'Chet'	5.5	W
Aristida purpurea	Purple threeawn	10	W
Bouteloua dactyloides	Buffalograss	244	B, C, W
Bouteloua curtipendula	Sideoats grama 'Vaughn'	12.45	W
Bouteloua gracilis	Blue grama	4.48	B, C, W
Eragrostis intermedia	Plains lovegrass	1.12	B, C, W
Eragrostis tef	Tef	5.6	B, C
Eragrostis trichodes	Sand lovegrass 'Bend'	3.4	W
Hilaria jamesii	Galleta 'Viva'	471	W
Hilaria rigida	Big galleta	195	B, C
Lippia nodiflora	Kurapia	17 628[b]	B, C
Muhlenbergia asperifolia	Alkali muhly	1.35	B, C
Schizachyrium scoparium	Little bluestem 'Cimarron'	5.5	W
Sporobolus airoides	Alkali sacaton	3.36	B, C, W
Sporobolus contractus	Spike dropseed	1.12	B, C
Sporobolus cryptandrus	Sand dropseed	1.12	B, C, W
Zinnia acerosa	Desert zinnia	2.47	B, C

[a]Field experiment locations and establishment year: B = Briarwood Country Club (Sun City West, AZ, USA), 2017; C = Camelback Golf Club (Scottsdale, AZ, USA), 2016; W = Wigwam Golf Club (Litchfield Park, AZ, USA), 2019.
[b]Small plugs planted.

Karsten Turfgrass Research Facility (Tucson, AZ, USA) provided seed for *Hilaria rigida* (big galleta) and *Sporobolus cryptandrus* (sand dropseed) in 2016 and 2017, a commercial vendor was the source of seeds in 2019, and Kurapia Inc. (South Pasadena, CA, USA) provided Kurapia plugs for all three sites.

During the May–June timing of each installation at each site, supplemental overhead irrigation was required to germinate and establish the seeded native grasses and Kurapia transplants in all three experiments. At Camelback GC, an overhead sprinkler system was installed to irrigate the experimental area and approximately 6.6 mm was applied daily during the first month of establishment. At Briarwood CC, an existing irrigation system was utilized and approximately 1.02 mm was applied daily during the establishment period. At Wigwam GC, an existing overhead irrigation system was utilized during the establishment period. At Wigwam GC, an additional fourth replicate was planted in August 2019 to coincide with monsoon rains, and more effective germination and emergence were incidentally observed as a result of the rains.

Once established and into subsequent years, the intention was to eliminate supplemental irrigation and depend on seasonal rains during the summer monsoons (i.e. July-September) and rain during the winter-spring season (i.e. November–April).

At all three locations, a starter fertilizer (0.56 kg nitrogen•ha^{-1} and 2.8 kg P_2O_5•ha^{-1}) was applied only once for all plantings at the time of seeding or soon after standing emergence. The experimental area at Camelback GC was mowed in the late spring to early summer (i.e. May–June) and then again in September or December after the grasses matured. At Briarwood CC and Wigwam GC, the native grasses were mowed once per year during the fall to winter season.

At two test locations in 2016 and 2017, *Eragrostis tef* (tef), an annual forage grain crop, demonstrated the best rate of germination and stand establishment among the grasses and Kurapia successfully established from transplanted plugs. *Eragrostis* intermedia (plains lovegrass), *Bouteloua gracilis* (blue grama), *H. rigida* (big galleta), *Sporobolus cryptandrus* (sand dropseed), *S. airoides* (alkali sacaton), and *S. contractus* (spike dropseed) showed adequate emergence, establishment, and survival. *B. dactyloides* (buffalograss) failed to emerge at Briarwood CC and was very slow to establish at Camelback GC and Wigwam GC, and *Zinnia acerosa* (desert zinnia) did not emerge at Briarwood CC or Camelback GC nor in a separate laboratory germination test.

At Wigwam GC, *Andropogon halii* (sand bluestem 'Chet'), *Schizachyrium scoparium* (*little bluestem*) ('Cimarron'), and *Aristida purpurea* (purple threeawn) successfully emerged following that later planting in late summer during the monsoon rain period. Also, purple three awn aggressively spread beyond the small planted plots, and buffalograss, the dropseeds, big galleta, and alkali sacaton all emerged and established successful stands.

A desirable aesthetic characteristic of plant materials is to maintain visual greenness year-round. Most native grasses become dormant and turn brown in the winter season when temperatures fall and frosts occur (K. Umeda, personal observation). Desert grasses in the low desert of Arizona may slow down in growth with the lack of rainfall and the onset of intense summer heat in May–June. Under mild winter conditions when hard freezes did not occur, plains lovegrass, alkali sacaton, and Kurapia retained acceptable green color throughout the winter. *Muhlenbergia asperifolia* (alkali muhly) and blue grama maintained some green color during the winter, and the dropseeds and big galleta did not retain their green color.

Initial seeding rates that were used ensured successful stand establishment of the grasses. After establishment, during subsequent years of growth, thick and dense stands began to thin among the bunch grasses. Shorter bunch grasses could be encouraged in areas adjacent to the fairways and gradually taller grasses could be transitioned further away, while colorful wildflowers could be integrated among the grasses. Sustaining long-term succession of native grasses and wildflowers to continuously provide a natural appearance will require monitoring and considerable labor resources to maintain and occasionally seed desirable plant materials, especially annual wildflowers (K. Umeda, personal observation). Perennial shrubs and volunteer desert trees may encroach upon those native grass sites and could require timely maintenance if those species interfere with play, aesthetics, or functionality of the desired turf and native grass areas.

On golf courses, players' pace of play is critical, and lost golf balls in deep grasses are not a desirable feature for native grasses in non-play areas. Kurapia has a prostrate growth habit that provides uniform and complete surface coverage. Buffalograss, gramas, galletas, and alkali muhly tended to provide full surface cover and depth where golf balls could be difficult to find. The alkali sacaton, purple three awn, dropseeds, and bluestems grow as bunch-type grasses where establishment spacing provides visible bare ground in between plants where golf balls could be potentially found.

Native grasses can be utilized effectively for golf courses if a natural desert grassland appearance is desired. Wildflowers can be integrated with the native grasses, and native desert-adapted shrubs and trees can provide perennial plant material to those areas. Irrigation water inputs will be reduced to only supplemental applications when severe drought occurs, and golf course labor can be directed away from those non-play areas to focus on those turf areas in play.

3.2.2 Camelback Golf Club – Ambiente Golf Course

The Ambiente Golf Course at Camelback Golf Club (Scottsdale, AZ, USA) is located in a floodplain through a heavily populated urbanized residential corridor. It was formerly turfed entirely with a total of 91 ha of common bermudagrass turf along with non-native desert-adapted pine trees along the fairways. It reopened in the fall of 2013 following an extensive renovation that reduced playable turfgrass areas to 32 ha. The drainage wash channel area on the golf course was deepened and strategically placed lakes were created and contoured to hold excess water during flooding events and also for irrigation water-holding purposes. The deepened channel prevented flood waters from rising to interfere with golf course playability. The ponds provided new habitats for fish and amphibians.

Few of the pine trees along the perimeter of the golf course adjacent to residential properties were preserved. Approximately, 20 ha of land adjacent to residences around the golf course perimeter and 28 ha of non-play areas in the drainage wash channel area were planted with native grasses and wildflowers (Table 2). During the initial winter following the establishment of those seed

Table 2 Wildflower and grass seed mixtures for the Ambiente Golf Course at Camelback Golf Club (Scottsdale, AZ, USA)

Scientific name	Common name	Plant family
Wildflowers		
Ambrosia deltoidea	Triangle-leaf bursage	Asteraceae
Baileya multiradiata	Desert marigold	Asteraceae
Calliandra eriophylla	Fairy duster	Fabaceae
Dyssodia spp.	Daisy family	Asteraceae
Encelia farinose	Brittlebush	Asteraceae
Eschscholzia californica	California poppy	Papaveraceae
Gaillardia pulchella	Firewheel	Asteraceae
Linaria maroccana	Baby snapdragon, annual toadflax	Plantaginaceae
Linum spp.	Flax	Linaceae
Lupinus spp.	Lupine	Fabaceae
Penstemon parryi	Parry's penstemon	Plantaginaceae
Ratibida columnifera	Prairie coneflower (Mexican hat)	Asteraceae
Sphaeralcea ambigua	Globe mallow	Malvaceae
Grasses		
Aristida purpurea	Purple three awn	Poaceae
Bouteloua curtipendula	Side oats grama	Poaceae
Hilaria rigida	Galleta	Poaceae
Sporobolus airoides	Alkali sacaton	Poaceae

mixes, approximately, 75% of the plant material yielded a high variability of colorful annual wildflowers. The visual aesthetics of those colorful wildflowers were considered highly favorable to the citizens of Paradise Valley (AZ, USA). Approximately 8 years after its establishment in 2013, the succession of plants has now converted to 5% wildflowers and 50% native grasses, with almost half of the ground surface area as bare ground. Volunteer mesquite and palo verde trees have emerged, and small emerged trees were selectively removed by hand to not interfere with golf course routing or play. Since the renovation of these planting areas, the plant community succession has created habitats for desert birds and wildlife and foraging opportunities for pollinators.

Typically, the low desert in the southwest USA is not characterized as a dense understory of annual or perennial grasses. Cactus, mesquite trees, and palo verde trees are not accustomed to fire regimes that are common in grassland areas of the plain states of the USA. Bare ground is typically observed between cactus, desert trees, and shrubs. Most desert native plants are supported by winter rains from November to April and summer monsoon rains from July through September, thus totaling 12–25 mm precipitation annually. Invasive weeds have infiltrated the desert region contributing to many desert wildfires. *Bromus rubens* (red brome) a winter annual grass, *Pennisetum ciliare* (buffelgrass) a perennial forage grass that has become weedy, and *Oncosiphon pilulifer* (stinknet) are examples of fire-carrying invasives spreading in the desert region.

The maintenance of the native grasses and wildflowers at Ambiente Golf Course has been fine-tuned to two mowings per year. The first mowing in late summer following summer growth resulting from the monsoon rains, and a second mowing after the spring bloom of wildflowers. However, the desired goal is to mow only once per year to further save on labor costs. Of note, selective herbicides have been applied in recent years when red brome and other *Bromus* or *Hordeum* spp. invaded the open spaces among the native species' plantings following winter rains.

The major highlight of the renovation project at Ambiente Golf Course was not only the reduction in turf area but also significant water conservation was achieved by irrigating only tees, fairways, and putting greens. Supplemental irrigation water is applied to the native species areas only during extreme drought periods. In 2020, records indicated that only 60.5 ML water were used to maintain the native grasses and wildflowers. In prior years, 211.7 ML water were used in 2015 and 71.8 ML water in 2017. Overall, 170.1 ML $year^{-1}$ were saved between the periods of 2013 and 2016 versus pre-renovation from 2000 to 2009 Also, the overall annual reduction in irrigation water consumption at Ambiente Golf Course was over 1 million m^3 in 2006 to just over 800 000 m^3 in 2020.

3.3 Subsurface drip irrigation for golf courses

Sprinkler systems are the most used irrigation systems for turf-dominated landscapes and golf courses, despite their inefficiencies (B. Leinauer, personal observation). Sprinkler overspray, overlap, wind drift, and evaporation all contribute to water losses that increase overall water consumption and/or decrease turf quality. Subsurface irrigation systems, however, are considered to irrigate more efficiently because water is applied directly to the rootzone, thereby avoiding problems such as overspray, runoff, and wind drift. Drip irrigation systems have been frequently used to irrigate trees, shrubs, flower beds, or vegetables but have received little acceptance for turfgrass irrigation. The benefits of subsurface drip systems (SDIs) have been documented for agricultural and horticultural crops as well as trees (Burt and Styles, 1999). They also offer a solution for lawns that are difficult to irrigate - such as narrow strips, slopes, or unusual, irregular-shaped areas, which is the case for many residential lawns - and bunker slopes and tee areas on golf courses.

Subsurface drip irrigation generally applies water from either a point or a line source. Both systems use a polyethylene pipe at depths that vary

Published by Burleigh Dodds Science Publishing Limited, 2023.

depending on the plant or ground cover which is irrigated (Burt and Styles, 1999). Line source systems seep water uniformly along the length of the line and are often referred to as soaker or porous hose lines. Point source systems, also called drip lines, are equipped with either non-pressure-compensating or pressure-compensating drip emitters spaced equally (i.e. 30 cm, 45 cm, 60 cm) and fitted into the pipes.

Advantages of SDI include the uninterrupted use of the turf area during irrigation, energy savings as a result of lower operating water pressure, no human exposure to irrigation water, reduced turf disease pressure, and water savings because irrigation is limited to only the turf area and is not affected by wind drift or evaporation (Camp, 1998) (Fig. 4). Arguments against the use of SDI include high installation costs, a perceived interference with regular maintenance, and a perceived inability to leach salts (Sevostianova et al., 2011a). The suitability of SDI for turf was first demonstrated 40 years ago, but

Figure 4. Cross-section of turfgrass rootzone with a drip line surrounded by roots and no visible roots penetrating the emitter (left); aerial image of sprinkler irrigated tee box illustrating the irrigated area greater than the actual turfgrass area, as indicated by the lush green vegetation beyond the tee area perimeter (top right); and a subsurface drip irrigated tee box with the irrigation zone strictly limited to the turfgrass tee area (bottom right).

the technology has never gained significant market acceptance (B. Leinauer, personal observation). However, SDI has recently begun to receive greater attention in the context of water conservation and environmental sustainability and has been mandated by some water agencies for narrow and irregularly shaped turf areas (California Department of Water Resources, 2009).

Studies investigating turfgrass establishment with subsurface irrigation systems have given conflicting results. Snyder et al. (1974) investigated emitter spacing and depth for drip irrigation under bermudagrass and reported that a maximum distance of 60 cm between emitters can be allowed without a reduction in turfgrass quality. Manufacturers of drip irrigation systems recommend an emitter spacing between 30 cm and 60 cm and a line depth of 10 cm to ensure uniform water distribution, regardless of the soil type into which the system is installed. Leinauer and Makk (2007) documented a delayed establishment of subsurface drip-irrigated creeping bentgrass in a sand-based USGA-type rootzone, when compared to sprinkler irrigation. Johnson (2007) compared the establishment of several cool-season turfgrasses under SDI and aboveground sprinkler irrigation and reported that successful establishment with SDI depended on the turfgrass species and a cultivar used. Leinauer et al. (2010a) used SDI in an establishment study and reported efficient and uniform water distribution at the surface with the drip system and successful turfgrass establishment. Schiavon et al. (2012) and Serena et al. (2014) demonstrated that bermudagrass and seashore paspalum can be established successfully from seed and sod when SDI is used in combination with saline water. Subirrigation systems may have some limitations in leaching salts from the rootzone. Particularly the fraction of the rootzone above the emitters (i.e. where most of the roots are accumulated) that receives water only through capillary rise may not be sufficiently flushed with water to leach out the salts. Two studies (Devitt and Miller, 1988; Choi and Suarez-Rey, 2004) have demonstrated that warm-season turfgrasses irrigated with either high saline water or with effluent water can be successfully grown using SDI; however, long-term salt accumulation around drip emitter has not been investigated.

Several reports have documented irrigation water savings at no loss in turf quality when subirrigated from an adjustable water table, or SDI, compared to aboveground sprinkler irrigated turf (Leinauer et al., 2004; Sevostianova et al., 2011a,b). In areas with irrigation water restrictions where the implementation of water conservation strategies is mandated, the use of subsurface irrigation technologies will likely to increase (Duncan et al., 2009; Leinauer et al., 2010b). Thus, applying water directly to the rootzone with SDI can result in improved water conservation and an efficient irrigation system.

Information is limited on the long-term performance of SDI systems in heavily trafficked and intensively managed turfgrass systems as often observed in certain areas of a golf course. Consequently, a multi-year field study was conducted to investigate the effects of different SDI systems and a standard pop-up sprinkler irrigation system on turfgrass performance and irrigation water consumption on golf course tee areas.

3.3.1 Materials and methods

The study was conducted at the Club at Las Campanas (Santa Fe, NM, USA) in an area with an arid climate at 2133 m elevation. The championship tee boxes used for the study were constructed with a 20-cm-deep sand rootzone overlaying a 10-cm-deep pea gravel layer. Creeping bentgrass was established to the existing tee boxes that were between 18 and 25 years old. The first SDIs were installed in May 2016 and included Toro Rootguard® DL2000™ (Toro Co., Bloomington, MN, USA) delivering 2.0 $L{\bullet}hr^{-1}$ and operating at 241 kPa and Rainbird XSF (Rainbird Inc., Azusa, CA, USA) with copper shield technology, delivering 2.3 $L{\bullet}hr^{-1}$ and also operating at 241 kPa. In May 2017, Netafim Techline CVXR (Netafim Irrigation Inc., Fresno, CA, USA), delivering 2.0 $L{\bullet}hr^{-1}$ and operating at 344 kPa, and Hunter Eco-Mat (Hunter Industries Inc., San Marcos, CA, USA), delivering 2.0 $L{\bullet}hr^{-1}$ and operating at 241 kPa, were added to the study. Each system was installed with 30 cm spacing between lines, except for one Toro and one Rainbird system, for which a spacing of 23 cm between lines was chosen. Emitters were spaced 30 cm apart and all lines were installed at rootzone depths of approximately 15 cm. A screen filter and a pressure regulator were installed on each tee box before the main valve, in addition to an air release valve and an automatic flush valve. Two tee boxes irrigated with Rain Bird 5004 rotor sprinklers (Rainbird Inc., Azusa, CA, USA) were designated as the controls. Each SDI system and the controls were equipped with water meters installed after the valves. Water use for each tee box was recorded and the turf surface area was calculated. The amount of water delivered to each tee box was reported relative to the irrigated area. However, during the investigative period, irrigation was scheduled manually and not standardized across tees or SDI systems. Therefore, comparisons between SDI products regarding water use must be made with caution.

Two approaches were used to install the SDI systems under the pre-existing turf. The first strategy was to remove the sod and, using a disk trencher, cut into the sand profile. After the system was installed, the sod was returned into place, followed by sand topdressing and irrigation using the overhead sprinkler system. The second approach was to trench directly into the existing turf. After the system was installed, the lines were re-compacted, and the turf was then sand topdressed, followed by irrigation with the overhead sprinkler system.

Turfgrass maintenance was conducted similarly among all other tee boxes at the golf course. This included irrigation at approximately 100% ET rate replacement, daily mowing, fertilization, sand topdressing, and verticutting as needed. Due to concerns about potential damage to the irrigation systems, SDI tee boxes were not aerified until fall 2020.

3.3.2 Results

With the help of the in-ground sprinkler system, turf stands recovered within a few months from post-installation injury or disruption. Spacing the drip lines 23 cm apart resulted in no different recovery and turf quality than a spacing of 30 cm. SDIs installed using the sod removal approach resulted in a faster recovery, and overall better turfgrass quality during the first months of the study (Fig. 5). On two occasions, drip lines were inadvertently installed at the incorrect depth, resulting in uneven turf establishment and ultimately led to the system being re-installed. After 1 year of installation, all SDI systems performed equally well based on turfgrass quality, with little to non-visible signs of differences between SDI systems.

Normalized difference vegetation indices were measured and did not show any differences in stress between the controls and the SDI-irrigated tee boxes. The irrigation water use by SDI-irrigated tee boxes was lower compared to tee boxes irrigated with traditional pop-up sprinklers (Table 3). Sprinkler-irrigated tee boxes received approximately 2–5 times more water compared to the SDI-irrigated tee boxes (Table 4). Over the four-year investigative period (i.e. 2017–2020), SDI-irrigated tee boxes used on average between 38% (2019) and 60% (2020) less water than the sprinkler-irrigated counterparts. The higher irrigation amounts used by sprinkler systems were due in part to overspray. Pop-up sprinklers irrigated beyond the tee box area and ended up irrigating the surrounding native vegetation. In contrast, the SDI systems delivered the water only to the designated area. The growth of the native vegetation surrounding tee boxes was significantly reduced in SDI-irrigated tee boxes (i.e. from not receiving any additional water), thereby reducing or eliminating the need for vegetation maintenance of those areas (Fig. 5).

Figure 5 Example of tee boxes 1 month after subsurface drip irrigation (SDI) installation. Left: trench lines were dug directly into the turf canopy and rootzone. Right: sod was removed, SDI was installed by trenching only into the rootzone and sod was returned in place.

Table 3 Irrigation water use (mm) of subsurface drip (Hunter, Netafim, Rainbird, Toro) systems and aboveground pop-up sprinkler (control) system irrigated on golf course tees in Santa Fe, NM (USA) during summer months (i.e. May-August) 2017-2020

	2017	2018	2019	2020
	mm	mm	mm	mm
Control[a]	2310	3050	1830	2570
Hunter[b,c]	-	640	890	610
Netafim[b,c]	-	860	1350	1350
Rainbird[b]	1070	1090	860	660
Toro[b]	990	1170	1450	1370

[a]Water use values represent the average of two tee boxes.
[b]Water use values represent the average of three tee boxes.
[c]Irrigation system was added later to the study, and therefore data not available for 2017.

Table 4 Reduction in water use (mm) when subsurface drip (Hunter, Netafim, Rainbird, Toro) systems as compared to aboveground pop-up sprinkler irrigation. Water-use values (calculated as the average of three tee boxes) represent percent water savings or reduction compared to sprinkler irrigation (calculated as the average of two tee boxes)

	2017	2018	2019	2020	Four-year average
		% water savings			
Hunter[a]	-	79	52	73	68
Netafim[a]	-	72	26	47	48
Rainbird	54	64	53	75	62
Toro	57	62	21	47	47

[a]Irrigation system was added later to the study, and therefore data not available for 2017.

4 Conclusion

Golf course managers in arid environments must manage every drop of water like it is their last. Water restrictions continue to tighten and water costs are increasing exponentially in many areas. While superintendents in arid environments utilize onsite weather data to schedule irrigation, current research and new technologies offer ways to optimize soil moisture consistency and reduce water inputs. Courses in California are converting fairways and roughs from cool-season grass to warm-season grasses, saving 25% or more annual water use. Despite high conversion costs, the return on investment is typically less than 10 years. Significant water savings have been achieved in southern Nevada, California, and Arizona by removing irrigated turf and replacing with desert landscape or native grasses. Portable soil moisture meters are widely used and inground soil moisture sensors are gaining popularity for use in large turf areas such as fairways and roughs. Early indications are that this technology may yield substantial water savings and deliver more consistent playing conditions. Subsurface drip irrigation has been shown to save 40-80% water compared to overhead irrigation and this technology is growing in popularity for bunker surrounds and teeing grounds.

Diligence, daily observation, constant adjustment of the irrigation system and practices, and a good overall turfgrass management plan are required to optimize soil moisture uniformity across a golf course. The work is never done as the weather is constantly changing, sun angle changes dramatically from winter to summer and sprinklers fall out of adjustment - to name just a few of the variables involved. Hopefully, the strategies provided in this chapter will prove useful to improve soil moisture consistency at your golf course. In summary, the golf industry is adapting to water-related challenges through adopting new technologies and building on strategies illuminated through water-related research.

5 Acknowledgements

The native grasses project at the University of Arizona (Tucson, AZ, USA) was supported in-part by the Specialty Crop Block Grant Program of the U.S. Department of Agriculture (Beltsville, MD, USA), the United States Golf Association (Liberty Corner, NJ, USA), and the Horticultural Research Institute (Washington, DC, USA). The USDA Plant Materials Center (Tucson, AZ, USA) and Kurapia Inc. (South Pasadena, CA, USA) provided plant materials to conduct these studies. We also thank Camelback Golf Club (Scottsdale, AZ, USA), Briarwood Country Club (Sun City West, AZ, USA), and Wigwam Golf Club (Litchfield Park, AZ, USA) for providing experimental sites. Dr. Worku Burayu of the University of Arizona Cooperative Extension (Maricopa County, AZ, USA) contributed significantly to the development and implementation of the field projects.

6 Where to look for further information

Research is needed in the broad area of water conservation for golf courses. Topics such as new irrigation technologies including inline drip irrigation for large turf areas, breeding warm season grasses that retain green color year-round in southern states and are drought tolerant and utilization of in-ground soil moisture sensors to schedule irrigation are just a few examples that need further exploration. The golf industry will look to the United States Golf Association (USGA) as the primary source of guidance and funding for researchers in the future. Other entities include the Golf Course Superintendents Association of America (GCSAA) who also award grants for turfgrass research. Additionally, federal agencies such as the United Stated Department of Agriculture (USDA) and the National Science Foundation (NSF) are interested in funding turfgrass research that addresses water conservation.

7 References

Burt, C. M. and Styles, S. W. 1999. *Drip and Micro Irrigation for Trees, Vines, and Row Crops. Irrig*. San Luis Obispo, CA Training and Research Center, California Polytechnic State University (Cal Poly).

California Department of Water Resources. 2009. Model water efficient landscape ordinance. *California Code of Regulations, Title Division, Waters 2. Department of Water Resources, Chapter (Vol. 23)*, p. 2.7. Available at: www.water.ca.gov/wateruseefficiency/docs/MWELO09-10-09.pdf.

Camp, C. R. 1998. Subsurface drip irrigation: A review. *Transactions of the ASAE. American Society of Agricultural Engineers* 41: 1353–1367.

Choi, C. Y. and Suarez-Rey, E. M. 2004. Subsurface drip irrigation for Bermudagrass with reclaimed water. *Transactions of the ASAE. American Society of Agricultural Engineers* 47: 1943–1951.

Devitt, D. A. and Miller, W. W. 1988. Subsurface drip irrigation of Bermudagrass with saline water. *Applied Agricultural Research* 3: 133–143.

Duncan, R. R., Carrow, R. N. and Huck, M. T. 2009. *Turfgrass and Landscape Irrigation Water Quality - Assessment and Management*. Boca Raton, FL: CRC Press.

Dyer, W., Bremer, D., Patrignani, A., Fry, J., Lavis, C. and Friell, J. 2020. Integrating canopy dynamics, soil moisture, and soil physical properties to improve irrigation scheduling in turfgrass. *Summary*. United States Golf Association Turfgrass and Environmental Research Program, pp. 218–223.

Fidanza, M., Kostka, S. and Bigelow, C. 2020. Communication of soil water repellency causes, problems, and solutions of intensively managed amenity turf from 2000 to 2020. *Journal of Hydrology and Hydromechanics* 68(4): 306–312.

Gross, P. 2015. When is it time to replace an irrigation system? *USGA Green Section Record* 53(5).

Johnson, C. 2007. Establishing cool- and warm-season turfgrasses using saline irrigation in combination with subsurface drip irrigation. M.S. Thesis. New Mexico State University, Las Cruces, NM.

Leinauer, B. and Makk, J. 2007. Establishment of golf greens under different construction types, irrigation systems, and rootzones. *USGA Turf and Environ. Research* 6(7): 1–7.

Leinauer, B., Sallenave, R. and Schulz, H. 2004. A comparison of construction types and their associated irrigation systems: Effect on turfgrass quality, drought avoidance, and irrigation water use. *Acta Horticulturae* 661(661): 123–129. doi: 10.17660/ActaHortic.2004.661.15.

Leinauer, B., Serena, M., Singh, D. and Singh, D. 2010a. Seed coating and seeding rate effects on turfgrass germination and establishment. *Hort Technology* 20(1): 179–185.

Leinauer, B., Sevostianova, E., Serena, M., Schiavon, M. and Macolino, M. 2010b. Irrigation water conservation for urban lawn areas. *Acta Horticulturae* 881: 487–492.

Moeller, A. 2012. Identify soil moisture status more accurately than ever before. *USGA Green Section Record* 50(9): 1–5.

Murphy, J. A. 2002. *Best Management Practices for Irrigating Golf Course Turf. Rutgers Cooperative Extension*. New Jersey Agricultural Experiment Station. Publication E278. Rutgers, The State University of New Jersey.

Saxena, P., Green, R., Vis, E. and Mellano, V. 2020. *Research Summary. Soil Moisture Sensor Irrigation Scheduling in Bermudagrass*. United States Golf Association Turfgrass and Environmental Research Program, pp. 204-207.

Schiavon, M., Leinauer, B., Serena, M., Sallenave, R. and Maier, B. 2012. Bermudagrass and seashore paspalum establishment from seed using differing irrigation methods and water qualities. *Agronomy Journal* 104(3): 706-714.

Serena, M. and Baird, J. 2021. *Turfgrass and Landscape Research Field Day Summaries, Evaluation of Products for Water Conservation on Bermudagrass Turf Using a Linear Gradient Irrigation System*. University of California-Riverside, pp. 14-18.

Serena, M., Leinauer, B., Schiavon, M., Maier, B. and Sallenave, R. 2014. Establishment and rooting response of Bermudagrass propagated with saline water and subsurface irrigation. *Crop Science* 54(2): 827-836.

Sevostianova, E., Leinauer, B., Sallenave, R., Karcher, D. and Maier, B. 2011a. Soil salinity and quality of sprinkler and drip irrigated cool-season turfgrasses. *Agronomy Journal* 103(5): 1503-1513. doi: 10.2134/agronj2011.0162.

Sevostianova, E., Leinauer, B., Sallenave, R., Karcher, D. and Maier, B. 2011b. Soil salinity and quality of sprinkler and drip irrigated warm-season grasses. *Agronomy Journal* 103(6): 1773-1784. doi: 10.2134/agronj2011.0163.

Snyder, G. H., Burt, E. O., Rogers, J. S. and Campbell, K. L. 1974. Theory and experimentation for turf irrigation from multiple subsurface point sources. *Proceedings of the Soil Crop Science* Society *Florida* 33: 37-41.

Straw, C., Friell, J., Schwab, R.. and Watkins, E. 2020. *Research Summary, Encouraging Adoption of Precision Irrigation Technology Through On-Course Application and Demonstration of Water Savings*. United States Golf Association Turfgrass and Environmental Research Program, pp. 208-217.

Whitlark, B. 2012. Water use efficiency techniques from the Southwestern United States USGA golf's Use of water: Challenges and opportunities. In *A USGA Summit on Golf Course Water Use; USGA*, Liberty Cornery, NJ.

Whitlark, B. 2014. Winning strategies to overcome adverse soil conditions. *USGA Green Section Record* 52(15): 1-9.

Whitlark, B. 2020. No drain, no gain. USGA Green Section Record. *West Regional Update*. Available at: https://www.usga.org/content/usga/home-page/course-care/green-section-record/58/22/no-drain--no-gain.html.

Whitlark, B. 2021. Five proven methods to improve moisture uniformity. *USGA Green Section Record* 59(12).

Young, J., Deb, S., Ritchie, G., Guo, W., Escamilla, E., Cantu, J. and McCall, D. 2019. *Research Summary, Enhancing Water Conservation With Remote Sensing Technology on Golf Courses*. United States Golf Association Turfgrass and Environmental Research Program, pp. 235-239.

Zoldoske, D. F. 2003. *Improving Golf Course Irrigation Uniformity: A California Case Study*. The Center for Irrigation Technology, California State University Fresno.

Zontek, S. and Kostka, S. 2012. Understanding the different wetting agent chemistries. *USGA Green Section Record* 50(15): 1-6.

Chapter 20

Considerations with soil testing in turfgrass

Cole Thompson, United States Golf Association, USA; Elizabeth Guertal, Auburn University, USA; Pauric McGroary, Waypoint Analytical, USA; Douglas Soldat, University of Wisconsin-Madison, USA; and Bryan G. Hopkins, Brigham Young University and Soil Science Society of America, USA

1 Introduction

Soil testing can be a valuable method to help turfgrass managers make fertilizer decisions and choosing the most appropriate soil test extractant is key. This depends on soil properties and the availability of correlation data for turfgrass species in the desired region. This chapter describes common extracts and demonstrates their efficacy for phosphorus (P) and potassium (K) extraction with three soil samples from the North American Proficiency Testing program (www.naptprogram.org) administered by the Soil Science Society of America (Madison, WI, USA). Fertilizer recommendations were made based on regional sufficiency levels from university soil-testing laboratories and the Minimum Levels for Sustainable Nutrition (MLSN) guidelines from the Asian Turfgrass Center (www.asianturfgrass.com; Bangkok, Thailand) and PACE Turf (www.paceturf; San Diego, CA). Sufficiency Levels for Available Nutrients (SLAN) or MLSN guidelines are considered the most appropriate for deciding how to fertilize the turf. However, recommendations based on an inappropriate extractant, calibration, saturated paste extraction, or ideal ratios of major exchangeable cations (i.e. Basic Cation Saturation Ratio; BCSR) are considered inappropriate.

http://dx.doi.org/10.19103/AS.2022.0110.22

2 Soil testing

Obtaining a soil test and resulting fertilizer recommendations is based on a three-step process of 'correlation,' 'calibration,' and 'interpretation' (Brown, 1987; Hopkins, 2020). To start, the soil sample is mixed with a liquid solution varying in composition and concentration, with the resultant solution then containing extracted soil nutrients. The concentration of nutrients in the solution varies with a wide range of factors, including extractant properties, sample shaking time, relative ratio of soil-to-solution, and shaking method (i.e. orbital, speed, etc.). After extraction and filtering, the supernatant is analyzed (another entire topic of research) for various nutrients of interest. The amount of extracted nutrient should correlate well with plant nutrient uptake, yield, or another factor of interest, such as turfgrass color or density. Calibration is the process of determining crop response, over a range of soil-test result values (Mitchell and Mylavarapu, 2014). Ideally, calibration must be conducted for different crops and in different soil types and geographic regions. In many cases, we are missing key calibration data for turfgrasses. This is true for newer turfgrass species of interest, such as seashore paspalum, or for commonly utilized turfgrass species that may be managed differently by geographic region, such as annual bluegrass. A final step in the soil-testing process is interpretation, where a fertilizer recommendation is provided based on soil-test results as related to soil-test calibration data.

First, a soil sample is removed (a typical sampling depth is around 8–10 cm for turfgrass) and sent to a soil-testing laboratory. In the laboratory, the soil is air dried, pulverized, and extracted with a solution known to remove a nutrient amount that correlates well with plant growth. Often, the same extractant is used to remove several nutrients, as a single extractant that works well for multiple nutrients is more efficient than having to analyze a soil with numerous extractants. Fertilizer recommendations are then made based on available, extractant-specific calibrations of nutrient concentrations to crop responses.

There are plant nutrients for which we lack this correlation, calibration, and interpretation. The most common of these is nitrogen (N), which is readily leached or lost to the atmosphere from the soil, thus making it difficult to correlate with season-long needs for turfgrass and other crops. To overcome this problem, fertilizer recommendations for N are based on years of crop response testing, determined from years of application of N to various crops across a wide range of soil types and climates. Other plant nutrients for which we lack widely useful calibrations include sulfur (S) and most of the micronutrients.

2.1 Soil test extractants

Knowledge of basic soil types allows a turfgrass manager to best manage their soil and rootzones, including activities that vary with soil type, such as cultivation, fertilization, and establishment. Obtaining a soil test is one of the first activities for the management of a turfed soil. Results of that soil test can vary with many factors, and soil type or rootzone is a major one. This subsection will discuss the common soil test extractants used in the USA and how soil-test results may vary with soil and extractant type.

2.1.1 Bray P1

The Bray P1 extractant is primarily used to measure soil P. The Bray P1 soil test, which was developed in 1945 (Bray and Kurtz, 1945), uses fluoride in dilute acid to extract a proportion of soil P. More commonly used in the north central USA than in the south or west, this extractant is reliable on neutral or acid pH soils (Mallarino, 1995) and will underestimate plant-available P in calcareous soils (alkaline pH soil with excess carbonates). Most soil-testing laboratories in the north central region of the USA (except North and South Dakota, where soils are largely calcareous) use the Bray P1 extractant (Frank et al., 2011). It is also listed as a suitable extractant for other acid to neutral pH soils, even those in the western USA (Gavlak et al., 2005). In acidic soils, extraction with the Bray P1 and Mehlich 3 extract give nearly identical P concentrations that are strongly correlated when the extracts are analyzed via a colorimetric method. If Mehlich 3 extracts are analyzed using Inductively Coupled Plasma (ICP) as the analytical equipment, P concentrations from the Mehlich 3 extract will be greater than those measured with the Bray 1 extracts. However, they are still very strongly correlated (Pittman et al., 2005).

2.1.2 Bray P2

The Bray P2 extractant is more strongly acidic than the Bray P1 extractant, which means it will extract more P from the soil. Some soil-test reports will provide both Bray P1 and Bray P2 results, with the Bray P2 accounting for less soluble forms of P. However, most fertilizer recommendations for P do not consider the Bray P2 extractant because it does not correlate as well as Bray P1 or Mehlich 3. In some work, soil P extracted by Bray P2 was about twice that extracted by the Mehlich 3 extract (Wang et al., 2004).

2.1.3 Mehlich 1

Mehlich 1 is also known as the double acid extractant. This method was originally introduced by Dr. Mehlich in 1953 to determine the availability of P, K, calcium (Ca), and magnesium (Mg) (Mylavarapu and Miller, 2014). Additionally, boron (B), sodium (Na), copper (Cu), zinc (Zn), and manganese (Mn) can be extracted using Mehlich 1. However, S cannot be determined by Mehlich 1 because the extractant itself contains sulfur (sulfuric acid; H_2SO_4). Mehlich 1 is not a suitable P extractant from neutral or alkaline soils, or where apatite is the predominant source of plant-available P. Alkalinity neutralizes the effectiveness of dilute acids, therefore P will be under-extracted in alkaline soils. In the southeastern region of the USA, five states use the Mehlich 1 extractant (Hanlon, 2007), primarily on acidic and sandy soils.

2.1.4 Mehlich 3

Mehlich 3 was introduced in 1984 (Mehlich, 1984). The Mehlich 3 test includes fluoride to prevent P from re-precipitating with aluminum (Al) (Zhang et al., 2014). A chelate (ethylenediaminetetraacetic acid; EDTA) was also added for better extraction of micronutrients. Additionally, Dr. Mehlich replaced hydrochloric acid and H_2SO_4 with nitric acid to help extract calcium phosphate and added acetic acid to buffer the solution to a pH of 2.5 and minimize the solution from being neutralized in alkaline soils. The Mehlich 3 extractant is considered to be widely effective and works well over a wide range of nutrients, soils, and soil pHs. As such, it often is considered a 'universal' extractant. It is specifically mentioned for use in the southeastern USA (Hanlon, 2007) and for the extraction of P in northeastern soils where P availability is largely controlled by aluminum phosphates (Wolf and Beegle, 2011). In addition to extracting soil P, it is often used to extract other nutrients, including K (Watson and Mullen, 2007), Ca and Mg (Wolf and Beegle, 2011), Na, and micronutrients (Sawyer and Mallarino, 1999).

In the southeastern USA, various states use Mehlich 1 for extractable P and K, and others use Mehlich 3 for those same nutrients. As such, the relationship between these two extractants has been studied, and various conversion equations between the two are available (Sikora et al., 2005). The various conversion equations are state-specific, or they were developed as part of the North American Proficiency Testing program (NAPT), which is a quality program for soil-testing laboratories to check their accuracy for the various analytical tests they perform (Soil and Plant Analysis Council, 2000).

2.1.5 Olsen (sodium bicarbonate)

Olsen (sodium bicarbonate) or the Olsen P test ('Olsen') is considered the primary extractant for use in neutral to alkaline soils (Olsen et al., 1954; Wolf and Baker, 2008). In the Pacific Northwest and Northern Great Plains regions of the USA, the extractant is also used to measure plant-available K, nitrate, and S.

The Mehlich 3, Bray P1, and Olsen extracts are three of the most widely used soil extractants, and so comparisons have been made between the relative pools of P (or other nutrients) that they remove (Hopkins and Hansen, 2019). In general, results from the use of the Bray P1 and Mehlich 3 tend to be similar in acid soils, and equations documenting their close linear relationship for soil P extraction (albeit in limited crops and soils) exist (Eckert and Watson, 1996). In calcareous soils (soil pH >7.4), the Olsen extractant is generally recommended as the Bray P1 and, possibly, the Mehlich 3 test will yield incorrectly low P values. The reverse is true in acid soils (soil pH <5.0), a situation where the Olsen extract is not generally recommended (Sawyer and Mallarino, 1999).

2.1.6 Morgan

The Morgan (sodium acetate, pH 4.8) and modified Morgan (ammonium acetate, pH 4.8) extractants are used primarily by soil-testing laboratories not only in the northeastern USA but also in a few other locations. The Morgan test was developed as a 'universal' extractant for acid soils, with the single extractant used for P, Ca, Mg, and K, and, with the use of the modified Morgan, for micronutrients.

The Morgan methods often extract a smaller amount of nutrients than other methods (e.g. Bray). There are equations that quantify the relationship between extracted nutrients via the Mehlich 3 and Morgan methods (Ketterings et al., 2001a,b, 2002). This extractant is not commonly used outside the northeastern USA.

2.1.7 Lancaster

Lancaster (i.e. the Mississippi soil test method) is a soil test extractant developed for both the acid and alkaline soils of Mississippi where it is used for P, K, Ca, and Mg analyses (Oldham, 2014; Hanlon, 2007). It is not widely used outside Mississippi.

2.1.8 Ammonium acetate (1 M NH_4OAc)

The ammonium acetate (1 M NH_4OAc) extractant may be used to determine K, Ca, Mg, and/or Na in soils, especially in calcareous soils (Warncke and Brown, 2013). Two issues with this extractant are that it is not effective in saline or sodic soils (a pretreatment is needed), and the method does not account for free carbonates or gypsum (Gavlak et al., 2005). As it is an extractant for exchangeable cations, results from NH_4OAc extraction are sometimes used for the determination of cation exchange capacity (CEC) (Warncke and Brown, 2013).

2.1.9 Diethylenetriaminepentaacetic acid (DTPA)

Diethylenetriaminepentaacetic acid (DTPA) is a chelating agent which is sometimes used for the extraction of micronutrients (e.g. Fe, Mn, Cu, Zn) from soil (Whitney, 1998). B is the exception and is usually quantified via a hot-water extraction. However, many soil-test laboratories that use a general extractant such as Mehlich 3 or ammonium acetate will not use DTPA (or hot water) for micronutrients, and micronutrients are measured from the general supernatant.

2.1.10 Monocalcium phosphate

Recently, plant responses to added S have become more commonly observed, and have been documented in cotton and corn. This response is likely related to the improvement of air quality in the USA, and the subsequent reductions in S deposition. Sulfur soil tests, and correlation and calibration to various crop responses, are not especially well developed, and so if an S soil test result is reported it is often due to a general extractant such as Mehlich 3. However, the use of monocalcium phosphate may be a better-calibrated extractant for S (Singh et al., 2011), although research in this area is limited.

2.1.11 Saturated paste extract

Saturated paste extract (SPE) is created by saturating a soil sample with distilled water and allowing it to sit for as many as 8 h, after which the supernatant is suctioned off for analysis (Warncke, 2014). SPE is often used for soilless potting mixes and media or to determine if soils in arid parts of the USA are saline, saline-sodic, or sodic.

Soilless media have very low to no mineral soil content, thus, there are few to no reactive minerals such as iron (Fe)- and Al-oxides on which other minerals adsorb. Therefore, most of the nutrients present in soilless media are soluble and found in the media extract. When mineral soil is added to these substrates, nutrients that adsorb to mineral soil are not extracted and SPE will underestimate the nutrient supplying power of the mix. This phenomenon sometimes emerges as a topic in the popular turfgrass science literature where SPE is incorrectly interpreted as a measure of plant-available nutrients. However, it is critical to reiterate that a SPE does not extract nutrients that have the potential to become plant available over a growing season (i.e. those on the CEC sites or less soluble forms). Rather, SPE only measures elemental intensities as they relate to and affect salinity, not the quantity that will affect the nutrition of a plant. This mild, water-based extract is thus not useful for determining soil nutrient levels over a period of plant growth and therefore is not well calibrated to fertilizer recommendations (Carrow et al., 2003).

When SPE is used to determine soil pH, electrical conductivity (ECe), solution concentrations of Ca^{2+}, Mg^{2+}, K^{+}, Na^{+}, Cl^{-}, B, HCO_3^{-}, CO_3^{2-}, SO_4, sodium adsorption ratio (SAR), or exchangeable sodium percentage (ESP) ratio, the soil must also be analyzed for nutrients using appropriate extractants (i.e. Olsen P, ammonium acetate, or DTPA) for typical soils of the western USA.

2.2 Soil test calibration and the basis for fertilizer recommendations

Thus, a soil test provides some measure of plant nutrient content in the soil. Soil test calibration places the soil-test results into categories, upon which the likelihood of a plant response is based (Dahnke and Olson, 1990). Typically, those categories are something like 'Very Low,' 'Low,' 'Medium,' 'High,' and 'Very High,' and the fertilizer nutrient would be recommended at any soil test result below 'Medium,' as a crop response to that added nutrient would likely be observed. The point at which a soil test result moves from 'Medium' to 'High' is determined, in general, by two different concepts: (1) SLAN or (2) BCSR (Eckert, 1987). There have been many adjustments and small alterations to these two fundamental concepts. It is important to know that each method exists because both are used in the turfgrass industry and either may be manipulated, or unintentionally misused with good intentions, to encourage increased use of fertilizers. Typically, the SLAN concept is used more widely for determining fertilizer recommendations for turfgrass.

There is newer research that proposes a new calibration method for K, P, Ca, Mg, and S in turfgrass: MLSN (Stowell and Woods, 2013; Woods et al., 2014). The method was created from a large data set (16163 soil samples collected voluntarily by turfgrass practitioners worldwide) from sites with good turfgrass performance at sample collection. The data were filtered to include only soil samples with pH 5.5–8.5 (to avoid the factors of Al toxicity and alkalinity hazard) and with total CEC < 6 $cmol_c \cdot kg^{-1}$ (to select for samples where turfgrass was performing well in low nutrient content soils). The resulting 3683 soil samples were fit to a log-logistic model to identify, for each nutrient, the concentration where 10% of soil samples from the data set would have a lower nutrient concentration (and 90% would have a greater nutrient concentration). In other words, the MLSN sufficiency levels represent the 10th percentile nutrient concentrations from the data - again, where the turfgrass at the site was performing well. These guidelines (i.e. to withhold fertilizer of each nutrient unless K < 37 $mg \cdot kg^{-1}$, P < 21 $mg \cdot kg^{-1}$, Ca < 348 $mg \cdot kg^{-1}$, Mg < 47 $mg \cdot kg^{-1}$, or S < 7 $mg \cdot kg^{-1}$) are becoming increasingly popular in the turfgrass industry.

Currently, most public soil test laboratories base their fertilizer recommendations upon the SLAN concept, which states that there is some given nutrient level in the soil above which the crop will not respond, even if more of that nutrient is supplied. For many of our crops that response is the yield of the edible portion of the plant, and, at some soil-test level, the sufficiency level concept indicates that adding more of a nutrient will no longer increase crop yield. Of course, for turfgrass, the response can be a variety of indicators, including color, density, clipping yield, ball roll, abiotic or biotic stresses, or some other measure of turfgrass performance.

The BCSR does not focus on given levels of nutrient sufficiency but instead focuses on ideal ratios or proportions of the major exchangeable cations (typically, K, Mg, and Ca) (Dahnke and Olson, 1990). The originally proposed ranges (from the 1940s) were 65–75% Ca, 10% Mg, 2.5–5% K, and 2–10% H, with approximate ratios of 7:1 Ca:Mg, 15:1 Ca:K, and 3:1 Mg:K (Bear et al., 1945). Such ratios have been adjusted, fixed, and studied over many years. However, the overwhelming conclusion by many soil fertility experts is that there is little evidence to support the use of such ratios for fertilizer recommendations (Eckert and McLean, 1981; Rehm and Sorensen, 1985; Kopittke and Menzies, 2007; Chaganti and Culman, 2017). Further, recommendations for nutrients other than K, Mg, and Ca - such as P - typically come from SLAN guidelines. Last, even if the ratio of the cations in the soil is thought to be 'optimum,' a nutrient deficiency could still exist, especially in sandy soils with low organic matter content - a common situation for constructed turfgrass soils and rootzones such as putting greens and athletic fields. A brief comparison of SLAN, MLSN, and BCSR is available in Table 1.

Table 1 A brief comparison of common methods for interpreting soil nutrient tests for fertilizer recommendations

	Sufficiency levels for available nutrients (SLAN)	Minimum levels for sustainable nutrition (MLSN)	Basic cation saturation ratio (BCSR)
Relationship with soil nutrient test results	Assumes there is a concentration for each nutrient above which turfgrass will not respond to additional fertilizer	Assumes there is a concentration for each nutrient above which turfgrass will not respond to additional fertilizer	Assumes there is an ideal ratio of major exchangeable cations (i.e. K, Ca, and Mg) in soil
Calibration and interpretation	Based on field correlations among nutrient levels and turfgrass performance over a range of soil nutrient concentrations	A calibration of SLAN ideology to determine minimum sufficiency levels from a large number of low CEC soil samples from good-performing turfgrasses (over a range of species and geographies)	Only used to balance the ratio of K, Ca, and Mg SLAN or MLSN would be used for other nutrients, such as P
Considerations	Local calibrations for specific turfgrass species and soils are best	The same sufficiency levels are used regardless of turfgrass species, location, or geography	K, Ca, and Mg fertilizer recommendations are made regardless of their absolute concentration in soil

3 A case study in sustainable soil test interpretation

3.1 Rational

Soil test extraction procedures and calibration for soil fertility recommendations vary greatly across the USA and can differ even between border states. Therefore, a soil sample submitted to different laboratories in neighboring states may produce different results and different soil fertilizer recommendations. This does not mean that one of the laboratories is producing incorrect results or recommendations, it is simply variation due to the extractant and/or calibration.

The objective of this case study is to highlight the influence of soil test extractions on soil test results as well as the impact of using different soil extractants with different soil test fertility recommendations for cool-season turfgrasses on putting greens (generally qualified as 'high maintenance') from around the USA.

3.2 Sample selection and description

To demonstrate the impact that soil test extraction methods have on results from different soils, three soils were selected from the NAPT (www.naptprogram.org), which is administered by the Soil Science Society of America. The soils selected from the database represented soils with low levels of P and K from three geographic regions (Mid-Atlantic, Midwest, and Southwest) to represent different combinations of physical and chemical properties of soils. The NAPT database is comprised of soil test results from public and private agricultural laboratories using common soil tests on homogenized soil samples collected from around the United States (NAPT, 2021). Consequently, the NAPT database allowed us to demonstrate the influence of different extraction methods on P and K results for the soils selected. Furthermore, by comparing NAPT data to state soil test laboratory recommendations in each region, we were able to compare the effects of interpretations of calibrations from each laboratory, which may vary based on turfgrass response to soil type and environmental conditions.

The chemical characteristics of the selected samples in Table 2 demonstrate the importance of understanding how extractants affect soil-test results. Because the relative extraction efficiency of the extractants used under different combinations of physical and chemical properties is known, the variability within samples in Table 2 is expected and hence regionally preferred extractants have been identified. It is critical for turfgrass managers to be familiar with the most suitable extractant(s) for their region to prevent misinterpretation of soil-test reports. This is less challenging when local laboratories are used because university soil-testing labs, for example, default to their regionally appropriate extractants and local turfgrass response calibrations. However, if soil samples

Table 2 Example of chemical characteristics of soils from Texas, Pennsylvania, and Illinois (USA) utilizing different soil extractants for phosphorus and potassium. Data were selected from the North American Proficiency Testing program (http://www.naptprogram.org/), which is administered by the Soil Science Society of America (Madison, WI, USA). Regionally common phosphorous (P) and potassium (K) extractants are shaded by the sample

	Texas	Pennsylvania	Illinois
	pH		
Soil pH (1:1)	7.1	6.0	5.8
	Extractable P ($mg \cdot kg^{-1}$)		
Bray P1 (1:10)	68	26	20
Bray P2 (1:10)	79[a]	32	26
Mehlich 1	59	12	16
Mehlich 3	77	36[a]	27[a]
True Morgan	27	2[b]	4[b]
Olsen	22[b]	12	13
Upper quartile	*78*	*33*	*26*
Median	*64*	*19*	*18*
Lower quartile	*26*	*10*	*11*
	Extractable K ($mg \cdot kg^{-1}$)		
Ammonium Acetate	41	97	113
Mehlich 1	35[b]	74[b]	74[b]
Mehlich 3	45	99	114[a]
Olsen	41	107[a]	105
Bray P1	49[a]	83	89
True Morgan	36	82	89
Upper quartile	*46*	*101*	*113*
Median	*41*	*90*	*97*
Lower quartile	*36*	*80*	*85*
	Cation Exchange Capacity ($cmol_c \cdot kg^{-1}$)		
Estimate CEC	4	11	13

[a]Maximum extractable P or K within a sample.
[b]Minimum extractable P or K within a sample.

are submitted to a distant laboratory, a turfgrass manager should request only the most appropriate extraction method (and, preferably, locally developed SLANs for the extractant) to reduce misinterpretations.

To illustrate the range and variability of extractants on our selected soil samples, some descriptive statistics are also provided in Table 2. Extractable nutrient concentrations are shaded for extractants most suitable for the region of sample origin. Median extractable P ranged from 18 $mg \cdot kg^{-1}$for the Illinois sample to 64 $mg \cdot kg^{-1}$ for the sample from Texas and varied by up to 57 $mg \cdot kg^{-1}$

for extractants within a sample. The Texas sample had the largest interquartile range (52 mg·kg^{-1}) and varied the most among extractants. Median extractable K ranged from 41 mg·kg^{-1} for the Texas sample to 97 mg·kg^{-1} for the sample from Illinois and varied by up to 40 mg·kg^{-1} for extractants within a sample. Interquartile ranges were smaller among K extractants within each sample and the Illinois sample varied the most with an interquartile range of 28 mg·kg^{-1}.

It is important to note that neither aligning with the median, extracting the minimum, nor extracting the maximum nutrient concentration indicates the best extractant for a sample because of the known relationships among extractants and soils. The nutrient extracted from a sample must correlate with turfgrass performance under the present soil conditions, which is the basis of selecting regionally preferred extractants and calibrating ranges to plant nutrient sufficiency levels. For example, the Mehlich 3 extractant is widely used throughout the USA and extracted the most (or nearly the most) P and K across our three samples. This is a meaningless fact unless the Mehlich 3 SLANs, and preferably local Mehlich 3 SLANs, are available for the origin of the sample. Alternately, turfgrass managers can use generalized SLANs such as the MLSN, which, because of relatively low sufficiency levels, often reduces fertilizer use compared to other recommendations.

3.3 Sufficiency levels and recommendations

The next step is to compare extractable P and K from each sample to sufficiency ranges for regionally appropriate extractions. Table 3 contains categories of P and K soil ratings correlated to extractants for state laboratories from select universities across the USA. MLSN sufficiency levels also are included. Note differences in sufficiency levels among laboratories using the same extractant, which demonstrates local calibration of the appropriate extractant and turfgrass performance.

3.3.1 Phosphorous

Beginning with the sample from Texas, the Mehlich 3 or Olsen extractants would be considered appropriate (77 or 22 mg·kg^{-1} extractable P, respectively) (Table 2). Because of the neutral pH of the Texas sample, the Mehlich 3 extractant is preferred over Olsen – especially since soil ratings from Texas A&M University ('Texas A&M'; College Station, TX, USA) used Mehlich 3. A Mehlich 3 level of 77 mg·kg^{-1} extractable P is considered 'High' (or 'Above Optimum') for Pennsylvania State University ('Penn State'; University Park, PA, USA) ratings (>75 mg·kg^{-1}), 'High' for University of Florida (Gainesville, FL, USA) ratings (>45 mg·kg^{-1}), 'High' for Texas A&M ratings (50–200 mg·kg^{-1}), and 'Sufficient' for MLSN ratings (>21 mg·kg^{-1}), resulting in the recommendation for 1 pound P_2O_5 per 1000 square feet (49 kg P_2O_5 per hectare) from the Penn State calibration, and no P fertilizer recommended from the University of Florida, Texas A&M, or MLSN calibrations (Tables 3 and 4). In this case, the Mehlich 3 soil calibrations from Penn State seem to overestimate the amount of P needed for a turf grown in Texas.

North Dakota State University ('North Dakota State', Fargo, ND, USA) and Utah State University ('Utah State', Logan, UT, USA) provide ratings based on the Olsen extractant. For 22 $mg{\cdot}kg^{-1}$, both North Dakota State and Utah State ratings for the Texas sample would be 'Medium' (13–28 and 16–30 $mg{\cdot}kg^{-1}$, respectively), resulting in recommendations for applications of 2 pounds P_2O_5 per 1000 square feet and 0 pounds P_2O_5 per 1000 square feet, respectively (98 kg P_2O_5 per hectare and 0 kg P_2O_5 per hectare, respectively) (Table 4). The former case again overestimating the amount of P fertilizer when compared to local recommendations. Though not typically recommended for Texas, ratings for Bray 1 and Mehlich 1 extractions were considered 'Very High' and 'High' based on calibrations from Kansas State University ('Kansas State', Manhattan, KS, USA) and Clemson University ('Clemson', Clemson, SC, USA), respectively, which also would result in no fertilizer recommendation. Note, however, that there are scenarios where mixing results and ratings from different extractants would, incorrectly, indicate the Texas sample was low in P. For example, the Olsen test result of 22 $mg{\cdot}kg^{-1}$ would yield 'Medium' soil ratings for all labs except Kansas State (Bray 1) and would be just above the sufficiency level indicated by the MLSN, the most frugal method.

Mehlich 3 is considered the most suitable extractant for Pennsylvania, for which the sample in this case study had 36 $mg{\cdot}kg^{-1}$ extractable P (Table 2). This result would receive a rating of 'Medium' (or 'Below Optimum'), 'Medium', 'Medium', and 'Sufficient' based on the Mehlich 3 soil ratings from the Penn State, the University of Florida, Texas A&M, and the MLSN, respectively, yielding recommendations for applications of 2.0 pounds P_2O_5 per 1000 square feet and 0.2 pounds P_2O_5 per 1000 square feet (98 kg P_2O_5 per hectare and 10 kg P_2O_5 per hectare) from Penn State and University of Florida laboratories, respectively (Table 4). Mehlich 3 calibrations from other labs do not indicate a need for P fertilizer. Since Penn State is the local lab in this case, it seems that P fertilizer is needed for our Pennsylvania sample and calibrations from other states have underestimated the amount of P needed by turf in Pennsylvania. Other extractants are not widely used or recommended for Pennsylvania but note that results from the Bray 1 extraction would yield fertilizer recommendations based on University of Wisconsin (Madison, WI, USA) and Kansas State calibrations, as would the Olsen extraction based on North Dakota State and Utah State calibrations and Utah State calibrations and the Mehlich 1 extraction based on Clemson calibrations.

The Bray P1, Bray P2, and Mehlich 3 extractants are most suitable for P extraction in Illinois, for which our sample had 20 $mg{\cdot}kg^{-1}$, 26 $mg{\cdot}kg^{-1}$, and 27 $mg{\cdot}kg^{-1}$ extractable P, respectively (Table 2). University of Wisconsin ratings

Table 3 General sufficiency levels, unless otherwise stated, for soil phosphorus (P) and potassium (K) from various labs and the Minimum Levels for Sustainable Nutrition (MLSN)

Soil rating level	Pennsylvania State University[a]	Clemson University	University of Florida	University of Wisconsin	University of Wisconsin	Kansas State University	Texas A&M University	North Dakota State University	Utah State University	MLSN
	Soil Extractable P ($mg \cdot kg^{-1}$)									
	Mehlich 3	Mehlich 1	Mehlich 3	Mehlich 3[b]	Bray P1[b]	Bray P1	Mehlich 3	Olsen P	Olsen P	Mehlich 3
Very low	-	-	-	0-5	0-12	0-5	-	0-6	<5	-
Low	-	<15	<25	6-10	13-25	6-10	0-20	7-12	5-15	-
Medium	<45	16-30	26-45	11-19	26-37	10-20	20-50	13-28	16-30	-
Sufficient	45-75	31-40	-	20-30	38-50	-	-	-	-	21
High	>75	41-60	>45	-	-	20-50	50-200	>29	31-50	--
Very High	-	-	-	>30	>50	>50	-	-	>50	-
Excessive	-	>60	-	-	-	-	-	-	-	-
	Soil Extractable K ($mg\ kg^{-1}$)									
	Mehlich 3	Mehlich 1	Mehlich 3	Bray P1[b]	Bray P1[c]	Ammonium acetate	Mehlich 3	Olsen	Ammonium acetate	Mehlich 3
Very low	-	-	-	0-50	0-20	0-40	-	0-40	<75	-
Low	-	<35	<35	51-100	21-40	41-175	0-75	41-175	76-125	-
Medium	<180	36-78	36-60	101-150	41-60	175-250	75-125	175-250	126-400	-
Sufficient	180-220	79-91	-	151-175	61-80	-	-	-	-	37
High	>220	92-118	>60	-	-	250-300	125-250	250-300	>400	-
Very high	-	-	-	>175	>80	>300	-	>300	-	-
Excessive	-	>118	-	-	-	-	-	-	-	-

[a]Pennsylvania State University uses 'below optimum,' 'optimum,' and 'above optimum' rating scale. The 'below optimum' corresponds to 'very low,' 'low,' and 'medium,' 'optimum' corresponds with 'sufficient,' 'high,' 'very high,' and 'excessive' corresponds with 'above optimum.'
[b]Interpretations specific to golf course turf.
[c]Interpretations specific to general turf areas, excluding golf turf.

Table 4 Phosphorus fertilizer recommendations from different laboratories utilizing frequently used soil extractants for soil samples from three locations. Soil sample data was selected from the North American Proficiency Testing program (http://www.naptprogram.org/), which is administered by the Soil Science Society of America (Madison, WI, USA)

		P_2O_5 pounds per 1000 square feet[a]									
Soil sample origin	Phosphorus soil extractant	Pennsylvania State University	Clemson University[b]	University of Florida	University of Wisconsin[c]	University of Wisconsin[d]	Kansas State University	Texas A&M University	North Dakota State University	Utah State University	MLSN
Texas	Bray P1 (1:10)	–[e]	–	–	0[f]	0	0	–	–	–	–
	Mehlich 1	–	0	–	–	–	–	–	–	–	–
	Mehlich 3	1	–	0	0	0	–	0	–	–	0
	Olsen	–	–	–	–	–	–	–	2	0	–
Pennsylvania	Bray P1 (1:10)	–	–	–	2.0	2.0	1	–	–	–	–
	Mehlich 1	–	4.6	–	–	–	–	–	–	–	–
	Mehlich 3	2	–	0.2	0	0	0	0	–	–	0
	Olsen	–	–	–	–	–	–	–	3	1	–
Illinois	Bray P1 (1:10)	–	–	–	2.0	3.5	2	–	–	–	–
	Mehlich 1	–	2.3	–	–	–	–	–	–	–	–
	Mehlich 3	2	–	0.2	0.5	1.0	–	0	–	–	0
	Olsen	–	–	–	–	–	–	–	2	1	–

[a]Soil fertility recommendations for the maintenance of high-quality, cool-season turfgrass.
[b]Creeping bentgrass putting greens.
[c]Sand-based greens.
[d]High-traffic turfgrass areas, push-up tees, and greens (excludes sand greens).
[e]Hyphens indicate that a lab does not use the extractant.
[f]To convert pounds per 1000 square feet to kg per hectare, multiply cell values by 48.8.

for sand-based putting greens and Kansas State ratings (both Bray P1) for the Illinois soil were 'Low' and 'Medium', respectively (Table 3), and would each result in recommendations for 2 pounds P_2O_5 per 1000 square feet (98 kg P_2O_5 per hectare) (Table 4). Mehlich 3 ratings were 'Medium' (or 'Below Optimum') for Penn State, 'Medium' for the University of Florida, 'Medium', for Texas A&M, and 'Sufficient' for the MLSN. Depending on the specific lab, these ratings result in a range of recommendations for P fertilizer from 0 pounds P_2O_5 per 1000 square feet (0 kg P_2O_5 per hectare) (Texas A&M and MSLN) to 2.5 pounds P_2O_5 per 1000 square feet (122 kg P_2O_5 per hectare) (Penn State). Though not appropriate for Illinois, interpretation based on Olsen extraction and calibrations from North Dakota State or Utah State yielded similar fertilizer recommendations in this case. The same is true for the Mehlich 1 extraction and calibration from Clemson. The University of Wisconsin calibrations are likely the closest to conditions from the Illinois sample, so Texas A&M and MLSN Mehlich-3 calibrations may be underestimating the amount of P fertilizer necessary for the Illinois sample.

3.3.2 Potassium

Extractable K levels were less extractant-dependent than extractable P for these samples, but K fertilizer recommendations were more variable than the P recommendations. Beginning with the sample from Texas, ammonium acetate, Mehlich-3, and Olsen extractants could be appropriate and yielded 41 $mg{\cdot}kg^{-1}$, 45 $mg{\cdot}kg^{-1}$, and 41 $mg{\cdot}kg^{-1}$ extractable P, respectively (Table 2). Kansas State and Utah State ratings for ammonium acetate were 'Low' (41–175 $mg{\cdot}kg^{-1}$) and 'Very low' (<75 $mg{\cdot}kg^{-1}$) for the Texas sample (Table 2), respectively, and recommend 4.0 and 2-3 pounds K_2O per 1000 square feet (195 K_2O kg per hectare and 98–146 K_2O kg per hectare) (Table 4). Mehlich-3-based soil ratings were 'Medium' (or 'Below Optimum') for Penn State (<180 $mg{\cdot}kg^{-1}$), 'Medium' for the University of Florida (36–60 $mg{\cdot}kg^{-1}$), and 'Low' for Texas A&M (0–75 $mg{\cdot}kg^{-1}$), yielding recommendations for 0.5 pounds K_2O per 1000 square feet (24 kg K_2O per hectare) (University of Florida) to 5 pounds K_2O per 1000 square feet (244 kg K_2O per hectare) from Penn State. More frugally, the MLSN rates the Texas sample as 'Sufficient' (>37 $mg{\cdot}kg^{-1}$) resulting in a recommendation for no K fertilizer. The Olsen rating from North Dakota State was 'Low' and yielded a recommendation for 4 pounds K_2O per 1000 square feet (195 kg K_2O per hectare). As the recommendation from Texas A&M was for 2.9 pounds K_2O per 1000 square feet (142 kg K_2O per hectare), recommendations above and below this level likely have over- and underestimated the amount of K fertilizer needed.

The Pennsylvania and Illinois samples had more extractable K than the Texas sample and were quite similar, resulting in identical fertilizer recommendations. The samples (Pennsylvania-Illinois) had 97–113 $mg{\cdot}kg^{-1}$, 99–114 $mg{\cdot}kg^{-1}$, and 83–89 $mg{\cdot}kg^{-1}$ extractable K-based and ammonium acetate, Mehlich 3, and Bray extractants, respectively (Table 2). The ammonium acetate ratings were 'Low' for Kansas State and Utah State calibrations (Table 3), yielding recommendations for 4.0 pounds K_2O per 1000 square feet and 2 pounds K_2O per 1000 square feet, respectively (195 kg K_2O per hectare and 98 kg K_2O per hectare, respectively) (Table 5). The Mehlich 3 ratings were 'Medium' (or 'Below Optimum') for Penn State (<180 $mg\cdot kg^{-1}$), 'High' for the University of Florida (>60 $mg\cdot kg^{-1}$), and 'Medium' for Texas A&M (75–125 $mg\cdot kg^{-1}$), yielding recommendations for 0 pounds K_2O per 1000 square feet (0 kg K_2O per hectare) (University of Florida) to 3 pounds K_2O per 1000 square feet (146 kg K_2O per hectare) from Penn State. Again, the MLSN rating was 'Sufficient.' The University of Wisconsin lab uses Bray P1 for K extraction and rated the Illinois and Pennsylvania soils as 'Low' for golf turf with a recommendation for 3 pounds K_2O per 1000 square feet (146 kg K_2O per hectare), but 'Very High' for general turf (non-golf) with no K_2O recommended, highlighting the potential importance of land use for soil test interpretations and fertilizer recommendations. As discussed for P, the Penn State and University of Wisconsin recommendations likely should be considered the most accurate for samples from Pennsylvania and Illinois, respectively.

3.4 Case study conclusion

This case study reinforces the numerous ways that turfgrass fertilizer recommendations can be inaccurate and result in too little or too much fertilizer use. We only discussed P and K for simplicity, but similar issues exist with soil test interpretation for other nutrients. When submitting soil samples for testing, rating, and recommendations, it is always best to use regionally appropriate extractants and SLANs from local turfgrass calibration data - hopefully with the exact turf species and soil properties as the origin of the sample. Because this calibration data does not exist for all nutrients for all turfgrass species on all soils and in all locations, or - even less inspiring - for all turf cultivars on all soils and in all locations, concessions are inevitable. At a minimum, a turfgrass manager should ask for a regionally appropriate extractant and seek SLANs based on that extractant that have been calibrated to a soil similar to what is at their site. If local SLANs are unavailable, the MLSN is a reasonable guide for tests from Mehlich 3 extractions and will, at worst, underpredict the amount of fertilizer needed. In these cases, inspection for visual deficiency symptoms validated by low rates of the suspected deficient nutrient can help a turf manager adjust MLSN guidelines to their site.

4 Conclusion and future trends

The axiom 'don't guess, soil test' is used to convey the idea that soil testing helps turfgrass managers follow precise guidelines to prevent the over-application

Table 5 Potassium fertilizer recommendations from different laboratories utilizing frequently used soil extractants for soil samples from three locations. Soil sample data was selected from the North American Proficiency Testing program (http://www.naptprogram.org/), which is administered by the Soil Science Society of America (Madison, WI, USA)

		K_2O pounds per 1000 square feet[a]									
Soil sample origin	Potassium soil extractant	Pennsylvania State University	Clemson University[b]	University of Florida	University of Wisconsin[c]	University of Wisconsin[d]	Kansas State University	Texas A&M University	North Dakota State University	Utah State University	MLSN
Texas	Ammonium acetate	-[e]	-	-	-	-	4[f]	-	-	2-3	-
	Mehlich 1	-	4.6	-	-	-	-	-	-	-	-
	Mehlich 3	5	-	0.5	-	-	-	2.9	-	-	0
	Bray P1	-	-	-	4	2	-	-	-	-	-
	Olsen	-	-	-	-	-	-	-	4	-	-
Pennsylvania	Ammonium acetate	-	-	-	-	-	4	-	-	2	-
	Mehlich 1	-	2.3	-	-	-	-	-	-	-	-
	Mehlich 3	3	-	0	-	-	-	0	-	-	0
	Bray P1	-	-	-	3	0	-	-	-	-	-
	Olsen	-	-	-	-	-	-	-	4	-	-
Illinois	Ammonium acetate	-	-	-	-	-	4	-	-	2	-
	Mehlich 1	-	2.3	-	-	-	-	-	-	-	-
	Mehlich 3	3	-	0	-	-	-	0	-	-	0
	Bray P1	-	-	-	3	0	-	-	-	-	-
	Olsen	-	-	-	-	-	-	-	4	-	-

[a]Soil fertility recommendations for the maintenance of high-quality, cool-season turfgrass.
[b]Creeping bentgrass putting greens.
[c]Recommendations specific to golf course turf.
[d]Recommendations specific to general turf areas, excluding golf turf.
[e]Hyphens indicate that a lab does not use the extractant.
[f]To convert pounds per 1000 square feet to kg per hectare, multiply cell values by 48.8.

of fertilizer. A recent survey of golf course superintendents found that the most common nutrient conservation technique was to fertilize based on soil test results, with 55% of respondents using this method (Gelernter et al., 2016). However, the same survey reported that golf course superintendents who used soil testing applied significantly more nutrients than golf courses that did not soil test. While soil testing can prevent over-application of nutrients, the situation where soil test recommendations result in more nutrient use than normally would occur appears to be at least as likely. While the phrase 'guess, don't soil test' will never catch on, the fertilizer recommendations from a soil test are only as good as the correlation, calibration, and interpretation work that was used to create the recommendations. Continued scrutiny of existing recommendations and additional research into soil testing for turfgrass should remain a high priority.

In the absence of good soil-testing data, recommendations tend to be conservative - meaning the recommendations will skew toward applying more nutrients than the turfgrass actually requires. However, over applications of nutrients because of soil testing are not simply because of conservative recommendations alone. Unfortunately, some recommendations are more spurious. While most soil-testing laboratories report the soil test values and compare against some research-based guidelines (which may or may not result in a fertilizer recommendation), sometimes a secondary party (often a consulting agronomist or fertilizer company) will send client's soil samples to a reputable laboratory but disregard the laboratory's recommendations and instead use their own interpretation and recommendations from the laboratory results. The quality of these recommendations varies. In some cases, the recommendations are drastically different from research-based results. For example, SPE (which is useful for pH and salt and sodium hazard evaluation) is inappropriately but commonly used to make nutrient recommendations. SPEs are unable to extract the nutrients on the cation exchange sites and are also unable to extract much P because of that nutrient's low solubility in soil solution. Therefore, the nutrient levels in the saturated paste are much lower than the levels extracted by conventional soil nutrient extractants, and the fertilizer or consulting agronomist may recommend applying nutrients with a justification that says that while the conventional soil test finds adequate nutrition, the nutrients are actually 'locked up' in the soil and the saturated paste test correctly shows that nutrients are low.

The most common justification for using SPE to make fertilizer recommendations is that by mixing water with soil, the saturated paste test is mimicking what the plant actually 'sees.' This sounds logical to a layperson but ignores fundamental knowledge about soil science and plant nutrition. Laboratories and secondary parties that follow scientific principles do not use this method for nutrient recommendations. Turfgrass managers using this method are wasting time, money, and likely contributing to non-point source pollution. Interpretations from secondary parties are not always negative, however. In fact, a secondary party can improve on a laboratory's recommendation by using local knowledge and incorporating the latest scientific results that laboratories may be slower to adopt. It is up to the end user to determine if the secondary party is reliable or not and use of the saturated paste test to make nutrient recommendations is an easy way to identify an unreliable interpretation.

In summary, soil testing can be a valuable method to help turfgrass managers make fertilizer decisions. There are many different methods for soil testing and choosing the most appropriate soil test extractant is key. Whether an extractant is appropriate depends on the physical and chemical properties of the soil or rootzone in question and the availability of data that correlates extracted nutrient concentrations to the performance of the managed turfgrass species in the desired region. Recommendations from this scenario are ideal and allow turf managers to judiciously apply fertilizer. However, while the database of soil test interpretations is growing, there are many gaps to be filled by soil test calibration research. There also are nutrients (e.g. N, S, and most micronutrients) for which we do not have much or, in some cases, any reliable soil test calibration data. Because of the innumerable soil nutrient calibration scenarios among soil types, geographies, and turfgrass species (or even varieties), it is likely that the perfect soil-testing database will never exist. This reality has advanced creative research and interpretation to define sufficiency levels for some nutrients in broad terms (e.g. the MLSN). These guidelines are straightforward and partially fill the void of needed soil test calibration research, which has normalized their use in recent years. Conversely, and without question, turfgrass managers should be skeptical of fertilizer recommendations based on an inappropriate extractant, calibration, or, especially, one based on SPE or ideal ratios or of major exchangeable cations (i.e. BCSR).

5 References

Bear, F. E., Prince, A. L. and Malcolm, J. L. 1945. Potassium needs of New Jersey soils. *Bull. J. Agric. Exp. Stn., New Brunswick* 721.

Bray, R. H. and Kurtz, L. T. 1945. Determination of total organic and available forms of phosphorus in soils. *Soil Sci.* 59(1):39–46. DOI: 10.1097/00010694-194501000-00006.

Brown, J. R. 1987. *Soil Testing: Sampling, Correlation, Calibration, and Interpretation: Proceedings* [SSSA Special Publication]. Madison, WI: Madison Book Company, p. 21.

Carrow, R. N., Stowell, L., Gelernter, W., Davis, S., Duncan, R. R. and Skorulski, J. 2003. Clarifying soil testing: I. Saturated paste and dilute extracts. *Golf Course Manag.* 71:81-85.

Chaganti, V. N. and Culman, S. W. 2017. Historical perspective of soil balancing theory and identifying knowledge gaps: A review. *Crop Forage Turfgrass Manage.* 3:1-7.

Dahnke, W. C. and Olson, R. A. 1990. Soil test correlation, calibration, and recommendation. In: *Soil Testing and Plant Analysis SSSA Book Series (No. 3)* (3rd ed.), Soil Science Society of America, Madison, Wisconsin.

Eckert, D. J. 1987. Soil test interpretations: Basic cation saturation ratios and sufficiency levels. In Brown, J. R. (Ed.), *SSSA Special Publication*. Madison, WI: Madison Book Company, Vol. 21, pp. 53-64.

Eckert, D. J. and McLean, E. O. 1981. Basic cation saturation ratios as a basis for fertilizing and liming agronomic crops: I. Growth chamber studies 1. *Agron. J.* 73(5):795-799.

Eckert, D. J. and Watson, M. E. 1996. Integrating the Mehlich-3 extractant into existing soil test interpretation schemes. *Commun. Soil Sci. Plant Anal.* 27(5-8):1237-1249.

Frank, K., Beegle, D. and Denning, J. 2011. Phosphorus. In *Recommended Chemical Soil Test Procedures for the North Central Region*. North Central Regional Research Publication No. 221 (Revised). Missouri Ag. Expn. Stn. SB (Vol. 1001).

Gavlak, R. G., Horneck, D. A. and Miller, R. O. 2005. Plant, soil and water reference methods for the Western Region. In *WREP* (3rd ed.), p. 125.

Gelernter, W. D., Stowell, L. J., Johnson, M. E. and Brown, C. D. 2016. Documenting trends in nutrient use and conservation practices on US golf courses. *Crop Forage & Turfgrass Manag.* 2(1):1-10.

Hanlon, E. A. 2007. Procedures used by state soil testing laboratories in the southern region of the United States. *SAAESD Southern Cooperative Series Bulletin, SCSB 190-D*. Available at: http://www.clemson.edu/agsrvlb/sera6.

Hopkins, B. G. 2020. Chapter 19: Developments in the use of fertilizers. In Rengel, Z. (Ed.), *Achieving Sustainable Crop Nutrition*. Cambridge: Burleigh Dodds Science Publishing, pp.555-588. Available at: www.bdspublishing.com.

Hopkins, B. G. and Hansen, N. C. 2019. Phosphorus management in high-yield systems. *J. Environ. Qual.* 48(5):1265-1280. DOI: 10.2134/jeq2019.03.0130.

Ketterings, Q. M., Bellows, B., Czymmek, K. J., Reid, W. S. and Wildman, R. 2001a. *Conversion Equations Part 1: Do Modified Morgan and Mehlich III P Have a Morgan Equivalent? What's Cropping Up?* Cornell University (Vol. 11):2-3.

Ketterings, Q. M., Bellows, B., Czymmek, K. J. and Reid, W. S. 2001b. *Conversion Equations Part 2: Do Mehlich-III K, Ca and Mg Have Morgan Equivalents? What's Cropping Up?* Cornell University (Vol. 11):4-5.

Ketterings, Q. M., Czymmek, K. J., Reid, W. S. and Wildman, R. F. 2002. Conversion of modified Morgan and Mehlich-III soil tests to Morgan soil test values. *Soil Sci.* 167(12):830-837.

Kopittke, P. M. and Menzies, N. W. 2007. A review of the use of the basic cation saturation ratio and the "ideal" soil. *Soil Sci. Soc. Am. J.* 71(2):259-265.

Mallarino, A. P. 1995. Comparison of Mehlich-3, Olsen, and Bray-P1 procedures for phosphorus in calcareous soils. *Proceedings of the Twenty-Fifth North Central Extension-Industry Soil Fertil Conference* 11:96-101.

Mehlich, A. 1984. Mehlich 3 soil test extractant: A modification of Mehlich 2 extractant. *Commun. Soil Sci. Plant Anal.* 15(12):1409-1416.

Mitchell, C. C. and Mylavarapu, R. 2014. Soil test correlation and calibration for recommendations. In USDA-SERA, I. (Ed.), *Soil Test Methods From the Southeastern United States*. Southern Cooperative Series Bulletin No. 419 (Vol. 6). Available at: http://www.clemson.edu/sera6/MethodsManualFinalSERA6.pdf.

Mylavarapu, R. and Miller, R. 2014. Mehlich-1. *Soil Test Methods from the Southeastern United States*. Southern Cooperador Series Bulletin No. 419. Southern Extension and Research Activity Information Exchange Group - 6.

North American Proficiency Testing Program (NAPT). 2021. Laboratory results data [Online]. Available at: https://www.naptprogram.org/content/laboratory-results (accessed 26 July 2021). Madison, WI: Soil Science Society of America.

Oldham, J. L. 2014. Lancaster. *Soil Test Methods from the Southeastern United States.* Southern Cooperador Series Bulletin No. 419 Southern Extension and Research Activity Information Exchange Group (Vol. 6).

Olsen, S. R., Cole, C. V., Watanabe, F. S. and Dean, L. A. 1954. Estimation of available phosphorus in soils by extraction with sodium bicarbonate. *USDA Circular* 939:1-19. Washington, DC: Government Printing Office.

Pittman, J. J., Zhang, H., Schroder, J. L. and Payton, M. E. 2005. Differences of phosphorus in Mehlich-3 extracts determined by colorimetric and spectroscopic methods. *Commun. Soil Sci. Plant Anal.* 36(11-12):1641-1659.

Rehm, G. W. and Sorensen, R. C. 1985. Effects of potassium and magnesium applied for corn grown on an irrigated sandy soil. *Soil Sci. Soc. Am. J.* 49(6):1446-1450.

Sawyer, J. E. and Mallarino, A. P. 1999. Differentiating and understanding the Mehlich 3, Bray and Olsen soil phosphorus tests. *Proceedings of the and Presentations. Agronomy Conference* (Vol. 12). Available at: http://lib.dr.iastate.edu/agron_conf/12.

Sikora, F. J., Mylavarapu, R. S., Hardy, D. H., Tucker, M. R. and Franklin, R. E. 2005. Conversion equations for soil test extractants: Mehlich-1 and Mehlich-3. *Southern Regional Fact Sheet. SERA-IEG-6*5.*

Singh, R., Bhumbla, D. K. and Keefer, R. F. 2011. Recommended soil sulfate-S tests. In *Recommended Soil Testing Procedures for the Northeastern United States* (3rd ed.). Northeastern Regional Publication No. 493. Agricultural Experiment Stations of CT, DE. Rehabilitation International, ME, MD, MA, NH, NJ, NY, PA, VA and WV.

Soil and Plant Analysis Council, Inc. 2000. Chapter 1: Introduction. In *Soil Analysis Handbook of Reference Methods*. Boca Raton, FL: CRC Press.

Stowell, L. and Woods, M. 2013. Minimum levels for sustainable nutrition (MLSN). *Appl. Turfgrass Sci.* 10(1). DOI: 10.2134/ATS-2013-0008BC.

Wang, J. J., Harrell, D. L., Henderson, R. E. and Bell, P. F. 2004. Comparison of soil-test extractants for phosphorus, potassium, calcium, magnesium, sodium, zinc, copper, manganese, and iron in Louisiana soils. *Commun. Soil Sci. Plant Anal.* 35(1-2):145-160.

Warncke, D. and Brown, J. R. 2013. Potassium and other basic cations. In *Recommended Chemical Soil Test Procedures for the North Central Region*. North Central Region Research Publication No. 221 (Revised). Missouri Ag. Exp. Stn. SB 1001.

Warncke, D. D. 2014. Test procedures for greenhouse root media. *Soil Test Methods From the Southeastern United States.* Southern Cooperador Series Bulletin No. 419 Southern Extension and Research Activity Information Exchange Group (Vol. 6).

Watson, M. and Mullen, R. 2007. Understanding soil tests for plant available phosphorus. *The Ohio State University School of Environment and Natural Resources Extension Fact Sheet*. Available at: http://ohioline.osu.edu.

Whitney, D. A. 1998. Micronutrients: Zinc, iron, manganese and copper. In *Recommended Chemical Soil Test Procedures for the North Central Region*. North Central Region Research Publication No. 221 (Revised). Missouri Ag. Exp. Stn. SB 1001.

Wolf, A. M. and Baker, D. E. 2008. Comparisons of soil test phosphorus by Olsen, Bray P1, Mehlich I and Mehlich III methods. *Commun. Soil Sci. Plant Anal.* 16(5):467-484. DOI: 10.1080/00103628509367620.

Wolf, A. M. and Beegle, D. 2011. Recommended soil tests for macro and micronutrients. In *Recommended Soil Testing Procedures for the Northeastern United States* (3rd ed.). Northeastern Regional Publication No. 493. Agricultural Experiment Stations of CT, DE. Rehabilitation International, ME, MD, MA, NH, NJ, NY, PA, VA and WV.

Woods, M., Stowell, L. and Gelernter, W. 2014. Just what the grass requires: Using minimum levels for sustainable nutrition. *Golf Course Manag*, 132–138.

Zhang, H., Hardy, D. H., Mylavarapu, R. and Wang, J. J. 2014. Mehlich-3. *Soil Test Methods From the Southeastern United States.* Southern Cooperador Series Bulletin No. 419 Southern Extension and Research Activity Information Exchange Group (Vol. 6).

Index

www.ingramcontent.com/pod-product-compliance
Ingram Content Group UK Ltd.
Pitfield, Milton Keynes, MK11 3LW, UK
UKHW021823100926
8556IPUK00003B/135

9 781801 460194